QUANTUM FIELD THEORY

QUANTUM FIELD THEORY

A Modern Introduction

MICHIO KAKU

Department of Physics
City College of the
   City University of New York

New York   Oxford
OXFORD UNIVERSITY PRESS
1993

Oxford University Press

Oxford   New York   Toronto
Delhi   Bombay   Calcutta   Madras   Karachi
Kuala Lumpur   Singapore   Hong Kong   Tokyo
Nairobi   Dar es Salaam   Cape Town
Melbourne   Auckland   Madrid

and associated companies in

Berlin   Ibadan

Library of Congress Cataloging-in-Publication Data
Kaku, Michio. Quantum field theory: a modern introduction/
Michio Kaku.
p.   cm.   Includes bibliographical references and index.
ISBN 0-19-507652-4
ISBN 0-19-509158-2 (pbk.)
1. Quantum field theory.   2. Gauge fields (Physics)
3. Standard model (Nuclear physics)
I. title   QC174.45.K34   1993   530.1'43 — dc20   92-27704

9 8 7 6 5 4 3

Printed in the United States of America
on acid-free paper

This book is dedicated to my parents.

Preface

In the 1960s, there were a number of classic books written on quantum field theory. Because of the phenomenal experimental success of quantum electrodynamics (QED), quantum field theory became a rigorous body of physical knowledge, as established as nonrelativistic quantum mechanics.

In the 1970s and 1980s, because of the growing success of gauge theories, it was clear that a typical 1-year course in quantum field theory was rapidly becoming obsolete. A number of advanced books appeared on various aspects of gauge theories, so often a 1-year course on quantum field theory became disjoint, with one book on QED being the basis of the first semester and one of several books on various aspects of gauge theories being the basis of the second semester.

Today, because of the success of the Standard Model, it is necessary to consolidate and expand the typical 1-year quantum field theory course. There is obviously a need for a book for the 1990s, one that presents this material in a coherent fashion and uses the Standard Model as its foundation in the same way that earlier books used QED as their foundation. Because the Standard Model is rapidly becoming as established as QED, there is a need for a textbook whose focus is the Standard Model.

As a consequence, we have divided the book into three parts, which can be used in either a two- or a three-semester format:

> I: Quantum Fields and Renormalization
>
> II: Gauge Theory and the Standard Model
>
> III: Non preturbative Methods and Unification

Part I of this book summarizes the development of QED. It provides the foundation for a first-semester course on quantum field theory, laying the basis for perturbation theory and renormalization theory. (However, one may also use it in the last semester of a three-semester course on quantum mechanics, treating it as the relativistic continuation of a course on nonrelativistic quantum mechanics.

In this fashion, students who are not specializing in high-energy physics will find Part I particularly useful, since perturbation theory and Feynman diagrams have now penetrated into all branches of quantum physics.)

In Part II, the Standard Model is the primary focus. This can be used as the basis of a second semester course on quantum field theory. Particular attention is given to the method of path integrals and the phenomenology of the Standard Model. This chapter is especially geared to students wanting an understanding of high-energy physics, where a working knowledge of the Standard Model is a necessity. It is hoped that the student will finish this section with an appreciation of the overwhelming body of experimental evidence pointing to the correctness of the Standard Model.

Because the experiments necessary to go beyond the Standard Model are rapidly becoming prohibitively expensive and time consuming, we are also aware that the development of physics into the next decade may become increasingly theoretical, and therefore we feel that an attempt should be made to explore the various theories that take us beyond the Standard Model.

Part III of this book, therefore, is geared to the students who wish to pursue more advanced material and can be used in one of two ways. A lecturer may want to treat a few of the chapters in Part III at the end of a typical two semester course on quantum field theory. Or, Part III can be used as the basis of a third semester course. We are providing a variety of topics so that the lecturer may pick and choose the chapters that are most topical and are of interest. We have written Part III to leave as much discretion as possible for the lecturer in using this material.

The approach that we have taken in our book differs from that taken in other books in several ways:

First, we have tried to consolidate and streamline, as much as possible in a coherent fashion, a large body of information in one book, beginning with QED, leading to the Standard Model, and ending on supersymmetry.

Second, we have emphasized the role of group theory, treating many of the features of quantum field theory as the byproduct of the Lorentz, Poincaré, and internal symmetry groups. Viewed in this way, many of the rather arbitrary and seemingly contrived conventions of quantum field theory are seen as a consequence of group theory. Group theory, especially in Part III, plays an essential role in understanding unification.

Third, we have presented three distinct proofs of renormalization theory. Most books, if they treat renormalization theory at all, only present one proof. However, because of the importance of renormalization theory to today's research, the serious student may find that a single proof of renormalization is not enough. The student may be ill prepared to handle research when renormalization theory is developed from an entirely different approach. As a consequence, we have presented three different proofs of renormalization theory so that the student can become fluent in at least two different methods. We have presented the original

Dyson/Ward proof in Chapter 7. In Part II, we also present two different proofs based on the BPHZ method and the renormalization group.

Fourth, we should caution the reader that experimental proof of nonperturbative quark confinement or of supersymmetry is absolutely nonexistent. However, since the bulk of current research in theoretical high-energy physics is focused on the material covered in Part III, this section should give the student a brief overview of the main currents in high-energy physics. Moreover, our attitude is to treat nonperturbative field theory and supersymmetries as useful theoretical "laboratories" in which to test many of our notions about quantum field theory. We feel that these techniques in Part III, if viewed as a rich, productive laboratory in which to probe the limits of field theory, will yield great dividends for the serious student.

We have structured the chapters so that they can be adapted in many different ways to suit different needs. In Part I, for example, the heart of the canonical quantization method is presented in Chapters 3–6. These chapters are essential for building a strong foundation to quantum field theory and Feynman diagrams. Although path integral methods today have proven more flexible for gauge theories, a student will have a much better appreciation for the rigor of quantum field theory by reading these chapters. Chapters 2 and 7, however, can be skipped by the student who either already understands the basics of group theory and renormalization, or who does not want to delve that deeply into the intricacies of quantum field theory.

In Part II, the essential material is contained in Chapters 8–11. In these chapters, we develop the necessary material to understand the Standard Model, that is, path integrals, gauge theory, spontaneous symmetry breaking, and phenomenology. This forms the heart of this section, and cannot be omitted. However, Chapters 12–14 should only be read by the student who wants a much more detailed presentation of the subtleties of quantum field theory (BRST, anomalies, renormalization group, etc.).

In Part III, there is great freedom to choose which material to study, depending on the person's interests. We have written Part III to give the greatest flexibility to different approaches in quantum field theory. For those want an understanding of quark confinement and nonperturbative methods, Chapters 15–17 are essential. The student wishing to investigate Grand Unified Theories should study Chapter 18. However, the student who wishes to understand some of the most exciting theoretical developments of the past decade should read Chapters 19–21.

Because of the wide and often confusing range of notations and conventions found in the literature, we have tried to conform, at least in the early chapters, to those appearing in Bjorken and Drell, Itzykson and Zuber, and Cheng and Li. We also choose our metric to be $g_{\mu\nu} = (+, -, -, -)$.

We have also included 311 exercises in this book, which appear after each chapter. We consider solving these exercises essential to an understanding of the material. Often, students complain that they understand the material but cannot do the problems. We feel that this is a contradiction in terms. If one cannot do the exercises, then one does not really fully understand the material.

In writing this book, we have tried to avoid two extremes. We have tried to avoid giving an overly tedious treatise of renormalization theory and the obscure intricacies of Feynman graphs. One is reminded of being an apprentice during the Middle Ages, where the emphasis was on mastering highly specialized, arcane techniques and tricks, rather than getting a comprehensive understanding of the field.

The other extreme is a shallow approach to theoretical physics, where many vital concepts are deleted because they are considered too difficult for the student. Then the student receives a superficial introduction to the field, creating confusion rather than understanding. Although students may prefer an easier introduction to quantum field theory, ultimately it is the student who suffers. The student will be totally helpless when confronted with research. Even the titles of the high-energy preprints will be incomprehensible.

By taking this intermediate approach, we hope to provide the student with a firm foundation in many of the current areas of research, without overwhelming the student in an avalanche of facts. We will consider the book a success if we have been able to avoid these extremes.

New York   M. K.
July 1992

Acknowledgments

I would especially like to thank J. D. Bjorken, who has made countless productive and useful suggestions in every chapter of this book. This book has greatly benefited from his careful reading and critical comments, which have significantly strengthened the presentation in a number of important places.

I would like to thank Dr. Bunji Sakita and Dr. Joseph Birman for their constant support during the writing of this book. I would like to thank A. Das, S. Samuel, H. Chang, D. Karabali, R. Ray, and C. Lee, for reading various chapters of the book and making valuable comments and numerous corrections that have greatly enhanced this book.

I would also like to thank my editor Jeffrey Robbins, who has skillfully guided the passage of the three books that I have written for him. I would also like to thank the National Science Foundation and CUNY–FRAP for partial support.

Contents

II   Gauge Theory and the Standard Model

Part I

Quantum Fields
and Renormalization

Chapter 1
Why Quantum Field Theory?

Anyone who is not shocked by the quantum theory does not understand it.

—N. Bohr

1.1 Historical Perspective

Quantum field theory has emerged as the most successful physical framework describing the subatomic world. Both its computational power and its conceptual scope are remarkable. Its predictions for the interactions between electrons and photons have proved to be correct to within one part in 10^8. Furthermore, it can adequately explain the interactions of three of the four known fundamental forces in the universe. The success of quantum field theory as a theory of subatomic forces is today embodied in what is called the Standard Model. In fact, at present, there is no known experimental deviation from the Standard Model (excluding gravity).

This impressive list of successes, of course, has not been without its problems. In fact, it has taken several generations of the world's physicists working over many decades to iron out most of quantum field theory's seemingly intractable problems. Even today, there are still several subtle unresolved problems about the nature of quantum field theory.

The undeniable successes of quantum field theory, however, were certainly not apparent in 1927 when P.A.M. Dirac[1] wrote the first pioneering paper combining quantum mechanics with the classical theory of radiation. Dirac's union of nonrelativistic quantum mechanics, which was itself only 2 years old, with the special theory of relativity and electrodynamics would eventually lay the foundation of modern high-energy physics.

Breakthroughs in physics usually emerge when there is a glaring conflict between experiment and theory. Nonrelativistic quantum mechanics grew out of the inability of classical mechanics to explain atomic phenomena, such as black body radiation and atomic spectra. Similarly, Dirac, in creating quantum field theory, realized that there were large, unresolved problems in classical electrodynamics that might be solved using a relativistic form of quantum mechanics.

In his 1927 paper, Dirac wrote: "...hardly anything has been done up to the present on quantum electrodynamics. The questions of the correct treatment of a system in which the forces are propagated with the velocity of light instead of instantaneously, of the production of an electromagnetic field by a moving electron, and of the reaction of this field on the electron have not yet been touched."

Dirac's keen physical intuition and bold mathematical insight led him in 1928 to postulate the celebrated Dirac electron theory. Developments came rapidly after Dirac coupled the theory of radiation with his relativistic theory of the electron, creating Quantum Electrodynamics (QED). His theory was so elegant and powerful that, when conceptual difficulties appeared, he was not hesitant to postulate seemingly absurd concepts, such as "holes" in an infinite sea of negative energy. As he stated on a number of occasions, it is sometimes more important to have beauty in your equations than to have them fit experiment.

However, as Dirac also firmly realized, the most beautiful theory in the world is useless unless it eventually agrees with experiment. That is why he was gratified when his theory successfully reproduced a series of experimental results: the spin and magnetic moment of the electron and also the correct relativistic corrections to the hydrogen atom's spectra. His revolutionary insight into the structure of matter was vindicated in 1932 with the experimental discovery of antimatter. This graphic confirmation of his prediction helped to erase doubts concerning the correctness of Dirac's theory of the electron.

However, the heady days of the early 1930s, when it seemed like child's play to make major discoveries in quantum field theory with little effort, quickly came to a halt. In some sense, the early successes of the 1930s only masked the deeper problems that plagued the theory. Detailed studies of the higher-order corrections to QED raised more problems than they solved. In fact, a full resolution of these question would have to wait several decades. From the work of Weisskopf, Pauli, Oppenheimer, and many others, it was quickly noticed that QED was horribly plagued by infinities. The early successes of QED were premature: they only represented the lowest-order corrections to classical physics. Higher-order corrections in QED necessarily led to divergent integrals.

The origin of these divergences lay deep within the conceptual foundation of physics. These divergences reflected our ignorance concerning the small-scale structure of space–time. QED contained integrals which diverged as $x \to 0$, or, in momentum space, as $k \to \infty$. Quantum field theory thus inevitably faced divergences emerging from regions of space–time and matter–energy beyond its regime of applicability, that is, infinitely small distances and infinitely large energies.

These divergences had their counterpart in the classical "self-energy" of the electron. Classically, it was known to Lorentz and others near the turn of the century that a complete description of the electron's self-energy was necessarily plagued with infinities. An accelerating electron, for example, would produce a radiation field that would act back on itself, creating absurd physical effects such as the breakdown of causality. Also, other paradoxes abounded; for example it would take an infinite amount of energy to assemble an electron.

Over the decades, many of the world's finest physicists literally brushed these divergent quantities under the rug by manipulating infinite quantities as if they were small. This clever sleight–of–hand was called *renormalization theory*, because these divergent integrals were absorbed into an infinite rescaling of the coupling constants and masses of the theory. Finally, in 1949, Tomonaga, Schwinger, and Feynman[2−4] penetrated this thicket of infinities and demonstrated how to extract meaningful physical information from QED, for which they received the Nobel Prize in 1965.

Ironically, Dirac hated the solution to this problem. To him, the techniques of renormalization seemed so abstruse, so artificial, that he could never reconcile himself with renormalization theory. To the very end, he insisted that one must propose newer, more radical theories that required no renormalization whatsoever.

Nevertheless, the experimental success of renormalization theory could not be denied. Its predictions for the anomalous magnetic moment of the electron, the Lamb shift, etc. have been tested experimentally to one part in 10^8, which is a remarkable degree of confirmation for the theory.

Although renormalized QED enjoyed great success in the 1950s, attempts at generalizing quantum field theory to describe the other forces of nature met with disappointment. Without major modifications, quantum field theory appeared to be incapable of describing all four fundamental forces of nature.[5] These forces are:

1. *The electromagnetic force*, which was successfully described by QED.

2. *The strong force*, which held the nucleus together.

3. *The weak force*, which governed the properties of certain decaying particles, such as the beta decay of the neutron.

4. *The gravitational force*, which was described classically by Einstein's general theory of relativity.

In the 1950s, it became clear that quantum field theory could only give us a description of one of the four forces. It failed to describe the other interactions for very fundamental reasons.

Historically, most of the problems facing the quantum description of these forces can be summarized rather succinctly by the following:

$$
\begin{aligned}
\alpha_{\text{em}} &\sim 1/137.0359895(61) \\[4pt]
\alpha_{\text{strong}} &\sim 14 \\[4pt]
G_{\text{weak}} &\sim 1.02 \times 10^{-5}/M_p^2 \\[4pt]
&\sim 1.16639(2) \times 10^{-5}\,\text{GeV}^{-2} \\[4pt]
G_{\text{Newton}} &\sim 5.9 \times 10^{-39}/M_p^2 \\[4pt]
&\sim 6.67259(85) \times 10^{-1}\text{m}^3\text{kg}^{-1}\text{s}^{-2}
\end{aligned}
\tag{1.1}
$$

where M_p is the mass of the proton and the parentheses represent the uncertainties.

Several crucial features of the various forces can be seen immediately from this chart. The fact that the coupling constant for QED, the "fine structure constant," is approximately 1/137 meant that physicists could successfully power expand the theory in powers of α_{em}. The power expansion in the fine structure constant, called "perturbation theory," remains the predominant tool in quantum field theory. The smallness of the coupling constant in QED gave physicists confidence that perturbation theory was a reliable approximation to the theory. However, this fortuitous circumstance did not persist for the other interactions.

1.2   Strong Interactions

In contrast to QED, the strongly interacting particles, the "hadrons" (from the Greek word *hadros*, meaning "strong"), have a large coupling constant, meaning that perturbation theory was relatively useless in predicting the spectrum of the strongly interacting particles. Unfortunately, nonperturbative methods were notoriously crude and unreliable. As a consequence, progress in the strong interactions was painfully slow.

In the 1940s, the first seminal breakthrough in the strong interactions was the realization that the force binding the nucleus together could be mediated by the exchange of π mesons:

$$
\begin{aligned}
\pi^- + p &\leftrightarrow n \\[4pt]
\pi^+ + n &\leftrightarrow p
\end{aligned}
\tag{1.2}
$$

Theoretical predictions by Yukawa[6] of the mass and range of the π meson, based on the energy scale of the strong interactions, led experimentalists to find the π

meson in their cosmic ray experiments. The π meson was therefore deduced to be the carrier of the nuclear force that bound the nucleus together.

This breakthrough, however, was tempered with the fact that, as we noted, the pion–nucleon coupling constant was much greater than one. Although the Yukawa meson theory as a quantum field theory was known to be renormalizable, perturbation theory was unreliable when applied to the Yukawa theory. Nonperturbative effects, which were exceedingly difficult to calculate, become dominant.

Furthermore, the experimental situation became confusing when so many "resonances" began to be discovered in particle accelerator experiments. This indicated again that the coupling constant of some unknown underlying theory was large, beyond the reach of conventional perturbation theory. Not surprisingly, progress in the strong interactions was slow for many decades for these reasons. With each newly discovered resonance, physicists were reminded of the inadequacy of quantum field theory.

Given the failure of conventional quantum field theory, a number of alternative approaches were investigated in the 1950s and 1960s. Instead of focusing on the "field" of some unknown constituent as the fundamental object (which is in principle unmeasurable), these new approaches centered on the S matrix itself. Borrowing from classical optics, Goldberger and his colleagues[7] assumed the S matrix was an analytic function that satisfied certain dispersion relations. Alternatively, Chew[8] assumed a type of "nuclear democracy"; that is, there were no fundamental particles at all. In this approach, one hoped to calculate the S matrix directly, without using field theory, because of the many stringent physical conditions that it satisfied.

The most successful approach, however, was the $SU(3)$ "quark" theory of the strongly interacting particles (the hadrons). Gell–Mann, Ne'eman, and Zweig,[9–11] building on earlier work of Sakata and his collaborators,[12,13] tried to explain the hadron spectrum with the symmetry group $SU(3)$.

Since quantum field theory was unreliable, physicists focused on the quark model as a strictly phenomenological tool to make sense out of the hundreds of known resonances. Composite combinations of the "up," "down," and "strange" quarks could, in fact, explain all the hadrons discovered up to that time. Together, these three quarks formed a representation of the Lie group $SU(3)$:

$$q_i = \begin{pmatrix} u \\ d \\ s \end{pmatrix} \tag{1.3}$$

The quark model could predict with relative ease the masses and properties of particles that were not yet discovered. A simple picture of the strong interactions was beginning to emerge: Three quarks were necessary to construct a *baryon*, such as a proton or neutron (or the higher resonances, such as the Λ, Ξ, Ω, etc.), while a quark and an antiquark were necessary to assemble a meson, such as the π meson or the K meson:

$$
\text{Hadrons} = \left\{ \begin{array}{lcl} \text{Baryons} & = & q_i q_j q_k \\ \text{Mesons} & = & \bar{q}_i q_j \end{array} \right. \tag{1.4}
$$

Ironically, one problem of the quark model was that it was too successful. The theory was able to make qualitative (and often quantitative) predictions far beyond the range of its applicability. Yet the fractionally charged quarks themselves were never discovered in any scattering experiment. Perhaps they were just a mathematical artifice, reflecting a deeper physical reality that was still unknown. Furthermore, since there was no quantum field theory of quarks, it was unknown what dynamical force held them together. As a consequence, the model was unable to explain why certain bound states of quarks (called "exotics") were not found experimentally.

1.3 Weak Interactions

Equation (1.1), which describes the coupling constants of the four fundamental forces, also reveals why quantum field theory failed to describe the weak interactions. The coupling constant for the weak interactions has the dimensions of inverse mass squared. Later, we will show that theories of this type are nonrenormalizable; that is, theories with coupling constants of negative dimension predict infinite amplitudes for particle scattering.

Historically, the weak interactions were first experimentally observed when strongly interacting particles decayed into lighter particles via a much weaker force, such as the decay of the neutron into a proton, electron, and antineutrino:

$$
n \rightarrow p + e^- + \bar{\nu} \tag{1.5}
$$

These light particles, such as the electron, its neutrino, the muon μ, etc., were called *leptons*:

$$
\text{Leptons} = e^\pm, \nu, \mu^\pm, \text{etc.} \tag{1.6}
$$

Fermi, back in the 1930s, postulated the form of the action that could give a reasonably adequate description of the lowest-order behavior of the weak interactions. However, any attempt to calculate quantum corrections to the Fermi theory floundered because the higher-order terms diverged. The theory was nonrenormalizable because the coupling constant had negative dimension.

Furthermore, it could be shown that the naive Fermi theory violated unitarity at sufficiently large energies.

The mystery of the weak interactions deepened in 1956, when Lee and Yang[14] theorized that parity conservation, long thought to be one of the fundamental principles of physics, was violated in the weak interactions. Their conjecture was soon proved to be correct by the careful experimental work of Wu and also Lederman and Garwin and their colleagues.[15,16]

Furthermore, more and more weakly interacting leptons were discovered over the next few decades. The simple picture of the electron, neutrino, and muon was shattered as the muon neutrino and the tau lepton were found experimentally. Thus, there was the unexplained embarrassment of three exact copies or "generations" of leptons, each generation acting like a Xerox copy of the previous one. (The solution of this problem is still unknown.)

There were some modest proposals that went beyond the Fermi action, which postulated the existence of a massive vector meson or W boson that mediated the weak forces. Buoyed by the success of the Yukawa meson theory, physicists postulated that a massive spin-one vector meson might be the carrier of the weak force. However, the massive vector meson theory, although it was on the right track, had problems because it was also nonrenormalizable. As a result, the massive vector meson theory was considered to be one of several phenomenological possibilities, not a fundamental theory.

1.4 Gravitational Interaction

Ironically, although the gravitational interaction was the first of the four forces to be investigated classically, it was the most difficult one to be quantized.

Using some general physical arguments, one could calculate the mass and spin of the gravitational interaction. Since gravity was a long-range force, it should be massless. Since gravity was always attractive, this meant that its spin must be even. (Spin-one theories, such as electromagnetism, can be both attractive and repulsive.) Since a spin-0 theory was not compatible with the known bending of starlight around the sun, we were left with a spin-two theory. A spin-two theory could also be coupled equally to all matter fields, which was consistent with the equivalence principle. These heuristic arguments indicated that Einstein's theory of general relativity should be the classical approximation to a quantum theory of gravity.

The problem, however, was that quantum gravity, as seen from Eq. (1.1), had a dimensionful coupling constant and hence was nonrenormalizable. This coupling constant, in fact, was Newton's gravitational constant, the first important universal physical constant to be isolated in physics. Ironically, the very success of Newton's early theory of gravitation, based on the constancy of Newton's constant, proved to be fatal for a quantum theory of gravity.

Another fundamental problem with quantum gravity was that, according to Eq. (1.1), the strength of the interaction was exceedingly weak, and hence very difficult to measure. For example, it takes the entire planet earth to keep pieces of paper resting on a tabletop, but it only takes a charged comb to negate gravity and pick them up. Similarly, if an electron and proton were bound in a hydrogen atom by the gravitational force, the radius of the atom would be roughly the size of the known universe. Although gravitational forces were weaker by comparison to the electromagnetic force by a factor of about 10^{-40}, making it exceedingly difficult to study, one could also show that a quantum theory of gravity had the reverse problem, that its natural energy scale was 10^{19} Gev. Once gravity was quantized, the energy scale at which the gravitational interaction became dominant was set by Newton's constant G_N. To see this, let r be the distance at which the gravitational potential energy of a particle of mass M equals its rest energy, so that $G_N M^2 / r = M c^2$. Let r also be the Compton wavelength of this particle, so that $r = \hbar / M c$. Eliminating M and solving for r, we find that r equals the Planck length, 10^{-33} cm, or 10^{19} GeV:

$$
\begin{aligned}
[\hbar G_N c^{-3}]^{1/2} &= 1.61605(10) \times 10^{-33} \, \text{cm} \\
[\hbar c G_N^{-1}]^{1/2} &= 1.221047(79) \times 10^{19} \, \text{GeV}/c^2
\end{aligned}
\tag{1.7}
$$

This is, of course, beyond the range of our instruments for the foreseeable future. So physicists were faced with the double problem: The classical theory of gravity was so weak that macroscopic experiments were difficult to perform in the laboratory, but the quantum theory of gravity dominated subatomic reactions at the incredible energy scale of 10^{19} GeV, which was far beyond the range of our largest particle accelerators.

Yet another problem arose when one tried to push the theory of gravity to its limits. Phenomenologically, Einstein's general relativity has proved to be an exceptionally reliable tool over cosmological distances. However, when one investigated the singularity at the center of a black hole or the instant of the Big Bang, then the gravitational fields became singular, and the theory broke down. One expected quantum corrections to dominate in those important regions of space–time. However, without a quantum theory of gravity, it was impossible to make any theoretical calculation in those interesting regions of space and time.

In summary, an enormous amount of information is summarized in Eq. (1.1). Some of the fundamental reasons why the development of quantum field theory was stalled in the 1950s are summarized in this chart.

1.5   Gauge Revolution

In the 1950s and 1960s, there was a large mass of experimental data for the
strong and weak interactions that was patiently accumulated by many experimental
groups. However, most of it could not be explained theoretically. There were
significant strides taken experimentally, but progress in theory was, by contrast,
painfully slow.

In 1971, however, a dramatic discovery was made by G. 't Hooft,[17] then a
graduate student. He reinvestigated an old theory of Yang and Mills, which was
a generalization of the Maxwell theory of light, except that the symmetry group
was much larger. Building on earlier pioneering work by Veltman, Faddeev,
Higgs, and others, 't Hooft showed that Yang–Mills gauge theory, even when
its symmetry group was "spontaneously broken," was renormalizable. With this
important breakthrough, it now became possible to write down renormalizable
theories of the weak interactions, where the W bosons were represented as gauge
fields.

Within a matter of months, a flood of important papers came pouring out. An
earlier theory of Weinberg and Salam[18,19] of the weak interactions, which was a
gauge theory based on the symmetry group $SU(2) \otimes U(1)$, was resurrected and
given serious analysis. The essential point, however, was that because gauge
theories were now known to be renormalizable, concrete numerical predictions
could be made from various gauge theories and then checked to see if they
reproduced the experimental data. If the predictions of gauge theory disagreed
with the experimental data, then one would have to abandon them, no matter how
elegant or aesthetically satisfying they were. Gauge theorists realized that the
ultimate judge of any theory was experiment.

Within several years, the agreement between experiment and the Weinberg–
Salam theory proved to be overwhelming. The data were sufficiently accurate to
rule out several competing models and verify the correctness of the Weinberg–
Salam model. The weak interactions went from a state of theoretical confusion to
one of relative clarity within a brief period of time. The experimental discovery
of the gauge bosons $W^{\pm}$ and Z in 1983 predicted by Weinberg and Salam was
another important vindication of the theory.

The Weinberg–Salam model arranged the leptons in a simple manner. It
postulated that the (left–handed) leptons could be arranged according to $SU(2)$
doublets in three separate generations:

$$
\begin{pmatrix} \nu_e \\ e \end{pmatrix}; \quad
\begin{pmatrix} \nu_\mu \\ \mu \end{pmatrix}; \quad
\begin{pmatrix} \nu_\tau \\ \tau \end{pmatrix}
\tag{1.8}
$$

The interactions between these leptons were generated by the intermediate vector bosons:

$$\text{Vector mesons}: \quad W_\mu^\pm, \qquad Z_\mu \tag{1.9}$$

(The remaining problem with the model is to find the Higgs bosons experimentally, which are responsible for spontaneous symmetry breaking, or to determine if they are composite particles.)

In the realm of the strong interactions, progress was also fairly rapid. The gauge revolution made possible Quantum Chromodynamics (QCD), which quickly became the leading candidate for a theory of the strong interactions. By postulating a new "color" $SU(3)$ symmetry, the Yang–Mills theory now provided a glue by which the quarks could be held together. [The $SU(3)$ color symmetry should not be confused with the earlier $SU(3)$ symmetry of Gell-Mann, Ne'eman, and Zweig, which is now called the "flavor" symmetry. Quarks thus have two indices on them; one index $a = u, d, s, c, t, b$ labels the flavor symmetry, while the other index labels the color symmetry.]

The quarks in QCD are represented by:

$$\begin{pmatrix} u^1 & u^2 & u^3 \\ d^1 & d^2 & d^3 \\ s^1 & s^2 & s^3 \\ c^1 & c^2 & c^3 \\ \vdots & \vdots & \vdots \end{pmatrix} \tag{1.10}$$

where the 1, 2, 3 index labels the color symmetry. QCD gave a plausible explanation for the mysterious experimental absence of the quarks. One could calculate that the effective $SU(3)$ color coupling constant became large at low energy, and hence "confined" the quarks permanently into the known hadrons. If this picture was correct, then the gluons condensed into a taffy-like substance that bound the quarks together, creating a string-like object with quarks at either end. If one tried to pull the quarks apart, the condensed gluons would resist their being separated. If one pulled hard enough, then the string might break and another bound quark–antiquark pair would be formed, so that a single quark cannot be isolated (similar to the way that a magnet, when broken, simply forms two smaller magnets, and not single monopoles).

The flip side of this was that one could also prove that the $SU(3)$ color coupling constant became small at large energies. This was called "asymptotic freedom," which was discovered in gauge theories by Gross, Wilczek, Politzer, and 't Hooft.[20−22] At high energies, it could explain the curious fact that the quarks acted as if they were described by a free theory. This was because the effective coupling constants decreased in size with rising energy, giving the appearance of a free theory. This explained the fact that the quark model worked much better than it was supposed to.

Gradually, a small industry developed around finding nonperturbative solutions to QCD that could explain the confinement of quarks. On one hand, physicists showed that two-dimensional "toy models" could reproduce many of the features that were required of a quantum field theory of quarks, such as confinement and asymptotic freedom. Many of these features followed from the exact solution of these toy models. On the other hand, a compelling description of the four dimensional theory could be achieved through Wilson's lattice gauge theory,[23] which gave qualitative nonperturbative information concerning QCD. In the absence of analytic solutions to QCD, lattice gauge theories today provide the most promising approach to the still-unsolved problem of quark confinement.

Soon, both the electro–weak and QCD models were spliced together to become the Standard Model based on the gauge group $SU(3) \otimes SU(2) \otimes U(1)$. The Standard Model was more than just the sum of its parts. The leptons in the Weinberg–Salam model were shown to possess "anomalies" that threatened renormalizability. Fortunately, these potentially fatal anomalies precisely cancelled against the anomalies coming from the quarks. In other words, the lepton and quark sectors of the Standard Model cured each other's diseases, which was a gratifying theoretical success for the Standard Model. As a result of this and other theoretical and experimental successes, the Standard Model was rapidly recognized to be a first-order approximation to the ultimate theory of particle interactions.

The spectrum of the Standard Model for the left-handed fermions is schematically listed here, consisting of the neutrino ν, the electron e, the "up" and "down" quarks, which come in three "colors," labeled by the index i. This pattern is then repeated for the other two generations (although the top quark has not yet been discovered):

$$\begin{pmatrix} \nu \\ e \end{pmatrix} \begin{pmatrix} u^i \\ d^i \end{pmatrix}; \quad \begin{pmatrix} \nu_\mu \\ \mu \end{pmatrix} \begin{pmatrix} c^i \\ s^i \end{pmatrix}; \quad \begin{pmatrix} \nu_\tau \\ \tau \end{pmatrix} \begin{pmatrix} t^i \\ b^i \end{pmatrix} \qquad (1.11)$$

In the Standard Model, the forces between the leptons and quarks were mediated by the massive vector mesons for the weak interactions and the massless gluons for the strong interactions:

$$\begin{cases} \text{Massive vector mesons}: & W^\pm, Z \\ \text{Massless gluons}: & A_\mu^a \end{cases} \qquad (1.12)$$

The weaknesses of the Standard Model, however, were also readily apparent. No one saw the theory as a fundamental theory of matter and energy. Containing at least 19 arbitrary parameters in the theory, it was a far cry from the original dream of physicists: a single unified theory with at most *one* undetermined coupling constant.

1.6   Unification

In contrast to the 1950s, when physicists were flooded with experimental data without a theoretical framework to understand them, the situation in the 1990s may be the reverse; that is, the experimental data are all consistent with the Standard Model. As a consequence, without important clues coming from experiment, physicists have proposed theories beyond the Standard Model that cannot be tested with the current level of technology. In fact, even the next generation of particle accelerators may not be powerful enough to rule out many of the theoretical models currently being studied. In other words, while experiment led theory in the 1950s, in the 1990s theory may lead experiment.

At present, attempts to use quantum field theory to push beyond the Standard Model have met with modest successes. Unfortunately, the experimental data at very large energies are still absent, and this situation may persist into the near future. However, enormous theoretical strides have been made that give us some confidence that we are on the right track.

The next plausible step beyond the Standard Model may be the GUTs (Grand Unified Theories), which are based on gauging a single Lie group, such as $SU(5)$ or $SO(10)$ (Fig. 1.1).

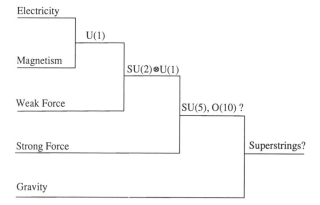

Figure 1.1. This chart shows how the various forces of nature, once thought to be fundamentally distinct, have been unified over the past century, giving us the possibility of unifying all known forces via quantum field theory.

According to GUT theory, the energy scale at which the unification of all three particle forces takes place is enormously large, about 10^{15} GeV, just below the Planck energy. Near the instant of the Big Bang, where such energies were found, the theory predicts that all three particle forces were unified by one GUT symmetry. In this picture, as the universe rapidly cooled down, the original GUT symmetry was broken down successively into the present–day symmetries of the Standard Model. A typical breakdown scheme might be:

$$O(10) \rightarrow SU(5) \rightarrow SU(3) \otimes SU(2) \otimes U(1) \rightarrow SU(3) \otimes U(1) \qquad (1.13)$$

Because quarks and electrons are now placed in the same multiplet, it also means that quarks can decay into leptons, giving us the violation of baryon number and the decay of the proton. So far, elaborate experimental searches for proton decay have been disappointing. The experimental data, however, are good enough to rule out minimal $SU(5)$; there are, however, more complicated GUT theories that can accomodate longer proton lifetimes.

Although GUT theories are a vast improvement over the Standard Model, they are also beset with problems. First, they cannot explain the three generations of particles that have been discovered. Instead, we must have three identical GUT theories, one for each generation. Second, it still has many arbitrary parameters (e.g., Higgs parameters) that cannot be explained from any simpler principle. Third, the unification scale takes place at energies near the Planck scale, where we expect gravitational effects to become large, yet gravity is not included in the theory. Fourth, it postulates that there is a barren "desert" that extends for twelve orders of magnitude, from the GUT scale down to the electro-weak scale. (The critics claim that there is no precedent in physics for such an extrapolation over such a large range of energy.) Fifth, there is the "hierarchy problem," meaning that radiative corrections will mix the two energy scales together, ruining the entire program.

To solve the last problem, physicists have studied quantum field theories that can incorporate "supersymmetry," a new symmetry that puts fermions and bosons into the same multiplet. Although not a single shred of experimental data supports the existence of supersymmetry, it has opened the door to an entirely new kind of quantum field theory that has remarkable renormalization properties and is still fully compatible with its basic principles.

In a supersymmetric theory, once we set the energy scale of unification and the energy scale of the low-energy interactions, we never have to "retune" the parameters again. Renormalization effects do not mix the two energy scales. However, one of the most intriguing properties of supersymmetry is that, once it is made into a local symmetry, it necessarily incorporates quantum gravity into the spectrum. Quantum gravity, instead of being an unpleasant, undesirable feature of quantum field theory, is necessarily an integral part of any theory with local supersymmetry.

Historically, it was once thought that all successful quantum field theories required renormalization. However, today supersymmetry gives us theories, like the $SO(4)$ super Yang–Mills theory and the superstring theory, which are *finite to all orders in perturbation theory*, a truly remarkable achievement. For the first time since its inception, it is now possible to write down quantum field theories that require no renormalization whatsoever. This answers, in some sense, Dirac's original criticism of his own creation.

Only time will tell if GUTs, supersymmetry, and the superstring theory can give us a faithful description of the universe. In this book, our attitude is that they are an exciting theoretical laboratory in which to probe the limits of quantum field theory. And the fact that supersymmetric theories can improve and in some cases solve the problem of ultraviolet divergences without renormalization is, by itself, a feature of quantum field theory worthy of study.

Let us, therefore, now leave the historical setting of quantum field theory and begin a discussion of how quantum field theory gives us a quantum description of point particle systems with an infinite number of degrees of freedom. Although the student may already be familiar with the foundations of classical mechanics and the transition to the quantum theory, it will prove beneficial to review this material from a slightly different point of view, that is, systems with an infinite number of degrees of freedom. This will then set the stage for quantum field theory.

1.7   Action Principle

Before we begin our discussion of field theory, for notational purposes it is customary to choose units so that:

$$c = 1$$
$$\hbar = 1 \tag{1.14}$$

(We can always do this because the definition of c and $\hbar = h/2\pi$ depends on certain conventions that grew historically in our understanding of nature. Imposing $c = 1$, for example, means that seconds and centimeters are to be treated on the same footing, such that exactly 299,792,458 meters is equivalent to 1 sec. Thus, the second and the centimeter are to be treated as if they were expressions of the same unit. Likewise, setting $\hbar = 1$ means that the erg × sec. is now dimensionless, so the erg and the second are inverses of each other. This also means that the gram is inversely related to the centimeter. With these conventions, the only unit that survives is the centimeter, or equivalently, the gram.)

In classical mechanics, there are two equivalent formulations, one based on Newton's equations of motion, and the other based on the action principle. At first, these two formalisms seem to have little in common. The first depends on iterating infinitesimal changes sequentially along a particle's path. The second depends on evaluating all possible paths between two points and selecting out the one with the minimum action. One of the great achievements of classical mechanics was the demonstration that the Newtonian equations of motion were equivalent to minimizing the action over the set of all paths:

$$\text{Equation of motion} \leftrightarrow \text{Action principle} \qquad (1.15)$$

However, when we generalize our results to the quantum realm, this equivalence breaks down. The Heisenberg Uncertainty Principle forces us to introduce probabilities and consider all possible paths taken by the particle, with the classical path simply being the most likely. Quantum mechanically, we know that there is a finite probability for a particle to deviate from its classical equation of motion. This deviation from the classical path is very small, on the scale determined by Planck's constant. However, on the subatomic scale, this deviation becomes the dominant aspect of a particle's motion. In the microcosm, motions that are in fact forbidden classically determine the primary characteristics of the atom. The stability of the atom, the emission and absorption spectrum, radioactive decay, tunneling, etc. are all manifestations of quantum behavior that deviate from Newton's classical equations of motion.

However, even though Newtonian mechanics fails within the subatomic realm, it is possible to generalize the action principle to incorporate these quantum probabilities. The action principle then becomes the only framework to calculate the probability that a particle will deviate from its classical path. In other words, the action principle is elevated into one of the foundations of the new mechanics.

To see how this takes place, let us first begin by describing the simplest of all possible classical systems, the nonrelativistic point particle. In three dimensions, we say that this particle has three degrees of freedom, each labeled by coordinates $q^i(t)$. The motion of the particle is determined by the Lagrangian $L(q^i, \dot{q}^i)$, which is a function of both the position and the velocity of the particle:

$$L = \frac{1}{2}m(\dot{q}^i)^2 - V(q) \qquad (1.16)$$

where $V(q)$ is some potential in which the particle moves. Classically, the motion of the particle is determined by minimizing the action, which is the integral of the

Lagrangian:

$$S = \int_{t_1}^{t_2} L(q^i, \dot{q}^i) \, dt \tag{1.17}$$

We can derive the classical equations of motion by minimizing the action; that is, the classical Newtonian path taken by the particle is the one in which the action is a minimum:

$$\delta S = 0 \tag{1.18}$$

To calculate the classical equations of motion, we make a small variation in the path of the particle $\delta q^i(t)$, keeping the endpoints fixed; that is, $\delta q^i(t_1) = \delta q^i(t_2) = 0$. To calculate δS, we must vary the Lagrangian with respect to both changes in the position and the velocity:

$$\delta S = \int_{t_1}^{t_2} dt \left(\frac{\delta L}{\delta q^i} \delta q^i + \frac{\delta L}{\delta \dot{q}^i} \delta \dot{q}^i \right) = 0 \tag{1.19}$$

We can integrate the last expression by parts:

$$\delta S = \int dt \left\{ \delta q^i \left[\frac{\delta L}{\delta q^i} - \frac{d}{dt} \frac{\delta L}{\delta \dot{q}^i} \right] + \frac{d}{dt} \left(\delta q^i \frac{\delta L}{\delta \dot{q}^i} \right) \right\} = 0 \tag{1.20}$$

Since the variation vanishes at the endpoints of the integration, we can drop the last term. Then the action is minimized if we demand that:

$$\frac{\delta L}{\delta q^i} = \frac{d}{dt} \frac{\delta L}{\delta \dot{q}^i} \tag{1.21}$$

which are called the Euler–Lagrange equations.

If we now insert the value of the Lagrangian into the Euler–Lagrange equations, we find the classical equation of motion:

$$m \frac{d^2 q^i}{dt^2} = -\frac{\partial V(q)}{\partial q^i} \tag{1.22}$$

which forms the basis of Newtonian mechanics.

Classically, we also know that there are two different formulations of Newtonian mechanics, the Lagrangian formulation, where the position q^i and velocity $\dot{q}^i$ of a point particle are the fundamental variables, and the Hamiltonian formulation, where we choose the position q^i and the momentum p^i to be the independent variables.

To make the transition from the Lagrangian to the Hamiltonian formulation of classical mechanics, we first define the momentum as follows:

$$p^i = \frac{\delta L}{\delta \dot{q}^i} \tag{1.23}$$

For our choice of the Lagrangian, we find that $p^i = m\dot{q}^i$. The definition of the Hamiltonian is then given by:

$$H(q^i, p^i) = p^i \dot{q}^i - L(q^i, \dot{q}^i) \tag{1.24}$$

With our choice of Lagrangian, we find that the Hamiltonian is given by:

$$H(q^i, p^i) = \frac{p^{i2}}{2m} + V(q) \tag{1.25}$$

In the transition from the Lagrangian to the Hamiltonian system, we have exchanged independent variables from $q^i, \dot{q}^i$ to q^i, p^i. To prove that $H(q^i, p^i)$ is not a function of $\dot{q}^i$, we can make the following variation:

$$
\begin{aligned}
\delta H &= p^i \delta \dot{q}^i + \delta p^i \dot{q}^i - \frac{\delta L}{\delta q^i} \delta q^i - \frac{\delta L}{\delta \dot{q}^i} \delta \dot{q}^i \\
&= \dot{q}^i \delta p^i - \frac{\delta L}{\delta q^i} \delta q^i \\
&= \delta p^i \frac{\delta H}{\delta p^i} + \delta q^i \frac{\delta H}{\delta q^i}
\end{aligned}
\tag{1.26}
$$

where we have used the definition of p^i to eliminate the dependence of the Hamiltonian on $\dot{q}^i$ and have used the chain rule for the last step. By equating the coefficients of the variations, we can make the following identification:

$$\dot{q}^i = \frac{\delta H}{\delta p^i}; \quad -\dot{p}^i = \frac{\delta H}{\delta q^i} \tag{1.27}$$

where we have used the equations of motion.

In the Hamiltonian formalism, we can also calculate the time variation of any field F in terms of the Hamiltonian:

$$
\begin{aligned}
\dot{F} &= \frac{\partial F}{\partial t} + \frac{\delta F}{\delta q^i} \dot{q}^i + \frac{\delta F}{\delta p^i} \dot{p}^i \\
&= \frac{\partial F}{\partial t} + \frac{\delta F}{\delta q^i} \frac{\delta H}{\delta p^i} - \frac{\delta F}{\delta p^i} \frac{\delta H}{\delta q^i}
\end{aligned}
\tag{1.28}
$$

If we define the Poisson bracket as:

$$\{A, B\}_{PB} = \sum_i \left(\frac{\partial A}{\partial p^i} \frac{\partial B}{\partial q^i} - \frac{\partial A}{\partial q^i} \frac{\partial B}{\partial p^i} \right) \tag{1.29}$$

then we can write the time variation of the field F as:

$$\dot{F} = \frac{\partial F}{\partial t} + \{H, F\}_{PB} \tag{1.30}$$

At this point, we have derived the Newtonian equations of motion by minimizing the action, reproducing the classical result. This is not new. However, we will now make the transition to the quantum theory, which treats the action as the fundamental object, incorporating both allowed and forbidden paths. The Newtonian equations of motion then specify the most likely path, but certainly not the only path. In the subatomic realm, in fact, the classically forbidden paths may dominate over the classical one.

There are many ways in which to make the transition from classical mechanics to quantum mechanics. (Perhaps the most profound and powerful is the *path integral* method, which we present in Chapter 8.) Historically, however, it was Dirac who noticed that the transition from classical to quantum mechanics can be achieved by replacing the classical Poisson brackets with commutators:

$$\{A, B\}_{PB} \rightarrow \frac{i}{\hbar}[A, B] \tag{1.31}$$

With this replacement, the Poisson brackets between canonical coordinates are replaced by:

$$[p^i, q^k] = -i\hbar \delta^{ik} \tag{1.32}$$

Quantum mechanics makes the replacement:

$$p^i = -i\hbar \frac{\partial}{\partial q^i}; \quad E = i\hbar \frac{\partial}{\partial t} \tag{1.33}$$

Because p^i is now an operator, the Hamiltonian also becomes an operator, and we can now satisfy Hamilton's equations by demanding that they vanish when acting on some function $\psi(q^i, t)$:

$$\left(-\frac{\hbar^2}{2m} \frac{\partial^2}{\partial q^{i2}} + V(q) \right) \psi = i\hbar \frac{\partial \psi}{\partial t} \tag{1.34}$$

This is the Schrödinger wave equation, which is the starting point for calculating the spectral lines of the hydrogen atom.

1.8 From First to Second Quantization

This process, treating the coordinates q^i and p^i as quantized variables, is called *first quantization*. However, the object of this book is to make the transition to *second quantization*, where we quantize *fields* which have an infinite number of degrees of freedom. The transition from quantum mechanics to quantum field theory arises when we make the transition from three degrees of freedom, described by x^i, to an infinite number of degrees of freedom, described by a field $\phi(x^i)$ and then apply quantization relations directly onto the fields.

Before we made the transition to quantum field theory in Chapter 3, let us discuss how we describe classical field theory, that is, classical systems with an infinite number of degrees of freedom. Let us consider a classical system, a series of masses arranged in a line, each connected to the next via springs (Fig. 1.2).

The Lagrangian of the system then contains two terms, the potential energy of each spring as well as the kinetic energy of the masses. The Lagrangian is therefore given by:

$$L = \sum_{r=1}^{N} \left[\frac{1}{2} m_r \, (\dot{x}_r)^2 - \frac{1}{2} k \, (x_r - x_{r+1})^2 \right] \tag{1.35}$$

Now let us assume that we have an infinite number of masses, each separated by a distance ϵ. In the limit $\epsilon \to 0$, the Lagrangian becomes:

$$
\begin{aligned}
L &= \sum_{r=1}^{N} \frac{\epsilon}{2} \left(\frac{m}{\epsilon} \dot{x}_r^2 - k \frac{(x_r - x_{r+1})^2}{\epsilon} \right) \\
&\to \int dx \, \frac{1}{2} \left[\mu \, \dot{\phi}(x)^2 - Y \left(\frac{\partial \phi(x)}{\partial x} \right)^2 \right]
\end{aligned} \tag{1.36}
$$

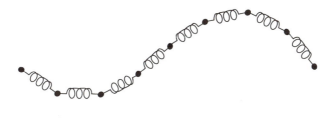

Figure 1.2. The action describing a finite chain of springs, in the limit of an infinite number of springs, becomes a theory with an infinite number of degrees of freedom.

where we have taken the limit via:

$$
\begin{aligned}
\epsilon &\rightarrow dx \\
\frac{m}{\epsilon} &\rightarrow \mu \\
k\epsilon &\rightarrow Y \\
x_r &\rightarrow \phi(x, t)
\end{aligned}
\tag{1.37}
$$

where $\phi(x, t)$ is the displacement of the particle located at position x and time t, where μ is the mass density along the springs, and Y is the Young's modulus.

If we use the Euler–Lagrange equations to find the equations of motion, we arrive at:

$$
\frac{\partial^2 \phi}{\partial x^2} - \frac{\mu}{Y} \frac{\partial^2 \phi}{\partial t^2} = 0
\tag{1.38}
$$

which is just the familiar wave equation for a one-dimensional system traveling with velocity $\sqrt{Y/\mu}$.

On one hand, we have proved the rather intuitively obvious statement that waves can propagate down a long, massive spring. On the other hand, we have made the highly nontrivial transition from a system with a finite number of degrees of freedom to one with an infinite number of degrees of freedom.

Now let us generalize our previous discussion of the Euler–Lagrange equations to include classical field theories with an infinite number of degrees of freedom. We begin with a Lagrangian that is a function of both the field $\phi(x)$ as well as its space–time derivatives $\partial_\mu \phi(x)$:

$$
L\left(\phi(x), \partial_\mu \phi(x)\right)
\tag{1.39}
$$

where:

$$
\partial_\mu = \left(\frac{\partial}{\partial t}, \frac{\partial}{\partial x^i}\right)
\tag{1.40}
$$

The action is given by a four dimensional integral over a Lagrangian density $\mathscr{L}$:

$$
\begin{aligned}
L &= \int d^3x\, \mathscr{L}(\phi, \partial_\mu \varphi) \\
S &= \int d^4x\, \mathscr{L} = \int dt\, L
\end{aligned}
\tag{1.41}
$$

integrated between initial and final times.

As before, we can retrieve the classical equations of motion by minimizing the action:

$$\delta S = 0 = \int d^4 x \left(\frac{\delta \mathscr{L}}{\delta \phi} \delta \phi + \frac{\delta \mathscr{L}}{\delta \partial_\mu \phi} \delta \partial_\mu \phi \right) \tag{1.42}$$

We can integrate by parts, reversing the direction of the derivative:

$$\delta S = \int d^4 x \left[\left(\frac{\delta \mathscr{L}}{\delta \phi} - \partial_\mu \frac{\delta \mathscr{L}}{\delta \partial_\mu \phi} \right) \delta \phi + \partial_\mu \left(\frac{\delta \mathscr{L}}{\delta \partial_\mu \phi} \delta \phi \right) \right] \tag{1.43}$$

The last term vanishes at the endpoint of the integration; so we arrive at the Euler–Lagrange equations of motion:

$$\partial_\mu \frac{\delta \mathscr{L}}{\delta \partial_\mu \phi} = \frac{\delta \mathscr{L}}{\delta \phi} \tag{1.44}$$

The simplest example is given by the scalar field $\phi(x)$ of a point particle:

$$\mathscr{L} = \frac{1}{2} \left(\partial_\mu \phi \partial^\mu \phi - m^2 \phi^2 \right) \tag{1.45}$$

Inserting this into the equation of motion, we find the standard Klein–Gordon equation:

$$\partial_\mu \partial^\mu \phi + m^2 \phi = 0 \tag{1.46}$$

where:

$$\partial^\mu = \left(\frac{\partial}{\partial t}, -\frac{\partial}{\partial x^i} \right) \tag{1.47}$$

One of the purposes of this book is to generalize the Klein–Gordon equation by introducing higher spins and higher interactions. To do this, however, we must first begin with a discussion of special relativity. It will turn out that invariance under the Lorentz and Poincaré group will provide the most important guide in constructing higher and more sophisticated versions of quantum field theory.

1.9   Noether's Theorem

One of the achievements of this formalism is that we can use the symmetries of the action to derive conservation principles. For example, in classical mechanics, if the Hamiltonian is time independent, then energy is conserved. Likewise, if the Hamiltonian is translation invariant in three dimensions, then momentum is conserved:

$$
\begin{aligned}
\text{Time independence} &\;\rightarrow\; \text{Energy conservation} \\
\text{Translation independence} &\;\rightarrow\; \text{Momentum conservation} \\
\text{Rotational independence} &\;\rightarrow\; \text{Angular momentum conservation} \quad (1.48)
\end{aligned}
$$

The precise mathematical formulation of this correspondence is given by Noether's theorem. In general, an action may be invariant under either an internal, isospin symmetry transformation of the fields, or under some space–time symmetry. We will first discuss the isospin symmetry, where the fields ϕ^α vary according to some small parameter $\delta\epsilon^\alpha$.

The action varies as:

$$
\begin{aligned}
\delta S &= \int d^4x \left(\frac{\delta \mathcal{L}}{\delta \phi^\alpha} \delta\phi^\alpha + \frac{\delta \mathcal{L}}{\delta \partial_\mu \phi^\alpha} \delta \partial_\mu \phi^\alpha \right) \\
&= \int d^4x \left(\partial_\mu \frac{\delta \mathcal{L}}{\delta \partial_\mu \phi^\alpha} \delta\phi^\alpha + \frac{\delta \mathcal{L}}{\delta \partial_\mu \phi^\alpha} \partial_\mu \delta\phi^\alpha \right) \\
&= \int d^4x\, \partial_\mu \left(\frac{\delta \mathcal{L}}{\delta \partial_\mu \phi^\alpha} \delta\phi^\alpha \right) \qquad (1.49)
\end{aligned}
$$

where we have used the equations of motion and have converted the variation of the action into the integral of a total derivative. This defines the *current* $J^{\mu\alpha}$:

$$
J^\mu_\alpha = \frac{\delta \mathcal{L}}{\delta \partial_\mu \phi^\beta} \frac{\delta\phi^\beta}{\delta\epsilon^\alpha} \qquad (1.50)
$$

If the action is invariant under this transformation, then we have established that the current is conserved:

$$
\partial_\mu J^\mu_\alpha = 0 \qquad (1.51)
$$

From this conserved current, we can also establish a conserved charge, given by the integral over the fourth component of the current:

$$
Q_\alpha \equiv \int d^3x\, J^0_\alpha \qquad (1.52)
$$

Now let us integrate the conservation equation:

$$0 = \int d^3x \, \partial_\mu J^\mu_\alpha = \int d^3x \, \partial_0 J^0_\alpha + \int d^3x \, \partial_i J^i_\alpha$$

$$= \frac{d}{dt} \int d^3x \, J^0_\alpha + \int dS_i \, J^i_\alpha = \frac{d}{dt} Q_\alpha + \text{surface term} \qquad (1.53)$$

Let us assume that the fields appearing in the surface term vanish sufficiently rapidly at infinity so that the last term can be neglected. Then:

$$\partial_\mu J^\mu_\alpha = 0 \rightarrow \frac{dQ_\alpha(t)}{dt} = 0 \qquad (1.54)$$

In summary, the symmetry of the action implies the conservation of a current J^μ_α, which in turn implies a conservation principle:

$$\text{Symmetry} \rightarrow \text{Current conservation} \rightarrow \text{Conservation principle} \qquad (1.55)$$

Now let us investigate the second case, when the action is invariant under the space–time symmetry of the Lorentz and Poincaré groups. Lorentz symmetry implies that we can combine familiar three-vectors, such as momentum and space, into four-vectors. We introduce the space–time coordinate x^μ as follows:

$$x^\mu \equiv (x^0, x^i) = (t, \mathbf{x}) \qquad (1.56)$$

where the time coordinate is defined as $x^0 = t$.

Similarly, the momentum three-vector $\mathbf{p}$ can be combined with energy to form the four-vector:

$$p^\mu = (p^0, p^i) = (E, \mathbf{p}) \qquad (1.57)$$

We will henceforth use Greek symbols from the middle of the alphabet μ, ν to represent four-vectors, and Roman indices from the middle of the alphabet i, j, k to represent space coordinates.

We will raise and lower indices by using the following *metric* $g_{\mu\nu}$ as follows:

$$A_\mu = g_{\mu\nu} A^\nu \qquad (1.58)$$

where:

$$g_{\mu\nu} = \begin{pmatrix} 1 & 0 & 0 & 0 \\ 0 & -1 & 0 & 0 \\ 0 & 0 & -1 & 0 \\ 0 & 0 & 0 & -1 \end{pmatrix} \qquad (1.59)$$

Now let us use this formalism to construct the current associated with making a translation:

$$x^\mu \rightarrow x^\mu + a^\mu \tag{1.60}$$

where a^μ is a constant. a^0 represents time displacements, and a^i represents space displacements. We will now rederive the result from classical mechanics that displacement in time (space) leads to the conservation of energy (momentum).

Under this displacement, a field $\phi(x)$ transforms as $\phi(x) \rightarrow \phi(x+a)$. For small a^μ, we can power expand the field in a power series in a^μ. Then the change in the field after a displacement is given by:

$$\delta\phi = \phi(x+a) - \phi(x) \sim \phi(x) + a^\mu \partial_\mu \phi(x) - \phi(x) = a^\mu \partial_\mu \phi(x) \tag{1.61}$$

Therefore, if we make a translation $\delta x^\mu = a^\mu$, then the fields transform, after making a Taylor expansion, as follows:

$$\begin{aligned}
\delta\phi &= a^\mu \partial_\mu \phi \\
\delta\partial_\mu \phi &= a^\nu \partial_\nu \partial_\mu \phi
\end{aligned} \tag{1.62}$$

The calculation of the current associated with translations proceeds a bit differently from the isospin case. There is an additional term that one must take into account, which is the variation of the Lagrangian itself under the space–time symmetry. The variation of our Lagrangian is given by:

$$\delta\mathscr{L} = a^\mu \partial_\mu \mathscr{L} = \frac{\delta\mathscr{L}}{\delta\phi}\delta\phi + \frac{\delta\mathscr{L}}{\delta\partial_\mu\phi}\delta\partial_\mu\phi \tag{1.63}$$

Substituting in the variation of the fields and using the equations of motion, we find:

$$\begin{aligned}
\delta\mathscr{L} &= a^\mu \partial_\mu \mathscr{L} \\
&= \frac{\delta\mathscr{L}}{\delta\phi}a^\nu \partial_\nu \phi + \frac{\delta\mathscr{L}}{\delta\partial_\mu\phi}a^\nu \partial_\nu \partial_\mu \phi \\
&= a^\nu \left(\partial_\mu \frac{\delta\mathscr{L}}{\delta\partial_\mu\phi}\partial_\nu \phi + \frac{\delta\mathscr{L}}{\delta\partial_\mu\phi}\partial_\mu \partial_\nu \phi \right) \\
&= a^\nu \partial_\mu \left(\frac{\delta\mathscr{L}}{\delta\partial_\mu\phi}\partial_\nu \phi \right)
\end{aligned} \tag{1.64}$$

Combining both terms under one derivative, we find:

$$\partial_\mu \left[\mathscr{L} \delta^\mu_\nu - \frac{\delta \mathscr{L}}{\delta \partial_\mu \phi} \partial_\nu \phi \right] a^\nu = 0 \qquad (1.65)$$

This then defines the *energy–momentum tensor* $T^\mu{}_\nu$:

$$T^\mu{}_\nu = \frac{\delta \mathscr{L}}{\delta \partial_\mu \phi} \partial_\nu \phi - \delta^\mu_\nu \mathscr{L} \qquad (1.66)$$

which is conserved:

$$\partial_\mu T^\mu{}_\nu = 0 \qquad (1.67)$$

If we substitute the Klein–Gordon action into this expression, we find the energy–momentum tensor for the scalar particle:

$$T_{\mu\nu} = \partial_\mu \phi \partial_\nu \phi - g_{\mu\nu} \mathscr{L} \qquad (1.68)$$

Using the equations of motion, we can explicitly check that this energy–momentum tensor is conserved.

By integrating the energy–momentum tensor, we can generate conserved currents, as we saw earlier. As the name implies, the conserved charges corresponding to the energy–momentum tensor give us energy and momentum conservation. Let us define:

$$P^\mu = (E, P^i) \qquad (1.69)$$

which combines the energy $P^0 = E$ and the momentum P^i into a single four-vector. We can show that energy and momentum are both conserved by making the following definition:

$$P^\mu \equiv \int d^3x (T^\mu_0)$$

$$\frac{d}{dt} P^\mu = 0 \qquad (1.70)$$

The conservation of energy–momentum is therefore a consequence of the invariance of the action under translations, which in turn corresponds to invariance under time and space displacements. Thus, we now have derived the result from classical Newtonian mechanics mentioned earlier in Eq. (1.48).

Next, let us generalize this discussion. We know from classical mechanics that invariance of the action under rotations generates the conservation of angular momentum. Now, we would like to derive the Lorentz generalization of this result from classical mechanics.

Rotations in three dimensions are described by:

$$\delta x^i = a^{ij} x^i \tag{1.71}$$

where a^{ij} is an anti-symmetric matrix describing the rotation (i.e., $a^{ij} = -a^{ji}$).

Let us now construct the generalization of this rotation, the current associated with Lorentz transformations. We define how a four–vector x^μ changes under a Lorentz transformation:

$$\text{Lorentz transformation}: \quad \begin{cases} \delta x^\mu = \epsilon^\mu{}_\nu x^\mu \\ \delta\phi(x) = \epsilon^\mu{}_\nu x^\nu \partial_\mu \phi(x) \end{cases} \tag{1.72}$$

where $\epsilon^\mu{}_\nu$ is an infinitesimal, antisymmetric constant matrix (i.e., $\epsilon^{\mu\nu} = -\epsilon^{\nu\mu}$).

Repeating the same steps with this new variation, we have:

$$
\begin{aligned}
\delta\mathscr{L} &= \epsilon^\mu{}_\nu x^\nu \partial_\mu \mathscr{L} \\
&= \frac{\delta\mathscr{L}}{\delta\phi}\delta\phi + \frac{\delta\mathscr{L}}{\delta\partial_\rho\phi}\delta\partial_\rho\phi \\
&= \partial_\rho\left(\frac{\delta\mathscr{L}}{\delta\partial_\rho\phi}\delta\phi\right) \\
&= \partial_\rho\left(\frac{\delta\mathscr{L}}{\delta\partial_\rho\phi}\epsilon^\mu{}_\nu x^\nu \partial_\mu\phi\right)
\end{aligned}
\tag{1.73}
$$

If we extract the coefficient of $\epsilon^\mu{}_\nu$ and put everything within the partial derivative ∂_ρ, we find:

$$\epsilon_{\mu\nu}\partial_\rho\left(\frac{\delta\mathscr{L}}{\delta\partial_\rho\phi}(\partial^\mu\phi x^\nu - \partial^\nu\phi x^\mu) - \delta^{\rho\mu}x^\nu\mathscr{L} + \delta^{\rho\nu}x^\mu\mathscr{L}\right) = 0 \tag{1.74}$$

This gives us the conserved current:

$$
\begin{aligned}
\mathscr{M}^{\rho,\mu\nu} &= T^{\rho\nu}x^\mu - T^{\rho\mu}x^\nu \\
\partial_\rho\mathscr{M}^{\rho,\mu\nu} &= 0
\end{aligned}
\tag{1.75}
$$

and the conserved charge:

$$
\begin{aligned}
M^{\mu\nu} &= \int d^3x\, \mathscr{M}^{0,\mu\nu} \\
\frac{d}{dt}M^{\mu\nu} &= 0
\end{aligned}
\tag{1.76}
$$

If we restrict our discussion to rotations in three dimensional space, then $\dot{M}^{ij} = 0$ corresponds to the conservation of angular momentum. If we take all the components of this matrix, however, we find that Lorentz transformations are an invariant of the action.

There is, however, a certain ambiguity in the definition of the energy–momentum tensor. The energy–momentum tensor is not a measurable quantity, but the integrated charges correspond to the physical energy and momentum, and hence are measurable.

We can exploit this ambiguity and add to the energy–momentum tensor a term:

$$\partial_\lambda E^{\lambda\mu\nu} \tag{1.77}$$

where $E^{\lambda\mu\nu}$ is antisymmetric in the first two indices:

$$E^{\lambda\mu\nu} = -E^{\mu\lambda\nu} \tag{1.78}$$

Because of this antisymmetry, this tensor satisfies trivially:

$$\partial_\lambda\partial_\mu E^{\lambda\mu\nu} = 0 \tag{1.79}$$

So we can make the replacement:

$$T^{\mu\nu} \rightarrow T^{\mu\nu} + \partial_\lambda E^{\lambda\mu\nu} \tag{1.80}$$

This new energy–momentum tensor is conserved, like the previous one. We can choose this tensor such that the new energy–momentum tensor is symmetric in μ and ν.

The addition of this extra tensor to the energy–momentum tensor does not affect the energy and the momentum, which are measurable quantities. If we take the integrated charge, we find that the contribution from $E^{\lambda\mu\nu}$ vanishes:

$$
\begin{aligned}
P^\mu &\rightarrow P^\mu + \int d^3x\, \partial_\lambda E^{\lambda 0\mu} \\
&= P^\mu + \int \partial_i E^{i0\mu} d^3x \\
&= P^\mu + \int_S E^{i0\mu} dS_i \\
&= P^\mu
\end{aligned}
\tag{1.81}
$$

Thus, the physical energy and momentum are not affected as long as this tensor vanishes sufficiently rapidly at infinity.

The purpose of adding this new term to the energy–momentum tensor is that the original one was not necessarily symmetric. This is a problem, since the conservation of angular momentum requires a symmetric energy–momentum tensor. For example, if we take the divergence of $\mathcal{M}^{\rho\mu\nu}$, we find that it does not vanish in general, but equals $T^{\mu\nu} - T^{\nu\mu}$. However, we can always choose $E^{\lambda\mu\nu}$ such that the energy–momentum tensor is symmetric, so angular momentum is conserved.

Yet another reason for requiring a symmetric energy–momentum tensor is that in general relativity, the gravitational field tensor, which is symmetric, couples to the energy–momentum tensor. By the equivalence principle, the gravitational field couples equally to all forms of matter via its energy–momentum content. Hence, when we discuss general relativity in Chapter 19, we will need a symmetric energy–momentum tensor.

In summary, in this chapter we have made the transition from a classical system with a finite number of degrees of freedom to a classical field theory with an infinite number of degrees of freedom. Instead of a one-particle, classical description of a point particle in terms of coordinates, we now have a classical formalism defined in terms of fields $\phi(x)$.

In constructing field theories, we clearly see that the study of symmetries plays a crucial role. In fact, over the past two decades, physicists have come to the realization that symmetries are perhaps the most powerful tool that we have in describing the physical universe. As a consequence, it will prove beneficial to begin a more systematic discussion of symmetries and group theory. With this foundation, many of the rather strange and seemingly arbitrary conventions of quantum field theory begin to take on an elegant and powerful form.

Therefore, in the next chapter we will discuss various symmetries that have been shown experimentally to describe the fundamental particles of nature. Then in Chapter 3 we will begin a formal introduction to the quantum theory of systems with an infinite number of degrees of freedom.

1.10   Exercises

1. Show that the Poisson brackets obey the Jacobi identity:

$$\{A, \{B, C\}\} + \{B, \{C, A\}\} + \{C, \{A, B\}\} = 0 \qquad (1.82)$$

2. A transformation from the coordinates p and q to the new set $P = P(q, p, t)$ and $Q = Q(q, p, t)$ is called *canonical* if Hamilton's equations in Eq. (1.27) are satisfied with the new variables when we introduce a new Hamiltonian $\tilde{H}(Q, P, t)$. Show that the Poisson brackets between P and Q are the same

as those between p and q. Thus, the Poisson brackets of the coordinates are preserved under a canonical transformation.

3. Show that Poisson's brackets of two arbitrary functions A and B are invariant under a canonical transformation.

4. Since the action principle must be satisfied in the new coordinates, then $p\dot{q} - H$ must be equal to $P\dot{Q} - \tilde{H}$ up to a total derivative, that is, up to some arbitrary function $\dot{F}$. Show that, without losing any generality, we can take F to be one of four functions, given by: $F_1(q, Q, t)$, $F_2(q, P, t)$, $F_3(p, Q, t)$ or $F_4(p, P, t)$.

5. If we choose $F = F_1(q, Q, t)$, then prove that $p = \partial F_1/\partial q$, $P = -\partial F/\partial Q$, and $\tilde{H} = H + \dot{F}_1$.

6. What are the analogous relations for the other three F_i functions?

Chapter 2
Symmetries and Group Theory

> *...although the symmetries are hidden from us, we can sense that they are latent in nature, governing everything about us. That's the most exciting idea I know: that nature is much simpler than it looks. Nothing makes me more hopeful that our generation of human beings may actually hold the key to the universe in our hands—that perhaps in our lifetimes we may be able to tell why all of what we see in this immense universe of galaxies and particles is logically inevitable.*
> —S. Weinberg

2.1 Elements of Group Theory

So far, we have only described the broad, general principles behind classical field theory. In this chapter, we will study the physics behind specific models. We must therefore impose extra constraints on our Lagrangian that come from group theory. These groups, in turn, are extremely important because they describe the symmetries of the subatomic particles found experimentally in nature. Then in the next chapter, we will make the transition from classical field theory to the quantum theory of fields.

The importance of symmetries is seen when we write down the theory of radiation. When we analyze Maxwell fields, we find that they are necessarily relativistic. Therefore, we can also say that that quantum field theory arises out of the marriage of group theory (in particular the Lorentz and Poincaré groups) and quantum mechanics. Roughly speaking, we have:

$$\text{Quantum field theory} = \begin{cases} \text{Group theory} \\ \text{Quantum mechanics} \end{cases} \tag{2.1}$$

In fact, once the group structure of a theory (including the specific representations) are fixed, we find that the S matrix is essentially unique, up to certain parameters specifying the interactions. More precisely, we will impose the following constraints in constructing a quantum field theory:

1. *We demand that all fields transform as irreducible representations of the Lorentz and Poincaré groups and some isospin group.*

2. *We demand that the theory be unitary and the action be causal, renormalizable, and an invariant under these groups.*

As simple as these postulates are, they impose enormous constraints on the theory. The first assumption will restrict the fields to be massive or massless fields of spin 0, 1/2, 1, etc. and will fix their isotopic representations. However, this constraint alone cannot determine the action, since there are invariant theories that are noncausal or nonunitary. (For example, there are theories with three or higher derivatives that satisfy the first condition. However, higher derivative theories have "ghosts," or particles of negative norm that violate unitarity, and hence must be ruled out.) The second condition, that the action obeys certain physical properties, then fixes the action up to certain parameters and representations, such as the various coupling constants found in the interaction.

Because of the power of group theory, we have chosen to begin our discussion of field theory with a short introduction to representation theory. We will find this detour to be immensely important; many of the curious "accidents" and conventions that occur in field theory, which often seem contrived and artificial, are actually byproducts of group theory.

There are three types of symmetries that will appear in this book.

1. *Space–time symmetries* include the Lorentz and Poincaré groups. These symmetries are noncompact, that is, the range of their parameters does not contain the endpoints. For example, the velocity of a massive particle can range from 0 to c, but cannot reach c.

2. *Internal symmetries* are ones that mix particles among each other, for example, symmetries like $SU(N)$ that mix N quarks among themselves. These internal symmetries rotate fields and particles in an abstract, "isotopic space," in contrast to real space–time. These groups are compact, that is, the range of their parameters is finite and contains their endpoints. For example, the rotation group is parametrized by angles that range between 0 and π or 2π. These internal symmetries can be either *global* (i.e., independent of space–time) or *local*, as in gauge theory, where the internal symmetry group varies at each point in space and time.

3. *Supersymmetry* nontrivially combines both space–time and internal symmetries. Historically, it was thought that space–time and isotopic symmetries

were distinct and could never be unified. "No-go theorems," in fact, were given to prove the incompatibility of compact and noncompact groups. Attempts to write down a nontrivial union of these groups with finite-dimensional unitary representations inevitably met with failure. Only recently has it become possible to unify them nontrivially and incorporate them into quantum field theories with supersymmetry, which manifest remarkable properties that were previously thought impossible. For example, certain supersymmetric theories are finite to all orders in perturbation theory, without the need for any renormalization. (However, since elementary particles with supersymmetry have yet to be discovered, we will only discuss this third class of symmetry later in the book in Chapters 20 and 21.)

2.2 $SO(2)$

We say that a collection of elements g_i form a *group* if they either obey or possess the following:

1. *Closure* under a multiplication operation; that is, if g_i and g_j are members of the group, then $g_i \cdot g_j$ is also a member of the group.

2. *Associativity* under multiplication; that is,

$$g_i \cdot (g_j \cdot g_k) = (g_i \cdot g_j) \cdot g_k \qquad (2.2)$$

3. *An identity element* **1**; that is, there exists an element **1** such that $g_i \cdot \mathbf{1} = \mathbf{1} \cdot g_i = g_i$.

4. *An inverse*; that is, every element g_i has an element g_i^{-1} such that $g_i \cdot g_i^{-1} = \mathbf{1}$.

There are many kinds of groups. A discrete group has a finite number of elements, such as the group of rotations that leave a crystal invariant. An important class of discrete groups are the parity inversion P, charge conjugation C, and time-reversal symmetries T. At this point, however, we are more interested in the *continuous* groups, such as the rotation and Lorentz group, which depend on a set of continuous angles.

To illustrate some of these abstract concepts, it will prove useful to take the simplest possible nontrivial example, $O(2)$, or rotations in two dimensions. Even the simplest example is surprisingly rich in content. Our goal is to construct the irreducible representations of $O(2)$, but first we have to make a few definitions. We know that if we rotate a sheet of paper, the length of any straight line on the paper is constant. If (x, y) describe the coordinates of a point on a plane, then this means that, by the Pythagorean theorem, the following length is an invariant under a rotation about the origin:

$$\text{Invariant}: \quad x^2 + y^2 \tag{2.3}$$

If we rotate the plane through angle θ, then the coordinates (x', y') of the same point in the new system are given by:

$$\begin{pmatrix} x' \\ y' \end{pmatrix} = \begin{pmatrix} \cos\theta & \sin\theta \\ -\sin\theta & \cos\theta \end{pmatrix} \begin{pmatrix} x \\ y \end{pmatrix} \tag{2.4}$$

We will abbreviate this by:

$$x^{i'} = O^{ij}(\theta)x^j \tag{2.5}$$

where $x^1 = x$ and $x^2 = y$. (For the rotation group, it makes no difference whether we place the index as a superscript, as in x^i, or as a subscript, as in x_i.)

For small angles, this can be reduced to:

$$\delta x = \theta y; \quad \delta y = -\theta x \tag{2.6}$$

or simply:

$$\delta x^i = \theta \epsilon^{ij} x^j \tag{2.7}$$

where ϵ^{ij} is antisymmetric and $\epsilon^{12} = -\epsilon^{21} = 1$. These matrices form a group; for example, we can write down the inverse of any rotation, given by $O^{-1}(\theta) = O(-\theta)$:

$$O(\theta)O(-\theta) = 1 = \begin{pmatrix} 1 & 0 \\ 0 & 1 \end{pmatrix} \tag{2.8}$$

We can also prove associativity, since matrix multiplication is associative.

The fact that these matrices preserve the invariant length places restrictions on them. To find the nature of these restrictions, let us make a rotation on the invariant distance:

$$
\begin{aligned}
x^{i'}x^{i'} &= O^{ij}x^j \, O^{ik}x^k \\
&= x^j \left\{ O^{ij} O^{ik} \right\} x^k \\
&= x^j x^j
\end{aligned}
\tag{2.9}
$$

that is, this is invariant if the O matrix is orthogonal;

$$O^{ij} O^{ik} = \delta^{jk} \tag{2.10}$$

or, more symbolically:

$$O^T 1 O = 1 \tag{2.11}$$

To take the inverse of an orthogonal matrix, we simply take its transpose. The unit matrix **1** is called the *metric* of the group.

The rotation group $O(2)$ is called the *orthogonal group* in two dimensions. The orthogonal group $O(2)$, in fact, can be defined as the set of all real, two-dimensional orthogonal matrices. Any orthogonal matrix can be written as the exponential of a single antisymmetric matrix τ:

$$O(\theta) = e^{\theta \tau} \equiv \sum_{n=0}^{\infty} \frac{1}{n!} (\theta \tau)^n \tag{2.12}$$

where:

$$\tau = \begin{pmatrix} 0 & 1 \\ -1 & 0 \end{pmatrix} \tag{2.13}$$

To see this, we note that the transpose of $e^{\theta \tau}$ is $e^{-\theta \tau}$:

$$O^T = \left(e^{\theta \tau} \right)^T = e^{-\theta \tau} = O^{-1} \tag{2.14}$$

Another way to prove this identity is simply to power expand the right-hand side and sum the series. We then see that the Taylor expansion of the cosine and sine functions re-emerge. After summing the series, we arrive at:

$$e^{\theta \tau} = \cos \theta \mathbf{1} + \tau \sin \theta = \begin{pmatrix} \cos \theta & \sin \theta \\ -\sin \theta & \cos \theta \end{pmatrix} \tag{2.15}$$

All elements of $O(2)$ are parametrized by one angle θ. We say that $O(2)$ is a one-parameter group; that is, it has dimension 1.

Let us now take the determinant of both sides of the defining equation:

$$\det \left(O O^T \right) = \det O \det O^T = (\det O)^2 = 1 \tag{2.16}$$

This means that the determinant of O is equal to ± 1. If we take $\det O = 1$, then the resulting subgroup is called $SO(2)$, or the special orthogonal matrices in two dimensions. The rotations that we have been studying up to now are members of $SO(2)$. However, there is also the curious subset where det $O = -1$. This subset consists of elements of $SO(2)$ times the matrix:

$$\begin{pmatrix} 1 & 0 \\ 0 & -1 \end{pmatrix} \tag{2.17}$$

This last transformation corresponds to a *parity* transformation:

$$x \rightarrow x$$
$$y \rightarrow -y \tag{2.18}$$

A parity transformation P takes a plane and maps it into its mirror image, and hence it is a discrete, not continuous, transformation, such that $P^2 = 1$.

An important property of groups is that they are uniquely specified by their multiplication law. It is easy to show that these two dimensional matrices O^{ij} can be multiplied in succession as follows:

$$O^{ij}(\theta)O^{jk}(\theta') = O^{ik}(\theta + \theta') \tag{2.19}$$

which simply corresponds to the intuitively obvious notion that if we rotate a coordinate system by an angle θ and then by an additional angle θ', then the net effect is a rotation of $\theta + \theta'$. In fact, *any* matrix $D(\theta)$ (not necessarily orthogonal or even 2×2 in size) that has this multiplication rule:

$$D(\theta)D(\theta') = D(\theta + \theta'); \quad D(\theta) = D(\theta + 2\pi) \tag{2.20}$$

forms a *representation* of $O(2)$, because it has the same multiplication table.

For our purposes, we are primarily interested in the transformation properties of fields. For example, we can calculate how a field $\phi(x)$ transforms under rotations. Let us introduce the operator:

$$L \equiv i\epsilon^{ij}x^i \frac{\partial}{\partial x^j} = i(x^1\partial^2 - x^2\partial^1) \tag{2.21}$$

Let us define:

$$U(\theta) \equiv e^{i\theta L} \tag{2.22}$$

Then we define a *scalar* field as one that transforms under $SO(2)$ as:

$$\text{Scalar}: \quad U(\theta)\phi(x)U^{-1}(\theta) = \phi(x') \tag{2.23}$$

(To prove this equation, we use the fact that:

$$e^A \, B \, e^{-A} = B + [A, B] + \frac{1}{2!}[A, [A, B]] + \frac{1}{3!}[A, [A, [A, B]]] + \cdots \quad (2.24)$$

Then we reassemble these terms via a Taylor expansion to prove the transformation law.)

We can also define a *vector* field $\phi^i(x)$, where the additional i index also transforms under rotations:

$$\text{Vector}: \quad U(\theta)\phi^i(x)U^{-1}(\theta) = O^{ij}(-\theta)\phi^j(x') \quad (2.25)$$

[For this relation to hold, Eq. (2.21) must contain an additional term that rotates the vector index of the field.] Not surprisingly, we can now generalize this formula to include the transformation property of the most arbitrary field. Let $\phi^A(x)$ be an arbitrary field transforming under some representation of $SO(2)$ labeled by some index A. Then this field transforms as:

$$U(\theta)\phi^A(x)U^{-1}(\theta) = \mathscr{D}^{AB}(-\theta)\phi^B(x') \quad (2.26)$$

where $\mathscr{D}^{AB}$ is some representation, either reducible or irreducible, of the group.

2.3 Representations of $SO(2)$ and $U(1)$

One of the chief goals of this chapter is to find the irreducible representations of these groups, so let us be more precise. If g_i is a member of a group G, then the object $D(g_i)$ is called a *representation* of G if it obeys:

$$D(g_i)D(g_j) = D(g_i g_j) \quad (2.27)$$

for all the elements in the group. In other words, $D(g_i)$ has the same multiplication rules as the original group.

A representation is called *reducible* if $D(g_i)$ can be brought into block diagonal form; for example, the following matrix is a reducible representation:

$$D(g_i) = \begin{pmatrix} D_1(g_i) & 0 & 0 \\ 0 & D_2(g_i) & 0 \\ 0 & 0 & D_3(g_i) \end{pmatrix} \quad (2.28)$$

where D_i are smaller representations of the group. Intuitively, this means $D(g_i)$ can be split up into smaller pieces, with each piece transforming under a smaller representation of the same group.

The principal goal of our approach is to find all irreducible representations of the group in question. This is because the basic fields of physics transform as irreducible representations of the Lorentz and Poincaré groups. The complete set of finite-dimensional representations of the orthogonal group comes in two classes, the *tensors* and *spinors*. (For a special exception to this, see Exercise 10.)

One simple way of generating higher representations of $O(2)$ is simply to multiply several vectors together. The product $A^i B^j$, for example, transforms as follows:

$$\left(A^{i'} B^{j'}\right) = \left[O^{i'i}(\theta)O^{j'j}(\theta)\right]\left(A^i B^j\right) \tag{2.29}$$

This matrix $O^{i'i}(\theta)O^{j'j}(\theta)$ forms a representation of $SO(2)$. It has the same multiplication rule as $O(2)$, but the space upon which it acts is 2×2 dimensional. We call any object that transforms like the product of several vectors a *tensor*.

In general, a tensor $T^{ijk\cdots}$ under $O(2)$ is nothing but an object that transforms like the product of a series of ordinary vectors:

$$\text{Tensor}: \quad (T')^{i_1,i_2,\cdots} = O^{i_1,j_1} O^{i_2,j_2} \cdots T^{j_1,j_2,\cdots} \tag{2.30}$$

The transformation of $T^{ijk\cdots}$ is identical to the transformation of the product $x^i x^j x^k \cdots$. This product forms a representation of $O(2)$ because the following matrix:

$$O^{i_1,i_2\cdots i_N;j_1,j_2,\cdots j_N}(\theta) \equiv O^{i_1,j_1}(\theta)O^{i_2,j_2}(\theta)\cdots O^{i_N,j_N}(\theta) \tag{2.31}$$

has the same multiplication rule as $SO(2)$.

The tensors that we can generate by taking products of vectors are, in general, reducible; that is, within the collection of elements that compose the tensor, we can find subsets that by themselves form representations of the group. By taking appropriate symmetric and antisymmetric combinations of the indices, we can extract irreducible representations (see Appendix).

A convenient method that we will use to create irreducible representations is to use two tensors under $O(2)$ that are actually constants: δ^{ij} and ϵ^{ij}, where the latter is the antisymmetric constant tensor and $\epsilon^{12} = -\epsilon^{21} = +1$.

Although they do not appear to be genuine tensors, it is easy to prove that they are. Let us hit them with the orthogonal matrix O^{ij}:

$$\begin{aligned}
\delta^{i'j'} &= O^{i'i}O^{j'j}\delta^{ij} \\
\epsilon^{i'j'} &= O^{i'i}O^{j'j}\epsilon^{ij}
\end{aligned} \tag{2.32}$$

We instantly recognize the first equation: it is just the definition of an orthogonal matrix, and so δ^{ij} is an invariant tensor. The second equation, however, is more difficult to see. Upon closer inspection, it is just the definition of the determinant of the O matrix, which is equal to one for $SO(2)$. Thus, both equations are satisfied by construction. Because the ϵ^{ij} transforms like a tensor only if the determinant of O is +1, we sometimes call it a *pseudotensor*. The pseudotensors pick up an extra minus one when they are transformed under parity transformations.

Using these two constant tensors, for example, we can immediately contract the tensor $A^i B^j$ to form two scalar combinations: $A^i B^i$ and $A^i \epsilon^{ij} B^j = A^1 B^2 - A^2 B^1$.

This process of symmetrizing and antisymmetrizing all possible tensor indices to find the irreducible representations is also aided by the simple identities:

$$\epsilon^{ij}\epsilon^{kl} = \delta^{ik}\delta^{jl} - \delta^{il}\delta^{jk}$$
$$\epsilon^{ij}\epsilon^{jk} = -\delta^{ik} \tag{2.33}$$

Finally, we can show the equivalence between $O(2)$ and yet another formulation. We can take a complex object $u = a + ib$, and say that it transforms as follows:

$$u' = U(\theta)u = e^{i\theta}u \tag{2.34}$$

The matrix $U(\theta)$ is called a *unitary matrix*, because:

$$\text{Unitary matrix}: \quad U \times U^\dagger = 1 \tag{2.35}$$

The set of all one-dimensional unitary matrices $U(\theta) = e^{i\theta}$ defines a group called $U(1)$. Obviously, if we make two such transformations, we find:

$$e^{i\theta}e^{i\theta'} = e^{i\theta+i\theta'} \tag{2.36}$$

We have the same multiplication law as $O(2)$, even though this construction is based on a new space, the space of complex one-dimensional numbers. We thus say that:

$$SO(2) \sim U(1) \tag{2.37}$$

This means that there is a correspondence between the two, even though they are defined in two different spaces:

$$e^{\tau(\theta)} \leftrightarrow e^{i\theta} \tag{2.38}$$

To see the correspondence between $O(2)$ and $U(1)$, let us consider two scalar fields ϕ^1 and ϕ^2 that transform infinitesimally under $SO(2)$ as in Eq. (2.6):

$$\delta\phi^i = \theta\epsilon^{ij}\phi^j \tag{2.39}$$

which is just the usual transformation rule for small θ. Because $SO(2) \sim U(1)$, these two scalar fields can be combined into a single complex scalar field:

$$\phi = \frac{1}{\sqrt{2}}(\phi^1 + i\phi^2) \tag{2.40}$$

Then the infinitesimal variation of this field under $U(1)$ is given by:

$$\delta\phi = -i\theta\phi \tag{2.41}$$

for small θ. Invariants under $O(2)$ or $U(1)$ can be written as:

$$\frac{1}{2}\phi^i\phi^i = \phi^*\phi \tag{2.42}$$

2.4 Representations of $SO(3)$ and $SU(2)$

The previous group $O(2)$ was surprisingly rich in its representations. It was also easy to analyze because all its elements commuted with each other. We call such a group an *Abelian group*. Now, we will study *non-Abelian groups*, where the elements do not necessarily commute with each other. We define $O(3)$ as the group that leaves distances in three dimensions invariant:

$$\text{Invariant}: \quad x^2 + y^2 + z^2 \tag{2.43}$$

where $x^{i'} = O^{ij}x^j$. Generalizing the previous steps for $SO(2)$, we know that the set of 3×3, real, and orthogonal matrices $O(3)$ leaves this quantity invariant. The condition of orthogonality reduces the number of independent numbers down to $9 - 6 = 3$ elements. Any member of $O(3)$ can be written as the exponential of an antisymmetric matrix:

$$O = \exp\left(i\sum_{i=1}^{3}\theta^i\tau^i\right) \tag{2.44}$$

where τ^i has purely imaginary elements. There are only three independent antisymmetric 3×3 matrices, so we have the correct counting of independent degrees of freedom. Therefore $O(3)$ is a three-parameter Lie group, parametrized by three angles.

These three antisymmetric matrices τ^i can be explicitly written as:

$$\tau^1 = \tau^x = -i \begin{pmatrix} 0 & 0 & 0 \\ 0 & 0 & 1 \\ 0 & -1 & 0 \end{pmatrix}; \quad \tau^2 = \tau^y = -i \begin{pmatrix} 0 & 0 & -1 \\ 0 & 0 & 0 \\ 1 & 0 & 0 \end{pmatrix} \qquad (2.45)$$

$$\tau^3 = \tau^z = -i \begin{pmatrix} 0 & 1 & 0 \\ -1 & 0 & 0 \\ 0 & 0 & 0 \end{pmatrix} \qquad (2.46)$$

By inspection, this set of matrices can be succinctly represented by the fully antisymmetric ϵ^{ijk} tensor as:

$$(\tau^i)^{jk} = -i\epsilon^{ijk} \qquad (2.47)$$

where $\epsilon^{123} = +1$. These antisymmetric matrices, in turn, obey the following properties:

$$[\tau^i, \tau^j] = i\epsilon^{ijk}\tau^k \qquad (2.48)$$

This is an example of a *Lie algebra* (not to be confused with the Lie group). The constants ϵ^{ijk} appearing in the algebra are called the *structure constants* of the algebra. A complete determination of the structure constants of any algebra specifies the Lie algebra, and also the group itself.

For small angles θ^i, we can write the transformation law as:

$$\delta x^i = \epsilon^{ijk}\theta^k x^j \qquad (2.49)$$

As before, we will introduce the operators:

$$L^i \equiv i\epsilon^{ijk}x^j\partial^k \qquad (2.50)$$

We can show that the commutation relations of L^i satisfy those of $SO(3)$. Let us construct the operator:

$$U(\theta^i) = e^{i\theta^i L^i} \qquad (2.51)$$

Then a scalar and a vector field, as before, transform as follows:

$$U(\theta^k)\phi(x)U^{-1}(\theta^k) = \phi(x')$$

$$U(\theta^k)\phi^i(x)U^{-1}(\theta^k) = (O^{-1})^{ij}(\theta^k)\phi^j(x') \tag{2.52}$$

For higher tensor fields, we must be careful to select out only the irreducible fields. The easiest way to decompose a reducible a tensor is to take various symmetric and anti-symmetric combinations of the indices. Irreducible representations can be extracted by using the two constant tensors, δ^{ij} and ϵ^{ijk}. In carrying out complicated reductions, it is helpful to know:

$$\epsilon^{ijk}\epsilon^{lmn} = \delta^{il}\delta^{jm}\delta^{kn} - \delta^{il}\delta^{jn}\delta^{km} + \delta^{im}\delta^{jn}\delta^{kl}$$

$$-\delta^{im}\delta^{jl}\delta^{kn} + \delta^{in}\delta^{jl}\delta^{km} - \delta^{in}\delta^{jm}\delta^{kl}$$

$$\epsilon^{ijk}\epsilon^{klm} = \delta^{il}\delta^{jm} - \delta^{im}\delta^{jl} \tag{2.53}$$

More generally, we can use the method of Young Tableaux described in the Appendix to find more irreducible representations.

As in the case of $O(2)$, we can also find a relationship between $O(3)$ and a unitary group. Consider the set of all unitary, 2×2 matrices with unit determinant. These matrices form a group, called $SU(2)$, which is called the special unitary group in two dimensions. This matrix has $8 - 4 - 1 = 3$ independent elements in it. Any unitary matrix, in turn, can be written as the exponential of a Hermitian matrix H, where $H = H^\dagger$:

$$U = e^{iH} \tag{2.54}$$

Again, to prove this relation, simply take the Hermitian conjugate of both sides: $U^\dagger = e^{-iH^\dagger} = e^{-iH} = U^{-1}$.

Since an element of $SU(2)$ can be parametrized by three numbers, the most convenient set is to use the Hermitian *Pauli spin matrices*. Any element of $SU(2)$ can be written as:

$$U = e^{i\theta^i\sigma^i/2} \tag{2.55}$$

where:

$$\sigma^x = \begin{pmatrix} 0 & 1 \\ 1 & 0 \end{pmatrix}; \quad \sigma^y = \begin{pmatrix} 0 & -i \\ i & 0 \end{pmatrix}; \quad \sigma^z = \begin{pmatrix} 1 & 0 \\ 0 & -1 \end{pmatrix} \tag{2.56}$$

where σ^i satisfy the relationship:

$$\left[\frac{\sigma^i}{2}, \frac{\sigma^j}{2}\right] = i\epsilon^{ijk}\frac{\sigma^k}{2} \tag{2.57}$$

We now have exactly the same algebra as $SO(3)$ as in Eq. (2.48). Therefore, we can say:

$$SO(3) \sim SU(2) \tag{2.58}$$

To make this correspondence more precise, we will expand the exponential and then recollect terms, giving us:

$$e^{i\sigma^j\theta^j/2} = \cos(\theta/2) + i(\sigma^k n^k)\sin(\theta/2) \tag{2.59}$$

where $\theta^i = n^i\theta$ and $(n^i)^2 = 1$. The correspondence is then given by:

$$e^{i\tau^j\theta^j} \leftrightarrow e^{i\sigma^j\theta^j/2} \tag{2.60}$$

where the left-hand side is a real, 3×3 orthogonal matrix, while the right-hand size is a complex, 2×2 unitary matrix. (The isomorphism is only local, i.e., within a small neighborhood of each of the parameters. In general, the mapping is actually one-to-two, and not one-to-one.) Even though these two elements exist in different spaces, they have the same multiplication law. This means that there should also be a direct way in which vectors (x, y, z) can be represented in terms of these spinors. To see the precise relationship, let us define:

$$h(\mathbf{x}) = \boldsymbol{\sigma} \cdot \mathbf{x} = \begin{pmatrix} z & x - iy \\ x + iy & z \end{pmatrix} \tag{2.61}$$

Then the $SU(2)$ transformation:

$$h' = UhU^{-1} \tag{2.62}$$

is equivalent to the $SO(3)$ transformation:

$$\mathbf{x}' = \mathbf{O} \cdot \mathbf{x} \tag{2.63}$$

2.5   Representations of $SO(N)$

By now, the generalization to $O(N)$ should be straightforward. (The representations of $SU(N)$, which are important when we discuss the quark model, are discussed in the Appendix.)

The essential feature of a rotation in N dimensions is that it preserves distances. Specifically, the distance from the origin to the point x^i, by the Pythagorean Theorem, is given by $\sqrt{(x^i)^2}$. Therefore, $x^i x^i$ is an invariant, where an N-dimensional rotation is defined by: $x'^i = O^{ij} x^j$.

The number of independent elements in each member of $O(N)$ is N^2 minus the number of constraints arising from the orthogonality condition:

$$N^2 - \frac{1}{2}N(N+1) = \frac{1}{2}N(N-1) \tag{2.64}$$

This is exactly the number of independent antisymmetric, $N \times N$ matrices, that is, we can parametrize the independent components within $O(N)$ by either orthogonal matrices or by exponentiating antisymmetric ones, (i.e., $O = e^A$).

Any orthogonal matrix can thereby be parametrized as follows:

$$O = \exp\left(i \sum_{i=1}^{N(N-1)/2} \theta^i \tau^i \right) \tag{2.65}$$

where τ^i are linearly independent, antisymmetric matrices with purely imaginary elements. They are called the *generators* of the group, and θ^i are the rotation angles or the *parameters* of the group.

Finding representations of $O(N)$ is complicated by the fact that the multiplication table for the parameters of $O(N)$ are quite complicated. However, we know from the Baker–Campbell–Hausdorff theorem that:

$$e^A e^B = e^{A+B+(1/2)[A,B]+\cdots} \tag{2.66}$$

where the ellipsis represent multiple commutators of A and B. If e^A and e^B are close together, then these elements form a group as long as the commutators of A and B form an algebra.

If we take the commutator of two antisymmetric matrices, then it is easy to see that we get another antisymmetric matrix. This means that the algebra created by commuting all antisymmetric matrices is a closed one. We will represent the algebra as follows:

$$[\tau^i, \tau^j] = i f^{ijk} \tau^k \tag{2.67}$$

As before, we say that the f^{ijk} are the structure constants of the group. (It is customary to insert an i in the definition of τ, so that an i appears in the commutator.)

For arbitrary N, it is possible to find an exact form for the structure constants. Let us define the generator of $O(N)$ as:

$$(M^{ij})_{ab} = -i(\delta_a^i \delta_b^j - \delta_a^j \delta_b^i) \tag{2.68}$$

Since the matrix is antisymmetric in i, j, there are $N(N-1)/2$ such matrices, which is the correct number of parameters for $O(N)$. The indices a, b denote the various matrix entries of the generator. If we commute these matrices, the calculation is rather easy because it reduces to contracting over a series of delta functions:

$$[M^{ij}, M^{lm}] = i\left(-\delta^{il} M^{jm} + \delta^{jl} M^{im} + \delta^{im} M^{jl} - \delta^{mj} M^{il}\right) \tag{2.69}$$

To define the action of $SO(N)$ on the fields, let us define the operator:

$$L^{ij} \equiv i(x^i \partial^j - x^j \partial^i) \tag{2.70}$$

It is easy to check that L^{ij} satisfies the commutation relations of $SO(N)$. Now construct the operator:

$$U(\theta^{ij}) = e^{i\theta^{ij} L^{ij}} \tag{2.71}$$

where θ^{ij} is antisymmetric. The structure constants of the theory f^{ijk} can also be thought of as a representation of the algebra. If we define:

$$\left(\tau^i\right)^{jk} = f^{ijk} \tag{2.72}$$

then τ^i as written as a function of the structure constants also forms a representation of the generators of $O(N)$. We call this the *adjoint* representation. This also means that the structure constant f^{ijk} is a constant tensor, just like δ^{ij}.

For our purposes, we will often be interested in how fields transform under some representation of $O(N)$. Without specifying the exact representation, we can always write:

$$\delta\phi^i = i\theta^a \left(\tau^a\right)_{ij} \phi^j \tag{2.73}$$

This simply means that we are letting ϕ^i transform under some unspecified representation of the generators of $O(N)$, labeled by the indices i, j.

Finally, one might wonder whether we can find more identities such as $O(2) \sim U(1)$ and $O(3) \sim SU(2)$. With a little work, we can show:

$$SO(4) \sim SU(2) \otimes SU(2); \quad SO(6) \sim SU(4) \tag{2.74}$$

One may then wonder whether there are higher sequences of identities between $O(N)$ and $SU(M)$. Surprisingly, the answer is no. To see that any correspondence would be nontrivial can be checked by simply calculating the number of parameters in $O(N)$ and $SU(M)$, which are quite different. These "accidents" between Lie groups only occur at low dimensionality and are the exception, rather than the rule.

2.6   Spinors

In general, there are two major types of representations that occur repeatedly throughout physics. The first, of course, are tensors, which transform like the product of vectors. Irreducible representations are then found by taking their symmetric and antisymmetric combinations.

However, one of the most important representations of $O(N)$ are the *spinor representations*. To explain the spinor representations, let us introduce N objects Γ^i, which obey:

$$\{\Gamma^i, \Gamma^j\} = 2\delta^{ij}$$
$$\{A, B\} \equiv AB + BA \tag{2.75}$$

where the brackets represent an anticommutator rather than a commutator. This is called a *Clifford algebra*. Then we can construct the spinor representation of the generators of $SO(N)$ as follows:

$$\text{Spinor representation}: \quad M^{ij} = \frac{i}{4}[\Gamma^i, \Gamma^j] \tag{2.76}$$

By inserting this value of M^{ij} into the definition of the algebra of $O(N)$, it satisfies the commutation relations of Eq. (2.69) and hence gives us a new representation of the group.

In general, we can find a spinorial matrix representation of $O(N)$ (for N even) that is $2^{N/2}$ dimensional and complex. The simplest spinor representation of $O(4)$, we will see, gives us the compact version of the celebrated Dirac matrices. For the odd orthogonal groups $O(N + 1)$ (where N is even), we can construct the spinors Γ^i from the spinors for the group $O(N)$. We simply add a new element to the old set:

$$\Gamma_{N+1} = \Gamma^1 \Gamma^2 \cdots \Gamma^N \tag{2.77}$$

Γ_{N+1} has the same anti-commutation relations as the other spinors, and hence we can form the generators of $O(N+1)$ out of the spinors of $O(N)$. As before, we can construct the transformation properties of the Γ^i. Let us define:

$$U(\theta^{ij}) = \exp\left(i\theta^{ij} M^{ij}\right) \tag{2.78}$$

where M^{ij} is constructed out of the spinors Γ^i.

Then it is easy to show that Γ^i satisfies the following identity:

$$U(\theta^{ij})\Gamma^k U^{-1}(\theta^{ij}) = (O^{-1})^{kl}(\theta^{ij})\Gamma^l \tag{2.79}$$

which proves that Γ^i transforms as a vector.

Finally, we should also mention that this spinorial representation of the group $O(N)$ is reducible. For example, we could have constructed two projection operators:

$$P_R = \frac{1 + \Gamma_{N+1}}{2}$$

$$P_L = \frac{1 - \Gamma_{N+1}}{2} \tag{2.80}$$

which satisfy the usual properties of projection operators:

$$
\begin{aligned}
P_L^2 &= P_L \\
P_R^2 &= P_R \\
P_R P_L &= 0 \\
P_L + P_R &= 1
\end{aligned}
\tag{2.81}
$$

With these two projection operators, the group splits into two self-contained pieces. (When we generalize this construction to the Lorentz group, these will be called the "left-handed" and "right-handed" Weyl representations. They will allow us to describe neutrino fields.)

2.7   Lorentz Group

Now that we have completed our brief, warm-up discussion of some compact Lie groups, let us tackle the main problem of this chapter, finding the representations of the noncompact Lorentz and Poincaré groups.

We define the Lorentz group as the set of all 4×4 real matrices that leave the following invariant:

$$\text{Invariant}: \quad s^2 = c^2 t^2 - x^i x^i = (x^0)^2 - (x^i)^2 = x^\mu g_{\mu\nu} x^\nu \tag{2.82}$$

The minus signs, of course, distinguish this from the group $O(4)$.

A Lorentz transformation can be parametrized by:

$$x'^\mu = \Lambda^\mu{}_\nu x^\nu \tag{2.83}$$

Inserting this transformation into the invariant, we find that the Λ matrices must satisfy:

$$g_{\mu\nu} = \Lambda^\rho{}_\mu g_{\rho\sigma} \Lambda^\sigma{}_\nu \tag{2.84}$$

which can be written symbolically as $g = \Lambda^T g \Lambda$. Comparing this with Eq. (2.11), we say that $g_{\mu\nu}$ is the *metric* of the Lorentz group. If the signs within the metric $g_{\mu\nu}$ were all the same sign, then the group would be $O(4)$. To remind ourselves that the signs alternate, we call the Lorentz group $O(3, 1)$:

$$\text{Lorentz group}: \quad O(3, 1) \tag{2.85}$$

where the comma separates the positive from the negative signs within the metric. In general, an orthogonal group that preserves a metric with M indices of one sign and N indices of another sign is denoted $O(M, N)$.

The minus signs in the metric create an important difference between $O(4)$ and the Lorentz group: The invariant distance s^2 can be negative or positive, while the invariant distance for $O(4)$ was always positive. This means that the x_μ plane splits up into distinct regions that cannot be connected by a Lorentz transformation. If x and y are two position vectors, then these regions can be labeled by the value of the invariant distance s^2:

$$(x - y)^2 > 0: \qquad \text{time-like}$$

$$(x - y)^2 = 0: \qquad \text{light-like}$$

$$(x - y)^2 < 0: \qquad \text{space-like} \tag{2.86}$$

Excluding this crucial difference, the representations of the Lorentz group, for the most part, bear a striking resemblance with those of $O(4)$. For example, as before, we can introduce the operator $L^{\mu\nu}$ in order to define the action of the Lorentz group on fields:

$$L^{\mu\nu} = x^{\mu}p^{\nu} - x^{\nu}p^{\nu} = i(x^{\mu}\partial^{\nu} - x^{\nu}\partial^{\mu}) \tag{2.87}$$

where $p_{\mu} = i\partial_{\mu}$. As before, we can show that this generates the algebra of the Lorentz group:

$$[L^{\mu\nu}, L^{\rho\sigma}] = ig^{\nu\rho}L^{\mu\sigma} - ig^{\mu\rho}L^{\nu\sigma} - ig^{\nu\sigma}L^{\mu\rho} + ig^{\mu\sigma}L^{\nu\rho} \tag{2.88}$$

Let us also define:

$$U(\Lambda) = \exp\left(i\epsilon^{\mu\nu}L_{\mu\nu}\right) \tag{2.89}$$

where, infinitesimally, we have:

$$\Lambda^{\mu\nu} = g^{\mu\nu} + \epsilon^{\mu\nu} + \cdots \tag{2.90}$$

Then the action of the Lorentz group on a vector field ϕ^{μ} can be expressed as:

$$\text{Vector}: \quad U(\Lambda)\phi^{\mu}(x)U^{-1}(\Lambda) = (\Lambda^{-1})^{\mu}{}_{\nu}\phi^{\nu}(x') \tag{2.91}$$

(where $U(\Lambda)$ contains an additional piece which rotates the vector index of ϕ^{μ}). Let us parametrize the Λ^{μ}_{ν} of a Lorentz transformation as follows:

$$x' = \frac{x + vt}{\sqrt{1 - v^2/c^2}}; \quad y' = y; \quad z' = z; \quad t' = \frac{t + vx/c^2}{\sqrt{1 - v^2/c^2}} \tag{2.92}$$

We can make several observations about this transformation from a group point of view. First, the velocity v is a parameter of this group. Since the velocity varies as $0 \leq v < c$, where v is strictly less than the speed of light c, we say the Lorentz group is *noncompact*; that is, the range of the parameter v does not include the endpoint c. This stands in contrast to the group $O(N)$, where the parameters have finite range, include the endpoints, and are hence compact.

We say that the three components of velocity v^x, v^y, v^z are the parameters of Lorentz boosts. If we make the standard replacement:

$$\gamma = \frac{1}{\sqrt{1 - v^2/c^2}} = \cosh\phi, \quad \beta\gamma = \sinh\phi; \quad \beta = v/c \tag{2.93}$$

then this transformation can be written as:

$$
\begin{pmatrix} x'^0 \\ x'^1 \\ x'^2 \\ x'^3 \end{pmatrix} = \begin{pmatrix} \cosh\phi & \sinh\phi & 0 & 0 \\ \sinh\phi & \cosh\phi & 0 & 0 \\ 0 & 0 & 1 & 0 \\ 0 & 0 & 0 & 1 \end{pmatrix} \begin{pmatrix} x^0 \\ x^1 \\ x^2 \\ x^3 \end{pmatrix} \tag{2.94}
$$

Let us rewrite this in terms of $M^{\mu\nu}$. Let us define:

$$
\begin{aligned}
J^i &= \frac{1}{2}\epsilon^{ijk} M^{jk} \\
K^i &= M^{0i}
\end{aligned} \tag{2.95}
$$

Written out explicitly, this becomes:

$$
K^x = K^1 = -i \begin{pmatrix} 0 & 1 & 0 & 0 \\ 1 & 0 & 0 & 0 \\ 0 & 0 & 0 & 0 \\ 0 & 0 & 0 & 0 \end{pmatrix} \tag{2.96}
$$

Then this Lorentz transformation can be written as:

$$
e^{iK^x v^x} = \cosh\phi + i \sinh\phi K^x \tag{2.97}
$$

Similarly, Lorentz boosts in the y and z direction are generated by exponentiating K^y and K^z:

$$
K^y = K^2 = -i \begin{pmatrix} 0 & 0 & 1 & 0 \\ 0 & 0 & 0 & 0 \\ 1 & 0 & 0 & 0 \\ 0 & 0 & 0 & 0 \end{pmatrix} \quad ; \quad K^z = K^3 = -i \begin{pmatrix} 0 & 0 & 0 & 1 \\ 0 & 0 & 0 & 0 \\ 0 & 0 & 0 & 0 \\ 1 & 0 & 0 & 0 \end{pmatrix} \tag{2.98}
$$

Unfortunately, it is easily checked that a boost in the x direction, followed by a boost in the y direction, does not generate another Lorentz boost. Thus, the three K matrices by themselves do not generate a closed algebra. To complete the algebra, we must introduce the generators of the ordinary rotation group $O(3)$ as well:

$$
J^x = J^1 = -i \begin{pmatrix} 0 & 0 & 0 & 0 \\ 0 & 0 & 0 & 0 \\ 0 & 0 & 0 & 1 \\ 0 & 0 & -1 & 0 \end{pmatrix} ; \quad J^y = J^2 = -i \begin{pmatrix} 0 & 0 & 0 & 0 \\ 0 & 0 & 0 & -1 \\ 0 & 0 & 0 & 0 \\ 0 & 1 & 0 & 0 \end{pmatrix}
$$

$$(2.99)$$

$$
J^z = J^3 = -i \begin{pmatrix} 0 & 0 & 0 & 0 \\ 0 & 0 & 1 & 0 \\ 0 & -1 & 0 & 0 \\ 0 & 0 & 0 & 0 \end{pmatrix} \qquad (2.100)
$$

The K and J matrices have the following commutation relations:

$$
\begin{aligned}
[K^i, K^j] &= -i\epsilon^{ijk} J^k \\
[J^i, J^j] &= i\epsilon^{ijk} J^k \\
[J^i, K^j] &= i\epsilon^{ijk} K^k
\end{aligned}
$$

$$(2.101)$$

Two pure Lorentz boosts generated by K^i taken in succession do *not* generate a third boost, but must generate a rotation generated by J^i as well. (Physically, this rotation, which arises after several Lorentz boosts, gives rise to the Thomas precession effect.)

By taking linear combinations of these generators, we can show that the Lorentz group can be split up into two pieces. We will exploit the fact that $SO(4) = SU(2) \otimes SU(2)$, and that the algebra of the Lorentz group is similar to the algebra of $SO(4)$ (modulo minus signs coming from the metric). By taking linear combinations of these generators, one can show that the algebra actually splits into two pieces:

$$
\begin{aligned}
A^i &= \frac{1}{2}(J^i + iK^i) \\
B^i &= \frac{1}{2}(J^i - iK^i)
\end{aligned}
$$

$$(2.102)$$

We then have $[A^i, B^j] = 0$, so the algebra has now split up into two distinct pieces, each piece generating a separate $SU(2)$.

If we change the sign of the metric so that we only have compact groups, then we have just proved that the Lorentz group, for our purposes, can be written as $SU(2) \otimes SU(2)$. This means that irreducible representations (j) of $SU(2)$, where $j = 0, 1/2, 1, 3/2$, etc., can be used to construct representations of the

Lorentz group, labeled by (j, j') (see Appendix). By simply pairing off two representations of $SU(2)$, we can construct all the representations of the Lorentz group. (The representation is spinorial when $j + j'$ is half-integral.) In this fashion, we can construct both the tensor and spinor representations of the Lorentz group.

We should also remark that not all groups have spinorial representations. For example, $GL(N)$, the group of all real $N \times N$ matrices, does not have any finite-dimensional spinorial representation. (This will have a great impact on the description of electrons in general relativity in Chapter 19.)

2.8   Representations of the Poincaré Group

Physically, we can generalize the Lorentz group by adding translations:

$$x'^{\mu} = \Lambda^{\mu}{}_{\nu} x^{\nu} + a^{\mu} \qquad (2.103)$$

The Lorentz group with translations now becomes the Poincaré group. Because the Poincaré group includes four translations in addition to three rotations and three boosts, it is a 10-parameter group. In addition to the usual generator of the Lorentz group, we must add the translation generator $p_{\mu} = i \partial_{\mu}$.

The Poincaré algebra is given by the usual Lorentz algebra, plus some new relations:

$$
\begin{aligned}
[L_{\mu\nu}, P_{\rho}] &= -i g_{\mu\rho} P_{\nu} + i g_{\nu\rho} P_{\mu} \\
[P_{\mu}, P_{\nu}] &= 0
\end{aligned}
\qquad (2.104)
$$

These relations mean that two translations commute, and that translations transform as a genuine vector under the Lorentz group.

To find the irreducible representations of a Lie group, we will often use the technique of simultaneously diagonalizing a subset of its generators. Let the *rank* of a Lie group be the number of generators that simultaneously commute among themselves. The rank and dimension of $O(N)$ and $SU(N)$ are given by:

Group	Dimension	Rank
$SO(N)$ (N even)	$(1/2)N(N-1)$	$N/2$
$SO(N)$ (N odd)	$(1/2)N(N-1)$	$(N-1)/2$
$SU(N)$	$N^2 - 1$	$N-1$

(2.105)

For example, the group $SO(3)$ has rank 1, so we can choose L_3 to be the generator to diagonalize. The group $SO(4)$, as we saw, can be re-expressed in terms of two $SU(2)$ subgroups, so that there are two generators that commute among themselves.

In addition, we have the *Casimir operators* of the group, that is, those operators that commute with all the generators of the algebra. For the group $O(3)$, we know, for example, that the Casimir operator is the sum of the squares of the generators: L_i^2. Therefore, we can simultaneously diagonalize L_i^2 and L_3. The representations of $SO(3)$ then correspond to the eigenvalues of these two operators.

For the Poincaré group, we first know that $P_\mu^2 = m^2$, or the mass squared, is a Casimir operator. Under Lorentz transformations, it transforms as a genuine scalar, and is hence invariant. Also, it is invariant under translations because all translations commute.

To find the other Casimir operator, we introduce:

$$W^\mu = \frac{1}{2}\epsilon^{\mu\nu\rho\sigma} P_\nu L_{\rho\sigma} \tag{2.106}$$

which is called the Pauli–Lubanski tensor (where $\epsilon^{0123} = +1$). Then the square of this tensor is a Casimir operator as well:

$$\text{Casimir operators} = \{P_\mu^2, W_\mu^2\} \tag{2.107}$$

All physical states in quantum field theory can be labeled according to the eigenvalue of these two Casimir operators (since the Casimir commutes with all generators of the algebra). However, the physical significance of this new Casimir operator is not immediately obvious. It cannot correspond to spin, since our intuitive notion of angular momentum, which we obtain from nonrelativistic quantum mechanics, is, strictly speaking, lost once we boost all particles by a Lorentz transformation. The usual spin operator is no longer a Casimir operator of the Lorentz group.

To find the physical significance of W_μ^2, let us therefore to go the rest frame of a massive particle: $P^\mu = (m, 0)$. Inserting this into the equation for the Pauli–Lubanski tensor, we find:

$$
\begin{aligned}
W_i &= -\frac{1}{2}m\epsilon_{ijk0}J^{jk} \\
&= -mL_i \\
W_0 &= 0
\end{aligned}
\tag{2.108}
$$

where L_i is just the usual rotation matrix in three dimensions. Thus, in the rest frame of a massive particle, the Pauli–Lubanski tensor is just the spin generator.

Its square is therefore the Casimir of $SO(3)$, which we know yields the spin of the particle:

$$W_i^2 = m^2 s(s+1) \tag{2.109}$$

where s is the spin eigenstate of the particle.

In the rest frame of the massive particle, we have $2s + 1$ components for a spin-s particle. This corresponds to a generalization of our intuitive understanding of spin coming from nonrelativistic quantum mechanics.

However, this analysis is incomplete because we have not discussed massless particles, where $P_\mu^2 = 0$. In general, the counting rule for massive spinning states breaks down for massless ones. For these particles, we have:

$$W_\mu^2 |p\rangle = W_\mu P^\mu |p\rangle = P_\mu P^\mu |p\rangle = 0 \tag{2.110}$$

The only way to satisfy these three conditions is to have W_μ and P^μ be proportional to each other; that is, $W_\mu |p\rangle = h P_\mu |p\rangle = 0$ on a massless state $|p\rangle$. This number h is called the *helicity*, and describes the number of independent components of a massless state.

Using the definition of W_μ for a massless state, we can show that h can be written as:

$$h = \frac{\mathbf{J} \cdot \mathbf{P}}{|\mathbf{P}|} \tag{2.111}$$

Because of the presence of $\epsilon^{\mu\nu\rho\sigma}$ in the definition of W_μ, the Pauli–Lubanski vector is actually a pseudovector. Under a parity transformation, the helicity h therefore transforms into $-h$. This means that massless states have two helicity states, corresponding to a state where W_μ is aligned parallel to the momentum vector and also aligned antiparallel to the momentum vector. Regardless of the spin of a massless particle, the helicity can have only *two* values, h and $-h$. There is thus an essential difference between massless and massive states in quantum field theory.

It is quite remarkable that we can label all irreducible representations of the Poincaré group (and hence all the known fields in the universe) according to the eigenvalues of these Casimir operators. A complete list is given as follows in terms of the mass m, spin s, and helicity h:

$$\begin{cases} P_\mu^2 > 0 : & |m, s\rangle, \ s = 0, 1/2, 1, 3/2, \cdots \\ P_\mu^2 = 0 : & |h\rangle, \ h = \pm s \\ P_\mu^2 = 0 : & s \text{ continuous} \\ P_\mu^2 < 0 : & \text{tachyon} \end{cases} \tag{2.112}$$

In nature, the physical spectrum of states seems to be realized only for the first two categories with $P^0 > 0$. The other states, which have continuous spins or tachyons, have not been seen in nature.

2.9 Master Groups and Supersymmetry

Although we have studied the representations of $O(3, 1)$ by following our discussion of the representations of $O(4)$, the two groups are actually profoundly different. One nontrivial consequence of the difference between $O(4)$ and $O(3, 1)$ is summarized by the following:

No-Go Theorem: *There are no finite-dimensional unitary representations of noncompact Lie groups. Any nontrivial union of the Poincaré and an internal group yields an S matrix which is equal to 1.*

This theorem has caused a certain amount of confusion in the literature. In the 1960s, after the success of the $SU(3)$ description of quarks, attempts were made to construct Master Groups that could nontrivially combine both the Poincaré group and the "internal" group $SU(3)$:

$$\text{Master group} \supset P \otimes SU(3) \qquad (2.113)$$

In this way, it was hoped to give a unified description of particle physics in terms of group theory. There was intense interest in groups like $SU(6, 6)$ or $\tilde{U}(12)$ that combined both the internal and space–time groups. Only later was the no-go theorem discovered, which seemed to doom all these ambitious efforts. Because of the no-go theorem, unitary representations of the particles were necessarily infinite dimensional: These groups possessed nonphysical properties, such as an infinite number of particles in each irreducible representation, or a continuous spectrum of masses for each irreducible representation. As a consequence, after a period of brief enthusiasm, the no-go theorem doomed all these naive efforts to build a Master Group for all particle interactions.

Years later, however, it was discovered that there was a loophole in the no-go theorem. It was possible to evade this no-go theorem (the most comprehensive version being the Coleman–Mandula theorem) because it made an implicit assumption: that the parameters θ_i of the Master Group were all c numbers. However, if the θ_i could be anticommuting, then the no-go theorem could be evaded.

This leads us to the super groups, and eventually to the superstring, which have revived efforts to build Master Groups containing all known particle interactions, including gravity. Although supersymmetry holds the promise of being a fundamental symmetry of physics, we will study these theories not because they have any immediate application to particle physics, but because they provide a fascinating laboratory in which one can probe the limits of quantum field theory.

In summary, the essential point is that quantum field theory grew out of the marriage between quantum mechanics and group theory, in particular the Lorentz and Poincaré group. In fact, it is one of the axioms of quantum field theory that the fundamental fields of physics transform as irreducible representations of these groups. Thus, a study of group theory goes to the heart of quantum field theory. We will find that the results of this chapter will be used throughout this book.

2.10   Exercises

1. By a direct calculation, show that M^{ij} given in Eq. (2.68) and the spinor representation given in Eq. (2.76) do, in fact, satisfy the commutation relations of the Lorentz algebra in Eq. (2.88) if we use the Lorentz metric instead of Kronecker delta functions.

2. Prove that, under a proper orthochronous Lorentz transformation (see Appendix for a definition), the sign of the time t variable (if we are in the forward light cone) does not change. Thus, the sign of t is an invariant in the forward light cone, and we cannot go backwards in time by using rotations and proper orthochronous Lorentz boosts.

3. Show that the proper orthochronous Lorentz group is, in fact, a group. Do the other branches of the Lorentz group also form a group?

4. For $O(3)$, show that the dimensions of the irreducible tensor representations are all positive odd integers.

5. For $O(3)$, show that:

$$\mathbf{3} \otimes \mathbf{3} \otimes \mathbf{3} = \mathbf{7} \oplus \mathbf{5} \oplus \mathbf{5} \oplus \mathbf{3} \oplus \mathbf{3} \oplus \mathbf{3} \oplus \mathbf{1} \qquad (2.114)$$

(see Appendix).

6. Prove that:

$$e^A F(B) e^{-A} = F\left(e^A B e^{-A}\right) \qquad (2.115)$$

where A and B are operators, and F is an arbitrary function. (Hint: use a Taylor expansion of F.)

7. Prove Eq. (2.24).

8. Prove that the Pauli–Lubanski vector is a genuine Casimir operator.

9. Prove that an element of a Clifford algebra Γ^μ transforms as a vector under the Lorentz group; that is, verify Eq. (2.79).

10. For $SO(2)$, show that the spin eigenvalue can be continuous. (Hint: examine $\exp(i\alpha\theta)$ under a complete rotation in θ if α is fractional.)

11. Prove that the Lorentz group can be written as $SL(2, C)$, the set of complex, 2×2 matrices with unit determinant. Show that this representation (as well as the other representations we have discussed in this chapter) is not unitary, and hence satisfies the no-go theorem. Unitarity, however, is required to satisfy the conservation of probability. Does the nonunitarity of these representations violate this important principle?

12. In three dimensions, using the contraction of two ϵ^{ijk} constant tensors into Kronecker deltas in Eq. (2.53), prove that the curl of the curl of a vector A is given by $\nabla^2 A - \nabla(\nabla \cdot A)$.

13. Prove that $SU(4)$ is locally isomorphic to $SO(6)$. (Hint: show the equivalence of their Lie algebras.)

14. Prove that there are two constants, δ^{ij} and $\epsilon^{ijklm\cdots}$, which transform as genuine tensors under $SO(N)$. To prove these constants are genuine tensors, act upon them with O^{ij}. Show that the tensor equation for δ^{ij} reduces to the definition of an orthogonal group. Prove that $\epsilon^{ijklm\cdots}$ satisfies:

$$O^{i_1 j_1} O^{i_2 j_2} \cdots O^{i_N j_N} \epsilon^{j_1 j_2 \cdots j_N} = A \, \epsilon^{i_1 i_2 \cdots i_N} \qquad (2.116)$$

What is the constant A?

15. Prove that:

$$L_{\mu\nu}^2; \quad \epsilon^{\mu\nu\rho\sigma} L_{\mu\nu} L_{\rho\sigma} \qquad (2.117)$$

are Casimir operators for the Lorentz group (but are not Casimir operators for the full Poincaré group).

16. Re-express:

$$\epsilon^{i_1 i_2 \cdots i_N} \epsilon^{j_1 j_2 \cdots j_N} \qquad (2.118)$$

entirely in terms of delta functions for $N = 4$ and 5. (Hint: check that the antisymmetry properties of the $\epsilon^{\mu\nu\cdots}$ tensor are satisfied for the product of delta functions.)

Chapter 3
Spin-0 and $\frac{1}{2}$ Fields

> *It is more important to have beauty in one's equations than to have them*
> *fit experiment . . . because the discrepancy may be due to minor features*
> *that are not properly taken into account and that will get cleared up with*
> *further developments of the theory . . . . It seems that if one is working*
> *from the point of view of getting beauty in one's equations, and if one has*
> *really a sound insight, one is on a sure line of progress.*
>
> —P.A.M. Dirac

3.1 Quantization Schemes

In the previous chapters, we presented the classical theory of fields and also the symmetries they obey. In this chapter, we now make the transition to the quantum theory of fields.

Symbolically, we may write:

$$\lim_{N \to \infty} \text{Quantum mechanics} = \text{Quantum field theory} \qquad (3.1)$$

where N is the number of degrees of freedom of the system. We will see that one important consequence of this transition is that quantum field theory describes *multiparticle* states, while ordinary quantum mechanics is based on a single-particle interpretation. We will find that second quantized systems are ideally suited to describing relativistic physics, since relativity introduces pair creation and annihilation and hence inevitably introduces multiparticle states.

In this chapter, we will develop the second quantization program for the irreducible representations of the Lorentz group for fields with spin 0 and $\frac{1}{2}$. We stress, however, that a number of different types of quantization schemes have been proposed over the decades, each with their own merits and drawbacks:

1. The most direct method is the *canonical quantization* program, which we will develop in this chapter. Canonical quantization closely mimics the development of quantum mechanics; that is, time is singled out as a special coordinate and manifest Lorentz invariance is sacrificed. The advantage of canonical quantization is that it quantizes only physical modes. Unitarity of the system is thus manifest. At the level of QED, the canonical quantization method is not too difficult, but the canonical quantization of more complicated theories, such as non-Abelian gauge theories, is often prohibitively tedious.

2. The *Gupta–Bleuler* or covariant quantization method will also be mentioned in this chapter. Contrary to canonical quantization, it maintains full Lorentz symmetry, which is a great advantage. The disadvantage of this approach is that ghosts or unphysical states of negative norm are allowed to propagate in the theory, and are eliminated only when we apply constraints to the state vectors.

3. The *path integral method* is perhaps the most elegant and powerful of all quantization programs. One advantage is that one can easily go back and forth between many of the other quantization programs to see the relationships between them. Although some of the conventions found in various quantization programs may seem a bit bizarre or contrived, the path integral approach is based on simple, intuitive principles that go to the very heart of the assumptions of quantum theory. The disadvantage of the path integral approach is that functional integration is a mathematically delicate operation that may not even exist in Minkowski space.

4. The Becchi–Rouet–Stora–Tyupin (BRST) approach is one of the most convenient and practical covariant approaches used for gauge theories. Like the Gupta-Bleuler quantization program, negative norm states or ghosts are allowed to propagate and are eliminated by applying the BRST condition onto the state vectors. All the information is contained in a single operator, making this a very attractive formalism. The BRST approach can be easily expressed in terms of path integrals.

5. Closely related to the BRST method is the Batalin–Vilkovisky (BV) quantization program, which has proved powerful enough to quantize the most complicated actions so far proposed, such as those found in string and membrane theories. The formalism is rather cumbersome, but it remains the only program that can quantize certain complex actions.

6. *Stochastic quantization* is yet another quantization program that preserves gauge invariance. One postulates a fictitious fifth coordinate, such that the physical system eventually settles down to the physical solution as the fifth coordinate evolves.

3.2 Klein–Gordon Scalar Field

Let us begin our discussion by quantizing the simplest possible relativistic field theory, the free scalar field. The theory was proposed independently by six different physicists.[1-6] The Lagrangian is given by:

$$\mathscr{L} = \frac{1}{2}(\partial_\mu \phi)^2 - \frac{1}{2}m^2\phi^2 \tag{3.2}$$

Historically, the quantization of the Klein–Gordon equation caused much confusion. Schrödinger, even before he postulated his celebrated nonrelativistic equation, considered this relativistic scalar equation but ultimately discarded it because of problems with negative probability and negative energy states. In any fully relativistic equation, we must obey the "mass-shell condition" $p_\mu^2 = E^2 - \mathbf{p}^2 = m^2$. This means that the energy is given by:

$$E = \pm\sqrt{\mathbf{p}^2 + m^2} \tag{3.3}$$

The energy can be negative, which is quite disturbing. Even if we banish the negative energy states by fiat, we find that interactions with other particles will reduce the energy and create negative energy states. This means that all positive energy states will eventually collapse into negative energy states, destabilizing the entire theory. One can show that even if we prepare a wave packet with only states of positive energy, interactions will inevitably introduce negative energy states. We will see, however, that the solution of these problems with negative probability and negative energy can be resolved once one quantizes the theory.

The canonical quantization program begins with fields ϕ and their conjugate momentum fields π, which satisfy equal time commutation relations among themselves. Then the time evolution of these quantized fields is governed by a Hamiltonian. Thus, we closely mimic the dynamics found in ordinary quantum mechanics. We begin by singling out time as a special coordinate and then defining the canonical conjugate field to ϕ:

$$\pi(\mathbf{x}, t) = \frac{\delta \mathscr{L}}{\delta \dot{\phi}(\mathbf{x}, t)} = \dot{\phi}(\mathbf{x}, t) \tag{3.4}$$

We can introduce the Hamiltonian as:

$$\mathscr{H} = \pi\dot{\phi} - \mathscr{L} = \frac{1}{2}\left[\pi^2 + (\nabla\phi)^2 + m^2\phi^2\right] \tag{3.5}$$

Then the transition from classical mechanics to quantum field theory begins when we postulate the commutation relations between the field and its conjugate momentum:

$$[\phi(\mathbf{x}, t), \pi(\mathbf{y}, t)] = i\delta^3(\mathbf{x} - \mathbf{y}) \tag{3.6}$$

(The right-hand side is proportional to $\hbar$, which we omit. This is the point where the quantum principle begins to emerge from the classical theory.) All other commutators (i.e., between π and itself, and ϕ and itself) are set equal to zero. [Although this expression looks non-relativistic, notice that x^μ and y^μ are separated by a space-like distance, $(x - y)^2 < 0$, which is preserved under a Lorentz transformation.]

Much of what follows is a direct consequence of this commutation relation. There are an infinite number of ways in which to satisfy this relationship, but our strategy will be to find a specific Fourier representation of this commutation relation in terms of plane waves. When these plane-wave solutions are quantized in terms of harmonic oscillators, we will be able to construct the multiparticle Hilbert space and also find a specific operator representation of the Lorentz group in terms of oscillators.

We first define the quantity:

$$k \cdot x \equiv k_\mu x^\mu = (Et - \mathbf{p} \cdot \mathbf{x}) \tag{3.7}$$

We want a decomposition of the scalar field where the energy k^0 is positive, and where the Klein–Gordon equation is explicitly obeyed. In momentum space, the operator $\partial_\mu^2 + m^2$ becomes $k^2 - m^2$. Therefore, we choose:

$$\phi(x) = \frac{1}{(2\pi)^{3/2}} \int d^4k \; \delta(k^2 - m^2)\theta(k_0)\left[A(k)e^{-ik\cdot x} + A^\dagger(k)e^{ik\cdot x}\right] \tag{3.8}$$

where θ is a step function [$\theta(k_0) = +1$ if $k_0 > 0$ and $\theta(k_0) = 0$ otherwise], and where $A(k)$ are operator-valued Fourier coefficients. It is now obvious that this field satisfies the Klein–Gordon equation. If we hit this expression with $(\partial_\mu^2 + m^2)$, then this pulls down a factor of $k^2 - m^2$, which then cancels against the delta function.

We can simplify this expression by integrating out dk^0 (which also breaks manifest Lorentz covariance). To perform the integration, we need to re-express the delta function. We note that a function $f(x)$, which satisfies $f(a) = 0$, obeys the relation:

$$\delta(f(x)) = \frac{\delta(x - a)}{|f'(a)|} \tag{3.9}$$

for x near a. (To prove this relation, simply integrate both sides over x, and then change variables. This generates a Jacobian, which explains the origin of f'.)

Since $k^2 = m^2$ has two roots, we find:

$$\delta(k^2 - m^2) = \frac{\delta(k^0 - \sqrt{\mathbf{k}^2 + m^2})}{2k^0} + \frac{\delta(k^0 + \sqrt{\mathbf{k}^2 + m^2})}{2|k^0|} \tag{3.10}$$

Putting this back into the integral, and using only the positive value of k^0, we find:

$$\int d^4k \, \delta(k^2 - m^2)\theta(k_0) = \int d^3k \int_0^\infty \frac{dk^0}{2k^0} \delta\left(k_0 - \sqrt{\mathbf{k}^2 + m^2}\right)$$

$$= \int \frac{d^3k}{2\omega_k}; \qquad \omega_k = \sqrt{\mathbf{k}^2 + m^2} \tag{3.11}$$

Now let us insert this expression back into the Fourier decomposition of $\phi(x)$. We now find:

$$\phi(x) = \int \frac{d^3k}{\sqrt{(2\pi)^3 2\omega_k}} \left[a(k)e^{-ikx} + a^\dagger(k)e^{ikx}\right] \tag{3.12}$$

$$= \int d^3k \left[a(k)e_k(x) + a^\dagger(k)e_k^*\right] \tag{3.13}$$

$$\pi(x) = \int d^3k \, i\omega_k \left[-a(k)e_k + a^\dagger(k)e_k^*\right] \tag{3.14}$$

where:

$$e_k(x) = \frac{e^{-ik\cdot x}}{\sqrt{(2\pi)^3 2\omega_k}} \tag{3.15}$$

where $A(k) = \sqrt{2\omega_k}a(k)$ and where k^0 appearing in $k \cdot x$ is now equal to ω_k. We can also invert these relations, solving for the Fourier modes $a(k)$ in terms of the original scalar field:

$$a(k) = i \int d^3x \, e_k^*(x) \overset{\leftrightarrow}{\partial_0} \phi(x)$$

$$a^\dagger(k) = -i \int d^3x \, e_k(x) \overset{\leftrightarrow}{\partial_0} \phi(x)$$

$$\tag{3.16}$$

where:

$$A \overset{\leftrightarrow}{\partial} B \equiv A\partial B - (\partial A)B \tag{3.17}$$

Because the fields satisfy equal-time canonical commutation relations, the Fourier modes must also satisfy commutation relations:

$$[a(k), a^\dagger(k')] = \delta^3(\mathbf{k} - \mathbf{k}') \tag{3.18}$$

and all other commutators are zero. To prove that this commutation relation is consistent with the original commutator, let us insert the Fourier expansion into the equal-time commutator:

$$
\begin{aligned}
\left[\phi(\mathbf{x}, t), \pi(\mathbf{x}', t)\right] &= \int \frac{d^3k}{\sqrt{(2\pi)^3 2\omega_k}} \int \frac{i\omega_{k'} d^3k'}{\sqrt{(2\pi)^3 2\omega_{k'}}} \Big[a(k)e^{-ik\cdot x} \\
&\qquad + a^\dagger(k)e^{ik\cdot x}, -a(k')e^{-ik'\cdot x'} + a^\dagger(k')e^{ik'\cdot x'} \Big] \\
&= \int \frac{d^3k}{\sqrt{(2\pi)^3 2\omega_k}} \int \frac{d^3k'}{\sqrt{(2\pi)^3 2\omega_{k'}}} i\omega_{k'} \\
&\qquad \times \delta^3(\mathbf{k} - \mathbf{k}') \left(e^{-ik\cdot x + ik'\cdot x'} + e^{ik\cdot x - ik'\cdot x'} \right) \\
&= i \int \frac{d^3k}{2(2\pi)^3} \left(e^{i\mathbf{k}\cdot(\mathbf{x}-\mathbf{x}')} + e^{-i\mathbf{k}\cdot(\mathbf{x}-\mathbf{x}')} \right) \\
&= i\delta^3(\mathbf{x} - \mathbf{x}') \tag{3.19}
\end{aligned}
$$

Thus, this is a consistent choice for the commutators. Now we can calculate the Hamiltonian in terms of these Fourier modes:

$$
\begin{aligned}
H &= \frac{1}{2} \int d^3x \left[\pi^2 + \partial_i\phi\partial_i\phi + m^2\phi^2 \right] \\
&= \frac{1}{2} \int d^3k\, \omega_k \left[a(k)a^\dagger(k) + a^\dagger(k)a(k) \right] \\
&= \int d^3k\, \omega_k \left[a^\dagger(k)a(k) + \frac{1}{2} \right] \tag{3.20}
\end{aligned}
$$

Similarly, we can calculate the momentum $\mathbf{P}$:

$$
\begin{aligned}
\mathbf{P} &= -\int \pi \nabla \phi\, d^3x \\
&= \frac{1}{2} \int d^3k\, \mathbf{k} \left[a^\dagger(k)a(k) + a(k)a^\dagger(k) \right]
\end{aligned}
$$

$$= \int d^3k \; \mathbf{k} \left[a^\dagger(k)a(k) + \frac{1}{2} \right] \tag{3.21}$$

(We caution that both the energy and momentum are actually divergent because of the factor $\frac{1}{2}$ appearing in the infinite sum. We will clarify this important point shortly.)

With these expressions, it is now easy to check that the operators P^μ and $M^{\mu\nu}$ generate translations and Lorentz rotations, as they should:

$$i[P_\mu, \phi] = \partial_\mu \phi$$

$$i[M^{\mu\nu}, \phi] = (x^\mu \partial^\nu - x^\nu \partial^\mu)\phi \tag{3.22}$$

If we exponentiate the generators of translations and Lorentz rotations, we can calculate how the field $\phi(x)$ transforms under the Poincaré group. Let us define:

$$U(\Lambda, a) = \exp\left(-\frac{i}{2}\epsilon_{\mu\nu}M^{\mu\nu} + ia_\mu P^\mu \right) \tag{3.23}$$

where $\Lambda_{\mu\nu} = g_{\mu\nu} + \epsilon_{\mu\nu} + \cdots$. Then it is straightforward to show:

$$U(\Lambda, a)\phi(x)U^{-1}(\Lambda, a) = \phi(\Lambda x + a) \tag{3.24}$$

This demonstrates that $\phi(x)$ transforms as a scalar field under the Poincaré group.

Now that we have successfully shown how to quantize the Klein–Gordon field, we must now calculate the eigenstates of the Hamiltonian to find the spectrum of states. Let us now define the "vacuum" state as follows:

$$a(k)|0\rangle = 0 \tag{3.25}$$

By convention, we call $a(k)$ an "annihilation" operator. We define a one-particle state via the "creation" operator as a Fock space:

$$a^\dagger(k)|0\rangle = |k\rangle \tag{3.26}$$

The problem with this construction, however, is that the energy associated with the vacuum state is formally infinite because of the presence of $1/2$ in the sum in Eq. (3.20). We will simply drop this infinite term, since infinite shifts in the Hamiltonian cannot be measured. Dropping the zero point energy in the expression for harmonic oscillators has a simple counterpart in x space. The zero-point energy emerged when we commuted creation and annihilation operators past each other. Dropping the zero-point energy is therefore equivalent to moving all creation operators to the left and annihilation operators to the right. This operation, in x space, can be accomplished by "normal ordering." Since the product of two or more fields at the same point is formally divergent, we can remove this divergence by the normal ordering operation, which corresponds to moving the part containing the creation operators to the left and the annihilation operators to the right. If we decompose $\phi = \phi^+ + \phi^-$, where $-$ ($+$) represent the creation (annihilation) part of an operator with negative (positive) frequency, then we define:

$$: \phi_1 \phi_2 := \phi_1^\dagger \phi_2^\dagger + \phi_1^- \phi_2^+ + \phi_1^- \phi_2^- + \phi_2^- \phi_1^\dagger \tag{3.27}$$

Then, by applying the normal ordering to the definition of the Hamiltonian, we can simply drop the $\frac{1}{2}$ appearing in Eq. (3.20). From now on, we assume that when two fields are multiplied at the same point in space–time, they are automatically normal ordered.

Once we have normal ordered the operators, we now have an explicitly positive Hamiltonian. In this fashion, we have been able to handle the question of negative energy states for the Klein–Gordon theory. (More subtleties concerning negative energy states will be discussed when we analyze the Dirac equation.)

One essential point in introducing these creation and annihilation operators is that we can write down the N-particle Fock space:

$$|k_1, k_2, \cdots, k_N\rangle = a^\dagger(k_1) a^\dagger(k_2) \cdots a^\dagger(k_N) |0\rangle \tag{3.28}$$

This is the chief distinguishing feature between first and second quantization. In first quantization, we quantized the x_i corresponding to a single particle. First quantized systems were hence inherently based on single-particle dynamics. In the second quantized formalism, by contrast, we quantize *multiparticle states*. To count how many particles we have of a certain momentum, we introduce the "number" operator:

$$N = \int d^3k \, a^\dagger(k) a(k) \tag{3.29}$$

The advantage of this number operator is that we can now calculate how many particles there are of a certain momentum. For example, let $|n(k)\rangle$ equal a state consisting of $n(k)$ identical particles with momentum k:

$$|n(k)\rangle = \frac{a^\dagger(k)^{n(k)}}{\sqrt{n(k)!}} |0\rangle \tag{3.30}$$

It is easy to show [by commuting $a(k)$ to the right, until they annihilate on the vacuum] that:

$$N|n(k)\rangle = n(k)|n(k)\rangle \tag{3.31}$$

that is, N simply counts the number of states there are at momentum k. Not surprisingly, a multiparticle state, consisting of many particles of different momenta, can be represented as a Fock space:

$$|n(k_1)n(k_2)\cdots n(k_m)\rangle = \prod_{i=1}^{m} \frac{\left(a^\dagger(k_i)\right)^{n(k_i)}}{\sqrt{n(k_i)!}}|0\rangle \qquad (3.32)$$

Then the number operator N acting on this multiparticle state just counts the number of particles present:

$$N|n(k_1)n(k_2)\cdots n(k_m)\rangle = \left(\sum_{i=1}^{m} n(k_i)\right)|n(k_1)n(k_2)\cdots n(k_m)\rangle \qquad (3.33)$$

Finally, it is essential to notice that the norm of these multiparticle states is positive. If we define $\langle k| \equiv \langle 0|a(k)$ and set $\langle 0|0\rangle = 1$, then the norm is given by $\langle k|k'\rangle = +\delta^3(\mathbf{k} - \mathbf{k}')$. The norm is positive because the appropriate sign appears in the commutation relation, Eq. (3.18). If the sign of the commutator had been reversed and the norm were negative, then we would have a negative norm state, or "ghost" state, which would give us negative probabilities and would violate unitarity. (For example, we would not be able to write the completeness statement $\mathbf{1} = \sum_n |n\rangle\langle n|$ which is used in unitarity arguments.) To preserve unitarity, it is essential that a physical theory be totally free of ghost states (or that they cancel completely). We will encounter this important question of ghosts repeatedly throughout this book.

3.3 Charged Scalar Field

We can generalize our discussion of the Klein–Gordon field by postulating the existence of several scalar fields. In particular, we can arrange two independent scalar fields into a single complex field:

$$\phi = \frac{1}{\sqrt{2}}(\phi_1 + i\phi_2) \qquad (3.34)$$

The action then becomes:

$$\mathscr{L} = \partial_\mu \phi^\dagger \partial^\mu \phi - m^2 \phi^\dagger \phi \qquad (3.35)$$

If we insert this decomposition into the action, then we find the sum of two independent actions for ϕ_1 and ϕ_2.

The quantization of this action proceeds as before by calculating the conjugate field and postulating the canonical commutation relations. The conjugate field is given by:

$$\pi = \frac{\delta \mathscr{L}}{\delta \dot{\phi}} = \dot{\phi}^{\dagger} \tag{3.36}$$

The commutation relations now read:

$$[\phi(\mathbf{x}, t), \pi(\mathbf{y}, t)] = i\delta^{3}(\mathbf{x} - \mathbf{y}) \tag{3.37}$$

We can always decompose this field in terms of its Fourier components:

$$\phi_{i}(x) = \int \frac{d^{3}k}{\sqrt{(2\pi)^{3}2\omega_{k}}} \left(a_{i}(k)e^{-ik \cdot x} + a_{i}^{\dagger}(k)e^{ik \cdot x} \right) \tag{3.38}$$

Then the canonical commutation relations can be satisfied if the Fourier components obey the following commutation relations:

$$[a_{i}(k), a_{j}^{\dagger}(k')] = \delta^{3}(\mathbf{k} - \mathbf{k}')\delta_{ij} \tag{3.39}$$

All other commutators vanish. We could also choose the decomposition:

$$a(k) = \frac{1}{\sqrt{2}}[a_{1}(k) + ia_{2}(k)]; \quad a^{\dagger}(k) = \frac{1}{\sqrt{2}}\left[a_{1}^{\dagger}(k) - ia_{2}^{\dagger}(k)\right]$$

$$b(k) = \frac{1}{\sqrt{2}}[a_{1}(k) - ia_{2}(k)]; \quad b^{\dagger}(k) = \frac{1}{\sqrt{2}}\left[a_{1}^{\dagger}(k) + ia_{2}^{\dagger}(k)\right] \tag{3.40}$$

For these operators, the new commutation relations read:

$$[a(k), a^{\dagger}(k')] = [b(k), b^{\dagger}(k')] = \delta^{3}(\mathbf{k} - \mathbf{k}') \tag{3.41}$$

All other commutators are zero. Now let us construct the symmetries of the action and the corresponding Noether currents. The action is symmetric under the following transformation:

$$\phi \rightarrow e^{i\theta}\phi; \quad \phi^{\dagger} \rightarrow e^{-i\theta}\phi^{\dagger} \tag{3.42}$$

which generates a $U(1)$ symmetry. Written out in components, we find, as in the previous chapter, the following $SO(2)$ transformation:

$$\begin{pmatrix} \phi_{1}' \\ \phi_{2}' \end{pmatrix} = \begin{pmatrix} \cos\theta & -\sin\theta \\ \sin\theta & \cos\theta \end{pmatrix} \begin{pmatrix} \phi_{1} \\ \phi_{2} \end{pmatrix} \tag{3.43}$$

This symmetry generates a Noether current, which equals:

$$J_\mu = i\phi^\dagger \partial_\mu \phi - i\partial_\mu \phi^\dagger \phi \tag{3.44}$$

Now let us calculate the charge Q corresponding to this current in terms of the quantized operators:

$$
\begin{aligned}
Q &= \int d^3x\, i(\phi^\dagger \dot{\phi} - \dot{\phi}^\dagger \phi) \\
&= \int d^3k \left[a^\dagger(k)a(k) - b^\dagger(k)b(k) \right] \\
&= N_a - N_b
\end{aligned}
\tag{3.45}
$$

where the number operator for the a and b oscillator is given by:

$$N_a = \int d^3k\, a^\dagger(k)a(k); \quad N_b = \int d^3k\, b^\dagger(k)b(k) \tag{3.46}$$

Historically, this conserved current caused a certain amount of confusion. If J^0 is considered to be the probability density of the wave function, then it can be negative, and hence negative probabilities creep into the theory. In fact, Schrödinger originally studied this equation as a candidate for the theory of the electron but abandoned it because of these negative probabilities. As a consequence, he later went on to write another equation that did not suffer from this problem, the celebrated nonrelativistic Schrödinger equation.

However, in 1934 Pauli and Weisskopf[7] finally gave the correct quantum interpretation of these negative probabilities.

First, because of the crucial minus sign appearing in front of the b oscillators, we will find it convenient to redefine the current J_μ as the current corresponding to the *electric charge*, rather than probability density, so that the a oscillators correspond to a positively charged particle and the b oscillators correspond to a negatively charged one. In this way, we can construct the quantum theory of charged scalar particles, where the minus sign appearing in the current is a desirable feature, rather than a fatal illness of the theory. In the next chapter, we will show how to couple this theory to the Maxwell field and hence rigorously show how this identification works.

Second, we will interpret the $b^\dagger$ oscillator as the creation operator for a new state of matter, antimatter. It was Dirac who originally grappled with these new states found in any relativistic theory and deduced the fact that a new form of matter, with opposite charge, must be given serious physical consideration. The discovery of the antielectron gave graphic experimental proof of this conjecture.

Third, we no longer have a simple, single-particle interpretation of the $\phi(x)$, which now contains both the matter and antimatter fields. Thus, we must abandon the strict single-particle interpretation for ϕ and reinterpret it as a field. We will see this unusual feature emerging again when we discuss the Dirac equation.

3.4   Propagator Theory

Now that we have defined the canonical commutation relations among the particle fields, we are interested in how these particles actually move in space–time. To define a propagator, and also anticipate interactions, let us modify the Klein–Gordon equation to include a source term $J(x)$:

$$(\partial_\mu^2 + m^2)\phi(x) = J(x); \partial_\mu^2 \equiv \partial_\mu \partial^\mu \tag{3.47}$$

To solve this equation, we use the standard theory of Green's functions. We first define a propagator that satisfies:

$$(\partial_\mu^2 + m^2)\Delta_F(x - y) = -\delta^4(x - y) \tag{3.48}$$

Then the solution of the interacting ϕ field is given by:

$$\phi(x) = \phi_0(x) - \int d^4x\, \Delta_F(x - y)J(y) \tag{3.49}$$

where $\phi_0(x)$ is any function that satisfies the Klein–Gordon equation without any source term. If we hit both sides of this expression with $(\partial_\mu^2 + m^2)$, then we find that it satisfies the original Klein–Gordon equation in the presence of a source term. As we know from the theory of Green's functions, the way to solve this equation is to take the Fourier transform:

$$\Delta_F(x - y) = \int \frac{d^4k}{(2\pi)^4} e^{-ik(x-y)} \Delta_F(k) \tag{3.50}$$

If we hit both sides of this equation with $(\partial_\mu^2 + m^2)$, then we can solve for $\Delta(k)$:

$$\Delta_F(k) = \frac{1}{k^2 - m^2} \tag{3.51}$$

At this point, however, we realize that there is an ambiguity in this equation. The integral over d^4k cannot be performed on the real axis, because the denominator diverges at $k_\mu^2 = m^2$. This same ambiguity, of course, occurs even in the classical

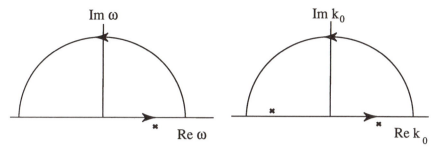

Figure 3.1. Contour integration for the Green's function. The contour on the left gives us the nonrelativistic retarded Green's functions, while the contour on the right gives us the Feynman prescription for a relativistic Green's function.

theory of wave equations, and is not specific to the Lorentz covariant theory. The origin of this ambiguity lies not in the mathematics, but in the physics, in the fact that we have yet to fix our boundary conditions.

For example, consider the Green's function for the Schrödinger equation:

$$\left(i\frac{\partial}{\partial t} - H_0\right) G_0(x - x') = \delta^4(x - x') \tag{3.52}$$

If we take the Fourier transform of this equation and solve for the Green's function, we find:

$$G_0(x - x') = \int \frac{d^4p}{(2\pi)^4} \frac{1}{\omega - \mathbf{p}^2/2m} e^{-ip\cdot(x-x')} \tag{3.53}$$

where $p^\mu = (\omega, \mathbf{p})$. This expression also suffers from an ambiguity, because the integration over ω is divergent.

Let us take the convention that we integrate over the real ω axis as in Figure 3.1, so that we integrate above the singularity. This can be accomplished by inserting a factor of $i\epsilon$ into the denominator, replacing $\omega - p^2/2m$ with $\omega - p^2/2m + i\epsilon$. Then the ω integration can be performed.

We simply convert the integration over the real axis into a contour integration over a complex variable ω. We are going to add the contour integral over the upper half plane (which vanishes) such that:

$$
\begin{aligned}
G_0(x - x') &= \int \frac{d^3p}{(2\pi)^3} e^{i\mathbf{p}\cdot(\mathbf{x}-\mathbf{x}')} \int_{-\infty}^{\infty} \frac{d\omega}{(2\pi)} \frac{e^{-i\omega(t-t')}}{\omega - \mathbf{p}^2/2m + i\epsilon} \\
&= -i \int \frac{d^3p}{(2\pi)^3} e^{i\mathbf{p}\cdot(\mathbf{x}-\mathbf{x}')-i(p^2/2m)(t-t')} \theta(t - t') \tag{3.54}
\end{aligned}
$$

where the θ function has been written as:

$$
\theta(t - t') = -\lim_{\epsilon \to 0} \frac{1}{2\pi i} \int_{-\infty}^{\infty} \frac{e^{-i\omega(t-t')} \, d\omega}{\omega + i\epsilon}
$$

$$
= \begin{cases} 1 & \text{if } t > t' \\ \\ 0 & \text{otherwise} \end{cases} \tag{3.55}
$$

which equals 1 for $t > t'$, and vanishes otherwise. (To prove this last relation, extend the contour integral into a semicircle in the complex ω plane, closing the contour in the lower half plane when $t > t'$. Then the contour integral picks up the pole at $\omega = -i\epsilon$.)

To see this a bit more explicitly, let us define:

$$
\phi_p(x) \equiv \frac{e^{-ip \cdot x}}{(2\pi)^{3/2}} \tag{3.56}
$$

Then the Green's function can be written as:

$$
G_0(x - x') = -i\theta(t - t') \int d^3p \, \phi_p(x)\phi_p^*(x') \tag{3.57}
$$

In this way, the $+i\epsilon$ insertion has selected out the *retarded* Green's function, which obeys the usual concept of causality. Taking the $-i\epsilon$ prescription would have given us the *advanced* Green's function, which would violate causality.

Finally, taking the d^3p integration (which is simply a Gaussian integral), we find the final result for the Green's function:

$$
G_0(x - x') = -i \left(\frac{m}{2\pi i(t - t')} \right)^{3/2} \exp \left\{ \frac{im|\mathbf{x} - \mathbf{x}'|^2}{2(t - t')} \right\} \theta(t - t') \tag{3.58}
$$

which is just the Green's function found in ordinary quantum mechanics.

Now that we have seen how various prescriptions for $i\epsilon$ give us various boundary conditions, let us apply this knowledge to the relativistic case and choose the following, unorthodox prescription:

$$
\Delta_F(k) = \frac{1}{k^2 - m^2 + i\epsilon} \tag{3.59}
$$

To see how this $i\epsilon$ prescription modifies the boundary conditions, we will find it useful to decompose this as:

$$
\frac{1}{k^2 - m^2 + i\epsilon} = \frac{1}{2k^0} \left(\frac{1}{k^0 - \sqrt{m^2 + \mathbf{k}^2} + i\epsilon} + \frac{1}{k^0 + \sqrt{m^2 + \mathbf{k}^2} - i\epsilon} \right) \tag{3.60}
$$

The integral over k^0 now picks up contributions from both terms. Performing the integration as before, we find:

$$
\begin{aligned}
\Delta_F(x - x') &= -i\theta(t - t') \int \frac{d^3k}{(2\pi)^3 2\omega_k} e^{-ik\cdot(x-x')} \\
&\quad -i\theta(t' - t) \int \frac{d^3k}{(2\pi)^3 2\omega_k} e^{ik\cdot(x-x')} \\
&= -i\theta(t - t') \int \frac{d^3k}{(2\pi)^3 2\omega_k} \phi_p(x)\phi_p^*(x') \\
&\quad -i\theta(t' - t) \int \frac{d^3k}{(2\pi)^3 2\omega_k} \phi_p(x)^* \phi_p(x')
\end{aligned}
\tag{3.61}
$$

We see the rather unusual feature of this prescription: positive energy solutions are carried forward in time, but negative energy solutions are carried *backwards* in time.

In classical physics, the usual solutions of the Maxwell theory give us retarded and advanced waves, and we eliminate the advanced waves by a choice of boundary conditions. However, in the quantum theory we are encountering a new type of propagator that, classically, makes no sense, with negative energy solutions going backwards in time. This propagator never appears in classical physics because it is complex and is hence forbidden.

Quantum mechanically, negative energy solutions are an inherent problem with any relativistic theory. Even if we ban them at the beginning, quantum interactions will inevitably re-create them later. However, as we saw in the previous section, these negative energy states can be reinterpreted. Feynman's approach to this problem was to assume that these negative energy states, because they are going backwards in time, appear as a new form of matter with positive energy going forwards in time, antimatter. Although matter going backwards in time seems to contradict causality, this poses no problem because one can show that, experimentally, a system where matter is going backwards in time is indistinguishable (if we reverse certain quantum numbers such as charge) from antimatter going forwards in time. For example, an electron placed in an electric field may move to the right; however, if it is moving backwards in time, it appears to move to the left. However, this is indistinguishable experimentally from a positively charged electron moving forwards in time to the left. In this way, we can interpret this theory as one in which everything (matter plus antimatter) has positive energy. (We will discuss this new reinterpretation further when we analyze the Dirac theory.)

The previous expression for the Green's function was written in terms of plane waves ϕ_p. However, we can replace the plane wave ϕ_p by the quantum field $\phi(x)$ if we take the vacuum expectation value of the product of fields. From the previous equation Eq. (3.61), we easily find:

$$i\Delta_F(x - x') = \langle 0|T\phi(x)\phi(x')|0\rangle \tag{3.62}$$

where T is called the time-ordered operator, defined as:

$$T\phi(x)\phi(x') = \begin{cases} \phi(x)\phi(x') & \text{if } t > t' \\ \phi(x')\phi(x) & \text{if } t' > t \end{cases} \tag{3.63}$$

T makes sure that the operator with the latest time component always appears to the left. This equation for Δ_F is our most important result for propagators. It gives us a bridge between the theory of propagators, in which scattering amplitudes are written in terms of $\Delta_F(x - x')$, and the theory of operators, where everything is written in terms of quantum field $\phi(x)$. The previous expression will be crucial to our discussion when we calculate the S matrix for QED.

Finally, we remark that our theory must obey the laws of causality. For our purposes, we will define microscopic causality as the statement that information cannot travel faster than the speed of light. For field theory, this means that $\phi(x)$ and $\phi(y)$ cannot interact with each other if they are separated by space-like distances. Mathematically, this means that the commutator between these two fields must vanish for space-like separations.

Repeating the earlier steps, we can show that this commutator equals:

$$\begin{aligned} [\phi(x), \phi(y)] &= i\Delta(x - y) \\ &= \int \frac{d^4k}{(2\pi)^3} \delta(k^2 - m^2)\epsilon(k_0)e^{-ik(x-y)} \\ &= \frac{i}{4\pi r}\frac{\partial}{\partial r} \begin{cases} J_0(m\sqrt{t^2 - r^2}) & t > r \\ 0 & -r < t < r \\ -J_0(m\sqrt{t^2 - r^2}) & t < -r \end{cases} \end{aligned} \tag{3.64}$$

where $\epsilon(k)$ equals $+1(-1)$ for positive (negative) k, $t = x^0 - y^0$, $r = |\mathbf{x} - \mathbf{y}|$, and J_0 is the Bessel function. (To prove this, convert the integral to radial coordinates, and then perform the k^0 and $|\mathbf{k}|$ integrations.)

With this explicit form for the commutator, we can easily show that, for space-like separations, we have:

$$\Delta(x - y) = 0 \text{ if } (x - y)^2 < 0 \tag{3.65}$$

This shows that our construction obeys microscopic causality.

3.5   Dirac Spinor Field

After Dirac considered the relativistic theory of radiation in 1927, he set out the next year to construct the relativistic theory of electrons. One severe limitation was the problem of negative probabilities. He started with the observation that the nonrelativistic Schrödinger equation did not have negative probabilities because it was linear in time, while the Klein–Gordon equation, being quadratic in time, did have negative probabilities.

Therefore Dirac tried to find a wave equation that was linear in time but still satisfied the relativistic mass-shell constraint:

$$p_\mu p^\mu = E^2 - \mathbf{p}^2 = m^2 \tag{3.66}$$

Dirac's original idea was to take the "square root" of the energy equation. In this way, he stumbled onto the spinorial representation of the Lorentz group discussed in Chapter 2. He began with a first-order equation:

$$i\frac{\partial \psi}{\partial t} = \left(-i\alpha_i \nabla^i + \beta m\right)\psi \tag{3.67}$$

where α_i and β are now constant matrices, not ordinary c numbers, which act on ψ, a column matrix.

By squaring the operator in front of the ψ field, we want to recover the mass-shell condition:

$$
\begin{aligned}
-\frac{\partial^2}{\partial t^2}\psi &= (-i\boldsymbol{\alpha}\cdot\nabla + \beta m)^2\,\psi \\
&= (-\nabla^2 + m^2)\psi
\end{aligned}
\tag{3.68}
$$

This is only possible if we demand that the matrices satisfy:

$$
\begin{aligned}
\{\alpha_i, \alpha_k\} &= 2\delta_{ik} \\
\{\alpha_i, \beta\} &= 0 \\
\alpha_i^2 &= \beta^2 = 1
\end{aligned}
\tag{3.69}
$$

To make the equations more symmetrical, we can then define $\gamma^0 = \beta$ and $\gamma^i = \beta\alpha^i$. Multiplying the wave equation by β, we then have the celebrated Dirac equation[8]:

$$(i\gamma^\mu \partial_\mu - m)\psi = 0 \tag{3.70}$$

where the γ^μ matrices satisfy:

$$\{\gamma^\mu, \gamma^\nu\} = 2g^{\mu\nu} \tag{3.71}$$

It is no accident that such a relativistic construction is possible. After all, in the previous chapter we studied spinor representations of $O(N)$ in Section 2.6 by defining a Clifford algebra, which is precisely the algebra formed by the γ^μ. Thus, what we are really constructing is the spin $\frac{1}{2}$ representation of the Lorentz group, that is, the spinors.

To calculate the behavior of this equation under the Lorentz group, let us define how spinors transform under some representation $S(\Lambda)$ of the Lorentz group:

$$\psi'(x') = S(\Lambda)\psi(x) \tag{3.72}$$

Then the Dirac equation transforms as follows:

$$[iS^{-1}(\Lambda)\gamma^\mu S(\Lambda)\partial_\mu - m]\psi = [i\gamma^\mu(\Lambda)^\nu_\mu \partial_\nu - m]\psi = 0 \tag{3.73}$$

where we have multiplied the transformed Dirac equation by $S^{-1}(\Lambda)$ on the left, and we have taken into account the transformed $\partial'_\mu = (\Lambda)^\nu_\mu \partial_\nu$. In order for the equation to be Lorentz covariant, we must therefore have the following relation:

$$S(\Lambda)\gamma^\mu S^{-1}(\Lambda) = (\Lambda^{-1})^\mu_\nu \gamma^\nu \tag{3.74}$$

which we first encountered in Section 2.6. To find an explicit representation for $S(\Lambda)$, let us introduce the following matrix:

$$\sigma_{\mu\nu} = \frac{i}{2}[\gamma_\mu, \gamma_\nu] \tag{3.75}$$

In Chapter 2, we saw that $(i/4)[\Gamma_\mu, \Gamma_\nu]$ are the generators of $O(N)$ in the spinor representation. Thus, the $\sigma_{\mu\nu}/2$ are the generators of the Lorentz group in this representation.

Thus, we can write a new Lorentz group generator that is the sum of the old generator $L_{\mu\nu}$ (which acts on the space–time coordinate) plus a new piece that also generates the Lorentz group but in the spinor representation:

$$M_{\mu\nu} = L_{\mu\nu} + \frac{1}{2}\sigma_{\mu\nu} \tag{3.76}$$

The $\sigma_{\mu\nu}$ also obey the following relation:

$$[\gamma^\mu, \sigma_{\alpha\beta}] = 2i(\delta^\mu_\alpha \gamma_\beta - \delta^\mu_\beta \gamma_\alpha) \tag{3.77}$$

which shows that the Dirac matrices transform as vectors under the spinor representation of the Lorentz group.

In terms of this new matrix, we can find an explicit representation of the $S(\Lambda)$ matrix:

$$S(\Lambda) = e^{-(i/4)\sigma_{\mu\nu}\omega^{\mu\nu}} \tag{3.78}$$

Now that we know how spinors transform under the Lorentz group, we would like next to construct invariants under the group. Let us take the Hermitian conjugate of the Dirac equation:

$$\psi^\dagger (i\gamma^{\dagger\mu}\overleftarrow{\partial}_\mu + m) = 0 \tag{3.79}$$

We will show shortly that there exists a representation of the Dirac matrices that satisfies:

$$
\begin{aligned}
(\gamma^0)^\dagger &= \gamma^0 \\
(\gamma^i)^\dagger &= -\gamma^i
\end{aligned}
\tag{3.80}
$$

where γ^i is anti-Hermitian and γ^0 is Hermitian. This can also be written as $\gamma^{\mu\dagger} = \gamma_\mu$. (It may be puzzling that we did not take a representation that was completely Hermitian. However, as we mentioned earlier, there are no finite-dimensional unitary representations of the Lorentz group. If a purely Hermitian representation of the Dirac matrices could be found, then we could construct the generators of the Lorentz group out of them that would be unitary, violating this theorem. Thus, we are forced to take non-Hermitian representations.)

Now let us define:

$$\bar{\psi} \equiv \psi^\dagger \gamma^0 \tag{3.81}$$

If we hit the conjugated equation of motion with γ^0, we can replace the $\gamma^\dagger$ with γ matrices, leaving us with:

$$\bar{\psi}(i\gamma^\mu\overleftarrow{\partial}_\mu + m) = 0 \tag{3.82}$$

Under a Lorentz transformation, this new field $\bar{\psi}$ obeys:

$$
\begin{aligned}
\bar{\psi}'(x') &= \bar{\psi}(x)\gamma^0 S(\Lambda)^\dagger \gamma^0 \\
&= \bar{\psi}(x)S^{-1}(x)
\end{aligned}
\tag{3.83}
$$

This is just what we need to form invariants and covariant tensors. For example, notice that $\bar{\psi}\psi$ is an invariant under the Lorentz group:

$$\bar{\psi}'(x')\psi'(x') = \bar{\psi}(x)S^{-1}(\Lambda)S(\Lambda)\psi = \bar{\psi}(x)\psi(x) \qquad (3.84)$$

Similarly, $\bar{\psi}\gamma^\mu\psi$ is a genuine vector under the Lorentz group. We find:

$$\bar{\psi}'(x')\gamma^\mu\psi'(x') = \bar{\psi}(x)S^{-1}\gamma^\mu S\psi = \bar{\psi}(x)\Lambda^\mu{}_\nu\gamma^\nu\psi(x) \qquad (3.85)$$

where we have used the fact that:

$$S^{-1}\gamma^\mu S = \Lambda^\mu{}_\nu\gamma^\nu \qquad (3.86)$$

which is nothing but the statement that the γ^μ transform as vectors under the spinor representation of the Lorentz group, as we saw in Chapter 2. (To prove this formula, take an infinitesimal Lorentz transformation. Then $S^{-1}\gamma^\mu S$ becomes proportional to the commutator between $\sigma_{\lambda\rho}$ and γ^μ, which is gives just another gamma matrix. If we then exponentiate this process for finite transformations, we find the previous equation, as desired.)

In the same manner, it is also straightforward to show that $\bar{\psi}\sigma^{\mu\nu}\psi$ transforms as a genuine antisymmetric second-rank tensor under the Lorentz group. To find other Lorentz tensors that can be represented as bilinears in the spinors, let us follow our discussion of Chapter 2 and introduce the matrix:

$$\gamma_5 = \gamma^5 = i\gamma^0\gamma^1\gamma^2\gamma^3 = -\frac{i}{4!}\epsilon_{\mu\nu\sigma\rho}\gamma^\mu\gamma^\nu\gamma^\sigma\gamma^\rho \qquad (3.87)$$

where $\epsilon^{\mu\nu\sigma\rho} = -\epsilon_{\mu\nu\sigma\rho}$ and $\epsilon^{0123} = +1$. Because γ_5 transforms like $\epsilon^{\mu\nu\rho\sigma}$, it is a pseudoscalar; that is, it changes sign under a parity transformation. Thus, $\bar{\psi}\gamma_5\psi$ is a pseudoscalar.

In fact, the complete set of bilinears, their transformation properties, and the number of elements within each tensor are given by:

$$
\begin{array}{rcl}
\text{Scalar}: & \bar{\psi}\psi & [1] \\
\text{Vector}: & \bar{\psi}\gamma^\mu\psi & [4] \\
\text{Tensor}: & \bar{\psi}\sigma^{\mu\nu}\psi & [6] \\
\text{Pseudovector}: & \bar{\psi}\gamma_5\gamma^\mu\psi & [4] \\
\text{Pseudoscalar}: & \bar{\psi}\gamma_5\psi & [1]
\end{array}
\qquad (3.88)
$$

There is a total of 16 independent components in this table. We can show that the following 16 matrices are linearly independent;

$$\Gamma^A = \{I, \gamma_\mu, \sigma_{\mu\nu}, \gamma_5\gamma_\mu, \gamma_5\} \tag{3.89}$$

where $(\Gamma^A)^2 = \pm 1$. To show that this set of 16 matrices forms a complete set, let us assume, for the moment, that a relation exists among them, so that:

$$\sum_A c_A \Gamma^A = 0 \tag{3.90}$$

where c_A are numbers. Then multiply this by Γ^B and take the trace. If $\Gamma^B = I$, we find that $c_I = 0$. If $\Gamma^B \neq I$, then we use the fact that that there exists Γ_C not equal to unity such that $\Gamma^A \Gamma^B = \Gamma^C$ if $A \neq B$. Taking the trace, we find that $c_B = 0$. Since B was arbitrary, this means that all coefficients are zero; so these 16 matrices must be linearly independent

Because γ^μ transforms as a vector under the Lorentz group, the following Lagrangian is invariant under the Lorentz group:

$$\mathcal{L} = \bar{\psi}(i\gamma^\mu \partial_\mu - m)\psi \tag{3.91}$$

This, in turn, is the Lagrangian corresponding to the Dirac equation. Variations of this equation by ψ or by $\bar{\psi}$ will generate the two versions of the Dirac equation.

Up to now, we have not said anything specific about the representation of the Dirac matrices themselves. In fact, a considerable number of identities can be derived for these matrices in four dimensions without ever mentioning a specific representation, such as:

$$
\begin{aligned}
\gamma^\mu \gamma_\mu &= 4 \\
\gamma^\rho \gamma^\mu \gamma_\rho &= -2\gamma^\mu \\
\gamma^\rho \gamma^\mu \gamma^\nu \gamma_\rho &= 4g^{\mu\nu} \\
\gamma^\rho \gamma^\mu \gamma^\nu \gamma^\sigma \gamma_\rho &= -2\gamma^\sigma \gamma^\nu \gamma^\mu
\end{aligned}
\tag{3.92}
$$

Some trace operations can also be defined:

$$
\begin{aligned}
\mathrm{Tr}\,(\gamma^5 \gamma^\mu) &= \mathrm{Tr}\,\sigma^{\mu\nu} = \mathrm{Tr}\,\gamma^\mu \gamma^\nu \gamma^5 = 0 \\
\mathrm{Tr}\,(\gamma^\mu \gamma^\nu) &= 4g^{\mu\nu} \\
\mathrm{Tr}\,(\gamma^\mu \gamma^\nu \gamma^\rho \gamma^\sigma) &= 4(g^{\mu\nu}g^{\rho\sigma} - g^{\mu\rho}g^{\nu\sigma} + g^{\mu\sigma}g^{\nu\rho}) \\
\mathrm{Tr}\,(\gamma^5 \gamma^\mu \gamma^\nu \gamma^\rho \gamma^\sigma) &= 4i\epsilon^{\mu\nu\rho\sigma}
\end{aligned}
\tag{3.93}
$$

In particular, this means:

$$\text{Tr}\,(\slashed{a}\,\slashed{b}\,\slashed{c}\,\slashed{d}) = 4\,[(a\cdot b)(c\cdot d) - (a\cdot c)(b\cdot d) + (a\cdot d)(b\cdot c)] \qquad (3.94)$$

where $\slashed{a} \equiv a_\mu \gamma^\mu$.

It is often convenient to find an explicit representation of the Dirac matrices. The most common representation of these matrices is the *Dirac representation*:

$$\gamma^0 = \begin{pmatrix} I & 0 \\ 0 & -I \end{pmatrix}; \quad \gamma^i = \begin{pmatrix} 0 & \sigma^i \\ -\sigma^i & 0 \end{pmatrix}$$

$$\beta = \begin{pmatrix} I & 0 \\ 0 & -I \end{pmatrix}; \quad \alpha^i = \begin{pmatrix} 0 & \sigma^i \\ \sigma^i & 0 \end{pmatrix} \qquad (3.95)$$

where σ^i are the familiar Pauli spin matrices. Then the spinor ψ is a complex-valued field with four components describing a massive, spin $\frac{1}{2}$ field.

Now let us try to decompose $\psi(x)$ into plane waves in order to begin canonical quantization. To do this, we need to find a set of independent basis spinors for ψ. We will make the obvious choice:

$$u_1(0) = \begin{pmatrix} 1 \\ 0 \\ 0 \\ 0 \end{pmatrix}; \quad u_2(0) = \begin{pmatrix} 0 \\ 1 \\ 0 \\ 0 \end{pmatrix}; \quad v_1(0) = \begin{pmatrix} 0 \\ 0 \\ 1 \\ 0 \end{pmatrix}; \quad v_2(0) = \begin{pmatrix} 0 \\ 0 \\ 0 \\ 1 \end{pmatrix}$$

$$(3.96)$$

The trick is to act upon these spinors with $S(\Lambda)$ in order to boost them up to momentum p. The momentum-dependent spinors are given by:

$$\begin{aligned} u_\alpha(p) &= S(\Lambda)u_\alpha(0) \\ v_\alpha(p) &= S(\Lambda)v_\alpha(0) \end{aligned} \qquad (3.97)$$

which can be shown to obey:

$$\begin{aligned} (\gamma\cdot p - m)u(p) &= 0 \\ (\gamma\cdot p + m)v(p) &= 0 \\ \bar{u}(p)(\gamma\cdot p - m) &= 0 \\ \bar{v}(p)(\gamma\cdot p + m) &= 0 \end{aligned} \qquad (3.98)$$

The Lorentz transformation matrix $S(\Lambda)$ is not difficult to construct if we set all rotations to zero, leaving us with only Lorentz boosts. Then the only generators we have are the K generators, which in turn are proportional to σ^i. Specifically, we have:

$$S(\Lambda) = \begin{pmatrix} \cosh(\phi/2) & \boldsymbol{\sigma} \cdot \mathbf{n} \sinh(\phi/2) \\ \boldsymbol{\sigma} \cdot \mathbf{n} \sinh(\phi/2) & \cosh(\phi/2) \end{pmatrix}$$

$$= \sqrt{\frac{E+m}{2m}} \begin{pmatrix} 1 & \frac{\boldsymbol{\sigma} \cdot \mathbf{p}}{E+m} \\ \frac{\boldsymbol{\sigma} \cdot \mathbf{p}}{E+m} & 1 \end{pmatrix}$$

where $\cosh(\phi/2) = [(E+m)/2m]^{1/2}$ and $\sinh(\phi/2) = [(E-m)/2m]^{1/2}$.

Applying $S(\Lambda)$ to the independent spinor basis, we easily find:

$$u_1(p) = \sqrt{\frac{E+m}{2m}} \begin{pmatrix} 1 \\ 0 \\ \frac{p_z}{E+m} \\ \frac{p_+}{E+m} \end{pmatrix}, \quad u_2(p) = \sqrt{\frac{E+m}{2m}} \begin{pmatrix} 0 \\ 1 \\ \frac{p_-}{E+m} \\ \frac{-p_z}{E+m} \end{pmatrix} \tag{3.99}$$

$$v_1(p) = \sqrt{\frac{E+m}{2m}} \begin{pmatrix} \frac{p_z}{E+m} \\ \frac{p_+}{E+m} \\ 1 \\ 0 \end{pmatrix}, \quad v_2(p) = \sqrt{\frac{E+m}{2m}} \begin{pmatrix} \frac{p_-}{E+m} \\ \frac{-p_z}{E+m} \\ 0 \\ 1 \end{pmatrix} \tag{3.100}$$

where $p_\pm = p_x \pm i p_y$.

Because of the particular decomposition we have chosen, the u spinors correspond to electrons with positive energy particles (moving forwards in time), while the v spinors correspond to electrons with negative energy (moving backwards in time).

Next, we would like to describe spinors of definite spin. In many experiments, we can produce polarized beams of electrons; so it becomes important to understand how to incorporate projection operators that can select definite spin.

This is not as simple as one might suspect, since the intuitive concept of spin is rooted in our notion of the rotation group, which is only a subgroup of the Lorentz group. Hence, the naive concept of spin and its eigenfunctions no longer applies for boosted systems.

In the rest frame, however, we know that the spin of a system can be described by a three-vector $\mathbf{s}$ that points in a certain direction; so we may introduce the four-vector s_μ which, in its rest frame, reduces to $s_\mu = (0, \mathbf{s})$. Then, by demanding that this transform as a four-vector, we can boost this spin vector by a Lorentz transformation. Since we define $\mathbf{s}^2 = 1$, this means that $s_\mu^2 = -1$. In the rest frame, we have $p_\mu = (m, 0)$; so we also have $p_\mu s^\mu = 0$, which must also hold in any boosted frame by Lorentz invariance. Thus, we now have two Lorentz-invariant conditions on the spin four-vector:

$$s_\mu^2 = -1$$
$$p_\mu s^\mu = 0 \tag{3.101}$$

Next, we would like to define a projection operator that selects out states of definite spin. Again, we will define the Lorentz-invariant projection operator by first examining the rest frame. At rest, we know that the operator $\boldsymbol{\sigma} \cdot \mathbf{s}$ serves as an operator that determines the spin of a system:

$$\boldsymbol{\sigma} \cdot \mathbf{s} u_\alpha(0) = u_\alpha(0)$$
$$\boldsymbol{\sigma} \cdot \mathbf{s} v_\alpha(0) = -v_\alpha(0) \tag{3.102}$$

For a spin $\frac{1}{2}$ system, the projection operator at rest can be written as:

$$P(\mathbf{s}) = \frac{1 \pm \boldsymbol{\sigma} \cdot \mathbf{s}}{2} \tag{3.103}$$

where the $+$ refers to the u spinor, and the $-$ refers to the v spinor. Our goal is to write a boosted version of this expression. Let us define the projection operator:

$$P(s) \equiv \frac{1 + \gamma_5 \not{s}}{2} \tag{3.104}$$

It is easy to show that, in the rest frame, this projection reduces to:

$$P(s) = \frac{1}{2} \begin{pmatrix} 1 + \boldsymbol{\sigma} \cdot \mathbf{s} & 0 \\ 0 & 1 - \boldsymbol{\sigma} \cdot \mathbf{s} \end{pmatrix} \tag{3.105}$$

Therefore, this operator reduces to the previous one, so this is the desired expression. The new eigenfunctions now have a spin s associated with them: $u(k, s)$. They satisfy:

$$P(s)u(k, s) = u(k, s)$$
$$P(s)v(k, s) = v(k, s)$$
$$P(-s)u(k, s) = P(-s)v(k, s) = 0 \tag{3.106}$$

These spinors are quite useful for practical calculations because they satisfy certain *completeness relations*. Any four-spinor can be written in terms of linear combinations of the four $u_\alpha(0)$ and $v_\beta(0)$ because they span the space of four-spinors. If we boost these spinors with $S(\Lambda)$, then $u_\alpha(p)$ and $v_\beta(p)$ span the space of all four-spinors satisfying the Dirac equation.

Likewise, $u_\alpha^T(0)v_\beta(0)$, etc. have 16 independent elements, which in turn span the entire space of 4×4 matrices. Thus, $\bar{u}_\alpha(p)v_\beta(p)$, etc. span the space of all 4×4 matrices that also satisfy the Dirac equation.

We first normalize our spinors with the following conventions:

$$\bar{u}(p, s)u(p, s) = 1$$
$$\bar{v}(p, s)v(p, s) = -1 \tag{3.107}$$

With these normalizations, we can show that these spinors obey certain completeness relations:

$$\sum_s u_\alpha(p, s)\bar{u}_\beta(p, s) - v_\alpha(p, s)\bar{v}_\beta(p, s) = \delta_{\alpha\beta} \tag{3.108}$$

For the particular representation we have chosen, we find:

$$u_\alpha(p, s)\bar{u}_\beta(p, s) = \left(\frac{\not{p} + m}{2m} \frac{1 + \gamma_5 \not{s}}{2}\right)_{\alpha\beta} \tag{3.109}$$

and:

$$v_\alpha(p, s)\bar{v}_\beta(p, s) = -\left(\frac{m - \not{p}}{2m} \frac{1 + \gamma_5 \not{s}}{2}\right)_{\alpha\beta} \tag{3.110}$$

If we sum over the helicity s, we have two projection operators:

$$[\Lambda_+(p)]_{\alpha\beta} = \sum_{\pm s} u_\alpha(p, s)\bar{u}_\beta(p, s) = \left(\frac{\not{p} + m}{2m}\right)_{\alpha\beta}$$
$$[\Lambda_-(p)]_{\alpha\beta} = -\sum_{\pm s} v_\alpha(p, s)\bar{v}_\beta(p, s) = \left(\frac{-\not{p} + m}{2m}\right)_{\alpha\beta} \tag{3.111}$$

These projection operators satisfy:

$$\Lambda_\pm^2 = \Lambda_\pm; \quad \Lambda_+\Lambda_- = 0; \quad \Lambda_+ + \Lambda_- = 1 \tag{3.112}$$

Because of the completeness relations, $\Lambda_\pm$ has a simple interpretation: It projects out the positive or negative energy solution.

3.6 Quantizing the Spinor Field

So far, we have only discussed the classical theory. To second quantize the Dirac field, we first calculate the momentum canonically conjugate to the spinor field:

$$\pi(x) = \frac{\delta \mathscr{L}}{\delta \dot{\psi}(x)} = i \psi^{\dagger} \tag{3.113}$$

Let us decompose the spinor field into its Fourier moments:

$$\psi(x) = \int \sqrt{\frac{m}{k_0}} \frac{d^3k}{\sqrt{(2\pi)^3}} \sum_{\alpha=1,2} \left[b_\alpha(k) u_\alpha(k) e^{-ikx} + d_\alpha^{\dagger}(k) v_\alpha(k) e^{ikx} \right]$$

$$\bar{\psi}(x) = \int \sqrt{\frac{m}{k_0}} \frac{d^3k}{\sqrt{(2\pi)^3}} \sum_{\alpha=1,2} \left[b_\alpha^{\dagger}(k) \bar{u}_\alpha(k) e^{ikx} + d_\alpha(k) \bar{v}_\alpha(k) e^{-ikx} \right]$$

$$\tag{3.114}$$

In terms of particles and antiparticles, this particular decomposition gives the following physical interpretation:

$$\psi(x) = \begin{cases} b(p)u(p)e^{-ip\cdot x} & \text{Annihilates positive energy electron} \\ d^{\dagger}(p)v(p)e^{+ip\cdot x} & \text{Creates positive energy positron} \end{cases} \tag{3.115}$$

(Having $d^{\dagger}$ create a positive energy positron can be viewed as annihilating a negative energy electron.)

Repeating the steps we took for the scalar particle, we invert these equations and solve for the Fourier moments in terms of the fields themselves:

$$b_\alpha(k) = \int d^3x \, \bar{U}_k^\alpha(x) \gamma^0 \psi(x)$$

$$b_\alpha^{\dagger}(k) = \int d^3x \, \bar{\psi}(x) \gamma^0 U_k^\alpha(x)$$

$$d_\alpha(k) = \int d^3x \, \bar{\psi}(x) \gamma^0 V_k^\alpha(x)$$

$$d_\alpha^{\dagger}(k) = \int d^3x \, \bar{V}_k^\alpha(x) \gamma^0 \psi(x) \tag{3.116}$$

where:

$$U_k(x) \equiv \sqrt{\frac{m}{k_0 (2\pi)^3}} u(k) e^{-ik\cdot x}$$

$$V_k(x) \equiv \sqrt{\frac{m}{k_0(2\pi)^3}} v(k)e^{ik\cdot x} \qquad (3.117)$$

Now let us insert the Fourier decomposition back into the expression for the Hamiltonian:

$$
\begin{aligned}
H &= \int d^3x \, (\pi\dot\psi - \mathscr{L}) \\
&= \int d^3x \, \bar\psi(i\gamma^0\partial_0\psi) \\
&= \int d^3k \, k_0 \sum_\alpha \left[b_\alpha^\dagger(k)b_\alpha(k) - d_\alpha(k)d_\alpha^\dagger(k) \right] \qquad (3.118)
\end{aligned}
$$

Here we encounter a serious problem. We find that the energy of the Hamiltonian can be negative. There is, however, an important way in which this minus sign can be banished. Let us define the canonical equal-time commutation relations of the fields and conjugate fields with *anticommutators*, instead of commutators:

$$\{\psi_i(\mathbf{x}, t), \psi_j^\dagger(\mathbf{y}, t)\} = \delta^3(\mathbf{x} - \mathbf{y})\delta_{ij} \qquad (3.119)$$

In order to satisfy the canonical anticommutation relations, the Fourier moments must themselves obey anticommutation relations given by:

$$
\begin{aligned}
\{b_\alpha(k), b_{\alpha'}^\dagger(k')\} &= \delta_{\alpha,\alpha'}\delta^3(\mathbf{k} - \mathbf{k}') \\
\{d_\alpha(k), d_{\alpha'}^\dagger(k')\} &= \delta_{\alpha,\alpha'}\delta^3(\mathbf{k} - \mathbf{k}') \qquad (3.120)
\end{aligned}
$$

Now, if we normal order the Hamiltonian, we must also drop the infinite zero point energy, and hence:

$$
\begin{aligned}
H &= \int d^3k \, k_0 \sum_\alpha [b_\alpha^\dagger(k)b_\alpha(k) + d_\alpha^\dagger(k)d_\alpha(k)] \\
\mathbf{P} &= \int d^3k \, \mathbf{k} \sum_\alpha [b_\alpha^\dagger(k)b_\alpha(k) + d_\alpha^\dagger(k)d_\alpha(k)] \qquad (3.121)
\end{aligned}
$$

Thus, the use of anticommutation relations and normal ordering nicely solves the problem of the Hamiltonian with negative energy eigenvalues.

Furthermore, the $d^\dagger$ operators can be interpreted as creation operators for antimatter (or annihilation operators for negative energy electrons). In fact, this was Dirac's original motivation for postulating antimatter in the first place. To see how this interpretation of these new states emerges, we first notice that the Dirac Lagrangian is invariant under:

$$\psi \to e^{i\Lambda}\psi; \quad \bar\psi \to \bar\psi e^{-i\Lambda} \qquad (3.122)$$

Therefore, there should be a conserved current associated with this symmetry. A direct application of Noether's theorem yields:

$$J^\mu = \bar{\psi}\gamma^\mu\psi, \quad \partial_\mu J^\mu = 0 \tag{3.123}$$

which is conserved, if we use the Dirac equation.

Classically, the conserved charge is positive definite since it is proportional to $\psi^\dagger\psi$. This was, in fact, an improvement over the classical Klein–Gordon equation, where the charge could be negative. However, once we quantize the system, the Dirac charge can also be negative. The quantized charge associated with this current is given by:

$$
\begin{aligned}
Q &= \int d^3x\, J^0 = \int d^3x\; : \psi^\dagger\psi : \\
&= \int d^3k \sum_\alpha [b_\alpha^\dagger(k)b_\alpha(k) - d_\alpha^\dagger(k)d_\alpha(k)]
\end{aligned}
\tag{3.124}
$$

This quantity can be negative, and hence cannot be associated with the probability density. However, we can, as in the Klein–Gordon case, interpret this as the current associated with the coupling to electromagnetism; so Q corresponds to the electric charge. In this case, the minus sign in Q is a desirable feature, because it means that $d^\dagger$ is the creation operator of antimatter, that is, a positron with opposite charge to the electron.

Again, this also means that we have to abandon the simple-minded interpretation of ψ as a single-electron wave function, since it now describes both the electron and the antielectron. The anticommutation relations also reproduce the Pauli Exclusion Principle found in quantum mechanics. Because $d_\alpha^\dagger(k)d_\alpha^\dagger(k) = 0$, only one particle can occupy a distinct energy state with definite spin. Thus, a multiparticle state is given by:

$$\prod_{i=1}^N d_{\alpha_i}^\dagger(k_i) \prod_{j=1}^M b_{\alpha_j}^\dagger(k_j)|0\rangle \tag{3.125}$$

with only one particle in any given quantum state. This is the first example of the *spin-statistics theorem*, that field theories defined with integer spin are quantized with commutators and are called *bosons*, while theories with half-integral spins are quantized with anticommutators and are called *fermions*. The existence of two types of statistics, one based on commutators (i.e., Bose–Einstein statistics) and one based on anticommutators (i.e., Fermi–Dirac), has been experimentally observed in a wide variety of physical situations and has been applied to explain the behavior of low-temperature systems and even white dwarf stars. Repeating the same steps that we used for the Klein–Gordon field, we can also compute the energy–momentum tensor and the angular momentum tensor from Noether's theorem. It is easy to show:

$$
\begin{aligned}
T^{\mu\nu} &= i\bar{\psi}\gamma^{\mu}\partial^{\nu}\psi \\
\mathscr{M}^{\lambda\mu\nu} &= i\bar{\psi}\gamma^{\lambda}\left(x^{\mu}\partial^{\nu} - x^{\nu}\partial^{\mu} - \frac{i}{2}\sigma^{\mu\nu}\right)\psi
\end{aligned}
\tag{3.126}
$$

(The Lagrangian term in the energy–momentum tensor can be dropped since the Dirac Lagrangian is zero if the equations of motion are obeyed.)

The conserved angular momentum tensor is therefore:

$$
\begin{aligned}
M^{\mu\nu} &= \int \mathscr{M}^{0\mu\nu}\, d^3x \\
&= \int d^3x\, i\psi^{\dagger}\left(x^{\mu}\partial^{\nu} - x^{\nu}\partial^{\mu} - \frac{i}{2}\sigma^{\mu\nu}\right)\psi
\end{aligned}
\tag{3.127}
$$

The crucial difference between these equations and the Klein–Gordon case is that the angular momentum tensor contains an extra piece, proportional to $\sigma^{\mu\nu}$, which represents the fact that the theory has nontrivial spin $\frac{1}{2}$.

It is then easy to complete this discussion by calculating how a quantized spinor field transforms under the Poincaré group:

$$
\begin{aligned}
i[P_{\mu}, \psi] &= \partial_{\mu}\psi \\
i[M^{\mu\nu}, \psi] &= \left(x^{\mu}\partial^{\nu} - x^{\nu}\partial^{\mu} - \frac{i}{2}\sigma^{\mu\nu}\right)\psi
\end{aligned}
\tag{3.128}
$$

With these operator identities, we can confirm that the quantized spinor field transforms as a spin $\frac{1}{2}$ field under the Poincaré group:

$$
U(\Lambda, a)\psi_{\alpha}(x)U^{-1}(\Lambda, a) = S^{-1}(\Lambda)_{\alpha\beta}\psi_{\beta}(\Lambda x + a)
\tag{3.129}
$$

As we mentioned earlier, one of our fundamental assumptions about quantum field theory is that it must be causal. Not surprisingly, the spin-statistics theorem is also intimately tied to the question of microcausality, that is, that no signals can propagate faster than the speed of light. From our field theory perspective, microcausality can be interpreted to mean that the commutator (anticommutator) of two boson (fermion) fields vanishes for spacelike separations:

$$
\begin{aligned}
[\phi(x), \phi(y)] &= 0 \quad \text{for } (x - y)^2 < 0 \\
\{\psi(x), \bar{\psi}(y)\} &= 0 \quad \text{for } (x - y)^2 < 0
\end{aligned}
\tag{3.130}
$$

To demonstrate the spin-statistics theorem, let us quantize bosons with anticom-mutators and arrive at a contradiction. Repeating earlier steps, we find, for large separations:

$$\langle 0|\{\phi(x), \phi(y)\}|0\rangle = \int \frac{d^3k}{(2\pi)^3 2\omega_k} \left(e^{-ik\cdot(x-y)} + e^{ik\cdot(x-y)}\right)$$

$$\sim \frac{\exp[-m\sqrt{|\mathbf{x}-\mathbf{y}|^2 - (x^0-y^0)^2}]}{|\mathbf{x}-\mathbf{y}|^2 - (x^0-y^0)^2} \qquad (3.131)$$

which clearly violates our original assumption of microcausality. (Likewise, one can prove that fermions quantized with commutators violates microcausality.)

Historically, anticommutators and antimatter were introduced by Dirac, who was troubled that his theory seemed riddled with negative energy states. Since physical systems prefer the state of lowest energy, there is a finite probability that all the electrons in nature would decay into these negative energy states, thereby creating a catastrophe. To solve the problem of negative energy states, Dirac was led to postulate a radically new interpretation of the vacuum (which is consistent with Feynman's interpretation, which we have chosen in this chapter). He postulated that the vacuum consisted of an infinite sea of filled negative energy states. Ordinary matter did not suddenly radiate an infinite amount of energy and cascade down to negative energy because the negative energy sea was completely filled. By the anticommutation relations, only one electron can occupy a negative energy state at a time; so an electron could not decay into the negative energy sea if it was already filled. In this way, electrons of positive energy could not cascade in energy down to negative energy states.

However, once in a while an electron may be knocked out of the negative energy sea, creating a "hole." This hole would act as if it were a particle. Dirac noticed that the absence of an electron of charge $-|e|$ and negative energy $-E$ is equivalent to the presence of a particle of positive charge $+|e|$ and positive energy $+E$. This hole then had positive charge and the same mass as the electron. All particles therefore had positive energy: Both the original positive energy electron as well as the absence of a negative energy electron possessed positive charge E (Fig. 3.2).

Dirac postulated that this hole would correspond to a new state of matter, an antielectron (although he initially considered the possibility that the hole was a proton). The vacuum was now elevated to an infinite storehouse of negative energy matter.

Dirac's hole theory meant that a new physical process was possible, pair production, where matter appeared out of the empty vacuum. Photons could knock an electron out of its negative energy sea, leaving us with an electron and its hole, that is, an electron and an antielectron.

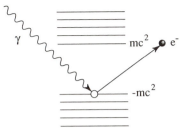

Figure 3.2. Dirac's "hole" picture. When a photon kicks an electron out of the infinite negative energy sea, it leaves a "hole" that behaves as if it had positive energy and positive charge (i.e., an anti-electron). This is pair production.

At first, Dirac's theory of an infinite sea of filled negative energy states was met with extreme skepticism. In the *Handbuch der Physik*, Pauli wrote: "Thus γ ray photons (at least two in order to satisfy the laws of conservation of energy and momentum) must be able to transform, by themselves, into an electron and an antielectron. We do not believe, therefore, that this explanation can be seriously considered." When Pauli's discouraging article appeared, Anderson had already observed the antielectron in cloud chamber photographs, verifying Dirac's conjecture. When confronted with the undeniable experimental verification of antimatter, Pauli later revised his opinion of Dirac's theory and made his famous remark, "... with his fine instinct for physical realities, he started his argument without knowing the end of it."

The interpretation that we have chosen in this chapter (that negative energy electrons going backwards in time are equivalent to positive energy antielectrons going forwards in time) is equivalent to Dirac's infinite negative energy sea. In fact, when we subtracted off an infinite constant in the Hamiltonian, this can be interpreted as subtracting off the energy of Dirac's infinite sea of negative energy states.

To see how these positive and negative energy states move in time, let us define the evolution of a wave function via a source as follows:

$$(i\gamma^\mu \partial_\mu - m)\psi(x) = J(x) \tag{3.132}$$

To solve this equation, we introduce the Dirac propagator by:

$$(i\gamma^\mu \partial_\mu - m)S_F(x - y) = \delta^4(x - y) \tag{3.133}$$

Then the solution to the wave equation is given by:

$$\psi(x) = \psi_0(x) + \int d^4y \, S_F(x - y)J(y) \tag{3.134}$$

where ψ_0 solves the homogeneous Dirac equation.

An explicit representation of the Dirac propagator can be obtained by using the Fourier transform:

$$S_F(x - y) = \int \frac{d^4 p}{(2\pi)^4} e^{-ip\cdot(x-y)} \frac{\gamma^\mu p_\mu + m}{p^2 - m^2 + i\epsilon} \tag{3.135}$$

which satisfies:

$$S_F(x - y) = (i\gamma^\mu \partial_\mu + m)\Delta_F(x - y) \tag{3.136}$$

where $\Delta_F(x - y)$ is the Klein–Gordon propagator.

As before, we can solve for the propagator by integrating over k^0. The integration is identical to the one found earlier for the Klein–Gordon equation, except now we have additional factors of $\not{p} + m$ in the numerator.

Integrating out the energy, we can write the Green's function in terms of plane waves. The result is almost identical to the expression found for the Klein–Gordon propagator, except for the insertion of gamma matrices:

$$
\begin{aligned}
S_F(x - x') &= -i \int \frac{d^3 p}{(2\pi)^3} \frac{m}{E} [\Lambda_+(p)e^{-ip\cdot(x-x')}\theta(t - t') \\
&\quad + \Lambda_-(p)e^{ip\cdot(x-x')}\theta(t' - t)] \\
&= \int d^3 p \left[-i\theta(t - t') \sum_{r=1}^{2} \psi_p^r(x)\bar\psi_p^r(x') \right. \\
&\qquad\qquad \left. + i\theta(t' - t) \sum_{r=3}^{4} \psi_p^r(x)\bar\psi_p^r(x') \right]
\end{aligned} \tag{3.137}
$$

where:

$$\psi_p^r(x) \equiv \sqrt{\frac{m}{E}}(2\pi)^{-3/2} w_r(\mathbf{p})e^{-i\epsilon^r p\cdot x} \tag{3.138}$$

where $\epsilon^r = (1, 1, -1, -1)$ and $w_1 = u_1$, $w_2 = u_2$, $w_3 = v_1$, and $w_4 = v_2$. Written in this fashion, the states with positive energy propagate forward in time, while the states with negative energy propagate backwards in time.

As in the Klein–Gordon case, we can now replace the plane waves ψ_p with the quantized spinor field $\psi(x)$ by taking the vacuum expectation value of the spinor fields:

$$iS_F(x - y)_{\alpha\beta} = \langle 0|T\psi_\alpha(x)\bar\psi_\beta(y)|0\rangle \tag{3.139}$$

which is one of the most important results of this section. We will use this expression repeatedly in our discussion of scattering matrices.

3.7 Weyl Neutrinos

In the previous chapter, we saw that the spinorial representation of the Lorentz group is actually reducible if we introduce projection operators P_L and P_R. Normally, we are not concerned with this because this spinorial representation is irreducible under the full Poincaré group for massive states.

However, there is a situation when the spinorial representation becomes reducible even under the Poincaré group, and that is when the fermion is massless. For example, we can take an imaginary representation of the γ matrices, which gives us *Majorana* spinors[9] (see Appendix). For our purposes, what is more interesting is taking the *Weyl* representation,[10] which gives us a representation of neutrinos.

If we take the representation:

$$\gamma^0 = \begin{pmatrix} 0 & -I \\ -I & 0 \end{pmatrix}; \quad \gamma^i = \begin{pmatrix} 0 & \sigma^i \\ -\sigma^i & 0 \end{pmatrix}; \quad \gamma_5 = \begin{pmatrix} I & 0 \\ 0 & -I \end{pmatrix} \quad (3.140)$$

then in this representation, we can write down two *chiral* operators:

$$P_R = \frac{1 + \gamma_5}{2} = \begin{pmatrix} I & 0 \\ 0 & 0 \end{pmatrix}$$

$$P_L = \frac{1 - \gamma_5}{2} = \begin{pmatrix} 0 & 0 \\ 0 & I \end{pmatrix}$$

To see how these projection operators affect the electron field, let us split ψ as follows:

$$\psi = \begin{pmatrix} \psi_R \\ \psi_L \end{pmatrix} \quad (3.141)$$

Then we have:

$$\frac{1 + \gamma_5}{2}\psi = \begin{pmatrix} \psi_R \\ 0 \end{pmatrix}$$

$$\frac{1 - \gamma_5}{2}\psi = \begin{pmatrix} 0 \\ \psi_L \end{pmatrix}$$

Because P_L and P_R commute with the Lorentz generators:

$$[P_{L,R}, \sigma^{\mu\nu}] = 0 \quad (3.142)$$

it means that the four-component spinor ψ actually spans a reducible representation of the Lorentz group. Contained within the complex Dirac representation are two distinct chiral representations of the Lorentz group. Although ψ_L and ψ_R separately form an irreducible representation of the Lorentz group, they do not form a representation of the Poincaré group for massive particles. To find an irreducible representation of the Poincaré group, we must impose $m = 0$.

The reason that these chiral fermions must be massless is because the mass term $m\bar{\psi}\psi$ in the Lagrangian is not invariant under these two separate Lorentz transformations. Because:

$$\bar{\psi}\psi = \bar{\psi}_L\psi_R + \bar{\psi}_R\psi_L \tag{3.143}$$

mass terms in the action necessarily mix these two distinct representations of the Lorentz group. Thus, this representation forces us to have massless fermions; that is, this is a theory of massless neutrinos.

The theory of massless neutrinos is therefore invariant under the following chiral transformation:

$$\begin{aligned} \psi &\rightarrow e^{i\gamma_5\Lambda}\psi \\ \bar{\psi} &\rightarrow \bar{\psi}e^{i\gamma_5\Lambda} \end{aligned} \tag{3.144}$$

This symmetry is violated by mass terms. In other words, the spinor representation of the Poincaré group (for zero-mass particles) is reducible, and we have the freedom to choose two-component rather than four-component spinor representations.

This will have important phenomenological implications later on when we consider the quark model in the limit of small quark masses. Then we can use the power of chiral symmetry in this limit to extract a large number of nontrivial relations among S matrix elements, called sum rules. In addition, the actual values of the masses of the quarks then give us a handle as to size of the violation of these chiral sum rules.

In summary, canonical quantization gives a rigorous formulation of a second quantized field theory capable of describing multiparticle states. There are other, more elegant quantization programs, but the canonical one is perhaps the most rigorous. We also saw that the Dirac equation emerged from a spinorial representation of the Lorentz group, which was developed in the previous chapter. One of the successes of the second quantized approach is that we have a physical interpretation for the negative energy states that inevitably occur in any relativistic formulation. In the next chapter, we will quantize a spin-one field and couple it to the Dirac electron theory, giving us quantum electrodynamics.

3.8 Exercises

1. Prove that the anticommutation relation of the Dirac spinors in Eq. (3.119) is satisfied if the harmonic oscillator states obey the anticommutation relation in Eq. (3.120).

2. Prove that:

$$
\begin{aligned}
\mathrm{Tr}\,(\not a_1 \not a_2 \cdots \not a_{2n}) \;=\;& a_1 \cdot a_2\,\mathrm{Tr}\,(\not a_3 \cdots \not a_{2n}) - a_1 \cdot a_3\,\mathrm{Tr}\,(\not a_2 \not a_4 \cdots \not a_{2n}) \\
&+ \cdots + a_1 \cdot a_{2n}\,\mathrm{Tr}\,(\not a_2 \cdots \not a_{2n-1})
\end{aligned}
\tag{3.145}
$$

3. Consider the 16 matrices Γ_A, where $A = S, T, V, P, A$. Show that $(\Gamma^A)^2 = \pm 1$. Show that each is traceless except for the scalar. Given Γ_A and Γ_B $(A \neq B)$, show that there exists Γ_C (not equal to unity) such that:

$$
\Gamma_A \Gamma_B = \Gamma_C
\tag{3.146}
$$

4. By inserting the Fourier decomposition of the fields in Eq. (3.64), prove explicitly that:

$$
\begin{aligned}
[\phi(x), \phi(y)] \;=\;& i\Delta(x - y) \\
\;=\;& \int \frac{d^4x}{(2\pi)^3}\,\delta(k^2 - m^2)\epsilon(k_0)e^{-ik(x-y)}
\end{aligned}
\tag{3.147}
$$

Then perform the integration over k, leaving us with a Bessel function. Then show that the commutator vanishes outside the light cone, thereby establishing the causality of the system.

5. Prove Eq. (3.22) by explicitly performing the commutation relations.

6. Prove that Eqs. (3.108) and (3.109) are obeyed by explicit computation.

7. Prove the following formula, due to Fierz:

$$
\sum_{D=S,V,T,A,P} c_D (\Gamma_D)_{\alpha\beta}(\Gamma_D)_{\gamma\delta} = \sum_{B=S,V,T,A,P} \tilde{c}_B(\Gamma_B)_{\alpha\delta}(\Gamma_B)_{\gamma\beta}
\tag{3.148}
$$

where:

$$
\begin{pmatrix} \tilde{c}_S \\ \tilde{c}_V \\ \tilde{c}_T \\ \tilde{c}_A \\ \tilde{c}_P \end{pmatrix} = \frac{1}{4} \begin{pmatrix} 1 & 4 & 12 & -4 & 1 \\ 1 & -2 & 0 & -2 & -1 \\ \frac{1}{2} & 0 & -2 & 0 & \frac{1}{2} \\ -1 & -2 & 0 & -2 & 1 \\ 1 & -4 & 12 & 4 & 1 \end{pmatrix} \begin{pmatrix} c_S \\ c_V \\ c_T \\ c_A \\ c_P \end{pmatrix}
\tag{3.149}
$$

(Hint: use the fact that the Γ_A matrices form a complete set of 16 matrices. Then treat the above expression as a matrix equation in order to power expand it in terms of the Γ_A.)

8. Use the Fierz identity to re-express $(\bar{\psi}_1 A \psi_2)(\bar{\psi}_3 B \psi_4)$ in terms of Dirac bilinears $(\bar{\psi}_1 C \psi_4)(\bar{\psi}_3 D \psi_2)$. Express the matrices C and D in terms of the matrices A and B.

9. The appearance of γ^0 within $\bar{\psi}\psi = \psi^\dagger \gamma^0 \psi$ seems to violate Lorentz invariance, since γ^0 is manifestly non-invariant and transforms as the zeroth component of a vector. So why is $\bar{\psi}\psi$ still a Lorentz invariant? Furthermore, $\bar{\psi}\sigma^{\mu\nu}\psi$ clearly forms a finite dimensional representation of the Lorentz group. But this seems to be a contradiction of our no-go theorem. Is this so? If not, then why not?

10. Let $u_i(p)$ be spinors which satisfy the Dirac equation. Then prove the Gordon formula:

$$\bar{u}(p_2)\gamma^\mu u(p_1) = \bar{u}(p_2) \left(\frac{p_2^\mu + p_1^\mu}{2m} + \frac{i\sigma^{\mu\nu}q_\nu}{2m} \right) u(p_1) \qquad (3.150)$$

where q_μ is the momentum transfer.

11. We define brackets to mean summing over all antisymmetric combinations of indices (see (A.12)):

$$\gamma^{\mu_1\mu_2\cdots\mu_N} \equiv \frac{1}{N!}\gamma^{[\mu_1}\gamma^{\mu_2}\cdots\gamma^{\mu_N]} \qquad (3.151)$$

In an arbitrary number of space–time dimensions, prove that:

$$\gamma^\mu\gamma^{\mu_1\mu_2\cdots\mu_N} = \gamma^{\mu\mu_1\mu_2\cdots\mu_N} + \sum_{i=1}^N (-1)^{i+1} g^{\mu\mu_i}\gamma^{\mu_1\cdots\hat{\mu}_i\cdots\mu_N} \qquad (3.152)$$

where $\hat{\mu}_i$ means that the μ_i index is to be deleted in the sum. Prove:

$$\begin{aligned}
\gamma^{\mu\nu}\gamma^{\mu_1\cdots\mu_N} &= \gamma^{\mu\nu\mu_1\cdots\mu_N} + \sum_i (-1)^{i+1} g^{\mu_i[\mu}\gamma^{\nu]\mu_1\cdots\hat{\mu}_i\cdots\mu_N} \\
&+ \sum_i (-1)^{i+j} g^{\mu_i\mu_j;\mu\nu}\gamma^{\mu_1\cdots\hat{\mu}_i\cdots\hat{\mu}_j\cdots\mu_N} \qquad (3.153)
\end{aligned}$$

where $g^{\mu\nu;\rho\sigma} = g^{\mu\rho}g^{\nu\sigma} - g^{\mu\sigma}g^{\nu\rho}$.

12. Based on the previous problem, derive a formula for:

$$\gamma^{\nu_1\nu_2\cdots\nu_N}\gamma^{\mu_1\mu_2\cdots\mu_M} \qquad (3.154)$$

Prove it by induction.

13. Prove, by direct computation, that the Hamiltonian and the charge can be written in terms of Dirac harmonic oscillators as in Eqs. (3.118) and (3.124).

14. The Feynman propagator in x space can actually be computed analytically. Set $m = 0$, use radial coordinates, and show that the Feynman propagator in x space can be written in terms of Bessel functions.

15. Consider a unitary transformation U, with $H' = UHU^\dagger$ and $\psi' = U\psi$, which changes the Dirac equation to:

$$i\frac{\partial \psi'}{\partial t} = H'\psi' \tag{3.155}$$

Let U be given by:

$$U = \sqrt{\frac{m + |E|}{2|E|}} + \frac{\beta \boldsymbol{\alpha} \cdot \mathbf{p}}{\sqrt{2|E|(m + |E|)}} \tag{3.156}$$

Show that U removes all coupling between the positive and negative energy parts of the Dirac equation. This is an example of the Foldy–Wouthuysen transformation.

16. Prove that the asymptotic behavior of the anticommutator appearing in Eq. (3.131) is correct. Repeat the same calculation to show that spinors quantized with commutators also violate the spin-statistics theorem.

17. Prove that the derivative of a theta function gives us a Dirac delta function:

$$\frac{\partial}{\partial t}\theta(t - t') = \delta(t - t') \tag{3.157}$$

From this, prove that:

$$(\partial_\mu^2 + m^2)T\phi(x)\phi(x') = -i\,\delta^4(x - x') \tag{3.158}$$

Show that this equation is compatible with the expression for Δ_F in terms of the time-ordered product of two scalar fields.

18. The Feynman propagator $\Delta_F(x - y)$ can be expressed as the vacuum expectation value of the time-ordered product of two fields. The time-ordering operator T appears to violate Lorentz invariance, since x^0 is singled out. Why is the expression still Lorentz invariant?

Chapter 4
Quantum Electrodynamics

*It was found that this equation gave the particle a spin of half a quantum.
And also gave it a magnetic moment. It gave us the properties that one
needed for an electron. That was really an unexpected bonus for me,
completely unexpected.*

—P.A.M. Dirac

4.1 Maxwell's Equations

Now that we have successfully quantized the free Dirac electron, we would like
to discuss the question of coupling the Dirac electron to a spin-one Maxwell
field A_μ. The resulting theory will be called *quantum electrodynamics*, which
is perhaps the most successful physical theory ever proposed. After several
decades of confusion, false starts, and frustration, QED has emerged as one of the
cornerstones of the quantum theory.

Our discussion of the massless, spin-one field begins with the classical equations of Maxwell:

$$\operatorname{div} \mathbf{E} = \rho$$

$$\operatorname{curl} \mathbf{B} - \frac{\partial \mathbf{E}}{\partial t} = \mathbf{j}$$

$$\operatorname{div} \mathbf{B} = 0$$

$$\operatorname{curl} \mathbf{E} + \frac{\partial \mathbf{B}}{\partial t} = 0 \tag{4.1}$$

The source, in turn, obeys a conservation equation:

$$\frac{\partial \rho}{\partial t} + \operatorname{div} \mathbf{j} = 0 \tag{4.2}$$

Because the divergence of a curl is equal to zero, and because the curl of
the gradient is equal to zero, we can replace the magnetic and electric field with

potentials A^0 and $\mathbf{A}$ as follows:

$$\mathbf{E} = -\nabla A^0 - \frac{\partial \mathbf{A}}{\partial t}; \quad \mathbf{B} = \text{curl } \mathbf{A} \tag{4.3}$$

The fact that these equations did not transform according to the standard Galilean transformation led to the discovery of special relativity. To see this relativistic invariance more clearly, let us define:

$$A^\mu = (A^0, \mathbf{A})$$
$$j^\mu = (\rho, \mathbf{j}) \tag{4.4}$$

Then the current conservation equation can be written:

$$\partial_\mu j^\mu = 0 \tag{4.5}$$

and Maxwell's equations can be summarized as:

$$\partial_\mu F^{\mu\nu} = j^\nu \tag{4.6}$$

where:

$$F_{\mu\nu} = \partial_\mu A_\nu - \partial_\nu A_\mu \tag{4.7}$$

or:

$$F^{\mu\nu} = \begin{pmatrix} 0 & -E^1 & -E^2 & -E^3 \\ E^1 & 0 & -B^3 & B^2 \\ E^2 & B^3 & 0 & -B^1 \\ E^3 & -B^2 & B^1 & 0 \end{pmatrix} \tag{4.8}$$

and:

$$F^{0i} = -E^i; \quad F^{ij} = -\epsilon^{ijk} B^k \tag{4.9}$$

We can now derive Maxwell's equations by writing down the following action:

$$\mathscr{L} = -\frac{1}{4} F_{\mu\nu} F^{\mu\nu} = \frac{1}{2}(\mathbf{E}^2 - \mathbf{B}^2) \tag{4.10}$$

If we insert this into the Euler-Lagrange equations of motion, we find that the equations of motion are given by:

$$\partial_\mu F^{\mu\nu} = 0 \tag{4.11}$$

which is just the classical Maxwell's equation with zero source.

A key consequence of this construction is that the Maxwell theory is invariant under a *local symmetry*, that is, one whose parameters are dependent on space–time:

$$\delta A_\mu = \partial_\mu \Lambda(x) \tag{4.12}$$

(A transformation whose parameters are constants is called a *global transformation*, like the isospin and Lorentz transformations discussed earlier.) If we apply successive gauge transformations, we find that they form a group with the simple addition law:

$$\Lambda_3 = \Lambda_1 + \Lambda_2 \tag{4.13}$$

This is the same group law we found for $U(1)$; so we see that Maxwell's equations are locally invariant under $U(1)$.

Under this transformation, the Maxwell tensor is an invariant:

$$\delta F_{\mu\nu} = 0 \tag{4.14}$$

so the Lagrangian is also invariant.

This also means that there is a large redundancy associated with the theory. The equations for A_μ are identical to the equations for $A'_\mu = A_\mu + \partial_\mu \Lambda$.

We also note that the naive energy–momentum tensor associated with Maxwell's theory has the wrong properties. It is neither symmetric, nor is it gauge invariant. A naive application of Noether's theorem gives us an energy–momentum tensor that equals:

$$T^{\mu\nu} = -F^{\mu\lambda}\partial^\nu A_\lambda + \frac{1}{4}g^{\mu\nu}F_{\rho\sigma}F^{\rho\sigma} \tag{4.15}$$

which is not symmetric. This means that there is no conserved angular momentum tensor. Worse, it is not even gauge invariant, since it is not written entirely in terms of the Maxwell tensor $F_{\mu\nu}$.

However, since the energy–momentum tensor is not a directly measurable quantity, we are free to add another tensor to it:

$$T^{\mu\nu} \rightarrow T^{\mu\nu} + \partial_\lambda(F^{\mu\lambda}A^\nu) \tag{4.16}$$

The resulting energy–momentum tensor is conserved, symmetric, and gauge invariant:

$$T^{\mu\nu} = F^{\mu\rho} F^{\nu}_{\ \rho} + \frac{1}{4} g^{\mu\nu} F_{\rho\sigma} F^{\rho\sigma} \tag{4.17}$$

It is important to note that the addition of this extra term does not affect the integrated charges, which are directly measurable. To see what conserved charges are associated with this energy–momentum tensor, we find:

$$
\begin{aligned}
T^{00} &= \frac{1}{2}(\mathbf{E}^2 + \mathbf{B}^2) \\
T^{i0} &= (\mathbf{E} \times \mathbf{B})^i
\end{aligned}
\tag{4.18}
$$

which we recognize as the energy density and the Poynting vector. Thus, this new energy–momentum tensor is a physically acceptable quantity and compatible with gauge invariance.

Gauge invariance is thus a guide to calculating the physical properties of field theory. As we shall see throughout this book, it is also absolutely important to maintain gauge invariance for QED, for several reasons:

1. The proof of renormalization, that we can extract a finite S matrix order by order from the quantum theory, is crucially dependent on gauge invariance.

2. The proof of unitarity, that there are no ghost states with negative norm in the theory, also depends on gauge invariance. (The longitudinal vibration modes of the Maxwell field are negative norm states, which can be eliminated by choosing a gauge, such as the Coulomb gauge.)

3. The proof that the theory is Lorentz invariant after we have fixed the gauge in a nonrelativistic fashion requires the use of gauge symmetry.

4.2 Relativistic Quantum Mechanics

The problem facing us now is to write down the action for the Dirac theory coupled to the Maxwell theory, creating the quantum theory of electrodynamics. The most convenient way is to use the electron current as the source for the Maxwell field. The electron current is given by $\bar{\psi}\gamma^\mu\psi$, and hence we propose the coupling:

$$e A_\mu \bar{\psi} \gamma^\mu \psi \tag{4.19}$$

We saw earlier that this current emerged because of the invariance of the Dirac action under the symmetry $\psi \to \exp(i\Lambda)\psi$. Let us promote Λ into a local gauge parameter, so that Λ is a function of space–time. We want an action invariant under:

$$\psi(x) \quad \to e^{ie\Lambda(x)}\psi(x)$$

$$A_\mu \quad \to A_\mu - \partial_\mu\Lambda(x) \qquad (4.20)$$

The problem with this transformation is that $\Lambda(x)$ is a function of space–time; therefore, the derivative of a spinor $\partial_\mu\psi$ is not a covariant object. It picks up an extraneous term $\partial_\mu\Lambda(x)$ in its transformation. To eliminate this extraneous term, we introduce the *covariant derivative*:

$$\partial_\mu \to D_\mu \equiv \partial_\mu + ieA_\mu \qquad (4.21)$$

The advantage of introducing the covariant derivative is that it transforms covariantly under a gauge transformation:

$$D_\mu\psi \quad \to \quad e^{ie\Lambda(x)}D_\mu\psi + (ie\partial_\mu\Lambda - ie\partial_\mu\Lambda)\psi$$

$$\to \quad e^{ie\Lambda(x)}D_\mu\psi \qquad (4.22)$$

This means that the following action is gauge invariant:

$$\mathscr{L} = \bar{\psi}(i\gamma^\mu D_\mu - m)\psi - \frac{1}{4}F_{\mu\nu}F^{\mu\nu} \qquad (4.23)$$

which we obtain by simply replacing ∂_μ by D_μ.

The coupling of the electron to the Maxwell field reproduces the coupling proposed earlier. In the limit of velocities small compared to the speed of light, the Dirac equation should reduce to a modified version of the Schrödinger equation. We are hence interested in checking the correctness of the Dirac equation to lowest order, to see if we can reproduce the nonrelativistic results and corrections to them. The Dirac equation of motion, in the presence of an electromagnetic potential, now becomes:

$$i\frac{\partial\psi}{\partial t} = \left[\boldsymbol{\alpha} \cdot (-i\boldsymbol{\nabla} - e\mathbf{A}) + \beta m + eA^0\right]\psi \qquad (4.24)$$

To find solutions of this equation, let us decompose this four-spinor into two smaller two-spinors:

$$\psi = \begin{pmatrix} \phi \\ \chi \end{pmatrix} = \begin{pmatrix} e^{-imt}\Phi \\ e^{-imt}\Psi \end{pmatrix} \qquad (4.25)$$

Then the Dirac equation can be decomposed as the sum of two two-spinor equations:

$$i\frac{\partial \phi}{\partial t} = \boldsymbol{\sigma} \cdot \boldsymbol{\pi} \chi + e A^0 \phi + m\phi$$

$$i\frac{\partial \chi}{\partial t} = \boldsymbol{\sigma} \cdot \boldsymbol{\pi} \phi + e A^0 \chi - m\chi \qquad (4.26)$$

where $\boldsymbol{\pi} = \mathbf{p} - e\mathbf{A}$. Next, we will eliminate χ. For small fields, we can make the approximation that $e A^0 \ll 2m$. Then we can solve the second equation as:

$$\chi \sim \frac{\boldsymbol{\sigma} \cdot \boldsymbol{\pi}}{2m} \phi \ll \phi \qquad (4.27)$$

In this approximation, the Dirac equation can be expressed as a Schrödinger-like equation, but with crucial spin-dependent corrections:

$$i\frac{\partial \Phi}{\partial t} = \left[\frac{(\boldsymbol{\sigma} \cdot \boldsymbol{\pi})^2}{2m} + e A^0 \right] \Phi$$

$$= \left[\frac{(\mathbf{p} - e\mathbf{A})^2}{2m} - \frac{e}{2m} \boldsymbol{\sigma} \cdot \mathbf{B} + e A^0 \right] \Phi \qquad (4.28)$$

where we have used the fact that :

$$(\boldsymbol{\sigma} \cdot \boldsymbol{\pi})^2 = \pi^2 - e\boldsymbol{\sigma} \cdot \mathbf{B} \qquad (4.29)$$

The previous equation gives the first corrections to the Schrödinger equation in the presence of a magnetic and electric field. Classically, we know that the energy of a magnetic dipole in a magnetic field is given by the dot product of the magnetic moment with the field:

$$E = -\boldsymbol{\mu} \cdot \mathbf{B} = -\frac{e\hbar}{2mc} \boldsymbol{\sigma} \cdot \mathbf{B} \qquad (4.30)$$

Since $\mathbf{S} = \hbar\boldsymbol{\sigma}/2$, the magnetic moment of the electron is therefore:

$$\boldsymbol{\mu} \equiv \frac{e}{mc}\frac{\hbar\boldsymbol{\sigma}}{2} = 2\left(\frac{e}{2mc}\right)\mathbf{S} \qquad (4.31)$$

Thus, the Dirac theory predicted that the electron should have a magnetic moment twice what one might normally expect, that is, twice the Bohr magneton. This was perhaps the first major success of the Dirac theory of the electron. Historically, it gave confidence to physicists that the Dirac theory was correct, even if it seemed to have problems with negative energy states.

Yet another classic result that gave credibility to this relativistic formulation was the splitting of the spectral lines of the hydrogen atom. To solve for corrections to the Schrödinger atom, we set **B** to zero, and take A^0 to be the Coulomb potential. The Dirac equation is now written as:

$$E\psi = \left(-i\boldsymbol{\alpha}\cdot\nabla + \beta m - \frac{Z\alpha}{r}\right)\psi = H\psi \tag{4.32}$$

where: $$\alpha = e^2/4\pi$$

and:

$$H = \begin{pmatrix} m - \frac{Z\alpha}{r} & \boldsymbol{\sigma}\cdot\mathbf{p} \\ \boldsymbol{\sigma}\cdot\mathbf{p} & -m - \frac{Z\alpha}{r} \end{pmatrix} \tag{4.33}$$

In hindsight, it turns out to be convenient to introduce a judicious ansatz for the solution of the Dirac equation. First, as in the Schrödinger case, we want to separate variables, so that we take $\psi \sim f(r)Y(\theta, \phi)$, where the radial function is explicitly separated off. Second, because ∇^2 contains the Casimir operator L_i^2 in the usual Schrödinger formalism, we choose $Y(\theta, \phi)$ to be the standard spherical harmonics, that is, eigenfunctions of the angular momentum operators. For the spinning case, however, what appears is:

$$\mathbf{J} = \mathbf{L} + \mathbf{S} = \mathbf{L} + \boldsymbol{\sigma}/2 \tag{4.34}$$

that is, we have a combination of orbital spin **L** and intrinsic spin **S**. Thus, our eigenfunctions for the Dirac case must be labeled by eigenvalues j, l, m.

Based on these arguments, we choose as our ansatz:

$$\psi_{jm}^l = \begin{pmatrix} i[G_{lj}(r)/r]\phi_{jm}^l \\ [F_{lj}(r)/r]\left(\sigma \cdot \hat{\mathbf{r}}\right)\phi_{jm}^l \end{pmatrix} \tag{4.35}$$

Inserting this ansatz into Eq. (4.32) and factoring out the angular part, we find that the radial part of our eigenfunctions obey:

$$\left(E - m + \frac{Z\alpha}{r}\right)G_{lj}(r) = -\frac{dF_{lj}(r)}{dr} \mp \left(j + \frac{1}{2}\right)\frac{F_{lj}(r)}{r}$$

$$\left(E + m + \frac{Z\alpha}{r}\right)F_{lj}(r) = \frac{dG_{lj}}{dr} \mp \left(j + \frac{1}{2}\right)\frac{G_{lj}(r)}{r} \tag{4.36}$$

where we use the $+(-)$ sign for $j = l + 1/2$ ($j = l - 1/2$). By power expanding this equation in r, this series of equations can be solved in terms of hypergeometric functions. These equations, when power expanded, give us the energy eigenstates of the hydrogen atom[1,2]:

$$E_{nj} = m \left[1 + \left(\frac{Z\alpha}{n - (j + 1/2) + \sqrt{(j + 1/2)^2 - Z^2\alpha^2}} \right)^2 \right]^{-1/2} \tag{4.37}$$

Experimentally, this formula correctly gave spin-dependent corrections to the Bohr formula to lowest order. In contrast to the usual Bohr formula, the energy is now a function of both the principal quantum number n as well as the total spin j. By power expanding, we can recover the usual nonrelativistic result to lowest order:

$$E_{nj} = m \left[1 - \frac{Z^2\alpha^2}{2n^2} - \frac{(Z^2\alpha^2)^2}{2n^4} \left(\frac{n}{j + 1/2} - 3/4 \right) + \cdots \right] \tag{4.38}$$

So far, we have only considered the Dirac electron interacting with a classical Coulomb potential. Therefore, it is not surprising that this formula neglects smaller quantum corrections in the hydrogen energy levels. In particular, the $2S_{1/2}$ and $2P_{1/2}$ levels have the same n and j values, so they should be degenerate according to the Dirac formula. However, experimentally these two levels are found to be split by a small amount, called the Lamb shift.

It was not until 1949, with the correct formulation of QED, that one could successfully calculate these small quantum corrections to the Dirac energy levels. The calculation of these quantum corrections in QED was one of the finest achievements of quantum field theory.

4.3   Quantizing the Maxwell Field

Because of gauge invariance, there are also complications when we quantize the theory. A naive quantization of the Maxwell theory fails for a simple reason: The propagator does not exist. To see this, let us write down the action in the following form:

$$\mathscr{L} = \frac{1}{2} A^\mu P_{\mu\nu} \partial^2 A^\nu \tag{4.39}$$

where:

$$P_{\mu\nu} = g_{\mu\nu} - \partial_\mu \partial_\nu / (\partial)^2 \tag{4.40}$$

The problem with this operator is that it is not invertible, and hence we cannot construct a propagator for the theory. In fact, this is typical of *any* gauge theory, not just Maxwell's theory. This also occurs in general relativity and in superstring theory. The origin of the noninvertibility of this operator is because $P_{\mu\nu}$ is a projection operator, that is, its square is equal to itself:

$$P_{\mu\nu} P^{\nu\lambda} = P_\mu^\lambda \tag{4.41}$$

and it projects out longitudinal states:

$$\partial^\mu P_{\mu\nu} = 0 \tag{4.42}$$

The fact that $P_{\mu\nu}$ is a projection operator, of course, goes to the heart of why Maxwell's theory is a gauge theory. This projection operator projects out any states with the form $\partial_\mu \Lambda$, which is just the statement of gauge invariance.

The solution to this problem is that we must break this invariance by choosing a gauge. Because we have the freedom to add $\partial_\mu \Lambda$ to A_μ, we will choose a specific value of Λ which will break gauge invariance. We are guaranteed this degree of freedom if the variation $\delta A_\mu = \partial_\mu \Lambda$ can be inverted. There are several ways in which we can fix the gauge and remove this infinite redundancy. We can, for example, place constraints directly on the gauge field A_μ, or we may add the following term to the action:

$$-\frac{1}{2\alpha} \left(\partial_\mu A^\mu\right)^2 \tag{4.43}$$

where α is arbitrary. We list some common gauges:

Coulomb gauge :	$\nabla_i A_i = 0$
Axial gauge :	$A_3 = 0$
Temporal gauge :	$A_0 = 0$
Landau gauge :	$\partial_\mu A^\mu = 0$
Landau gauge :	$\alpha = 0$
Feynman gauge :	$\alpha = 1$
Unitary gauge :	$\alpha = \infty$

$$\tag{4.44}$$

(Notice that there are two equivalent ways to represent the Landau gauge.) Each time we fix the constraint by restricting the gauge field A_μ, we must check that there exists a choice of Λ such that this gauge condition is possible. For example, if we set $A_3 = 0$, then we must show that:

$$A_3' = 0 = A_3 + \partial_3 \Lambda \tag{4.45}$$

so that:

$$\Lambda = -\int^x dx^3 A_3 \tag{4.46}$$

In the Coulomb gauge, one extracts out the longitudinal modes of the field from the very start. To show that the gauge degree of freedom allows us to make this choice, we write down:

$$\nabla \cdot \mathbf{A}' = \nabla \cdot (\mathbf{A} + \nabla\Lambda) = 0 \tag{4.47}$$

Solving for Λ, we find:

$$\Lambda = -\frac{1}{\nabla^2} \nabla \cdot \mathbf{A} = -\int \frac{d^3 x'}{4\pi |\mathbf{x} - \mathbf{x}'|} \nabla' \cdot A(x') \tag{4.48}$$

Likewise, the Landau gauge choice means that we can find a Λ such that:

$$\Lambda = -\frac{1}{\partial^2} \partial_\mu A^\mu \tag{4.49}$$

To begin the process of canonical quantization, we will take the Coulomb gauge in which only the physical states are allowed to propagate. Let us first calculate the canonical conjugate to the various fields. Since $\dot{A}_0$ does not occur in the Lagrangian, this means that A_0 does not appear to propagate, which is a sign that there are redundant modes in the action.

The other modes, however, have canonical conjugates:

$$\pi^0 \;=\; \frac{\delta \mathscr{L}}{\delta \dot{A}_0} = 0$$

$$\pi^i \;=\; \frac{\delta \mathscr{L}}{\delta \dot{A}_i} = -\dot{A}^i - \partial_i A_0 = E^i \tag{4.50}$$

We write the Lagrangian as:

$$\mathscr{L} = -\frac{1}{4} F_{\mu\nu}^2 = \frac{1}{2} F_{0i}^2 - \frac{1}{4} F_{ij}^2 \tag{4.51}$$

We now introduce the independent E_i field via a trick. We rewrite the action as:

$$\mathscr{L} = -\frac{1}{2} E_i^2 - E_i F_{0i} - \frac{1}{4} F_{ij}^2 \tag{4.52}$$

By eliminating E_i by its equation of motion, we find $E_i = -F_{0i}$. By inserting this value back into the Lagrangian, we find the original Lagrangian back again.

Then we can write the Lagrangian for QED as:

$$
\begin{aligned}
\mathscr{L} &= -\mathbf{E} \cdot \dot{\mathbf{A}} - A_0 \nabla \cdot \mathbf{E} - \frac{1}{2}\mathbf{E}^2 - \frac{1}{4}F_{ij}^2 + \bar{\psi}(i \not{D} - m)\psi \\
&= -\mathbf{E} \cdot \dot{\mathbf{A}} - \frac{1}{2}\left(\mathbf{E}^2 + \mathbf{B}^2\right) + \mathscr{L}(\psi, \mathbf{A}) - A_0 \left(\nabla \cdot \mathbf{E} - e\bar{\psi}\gamma^0\psi\right) \quad (4.53)
\end{aligned}
$$

A_0 is a Lagrange multiplier. If we solve for the equation of motion of this field, we find that there is the additional constraint:

$$
\text{Gauss's Law}: \quad \nabla \cdot \mathbf{E} = \rho = e\bar{\psi}\gamma_0\psi \qquad (4.54)
$$

Thus, Gauss's Law emerges only after solving for the equation of motion of A_0. If we now count the independent degrees of freedom, we find we have only two degrees left, which correspond to the two independent transverse helicity states. Of the original four components of A_μ, we see that A_0 can be eliminated by its equation of motion, and that we can gauge away the longitudinal mode, therefore leaving us with $4 - 2$ degrees of freedom, which is precisely the two helicity states predicted in Section 2.8 from group-theoretical arguments alone for massless representations of the Poincaré group.

Intuitively, this means that a photon moving in the z direction can vibrate in the x and y direction, but not the z direction or the timelike direction. This corresponds to the intuitive understanding of transverse photons. In the Coulomb gauge, we can reduce all fields to their transverse components by eliminating their longitudinal components. Let us separate out the transverse and longitudinal parts as follows: $\mathbf{E} = \mathbf{E}_T + \mathbf{E}_L$, where $\nabla \cdot \mathbf{E}_T = 0$ and $\nabla \cdot \mathbf{E}_L = \rho$. Let us now solve for $\mathbf{E}_L$ in terms of ρ. Then we have:

$$
\mathbf{E}_L = \nabla \frac{1}{\nabla^2} \rho \qquad (4.55)
$$

If we insert this back into the Lagrangian, then we find that all longitudinal contributions cancel, leaving only the transverse parts, except for the piece:

$$
\begin{aligned}
\frac{1}{2}\mathbf{E}_L^2 &= \frac{1}{2}\rho\frac{1}{\nabla^2}\rho \\
&= \frac{e^2}{8\pi}\int d^3x d^3y \, \frac{\psi^\dagger(\mathbf{x}, t)\psi(\mathbf{x}, t)\psi^\dagger(\mathbf{y}, t)\psi(\mathbf{y}, t)}{|\mathbf{x} - \mathbf{y}|} \qquad (4.56)
\end{aligned}
$$

This last term is called the "instantaneous four-fermion Coulomb term," which seems to violate special relativity since this interaction travels instantly across space. However, we shall show at the end of this section that this term precisely cancels against another term in the propagator, so Lorentz symmetry is restored.

If we impose canonical commutation relations, we find a further complication. Naively, we might want to impose:

$$[A_i(\mathbf{x}, t), \pi^j(\mathbf{y}, t)] = -i\delta_{ij}\delta^3(\mathbf{x} - \mathbf{y}) \tag{4.57}$$

However, this cannot be correct because we can take the divergence of both sides of the equation. The divergence of A_i is zero, so the left-hand side is zero, but the right-hand side is not. As a result, we must modify the canonical commutation relations as follows:

$$[A_i(\mathbf{x}, t), \pi^j(\mathbf{y}, t)] = -i\tilde{\delta}_{ij}(\mathbf{x} - \mathbf{y}) \tag{4.58}$$

where the right-hand side must be transverse; that is:

$$\tilde{\delta}_{ij} = \int \frac{d^3k}{(2\pi)^3} e^{i\mathbf{k}\cdot(\mathbf{x}-\mathbf{x}')} \left(\delta_{ij} - \frac{k_i k_j}{\mathbf{k}^2}\right) \tag{4.59}$$

As before, our next job is to decompose the Maxwell field in terms of its Fourier modes, and then show that they satisfy the commutation relations. However, we must be careful to maintain the transversality condition, which imposes a constraint on the polarization vector. The decomposition is given by:

$$A(x) = \int \frac{d^3k}{\sqrt{(2\pi)^3 2k_0}} \sum_{\lambda=1}^{2} \varepsilon^\lambda(k) \left[a^\lambda(k)e^{-ik\cdot x} + a^{\lambda\dagger}(k)e^{ik\cdot x}\right] \tag{4.60}$$

In order to preserve the condition that A is transverse, we take the divergence of this equation and set it to zero. This means that we must impose:

$$\varepsilon^\lambda \cdot \mathbf{k} = 0$$
$$\varepsilon^\lambda(k) \cdot \varepsilon^{\lambda'}(k) = \delta^{\lambda\lambda'} \tag{4.61}$$

(The simplest way of satisfying these transversality conditions is to take the momentum along the z direction, and keep the polarization vector totally in the transverse directions, i.e., in the x and y directions. However, we will keep our discussion as general as possible.) By inverting these relations, we can solve for the Fourier moments in terms of the fields:

$$a^\lambda(k) = i \int \frac{d^3x}{\sqrt{(2\pi)^3 2k_0}} e^{ik\cdot x} \overset{\leftrightarrow}{\partial_0} \varepsilon^\lambda(k) \cdot A(x)$$
$$a^{\dagger\lambda}(k) = -i \int \frac{d^3x}{\sqrt{(2\pi)^3 2k_0}} e^{-ik\cdot x} \overset{\leftrightarrow}{\partial_0} \varepsilon^\lambda(k) \cdot A(x) \tag{4.62}$$

In order to satisfy the canonical commutation relations among the fields, we must impose the following commutation relations among the Fourier moments:

$$[a^\lambda(k), a^{\dagger\lambda'}(k')] = \delta^{\lambda,\lambda'}\delta^3(\mathbf{k} - \mathbf{k'}) \tag{4.63}$$

(An essential point is that the sign of the commutation relations gives us positive norm states. There are no negative norm states, or ghosts, in this construction in the Coulomb gauge.) Let us now insert this Fourier decomposition into the expression for the energy and momentum:

$$
\begin{aligned}
H &= \frac{1}{2}\int d^3x\,(\mathbf{E}^2 + \mathbf{B}^2) \\[2mm]
&= \sum_{\lambda=1}^{2}\int d^3k\,\omega[a^{\dagger\lambda}(k)a^\lambda(k)] \\[2mm]
P &= \int d^3x\,(:\mathbf{E}\times\mathbf{B}:) \\[2mm]
&= \int d^3k\,\mathbf{k}\sum_{\lambda=1}^{2}a^{\lambda\dagger}(k)a^\lambda(k)
\end{aligned}
\tag{4.64}
$$

After normal ordering, once again the energy is positive definite. Finally, we wish to calculate the propagator for the theory. Again, there is a complication because the field is transverse. The simplest way to construct the propagator is to write down the time-ordered vacuum expectation value of two fields. The calculation is almost identical to the one for scalar and spinor fields, except we have the polarization tensor to insert:

$$
\begin{aligned}
iD_F^{\mathrm{tr}}(x-x')_{\mu\nu} &= \langle 0|TA_\mu(x)A_\nu(x')|0\rangle \\[2mm]
&= i\int\frac{d^4k}{(2\pi)^4}\frac{e^{-ik\cdot(x-x')}}{k^2+i\epsilon}\sum_{\lambda=1}^{2}\epsilon_\mu^\lambda(k)\epsilon_\nu^\lambda(k)
\end{aligned}
\tag{4.65}
$$

The previous expression is not Lorentz invariant since we are dealing with transverse states. This is a bit troubling, until we realize that Green's functions are off-shell objects and are not measurable. However, these Lorentz violating terms should vanish in the full S matrix. To see this explicitly, let us choose a new orthogonal basis of four vectors, given by $\epsilon_\mu^1(k)$, $\epsilon_\mu^2(k)$, k^μ, and a new vector $\eta^\mu = (1, 0, 0, 0)$. Any tensor can be power expanded in terms of this new basis. Therefore, the sum over polarization vectors appearing in the propagator can always be expanded in terms of the tensors $g_{\mu\nu}$, $\eta_\mu\eta_\nu$, $k_\mu k_\nu$, and $k_\mu\eta_\nu$. We can calculate the coefficients of this expansion by demanding that both sides of the equation be transverse. Then it is easy to show:

$$\sum_{\lambda=1}^{2} \epsilon_{\mu}^{\lambda}(k)\epsilon_{\nu}^{\lambda}(k) = -g_{\mu\nu} - \frac{k_{\mu}k_{\nu}}{(k \cdot \eta)^2 - k^2}$$

$$+ \frac{(k \cdot \eta)(k_{\mu}\eta_{\nu} + k_{\nu}\eta_{\mu})}{(k \cdot \eta)^2 - k^2} - \frac{k^2\eta_{\mu}\eta_{\nu}}{(k \cdot \eta)^2 - k^2} \qquad (4.66)$$

Fortunately, the noninvariant terms involving η can all be dropped. The terms proportional to k_{μ} vanish when inserted into a scattering amplitude. This is because the propagator couples to two currents, which in turn are conserved by gauge invariance. (To see this, notice that the theory is invariant under $\delta A_{\mu} = \partial_{\mu}\Lambda$. In a scattering amplitude, this means that adding k_{μ} to the polarization vector ϵ_{μ} cannot change the amplitude. Thus, k_{μ} terms in the propagator do not couple to the rest of the diagram. This will be discussed more in detail when we study the Ward identities.)

If we drop terms proportional to k_{μ} in the propagator, we are left with:

$$D_F^{\text{tr}}(x - x')_{\mu\nu} = -g_{\mu\nu}\Delta_F(x - x'; m = 0) - \eta_{\mu}\eta_{\nu}\frac{\delta(t - t')}{4\pi |\mathbf{x} - \mathbf{x}'|} \qquad (4.67)$$

The first term is what we want since it is covariant. The second term proportional to $\eta_{\mu}\eta_{\nu}$ is called the instantaneous Coulomb term. In any Feynman diagram, it occurs between two currents, creating $(\psi^{\dagger}\psi)\nabla^{-2}(\psi^{\dagger}\psi)$. This term precisely cancels the Coulomb term found in the Hamiltonian when we solved for $\mathbf{E}_L$ in Eq. (4.56).

As expected, we therefore find that although the Green's function is gauge dependent (and possesses terms that travel instantly across space), the S matrix is Lorentz invariant and causal.

4.4 Gupta–Bleuler Quantization

The advantage of the canonical quantization method in the Coulomb gauge is that we always work with transverse states. Thus, all states have positive norm:

$$\langle 0|a^{\lambda}(k)a^{\lambda' \dagger}(k')|0\rangle = \delta_{\lambda\lambda'}\delta^3(\mathbf{k} - \mathbf{k}') \qquad (4.68)$$

However, the canonical quantization method, although it is guaranteed to yield a unitary theory, is cumbersome because Lorentz invariance is explicitly broken. For higher spin theories, the loss of Lorentz invariance multiplies the difficulty of any calculation by several times.

There is another method of quantization, called the Gupta–Bleuler[3,4] quantization method or covariant method, which keeps manifest Lorentz invariance and simplifies any calculation. There is, however, a price that must be paid, and that is the theory allows negative norm states, or ghosts, to propagate. The resulting theory is manifestly Lorentz invariant with the presence of these ghosts, but the theory is still self-consistent because we remove these ghost states by hand from the physical states of the theory. We begin by explicitly breaking gauge invariance by adding a noninvariant term into the action:

$$\mathcal{L} = -\frac{1}{4} F_{\mu\nu}^2 - \frac{1}{2\alpha}(\partial_\mu A^\mu)^2 \tag{4.69}$$

for arbitrary α.

Then the action now reads:

$$\mathcal{L} \sim \frac{1}{2} A^\mu \tilde{P}_{\mu\nu} \partial^2 A^\nu \tag{4.70}$$

where:

$$\tilde{P}_{\mu\nu} = g_{\mu\nu} - \left(1 - \alpha^{-1}\right) \partial_\mu \partial_\nu / \partial^2 \tag{4.71}$$

Now that we have explicitly broken the gauge invariance, this operator is no longer a projection operator and hence can be inverted to find the propagator:

$$D_{\mu\nu} = -(\tilde{P}^{-1})_{\mu\nu}/\partial^2 = -\left[g_{\mu\nu} - (1 - \alpha)\partial_\mu \partial_\nu/\partial^2\right]/\partial^2 \tag{4.72}$$

This propagator explicitly propagates ghost states that violate unitarity. The D_{00} component of the propagator occurs with the wrong sign, and hence represents a ghost state. For our purposes, we will take the gauge $\alpha = 1$, so that the equation of motion now reads:

$$\partial^2 A_\mu = 0 \tag{4.73}$$

In this gauge, we find that A_0 is no longer a redundant Lagrange multiplier, but a dynamical field and hence has a canonical conjugate to it. The conjugate field of A_μ is now a four-vector:

$$\pi_\mu = \frac{\delta \mathcal{L}}{\delta \dot{A}^\mu} = \dot{A}_\mu \tag{4.74}$$

Then the covariant canonical commutation relations read:

$$[A_\mu(x), \pi^\nu(x')] = i\delta_\mu^\nu \delta^3(\mathbf{x} - \mathbf{x}') \tag{4.75}$$

As usual, we can decompose the field in terms of the Fourier moments:

$$A_\mu(x) = \int \frac{d^3k}{2\omega_k} \left[a^\lambda(k)\epsilon_\mu^\lambda(k)e^{-ik\cdot x} + a^{\dagger\lambda}(k)\epsilon_\mu^\lambda(k)e^{ik\cdot x} \right] \qquad (4.76)$$

The difference now is that the ϵ_μ vector is a genuine four-vector. In order for the canonical commutation relations to be satisfied, we necessarily choose the following commutation relations among the oscillators:

$$[a^\lambda(k), a^{\dagger\lambda'}(k')] = -g^{\lambda\lambda'}\delta^3(\mathbf{k} - \mathbf{k'}) \qquad (4.77)$$

The presence of the metric tensor in the commutation relation signals that the norm of the states may be negative; that is, a nonphysical, negative norm ghost is present in the theory. The norm of the state $a^{\dagger\lambda}(k)|0\rangle$ can now be negative. This is the price we pay for having a Lorentz covariant quantization scheme.

It is straightforward to prove that the propagator is now given by:

$$\langle 0|T A_\mu(x)A_\nu(y)|0\rangle = -ig_{\mu\nu}\Delta_F(x - y) \qquad (4.78)$$

This can be proved by explicitly inserting the operator expression for $A_\mu(x)$. The important aspect of this propagator is that it contains the metric $g_{\mu\nu}$, which alternates in sign. Hence, it propagates ghost states. Since ghosts now propagate in the theory, we must be careful how we remove them. If we take the condition $\partial_\mu A^\mu|\Psi\rangle = 0$, we find that this condition is too stringent; it has no solutions at all. The Gupta–Bleuler formalism is based on the observation that a weaker condition is required:

$$\left(\partial_\mu A^\mu\right)^{(+)}|\Psi\rangle = 0 \qquad (4.79)$$

where we only allow the destruction part of the constraint to act on physical states. In momentum space, this is equivalent to the condition that $k^\mu a_\mu(k)|\Psi\rangle = 0$. This guarantees that, although ghosts are allowed to circulate in the system, they are explicitly removed from all physical states of the theory.

(We can also quantize the massive vector field in much the same way. The quantization is almost identical to the one presented before, but now the counting of physical states is different. We recall from our discussion of the Poincaré group that a massless field only has two helicity components. However, the massive vector field has 3 components.)

4.5 *C*, *P*, and *T* Invariance

Although we have analyzed the behavior of quantum fields under continuous isospin and Lorentz symmetry, we must also investigate their behavior under discrete symmetries, such as that generated by parity, charge conjugation, and time-reversal symmetry:

$$
\begin{aligned}
P: &\quad \mathbf{x} \to -\mathbf{x} \\
C: &\quad e \to -e \\
T: &\quad t \to -t
\end{aligned}
\tag{4.80}
$$

Classically, we know that the laws of physics are invariant under these transformations. For example, we find that both Newton's and Maxwell's equations are invariant under these transformations.

The easiest way in which to calculate how the A_μ field transforms is to examine Maxwell's equation, especially the source term j^μ. Under a parity transformation, the electric charge distribution ρ does not change, but $\mathbf{j} \to -\mathbf{j}$ because we are reversing the direction of the electric current. Thus:

$$
\mathscr{P} j^\mu \mathscr{P}^{-1} = j_\mu
\tag{4.81}
$$

Under a charge conjugation, a positive electric current turns into a negative one, so that:

$$
\mathscr{C} j^\mu \mathscr{C}^{-1} = -j^\mu
\tag{4.82}
$$

Then under a time reversal, once again ρ remains the same, but $\mathbf{j}$ reverses sign (because the current reverses direction), so:

$$
\mathscr{T} j^\mu \mathscr{T}^{-1} = j_\mu
\tag{4.83}
$$

Then the transformation of A_μ can be found immediately. One way is to observe that $j^\mu A_\mu$ appears in the action and is an invariant. Then we can read off the transformation properties of A_μ. Another way in which to derive the transformation properties of A_μ is to use Maxwell's equations, $\partial_\mu F^{\mu\nu}(A) = j^\nu$. Knowing the transformation properties of ∂_μ and j^μ, then it is easy to solve for the transformation properties of A_μ. We summarize our results as follows:

	∂_μ	j^μ	A_μ
P	∂^μ	j_μ	A^μ
C	∂_μ	$-j^\mu$	$-A_\mu$
T	$-\partial^\mu$	j_μ	A^μ

$$(4.84)$$

Similarly, we can read off the transformation properties of ψ, using the Dirac equation and also the fact that $j^\mu = e\bar{\psi}\gamma^\mu\psi$, although the calculation is considerably longer. With a bit of work, we can summarize how Dirac bilinear scalars, vectors, axial vectors, etc. transform under these discrete transformations, including $CPT = \Theta$:

	$\bar{\psi}\psi$	$\bar{\psi}\gamma_5\psi$	$\bar{\psi}\gamma^\mu\psi$	$\bar{\psi}\gamma^\mu\gamma_5\psi$	$\bar{\psi}\sigma^{\mu\nu}\psi$
C	S	P	$-V^\mu$	A^μ	$-T^{\mu\nu}$
P	S	$-P$	V_μ	$-A_\mu$	$T_{\mu\nu}$
T	S	$-P$	V_μ	A_μ	$-T_{\mu\nu}$
Θ	S	P	$-V^\mu$	$-A^\mu$	$T^{\mu\nu}$

$$(4.85)$$

To prove this, let us examine each transformation separately.

4.5.1 Parity

When the Dirac field couples to the Maxwell field, we want the combination $j^\mu A_\mu$ appearing in the action to be parity conserving.

To find an explicit form for the operator $\mathscr{P}$ for the Dirac field, we will find it convenient to recall that any element of $O(3, 1)$ (which includes parity operations with $\det O = -1$) has the following effect on the Dirac matrix:

$$S(\Lambda)^{-1}\gamma^\mu S(\Lambda) = \Lambda^\mu_\nu \gamma^\nu \qquad (4.86)$$

For our case, we can set the matrix Λ to be our parity operator if we specify that it reverses the sign of $\mathbf{x}$:

$$\Lambda^{\mu}{}_{\nu} = \begin{pmatrix} 1 & 0 & 0 & 0 \\ 0 & -1 & 0 & 0 \\ 0 & 0 & -1 & 0 \\ 0 & 0 & 0 & -1 \end{pmatrix} \tag{4.87}$$

which is equal to the metric tensor $g_{\mu\nu}$. Thus, we want a solution of the following equation:

$$\mathscr{P}^{-1}\gamma^{\nu}\mathscr{P} = g^{\nu\nu}\gamma^{\nu} = -(-1)^{\delta_{\nu,0}}\gamma^{\nu} \tag{4.88}$$

An explicit solution of this equation is simply given by:

$$\mathscr{P} = e^{i\phi}\gamma^0 \tag{4.89}$$

where $e^{i\phi}$ is an irrelevant phase factor.

Thus, the action of the parity transformation on a spinor field is given by:

$$\text{Parity}: \quad \psi'(-\mathbf{x}, t) = S(\Lambda)\psi = e^{i\phi}\gamma^0\psi(\mathbf{x}, t) \tag{4.90}$$

This also means that the transformation of the $\bar{\psi}$ field is given by:

$$\bar{\psi}'(-\mathbf{x}, t) = \bar{\psi}(\mathbf{x}, t)\gamma^0 e^{-i\phi} \tag{4.91}$$

From this, we can calculate how the various bilinear combinations transform under the parity operation in Eq. (4.85).

4.5.2 Charge Conjugation

Charge conjugation is easily studied by taking the Dirac equation and then reversing the sign of the electric charge. If we let ψ_c represent the Dirac field that has the opposite charge as ψ, then we have:

$$\begin{aligned} (i\,\not\partial - e\,\not\!A - m)\psi &= 0 \\ (i\,\not\partial + e\,\not\!A - m)\psi_c &= 0 \end{aligned} \tag{4.92}$$

In order to find the relationship between ψ with charge e and ψ_c with charge $-e$, let us take the complex conjugate and then the transpose of the first equation. Then we find:

$$\gamma^{\mu T}(-i\partial_{\mu} - eA_{\mu})\left(\gamma^{0T}\psi^*\right) = 0 \tag{4.93}$$

It can be shown that for any representation of the Dirac algebra, there exists a matrix C that satisfies:

$$C\gamma_\mu^T C^{-1} = -\gamma_\mu \tag{4.94}$$

Now let us compare the previous equation with the equation for the ψ_c field. We have an exact correspondence (up to a phase) if we set:

$$\psi_c = e^{i\phi} C(\gamma^{0T}\psi^*) = e^{i\phi} C\bar{\psi}^T \tag{4.95}$$

So far, we have not specified the representation of the Dirac matrices. There is more than one solution of the equation for C. In the Dirac representation, we find the following solution for the C matrix as:

$$C = i\gamma^2\gamma^0 = \begin{pmatrix} 0 & -i\sigma^2 \\ -i\sigma^2 & 0 \end{pmatrix} \tag{4.96}$$

which satisfies the following additional constraints:

$$-C = C^{-1} = C^T = C^\dagger \tag{4.97}$$

For us, however, the important feature of the C matrix is that it allows us to identify the particle–antiparticle structure of the Dirac field.

Applying the C matrix to the particle field, we obtain the antiparticle field:

$$\psi = \begin{pmatrix} 0 \\ 0 \\ 0 \\ 1 \end{pmatrix}; \quad \psi_c = e^{i\phi} \begin{pmatrix} 1 \\ 0 \\ 0 \\ 0 \end{pmatrix} \tag{4.98}$$

which justifies our earlier statement that the Dirac equation contains both the particle and antiparticle fields, with the charges as well as spins reversed.

Now let us compute how the current j^μ transforms under the charge conjugation operation:

$$j_c^\mu = \bar{\psi}_c\gamma^\mu\psi_c = \psi^T C\gamma_\mu C\bar{\psi}^T = \psi^T \gamma_\mu^T \bar{\psi}^T = -\bar{\psi}\gamma_\mu\psi \tag{4.99}$$

The last minus sign is important: Because ψ is an anticommuting field, we pick up an extra minus sign when we move one spinor past another.

Now let us try to determine how the combined Dirac and Maxwell system transforms under charge conjugation. The Dirac equation is left invariant if we make the simultaneous change:

$$\psi \quad \rightarrow \quad \psi_c = e^{i\phi} C \bar{\psi}^T$$

$$j^\mu \quad \rightarrow \quad j^\mu_c = -j^\mu$$

$$A_\mu \quad \rightarrow \quad -A_\mu \tag{4.100}$$

(We have changed the sign of the Maxwell field, which simulates the change of the charge *e* that we made earlier.)

Thus, the combination $j_\mu A^\mu$ is invariant under charge conjugation, as desired.

4.5.3   Time Reversal

Finally, we analyze the effect of making the transformation $t \rightarrow -t$. We wish to find an explicit representation of the operator $\mathscr{T}$ in terms of harmonic oscillators. This can be done in several ways. We can represent the time-reversal operator as $S(\Lambda)$ acting on a spinor where $\Lambda_{\mu\nu} = -g_{\mu\nu}$. Or, we can write down the Dirac equation with a time reversal and try to retransform the equation back into the usual Dirac form.

Either way, we find the same result:

$$\mathscr{T} \psi(\mathbf{x}, t) \mathscr{T}^{-1} = e^{i\phi} T \psi(\mathbf{x}, -t) \tag{4.101}$$

where:

$$T \gamma^\mu T^{-1} = \gamma^T_\mu = \gamma^{\mu*} \tag{4.102}$$

An explicit representation of the *T* matrix is given by:

$$T = i\gamma^1 \gamma^3 \tag{4.103}$$

where:

$$T = T^\dagger = T^{-1} = -T^* \tag{4.104}$$

We should also mention that the $\mathscr{T}$ operator is unusual because it is antiunitary. For example, consider the time evolution equation:

$$[H, \phi(x)] = -i \frac{\partial \phi}{\partial t} \tag{4.105}$$

Let us make a time-reversal transformation on this equation. If we reverse the sign of t in this equation, we find:

$$[H, \phi(x')] = +i \frac{\partial \phi}{\partial t} \tag{4.106}$$

However, this has the net effect of transforming $H \rightarrow -H$, which is illegal (because then we would have negative energy states). This difficulty did not happen for parity or charge conjugation because $\partial / \partial t$ did not change for those symmetries.

Correspondingly we wish that $\mathscr{T}$ would reverse the exponent appearing in the time evolution operator:

$$\mathscr{T} e^{iH(t_1 - t_2)} \mathscr{T}^{-1} = e^{+iH(t_2 - t_1)} \tag{4.107}$$

However, this is impossible if the Hamiltonian commutes with $\mathscr{T}$.

This means that the operator $\mathscr{T}$ must be antiunitary:

$$\langle \mathscr{T} \phi | \mathscr{T} \psi \rangle = \langle \psi | \phi \rangle \tag{4.108}$$

Or this operator contains yet another operator that can take the complex conjugate of any c number. If we postulate an operator that can reverse $i \rightarrow -i$, then $\mathscr{T}$ can commute with the Hamiltonian yet still reverse the sign of t.

4.6 CPT **Theorem**

In nature, these discrete symmetries are violated. Parity is maximally violated by the weak interactions, and the combination CP is violated in K meson decays. However, there is a remarkable theorem that states that any quantum field theory is invariant under the combined operation of CPT[5-7] under very general conditions.

The theorem states that the Hamiltonian $\mathscr{H}$ is invariant under CPT:

$$(CPT) \mathscr{H}(x) (CPT)^{-1} = \mathscr{H}(x') \tag{4.109}$$

if the following two conditions are met:

1. The theory must be local, possess a Hermitian Lagrangian, and be invariant under proper Lorentz transformations.

2. The theory must be quantized with commutators for integral spin fields and quantized with anticommutators for half-integral spin fields (i.e., the usual spin-statistics connection).

Thus, although quantum field theories are easily written down that violate these discrete symmetries separately, any quantum theory obeying these very general features must be invariant under CPT.

Various proofs of this powerful theorem have been proposed over the years. We will not review the results from axiomatic field theory, which are the most rigorous. Rather than give a detailed proof, we will show that the CPT theorem is satisfied for the spin 0, 1/2, and 1 fields that we have so far investigated.

We first note that it is easy to show that the CPT operation changes any quantum field theory obeying these two assumptions in the following way:

1. The coordinates change as follows:

$$x^\mu \rightarrow -x^\mu$$
$$\partial_\mu \rightarrow -\partial_\mu \qquad (4.110)$$

2. The Maxwell field transforms as follows:

$$\mathscr{P} A_\mu(x) \mathscr{P}^{-1} = A^\mu(x')$$
$$\mathscr{C} A_\mu(x) \mathscr{C}^{-1} = -A_\mu(x)$$
$$\mathscr{T} A_\mu(x) \mathscr{T}^{-1} = A^\mu(x') \qquad (4.111)$$

where x' refers to the $\mathscr{P}$ or the $\mathscr{T}$ transformed variable. Thus, under the combined CPT, we have:

$$A_\mu(x) \rightarrow -A_\mu(-x) \qquad (4.112)$$

3. A Dirac spinor transforms under CPT as:

$$\psi_\alpha \rightarrow -i\psi(-x)_\beta (\gamma_5 \gamma_0)_{\beta\alpha} \qquad (4.113)$$

4. A Dirac bilinear changes as follows:

$$\bar{\psi}(x) O \psi(x) \rightarrow (-1)^k \bar{\psi}(x') O \psi(x') \qquad (4.114)$$

where k denotes the number of Lorentz indices appearing in the matrix O. (This assumed the spin-statistics connection, since we had to push one spinor past another.)

5. Any even-rank tensor transforms into its Hermitian conjugate, and all odd-rank tensors transform into the negative of their Hermitian conjugates.

6. All c numbers appearing in the theory are complex conjugated.

The sketch of the proof now proceeds as follows. First, we observe that the Lagrangian is invariant under CPT:

$$(CPT)\mathscr{L}(x)(CPT)^{-1} = \mathscr{L}(-x) \tag{4.115}$$

This is because the Lagrangian is a contraction of many tensors of different rank, but the sum of all the ranks must be even since $\mathscr{L}$ is a Lorentz scalar. By (5), this means that the Lagrangian is transformed into its Hermitian conjugate. But since the Lagrangian is Hermitian by the first assumption, we find that $\mathscr{L}$ must be CPT invariant.

One possible loophole in this construction is if the Lagrangian contains an infinite number of derivatives. Then condition (1) would be difficult to obey. This loophole is closed by the assumption that $\mathscr{L}$ is local. [Nonlocal theories containing terms like $\phi(x)\phi(y)$ can be power expanded as follows:

$$\phi(x)\phi(y) = \phi(x)e^{(y-x)^\mu \partial_\mu}\phi(x) \tag{4.116}$$

Nonlocal theories therefore contain an infinite number of derivatives, which are excluded from our discussion by the first assumption.]

Now that the Lagrangian is invariant, let us analyze the transformation property of the Hamiltonian, which is defined as:

$$\mathscr{H} = -\mathscr{L} + \sum_r :\pi_r(x)\dot{\phi}_r(x): \tag{4.117}$$

where the sum r is over both bosons and fermions.

To calculate the transformation of the Hamiltonian under CPT, we recall the transformation properties of the boson and fermion fields under CPT:

$$
\begin{aligned}
(CPT)\phi_r(x)(CPT) &= e^{i\theta_r}\phi_r(-x) \\
(CPT)\dot{\phi}_r(x)(CPT) &= -e^{i\theta_r}\dot{\phi}_r(-x) \\
(CPT)\pi_r(x)(CPT) &= -e^{-i\theta_r}\pi_r(-x)
\end{aligned}
\tag{4.118}
$$

To show that the Hamiltonian is invariant, we observe that the equal-time commutation relations for the bosons and fermions are preserved if the spin-statistics relation holds:

Boson : $[\pi_r(\mathbf{x}, t), \phi_s(\mathbf{x}', t)] = -i\delta^3(\mathbf{x} - \mathbf{x}')\delta_{rs}$

$$\text{Fermion :} \qquad \{\pi_r(\mathbf{x}, t), \phi_s(\mathbf{x}', t)\} = i\delta^3(\mathbf{x} - \mathbf{x}')\delta_{rs} \qquad (4.119)$$

Then we have all the identities necessary to show that:

$$(CPT)\mathcal{H}(x)(CPT)^{-1} = \mathcal{H}(-x) \qquad (4.120)$$

which completes the proof.

In summary, we have seen how to quantize the Maxwell spin-one field via the canonical method. The formulation is a bit clumsy because we must quantize the theory in the Coulomb gauge to eliminate the ghost states with negative norm. The Gupta–Bleuler formulation is more convenient because it is Lorentz covariant, but we must apply the ghost-killing constraint on the states. The reason for this complication is that the Maxwell theory is a gauge theory with the group $U(1)$.

In Chapter 5, we will analyze how to calculate scattering cross sections for QED with quantum field theory.

4.7   Exercises

1. Prove that the radial functions G_{lj} and F_{lj} obey Eq. (4.36) starting with the Dirac equation for an electron in a Coulomb field.

2. There are no finite-dimensional unitary representations of the Lorentz group, but what about a vector field A_μ, which forms a finite-dimensional representation space for the Lorentz group. Does this violate the no-go theorem? Why not?

3. There are very few exactly solvable problems involving the classical Dirac equation. One of them is the electron in a Coulomb potential. Another is the electron in a constant magnetic field. Solve the Dirac equation in this case, and show that:

$$E = \sqrt{m^2 + p_z^2 + 2n|eB|} \qquad (4.121)$$

where $n = 0, 1, 2, \ldots$.

4. Prove that:

$$\mathscr{P}b(p, s)\mathscr{P}^{-1} = b(-p, s); \quad \mathscr{P}d^\dagger(p, s)\mathscr{P}^{-1} = -d^\dagger(-p, s) \qquad (4.122)$$

Prove that an explicit operator representation for $\mathscr{P}$ is given by:

$$\mathscr{P} = \exp\left(-\frac{i\pi}{2}\int d^3p\left[b^\dagger(p, s)b(p, s) - b^\dagger(p, s)b(-p, s)\right.\right.$$

$$+d^\dagger(p, s)d(p, s) + d^\dagger(p, s)d(-p, s)\Big]\Big) \qquad (4.123)$$

5. Prove that:

$$\mathscr{C}b(p, s)\mathscr{C}^{-1} = d(p, s)e^{i\phi}$$
$$\mathscr{C}d^\dagger(p, s)\mathscr{C}^{-1} = b^\dagger(p, s)e^{i\phi} \qquad (4.124)$$

Prove that this operation, in turn, can be generated by the operator expression:

$$\mathscr{C} = \exp\left(\frac{i\pi}{2}\int d^3p \sum_s \left[b^\dagger(p, s) - d^\dagger(p, s)\right]\left[b(p, s) - d(p, s)\right]\right) \qquad (4.125)$$

(for $\varphi = 0$).

6. Prove Eq. (4.66).

7. Show that:

$$(CPT)U(t, t_0)(CPT)^{-1} = U(-t, -t_0) \qquad (4.126)$$

where U is the operator e^{-iHt} for constant H, so therefore:

$$(CPT)S(CPT)^{-1} = S^\dagger \qquad (4.127)$$

where S is the S matrix.

8. Consider the magnetic field created by electrons both moving in a straight wire and circulating in a solenoid. Also consider the electric field created by a stationary electron. Draw the diagrams for these system when a C, P, and T transformation is applied to them. Show that we reproduce the results in Eq. (4.84).

9. Choose the $A_0 = 0$ gauge. Since A_0 is no longer a Lagrange multiplier, this means that Gauss's Law can no longer be applied; therefore, we cannot impose $\nabla \cdot \mathbf{E} = 0$. This means we cannot reduce the system to the Coulomb gauge. Resolve this paradox. [Hint: show that Gauss's Law commutes with the Hamiltonian, and then use Eqs. (1.30) and (1.31).]

10. Given a Lagrangian:

$$\mathscr{L} \sim \theta F^a_{\mu\nu} F^a_{\sigma\rho} \epsilon^{\mu\nu\sigma\rho} \qquad (4.128)$$

what symmetries among C, P, T are broken?

11. Consider the coupling: $\bar{\psi}\sigma_{\mu\nu}\psi F^{\mu\nu}$. Take the nonrelativistic limit of this expression, and show how this relates to the magnetic moment of the electron.

12. Prove that the inversion of the operator in Eq. (4.71) gives us the propagator in Eq. (4.72).

13. Prove that the canonical commutation relation in Eq. (4.58) is satisfied if we take the commutation relations Eq. (4.63) among the harmonic oscillator states. Show that transversality is preserved.

14. It is often convenient to describe gauge theory in the mathematical language of *forms* (especially when working in higher dimensions). Let the infinitesimal dx^{μ} be antisymmetric under an operation we call $\wedge$; that is, $dx^{\mu} \wedge dx^{\nu} = -dx^{\nu} \wedge dx^{\mu}$. Define the operator $d \equiv dx^{\mu}\partial_{\mu}$. Prove that d is nilpotent, that is, $d^2 = 0$. Define the one-form $A \equiv A_{\mu}dx^{\mu}$. Define a p-form as $\omega = \omega_{\mu_1\mu_2\cdots\mu_p}dx^{\mu_1} \wedge dx^{\mu_2} \wedge \cdots \wedge dx^{\mu_p}$. Show that $dA = F$, where $F = F_{\mu\nu}dx^{\mu} \wedge dx^{\nu}$, where A is the vector potential and F is the Maxwell tensor.

15. In n dimensions, define the Hodge operator $*$, which generates a duality transformation:

$$*(dx^{\mu_1} \wedge dx^{\mu_2} \wedge \cdots \wedge dx^{\mu_p}) = \frac{1}{(n-p)!}\epsilon_{\mu_1\mu_2\cdots\mu_n}dx^{\mu_{p+1}} \wedge dx^{\mu_{p+2}} \wedge \cdots \wedge dx^{\mu_n}$$

$$(4.129)$$

for $p < n$. Show that Maxwell's equations can now be summarized as:

$$dF = 0; \quad d * F = J \tag{4.130}$$

Show that the Bianchi identity is a consequence of $d^2 = 0$.

16. Define the operator δ as:

$$\delta \equiv (-1)^{np+n+1} * d* \tag{4.131}$$

Then show that $\delta^2 = 0$ and that the Laplacian is given by:

$$\Delta = (d+\delta)^2 = d\delta + \delta d \tag{4.132}$$

17. Prove that $T^{\mu\nu}$ in Eq. (4.17) is conserved. (Hint: use the Bianchi identity.)

Chapter 5
Feynman Rules
and LSZ Reduction

> *The reason Dick's physics was so hard for ordinary people to grasp was
> that he did not use equations. The usual way theoretical physics was done
> since the time of Newton was to begin by writing down some equations
> and then to work hard calculating solutions of the equations .... He had
> a physical picture of the way things happen, and the picture gave him the
> solution directly with a minimum of calculation. It was no wonder that
> people who had spent their lives solving equations were baffled by him.
> Their minds were analytical; his was pictorial.*
>
> —F. Dyson on R. Feynman

5.1 Cross Sections

So far, our discussion has been rather formal, with no connection to experiment.
This is because we have been concentrating on Green's functions, which are
unphysical; that is, they describe the motion of "off-shell" particles where $p_\mu^2 \neq m^2$. However, the physical world that we measure in our laboratories is on-shell.
To make the connection to experiment, we need to rewrite our previous results in
terms of numbers that can be measured in the laboratory, such as decay rates of
unstable particles and scattering cross sections. There are many ways in which
to define the cross section, but perhaps the simplest and most intuitive way is to
define it as the effective "size" of each particle in the target:

$$\text{Cross section} = \text{Effective size of target particle} \tag{5.1}$$

The cross section is thus the effective area of each target particle as seen by an
incoming beam. Cross sections are often measured in terms of "barns." (One barn
is 10^{-24} cm^2.) A nucleon is about one fermi, or 10^{-13} cm across. Its area is
therefore about 10^{-26} cm^2, or about 0.01 barns. Thus, by giving the cross section

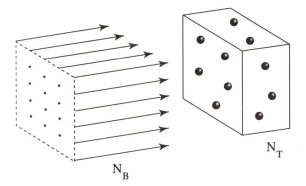

Figure 5.1. A target with N_T nuclei is bombarded with a beam with N_B particles. The cross section σ is the effective size in cm^2 of each nuclei as seen by the beam.

of a particle in a certain reaction, we can immediately calculate the effective size of that particle in relationship to a nucleon.

To calculate the cross section in terms of the rate of collisions in a scattering experiment, let us imagine a thin target with N_T particles in it, each particle with effective area σ or cross section. As seen from an incoming beam, the total amount of area taken up by these particles is therefore $N_T\sigma$. If we aim a beam of particles at the target with area A, then the chance of hitting one of these particles is equal to the total area that these target particles occupy ($N_T\sigma$) divided by the area A:

$$\text{Chance of hitting a particle} = \frac{N_T\sigma}{A} \tag{5.2}$$

Let us say we fire a beam containing N_B particles at the target. Then the number of particles in the beam that are absorbed or deflected is N_B times the chance of being hit. Thus, the number of scattering events is given by:

$$\text{Number of events} = N_B \frac{N_T\sigma}{A} \tag{5.3}$$

or simply:

$$\sigma = \left(\frac{\text{number of events}}{N_B N_T}\right) A \tag{5.4}$$

This reconfirms that the cross section has dimensions of area (Fig. 5.1).

In actual practice, a more convenient way of expressing the cross section is via the flux of the incoming beam, which is equal to ρv. If the beam is moving at velocity v toward a stationary target, then the number of particles in the beam N_B is equal to the density of the beam ρ times the volume. If the beam is a pulse that is turned on for t seconds, then the volume of the beam is vtA. Therefore, $N_B = \rho vtA$. The cross section can therefore be written as:

$$
\begin{aligned}
\sigma &= \frac{\text{number of events}/t}{(\rho vtA)N_T/t} A \\
&= \frac{\text{number of events}/t}{\rho v} \\
&= \frac{\text{transition rate}}{\text{flux}}
\end{aligned}
\tag{5.5}
$$

where we have normalized N_T to be 1 and the transition rate is the number of scattering events per second. The cross section is therefore equal to the transition rate divided by the flux of the beam. (More precisely, the cross section is equal to the transition probability per scatterer in the target and per unit incident flux.) This is usually taken as the starting point in discussions of the cross section, but unfortunately it is rather obscure and does not reveal its intuitive meaning as the effective size of the target particle.

(We will be calculating the cross section in collinear Lorentz frames, i.e., where the incoming beam and target move along the same axis. Two common collinear frames are the laboratory frame and the center-of-mass frame. If we make a Lorentz transformation to any other collinear frame, the cross section is an invariant, since a Lorentz contraction does not affect the cross section if we make a boost along this axis. However, the cross section is *not* a true Lorentz invariant. In fact, it transforms like an area under arbitrary Lorentz transformations.)

The next problem is to write the transition rate appearing in the cross section in terms of the S matrix. We must therefore calculate the probability that a collection of particles in some initial state i will decay or scatter into another collection of particles in some final state j. From ordinary nonrelativistic quantum mechanics, we know that the cross section σ can be calculated by analyzing the properties of the scattered wave. Using classical wave function techniques dating back to Rayleigh, we know that a plane wave e^{ikz} scattering off a stationary, hard target is given by:

$$
e^{ikz} + \frac{f(\theta)}{r} e^{ikr}
\tag{5.6}
$$

where the term with e^{ikr} represents the scattered wave, which is expanding radially from the target. Therefore $|f(\theta)|^2$ is proportional to the probability that a particle scatters into an angle θ.

More precisely, the differential cross section is given by the square of $f(\theta)$:

$$
\frac{d\sigma}{d\Omega} = |f(\theta)|^2
\tag{5.7}
$$

where the solid angle differential is given by:

$$\int d\Omega = \int_0^{2\pi} d\phi \int_{-1}^{1} d\cos\theta \tag{5.8}$$

and the total cross section is given by:

$$\int d\Omega \, |f(\theta)|^2 = \int d\Omega \frac{d\sigma}{d\Omega} = \sigma \tag{5.9}$$

For our purposes, however, this formulation is not suitable because it is inherently nonrelativistic. To give a relativistic formulation, let us start at the beginning. We wish to describe the scattering process that takes us from an initial state consisting of a collection of free, asymptotic states at $t \to -\infty$ to a final state $|f\rangle$ at $t \to \infty$. To calculate the probability of taking us from the initial state to the final state, we introduce the S matrix:

$$
\begin{aligned}
S_{fi} &= \langle f|S|i\rangle \\
&= \delta_{fi} - i(2\pi)^4 \delta^4(P_f - P_i)\mathcal{T}_{fi} \tag{5.10}
\end{aligned}
$$

where δ_{fi} symbolically represents the particles not interacting at all, and $\mathcal{T}_{fi}$ is called the transition matrix, which describes non-trivial scattering.

One of the fundamental constraints coming from quantum mechanics is that the S matrix is unitary:

$$\sum_f S_{fi}^* S_{fk} = \delta_{ik} \tag{5.11}$$

By taking the square of the S matrix, we can calculate the transition probabilities. The probability that the collection of states i will make the transition to the final states f is given by:

$$P_{fi} = S_{fi}^* S_{fi} \tag{5.12}$$

Likewise, the total probability that the initial states i will scatter into all possible final states f is given by:

$$P_{\text{total}} = \sum_f S_{fi}^* S_{fi} \tag{5.13}$$

Now we must calculate precisely what we mean by $\sum_f$. We begin by defining our states within a box of volume V:

$$|\mathbf{p}\rangle = \sqrt{(2\pi)^3 2E_p/V} a^\dagger(p)|0\rangle \quad \text{(bosons)}$$

$$|\mathbf{p}\rangle = \sqrt{(2\pi)^3(E_p/mV)}a^\dagger(p)|0\rangle \quad \text{(fermions)} \qquad (5.14)$$

Our states are therefore normalized as follows:

$$\langle\mathbf{p}|\mathbf{p}'\rangle = (2\pi)^3 2E_p\delta^3(\mathbf{p}-\mathbf{p}')/V \quad \text{(bosons)}$$

$$\langle\mathbf{p}|\mathbf{p}'\rangle = (2\pi)^3\frac{E_p}{m}\delta^3(\mathbf{p}-\mathbf{p}')/V \quad \text{(fermions)} \qquad (5.15)$$

With this normalization, the unit operator (on single paricle states) can be expressed as:

$$1 = \int \frac{Vd^3p}{(2\pi)^3 2E_p}|\mathbf{p}\rangle\langle\mathbf{p}| \quad \text{(bosons)}$$

$$1 = \int \frac{Vd^3p}{(2\pi)^3}\frac{m}{E_p}|\mathbf{p}\rangle\langle\mathbf{p}| \quad \text{(fermions)} \qquad (5.16)$$

To check our normalizations, we can let the number one act on an arbitrary state $|\mathbf{q}\rangle$, and we see that it leaves the state invariant. This means, however, that we have an awkward definition of the number of states at a momentum $\mathbf{p}$. With this normalization, we find that:

$$\langle\mathbf{p}|\mathbf{p}\rangle = (2\pi)^3 2E_p\delta^3(0)/V \qquad (5.17)$$

which makes no sense. However, we will interpret this to mean that we are actually calculating particle densities inside a large but finite box of size L and volume V; that is, we define:

$$\delta^3(\mathbf{p}) = \lim_{L\to\infty}\left(\frac{1}{(2\pi)^3}\int\int\int_{-L/2}^{L/2} dxdydz\; e^{-i\mathbf{p}\cdot\mathbf{r}}\right) \qquad (5.18)$$

This implies that we take the definition:

$$\delta^3(0) = \frac{V}{(2\pi)^3}$$

We will let the volume of the box V tend to infinity only at the end of the calculation. The origin of this problem is that we have been dealing with plane waves, rather than wave packets that are confined to a specific region of space and time. The price we pay for these nonlocalized plane waves is that we must carefully divide out infinite quantities proportional to the volume of space and time. (A more careful analysis would use wave packets that are completely localized in space and time; i.e., they have an envelope that restricts most of the wave packet to a definite region of space and time. This analysis with wave packets is somewhat more complicated, but yields precisely the same results.)

Our task is now to calculate the scattering cross section $1 + 2 \rightarrow 3 + 4 + \cdots$ and the rate of decay of a single particle $1 \rightarrow 2 + 3 \cdots$.

We must now define how we normalize the sum over final states. We will integrate over all momenta of the various final states, and sum over all possible final states. For each final state, we will integrate over the final momentum in a Lorentz covariant fashion. We will use:

$$\int \frac{d^4 p}{(2\pi)^4} \delta^4(p^2 - m^2)\theta(p_0) = \int \frac{d^3 p}{(2\pi)^3 2E_p} \tag{5.19}$$

The density of states dN_f, that is, the number of states within $\mathbf{p}$ and $\mathbf{p} + \delta\mathbf{p}$, is:

$$dN_f = \prod_{i=1}^{N_f} \frac{V d^3 p}{(2\pi)^3} \tag{5.20}$$

As before, the differential cross section $d\sigma$ is the number of transitions per unit time per unit volume divided by the flux J of incident particles:

$$
\begin{aligned}
d\sigma &= \frac{\text{transitions per second per cm}^3}{\text{incident flux}} \\
&= \left(\frac{|S_{fi}|^2 dN_f}{VT}\right)\frac{1}{J} \tag{5.21}
\end{aligned}
$$

We also know that the transition rate per unit volume (within a momentum-space interval) is given by:

$$
\begin{aligned}
\text{Transition rate per cm}^3 &= \frac{|S_{fi}|^2 dN_f}{VT} \\
&= \frac{(2\pi)^8 |\mathscr{T}_{fi}|^2 \delta^4(P_f - P_i)\delta^4(0)}{VT} dN_f \\
&= (2\pi)^4 \delta^4(P_f - P_i)|\mathscr{T}_{fi}|^2 dN_f \tag{5.22}
\end{aligned}
$$

where $(2\pi)^4 \delta^4(0) = VT$.

To calculate the incident flux, we will first take a collinear frame, such as the laboratory frame or center-of-mass frame. The incident flux J equals the product of the density of the initial state $(1/V)$ and the relative velocity $v = |v_1 - v_2|$, where $v_1 = |\mathbf{p}_1|/E_1$:

$$J = |v_1 - v_2|/V \tag{5.23}$$

In the center-of-mass frame, where $\mathbf{p}_1 = -\mathbf{p}_2$, we have:

$$
\begin{aligned}
V(2E_1)(2E_2)J & = (2E_1)(2E_2)|v_1 - v_2| = 2E_1 2E_2 \left| \frac{\mathbf{p}_1}{E_1} - \frac{\mathbf{p}_2}{E_2} \right| \\
& = 4|\mathbf{p}_1 E_2 - \mathbf{p}_2 E_1| \\
& = 4|\mathbf{p}_1|(E_1 + E_2) = 4[(p_1 \cdot p_2)^2 - m_1^2 m_2^2]^{1/2} \quad (5.24)
\end{aligned}
$$

(The last step is a bit deceiving, since it appears as if the flux is a Lorentz invariant in all frames. The last equality only holds if the two particles are collinear. The last step is not necessarily true for arbitrary Lorentz frames in which the two particles are not collinear. To see this, we can also write the flux for a beam moving in the z direction as:

$$
J \sim \epsilon_{xy\mu\nu} p_1^\mu p_2^\nu \quad (5.25)
$$

The flux now transforms as the x, y component of an antisymmetric second-rank tensor; that is, as an inverse area. Since the cross section is an area as seen by a particle in the beam, we expect it should transform as an area under an arbitrary Lorentz transformation. From now on, we assume that all Lorentz frames are collinear, so we can drop this distinction.)

The final formula for bosons for the differential cross section for $1 + 2 \rightarrow 3 + 4 \cdots$ is therefore given by:

$$
d\sigma = \frac{(2\pi)^4 |\mathcal{M}_{fi}|^2 \delta^4(P_f - P_i)}{4 \left[(p_1 \cdot p_2)^2 - m_1^2 m_2^2 \right]^{1/2}} \prod_{i=3}^N \frac{d^3 p_i}{(2\pi)^3 2E_{p_i}} \quad (5.26)
$$

in a collinear frame where $\mathcal{T}_{fi} = \prod_{i=1}^N (2E_{p_i} V)^{-1/2} \mathcal{M}_{fi}$. Notice that all factors of V in the S matrix have precisely cancelled against other factors coming from dN_f and the flux.

[We should note that other normalization conventions are possible. For example, we can always change the fermion normalization such that the $2m$ factor appearing in Eqs. (3.109) and (3.110) disappears. Then, with this new normalization, Eq. (5.26) works for both fermions and bosons. The advantage of this is that we can then use Eq. (5.26) without having to make the distinction between boson and fermion normalizations.]

Finally, we will use this formalism to compute the probability of the decay of a single particle. The decay probability is given by:

$$
\begin{aligned}
P_{\text{total}} & = \sum_f \int dN_f |S_{fi}|^2 \\
& = \sum_f \int dN_f |\mathcal{T}_{fi}|^2 \left[(2\pi)^4 \delta^4(P_f - P_i) \right]^2 \quad (5.27)
\end{aligned}
$$

where we have taken $\delta_{fi} = 0$ for a decay process. The last expression, unfortunately, is singular because of the delta function squared. As before, however, we assume that all our calculations are being performed in a large but finite box of volume V over a large time interval T. We thus reinterpret one of the delta functions as:

$$\delta^4(0) = \frac{VT}{(2\pi)^4} \qquad (5.28)$$

We now define the decay rate Γ of an unstable particle as the transition probability per unit volume of space and time:

$$\Gamma = \frac{\text{transition probability}}{\text{sec} \times \text{volume}}$$

$$= \frac{P_{\text{total}}}{VT(2E_i)} \qquad (5.29)$$

The final result for the decay rate is given by:

$$\Gamma = \frac{(2\pi)^4}{2E_i} \int \prod_{j=1}^{N_f} \frac{d^3 p_j}{(2\pi)^3 2E_j} |\mathcal{M}_{fi}|^2 \delta^4(P_f - P_i) \qquad (5.30)$$

The lifetime of the particle τ is then defined as the inverse of the decay rate:

$$\tau = \frac{1}{\Gamma} \qquad (5.31)$$

5.2   Propagator Theory and Rutherford Scattering

Historically, calculations in QED were performed using two seemingly independent formulations. One formulation was developed by Schwinger[1] and Tomonaga[2] using a covariant generalization of operator methods developed in quantum mechanics. However, the formulation was exceedingly difficult to calculate with and was physically opaque. The second formulation was developed by Feynman[3] using the propagator approach. Feynman postulated a list of simple "rules" from which one could pictorially setup the calculation for scattering matrices of arbitrary complexity. The weakness of Feynman's graphical methods, however, was that they were not rigorously justified. Later, Dyson demonstrated the equivalence of these two formulations by deriving Feynman's rules from the interaction picture.

In this chapter, we first follow Feynman to show how the propagator method gives us a rapid, convenient method of calculating the lowest order terms in the scattering matrix. Then we will develop the Lehmann–Symanzik–Zimmermann (LSZ) reduction formalism, in which one can develop the Feynman rules for diagrams of arbitrary complexity.

At this point, we should emphasize that the Green's functions that appear in the propagator approach are "off-shell"; that is, they do not satisfy the mass-shell condition $p_\mu^2 = m^2$. Neither do they obey the usual equations of motion. The Green's functions describe virtual particles, not physical ones. As we saw in the previous chapter, the Green's function develops a pole in momentum space at $p_\mu^2 = m^2$. However, there is no violation of cherished physical principles because the Green's functions are not measurable quantities. The only measurable quantity is the S matrix, where the external particles obey the mass-shell condition.

To begin calculating cross sections, let us review the propagator method in ordinary quantum mechanics, where we wish to solve the equation:

$$\left(i \frac{\partial}{\partial t} - H \right) \psi = 0 \qquad (5.32)$$

We assume that the true Hamiltonian is split into two pieces: $H = H_0 + H_I$, where the interaction piece H_I is small. We wish to solve for the propagator $G(\mathbf{x}, t; \mathbf{x}', t')$:

$$\left(i \frac{\partial}{\partial t} - H_0 - H_I \right) G(\mathbf{x}, t; \mathbf{x}', t') = \delta^3(\mathbf{x} - \mathbf{x}')\delta(t - t') \qquad (5.33)$$

If we could solve for the Green's function for the interacting case, then we can use Huygen's principle to solve for the time evolution of the wave. We recall that Huygen's principle says that the future evolution of a wave front can be determined by assuming that every point along a wave front is an independent source of an infinitesimal wave. By adding up the contribution of all these small waves, we can determine the future location of the wave front. Mathematically, this is expressed by the equation:

$$\psi(\mathbf{x}, t) = \int d^3 x' G(\mathbf{x}, t; \mathbf{x}', t')\psi(\mathbf{x}', t'); \qquad t > t' \qquad (5.34)$$

Our next goal, therefore, is to solve for the complete Green's function G, which we do not know, in terms of the free Green's function G_0, which is well understood. To find the propagator for the interacting case, we have to power expand in H_I. We will use the following formula for operators A and B:

$$\frac{1}{A + B} = \frac{1}{A(1 + A^{-1}B)}$$

$$= \frac{1}{1 + A^{-1}B} \frac{1}{A}$$

$$= (1 - A^{-1}B + A^{-1}BA^{-1}B + \cdots)A^{-1}$$

$$= A^{-1} - A^{-1}BA^{-1} + A^{-1}BA^{-1}BA^{-1} + \cdots \qquad (5.35)$$

Another way of writing this is:

$$\frac{1}{A+B} = A^{-1} - \frac{1}{A+B}BA^{-1} \qquad (5.36)$$

Now let:

$$A = -H_0 + i\frac{\partial}{\partial t}$$

$$B = -H_I$$

$$G = 1/(A+B)$$

$$G_0 = 1/A \qquad (5.37)$$

Then we have the symbolic identities:

$$G = G_0 + GH_I G_0$$

$$G = G_0 + G_0 H_I G_0 + G_0 H_I G_0 H_I G_0 + \cdots \qquad (5.38)$$

More explicitly, we can recursively write this as:

$$G(\mathbf{x}, t; \mathbf{x}', t') = G_0(\mathbf{x}, t; \mathbf{x}', t') + \int dt_1 \int d^3x_1 \, G(\mathbf{x}, t; \mathbf{x}_1, t_1)$$

$$\times H_I(\mathbf{x}_1, t_1)G_0(\mathbf{x}_1, t_1; \mathbf{x}', t') \qquad (5.39)$$

If we power expand this expression, we find:

$$G(\mathbf{x}, t; \mathbf{x}', t') = G_0(\mathbf{x}, t; \mathbf{x}', t') + \int dt_1 \int d^3x_1$$

$$\times G_0(\mathbf{x}, t; \mathbf{x}_1, t_1)H_I(\mathbf{x}_1, t_1)G_0(\mathbf{x}_1, t_1; \mathbf{x}', t')$$

$$+ \int_{-\infty}^{\infty} dt_1 \, dt_2 \, d^3x_1 \, d^3x_2 G_0(\mathbf{x}, t; \mathbf{x}_1, t_1)H_I(\mathbf{x}_1, t_1)$$

$$\times G_0(\mathbf{x}_1, t_1; \mathbf{x}_2, t_2)H_I(\mathbf{x}_2, t_2)G_0(\mathbf{x}_2, t_2; \mathbf{x}', t') + \cdots (5.40)$$

Using Huygen's principle, we can power expand for the time evolution of the wave function:

$$\psi(\mathbf{x}, t) = \psi_0(\mathbf{x}, t) + \int d^4x_1 \, G_0(\mathbf{x}, t; \mathbf{x}_1, t_1)\psi_0(\mathbf{x}_1, t_1)$$

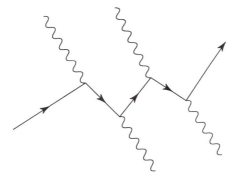

Figure 5.2. In the propagator approach, perturbation theory can be pictorially represented as a particle interacting with a background potential at various points along its trajectory.

$$+ \int d^4x_1 \, d^4x_2 \; G_0(\mathbf{x}, t; \mathbf{x}_1, t_1) H_I(\mathbf{x}_1, t_1) G_0(\mathbf{x}_1, t_1; \mathbf{x}_2, t_2) \psi_0(\mathbf{x}_2, t_2)$$

$$+ \cdots + \int d^4x_1 \cdots d^4x_n \; G_0(\mathbf{x}, t; \mathbf{x}_1, t_1) H_I(\mathbf{x}_1, t_1) \cdots$$

$$\times \; G_0(\mathbf{x}_{n-1}, t_{n-1}; \mathbf{x}_n, t_n) \psi_0(\mathbf{x}_n, t_n) + \cdots \tag{5.41}$$

In Figure 5.2, we have a pictorial representation of this diagram.

Let us now solve for the S matrix in lowest order. We postulate that at infinity, there are free plane waves given by $\phi = e^{-ik \cdot x}/(2\pi)^{3/2}$. We want to calculate the transition probability that a wave packet starts out in a certain initial state i, scatters off the potential, and then re-emerges as another free plane wave, but in a different final state f. To lowest order, the transition probability can be calculated by examining Huygen's principle:

$$\psi(\mathbf{x}, t) = \phi_i(\mathbf{x}, t) + \int d^4x' \; G_0(\mathbf{x}, t; \mathbf{x}', t') H_I(\mathbf{x}', t') \phi_i(\mathbf{x}', t') + \cdots \tag{5.42}$$

To extract the S matrix, multiply this equation on the left by ϕ_f^* and integrate. The first term on the right then becomes δ_{ij}. Using the power expansion of the Green's function, we can express the Green's function G_0 in terms of these free fields. After integration, we find:

$$S_{fi} = \delta_{fi} + i \int d^4x \; \phi_f^*(x') H_I(x') \phi_i(x') + \cdots \tag{5.43}$$

Therefore, the transition matrix is proportional to the matrix element of the potential H_I. We will now generalize this exercise to the problem in question: the calculation in QED of the scattering of an electron due to a stationary Coulomb

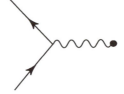

Figure 5.3. An electron scatters off a stationary Coulomb field in lowest order perturbation theory. This reproduces the Rutherford scattering cross section in the nonrelativistic limit.

potential. Our calculation should be able to reproduce the old Rutherford scattering amplitude to lowest order in the nonrelativistic limit and give higher-order quantum corrections to it. Our starting point is the Dirac electron in the presence of an external, classical Coulomb potential (Fig. 5.3).

The interacting Dirac equation reads:

$$(i\gamma^{\mu}\partial_{\mu} - m)\psi(x) = e\,\mathcal{A}(x)\psi(x) \tag{5.44}$$

Since we are only working to lowest order and are treating the potential A_{μ} as a classical potential, we can solve this equation using only propagator methods. The solution of this equation, as we have seen, is given by:

$$\psi(x) = \psi_0(x) + e \int d^4y\, S_F(x - y)\,\mathcal{A}(y)\psi(y) \tag{5.45}$$

where ψ_0 is a solution of the free, homogeneous Dirac equation. To calculate the scattering matrix, it is convenient to insert the expansion of the Feynman propagator $S_F(x - x')$ in terms of the time-ordered function $\theta(t - t')$, as in Eq. (3.137). Then we find:

$$\psi(x) = \psi_i(x) - ie \int d^4y\, i\theta(t - t') \int d^3p \sum_{r=1}^{2} \bar{\psi}_p^r(x)\psi_p^r(y)\,\mathcal{A}(y)\psi(y) \tag{5.46}$$

for $t \to \infty$. We now wish to extract from this expression the amplitude that the outgoing wave $\psi(x)$ will be scattered in the final state, given by $\psi_f(x)$. This is easily done by multiplying both sides of the equation by $\bar{\psi}_f$ and integrating over all space-time. The result gives us the S matrix to lowest order:

$$S_{fi} = \delta_{fi} - ie \int d^4x\, \bar{\psi}_f(x)\,\mathcal{A}(x)\psi_i(x) \tag{5.47}$$

Now we insert the expression for the vector potential, which corresponds to an electric potential A_0 given by the standard $1/r$ Coulomb potential:

$$A_0(x) = -\frac{Ze}{4\pi |\mathbf{x}|} \tag{5.48}$$

whose Fourier transform is given by:

$$\int \frac{d^3x}{|\mathbf{x}|} e^{i\mathbf{q}\cdot\mathbf{x}} = \frac{4\pi}{|\mathbf{q}|^2} \tag{5.49}$$

Inserting the plane-wave expression for the fermion fields into the expression for the scattering matrix in Eq. (5.47) and performing the integration over x, we have:

$$
\begin{aligned}
S_{fi} &= \frac{iZe^2}{4\pi V} \int d^4x \sqrt{\frac{m}{E_f}} \bar{u}(p_f, s_f) e^{ip_f \cdot x} \left[\gamma^0 \frac{1}{|\mathbf{x}|} \right] \sqrt{\frac{m}{E_i}} u(p_i, s_i) e^{-ip_i \cdot x_i} \\
&= \frac{iZe^2}{V} \sqrt{\frac{m^2}{E_f E_i}} \frac{\bar{u}(p_f, s_f)\gamma^0 u(p_i, s_i)}{|\mathbf{q}|^2} 2\pi \delta(E_f - E_i)
\end{aligned}
\tag{5.50}
$$

We recall that $Vd^3p_f/(2\pi)^3$ is the number of final states contained in the momentum interval d^3p_f. Multiply this by $|S_{fi}|^2$ and we have the probability of transition per particle into these states. [We recall that squaring the S matrix give us divergent quantities like $\delta(0)$, which is due to the fact that we have not rigorously localized the wave packets. We set $2\pi \delta(0) = T$, where we localize the scattering process in a box of size V and duration T.]

If we divide by T, this give us the rate R of transitions per unit time into this momentum interval. Finally, if we divide the rate of transitions by the flux of incident particles $|\mathbf{v}_i|/V$, this gives us the differential cross section:

$$d\sigma = |S_{fi}|^2 \frac{Vd^3p_f}{(2\pi)^3} \frac{1}{VT|\mathbf{v}_i|/V} \tag{5.51}$$

To calculate the differential cross section per unit of solid angle, we must decompose the momentum volume element:

$$d^3p = d\Omega p^2 \, dp \tag{5.52}$$

Using the fact that $p_f \, dp_f = E_f \, dE_f$, we have the result:

$$
\begin{aligned}
\frac{d\sigma}{d\Omega} &= \frac{4Z^2\alpha^2 m^2}{|\mathbf{q}|^4} \sum_{\text{spins}} \frac{1}{2} |\bar{u}(p_f, s_f)\gamma^0 u(p_i, s_i)|^2 \\
&= \frac{4Z^2\alpha^2 m^2}{|\mathbf{q}|^4} \frac{1}{2} \text{Tr} \left(\gamma^0 \frac{\not{p}_i + m}{2m} \gamma^0 \frac{\not{p}_f + m}{2m} \right)
\end{aligned}
\tag{5.53}
$$

In the last step, we have used the fact that the summation of spins can be written as:

$$
\begin{aligned}
\left|\bar{u}_f \Gamma u_i\right|^2 &= \left(\bar{u}_f \Gamma u_i\right) \left(u_i^\dagger \Gamma^\dagger \bar{u}_f^\dagger\right) \\
&= \left(\bar{u}_f \Gamma u_i\right) \left(\bar{u}_i \gamma_0 \Gamma^\dagger \gamma_0 u_f\right) \\
&= \bar{u}_{f,\alpha} \Gamma_{\alpha,\beta} u_{i,\beta} \bar{u}_{i,\gamma} \left(\gamma_0 \Gamma^\dagger \gamma_0\right)_{\gamma,\delta} u_{f,\delta} \\
&= \left(\frac{\not{p}_f + m}{2m}\right)_{\delta,\alpha} \Gamma_{\alpha,\beta} \left(\frac{\not{p}_i + m}{2m}\right)_{\beta,\gamma} \left(\gamma_0 \Gamma^\dagger \gamma_0\right)_{\gamma,\delta} \\
&= \operatorname{Tr}\left\{\frac{\not{p}_f + m}{2m} \Gamma \frac{\not{p}_i + m}{2m} \gamma_0 \Gamma^\dagger \gamma_0\right\}
\end{aligned}
\tag{5.54}
$$

where we have use the fact that the sum over spins in Eq. (3.109) gives us:

$$
\sum_{\text{spins}} u_\beta(p, s) \bar{u}_\alpha(p, s) = \left(\frac{\not{p}_i + m}{2m}\right)_{\beta\alpha}
\tag{5.55}
$$

The last trace can be performed, since only the trace of even numbers of Dirac matrices survives:

$$
\operatorname{Tr} \gamma^0 \not{p}_i \gamma^0 \not{p}_f = 4(2E_i E_f - p_i \cdot p_f)
\tag{5.56}
$$

Finally, we need some kinematical information. If θ is the angle between p_f and p_i, then:

$$
\begin{aligned}
p_i \cdot p_f &= m^2 + 2\beta^2 E^2 \sin^2(\theta/2) \\
|\mathbf{q}|^2 &= 4|\mathbf{p}|^2 \sin^2(\theta/2)
\end{aligned}
\tag{5.57}
$$

We then obtain the Mott cross section[4]:

$$
\frac{d\sigma}{d\Omega} = \frac{Z^2 \alpha^2}{4|\mathbf{p}|^2 \sin^4(\theta/2)} \left[1 - \beta^2 \sin^2\left(\frac{\theta}{2}\right)\right]
\tag{5.58}
$$

In the nonrelativistic limit, as $\beta \to 0$, we obtain the celebrated Rutherford scattering formula.

5.3   LSZ Reduction Formulas

Up to now, we have made the approximation that certain fields, such as the electromagnetic field, were classical. The development of scattering theory in this approximation was rather intuitive.

In this chapter, we would like to introduce a more convenient method, the LSZ reduction formalism,[5] from which we can derive scattering amplitudes to all orders in perturbation theory. The LSZ method gives us a simple derivation of Feynman's rules, which were originally derived from a more intuitive approach using propagator theory.

The LSZ approach begins with the physical S matrix, making as few assumptions as possible. We start by defining the "in" and "out" states, which are free particle states at asymptotic times; that is, $t = -\infty$ and ∞, respectively. We choose to distinguish them from the intermediate states, which are defined off-shell and are interacting. Our goal is to express the interacting S matrix, defined in terms of the unknown interacting field $\phi(x)$, in terms of these free asymptotic states. (We caution that in certain theories, such as QCD, the asymptotic states are bound and do not correspond to free states.)

The S matrix is defined as the matrix element of the transition from one asymptotic set of states to another. Let f denote a collection of free asymptotic states at $t = \infty$, while i refers to another collection of asymptotic states at $t = -\infty$. Then the S matrix describes the scattering of the i states into the f states:

$$S_{fi} = \text{out}\langle f | i \rangle_{\text{in}} \tag{5.59}$$

We postulate the existence of an operator S that converts asymptotic states at $t = \infty$ to states at $t = -\infty$:

$$
\begin{aligned}
|f\rangle_{\text{in}} &= S|f\rangle_{\text{out}} \\
S_{fi} &= \text{out}\langle f | S | i \rangle_{\text{out}} = \text{in}\langle f | S | i \rangle_{\text{in}}
\end{aligned}
\tag{5.60}
$$

For these asymptotic states, we also have asymptotic fields ϕ_{in} and ϕ_{out} that are free fields. Thus, we can use the machinery developed in Chapter 4 to describe these asymptotic free states. In particular, we can use Eqs. (3.15) and (3.16) to define the state vector as the vacuum state multiplied by creation operators:

$$
\begin{aligned}
|q\rangle_{\text{in}} &= a_{\text{in}}^\dagger(q)|0\rangle_{\text{in}} \\
&= -i \int d^3x \, e_q(x) \overleftrightarrow{\partial}_0 \, \phi_{\text{in}}(x)|0\rangle_{\text{in}}
\end{aligned}
\tag{5.61}
$$

The goal of the LSZ method is to reduce all expressions involving the full, interacting field $\phi(x)$ (which has matrix elements that cannot be easily computed) into simpler expressions involving the free, asymptotic fields ϕ_{in} and ϕ_{out}. The S matrix, written as a transition matrix, is useless to us at this moment. The goal of the LSZ approach, therefore, is to continue this process until we have gradually extracted out of the S matrix the entire set of fields contained within the asymptotic states. Then we can use the machinery developed in the previous chapter to manipulate and reduce these fields.

There is, however, a subtle point we should mention. Naively, one might expect that the interacting field $\phi(x)$, taken at infinitely negative or positive times, should smoothly approach the value of the free asymptotic fields, so that:

$$x^0 \to -\infty; \quad \phi(x) \to Z^{1/2}\phi(x)_{in} \tag{5.62}$$

where the factor $Z^{1/2}$ arises because of renormalization effects (which will be eliminated in Chapter 7). However, this naive assumption is actually incorrect. If we take this "strong" assumption, then it can be shown that the S matrix becomes trivial and no scattering takes place. We must therefore take the "weak" assumption, that the matrix elements of the two fields $\phi(x)$ and ϕ_{in} approach each other at infinitely negative times; that is:

$$x^0 \to -\infty; \quad \langle f|\phi(x)|i\rangle \to Z^{1/2}\langle f|\phi_{in}(x)|i\rangle \tag{5.63}$$

For the moment, however, we will simply ignore the complications that arise due to this. We will return to the question of evaluating Z later.

Let us now take an arbitrary S matrix element for the scattering of m particles with momenta q_i into n particles with momenta p_j. We first extract the field $\phi_{in}(q_1)$ from the asymptotic "in" state:

$$_{out}\langle p_1, p_2 \cdots p_n|q_1, q_2, \cdots q_m\rangle_{in} = {}_{out}\langle p_1, p_2, \cdots p_n|a_{in}^\dagger(q_1)|q_2, q_3, \cdots q_m\rangle_{in}$$

$$= -i \lim_{t\to-\infty} \int d^3x \, e_{q_1}(x) \overset{\leftrightarrow}{\partial_0} \, {}_{out}\langle p_1, p_2, \cdots, p_n|\phi(x)_{in}|q_2, q_3, \cdots, q_m\rangle_{in} \tag{5.64}$$

Next, we wish to convert the three-dimensional integral $\int d^3x$ into a four dimensional one. We use the identity:

$$\left(\lim_{t\to\infty} - \lim_{t\to-\infty}\right)\int d^3x \, A(x, t) = \int_{-\infty}^{\infty} dt \frac{\partial}{\partial t} \int d^3x \, A(x, t) \tag{5.65}$$

Since we already have a term at $t \to -\infty$, we add and subtract the same term as $t \to \infty$. This therefore gives us an integral over four dimensional space–time, plus a term at $t \to \infty$:

$$_{\text{out}}\langle p_1, p_2, \cdots, p_n | q_1, q_2, \cdots, q_m \rangle_{\text{in}} =$$

$$+ iZ^{-1/2} \int d^4x \, \partial_0 \left[e_{q_1}(x) \overset{\leftrightarrow}{\partial}_0 \,_{\text{out}}\langle p_1, p_2, \cdots | \phi(x) | q_2, \cdots \rangle_{\text{in}} \right]$$

$$- iZ^{-1/2} \lim_{t \to \infty} \int d^3x \left[e_{q_1}(x) \overset{\leftrightarrow}{\partial}_0 \,_{\text{out}}\langle p_1, p_2, \cdots | \phi(x) | q_2, \cdots \rangle_{\text{in}} \right]$$

$$(5.66)$$

This last term, in turn, can be written as the creation operator of an "out" state:

$$-iZ^{-1/2} \lim_{t \to \infty} \int d^3x \left[e_{q_1}(x) \overset{\leftrightarrow}{\partial}_0 \,_{\text{out}}\langle p_1, p_2, \cdots | \phi(x) | q_2, \cdots \rangle_{\text{in}} \right]$$

$$= \,_{\text{out}}\langle p_1, p_2, \cdots | a_{\text{out}}^\dagger(q_1) | q_2, \cdots \rangle_{\text{in}} \qquad (5.67)$$

The price of converting a three-dimensional integral to a four-dimensional one is that we have now generated a new term at $t \to \infty$, which is the matrix element of an "out" operator $a_{\text{out}}^\dagger(q_1)$. Because this "out" operator is an annihilation operator if it acts to the left, in general it gives us zero. The only exception is if there is, within the collection of "out" states, precisely the same state with momentum q_1. Thus, the matrix element vanishes unless, for example, the ith state with momentum p_i has exactly the same momentum as q_1:

$$_{\text{out}}\langle p_1, p_2 \cdots | a_{\text{out}}^\dagger(q_1) | q_2, q_3, \cdots \rangle_{\text{in}} = \sum_{i=1}^{n} 2p_i^0 (2\pi)^3 \delta^3(p_i - q_1)$$

$$\times \,_{\text{out}}\langle p_1, \cdots \hat{p}_i \cdots, p_n | q_2, \cdots q_m \rangle_{\text{in}} \qquad (5.68)$$

where the caret over a variable means that we delete that particular variable. This term is called a "disconnected graph," because one particle emerges unaffected by the scattering process, and is hence disconnected from the rest of the particles.

Now, we come to a key step in the calculation. So far, the expressions are noncovariant because of the presence of two time derivatives. We will now convert the time derivatives in the reduced matrix element into a fully covariant object, thus restoring Lorentz invariance. This is possible because the operator $\partial_0^2 - \partial_i^2 + m^2$ annihilates the plane wave $\exp(-iq_1 \cdot x)$. By integrating by parts, we will be able to convert the various time derivatives into a fully covariant ∂_μ^2:

$$\int d^4x \, \partial_0 \left[e_{q_1}(x) \overset{\leftrightarrow}{\partial}_0 \,_{\text{out}}\langle p_1, \cdots | \phi(x) | q_2, \cdots \rangle_{\text{in}} \right]$$

$$= - \int d^4x \left[\partial_0^2 e_{q_1}(x) \right] \, _{\text{out}}\langle p_1, \cdots | \phi(x) | q_2, \cdots \rangle_{\text{in}}$$

$$+ \int d^4x \, e_{q_1}(x) \, \partial_0^2 \, _{\text{out}}\langle p_1, \cdots | \phi(x) | q_2, \cdots \rangle_{\text{in}}$$

$$= \int d^4x \left[(-\partial_i^2 + m^2) e_{q_1}(x) \right] \, _{\text{out}}\langle p_1, \cdots | \phi(x) | q_2, \cdots \rangle_{\text{in}}$$

$$+ \int d^4x \, e_{q_1}(x) \, \partial_0^2 \, _{\text{out}}\langle p_1, \cdots | \phi(x) | q_2, \cdots \rangle_{\text{in}}$$

$$= \int d^4x \, e_{q_1}(x)(\partial_\mu^2 + m^2) \, _{\text{out}}\langle p_1 \cdots | \phi(x) | q_2, \cdots \rangle_{\text{in}} \qquad (5.69)$$

Collecting our results, we find:

$$_{\text{out}}\langle p_1, p_2, \cdots, p_n | q_1, q_2, \cdots, q_m \rangle_{\text{in}} = \text{disconnected graph}$$

$$+ iZ^{-1/2} \int d^4x \, e_{q_1}(x) \, (\partial_\mu^2 + m^2) \, _{\text{out}}\langle p_1, p_2, \cdots | \phi(x) | q_2, \cdots \rangle_{\text{in}}$$

$$(5.70)$$

We have now completed the first step of the LSZ program. By extracting the state with momentum q_1 from the "in" states, we now have reduced an abstract S matrix element into the matrix element of a field ϕ. Our goal, obviously, is to continue this process until all the asymptotic fields are extracted from the asymptotic states.

A small complication emerges when we extract out a second field from the "out" state with momentum p_1. We find that we must adopt the time-ordered product of two fields (otherwise, we cannot make the transition from time derivatives to the Lorentz covariant derivatives). Repeating the identical steps as before, and including this important feature of time ordering, we find:

$$_{\text{out}}\langle p_1, \cdots | \phi(x_1) | q_2, \cdots \rangle_{\text{in}} = \, _{\text{out}}\langle p_2, \cdots | \phi(x_1) a_{\text{in}}(p_1) | q_2, \cdots \rangle_{\text{in}}$$

$$+ iZ^{-1/2} \int d^4y_1 \, e_{p_1}^*(y_1) \, (\partial_\mu^2 + m^2)_{y_1} \, _{\text{out}}\langle p_2, \cdots | T \phi(y_1) \phi(x_1) | q_2 \cdots \rangle_{\text{in}}$$

$$(5.71)$$

As before, the a_{in} operator now acts to the right, where it is an annihilation operator that destroys a state with exactly momentum p_1, generating a disconnected graph. Completing all steps, we now have the twice-reduced matrix element:

$$_{\text{out}}\langle p_1, \cdots | q_1 \cdots \rangle_{\text{in}} = \text{disconnected graphs}$$

$$+ \left(iZ^{-1/2}\right)^2 \int d^4x_1 \, d^4y_1 \, e_{p_1}^*(y_1) e_{q_1}(x_1) \, (\partial_\mu^2 + m^2)_{y_1} (\partial_\mu^2 + m^2)_{x_1}$$

$$\times \;_{\text{out}}\langle p_2, \cdots | T\phi(y_1)\phi(x_1)|q_2, \cdots \rangle_{\text{in}} \tag{5.72}$$

It is now straightforward to apply this reduction process to all fields contained within the asymptotic state vectors. After each reduction, the only complication is that we generate disconnected graphs and we must be careful with time ordering. The final result is:

$$_{\text{out}}\langle p_1, p_2, \cdots, p_n | q_1, q_2, \cdots, q_m \rangle_{\text{in}} = \text{disconnected graphs}$$

$$+ \left(iZ^{-1/2}\right)^{n+m} \int d^4y_1 \cdots d^4x_m \prod_{i=1}^n \prod_{j=1}^m e_{p_i}^*(y_i) e_{q_j}(x_j)$$

$$\times \, (\partial_\mu^2 + m^2)_{y_1} \cdots (\partial_\mu^2 + m^2)_{x_m} \langle 0|T\phi(y_1) \cdots \phi(x_m)|0\rangle \tag{5.73}$$

(In general, we can choose momenta so that the disconnected graphs are zero. For values of the momenta where the disconnected graphs are non-zero, we can use this formalism to reduce them out as well.)

5.4 Reduction of Dirac Spinors

It is now straightforward to apply this formalism to the reduction of fermionic S matrices. Following the steps of the bosonic case, we write the creation and annihilation operators in terms of the original Dirac field. Then we write the asymptotic condition as:

$$x^0 \to -\infty; \qquad \langle f|\psi(x)|i\rangle \to Z_2^{-1/2} \langle f|\psi_{\text{in}}(x)|i\rangle \tag{5.74}$$

Next, we write the S matrix and reduce out one creation operator using Eq. (3.116):

$$_{\text{out}}\langle f|k, i\rangle_{\text{in}} \;=\; _{\text{out}}\langle f|b_{\text{in}}^\dagger(k, \epsilon)|i\rangle_{\text{in}}$$

$$=\; -i \lim_{t \to -\infty} \int d^3x \;_{\text{out}}\langle f|\bar{\psi}(x)\gamma^0|i\rangle_{\text{in}} U_k(x, \epsilon)$$

$$=\; _{\text{out}}\langle f|b_{\text{out}}^\dagger(k, \epsilon)|i\rangle_{\text{in}}$$

$$-\, iZ_2^{-1/2} \int d^3x \left[_{\text{out}}\langle f|\bar{\psi}(x)|i\rangle_{\text{in}} (-i)\overleftarrow{\partial}_0 \gamma^0 U_k(x, \epsilon) \right.$$

$$+ \, _{\text{out}}\langle f | \bar{\psi}(x) | i \rangle_{\text{in}} (-i) \overset{\rightarrow}{\partial}_0 \gamma^0 U_k(x, \epsilon) \Bigg] \qquad (5.75)$$

As before, we must convert the $\gamma^0 \partial_0$ into the covariant $\gamma^\mu \partial_\mu$. This is accomplished because we can replace the time derivative with a space derivative (because the spinor $u(k, \epsilon)e^{-ik \cdot x}$ satisfies the Dirac equation). In this way, we can now write down four different types of reduction formulas, depending on which creation or annihilation process we are analyzing.

The reduction formulas after a single reduction now read as follows:

$$_{\text{out}}\langle f | b_{\text{in}}^\dagger(k, \epsilon) | i \rangle_{\text{in}} = -i Z_2^{-1/2} \int d^4x \, _{\text{out}}\langle f | \bar{\psi}(x) | i \rangle_{\text{in}} (-i \overset{\leftarrow}{\not{\partial}} - m) U_k(x, \epsilon)$$

$$_{\text{out}}\langle f | d_{\text{in}}^\dagger(k, \epsilon) | i \rangle_{\text{in}} = +i Z_2^{-1/2} \int d^4x \, \bar{V}_k(x, \epsilon)(i \overset{\rightarrow}{\not{\partial}} - m) \, _{\text{out}}\langle f | \psi(x) | i \rangle_{\text{in}}$$

$$_{\text{out}}\langle f | b_{\text{out}}(k, \epsilon) | i \rangle_{\text{in}} = -i Z_2^{-1/2} \int d^4x \, \bar{U}_k(x, \epsilon)(i \overset{\rightarrow}{\not{\partial}} - m) \, _{\text{out}}\langle f | \psi(x) | i \rangle_{\text{in}}$$

$$_{\text{out}}\langle f | d_{\text{out}}(k, \epsilon) | i \rangle_{\text{in}} = +i Z_2^{-1/2} \int d^4x \, _{\text{out}}\langle f | \bar{\psi}(x) | i \rangle_{\text{in}} (-i \overset{\leftarrow}{\not{\partial}} - m) V_k(x, \epsilon)$$

$$(5.76)$$

where we have dropped the disconnected graph.

Making successive reductions, until all creation and annihilation operators are reduced out, is also straightforward. As before, we find that we must take time-ordered matrix elements of the various fields. Let us take the matrix element between incoming particles and outgoing particles. The incoming particles are labeled by $p_1, p_2, \ldots$, while the incoming antiparticles are labeled by $p_1', p_2', \ldots$. The outgoing particles are labeled by $q_1, q_2, \ldots$, while the outgoing antiparticles are labeled by $q_1', q_2', \ldots$.

The matrix element, after reduction, yields the following:

$$_{\text{out}}\langle 0 | b_{\text{out}}(q_1) \cdots d_{\text{out}}(q_1') \cdots b_{\text{in}}^\dagger(p_1) \cdots d_{\text{in}}^\dagger(p_1') \cdots | 0 \rangle_{\text{in}}$$

$$= (-i Z_2^{-1/2})^{n/2} (i Z_2^{-1/2})^{-n'/2} \int d^4x_1 \cdots d^4x_1' \cdots d^4y_1 \cdots d^4y_1' \cdots$$

$$\times \bar{U}_{q_1}(y_1)(i \overset{\rightarrow}{\not{\partial}} - m)_{y_1} \cdots \bar{V}_{p_1'}(x_1')(i \gamma^\mu \overset{\rightarrow}{\not{\partial}} - m)_{x_1'} \cdots$$

$$\times \langle 0 | T \left[\bar{\psi}(y_1') \cdots \psi(y_1) \cdots \bar{\psi}(x_1) \cdots \psi(x_1') \cdots \right] | 0 \rangle$$

$$\times (-i \overset{\leftarrow}{\not{\partial}} - m)_{x_1} U_{p_1}(x_1) \cdots (-i \overset{\leftarrow}{\not{\partial}} - m)_{y_1'} V_{q_1'}(y_1') \cdots \qquad (5.77)$$

where we have discarded the disconnected graphs. In this way, we can reduce out even the most complex scattering processes in terms of the reduction formulas.

5.5   Time Evolution Operator

The LSZ reduction formulas have been able to convert the abstract S matrix element into the product of vacuum expectation values of the fully interacting fields. No approximations have been made. However, we still do not know how to take matrix elements of the interacting fields. Hence, we cannot yet extract out numbers out of these matrix elements. The problem is that everything is written in terms of the fully interacting fields, of which we know almost nothing. The key is now to make an approximation to the theory by power expanding in the coupling constant, which is of the order of 1/137 for QED. We begin by splitting the Hamiltonian into two distinct pieces:

$$H = H_0 + H_I \tag{5.78}$$

where H_0 is the free Hamiltonian and H_I is the interacting part. For the ϕ^4 theory, for example, the interacting part would be:

$$H_I = \int d^3x \, \mathcal{H}_I$$
$$\mathcal{H}_I = \frac{\lambda}{4!}\phi^4 \tag{5.79}$$

At this point, it is useful to remind ourselves from ordinary quantum mechanics that there are several "pictures" in which to describe this time evolution. In the *Schrödinger picture*, we recall, the wave function $\psi(\mathbf{x}, t)$ and state vector are functions of time t, but the operators of the theory are constants in time. In the *Heisenberg picture*, the reverse is true; that is, the wave function and state vectors are constants in time, but the time evolution of the operators and dynamical variables of the theory are governed by the Hamiltonian:

$$\phi(\mathbf{x}, t) = e^{iHt}\phi(\mathbf{x}, 0)e^{-iHt} \tag{5.80}$$

In the LSZ formalism, we will find it convenient to define yet another picture, which resembles the interaction picture. In this new picture, we need to find a unitary operator $U(t)$ that takes us from the fully interacting field $\phi(x)$ to the free, asymptotic "in" states:

$$\phi(t, \mathbf{x}) = U^{-1}(t)\phi_{\text{in}}(t, \mathbf{x})U(t) \tag{5.81}$$

where $U(t) \equiv U(t, -\infty)$ is a time evolution operator, which obeys:

$$
\begin{aligned}
U(t_1, t_2)U(t_2, t_3) &= U(t_1, t_3) \\
U^{-1}(t_1, t_2) &= U(t_2, t_1) \\
U(t, t) &= 1
\end{aligned}
\tag{5.82}
$$

Because we now have two totally different types of scalar fields, one free and the other interacting, we must also be careful to distinguish the Hamiltonian written in terms of the free or the interacting fields. Let $H(t)$ be the fully interacting Hamiltonian written in terms of the interacting field, and let $H_0(\phi_{\mathrm{in}})$ represent the free Hamiltonian written in terms of the free asymptotic states. Then the free field ϕ_{in} and the interacting field satisfy two different equations of motion:

$$
\begin{aligned}
\frac{\partial}{\partial t}\phi(t, \mathbf{x}) &= i[H(t), \phi(t, \mathbf{x})] \\
\frac{\partial}{\partial t}\phi_{\mathrm{in}}(t, \mathbf{x}) &= i[H_0^{\mathrm{in}}, \phi_{\mathrm{in}}(t, \mathbf{x})]
\end{aligned}
\tag{5.83}
$$

To solve for $U(t)$, we need to extract a few more identities. If we differentiate the expression $UU^{-1} = 1$, we find:

$$
\left[\frac{d}{dt}U(t)\right]U^{-1}(t) + U(t)\frac{d}{dt}U^{-1}(t) = 0
\tag{5.84}
$$

Now let us take the derivative of ϕ_{in} and use the identities that we have written down:

$$
\begin{aligned}
\frac{\partial}{\partial t}\phi_{\mathrm{in}}(t, \mathbf{x}) &= \frac{\partial}{\partial t}\left[U(t)\phi(t, \mathbf{x})U^{-1}\right] \\
&= \dot{U}(t)\phi(t, \mathbf{x})U^{-1} + U(t)\dot{\phi}(t, \mathbf{x})U^{-1}(t) + U(t)\phi(t, \mathbf{x})\dot{U}^{-1}(t) \\
&= \dot{U}(t)\left(U^{-1}\phi_{\mathrm{in}}U\right)U^{-1} + U(t)[iH(\phi, \pi), \phi]U^{-1} \\
&\quad + UU^{-1}\phi_{\mathrm{in}}U\dot{U}^{-1} \\
&= \dot{U}U^{-1}\phi_{\mathrm{in}} + iU[H(\phi, \pi), \phi]U^{-1} + \phi_{\mathrm{in}}U\dot{U}^{-1} \\
&= \left[\dot{U}U^{-1} + iH(\phi_{\mathrm{in}}, \pi_{\mathrm{in}}), \phi_{\mathrm{in}}\right]
\end{aligned}
\tag{5.85}
$$

The last expression, in turn, must equal $i[H_0^{\mathrm{in}}, \phi_{\mathrm{in}}]$. This means that the following expression commutes with every "in" operator, and hence must be a c number:

$$
\dot{U}U^{-1} + i\left[H(\phi_{\mathrm{in}}, \pi_{\mathrm{in}}) - H_0^{\mathrm{in}}\right] = c\text{ number}
\tag{5.86}
$$

(This c number, one can show, does not contribute to the S matrix.) Thus, the $U(t)$ operator satisfies the following:

$$i \frac{\partial}{\partial t} U(t, t_0) = H_I(t) U(t, t_0) \tag{5.87}$$

where:

$$H_I(t) \equiv H(\phi_{\text{in}}, \pi_{\text{in}}) - H_0^{\text{in}} \tag{5.88}$$

that is, $H_I(t)$ is defined to be the interaction Hamiltonian defined only with free, asymptotic fields. Since $H_I(t)$ does not necessarily commute with $H_I(t')$ at different times, the integration of the previous equation is a bit delicate. However, one can show that:

$$
\begin{aligned}
U(t) &= U(t, -\infty) \\
&= T \exp\left(-i \int_{-\infty}^{t} dt_1 \, H_I(t_1)\right) \\
&= T \exp\left(-i \int_{-\infty}^{t} dt_1 \int d^3 x_1 \, \mathcal{H}_I(\mathbf{x}_1, t_1)\right) \tag{5.89}
\end{aligned}
$$

where the operator T means that, as we integrate over t_1, we place the exponentials sequentially in time order. To prove this expression, we simply insert it into Eq. (5.87). Written in this form, however, this expression is not very useful. We will find it much more convenient to power expand the exponential in a Taylor series, so we have:

$$
\begin{aligned}
U(t) &= 1 + \sum_{n=1}^{\infty} \frac{(-i)^n}{n!} \int_{-\infty}^{t} d^4 x_1 \int_{-\infty}^{t} d^4 x_2 \cdots \\
&\quad \times \int_{-\infty}^{t} d^4 x_n \, T\left[\mathcal{H}_I(x_1) \mathcal{H}_I(x_2) \cdots \mathcal{H}_I(x_n)\right] \tag{5.90}
\end{aligned}
$$

(See Exercise 10 concerning the change in the upper limits of integration.) Now that we have an explicit solution for $U(t)$, let us decompose the interacting Green's function. If we take the matrix element of a series of interacting ϕ fields, we will use the $U(t)$ operator to convert the entire expression to free, asymptotic fields. To do this, let us first choose a sequence of space–time points x_i^{μ}, time ordered such that $x_1^0 > x_2^0 > \cdots x_n^0$, so we can drop the time ordering operator T.

Then we can replace all interacting fields with free fields by making the conversion $\phi = U^{-1} \phi_{\text{in}} U$ everywhere:

$$\langle 0| T\left(\phi(x_1)\phi(x_2) \cdots \phi(x_n)\right)|0\rangle = \langle 0|\phi(x_1) \cdots \phi(x_n)|0\rangle$$

$$= \langle 0|U^{-1}(t_1)\phi_{\text{in}}(x_1)U(t_1)U^{-1}(t_2)\phi_{\text{in}}(x_2)U(t_2)$$

$$\cdots U(t_{n-1})U^{-1}(t_n)\phi_{\text{in}}(x_n)U(t_n)|0\rangle$$

$$= \langle 0|U^{-1}(t_1)\phi_{\text{in}}(x_1)U(t_1,t_2)\phi_{\text{in}}(x_2)\cdots U(t_{n-1},t_n)\phi_{\text{in}}(x_n)U(t_n)|0\rangle$$

$$= \langle 0|U^{-1}(t)U(t,t_1)\phi_{\text{in}}(x_1)\cdots\phi_{\text{in}}(x_n)U(t_n,-t)U(-t)|0\rangle$$

$$= \langle 0|U^{-1}(t)T\left[\phi_{\text{in}}(x_1)\phi_{\text{in}}(x_2)\cdots\phi_{\text{in}}(x_n)U(t,-t)\right]U(-t)|0\rangle$$

$$= \left\langle 0\left|U^{-1}(t)T\left[\phi_{\text{in}}(x_1)\cdots\phi_{\text{in}}(x_n)\exp\left(-i\int_{-t}^{t}dt'H_I(t')\right)\right]U(-t)\right|0\right\rangle$$

$$(5.91)$$

where t is an arbitrarily long time, much greater than t_1, and $-t$ is much less than t_n. We will later set $t = \infty$. In the last line, since we have restored the time ordering operator T, we are allowed to move all U's around inside the matrix element. We can thus combine them all into one large $U(t, -t)$, which in turn can be expanded as a function of the interacting Lagrangian (defined strictly in terms of free, asymptotic fields).

There is now one last step that must be performed. We still have the term $U(-t)|0\rangle$ and $\langle 0|U^{-1}(t)$ to eliminate in the limit as $t \rightarrow \infty$. In general, since we assume that the vacuum is stable for this theory, we know that the vacuum is an eigenstate of the U operator, up to some phase.

Since $U(t) \equiv U(t, -\infty)$, we can set:

$$\lim_{t\to\infty} U(-t)|0\rangle = |0\rangle \qquad (5.92)$$

where we have taken the limit so that $U(-\infty, -\infty) = 1$. However, for the other state $\langle 0|U^{-1}(t)$, we must be a bit more careful, since the limit gives us $U^{-1}(\infty) = U(-\infty, +\infty)$. Since the vacuum is stable, this means that the vacuum at $t = -\infty$ remains the vacuum at $t = +\infty$, modulo a possible phase λ. We thus find:

$$\lim_{t\to\infty} \langle 0|U^{-1}(t) = \lambda\langle 0| \qquad (5.93)$$

By hitting the equation with $|0\rangle$, the phase λ is equal to:

$$\lambda = \lim_{t\to\infty} \langle 0|U^{-1}(t)|0\rangle$$

$$= \lim_{t\to\infty} \langle 0|U(t)|0\rangle^{-1}$$

$$= \left(\left\langle 0 \middle| T \exp\left(i \int d^4x \, \mathscr{L}_I(\phi_{\text{in}}) \right) \middle| 0 \right\rangle \right)^{-1} \tag{5.94}$$

The phase λ thus gives us the contribution of vacuum-to-vacuum graphs (i.e., graphs without any external legs). Putting everything together, we now have an expression for the Green's function defined with interacting fields:

$$G(x_1, x_2, \cdots, x_n) = \langle 0| T \, [\phi(x_1)\phi(x_2)\cdots\phi(x_n)] \, |0\rangle \tag{5.95}$$

This interacting Green's function, written in terms of free fields, becomes:

$$G(x_1, x_2, \cdots, x_n)$$
$$= \frac{\langle 0| T \, \phi_{\text{in}}(x_1) \cdots \phi_{\text{in}}(x_n) \exp\left\{ i \int d^4x \, \mathscr{L}_I(\phi_{\text{in}}) \right\} |0\rangle}{\langle 0| T \exp\left\{ i \int d^4x \, \mathscr{L}_I \, [\phi_{\text{in}}(x)] \right\} |0\rangle} \tag{5.96}$$

Using the formalism that we have constructed, we can rewrite the previous matrix element entirely in terms of the asymptotic "in" fields by a power expansion of the exponential:

$$G(x_1, x_2, \cdots, x_n) = \sum_{m=0}^{\infty} \frac{(-i)^m}{m!} \int_{-\infty}^{\infty} d^4y_1 \cdots d^4y_m \left\langle 0 \middle| T \middle[\phi(x_1)\phi(x_2) \right.$$
$$\left. \cdots \phi(x_n) \mathscr{H}_I(y_1) \mathscr{H}_I(y_2) \cdots \mathscr{H}_I(y_m) \middle] |0\rangle_c \right. \tag{5.97}$$

where the subscript c refers to connected diagrams only. (From now on, we will drop the "in" subscript on all fields. However, we must remind the reader that we have made the transition from the Heisenberg picture to this new picture where all fields are free.)

The next step is to actually evaluate the time ordered product of an arbitrary number of free fields. To do this, we appeal to *Wick's theorem*.

5.6 Wick's Theorem

We begin our discussion by defining the "normal ordering" of operators. In general, unlike the classical situation, the product of two quantum fields taken at the same point is singular:

$$\lim_{x \to y} \phi(x)\phi(y) = \infty \tag{5.98}$$

(In fact, in Chapter 14, we will investigate precisely how divergent this expression is.) Unfortunately, our action consists of fields multiplied at the same point; so the transition to the quantum theory is actually slightly ambiguous. To render our expressions rigorous, we recall the normal ordered product of two fields introduced in Eq. (3.27). Let us decompose the scalar field into its creation operators [labeled by a $-$ sign, and annihilation operators, labeled by a $+$ sign: $\phi(x) = \phi^+(x) + \phi(x)^-$]. Then the normal ordered product of these fields simply rearranges the creation and annihilation parts such that the creation operators always appear on the left, and annihilation operators always appear on the right. The normal ordered product is defined as:

$$: \phi(x)\phi(y) : \quad \equiv \quad \phi(x)^+\phi^+(y) + \phi(x)^-\phi^+(y) + \phi(y)^-\phi^+(x)$$
$$+ \quad \phi(x)^-\phi^-(y) \tag{5.99}$$

(The normal ordered product of two fields is no longer a local object, since we have split up the components within the fields and reshuffled them.)

One consequence of normal ordering is that the vacuum expectation value of any normal ordered product vanishes, since annihilation operators always appear on the right. It vanishes because:

$$\phi^+|0\rangle = \langle 0|\phi^- = 0 \tag{5.100}$$

Similarly, one can define normal ordered products for more complicated products of fields. Now we can proceed to find the relationship between normal ordered products and time ordered products. It is easy to prove the following identity, or Wick's theorem for two fields:

$$T\left[\phi(x_1)\phi(x_2)\right] =: \phi(x_1)\phi(x_2) : + c \quad \text{number} \tag{5.101}$$

The only difference between the time ordered and normal ordered products is that we have reshuffled the various annihilation and creation parts of the fields. Each time we commute parts of the fields past each other, we pick up c-number expressions. To find what this c-number expression is, we now simply take the vacuum expectation value of both sides. Since the vacuum expectation value of normal ordered products is zero, we find:

$$T\left[\phi(x_1)\phi(x_2)\right] =: \phi(x_1)\phi(x_2) : +\langle 0|T\left[\phi(x_1)\phi(x_2)\right]|0\rangle \tag{5.102}$$

If we have three fields, then Wick's theorem[6] reads:

$$T\left(\phi(x_1)\phi(x_2)\phi(x_3)\right) \quad = \quad : \phi(x_1)\phi(x_2)\phi(x_3) :$$

$$+ \langle 0|T \, [\phi(x_1)\phi(x_2)] \,|0\rangle \phi(x_3)$$

$$+ \langle 0|T \, [\phi(x_2)\phi(x_3)] \,|0\rangle \phi(x_1)$$

$$+ \langle 0|T \, [\phi(x_3)\phi(x_1)] \,|0\rangle \phi(x_2) \qquad (5.103)$$

The last three terms can be succinctly summarized by the sum $\sum_{\text{perm}}$, which means summing over all permutations of x_1, x_2, x_3. To prove this expression with three scalar fields, we take the original identity with just two scalar fields, and multiply both sides of the equation on the right by $\phi(x_3)$, which has the earliest time. Then we merge $\phi(x_3)$ into the normal ordered product. Merging $\phi^+(x_3)$ into the normal ordered product is easy, since the annihilation operators are on the right, anyway. However, merging $\phi^-(x_3)$ into the normal ordered product is more difficult, since $\phi^-(x_3)$ must move past $\phi^+(x_1)$ and $\phi^+(x_2)$. Each time it moves past one of these terms, we pick up a c-number expression, which is equal to the time ordered product; that is:

$$\langle 0|\phi^+(x_2)\phi^-(x_3)|0\rangle \;=\; \langle 0|\phi(x_2)\phi(x_3)|0\rangle$$

$$\;=\; \langle 0|T \, [\phi(x_2)\phi(x_3)] \,|0\rangle \qquad (5.104)$$

In this way, we pick up all the terms in the Wick identity. Likewise, for four fields, we have:

$$\langle 0|T \, [\phi(x_1)\cdots\phi(x_4)] \,|0\rangle =: \phi(x_1)\cdots\phi(x_4):$$

$$+ \sum_{\text{perm}} \langle 0|T \, [\phi(x_1)\phi(x_2)] \,|0\rangle \; :\phi(x_3)\phi(x_4):$$

$$+ \sum_{\text{perm}} \langle 0|T \, [\phi(x_1)\phi(x_2)] \,|0\rangle \langle 0|T \, [\phi(x_3)\phi(x_4)] \,|0\rangle \qquad (5.105)$$

By now, it should be obvious that the time ordered product of n fields can be written in terms of sum of normal ordered products. For the general n-point case (n even), Wick's theorem reads:

$$T \, [\phi(x_1)\phi(x_2)\cdots\phi(x_n)] =: \phi(x_1)\phi(x_2)\cdots\phi(x_n):$$

$$+ \sum_{\text{perm}} \langle 0|T \, [\phi(x_1)\phi(x_2)] \,|0\rangle \; :\phi(x_3)\cdots\phi(x_n):$$

$$+ \sum_{\text{perm}} \langle 0|T \, [\phi(x_1)\phi(x_2)] \,|0\rangle \langle 0|T \, [\phi(x_3)\phi(x_4)] \,|0\rangle \; :\phi(x_5)\cdots\phi(x_n):$$

$$\vdots$$

$$+ \sum_{\text{perm}} \langle 0|T \, (\phi(x_1)\phi(x_2)) \,|0\rangle \cdots \langle 0|T \, (\phi(x_{n-1})\phi(x_n)) \,|0\rangle \qquad (5.106)$$

For n odd, the last line reads:

$$\sum_{\text{perm}} \langle 0|T\,[\phi(x_1)\phi(x_2)]\,|0\rangle \cdots \langle 0|T\,[\phi(x_{n-2})\phi(x_{n-1})]\,|0\rangle\phi(x_n) \qquad (5.107)$$

These formulas are proved by induction on n. If we assume that they hold for $n-1$, then we multiply the entire formula on the right by $\phi(x_n)$, which has the earliest time. Then, by merging $\phi(x_n)$ with the rest of the products, we find that the formula now holds for n. The proof is straightforward, and runs the same as for the case described earlier. Merging $\phi^+(x_n)$ poses no problem, since annihilation operators are on the right anyway. However, each time $\phi^-(x_n)$ moves past $\phi^+(x_j)$, we pick up a commutator, which is equal to the vacuum expectation of the time ordered product of the two fields. In this way, it is easy to show that we pick up the nth Wick's theorem.

If we take the vacuum expectation value of both sides of the equation, then Wick's theorem for vacuum expectation values reads:

$$\langle 0|T[\phi(x_1)\phi(x_2) \quad \cdots \phi(x_n)]|0\rangle = \sum_{\text{perm}} \langle 0|T\,[\phi(x_1)\phi(x_2)]\,|0\rangle$$

$$\cdots \langle 0|T\,[\phi(x_{n-1})\phi(x_n)]\,|0\rangle \qquad (5.108)$$

The generalization to fermionic fields is also straightforward. [The only complication is that we pick up extra minus signs because of the anti-commutation properties of fermionic fields. For more complicated products, we must always insert (-1) whenever fermion fields move past each other.] We find:

$$T\,[\psi(x)\bar{\psi}(y)] =: \psi(x)\bar{\psi}(y) : +\langle 0|T\,[\psi(x)\bar{\psi}(y)]\,|0\rangle \qquad (5.109)$$

Now insert Eqs. (5.97) and (5.108) into Eq. (5.73). This gives us a complete reduction of the S matrix (written as a function of interacting fields) in terms of Green's functions of free fields, as desired. In the last step, we can eliminate the $(\partial_\mu^2 + m^2)$ factors appearing in Eq. (5.73), because they act on two-point Green's functions and become delta functions:

$$(\partial_\mu^2 + m^2)_x \langle 0|T\phi(x)\phi(y)|0\rangle \;=\; -i\delta^4(x-y)$$

$$(i\,\slashed{\partial} - m)_x \langle 0|T\psi(x)\bar{\psi}(y)|0\rangle \;=\; i\delta^4(x-y) \qquad (5.110)$$

In this fashion, we have how completely reduced the S matrix element into sums of products of two-point Green's functions and certain vertex elements. Although Wick's theorem seems a bit tedious, in actual practice the decomposition proceeds rapidly. To see this, we will analyze the four-point function taken to first order in with an interaction given by $-\lambda\phi^4/4!$. We want to expand:

$$
\begin{aligned}
G(x_1, \cdots x_4) &= -\frac{i\lambda}{4!} \int d^4 y \langle 0|T\left[\phi(x_1) \cdots \phi(x_4)\phi(y)^4\right]|0\rangle + \cdots \\
&= (-i\lambda) \int d^4 y \prod_{i=1}^{4} \langle 0|T\left[\phi(x_i)\phi(y)\right]|0\rangle + \cdots \\
&= (-i\lambda) \int d^4 y \prod_{i=1}^{4} \left[i\Delta_F(x_i - y)\right] + \cdots \quad (5.111)
\end{aligned}
$$

where we have used Wick's theorem. The 4! term has disappeared, because there are 4! ways in which four external legs at x_i can be connected to the four fields contained within ϕ^4.

Another example of this decomposition is given by the four-point function taken to second order:

$$
\begin{aligned}
G(x_1, \cdots, x_4) &= \left(\frac{-i\lambda}{4!}\right)^2 \frac{1}{2!} \int d^4 y_1 \int d^4 y_2 \langle 0|T\left\{\phi(x_1)\cdots\phi(x_4)\right. \\
&\quad \times \left.[\phi(y_1)^4][\phi(y_2)^4]\right\}|0\rangle \quad (5.112)
\end{aligned}
$$

The expansion, via Wick's theorem, is straightforward:

$$
\begin{aligned}
G(x_1, \cdots x_4) &= \frac{(-i\lambda)^2}{2!} \int d^4 y_1 \int d^4 y_2 \left\{[i\Delta_F(y_1 - y_2)]^2\Delta_A\right. \\
&\quad \left. + [i\Delta_F(y_1 - y_1)][i\Delta_F(y_1 - y_2)]\Delta_B\right\} \quad (5.113)
\end{aligned}
$$

where:

$$
\begin{aligned}
\Delta_A &= \Delta_F(x_1 - y_1)\Delta_F(x_2 - y_1)\Delta_F(x_3 - y_2)\Delta_F(x_4 - y_2) \\
&\quad + \Delta_F(x_1 - y_1)\Delta_F(x_3 - y_1)\Delta_F(x_2 - y_2)\Delta_F(x_4 - y_2) \\
&\quad + \Delta_F(x_1 - y_1)\Delta_F(x_4 - y_1)\Delta_F(x_2 - y_2)\Delta_F(x_3 - y_2) \quad (5.114)
\end{aligned}
$$

$$
\begin{aligned}
\Delta_B &= \Delta_F(x_1 - y_1)\Delta_F(x_2 - y_2)\Delta_F(x_3 - y_2)\Delta_F(x_4 - y_2) \\
&\quad + \Delta_F(x_1 - y_2)\Delta_F(x_2 - y_1)\Delta_F(x_3 - y_2)\Delta_F(x_4 - y_2) \\
&\quad + \Delta_F(x_1 - y_2)\Delta_F(x_2 - y_2)\Delta_F(x_3 - y_1)\Delta_F(x_4 - y_2) \\
&\quad + \Delta_F(x_1 - y_2)\Delta_F(x_2 - y_2)\Delta_F(x_3 - y_1)\Delta_F(x_4 - y_2) \quad (5.115)
\end{aligned}
$$

These are shown graphically in Figure 5.4.

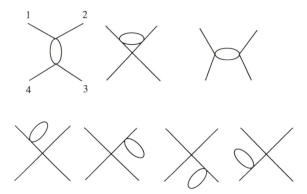

Figure 5.4. The Feynman diagrams corresponding to the Wick decomposition of ϕ^4 theory to second order.

Finally, we will convert from x-space to p-space by using Eqs. (3.50) and (3.135). When we perform the x integrations, we obtain a delta function at each vertex, which represents the conservation of momentum. When all x-space integrations are performed, we are left with one delta function representing overall momentum conservation. We are also left with a momentum integration for each internal loop. Then all Feynman's rules can be represented entirely in momentum space.

5.7   Feynman's Rules

From this, we can extract graphical rules by Feynman rules, by which we can almost by inspection construct Green's functions of arbitrary complexity. With an interaction Lagrangian given by $-\lambda\phi^4/4!$, $\mathcal{M}_{fi}$ appearing in Eq. (5.26) can be calculated as follows:

1. Draw all possible connected, topologically distinct diagrams, including loops, with n external legs. Ignore vacuum-to-vacuum graphs.

2. For each internal line, associate a propagator given by:

$$- - - \!\!\!\succ - - \qquad i\Delta_F(p) = \frac{i}{p^2 - m^2 + i\epsilon} \qquad\qquad (5.116)$$
$$p$$

3. For each vertex, associate the factor $-i\lambda$.

4. For each internal momentum corresponding to an internal loop, associate an integration factor:

$$\int \frac{d^4 p}{(2\pi)^4} \tag{5.117}$$

5. Divide each graph by an overall symmetry factor S corresponding to the number of ways one can permute the internal lines and vertices, leaving the external lines fixed.

6. Momentum is conserved at each vertex.

The symmetry factor S is easily calculated. For the four-point function given above, the $1/4!$ coming from the interaction Lagrangian cancels the 4! ways in which the four external lines can be paired off with the four scalar fields appearing in ϕ^4, so $S = 1$. Now consider the connected two-point diagram at second order, which is a double-loop diagram (which has the topology of the symbol ϕ). There are 4 ways in which each external leg can be connected to each vertex. There are 3×2 ways in which the internal vertices can be paired off. So this gives us a factor of $1/S = (1/4!)(1/4!) \times (4 \times 4) \times (3 \times 2) = 1/3!$, so $S = 6$.

For QED, the Feynman's rules are only a bit more complicated. The interaction Hamiltonian becomes:

$$\mathcal{H}_I = -ie\bar{\psi}\gamma^\mu\psi A_\mu \tag{5.118}$$

As before, the power expansion of the interacting Lagrangian will pull down various factors of $\mathcal{H}_I$. Then we use Wick's theorem to pair off the various fermion and vector meson lines to form propagators and vertices.

There are only a few differences that we must note. First, when contracting over an internal fermion loop, we must flip one spinor past the others to perform the trace and Wick decomposition. This means that there must be an extra -1 factor inserted into all fermion loop integrations.

Second, various vector meson propagators in different gauges may be used, but all the terms proportional to p_μ or p_ν vanish because of gauge invariance (which will be discussed more in detail later).

Thus, the Feynman's rules for QED become:

1. For each internal fermion line, associate a propagator given by:

$$iS_F(p) = \frac{i}{\not{p} - m + i\epsilon} = \frac{i(\not{p} + m)}{p^2 - m^2 + i\epsilon} \qquad \xrightarrow{} \atop p \tag{5.119}$$

2. For each internal photon line, associate a propagator:

$$i D_F(p)_{\mu\nu} = -\frac{i g_{\mu\nu}}{p^2 + i\epsilon}$$

$$(5.120)$$

3. At each vertex, place a factor of:

$$- i e \gamma_\mu$$

$$(5.121)$$

4. Insert an additional factor of -1 for each closed fermion loop.

5. For each internal loop, integrate over:

$$\int \frac{d^4 q}{(2\pi)^4} \tag{5.122}$$

6. A relative factor -1 appears between graphs that differ from each other by an interchange of two identical external fermion lines.

7. Internal fermion lines appear with arrows in both clockwise and counter-clockwise directions. However, diagrams that are topologically equivalent are counted only once.

8. External electron and positron lines entering a graph appear with factors $u(p, s)$ and $\bar{v}(p, s)$, respectively. External electron and positron lines leaving a graph appear with factors $\bar{u}(p, s)$ and $v(p, s)$, respectively. The direction of the positron lines is taken to be opposite of the electron lines, so that incoming positrons have momenta leaving the diagram.

Likewise, we can calculate Feynman's rules for any of the actions that we have investigated earlier.

For example, for charged scalar electrodynamics, with the additional term in the Lagrangian:

$$\mathcal{L} = D_\mu \phi^\dagger D^\mu \phi - m^2 \phi^\dagger \phi \tag{5.123}$$

one has the following interaction Hamiltonian:

$$\mathcal{H}_I = -i e \phi^\dagger \left(\overleftarrow{\partial}_\mu - \overrightarrow{\partial}_\mu \right) \phi A^\mu - e^2 A_\mu^2 \phi^\dagger \phi \tag{5.124}$$

The Feynman rules are as follows:

1. For each scalar-scalar-vector vertex, insert the factor:

$$-ie(p+p')_\mu \qquad (5.125)$$

where p and p' are the momenta for the scalar line.

2. Insert a factor of:

$$2ie^2 g_{\mu\nu} \qquad (5.126)$$

for each "seagull" graph.

3. Insert an additional factor of $1/2$ for each closed loop with only two photon lines.

In summary, we have seen that, historically, there were two ways in which to quantize QED. The first method, pioneered by Feynman, was the propagator approach, which was simple, pictorial, but not very rigorous. The second was the more conventional operator approach of Schwinger and Tomonaga. In this chapter, we have presented the LSZ approach, which is perhaps the most convenient method for deriving the Feynman rules for any quantum field theory.

With Feynman rules, one can almost, by inspection, write down the perturbation expansion for any quantum field theory. In the next chapter, we will use these rules to calculate higher-order interactions in QED.

5.8 Exercises

1. Set up the reduction formulas for a massless and massive vector meson. Derive the counterpart of Eqs. (5.73) and (5.77).

2. Write down the Feynman rules for a massive pseudoscalar field interacting with a Dirac electron via the interaction $\bar{\psi}\gamma_5\psi\phi$.

3. Write down the Feynman rules for a massive pseudovector field interacting with the Dirac field via the term $\bar{\psi}\gamma_5\gamma^\mu\psi A_\mu$.

4. In the "old-fashioned" noncovariant canonical approach to QED in the Coulomb gauge, one derived the scattering matrix by solving the Schrödinger

equation with a second-quantized Hamiltonian, in which the interactions were ordered in time. Draw the complete set of noncovariant diagrams that are necessary to describe electron–electron scattering in QED to the one-loop level. Sketch how several noncovariant diagrams can be summed to produce a single covariant Feynman diagram.

5. Prove:

$$\frac{(-1)^{N-1}}{(N-M)!}\epsilon^{\mu_1\mu_2\cdots\mu_M\sigma_1\sigma_2\cdots\sigma_{N-M}}\epsilon_{\nu_1\nu_2\cdots\nu_M\sigma_1\sigma_2\cdots\sigma_{N-M}} = \delta^{\mu_1\mu_2\cdots\mu_M}_{\nu_1\nu_2\cdots\nu_M} \qquad (5.127)$$

where:

$$\delta^{\mu_1\cdots\mu_M}_{\mu_1\cdots\mu_M} = \delta^{\mu_1}_{[\mu_1}\delta^{\mu_2}_{\nu_2}\cdots\delta^{\mu_M}_{\nu_M]} \qquad (5.128)$$

6. Show the equivalence of the two expressions for the flux J in Eqs. (5.24) and (5.25) and show the equality only holds in collinear frames. [Hint: square Eq. (5.25), and then expand the product of two $\epsilon^{\mu\nu\alpha\beta}$ tensors in terms of delta functions.]

7. In Compton scattering, a photon scatters off an electron. Show that the relative velocity $|v_1 - v_2|$ appearing in the cross-section formula can exceed the speed of light. Is this a violation of relativity? Why or why not?

8. Prove Furry's theorem[9], which states that a Feynman loop diagram containing an odd number of external photon lines vanishes. (Hint: Show that a fermion loop with an odd number of legs cancels against another fermion loop with the arrows reversed, or use the fact that QED is invariant under charge conjugation. The fields transform as: $\psi \to \psi^c = C\bar{\psi}^T$ and $A_\mu \to -A_\mu$. Show that these diagrams are odd under C and hence not allowed.)

9. The factor λ in Eq. (5.94) contains Feynman graphs with no external legs. For QED, draw all such diagrams up to the second-loop level. Do not solve.

10. If we power expand Eq. (5.89) in a Taylor series, we find:

$$U(t, t') = 1 + \sum_{n=1}^{\infty}(-i)^n \int_{t'}^{t} dt_1 \int_{t'}^{t_1} dt_2 \cdots \int_{t'}^{t_{n-1}} dt_n$$
$$\times T\left(\mathcal{H}_I(t_1)\cdots\mathcal{H}_I(t_n)\right) \qquad (5.129)$$

Prove this, and then show that it equals Eq. (5.90). (Hint: take the lowest order, and on a graph, draw the integration regions for t_1 and t_2. Show that the identity holds for this case, and then generalize to the arbitrary case.)

11. Prove that the time-ordered product in Eq. (5.73) always emerges when we make the LSZ reduction of more than one field from the asymptotic states.

12. Let us define the following Green's function:

$$\Delta'(x, x') \equiv -i \langle 0|[\phi(x), \phi(x')]|0\rangle \tag{5.130}$$

Insert a complete set of intermediate states within the commutator:

$$1 = \sum_n |n\rangle\langle n| = \int d^4q \delta^4(p_n - q) \tag{5.131}$$

Prove that the commutator becomes:

$$\Delta'(x - x') = \int_0^\infty d\sigma^2 \, \rho(\sigma^2)\Delta(x - x', \sigma) \tag{5.132}$$

where:

$$
\begin{aligned}
\rho(q) &= \rho(q^2)\theta(q_0) \\
&= (2\pi)^3 \sum_n \delta^4(p_n - q)|\langle 0|\phi(0)|n\rangle|^2 \tag{5.133}
\end{aligned}
$$

This is the Källén–Lehmann spectral representation[8,9]. [Since σ appears in this formula as a mass, this formula states that the complete, interacting value of Δ' is equal to the integral over all possible free Δ' with arbitrary mass σ, weighted by the unknown function $\rho(\sigma^2)$.]

13. Let us separate out the one-particle contribution to the smeared average. Then we find:

$$\Delta'(x - x') = Z\Delta(x - x'; m) + \int_{m^2}^\infty d\sigma^2 \, \rho(\sigma^2)\Delta(x - x'; \sigma) \tag{5.134}$$

Also, m^2 is the lowest mass squared that contributes to the continuum above the one-particle contribution. For example, for pions, $m^2 = 4m_\pi^2$. Take the time derivative of both sides, reducing the Green's functions to delta functions. Then show:

$$1 = Z + \int_{m^2}^\infty \rho(\sigma^2) \, d\sigma^2 \tag{5.135}$$

which implies:

$$0 \le Z < 1 \tag{5.136}$$

When $Z = 1$, this means that the ρ function is zero, so the theory has collapsed into a free theory. For an interacting theory, we must have $Z < 1$.

14. Renormalization constants are usually thought to be infinite quantities, yet we have just shown that Z is less than one. Is there a contradiction?

15. For ϕ^4 theory, prove that the symmetry factor S equals:

$$S = c \prod_{n=2,3,\ldots} 2^b (n!)^{a_n} \tag{5.137}$$

where a_n equals the number of pairs of vertices that are connected by n identical lines, b is the number of lines that connect a vertex with itself, and c is the number of permutations of vertices that leave the diagram invariant when the external lines are fixed. What is S for QED?

Chapter 6
Scattering Processes and the S Matrix

I was sort of half-dreaming, like a kid would ... that it would be funny if these funny pictures turned out to be useful, because the damned Physical Review would be full of these odd-looking things. And that turned out to be true.

—R. Feynman

6.1 Compton Effect

Now that we have derived the Feynman rules for various quantum field theories, the next step is to calculate cross sections for elementary processes involving photons, electrons, and antielectrons. At the lowest order, these cross sections reproduce classical results found with earlier methods. However, the full power of the quantum field theory will be seen at higher orders, where we calculate radiative corrections to the hydrogen atom that have been verified to great accuracy. In the process, we will solve the problem of the electron self-energy, which completely eluded earlier, classical attempts by Lorentz and others.

At the end of this chapter, we will also investigate the S matrix itself. Rather than appeal to perturbation theory and summing Feynman diagrams, we will impose mathematical constraints directly on the S matrix, like unitarity and analyticity, to obtain nontrivial constraints on π-nucleon scattering. These results hold without ever appealing to any perturbative power expansion.

The material covered in this chapter is fairly standard, and the reader is urged to consult other excellent texts for other details, such as Bjorken and Drell, Itzykson and Zuber, and Mandl and Shaw.

To begin our discussion, we will divide Feynman diagrams into two types, "trees" and "loops," on the basis of their topology. Loop diagrams, as their name suggests, have closed loops in them. Tree diagrams have no loops; that is, they only have branches. In a scattering process, we will see that the sum over tree diagrams is finite and reproduces the classical result. The loop diagrams, by contrast, are usually divergent and are purely quantum-mechanical effects.

We start by analyzing the lowest-order terms in the scattering matrix for four particles or fields. To this order, we find only tree diagrams and no loops. Thus, we should be able to reproduce generalizations of classical and nonrelativistic physics. (Interestingly enough, we will find that negative energy states cannot be omitted in these calculations, even in the nonrelativistic limit. Although these negative energy states are purely a byproduct of relativity, if we drop them, then we will fail to reproduce the classical and nonrelativistic results.)

In this chart, we will summarize the scattering processes that we will analyze in the first part of this chapter:

Compton scattering:	$e^- + \gamma \rightarrow e^- + \gamma$
Pair annihilation:	$e^- + e^- \rightarrow \gamma + \gamma$
Møller scattering:	$e^- + e^- \rightarrow e^- + e^-$
Bhabha scattering:	$e^- + e^+ \rightarrow e^- + e^+$
Bremsstrahlung:	$e^- + N \rightarrow e^- + N + \gamma$
Pair creation:	$\gamma + \gamma \rightarrow e^- + e^+$

There is a reason for writing these scattering processes in this particular order. If we take the Feynman diagrams for Compton scattering and rotate them by 90 degrees, we find that they turn into the Feynman diagrams for pair annihilation. This is called the *substitution rule*, where we take the process:

$$1 + 2 \rightarrow 3 + 4 \tag{6.1}$$

and convert it into:

$$1 + \bar{3} \rightarrow \bar{2} + 4 \tag{6.2}$$

Using the substitution rule, we can group these scattering processes into pairs:

Compton effect	$\leftrightarrow$	Pair annihilation
Møller scattering	$\leftrightarrow$	Bhabha scattering
Bremsstrahlung	$\leftrightarrow$	Pair creation

There are several advantages to using this symmetry. At the superficial level, this means that we can, almost by inspection, convert the scattering amplitude of one process into that of the other, thereby saving a considerable amount of time. At a deeper level, it signals the fact that the S matrix obeys a new kind of symmetry, called *crossing symmetry*. If we treat the S matrix as an analytic

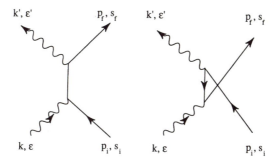

Figure 6.1. Compton scattering: a photon of momentum k scatters elastically off an electron of momentum p_i.

function of the energy variables, then crossing symmetry relates different analytic regions of the S matrix to each other in a non-trivial way.

To begin, the first process we will examine is Compton scattering, which occurs when an electron and a photon collide and scatter elastically. Historically, this process was crucial in confirming that electromagnetic radiation had particlelike properties, that is, that the photon was acting like a particle in colliding with the electron. (We will then, using the substitution rule, derive the Feynman amplitude for pair annihilation.)

We will assume that the electron has momentum p_i before the collision and p_f afterwards. The photon has momentum k before and k' afterwards. The reaction can be represented symbolically as:

$$\gamma(k) + e(p_i) \rightarrow \gamma(k') + e(p_f) \tag{6.3}$$

By energy–momentum conservation, we also have:

$$k + p_i = k' + p_f \tag{6.4}$$

Compton scattering, to lowest order, is shown in Figure 6.1.

We normalize the wave function of the photon by:

$$A_\mu(x) = \frac{1}{\sqrt{2kV}} \epsilon_\mu (e^{-ik \cdot x} + e^{ik \cdot x}) \tag{6.5}$$

To lowest order, the S matrix is:

$$S_{fi} = \frac{e^2}{V^2} (2\pi)^4 \delta^4 (p_f + k' - p_i - k)$$

$$\times \sqrt{\frac{m}{E_f}} \bar{u}(p_f, s_f) \left(\frac{(-i\,\rlap{/}{\epsilon}')}{\sqrt{2k'}} \frac{i}{\rlap{/}{p}_i + \rlap{/}{k} - m} \frac{(-i\,\rlap{/}{\epsilon})}{\sqrt{2k}} \right.$$

$$+ \frac{(-i\,\rlap{/}{\epsilon})}{\sqrt{2k}} \frac{i}{\rlap{/}{p}_i - \rlap{/}{k}' - m} \frac{(-i\,\rlap{/}{\epsilon}')}{\sqrt{2k'}} \left) \sqrt{\frac{m}{E_i}} u(p_i, s_i) \right. \qquad (6.6)$$

The differential cross section $\cdot$, found in several steps. First, we square the S matrix, which gives us a divergent result. We divide by the singular quantity $(2\pi)^4 \delta(0)$ and obtain the rate of transitions. We divide by the flux $|\mathbf{v}|/V$, divide by the number of particles per unit volume $1/V$, multiply by the phase factor for outgoing particles $[V^2/(2\pi)^6]\, d^3p_f\, d^3k'$. This give us the differential cross section:

$$d\sigma = \frac{|S_{fi}|^2}{(2\pi)^4 \delta(0)} \frac{1}{|\mathbf{v}|} \frac{V^2 d^3 p_f d^3 k'}{(2\pi)^6}$$

$$= \frac{e^4}{(2\pi)^2 |\mathbf{v}|} \frac{m}{E_i} \frac{1}{2k} \int \sum_{\text{spins}} |\bar{u}_f \Gamma u_i|^2$$

$$\times \delta^4(p_i + k - p_f - k') \frac{m}{E_f} \frac{d^3 p_f}{} \frac{d^3 k'}{2k'} \qquad (6.7)$$

where:

$$\Gamma = \frac{\rlap{/}{\epsilon}' \, \rlap{/}{\epsilon} \, \rlap{/}{k}}{2 p_i \cdot k} + \frac{\rlap{/}{\epsilon} \rlap{/}{\epsilon}' \, \rlap{/}{k}'}{2 p_i \cdot k'} \qquad (6.8)$$

To reduce out the spins, we will once again use the convenient formula given in Eq. (5.54):

$$\sum_{\text{spins}} |\bar{u}(p_f, s_f) \Gamma u(p_i, s_i)|^2 = \text{Tr}\left(\Gamma \frac{\rlap{/}{p}_i + m}{2m} \gamma^0 \Gamma^\dagger \gamma^0 \frac{\rlap{/}{p}_f + m}{2m} \right) \qquad (6.9)$$

Although this calculation looks formidable, we can perform the trace of up to eight Dirac matrices by reducing it to the trace of six, and then four Dirac matrices, etc. We will use the formula:

$$\text{Tr}(\rlap{/}{k}_1 \, \rlap{/}{k}_2 \, \rlap{/}{k}_3 \cdots \rlap{/}{k}_{2n}) = k_1 \cdot k_2 \text{Tr}(\rlap{/}{k}_3 \cdots \rlap{/}{k}_{2n}) - k_1 \cdot k_3 \text{Tr}(\rlap{/}{k}_2 \, \rlap{/}{k}_4 \cdots \rlap{/}{k}_{2n})$$

$$+ \cdots + k_1 \cdot k_{2n} \text{Tr}(\rlap{/}{k}_2 \, \rlap{/}{k}_3 \cdots \rlap{/}{k}_{2n-1}) \qquad (6.10)$$

The problem simplifies enormously because we can eliminate entire groups of terms every time certain dot products appear, since:

$$k^2 = k'^2 = \epsilon \cdot k = \epsilon' \cdot k' = 0 \qquad (6.11)$$

We can also simplify the calculation by using:

$$\epsilon^2 = \epsilon'^2 = -1; \quad p_i^2 = p_f^2 = m^2 \tag{6.12}$$

In short, each trace consists of collecting the complete set of all possible pairs of dot products of vectors, most of which vanish. Dividing the factors into smaller pieces, we now find:

$$\sum_{\text{spins}} |\bar{u}_f \Gamma u_i|^2 = \sum_{i=1}^{4} T_i \tag{6.13}$$

where:

$$
\begin{aligned}
T_1 &= \text{Tr}\left[\not{\epsilon}' \not{\epsilon} \not{k}(\not{p}_i + m) \not{k} \not{\epsilon} \not{\epsilon}'(\not{p}_f + m)\right] \\
&= 2p_i \cdot k \, \text{Tr}\left(\not{\epsilon}' \not{\epsilon} \not{k} \not{\epsilon} \not{\epsilon}' \not{p}_f \right) \\
&= 2p_i \cdot k \, \text{Tr}\left(\not{\epsilon}' \not{k} \not{\epsilon}' \not{p}_f \right) \\
&= 8p_i \cdot k[2(\epsilon' \cdot k)^2 + k' \cdot p_i] \tag{6.14}
\end{aligned}
$$

and:

$$
\begin{aligned}
T_2 &= \text{Tr}\left(\not{\epsilon}' \not{\epsilon} \not{k}(\not{p}_i + m) \not{k}' \not{\epsilon}' \not{\epsilon}(\not{p}_f + m)\right) \\
&= 2k \cdot p_i \, \text{Tr}\left(\not{k}' \not{\epsilon}' \not{\epsilon} \not{\epsilon}' \not{\epsilon} \not{p}_i \right) + 8(k' \cdot \epsilon)^2 k \cdot p_i - 8(k \cdot \epsilon')^2 p_i \cdot k' \\
&= 8k \cdot p_i \, k' \cdot p_f[2(\epsilon' \cdot \epsilon)^2 - 1] + 8(k' \cdot \epsilon)^2 k \cdot p_i - 8(k \cdot \epsilon')^2 k' \cdot p_i
\end{aligned}
$$

$$\tag{6.15}$$

We also have T_3 equal to T_2, and T_4 can be obtained from T_3 if we make the substitution: $(\epsilon, k) \rightarrow (\epsilon', -k')$. Since the calculation is Lorentz invariant, we can always take a specific Lorentz frame. We lose no generality by letting the electron be at rest, and let the incoming photon lie along the z axis. Let the outgoing photon scatter within the $y - z$ plane, making an angle θ with the z axis (Fig. 6.2).

Then the specific parametrization is given by:

$$
\begin{aligned}
p_{i,\mu} &= (m, 0, 0, 0) \\
k_\mu &= k(1, 0, 0, 1) \\
k'_\mu &= k'(1, 0, \sin\theta, \cos\theta) \\
p_{f,\mu} &= (E, 0, -k'\sin\theta, k - k'\cos\theta) \tag{6.16}
\end{aligned}
$$

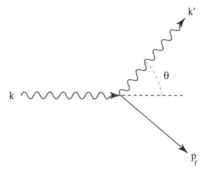

Figure 6.2. Compton scattering in the laboratory frame, where the electron is at rest.

It is important to notice that the only independent variables in this scattering are k and θ. All other variables can be expressed in terms of these two variables. For example, we can solve for k' and E in terms of the independent variables k and θ:

$$k' = \frac{k}{1 + (k/m)(1 - \cos\theta)}$$

$$E = m + k - k' \tag{6.17}$$

Adding all four contributions, we now have:

$$\sum_{\text{spins}} |\bar{u}(p_f, s_f)\Gamma u(p_i, s_i)|^2 = \frac{1}{2m^2}\left(\frac{k'}{k} + \frac{k}{k'} + 4(\epsilon' \cdot \epsilon)^2 - 2\right) \tag{6.18}$$

We now must integrate over the momenta of the outgoing photon k' and electron p_f. Since the only independent variables in the problem are given by k and θ, all integrations are easy, except for the Jacobian, which arises when we change variables and integrate over Dirac delta functions. Thus, the integration over d^3p_f is trivial because of momentum conservation; it simply sets the momenta to be the values given above. That leaves one complication, the integration over the time components dp_{f0}. However, this integration can be rewritten in a simple fashion:

$$\frac{1}{2E_f} = \int dp_{f0}\,\delta(p_f^2 - m^2)\theta(p_{f,0}) \tag{6.19}$$

The integration over $p_{f,0}$ in the integral just sets its value to be the on-shell value. Finally, this last delta function can be removed because of the integration over k'. The only tricky part is to extract from this last integration the measure when we integrate over k'.

This last delta function can be written as:

$$
\begin{aligned}
\delta\left([k + p_i - k']^2 - m^2\right) &= \delta\left(2m(k - k') - 2kk'(1 - \cos\theta)\right) \\
&= \frac{\delta(k' - k'(k))}{(2mk/k')}
\end{aligned}
\tag{6.20}
$$

where $k'(k)$ is the value given in Eq. (6.17). Putting all integration factors together, we now have:

$$
\delta^4(p_i + k - k' - p_f)\frac{d^3 p_f}{2E_f}\, d^3 k' = \frac{k'^2\, d\Omega}{(2mk/k')}
\tag{6.21}
$$

Inserting all expressions into the cross section, we obtain the Klein–Nishina formula[1]:

$$
\frac{d\sigma}{d\Omega} = \frac{\alpha^2}{4m^2}\left(\frac{k'}{k}\right)^2\left(\frac{k'}{k} + \frac{k}{k'} + 4(\epsilon \cdot \epsilon')^2 - 2\right)
\tag{6.22}
$$

If we take the low-energy limit $k \to 0$, then the Klein–Nishina formula reduces to the Thompson scattering formula:

$$
\frac{d\sigma}{d\Omega} = \frac{\alpha^2}{m^2}(\epsilon \cdot \epsilon')^2
\tag{6.23}
$$

If the initial and final photon are unpolarized, we can average over the initial and final polarizations ϵ and ϵ'. In the particular parametrization that we have chosen for our momenta, we can choose our polarizations ϵ and ϵ', such that they are purely transverse and perpendicular to the momenta p_i and p_f, respectively:

$$
\begin{aligned}
\epsilon^{(1)} &= (0, 1, 0, 0) \\
\epsilon^{(2)} &= (0, 0, 1, 0) \\
\epsilon'^{(1)} &= (0, 1, 0, 0) \\
\epsilon'^{(2)} &= (0, 0, \cos\theta, -\sin\theta)
\end{aligned}
\tag{6.24}
$$

It is easy to check that these polarization vectors satisfy all the required properties. Then the sum over these polarization vectors is easy to perform:

$$
\sum_{\text{spins}}(\epsilon \cdot \epsilon')^2 = \sum_{i,j}(\epsilon^{(i)} \cdot \epsilon'^{(j)})^2 = 1 + \cos^2\theta
\tag{6.25}
$$

The averaged cross section is given by:

$$\frac{d\sigma}{d\Omega}\bigg|_{av} = \frac{\alpha^2}{2m^2}\left(\frac{k'}{k}\right)^2\left(\frac{k'}{k}+\frac{k}{k'}-\sin^2\theta\right) \tag{6.26}$$

The integral over θ is straightforward. Let us define $z = \cos\theta$. The integration yields:

$$\begin{aligned}
\sigma &= \frac{\pi\alpha^2}{m^2}\int_{-1}^{1}dz\left(\frac{1}{[1+a(1-z)]^3}\right.\\
&\quad + \left.\frac{1}{1+a(1-z)}-\frac{1-z^2}{[1+a(1-z)]^2}\right)\\
&= \left(\frac{8\pi\alpha^2}{3m^2}\right)(3/4)\left[\frac{1+a}{a^3}\left(\frac{2a(1+a)}{1+2a}-\log(1+2a)\right)\right.\\
&\quad + \left.\frac{\log(1+2a)}{2a}-\frac{1+3a}{(1+2a)^2}\right] \tag{6.27}
\end{aligned}$$

where $a = k/m$. For small energies, this reduces to the usual Thompson total cross section:

$$\lim_{k\to 0}\sigma = \sigma_{\text{Thompson}} = \frac{8\pi\alpha^2}{3m^2} = 0.665\times 10^{-24}\text{ cm}^2 \tag{6.28}$$

For high energies, the logarithm starts to dominate the cross section:

$$\lim_{k\to\infty}\sigma = \frac{\pi\alpha^2}{km}\left[\log\frac{2k}{m}+\frac{1}{2}+O\left(\frac{m}{k}\ln\frac{k}{m}\right)\right] \tag{6.29}$$

6.2 Pair Annihilation

The Feynman graphs for pair annihilation of an electron and position into two gamma rays is shown in Figure 6.3. Pair annihilation is represented by the process:

$$e^-(p_1) + e^+(p_2) \to \gamma(k_1) + \gamma(k_2) \tag{6.30}$$

However, notice that we can obtain this diagram if we simply rotate the diagram for Compton scattering in Figure 6.1. Thus, by a subtle redefinition of the various momenta, we should be able to convert the Compton scattering amplitude, which we have just calculated, into the amplitude for pair annihilation. This is the

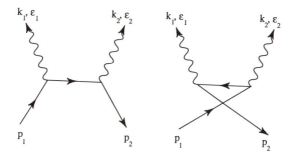

Figure 6.3. Pair annihilation: an electron of momentum p_1 annihilates with a positron of momentum p_2 into two photons.

substitution rule to which we referred earlier. For example, the S matrix now yields, to lowest order:

$$
S_{fi} = -e^2 (2\pi)^4 \delta^4 (k_1 + k_2 - p_1 - p_2) \frac{1}{\sqrt{2k_1 2k_2}}
$$

$$
\times \sqrt{\frac{m}{E_2}} \bar{v}(p_2, s_2) \left(\not{\epsilon}_2 \frac{1}{\not{p}_1 - \not{k}_1 - m} \not{\epsilon}_1 + \not{\epsilon}_1 \frac{1}{\not{p}_1 - \not{k}_2 - m} \not{\epsilon}_2 \right) \sqrt{\frac{m}{E_1}} u(p_1, s_1)
$$

$$(6.31)$$

where we have made the substitutions:

$$
\begin{aligned}
(k, \epsilon) &\rightarrow (-k_1, \epsilon_1) \\
(k', \epsilon') &\rightarrow (k_2, \epsilon_2) \\
u(p_i, s_i) &\rightarrow u(p_1, s_1) \\
\bar{u}(p_f, s_f) &\rightarrow \bar{v}(p_2, s_2)
\end{aligned}
\qquad (6.32)
$$

We will, as usual, take the Lorentz frame where the electron is at rest. Then our momenta become, as in Figure 6.4:

$$
\begin{aligned}
p_{1\mu} &= (m, 0, 0, 0) \\
p_{2\mu} &= (E, 0, 0, |\mathbf{p}|) \\
k_{1\mu} &= k_1(1, 0, \sin\theta, \cos\theta) \\
k_{2\mu} &= (k_2, 0, -k_1 \sin\theta, |\mathbf{p}| - k_1 \cos\theta)
\end{aligned}
\qquad (6.33)
$$

There are only two independent variables in this process, $|\mathbf{p}|$ and θ. All other variables can be expressed in terms of them. For example, we can easily show

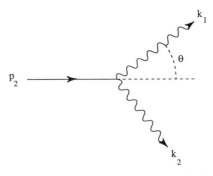

Figure 6.4. Pair annihilation in the laboratory frame, where the electron is at rest.

that:

$$k_1 = \frac{m(m+E)}{m+E-|\mathbf{p}|\cos\theta}$$

$$k_2 = \frac{(m+E)(E-|\mathbf{p}|\cos\theta)}{m+E-|\mathbf{p}|\cos\theta} \qquad (6.34)$$

We also have:

$$m+E = k_1 + k_2$$

$$k_1 \cdot k_2 = m(m+E) = k_1(m+E-|\mathbf{p}|\cos\theta) \qquad (6.35)$$

When we contract over the Dirac matrices, the calculation proceeds just as before, except that we want to evaluate $|\bar{v}\Gamma u|^2$. We have to use Eq. (3.111):

$$\sum_{\text{spins}} v(p,s)\bar{v}(p,s) = \frac{\slashed{p}-m}{2m} \qquad (6.36)$$

The trace becomes:

$$\text{Trace} = -\frac{1}{2m^2}\left(\frac{k_2}{k_1}+\frac{k_1}{k_2} - 4(\epsilon_1 \cdot \epsilon_2)^2 + 2\right) \qquad (6.37)$$

The integration over d^3k_1 and d^3k_2 also proceeds as before. The integrations over the delta functions are straightforward, except that we must be careful when picking up a measure term when we make a transformation on a Dirac delta function. When this additional measure term is inserted, the differential cross section becomes:

$$\frac{d\sigma}{d\Omega} = \frac{\alpha^2(m+E)}{8|\mathbf{p}|(m+E-|\mathbf{p}|\cos\theta)^2}\left(\frac{k_2}{k_1}+\frac{k_1}{k_2} - 4(\epsilon_1 \cdot \epsilon_2)^2 + 2\right) \qquad (6.38)$$

The total cross section is obtained by summing over photon polarizations. As before, we can take a specific set of polarizations which are transverse to p_1 and p_2:

$$
\begin{aligned}
\epsilon_1^{(1)} &= 0, 1, 0, 0) \\
\epsilon_1^{(2)} &= (0, 0, \cos\theta, -\sin\theta) \\
\epsilon_2^{(1)} &= (0, 1, 0, 0) \\
\epsilon_2^{(2)} &= (0, 0, |\mathbf{p}| - k_1\cos\theta, k_1\sin\theta)
\end{aligned}
\tag{6.39}
$$

Then we can sum over all polarizations. The only difficult sum involves:

$$
\begin{aligned}
\epsilon_1^{(2)} \cdot \epsilon_2^{(2)} &= \mathbf{k}_1 \cdot \mathbf{k}_2 \\
&= 1 - \frac{(k_1 + k_2)^2}{2k_1 k_2} \\
&= 1 - \frac{2m(m + E)}{2k_1 k_2} \\
&= 1 - m\left(\frac{1}{k_1} + \frac{1}{k_2}\right)
\end{aligned}
\tag{6.40}
$$

Then the sum over spins can be written as:

$$
\sum_{i,j}(\epsilon_1^{(i)} \cdot \epsilon_2^{(j)})^2 = 1 + \left[1 - \left(\frac{m}{k_1} + \frac{m}{k_2}\right)\right]^2
\tag{6.41}
$$

The only integration left is the one over the solid angle, which leaves us with ($\gamma = E_2/m$):

$$
\sigma = \frac{\pi\alpha^2}{m^2(1+\gamma)}\left(\frac{\gamma^2 + 4\gamma + 1}{\gamma^2 - 1}\log\left(\gamma + \sqrt{\gamma^2 - 1}\right) - \frac{\gamma + 3}{\sqrt{\gamma^2 - 1}}\right)
\tag{6.42}
$$

which is a result first obtained by Dirac.[2]

6.3 Møller Scattering

Next, we investigate electron–electron scattering. To lowest order, this scattering amplitude contains two graphs, as in Figure 6.5. This scattering is represented by:

$$
e(p_1) + e(p_2) \rightarrow e(p_1') + e(p_2')
\tag{6.43}
$$

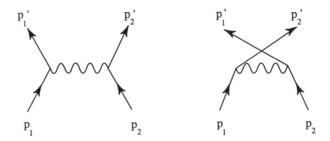

Figure 6.5. Møller scattering of two electrons with momenta p_1 and p_2.

By a straightforward application of the formulas for differential cross sections, we find:

$$d\sigma = \frac{e^4 m^4}{[(p_1 \cdot p_2)^2 - m^4]^{1/2}} (2\pi)^4 \delta^4(p_1' + p_2' - p_1 - p_2)$$

$$\times \int \frac{d^3 p_1'}{(2\pi)^3 E_1'} \frac{d^3 p_2'}{(2\pi)^3 E_2'} |\mathcal{M}_{fi}|^2 \qquad (6.44)$$

where, using a straightforward application of Feynman's rules for these two diagrams, we can compute $|\mathcal{M}_{fi}|^2$:

$$|\mathcal{M}_{fi}|^2 = \frac{1}{4} \Bigg[\mathrm{Tr}\left(\gamma_\nu \frac{\not{p}_1 + m}{2m} \gamma_\sigma \frac{\not{p}_1' + m}{2m} \right)$$

$$\times \mathrm{Tr}\left(\gamma^\nu \frac{\not{p}_2 + m}{2m} \gamma^\sigma \frac{\not{p}_2' + m}{2m} \right) \frac{1}{[(p_1' - p_1)^2]^2}$$

$$- \mathrm{Tr}\left(\gamma_\nu \frac{\not{p}_1 + m}{2m} \gamma_\sigma \frac{\not{p}_2' + m}{2m} \gamma^\nu \frac{\not{p}_2 + m}{2m} \gamma^\sigma \frac{\not{p}_1' + m}{2m} \right)$$

$$\times \frac{1}{(p_1' - p_1)^2 (p_2' - p_2)^2} + (p_1' \leftrightarrow p_2') \Bigg] \qquad (6.45)$$

Since the trace is over only four Dirac matrices, taking the trace is not hard to do:

$$\frac{1}{2m^4} \Bigg(\frac{(p_1 \cdot p_2)^2 + (p_1 \cdot p_2')^2 + 2m^2(p_1 \cdot p_2' - p_1 \cdot p_2)}{[(p_1' - p_1)^2]^2}$$

$$+ \frac{(p_1 \cdot p_2)^2 + (p_1 \cdot p_1')^2 + 2m^2(p_1 \cdot p_1' - p_1 \cdot p_2)}{[(p_2' - p_1)^2]^2}$$

$$+ 2\frac{(p_1 \cdot p_2)^2 - 2m^2 p_1 \cdot p_2}{(p_1' - p_1)^2 (p_2' - p_1)^2} \Bigg) \qquad (6.46)$$

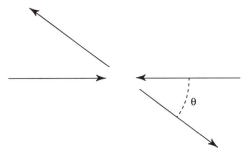

Figure 6.6. Møller scattering in the center-of-mass frame.

The kinematics is illustrated in Figure 6.6. Without any loss of generality, we can choose the center-of-mass frame, where the electron momenta p_1 and p_2 lie along the z axis:

$$p_{1,\mu} = (E, 0, 0, |\mathbf{p}|)$$

$$p_{2,\mu} = (E, 0, 0, -|\mathbf{p}|)$$

$$p'_{1,\mu} = (E, 0, |\mathbf{p}| \sin\theta, |\mathbf{p}| \cos\theta)$$

$$p'_{2,\mu} = (E, 0, -|\mathbf{p}| \sin\theta, -|\mathbf{p}| \cos\theta) \tag{6.47}$$

The only independent variables are θ and $|\mathbf{p}|$. In terms of this parametrization, we easily find:

$$p_1 \cdot p_2 = 2E^2 - m^2$$

$$p_1 \cdot p'_1 = E^2(1 - \cos\theta) + m^2 \cos\theta$$

$$p_1 \cdot p'_2 = E^2(1 + \cos\theta) - m^2 \cos\theta \tag{6.48}$$

Then the entire cross section can be written in terms of these independent variables. We finally obtain the Møller formula[3] in the center-of-mass frame:

$$\frac{d\sigma}{d\Omega} = \frac{\alpha^2(2E^2 - m^2)^2}{4E^2(E^2 - m^2)^2} \left[\frac{4}{\sin^4\theta} - \frac{3}{\sin^2\theta} + \frac{(E^2 - m^2)^2}{(2E^2 - m^2)^2} \left(1 + \frac{4}{\sin^2\theta} \right) \right] \tag{6.49}$$

In the relativistic limit, as $E \to \infty$, this formula reduces to:

$$\frac{d\sigma}{d\Omega} = \frac{\alpha^2}{E^2} \left(\frac{4}{\sin^4\theta} - \frac{2}{\sin^2\theta} + \frac{1}{4} \right) \tag{6.50}$$

For the low-energy, nonrelativistic result, we find:

$$\frac{d\sigma}{d\Omega} = \frac{\alpha^2}{m^2} \frac{1}{4v^2} \left(\frac{4}{\sin^4\theta} - \frac{3}{\sin^2\theta} \right) \tag{6.51}$$

6.4 Bhabha Scattering

To calculate the cross section for electron–positron scattering (Fig. 6.7), we can use the substitution rule. By rotating the diagram for Møller scattering, we find the Feynman diagrams for Bhabha scattering.[4] The only substitutions we must make are:

$$u(p_1) \quad \rightarrow \quad u(p_1)$$
$$u(p_1') \quad \rightarrow \quad u(p_1')$$
$$u(p_2) \quad \rightarrow \quad v(-p_2)$$
$$u(p_2') \quad \rightarrow \quad v(-p_2') \tag{6.52}$$

for the process:

$$e^-(p_1) + e^+(-p_2) \rightarrow e^-(p_1') + e^+(-p_2') \tag{6.53}$$

The calculation and the traces are performed exactly as before with these simple substitutions. We merely quote the final result, due to Bhabha (1935):

$$\frac{d\sigma}{d\Omega} = \frac{\alpha}{2E^2} \left(\frac{5}{4} - \frac{8E^4 - m^4}{E^2(E^2 - m^2)(1 - \cos\theta)} + \frac{(2E^2 - m^2)^2}{2(E^2 - m^2)^2(1 - \cos\theta)^2} \right.$$
$$+ [16E^4]^{-1} \left[2E^4(-1 + 2\cos\theta + \cos^2\theta) \right.$$
$$+ \left. \left. 4E^2m^2(1 - \cos\theta)(2 + \cos\theta) + 2m^4\cos^2\theta \right] \right) \tag{6.54}$$

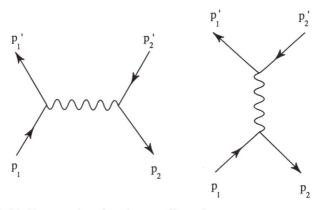

Figure 6.7. Bhabha scattering of an electron off a positron.

In the relativistic limit, we have:

$$\frac{d\sigma}{d\Omega} = \frac{\alpha^2}{8E^2}\left(\frac{1+\cos^4\theta/2}{\sin^4\theta/2}\right.$$

$$\left. + \frac{1}{2}(1+\cos^2\theta) - 2\frac{\cos^4\theta/2}{\sin^2\theta/2}\right) \tag{6.55}$$

In the nonrelativistic limit, we find:

$$\frac{d\sigma}{d\Omega} = \frac{\alpha^2}{m^2}\frac{1}{16v^4\sin^4\theta/2} \tag{6.56}$$

6.5 Bremsstrahlung

Bremsstrahlung is the process by which radiation is emitted from an electron as it moves past a nucleus (Fig. 6.8). Momentum conservation gives us:

$$p_i + q = k + p_f \tag{6.57}$$

Classically, one can calculate the radiation emitted by a moving charge as it accelerates past a proton. (*Bremsstrahlung* means "braking radiation.") However, unlike the previous scattering processes, which agree to first order with the experimental data, we find a severe problem with this amplitude, which is the *infrared divergence*. The quantum field theory calculation, to lowest order, reproduces the classical result, including the unwanted infrared divergence, which has its roots in the classical theory.

Although the infrared divergence first arose (in another form) in the classical theory, the final resolution of this problem comes when we take into account higher

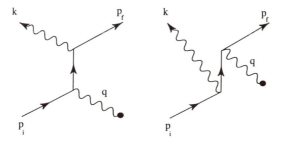

Figure 6.8. Bremsstrahlung, or the radiation emitted by an electron scattering in the presence of a nucleus.

quantum loop corrections to the scattering amplitude. The scattering matrix, using Feynman's rules, is:

$$S_{fi} = -\frac{Ze^3}{V^{3/2}} 2\pi \delta(E_f + k - E_i) \sqrt{\frac{m^2}{E_f E_i}} \bar{u}(p_f, s_f) \left(\frac{(-i \not\epsilon)}{\sqrt{2k}} \frac{i}{\not p_f + \not k - m} \frac{(-i\gamma_0)}{|\mathbf{q}|^2} \right.$$

$$\left. + \frac{(-i\gamma_0)}{|\mathbf{q}|^2} \frac{i}{\not p_i - \not k - m} \frac{(-i \not\epsilon)}{\sqrt{2k}} \right) u(p_i, s_i) \tag{6.58}$$

The differential cross section now becomes:

$$d\sigma = \frac{mZ^2 e^6}{E_i |\mathbf{v}_i|} \int 2\pi \delta(E_f + \omega - E_i) |\bar{u}_f \Gamma u_i|^2$$

$$\times \frac{1}{|\mathbf{q}|^4} \frac{m \, d^3 p_f}{E_f} \frac{d^3 k}{2\omega (2\pi)^6} \tag{6.59}$$

where $\omega = k_0$ and where:

$$\Gamma = \not\epsilon \frac{1}{\not p_f + \not k - m} \gamma^0 + \gamma^0 \frac{1}{\not p_i - \not k - m} \not\epsilon \tag{6.60}$$

The trace we wish to calculate is:

$$T = (2^{-5} m^{-2})(T_1 + T_2 + T_3)$$

$$= \frac{1}{2} \sum_\epsilon \text{Tr} \left[\left(\not\epsilon \frac{\not p_f + \not k + m}{2 p_f \cdot k} \gamma^0 - \gamma^0 \frac{\not p_i - \not k + m}{2 p_i \cdot k} \not\epsilon \right) \left(\frac{\not p_i + m}{2m} \right) \right.$$

$$\left. \times \left(\gamma^0 \frac{\not p_f + \not k + m}{2 p_f \cdot k} \not\epsilon - \not\epsilon \frac{\not p_i - \not k + m}{2 p_i \cdot k} \gamma^0 \right) \left(\frac{\not p_f + m}{2m} \right) \right] \tag{6.61}$$

The traces involved in the calculation yield:

$$T_1 = (p_f \cdot k)^{-2} \sum_\epsilon \text{Tr} \left[\not\epsilon (\not p_f + \not k + m) \gamma^0 (\not p_f + \not k + m) \not\epsilon (\not p_f + m) \right]$$

$$= 8(p_f \cdot k)^{-2} \sum_\epsilon [2(\epsilon \cdot p_f)^2 (m^2 + 2 p_i^0 p_f^0 + 2 p_i^0 \omega - p_i \cdot p_f - p_i \cdot k)$$

$$+ 2\epsilon \cdot p_f \, \epsilon \cdot p_i \, k \cdot p_f + 2 p_i^0 \omega \, k \cdot p_f - p_i \cdot k \, p_j \cdot k] \tag{6.62}$$

and:

$$T_2 = T_1(p_i \leftrightarrow -p_f) \tag{6.63}$$

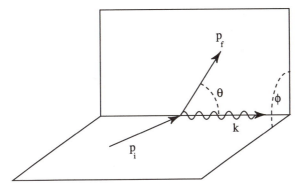

Figure 6.9. Bremsstrahlung, where the emitted photon is in the z direction and the emitted electron is in the $y - z$ plane. The incoming electron is in a plane rotated by angle ϕ from the $y - z$ plane.

and lastly:

$$
\begin{aligned}
T_3 \;=\; & -(p_f \cdot k)^{-1}(p_i \cdot k)^{-1}\mathrm{Tr}\left\{ \gamma^0 (\slashed{p}_i - \slashed{k} + m)\,\slashed{\epsilon}(\slashed{p}_i + m)\gamma^0 \right. \\
& \left. \times(\slashed{p}_f + \slashed{k} + m)\,\slashed{\epsilon}(\slashed{p}_f + m) + (p_i \leftrightarrow -p_f) \right\} \\
\;=\; & 16(p_f \cdot k)^{-1}(p_i \cdot k)^{-1}\sum_\epsilon \left[\epsilon \cdot p_i \epsilon \cdot p_f(p_i \cdot k - p_f \cdot k \right. \\
& +2p_i \cdot p_f - 4p_i^0 \cdot p_f^0 - 2m^2) \\
& +(\epsilon \cdot p_f)^2 k \cdot p_i - (\epsilon \cdot p_i)^2 k \cdot p_f + p_i \cdot k p_f \cdot k - m^2\omega^2 \\
& \left. +\omega(\omega p_i \cdot p_f - p_i^0 p_f \cdot k - p_f^0 p_i \cdot k) \right]
\end{aligned}
\tag{6.64}
$$

The parametrization of the momenta is a bit complicated, since the reaction does not take place in a plane. In Figure 6.9, we place the emitted photon momenta in the z direction, the emitted electron momenta p_f in the $y - z$ plane, and the incoming electron momenta p_i in a plane that is rotated by an angle ϕ from the $y - z$ plane.

The specific parametrization is equal to:

$$
\begin{aligned}
k_\mu \;&=\; \omega(1, 0, 0, 1) \\
p_{f,\mu} \;&=\; (E_f, 0, p_f \sin\theta_f, p_f \cos\theta_f) \\
p_{i,\mu} \;&=\; (E_i, p_i \sin\theta_i \sin\phi, p_i \sin\theta_i \cos\phi, p_i \cos\theta_i)
\end{aligned}
\tag{6.65}
$$

where $p_i = |\mathbf{p}_i|$ and $p_f = |\mathbf{p}_f|$. Now we must calculate the sum over transverse photon polarizations. Since k_μ points in the z direction, we can choose:

$$
\begin{aligned}
\epsilon^{(1)} &= (0, 1, 0, 0) \\
\epsilon^{(2)} &= (0, 0, 1, 0)
\end{aligned}
\tag{6.66}
$$

which satisfies all the desired properties of the polarization tensor. With this choice of parametrization for the momenta and polarizations, we easily find:

$$
\sum_i (\epsilon^{(i)} \cdot p_f)^2 = p_f^2 \sin^2 \theta_f; \quad \sum_i (\epsilon^{(i)} \cdot p_i)^2 = p_i^2 \sin^2 \theta_i
$$

$$
\sum_i (\epsilon^{(i)} \cdot p_f)(\epsilon^{(i)} \cdot p_i) = p_i p_f \sin \theta_f \sin \theta_i \cos \phi
\tag{6.67}
$$

It is now a simple matter to collect everything together, and we now have the Bethe–Heitler formula[5] (1934), which was first computed without using Feynman's rules:

$$
\begin{aligned}
d\sigma &= \frac{Z^2 \alpha^3}{(2\pi)^2} \frac{p_f}{p_i q^4} \frac{d\omega}{\omega} d\Omega_\gamma d\Omega_e \\
&\times \left(\frac{p_f^2 \sin^2 \theta_f}{(E_f - p_f \cos \theta_f)^2}(4E_i^2 - q^2) + \frac{p_i^2 \sin \theta_i}{(E_i - p_i \cos \theta_i)^2}(4E_f^2 - q^2) \right. \\
&\quad + 2\omega^2 \frac{p_i^2 \sin^2 \theta_i + p_f^2 \sin^2 \theta_f}{(E_f - p_f \cos \theta_f)(E_i - p_i \sin \theta_i)} \\
&\quad - 2 \frac{p_f p_i \sin \theta_i \sin \theta_f \cos \phi}{(E_f - p_f \cos \theta_f)(E_i - p_i \cos \theta_i)} \\
&\quad \left. \times (4 E_i E_f - q^2 + 2\omega^2) \right)
\end{aligned}
\tag{6.68}
$$

Now let us make the approximation that $\omega \to 0$. In the soft bremsstrahlung limit, we find a great simplification, and the differential cross section becomes the one found by classical methods:

$$
\frac{d\sigma}{d\Omega} \sim \left(\frac{d\sigma}{d\Omega} \right)_{\text{elastic}} \frac{e^2 \, d^3 k}{2\omega (2\pi)^3} \sum_\epsilon \left(\frac{\epsilon \cdot p_f}{k \cdot p_{-f}} - \frac{\epsilon \cdot p_i}{k \cdot p_i} \right)^2
\tag{6.69}
$$

Here the infrared divergence[6–8] appears for the first time. This problem was first correctly analyzed by Bloch and Nordsieck.[8] The integral $d^3 k / \omega$ is divergent for small ω, and therefore the amplitude for soft photon emission makes no sense. This is rather discouraging, and revealed the necessity of properly adding all quantum corrections. The resolution of this question only comes when the one-loop vertex corrections are added in properly, which we will discuss later in this chapter. For now, it is important to understand where the divergence comes from and its general form.

The infrared divergence always emerges whenever we have massless particles in a theory that can be emitted from an initial or final leg that is on the mass shell. For example, whenever we emit a soft photon of momentum k from an on-shell electron with momentum p, we find that the propagator just before the emission is given by:

$$\frac{1}{(p-k)^2 - m^2} \sim \frac{1}{-2p \cdot k} \sim \infty \tag{6.70}$$

Because $p^2 = m^2$ and k is small, we find that an integration over momentum k inevitably produces an infrared divergence. In order to quantify this infrared divergence, let us perform the integration over the momentum k, separating out the angular part $d\Omega$ from d^3k. To parametrize the divergence, we will regulate the integral by allowing the photon to have a small but finite mass μ. (This is, of course, a bit delicate since we are breaking gauge invariance by having massive photons, but one can show that, at this order of approximation, there are no problems.) We will integrate k from μ to some energy E given by the sensitivity of the detector. Expanding out the expression in the square, the amplitude now becomes:

$$\frac{d\sigma}{d\Omega} = \left(\frac{d\sigma}{d\Omega}\right)_0 \frac{\alpha}{4\pi^2} \int_\mu^E k\, dk \int d\Omega \left(\frac{2p_f \cdot p_i}{(k \cdot p_f)(k \cdot p_i)} - \frac{m^2}{(k \cdot p_f)^2} - \frac{m^2}{(k \cdot p_i)^2}\right) \tag{6.71}$$

In our approximation, we can perform the angular integral over the last three terms in the large parentheses.

Let us calculate the last two terms appearing on the right-hand side of the equation. We use the fact that:

$$k \cdot p_f \sim E(1 - \cos\theta_f) \tag{6.72}$$

Because $d\Omega = 2\pi d(\cos\theta)$, we can trivially integrate the last two terms:

$$\int \frac{d\Omega}{4\pi} \frac{m^2}{(k \cdot p_f)^2} = \frac{m^2}{2E^2} \int_{-1}^{+1} d(\cos\theta) \frac{1}{(1 - \beta\cos\theta)^2} = 1 \tag{6.73}$$

The integral over the first term is also easy to perform. We have to use the fact that:

$$\frac{2p_f \cdot p_i}{(k \cdot p_f)(k \cdot p_i)} \sim \frac{2E^2(1 - \beta^2\cos\theta)}{E(1 - \beta\cos\theta_i)E(1 - \beta\cos\theta_f)} \tag{6.74}$$

We will now introduce the Feynman parameter trick, which is often used to evaluate Feynman integrals:

$$\frac{1}{ab} = \int_0^1 \frac{dx}{[ax + b(1 - x)]^2} \tag{6.75}$$

By introducing a new variable x, we are able to perform the angular integration. We find:

$$\int \frac{d\Omega}{4\pi} \frac{2p_f \cdot p_f}{(k \cdot p_f)(k \cdot p_i)} \sim 2(1 - \beta^2 \cos^2 \theta) \int_0^1 dx$$

$$\times \int d(\cos \theta)(1/2) \frac{1}{[1 - \beta \cos \theta_f x - \beta \cos \theta_i (1 - x)]^2}$$

$$= 2(1 - \beta^2 \cos \theta) \int_0^1 \frac{dx}{1 - \beta^2 + 4\beta^2 \sin^2(\theta/2)x(1 - x)}$$

$$= \begin{cases} 2[1 + \frac{4}{3}\beta^2 \sin^2(\theta/2] + O(\beta^4) & \text{if } \beta \ll 1 \\ 2 \log(-q^2/m^2) + O(m^2/q^2) & \text{if } m^2/q^2 \ll 1 \end{cases} \tag{6.76}$$

Inserting this value back into the previous expression, we find that the final soft bremsstrahlung cross section is then given by:

$$\frac{d\sigma}{d\Omega} = \left(\frac{d\sigma}{d\Omega}\right)_0 \frac{\alpha}{\pi} \ln \frac{E^2}{\mu^2} \begin{cases} \frac{4}{3}\beta^2 \sin^2 \frac{\theta}{2} + O(\beta^4) & \beta \ll 1 \\ \log(-q^2/m^2) - 1 + O(m^2/q^2) & \beta \sim 1 \end{cases} \tag{6.77}$$

Although this formula agrees well with experiment at large photon momenta, this amplitude is clearly divergent if we let the fictitious mass of the photon μ go to zero. Thus, the infrared divergence occurs because we have massless photons present in the theory.

We should mention that the infrared problem arose (in another form) in classical physics, before the advent of quantum mechanics. The essential point is that, even at the classical level, we have the effects due to the long-range Coulomb field. If one were to calculate the radiation field created by a particle being accelerated by a stationary charge, one would find a similar divergence using only classical equations. If one tries to divide the energy by k_0 to calculate the number of photons emitted by bremsstrahlung, it turns out to be proportional to the result presented above. Thus, as the momentum of the emitted photon goes to zero, the number of emitted photons becomes infinite. (Classically, the infrared divergence appears in the *number* of emitted photons, not the emitted *energy*.)

Quantum field theory gives us a novel, but rigorous, solution to the infrared problem, which goes to the heart of the measurement process and the quantum

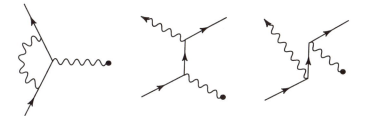

Figure 6.10. The infrared divergence cancels if we add the contributions of two different physical processes. These diagrams can be added together because the resolution of any detector is not sensitive enough to select out just one process.

theory. To this order of approximation, we have to add the contribution of *two* different physical processes to find the cross section of electrons scattering off protons or other charged particles (Fig. 6.10).

The first diagram describes the bremsstrahlung amplitude for the emission of an electron and a photon. The divergence of this amplitude is classical. Second, we have to sum over a purely quantum-mechanical effect, the radiative one-loop corrections of the electron elastically colliding off the charged proton. This may seem strange, because we are adding the cross sections of two different physical processes together, one elastic and one inelastic, to cancel the infrared divergence. However, this makes perfect sense from the point of view of the measuring process.

The essential point is to observe that our detectors cannot differentiate the presence of pure electrons from the presence of electrons accompanied by sufficiently soft photons. This is not just a problem of having crude measuring devices. No matter how precise our measuring apparatus may become, it can never be perfect; there will always be photons with momenta sufficiently close to zero that will sneak past them. Therefore, from an experimental point of view, our measuring apparatus cannot distinguish between these two types of processes and we must necessarily add these two diagrams together. Fortunately, we get an exact cancellation of the infrared divergences when these two scattering amplitudes are added together.

A full discussion of this cancellation, however, cannot be described until we discuss one-loop corrections to scattering amplitudes. Therefore, in section 6.7, we will prove that the bremsstrahlung amplitude, given by:

$$\frac{d\sigma}{d\Omega} \sim \left(\frac{d\sigma}{d\Omega}\right)_0 \frac{\alpha}{\pi} \log \frac{E^2}{\mu^2} \log \frac{-q^2}{m^2} \tag{6.78}$$

must be added to the one-loop vertex correction in order to yield a convergent integral. We will return to this problem at that time.

Finally, we note that, by the substitution rule, we can show the relationship between bremsstrahlung and pair annihilation. Once again, by rotating the diagram around, we can convert bremsstrahlung into pair creation.

This ends our discussion of the tree-level, lowest-order scattering matrix. Although we have had great success in reproducing and extending known classical results, there are immense difficulties involved in extending quantum field theory beyond the tree level. When loop corrections are calculated, we find that the integrals diverge in the ultraviolet region of momentum space. In fact, it has taken over a half century, involving the combined efforts of several generations of physicists, to resolve many of the difficulties of renormalization theory.

We now turn to a detailed calculation of single-loop radiative corrections. Although the calculations are often long and tedious, involving formally divergent quantities, the final conclusions are simple and show that the various infinities can be consistently absorbed into a redefinition of the physical constants of the theory, such as the electric charge and electron mass. Most important, the agreement with experiment is astonishing.

We will begin our discussion of radiative corrections by first examining the self-energy correction to the photon propagator, called the vacuum polarization graph. We will show that the divergence of this graph can be absorbed into a renormalization of the electric charge.

Then, we will calculate the single-loop correction to the electron–photon vertex and show that this leads to corrections to the magnetic moment of the electron. The theoretical value of the anomalous magnetic moment will agree with experiment to one part in 10^8. After that, we will show that the radiative correction to the vertex function is also infrared divergent. Fortunately, the sign of this infrared divergence is opposite the sign found in the bremsstrahlung amplitude. When added together, we will find that the two cancel exactly, giving us a quantum mechanical resolution of the infrared problem.

And finally, we will close this chapter by analyzing the Lamb shift between the energy levels of the $2S_{1/2}$ and $2P_{1/2}$ orbitals of the hydrogen atom. The calculation is rather intricate, because the hydrogen atom is a bound state, and also there are various contributions coming from the vertex correction, the anomalous magnetic moment, the self-energy of the electron, the vacuum polarization graph, etc. However, when all these contributions are added, we find agreement with experiment to within one part in 10^6.

6.6   Radiative Corrections

The simplest higher-order radiative correction is the vacuum polarization graph, shown in Figure 6.11. This graph is clearly divergent. For large momenta,

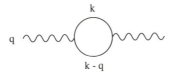

Figure 6.11. First-loop correction to the photon propagator, the vacuum polarization graph, which gives us a correction to the coupling constant and contributes to the Lamb shift.

the Feynman propagators of the two electrons give us two powers of p in the denominator, while the overall integration over $d^4 p$ gives us four powers of p in the numerator. So this graph diverges quadratically in the ultraviolet region of momentum space:

$$\Pi_{\mu\nu,m} = -e^2 \, \text{Tr} \int \frac{d^4 k}{(2\pi)^4} \gamma_\mu \frac{1}{\not{k} - m + i\epsilon} \gamma_\nu \frac{1}{\not{k} - \not{q} - m + i\epsilon} \qquad (6.79)$$

We will perform this integration via the Pauli–Villars method,[9] although the dimensional regularization method, which we will present in the next chapter, is significantly simpler. The Pauli–Villars method replaces this divergent integral with a convergent one by assuming that there are fictitious fermions with mass M in the theory with ghost couplings. At the end of the calculation, these fictitious particles will decouple if we take the limit as their masses tend to infinity. Therefore, M gives us a convenient way of cutting off the divergences of the self-energy correction.

The graph then becomes modified as follows:

$$\tilde{\Pi}_{\mu\nu} = \Pi_{\mu\nu,m} - \Pi_{\mu\nu,M} \qquad (6.80)$$

The most convenient way in which to perform the integration is to add additional auxiliary variables. This allows us to reverse the order of integration. We can then perform the integration over the momenta, and save the integration over the auxiliary variables to the very end. We will use:

$$\frac{i}{k^2 - m^2 + i\epsilon} = \int_0^\infty d\alpha \, e^{i\alpha(k^2 - m^2 + i\epsilon)} \qquad (6.81)$$

Inserting this expression for the electron propagators and performing the trace, we find:

$$\Pi_{\mu\nu,m} = 4e^2 \int_0^\infty d\alpha_1 \int_0^\infty d\alpha_2 \int \frac{d^4 k}{(2\pi)^4}$$
$$\times \exp\left\{ i\alpha_1 (k^2 - m^2 + i\epsilon) + i\alpha_2 [(k - q)^2 - m^2 + i\epsilon] \right\}$$

$$\times \left[k_\mu (k-q)_\nu + k_\nu (k-q)_\mu - g_{\mu\nu}(k^2 - k \cdot q - m^2) \right] \quad (6.82)$$

With the insertion of these auxiliary variables, we can now perform the integration over $d^4 k$. First, we shift momenta and complete the square:

$$p = k - \frac{q\alpha_2}{\alpha_1 + \alpha_2} \quad (6.83)$$

Then we use the fact that:

$$\int \frac{d^4 p}{(2\pi)^4} e^{ip^2(\alpha_1+\alpha_2)} = \frac{1}{16\pi^2 i (\alpha_1 + \alpha_2)^2}$$

$$\int \frac{d^4 p}{(2\pi)^4} p_\mu p_\nu e^{ip^2(\alpha_1+\alpha_2)} = \frac{ig_{\mu\nu}}{32\pi^2 i (\alpha_1 + \alpha_2)^3} \quad (6.84)$$

Putting everything back into $\Pi_{\mu\nu}$, we have:

$$\Pi_{\mu\nu,m} = (g_{\mu\nu} q^2 - q_\mu q_\nu)\Pi_1 + g_{\mu\nu}\Pi_2 \quad (6.85)$$

where:

$$\begin{aligned} \Pi_1 &= -i\frac{2\alpha}{\pi} \int_0^\infty \int_0^\infty \frac{\alpha_1\alpha_2 \, d\alpha_1 \, d\alpha_2}{(\alpha_1 + \alpha_2)^4} e^{f(\alpha_1,\alpha_2)} \\ \Pi_2 &= -i\frac{\alpha}{\pi} \int_0^\infty \int_0^\infty i \frac{d\alpha_1 \, d\alpha_2}{(\alpha_1 + \alpha_2)^3} [f(\alpha_1, \alpha_2) - 1] \, e^{f(\alpha_1,\alpha_2)} \end{aligned} \quad (6.86)$$

and where:

$$f(\alpha_1, \alpha_2) = iq^2 \frac{\alpha_1\alpha_2}{\alpha_1 + \alpha_2} - i(m^2 - i\epsilon)(\alpha_1 + \alpha_2) \quad (6.87)$$

There is a similar expression for $\Pi_{\mu\nu,M}$. We notice that Π_2 diverges quadratically, which is bad. However, since we have carefully regularized this integral using the Pauli–Villars method, the integral is finite for fixed M, and we are free to manipulate this expression. We can then show that Π_2 vanishes. This rather remarkable fact can be proved using simple scaling arguments. Consider integrals of the following type:

$$\int_0^\infty \frac{dx}{x} f(x) e^{f(x)} = \frac{\partial}{\partial\rho} \int_0^\infty \frac{dx}{x} e^{\rho f(x)} \bigg|_{\rho=1} \quad (6.88)$$

Assume that $f(x/\rho) = f(x)/\rho$. After this simple rescaling, this integral equals:

$$\frac{\partial}{\partial\rho} \int_0^\infty \frac{dx}{x} e^{f(x)} \bigg|_{\rho=1} = 0 \quad (6.89)$$

since all dependence on ρ has vanished. With a few modifications, this argument can be used to show that $\Pi_2 = 0$. Thus, the vacuum polarization graph is only logarithmically divergent.

To perform the integrations in Π_1, we use one last identity:

$$1 = \int_0^\infty \frac{d\rho}{\rho} \, \delta \left(1 - \frac{\alpha_1 + \alpha_2}{\rho} \right) \tag{6.90}$$

Inserting this into the expression for Π_1 and rescaling α_i, we find:

$$
\begin{aligned}
\Pi_1 &= -i \frac{2\alpha}{\pi} \int_0^\infty \int_0^\infty \frac{\alpha_1 \alpha_2 \, d\alpha_1 \, d\alpha_2}{(\alpha_1 + \alpha_2)^4} \int_0^\infty \frac{d\rho}{\rho} \\
&\quad \times \delta \left(1 - \frac{\alpha_1 + \alpha_2}{\rho} \right) e^{f(\alpha_1, \alpha_2)} \\
&= -\frac{2i\alpha}{\pi} \int_0^\infty \int_0^\infty d\alpha_1 \, d\alpha_2 \, \alpha_1 \alpha_2 \, \delta(1 - \alpha_1 - \alpha_2) \\
&\quad \times \int_0^\infty \frac{d\rho}{\rho} \exp \left[i\rho(q^2 \alpha_1 \alpha_2 - m^2 + i\epsilon) \right] \tag{6.91}
\end{aligned}
$$

As expected, this integral is logarithmically divergent. At this point, we now use the Pauli–Villars regulator, which lowers the divergence of the theory. To perform the tricky ρ integration, we use the fact that $m^2 - \alpha_1 \alpha_2 q^2$ is positive, so we can rotate the contour integral of ρ in the complex plane by -90 degrees. Using integration by parts and rescaling, we have the following identity:

$$\int_\epsilon^\infty \frac{d\rho}{\rho} e^{-a\rho} = \log \rho e^{-a\rho} \Big|_\epsilon^\infty - \int_{a\epsilon}^\infty \log a e^{-\rho} \, d\rho + \int_{a\epsilon}^\infty \log \rho e^{-a\rho} \, d\rho \tag{6.92}$$

where $a(m) = m^2 - \alpha_1 \alpha_2 q^2$. The dangerous divergence comes from the last term. However, since the last term is independent of a, it cancels against the same term, with a minus sign, coming from the Pauli–Villars contribution. Thus, we have the identity:

$$
\begin{aligned}
\lim_{\epsilon \to \infty} \int_\epsilon^\infty \frac{d\rho}{\rho} \left(e^{-a(m)\rho} - e^{-a(M)\rho} \right) &= -\log a(m) + \log a(M) \\
&= -\log \left(1 - \frac{\alpha_1 \alpha_2 q^2}{m^2} \right) + \log \frac{M^2}{m^2} \tag{6.93}
\end{aligned}
$$

for large but finite M. We can now take the limit at $\epsilon \to 0$ and M becomes large. Then:

$$\tilde{\Pi}_1 = -\frac{i\alpha}{3\pi} \log \frac{M^2}{m^2} + \frac{2i\alpha}{\pi} \int_0^\infty d\alpha_1 \, \alpha_1(1 - \alpha_1) \log \left(1 - \alpha_1(1 - \alpha_1) \frac{q^2}{m^2} \right) \tag{6.94}$$

If we perform the last and final integration over α_1, we arrive at:

$$\tilde{\Pi}_1 = \frac{i\alpha}{3\pi} \left\{ -\log \frac{M^2}{m^2} + \frac{1}{3} + 2\left(1 + \frac{2m^2}{q^2}\right)\left[\left(\frac{4m^2}{q^2} - 1\right)^{1/2}\right.\right.$$

$$\left.\left. \times \operatorname{arccot}\left(\frac{4m^2}{q^2} - 1\right)^{1/2} - 1\right]\right\} \tag{6.95}$$

This is our final result. After a long calculation, we find a surprisingly simple result that has a physical interpretation. We claim that the logarithmic divergence can be cancelled against another logarithmic divergence coming from the bare electric charge e_0. In fact, we will simply *define* the divergence of the electric charge so that it precisely cancels against the logarithmic divergence of Π_1.

To lowest order, we find that we can add the usual photon propagator $D_{\mu\nu}$ to the one-loop correction, leaving us with a revised propagator:

$$-i\frac{g_{\mu\nu}}{q^2}\left(1 - \frac{\alpha}{3\pi}\log\frac{M^2}{m^2} - \frac{\alpha}{15\pi}\frac{q^2}{m^2}\right) \tag{6.96}$$

in the limit as $q^2 \to 0$. This leaves us with the usual theory, except that the photon propagator is multiplied by a factor:

$$-i\frac{g_{\mu\nu}}{q^2} \to Z_3(-i)\frac{g_{\mu\nu}}{q^2} \tag{6.97}$$

where:

$$Z_3 \sim 1 - \frac{\alpha}{3\pi}\log\frac{M^2}{m^2} \tag{6.98}$$

Now let us absorb this divergence into the coupling constant e_0. We are then left with the usual theory with an extra (infinite) factor Z_3 multiplying each propagator. Since the photon propagator is connected to two electron vertices, with coupling e_0, we can absorb Z_3 into the coupling constant, so we have, to lowest order:

$$e = \sqrt{Z_3}e_0 \tag{6.99}$$

where e is called the renormalized electric charge. Since the infinity coming from Z_3 cancels (by construction) against the infinity coming from the bare electric charge, the renormalized electric charge e is finite. (Other renormalization constants will be discussed in Chapter 7.)

Although this works at the lowest level, it remains to be seen whether we can extend this procedure to all orders in perturbation theory. This will be further discussed in the next chapter. Next, we turn to the calculation of the single-loop correction to the vertex function, which gives us the anomalous magnetic moment of the electron.

6.7 Anomalous Magnetic Moment

We recall that in Chapter 4 we derived the magnetic moment of the electron by analyzing the coupling of the electron to the vector potential. At the tree level, we know that the coupling of an electron to the photon is given by $A^\mu \bar{u} \gamma_\mu u$, which in turn gives us a gyromagnetic ratio of $g = 2$. However, the experimentally observed value differed from this predicted value by a small but important amount. Schwinger's original calculation[10] of the anomalous magnetic moment of the electron helped to establish QED as the correct theory of electrons and photons.

To calculate the higher-order corrections to the magnetic moment of the electron, we will use the Gordon identity:

$$\bar{u}(p')\gamma_\mu u(p) = \frac{1}{2m}\bar{u}(p')\left[(p+p')_\mu + i\sigma_{\mu\nu}q^\nu\right]u(p) \qquad (6.100)$$

(To prove this, we simply use the Dirac equation repeatedly on the left and right spinors, which are on-shell.) The magnetic moment of the electron comes from the second term $\bar{u}\sigma_{\mu\nu}q^\nu u A^\mu$. To see this, we take the Fourier transform, so q^ν becomes ∂^ν, and the coupling becomes $\bar{u}\sigma_{\mu\nu}F^{\mu\nu}u$. The magnetic field B_i is proportional to $\epsilon_{ijk}F^{jk}$, so this coupling term in the rest frame now becomes $\bar{u}\sigma_i B_i u$, where we use the Dirac representation of the Dirac matrices. Since σ_i is proportional to the spin of the electron, which in turn is proportional to the magnetic moment of the electron, the coupling becomes $\boldsymbol{\mu} \cdot \mathbf{B}$. This is the energy of a magnet with moment $\boldsymbol{\mu}$ in a magnetic field $\mathbf{B}$.

In this section, we will calculate the one-loop vertex correction, which gives us a correction to the electron–photon coupling given by (to lowest order in α):

$$\bar{u}(p')\left[\frac{(p+p')_\mu}{2m} + \left(1 + \frac{\alpha}{2\pi}\right)\frac{i\sigma_{\mu\nu}q^\nu}{2m}\right]u(p) \qquad (6.101)$$

for the process given in Figure 6.12.

Notice that the $\bar{u}\sigma_{\mu\nu}q^\nu u$ term is modified by the one-loop correction, so that the g of the electron becomes:

$$\frac{g}{2} = 1 + \frac{\alpha}{2\pi} \qquad (6.102)$$

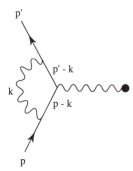

Figure 6.12. First-loop correction to the electron vertex function, which contributes to the anomalous electron magnetic moment.

Thus, QED predicts a correction to the moment of the electron. To show this, we will begin our calculation with the one-loop vertex correction:

$$\Lambda_\mu(p', p) = (-ie)^2 \int \frac{d^4k}{(2\pi)^4} \frac{(-i)}{k^2 - \mu^2 + i\epsilon} \gamma_\nu \frac{i}{\not{p}' - \not{k} - m + i\epsilon}$$

$$\times \gamma_\mu \frac{i}{\not{p} - \not{k} - m + i\epsilon} \gamma^\nu \tag{6.103}$$

Anticipating that the integral is infrared divergent, we have added μ, the fictitious mass of the photon. The integral is also divergent in the ultraviolet region, so we will use the Pauli–Villars cutoff method later to isolate the divergence.

Throughout this calculation, we tacitly assume that we have sandwiched this vertex between two on-shell spinors, so we can use the Gordon decomposition and the mass-shell condition. Our goal is to write this expression in the form:

$$\Lambda_\mu \sim \gamma_\mu F_1(q^2) + \frac{i\sigma_{\mu\nu}q^\nu}{2m} F_2(q^2) \tag{6.104}$$

sandwiched between $\bar{u}(p')$ and $u(p)$, where F_1 and F_2 are the form factors that measure the deviation from the simple γ_μ vertex. We will calculate explicit forms for these two form factors. (We will find that F_1 cancels against the infrared divergence found in the bremsstrahlung calculation, giving us a finite result. We will show that F_2 gives us a correction to the magnetic moment of the electron.)

We begin with the Feynman parameter trick, generalizing Eq. (6.75). The following can be proved by induction:

$$\frac{1}{a_1 \cdots a_n} = (n-1)! \int_0^\infty \int_0^\infty \int_0^\infty \prod_{i=1}^n dz_i \frac{\delta\left(1 - \sum_i z_i\right)}{\Delta^n} \tag{6.105}$$

where: $\Delta = \sum_i a_i z_i$. For our purposes, we want:

$$
\begin{aligned}
a_1 &= k^2 - \mu^2 + i\epsilon \\
a_2 &= (p' - k)^2 - m^2 + i\epsilon \\
a_3 &= (p - k)^2 - m^2 + i\epsilon
\end{aligned}
\tag{6.106}
$$

Therefore:

$$
\Lambda_\mu(p', p) = 2ie^2 \int \frac{d^4k}{(2\pi)^4} \int_0^\infty dz_1\, dz_2\, dz_3\, \delta\,(1 - z_1 - z_2 - z_3) \left(\frac{N_\mu(k)}{\Delta^3} \right)
\tag{6.107}
$$

where:

$$
N_\mu(k) = \gamma_\nu \left(p' - \slashed{k} + m \right) \gamma_\mu \left(\slashed{p} - \slashed{k} + m \right) \gamma^\nu
\tag{6.108}
$$

Next, we would like to perform the integration by completing the square:

$$
\begin{aligned}
\Delta &= \sum_{i=1}^3 a_i z_i = (k - p'z_2 - pz_3)^2 - \Delta_0 \\
\Delta_0 &= m^2(1 - z_1)^2 + \mu^2 z_1 - q^2 z_2 z_3 - i\epsilon \\
&\quad - [(p')^2 - m^2]^2 z_2(1 - z_2) - (p^2 - m^2)z_3(1 - z_3)
\end{aligned}
\tag{6.109}
$$

This allows us to make the shift in integration:

$$
k \to k + p'z_2 + pz_3
\tag{6.110}
$$

Therefore, after the shift, we have:

$$
\begin{aligned}
\Lambda_\mu(p', p) &= \frac{2ie^2}{(2\pi)^4} \int_0^1 dz_1\, dz_2\, dz_3\, \delta(1 - z_1 - z_2 - z_3) \\
&\quad \times \int_{-\infty}^\infty d^4k\, \frac{N_\mu(k + p'z_2 + pz_3)}{(k^2 - \Delta_0 + i\epsilon)^3}
\end{aligned}
\tag{6.111}
$$

By power counting, the integral diverges. This is why we must subtract off the contribution of the Pauli–Villars field, which has mass Λ. Let us expand:

$$
N_\mu(k + p'z_2 + pz_3) = -k^2 \gamma_\mu + 2k_\mu \slashed{k} + A_\mu(k) + N_\mu(p'z_2 + pz_3)
\tag{6.112}
$$

where $A_\mu(k)$ is linear in the k_μ variable. (This term can be dropped, since its integral over d^4k vanishes.)

Therefore, the leading divergence behaves like:

$$\int d^4k \left(\frac{k_\mu k_\nu}{\left(k^2 - \Delta_0\right)^3} - \frac{k_\mu k_\nu}{\left(k^2 - \Delta_\Lambda\right)^3} \right) \tag{6.113}$$

where Δ_Λ represents the Pauli–Villars contribution, where μ is replaced by Λ. With this insertion, the integral converges for finite but large Λ. To perform this integration, we must do an analytic continuation of the previous equation. We know that (see Appendix):

$$\int d^4k \frac{k_\mu k_\nu}{\left(k^2 - \Delta\right)^\alpha} = \frac{i\pi^2}{2\Gamma(\alpha)(-\Delta)^{\alpha-3}} \Gamma(\alpha - 3) g_{\mu\nu} \tag{6.114}$$

Now let us take the limit at $\alpha \to 3$. This expression, of course, diverges, but the Pauli–Villars term subtracts off the divergence. If we let $\alpha - 3 = \epsilon$, then we have:

$$
\begin{aligned}
\lim_{\alpha \to 3} \int d^4k & \left(\frac{k^2}{\left(k^2 - \Delta_0\right)^\alpha} - \frac{k^2}{\left(k^2 - \Delta_\Lambda\right)^\alpha} \right) \\
&= \lim_{\epsilon \to 0} i\pi^2 \Gamma(\epsilon) \left(\frac{1}{(-\Delta_0)^\epsilon} - \frac{1}{(-\Delta_\Lambda)^\epsilon} \right) \\
&= \lim_{\epsilon \to 0} i\pi^2 \frac{1}{\epsilon} \left(e^{\epsilon \log(-\Delta_0)} - e^{\epsilon \log(-\Delta_\Lambda)} \right) \\
&= -i\pi^2 \log \left(\frac{\Delta_\Lambda}{\Delta_0} \right) \\
&= -i\pi^2 \log \left(\frac{z_1 \Lambda^2}{\Delta_0} \right) \tag{6.115}
\end{aligned}
$$

where we drop terms like Λ^{-n}. Because this expression is sandwiched between two on-shell spinors, we can also reduce the term:

$$
\begin{aligned}
N_\mu(p'z_2 + pz_3) = {}& -\gamma_\mu \left[2m^2(1 - 4z_1 + z_1^2) + 2q^2(1 - z_2)(1 - z_3) \right] \\
& -2mz_1(1 - z_1) \left[\slashed{q}, \gamma_\mu \right] \tag{6.116}
\end{aligned}
$$

At this point, all integrals can be evaluated. Putting everything together, and dropping all terms of order Λ^{-1} or less, we now have:

$$\Lambda_\mu(p', p) = \frac{\alpha}{2\pi} \int_0^\infty dz_1 \, dz_2 \, dz_3 \, \delta(1 - z_1 - z_2 - z_3) \left[\log \left(\frac{z_1 \Lambda^2}{\Delta_0} \right) \right] \gamma_\mu$$

$$+ \quad \gamma_\mu \frac{m^2(1 - 4z_1 + z_1^2) + q^2(1 - z_2)(1 - z_3)}{\Delta_0} + \frac{im\sigma_{\mu\nu}q^\nu z_1(1 - z_1)}{\Delta_0} \Bigg] (6.117)$$

Now let us compare this expression with the form factors F_1 and F_2 appearing in Eq. (6.104). It is easy to read off:

$$F_1(q^2) \quad \rightarrow \quad 1 + \frac{\alpha}{2\pi} \int_0^\infty dz_1\, dz_2\, dz_3\, \delta\,(1 - z_1 - z_2 - z_3)$$

$$\times \left(-\log \Delta_0 + \frac{m^2(1 - 4z_1 + z_1^2) + q^2(1 - z_2)(1 - z_3)}{\Delta_0} - (q^2 = 0) \right)$$

$$(6.118)$$

[We have deliberately subtracted off the integrand defined at $q^2 = 0$ in order to maintain the constraint $F_1(q^2 = 0) = 1$, which preserves the correct normalization of the vertex function. This extra subtraction term comes from the fact that we have separated out the divergent logarithmic part, which is absorbed in an infinite constant called Z_1 which renormalizes the vertex function. This point will be discussed in more detail in Chapter 7.]

The value of F_2 can similarly be read off:

$$F_2(q^2) = \frac{\alpha}{2\pi} \int_0^\infty dz_1\, dz_2\, dz_3\, \delta\,(1 - z_1 - z_2 - z_3) \left(\frac{2m^2 z_1(1 - z_1)}{\Delta_0} \right) \quad (6.119)$$

The calculation for $F_2(q^2)$ is a bit easier, since there are no ultraviolet or infrared divergences. Because of this, we can set $\mu = 0$. Let us choose new variables:

$$p \cdot p' = m^2 \cosh\theta; \quad q^2 = -4m^2 \sinh^2(\theta/2) = 2m^2(1 - \cosh\theta) \quad (6.120)$$

where $q^2 = (p - p')^2$. Then we have:

$$F_2(q^2) \quad = \quad \frac{\alpha}{\pi} \int_0^1 dz_2\, dz_3\, \theta(1 - z_2 - z_3) \frac{z_1 + z_3 - (z_2 + z_3)^2}{z_2^2 + z_3^2 + 2z_2 z_3 \cosh\theta}$$

$$= \quad \frac{\alpha}{2\pi} \int_0^1 d\beta \frac{1}{\beta^2 + (1 - \beta)^2 + 2\beta(1 - \beta)\cosh\theta} \quad (6.121)$$

This leaves us with the exact result:

$$F_2(q^2) = \frac{\alpha}{2\pi} \frac{\theta}{\sinh\theta} \quad (6.122)$$

We are especially interested in taking the limit as $|q^2| \rightarrow 0$ and $\theta \rightarrow 0$:

$$F_2(0) = \frac{\alpha}{2\pi} \quad (6.123)$$

This gives us the correction to the magnetic moment of the electron, as in Eq. (6.101):

$$\frac{g-2}{2} = \frac{\alpha}{2\pi} \sim 0.0011614 \tag{6.124}$$

This is only a first-order calculation, yet already we are very close to the experimental value. Since the calculation was originally performed by Schwinger, the calculation since has been taken to α^3 order (where there are 72 Feynman diagrams). The theoretical value to this order is given by:

$$
\begin{aligned}
a_{\text{th.}} &\equiv \frac{1}{2}(g-2) \\
&= 0.5\left(\frac{\alpha}{\pi}\right) - 0.32848 \left(\frac{\alpha}{\pi}\right)^2 + 1.49 \left(\frac{\alpha}{\pi}\right)^3 + \cdots
\end{aligned} \tag{6.125}
$$

The final results for both the theoretical and experimental values are[11]:

$$
\begin{aligned}
a_{\text{th}} &= 0.001159652411(166) \\
a_{\text{expt}} &= 0.001159652209(31)
\end{aligned} \tag{6.126}
$$

where the estimated errors are in parentheses.

The calculation agrees to within one part in 10^8 for a and to one part in 10^9 for g, which is graphic vindication of QED. (To push the calculation to the fourth order involves calculating 891 diagrams and 12,672 diagrams at the fifth order.)

6.8 Infrared Divergence

The calculation for F_1 is much more difficult. However, it will be very important in resolving the question of the infrared divergence, which we found in the earlier discussion of bremsstrahlung. We will find that the infrared divergence coming from the bremsstrahlung graph and F_1 cancel exactly. Although the calculation of F_1 is difficult, one can can extract useful information from the integral by taking the limit as μ becomes small. Then F_1 integration in Eq. (6.118) splits up into four pieces:

$$F_1 = \sum_{i=1}^{4} P_i + \cdots \tag{6.127}$$

where the ellipsis represents constant terms.

After changing variables, each of the P_i pieces can be exactly evaluated:

$$P_1 = -\frac{\alpha}{\pi} \int_0^1 dz_2 \int_0^{1-z_2} dz_3 \frac{\cosh\theta}{z_2^2 + z_3^2 + 2z_2z_3\cosh\theta + (\mu^2/m^2)(1 - z_2 - z_3)}$$

$$= \frac{\alpha}{\pi}\theta\coth\theta\log\frac{\mu}{m} - \frac{2\alpha}{\pi}\coth\theta\int_0^{\theta/2} d\phi\,\phi\tanh\phi \qquad (6.128)$$

where:

$$1 - 2z_2 = \frac{\tanh\phi}{\tanh(\theta/2)} \qquad (6.129)$$

The others are given by:

$$P_2 = \frac{\alpha}{\pi}\cosh\theta\int_0^1 dz_2 \int_0^{1-z_2} dz_3\frac{z_2 + z_3}{z_2^2 + z_3^2 + 2z_2z_3\cosh\theta}$$

$$= \frac{\alpha}{\pi}\theta\coth\theta$$

$$P_3 = \frac{2\alpha}{\pi}(1 - \cosh\theta)\int_0^1 dz_2 \int_0^{1-\alpha_2} dz_3\frac{z_2z_3}{z_2^2 + z_3^2 + 2z_2z_3\cosh\theta}$$

$$= \frac{\alpha}{2\pi}\left(\frac{\theta}{\sinh\theta} - 1\right)$$

$$P_4 = -\frac{\alpha}{2\pi}\int_0^1 dz_2 \int_0^{1-z_2} dz_3\log\left[(z_2 + z_3)^2 + 4z_2z_3\sinh^2(\theta/2)\right]$$

$$= -\frac{\alpha}{2\pi}\frac{\theta/2}{\tanh(\theta/2)} + \text{const.} \qquad (6.130)$$

Thus, adding the pieces together, we find:

$$F_1(q^2) = \frac{\alpha}{\pi}\left[\left(\log\frac{\mu}{m} + 1\right)(\theta\coth\theta - 1)\right.$$

$$\left. - 2\coth\theta\int_0^{\theta/2} d\phi\,\phi\tanh\phi - \frac{\theta}{4}\tanh\frac{\theta}{2}\right] \qquad (6.131)$$

For $|q^2| \ll m^2$, we find using Eqs. (6.104), (6.120), (6.122), and (6.131):

$$\gamma_\mu + \Lambda_\mu^c(p', p) \sim \gamma_\mu\left[1 + \frac{\alpha}{3\pi}\frac{q^2}{m^2}\left(\log\frac{m}{\mu} - \frac{3}{8}\right)\right] + \frac{\alpha}{8\pi m}[\slashed{q}, \gamma_\mu] \qquad (6.132)$$

For $|q^2| \gg m^2$, we find:

$$\gamma_\mu + \Lambda_\mu^c(p', p) \sim \gamma_\mu \left\{ 1 - \frac{\alpha}{\pi} \log \frac{m}{\mu} \left[\log \left(\frac{-q^2}{m^2} \right) - 1 + O\left(m^2/q^2 \right) \right] \right\} \quad (6.133)$$

Plugging all this into the cross-section formula, we now find our final result:

$$\left(\frac{d\sigma}{d\Omega} \right)_\mu = \left(\frac{d\sigma}{d\Omega} \right)_0 \left[1 - \frac{2\alpha}{\pi} \log \frac{m}{\mu} \chi(q^2) \right] \quad (6.134)$$

where:

$$\chi(q^2) = \begin{cases} -\frac{1}{3} q^2/m^2 & \text{if } -q^2 \ll m^2 \\ \log -q^2/m^2 - 1 & \text{if } -q^2 \gg m^2 \end{cases} \quad (6.135)$$

Now we come to the final step, the comparison of the bremsstrahlung amplitude in Eq. (6.77) and the vertex correction for electron scattering. Although they represent different physical scattering processes, they must be added because there is always an uncertainty in our measuring equipment in measuring soft photons. Comparing the two amplitudes, we recall that the bremsstrahlung amplitude in Eq. (6.78) was given by:

$$\frac{d\sigma}{d\Omega} = \left(\frac{d\sigma}{d\Omega} \right)_0 \frac{\alpha}{\pi} \log \frac{-q^2}{m^2} \log \frac{E^2}{\mu^2} \quad (6.136)$$

while the vertex correction graph yields:

$$\frac{d\sigma}{d\Omega} = \left(\frac{d\sigma}{d\Omega} \right)_0 \left(1 - \frac{\alpha}{\pi} \log \frac{-q^2}{m^2} \log \frac{-q^2}{\mu^2} \right) \quad (6.137)$$

Clearly, when these two amplitudes are added together, we find a finite, convergent result independent of μ^2, as desired. The cancellation of infrared divergences to all orders in perturbation theory is a much more involved process. However, there are surprising simplifications that give a very simple result for this calculation (see Appendix).

6.9   Lamb Shift

Two of the great accomplishments of QED were the determination of the anomalous magnetic moment of the electron, which we discussed earlier, and the Lamb shift.[12] The fact that these two effects could not be explained by ordinary quantum

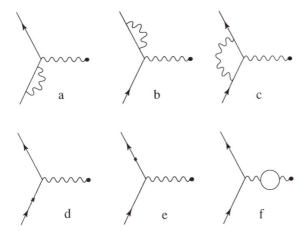

Figure 6.13. The various higher-order graphs that contribute to the Lamb shift: (a) and (b) the electron self-energy graphs; (c) the vertex correction; (d) and (e) the electron mass counterterm; (f) the photon self-energy correction.

mechanics, and the fact that the QED result was so accurate, helped to convince the skeptics that QED was the correct theory of the electron–photon system.

In 1947, Lamb and Retherford demonstrated that the $2S_{1/2}$ and the $2P_{1/2}$ energy levels of the hydrogen atom were split; the $2P_{1/2}$ energy level was depressed more than 1000 MHz below the $2S_{1/2}$ energy level. (The original Dirac electron in a classical Coulomb potential, as we saw earlier in Section 4.2, predicted that the energy levels of the hydrogen atom should depend only on the principal quantum number n and the total spin j, so these two levels should be degenerate.)

The calculation of the Lamb shift is rather intricate, because we are dealing with the hydrogen atom as a bound-state problem, and also because we must sum over all radiative corrections to the electron interacting with a Coulomb potential that modify the naive $\bar{u}\gamma_0 u A^0$ vertex. These corrections include the vertex correction, the anomalous magnetic moment, the self-energy of the electron, the vacuum polarization graph, and even infrared divergences (Fig. 6.13).

The original nonrelativistic bound-state calculation of Bethe,[13] which ignored many of these subtle higher-order corrections, could account for about 1000 MHz of the Lamb shift, but only a fully relativistic quantum treatment could calculate the rest of the difference. Because of the intricate nature of the calculation, we will only sketch the highlights of the calculation. To begin the discussion, we first see that the vacuum polarization graph can be attached to the photon line, changing the photon propagator to:

$$D_{\mu\nu} = -\frac{g_{\mu\nu}}{k^2}\left(1 - \frac{e^2}{60\pi^2}\frac{k^2}{m^2} + O(k^4)\right) \qquad (6.138)$$

This, of course, translates into a shift in the effective coupling of an electron to the Coulomb potential.[14,15] Analyzing the zeroth component of this propagator, we see that the coupling of the electron to the Coulomb potential changes as follows:

$$ie_0^2 \frac{\bar{u}\gamma_0 u}{q^2} \rightarrow ie^2 \frac{\bar{u}\gamma_0 u}{q^2}\left(1 - \frac{\alpha q^2}{15\pi m^2} + O(\alpha^2)\right) \tag{6.139}$$

To convert this back into x space, let us take the Fourier transform. We know that the Fourier transform of $1/q^2$ is proportional to $1/r$. This means that the static Coulomb potential that the electron sees is given by:

$$\left(1 - \frac{\alpha}{15\pi m^2}\nabla^2\right)\frac{e^2}{4\pi r} = \frac{e^2}{4\pi r} + \frac{e^2}{60\pi^2 m^2}\delta^3(\mathbf{x}) \tag{6.140}$$

meaning that there is a correction to Coulomb's law given by QED. This correction, in turn, shifts the energy levels of the hydrogen atom. We know from ordinary nonrelativistic quantum mechanics that, by taking matrix elements of this modified potential between hydrogen wave functions, we can calculate the first-order correction to the energy levels of the hydrogen atom due to the vacuum polarization graph.

Now let us generalize this discussion to include the other corrections to the calculation of the Lamb shift. Our method is the same: calculate the corrections to the vertex function $\bar{u}\gamma_\mu u$, take the zeroth component, and then take the low-energy limit. In Eqs. (6.104), (6.132), and (6.133), we saw how radiative corrections modified the vertex function with additional form factors $F_1(q^2)$ and $F_2(q^2)$. If we add the various contributions to the vertex correction, we find:

$$\begin{aligned}\bar{u}\gamma_\mu u \quad \rightarrow \quad &\bar{u}\left\{\gamma_\mu\left[1 - \frac{\alpha q^2}{3\pi m^2}\left(\log\frac{m}{\mu} - \frac{3}{8} - \frac{1}{5}\right)\right]\right.\\ &\left. + \frac{i\alpha}{4\pi m}\sigma_{\mu\nu}q^\nu\right\}u\end{aligned} \tag{6.141}$$

(For example, the vacuum polarization graph contributes the factor $-1/5$ to the vertex correction. The logarithm term comes from the vertex correction, and the μ term is eventually cancelled by the infrared correction.) Now take the low-energy limit of this expression. The $\sigma_{\nu\mu}q^\nu$ term reduces down to a spin-orbit correction, and we find the effective potential given by:

$$\Delta V_{\text{eff}} \sim \frac{4\alpha^2}{3m^2}\left(\log\frac{m}{\mu} - \frac{3}{8} - \frac{1}{5} + \frac{3}{8}\right)\delta^3(\mathbf{x}) + \frac{\alpha^2}{4\pi m^2 r^3}\boldsymbol{\sigma}\cdot\mathbf{L} \tag{6.142}$$

By taking the matrix element of this potential between two hydrogen wave functions, we can calculate the energy split due to this modified potential. The vertex correction, for example, gives us a correction of 1010 MHz. The anomalous magnetic moment of the electron contributes 68 MHz. And the vacuum polarization graph, calculated earlier, contributes -27.1 MHz. Adding these corrections together, we find, to the lowest loop level, that we arrive at the Lamb shift to within 6 MHz accuracy.

Since then, higher-order corrections have been calculated, so the difference between experiment and theory has been reduced to 0.01 MHz. Theoretically, the $2S_{1/2}$ level is above the $2P_{1/2}$ energy level by 1057.864 ± 0.014 MHz. The experimental result is 1057.862 ± 0.020 MHz. This is an excellent indicator of the basic correctness of QED.

6.10 Dispersion Relations

So far, we have been discussing the scattering matrix from the point of view of perturbation and renormalization theory, that is, as a sum of increasingly complicated divergent Feynman diagrams. However, the calculations and the renormalization procedures rapidly become extraordinarily tedious and complicated, involving hundreds of Feynman diagrams at the fourth order.

There is another approach one may take to the scattering matrix that completely avoids the complicated and often counterintuitive operations used in renormalization theory. Following Heisenberg and Chew,[16] one may also take the approach that the S matrix by itself satisfies so many stringent physical requirements that perhaps the S matrix is uniquely determined. This approach avoids the formidable apparatus that one must introduce in order to renormalize even the simplest quantum field theory. Although this approach does not give a systematic method to construct the S matrix for various physical processes, it does give us rigid constraints that are sometimes strong enough to solve for certain properties of the S matrix, such as sum rules and dispersion relations.

The S matrix, for example, should satisfy the following properties:

1. Unitarity or conservation of probability.

2. Analyticity in the various energy variables.

3. Lorentz invariance.

4. Crossing symmetry.

These, in turn, impose an enormous number of constraints on the S matrix, independent of perturbation theory, that may give clues to the final answer. One of the great successes of this approach was the use of *dispersion relations*[17] to calculate constraints on the pion–nucleon cross sections. It gave dramatic verification of the power of analyticity in deriving nontrivial relations among the cross sections. These relations were used extensively in strong interaction physics where a successful, renormalizable field theory eluded physicists for many decades.

In classical optics, we know that the imposition of causality on the propagation of a wave front is sufficient for us to write down a dispersion relation. Classically, we find that causality implies that the S matrix found in optics is analytic. Following this analogy to classical optics, we can show that the microscopic causality of the Green's functions is sufficient to prove that the S matrix is analytic and hence satisfies certain dispersion relations.

We know that functions $f(z)$ analytic in the upper half plane obey very stringent constraints; they can be written as Cauchy contour integrals:

$$f(z) = \frac{1}{2\pi i} \int_C \frac{f(u)}{u-z} du = \frac{1}{2\pi i} \int_{-\infty}^{\infty} \frac{f(u)}{u-z} dz \tag{6.143}$$

where the contour C is a large closed semicircle sitting on the real axis in the upper half complex plane. In the limit that f vanishes sufficiently rapidly at infinity, we can drop the integral around the outer semicircle. Notice that z sits in the upper half plane. If we slowly let z approach the real axis, then we must take the principal part of the integral:

$$
\begin{aligned}
f(z) &= \frac{1}{2\pi i} \left(\int_{-\infty}^{z-\epsilon} + \int_{z+\epsilon}^{\infty} \right) \frac{f(u)}{u-z} du + \frac{1}{2}\frac{1}{2\pi i} \int_{C'} \frac{f(u)}{u-z} du \\
&= \frac{1}{\pi i} P \int_{-\infty}^{\infty} \frac{f(u)}{u-z} du \tag{6.144}
\end{aligned}
$$

where C' is a circular contour integral taken infinitesimally around the point z.

Taking the real part of both sides, we then have:

$$\mathrm{Re}\, f(z) = \frac{1}{\pi} P \int_{-\infty}^{\infty} du \frac{\mathrm{Im}\, f(u)}{u-z} \tag{6.145}$$

We can therefore convert this formula to an integral from 0 to ∞:

$$
\begin{aligned}
\mathrm{Re}\, f(z) &= \frac{1}{\pi} P \int_{-\infty}^{0} du \frac{\mathrm{Im}\, f(u)}{u-z} + \frac{1}{\pi} \int_{0}^{\infty} du \frac{\mathrm{Im}\, f(u)}{u-z} \\
&= \frac{1}{\pi} P \int_{0}^{\infty} du \frac{\mathrm{Im}\, f(u)}{u+z} + \frac{1}{\pi} P \int_{0}^{\infty} du \frac{\mathrm{Im}\, f(u)}{z-u} \\
&= \frac{1}{\pi} P \int_{0}^{\infty} 2u \, du \frac{\mathrm{Im}\, f(u)}{u^2-z^2} \tag{6.146}
\end{aligned}
$$

where we assume $\mathrm{Im} f(z) = -\mathrm{Im} f(z)$.

In ordinary nonrelativistic quantum mechanics, we know that the scattering amplitude of a wave scattering off a stationary, hard sphere is given by a function $f(p)$, which in turn is given as a function of the cross section via the optical theorem:

$$\text{Im } f(p) = \frac{p}{4\pi}\sigma_{\text{tot}}(p) \tag{6.147}$$

Then we have the forward dispersion relation:

$$\text{Re } f(p) = \frac{1}{2\pi^2} P \int_0^\infty dq \frac{q^2\sigma_{\text{tot}}(q)}{q^2 - p^2} \tag{6.148}$$

If often turns out that f does not vanish fast enough to make the above relation valid. For example, for photon–proton scattering (see Section 6.1), we have, for zero frequency:

$$f(0) = -\frac{\alpha}{M} \tag{6.149}$$

which is the Thompson amplitude. Inserting this into the dispersion relation, we have:

$$-\frac{\alpha}{M} = \frac{1}{2\pi^2} \int_0^\infty \sigma_{\text{tot}}(q)dq \tag{6.150}$$

But this is obviously incorrect, because the right-hand side is positive, while the left-hand side is not. The error we have made is assuming that f vanishes at infinity. Instead, we will write a dispersion relation for $f(p)/p$, which does vanish at infinity. Repeating the same steps as before, we find the following dispersion relation:

$$\frac{f(p)}{p} = -\frac{\alpha}{pM} + \frac{1}{2\pi i} \lim_{\epsilon \to 0} \int dq \frac{f(q)}{q(q - p - i\epsilon)} \tag{6.151}$$

Taking the real part, we obtain our final result:

$$\begin{aligned}
\text{Re } f(p) &= -\frac{\alpha}{M} + \frac{p^2}{\pi} P \int_0^\infty 2q \, dq \frac{\text{Im } f(q)}{q^2(q^2 - p^2)} \\
&= -\frac{\alpha}{M} + \frac{p^2}{2\pi^2} P \int_0^\infty dq \frac{\sigma_{\text{tot}}(q)}{q^2 - p^2} \tag{6.152}
\end{aligned}$$

Now that we have shown how to use the power of analyticity to write dispersion relations, let us combine this with the added condition of unitarity to obtain conditions on the S matrix. We know that the scattering matrix satisfies:

$$\sum_n S_{nf}^* S_{ni} = \delta_{fi} \tag{6.153}$$

By subtracting off the delta function that describes when the particles do not interact with each other, the S matrix can be written in terms of a $\mathcal{T}$ matrix:

$$S_{fi} = \delta_{fi} - i(2\pi)\delta^4(\sum p_f - \sum p_i)\mathcal{T}_{fi} \tag{6.154}$$

Inserting the expression for S in terms of $\mathcal{T}$, and taking the case of forward scattering, $f = i$, we have:

$$\mathcal{T}_{ii} - \mathcal{T}_{ii}^* = -i(2\pi)^4 \sum_n \delta^4(P_i - P_n)\mathcal{T}_{ni}^* \mathcal{T}_{ni} \tag{6.155}$$

Earlier, in Eqs. (5.21) to (5.26) we derived:

$$\sigma_{\text{tot}} = \frac{(2\pi)^6}{v_{\text{lab}}} \sum_n (2\pi)^4 \delta^4(P_i - P_n)|\mathcal{T}_{ni}|^2 \tag{6.156}$$

where the sum over n includes integrating over final states. Inserting this into the unitarity condition, we find:

$$\text{Im}\,\mathcal{T}_{ii} = -\frac{1}{2} v_{\text{lab}} \frac{\sigma_{\text{tot}}(\omega)}{(2\pi)^6} \tag{6.157}$$

We now wish to apply this formalism to calculate dispersion relations for pion–nucleon scattering.[18] We first need to define the kinematics of the collision. We begin with the scattering of a pion of momentum q_1 off a proton of momentum p_1, producing a pion of momentum q_2 and a proton of momentum p_2.

Let us define the following *Mandelstam variables*:

$$\begin{aligned}
s &= (q_1 + p_1)^2 \\
t &= (q_2 - q_1)^2 \\
u &= (q_1 - p_2)^2
\end{aligned} \tag{6.158}$$

For the case of forward scattering, the most convenient variable is the laboratory energy ω of the pion:

$$\omega = \frac{p_1 \cdot q_1}{M} = \frac{s - M^2 - m_\pi^2}{2M} \tag{6.159}$$

By Lorentz invariance, the matrix $\mathcal{M}$ appearing in the transition matrix is given by:

$$\mathcal{M}(q_i, p_i) = \bar{u}(p_2)\left(A(s, t) + \frac{1}{2}(\not{q}_1 + \not{q}_2)B(s, t)\right)u(p_1) \qquad (6.160)$$

where nothing is known about the form factors A and B.

In the limit of forward scattering, these two form factors A and B merge into one unknown function:

$$\mathcal{T}_{ii} = \frac{1}{(2\pi)^6 2\omega}\mathcal{M}(q_1, p_1) = \frac{1}{(2\pi)^6 2\omega}4\pi\,\bar{u}(p_1)T(\omega)u(p_1)$$

$$4\pi\,T(\omega) = A(s, 0) + \omega B(s, 0) \qquad (6.161)$$

It can be shown that $T(\omega)$ is an analytic function of ω. Then we can use crossing relations to relate the π^+–nucleon scattering amplitude to the π^-–nucleon scattering amplitude:

$$T(\omega) = T^{\pi^+ p}(\omega) = T^{\pi^- p}(-\omega) \qquad (6.162)$$

Written in this fashion, we can write the unitarity condition as:

$$\operatorname{Im} T^{\pi^+ p}(-\omega) = \frac{1}{2i}\left[T^{\pi^+ p}(-\omega + i\epsilon) - T^{\pi^+ p}(-\omega - i\epsilon)\right]$$

$$= \frac{k}{4\pi}\sigma_{\text{tot}}^{\pi^- p}(\omega) \qquad (6.163)$$

where $k = \sqrt{\omega^2 - m_\pi^2}$.

Let us introduce the functions $T^{\pm}$:

$$T^{\pm} = \frac{1}{2}[T(\omega) \pm T(-\omega)] = \frac{1}{2}\left[T^{\pi^+ p}(\omega) \pm T^{\pi^- p}(\omega)\right] \qquad (6.164)$$

The point of this discussion is to use dispersion relations to write down nontrivial constraints for T that can then be compared with experiment. To write down the dispersion relations, we must deduce the analytic structure of T, which can be done by an analysis of Feynman diagrams. First, if we analyze four-point Feynman diagrams, there is a pole corresponding to the exchange of a single particle. This pole occurs at $\omega_B = m_\pi^2/2M$, which is due to the propagator. The intermediate particle corresponding to this pole is off-shell, and the value of ω is below the energy at which physical scattering takes place. Second, the intermediate states become on-shell in the region $[m_\pi, \infty]$ or $[-\infty, -m_\pi]$. For these values of the energy, real scattering takes place with real intermediate states. Analytically, this means that there is a Riemann cut in the real ω axis.

Putting these two facts together, we can now write down the dispersion relation for T, consisting of the contour integral and also the pole term:

$$T(\omega) = \frac{1}{\pi} \int_{-\infty}^{-m_\pi} d\omega \frac{\mathrm{Im}\, T(\omega')}{\omega' - \omega - i\epsilon} + \frac{1}{\pi} \int_{m_\pi}^{\infty} d\omega \frac{\mathrm{Im}\, T(\omega')}{\omega' - \omega - i\epsilon} + \frac{2f^2}{\omega - \omega_B} \quad (6.165)$$

where the f term comes from the pole contribution, and:

$$f = \left(\frac{g^2}{4\pi}\right) \frac{m_\pi^2}{4M^2} \quad (6.166)$$

Inserting the unitarity condition, we can rewrite the dispersion relation as:

$$\mathrm{Re}\, T^-(\omega)/\omega = \frac{1}{4\pi^2} P \int_{m_\pi}^{\infty} d\omega' \frac{\sqrt{\omega'^2 - m_\pi^2}}{\omega'^2 - \omega^2} \left[\sigma_{\mathrm{tot}}^{\pi^+ p}(\omega') - \sigma_{\mathrm{tot}}^{\pi^- p}(\omega')\right]$$

$$- \frac{2f^2}{\omega^2 - \omega_B^2} \quad (6.167)$$

This is our final result, which agrees well with experiment.

In summary, we have seen that a straightforward application of Feynman's rules allows us to calculate, to lowest order, the interactions of electrons and photons. In this approximation, we find good agreement with earlier, classical results and also experimental data. At higher orders, we find disastrous divergences that require careful renormalizations of the coupling constant and the masses. However, at the one-loop level, QED predicts results for the Lamb shift and the anomalous magnetic moment of the electron that are in excellent agreement with experimental data. This success, in fact, convinced the scientific community of the correctness of QED.

We will now turn to the conclusion of Part I, which is a proof that QED is finite to any order in perturbation theory.

6.11 Exercises

1. Prove the Feynman parameter formula in Eqs. (6.75) and (6.105). (Hint: use induction.)

2. Assume that electrons interact with massless neutrinos via the interaction Lagrangian $g A_\mu \bar{\psi}_e \gamma^\mu (1 - \gamma_5) \psi_\nu$, where A_μ is a massive vector field and ψ_ν is the neutrino field. To second order in the coupling constant, draw all Feynman diagrams for the four-point function $e^- + \nu \rightarrow e^- + \nu$ and write down the corresponding expression for the scattering formula. Do not solve.

3. Draw all the Feynman diagrams that appear in the electron–photon vertex function as well as the photon propagator to the third-loop order. Do not solve.

4. Explicitly reduce out the traces in Eq. (6.45).

5. Fill in the missing steps necessary to prove Eqs. (6.49) and (6.68).

6. Consider the scattering of an electron off a stationary spin-0 particle, using the Yukawa interaction $\bar{\psi}\psi\phi$ to lowest order. Compute the matrix element for this scattering. How will the formula differ from the standard Coulomb formula?

7. Compute the scattering matrix to lowest order of an electron with a stationary neutron if the interaction is given by $g\bar{\psi}\sigma^{\mu\nu}\psi F_{\mu\nu}$. Reduce out all gamma matrices. How does this result compare with the Coulomb formula?

8. Show by symmetry arguments that the most general coupling between the proton and the electromagnetic current is given by:

$$\bar{u}_{p'}\left(\gamma_\mu F_1(q^2) + i\frac{\kappa}{2m_p}\sigma_{\mu\nu}q^\nu F_2(q^2)\right)u_p \tag{6.168}$$

where κ is the anomalous magnetic moment of the proton in units of $e/2m_p$, and $F_1(q^2)$ and $F_2(q^2)$ are form factors and functions of the momentum transferred squared.

9. From this, derive the Rosenbluth formula for electrons scattering off a stationary proton target:

$$\left(\frac{d\sigma}{d\Omega}\right)_{lab} = \frac{\alpha^2\cos^2(\theta/2)}{4E^2\sin^4(\theta/2)}\left[1 + 2E\sin^2(\theta/2)/m_p\right]^{-1}$$

$$\times\left\{|F_1(q^2)|^2 + \frac{q^2}{4m_p^2}\left[2|F_1(q^2) + \kappa F_2(q^2)|^2\tan^2(\theta/2)\right.\right.$$

$$+\kappa^2|F_2(q^2)|^2\Big]\Big\} \tag{6.169}$$

where:

$$q^2 = \frac{4E^2\sin^2(\theta/2)}{1 + (2E/m_p)\sin^2(\theta/2)} \tag{6.170}$$

and where we have assumed that the laboratory energy of the electrons E is much larger than m_e.

10. Prove that $\Pi_2 = 0$ using the identities in Eqs. (6.88) and (6.89).

11. Prove Eqs. (6.62) and (6.64). Fill in the missing steps.

12. Prove Eqs. (6.17) and (6.34).

13. Show that the Mandelstam variables obey the following relationship:

$$s + t + u = \sum_{i=1}^{4} m_i^2 \tag{6.171}$$

where m_i are the masses of the external particles. Show that a four-point scattering amplitude has poles for various values of s, t, and u. Show how the scattering amplitude changes under crossing symmetry as a function of the Mandelstam variables. For example, prove Eq. (6.162).

14. Consider a higher derivative theory with a Lagrangian given by $\phi[\partial_\mu^2 - (\partial_\mu^2)^2/m^2]\phi$. Calculate its propagator and show that it corresponds to a Pauli–Villars-type propagator; that is, it propagates a negative norm ghost. Analyze its ultraviolet divergences, if it has any, if we add the interaction term ϕ^4. Show that the Lagrangian can be written in canonical form; that is, $\mathscr{L} = p\dot{q} - \mathscr{H}$. (Hint: add in auxiliary fields in order to absorb the large number of time derivatives in the action.)

15. Take an arbitrary Feynman graph. Using the Feynman parameter formula, prove[19,20]:

$$
\begin{aligned}
I &= \int \frac{d^4 q_1 \cdots d^4 q_k}{\prod_{i=1}^{n}(p_i^2 - m_i^2)} \\
&= \int d^4 q_1 \cdots d^4 q_k \int_0^1 d\alpha_1 \cdots d\alpha_n \delta\left(1 - \sum_{j=1}^{n} \alpha_j\right) \\
&\quad \times \left(\sum_j (k_j^2 - m_j^2)\alpha_j + 2\sum_{j,s} k_j \alpha_j \eta_{js} \cdot q_s + \sum_{j,s,t} \alpha_j \eta_{js}\eta_{jt} q_s \cdot q_t\right)^{-n}
\end{aligned}
\tag{6.172}
$$

where:

$$p_j = k_j + \sum_{s=1}^{k} \eta_{js} q_s \tag{6.173}$$

and η_{js} equals ± 1 or zero, depending on the placement of the jth and sth lines. If the j line lies in the sth loop, then $\eta = +1$ (-1) if p_j and q_s are parallel (antiparallel). Otherwise, η equals zero.

16. Prove that the previous integral can be written as:

$$\int_0^\infty d\alpha_1 \cdots d\alpha_n \frac{\delta\left(1 - \sum_{j=1}^n \alpha_j\right)}{\Delta^2 \left[\sum_{j=1}^n (k_j^2 - m_j^2)\alpha_j\right]^{n-2k}} \tag{6.174}$$

where:

$$\Delta = \det \left(\sum_{j=1}^n \eta_{js} \eta_{jt} \alpha_j\right)_{st} \tag{6.175}$$

(Hint: choose the k_i in order to set the cross-terms to zero:

$$\sum_{j=1}^n k_j \alpha_j \eta_{js} = 0 \tag{6.176}$$

Then diagonalize the integration over q_i.)

17. In the previous problem, we can make an analogy between a Feynman integral and Kirchhoff's laws found in electrical circuits. Show that we can make the analogy:

$$
\begin{aligned}
k &\rightarrow \text{current} \\
\alpha &\rightarrow \text{resistance} \\
k\alpha &\rightarrow \text{voltage} \\
k^2\alpha &\rightarrow \text{power}
\end{aligned} \tag{6.177}
$$

Show that the statement:

$$\sum_{\text{loop}} k_j \alpha_j = 0 \tag{6.178}$$

for momenta taken around a closed loop corresponds to the statement that the voltage around a closed loop is zero. Second, we also have the equation:

$$\sum_{\text{vertex}} k_j = 0 \tag{6.179}$$

which is the statement of current conservation. Because the Feynman parameters α are strictly real numbers, show that capacitors and inductors are not allowed in the circuit. In this way, we can intuitively analyze the analytic structure of a Feynman graph.

Chapter 7
Renormalization of QED

The war against infinities was ended. There was no longer any reason to fear the higher approximations. The renormalization took care of all infinities and provided an unambiguous way to calculate with any desired accuracy any phenomenon resulting from the coupling of electrons with the electromagnetic field .... It is like Hercules' fight against Hydra, the many-headed sea monster that grows a new head for every one cut off. But Hercules won the fight, and so did the physicists.

—V. Weisskopf

7.1 The Renormalization Program

One of the serious complications found in quantum field theory is the fact that the theory is naively divergent. When higher-order corrections are calculated for QED or ϕ^4 theory, one finds that the integrals diverge in the ultraviolet region, for large momentum p.

Since the birth of quantum field theory, several generations of physicists have struggled to renormalize it. Some physicists, despairing of ever extracting meaningful information from quantum field theory, even speculated that the theory was fundamentally sick and must be discarded. In hindsight, we can see that the divergences found in quantum field theory were, in some sense, inevitable. In the transition from quantum mechanics to quantum field theory, we made the transition from a finite number of degrees of freedom to an infinite number. Because of this, we must continually sum over an infinite number of internal modes in loop integrations, leading to divergences. The divergent nature of quantum field theory then reflects the fact that the ultraviolet region is sensitive to the infinite number of degrees of freedom of the theory. Another way to see this is that the divergent graphs probe the extremely small distance region of space-time, or, equivalently, the high-momentum region. Because almost nothing is known about the nature of physics at extremely small distances or momenta, we are disguising our ignorance of this region by cutting off the integrals at small distances.

Since that time, there have been two important developments in renormalization theory. First was the renormalization of QED via the covariant formulation developed by Schwinger and Tomonaga and by Feynman (which were shown to be equivalent by Dyson[1]). This finally showed that, at least for the electromagnetic interactions, quantum field theory was the correct formalism. Subsequently, physicists attacked the problem of the strong and weak interactions via quantum field theory, only to face even more formidable problems with renormalization that stalled progress for several decades.

The second revolution was the proof by 't Hooft that spontaneously broken Yang–Mills theory was renormalizable, which led to the successful application of quantum field theory to the weak interactions and opened the door to the gauge revolution.

There have been many renormalization proposals made in the literature, but all of them share the same basic physical features. Although details vary from scheme to scheme, the essential idea is that there is a set of "bare" physical parameters that are divergent, such as the coupling constants and masses. However, these bare parameters are unmeasurable. The divergences of these parameters are chosen so that they cancel against the ultraviolet infinities coming from infinite classes of Feynman diagrams, which probe the small-distance behavior of the theory. After these divergences have been absorbed by the bare parameters, we are left with the physical, renormalized, or "dressed" parameters that are indeed measurable.

Since there are a finite number of such physical parameters, we are only allowed to make a finite number of such redefinitions. Renormalization theory, then, is a set of rules or prescriptions where, after a finite number of redefinitions, we can render the theory finite to any order.

(If, however, an infinite number of redefinitions were required to render all orders finite, then *any* quantum field theory could be "renormalized." We could, say, find a different rule or prescription to cancel the divergences for each of the infinite classes of divergent graphs. Unless there is a well-defined rule that determines how this subtraction is to be carried out to all orders, the theory is not well defined; it is infinitely ambiguous.)

We should stress that, although the broad features of the renormalization program are easy to grasp, the details may be quite complicated. For example, solving the problem of "overlapping divergences" requires detailed graphical and combinatorial analysis and is perhaps the most important complication of renormalization theory. Due to these tedious details, there have been several errors made in the literature concerning renormalization theory. (For example, the original Dyson/Ward proof of the renormalization of QED actually breaks down at the 14th order[2,3] because of overlapping diagrams. The original claims of renormalization were thus incomplete. However, the proof can presumably be patched up.)

Mindful of these obscure complications, which tend to conceal the relatively simple essence of renormalization theory, in this chapter we will first try to approach the problem of renormalization from a schematic point of view, and then present the details later. Instead of presenting the complications first, which may be quite involved, we will discuss the basic components of renormalization theory, which occur in four essential steps:

1. *Power counting*

   By simply counting the powers of p in any Feynman graph, we can, for large p, tell whether the integral diverges by calculating the degree of divergence of that graph: each boson propagator contributes p^{-2}, each fermion propagator contributes p^{-1}, each loop contributes a loop integration with p^4, and each vertex with n derivatives contributes at most n powers of p. If the overall power of p; that is, the degree of divergence D, is 0 or positive, then the graph diverges. By simple power counting arguments, we can then calculate rather quickly whether certain theories are hopelessly nonrenormalizable, or whether they can be potentially renormalized.

2. *Regularization*

   Manipulating divergent integrals is not well defined, so we need to cutoff the integration over $d^4 p$. This formally renders each graph finite, order by order, and allows us to reshuffle the perturbation theory. At the end of the calculation, after we have rearranged the graphs to put all divergent terms into the physical parameters, we let the cutoff tend to infinity. We must also show that the resulting theory is independent of the regularization method.

3. *Counterterms or multiplicative renormalization*

   Given a divergent theory that has been regularized, we can perform formal manipulations on the Feynman graphs to any order. Then there are at least two equivalent ways in which to renormalize the theory:

   First, there is the method of *multiplicative renormalization*, pioneered by Dyson and Ward for QED, where we formally sum over an infinite series of Feynman graphs with a fixed number of external lines. The divergent sum is then absorbed into a redefinition of the coupling constants and masses in the theory. Since the bare masses and bare coupling constants are unmeasurable, we can assume they are divergent and that they cancel against the divergences of corresponding Feynman graphs, and hence the theory has absorbed all divergences at that level.

   Second, there is the method of *counterterms*, pioneered by Bogoliubov, Parasiuk, Hepp, and Zimmerman (BPHZ), where we add new terms directly to the action to subtract off the divergent graphs. The coefficients of these counterterms are chosen so that they precisely kill the divergent graphs. In a renormalizable theory, there are only a finite number of counterterms needed to render the theory finite to any order. Furthermore, these counterterms are proportional to terms in the original action. Adding the original action with the counterterms gives us a renormalization of the masses and coupling constants in the action. These two methods are therefore equivalent; that is, by adding counterterms to the action, they sum up, at the end of the calculation, to give a multiplicative rescaling of the physical parameters appearing in the action.

These methods then give us simple criteria that are necessary (but not sufficient) to prove that a theory is renormalizable:

a. The degree of divergence D of any graph must be a function only of the number of external legs; that is, it must remain constant if we add more internal loops. This allows us to collect all N-point loop graphs into one term. (For super-renormalizable theories, the degree of divergence actually decreases if we add more internal loops).

b. The number of classes of divergent N-point graphs must be finite. These divergences must cancel against the divergences contained within the bare parameters.

4. *Induction*

The last step in the proof of renormalizability is to use an induction argument. We assume the theory is renormalizable at the nth order in perturbation theory. Then we write down a recursion relation that allows us to generate the $n + 1$st-order graphs in terms of the nth-order graphs. By proving the $n + 1$st-order graphs are all finite, we can prove, using either multiplicative or counterterm renormalization, that the entire perturbation theory, order by order, is finite. Since there are various recursion relations satisfied by field theory (e.g., Schwinger–Dyson equations, renormalization group equations, etc.), there are also a variety of renormalization programs. However, all induction proofs ultimately rely on *Weinberg's theorem* (which states that a Feynman graph converges if the degree of divergence of the graph and all its subgraphs is negative).

7.2 Renormalization Types

Based on simple power counting of graphs, we can begin to catalog a wide variety of different field theories on the basis of their renormalizability. We will group quantum field theories into four distinct categories:

1. Nonrenormalizable theories.

2. Renormalizable theories.

3. Super-renormalizable theories.

4. Finite theories.

7.2.1 Nonrenormalizable Theories

To determine the degree of divergence of any graph, we need to know the dimension of the various fields and coupling constants. We can determine the dimension of a field by analyzing the behavior of the propagator at large momenta, or by analyzing the free action. We demand that the action has the dimension of $\hbar$, (i.e., has zero dimension). Since the volume element $d^4 x$ has dimension cm^4, this means that the Lagrangian must have dimension cm^{-4}.

For example, the Klein–Gordon action contains the term $(\partial_\mu \phi)^2$. Since the derivative has dimension cm^{-1}, it means that ϕ also has dimension cm^{-1}. The mass m has dimensions cm^{-1} in our units, so that $m^2 \phi^2$ has the required dimension (cm)$^{-4}$. By the same reasoning, the massless Maxwell field also has dimension (cm)$^{-1}$. The Dirac field, however, has dimension cm$^{-3/2}$, so that the term $\bar{\psi} i \partial \psi$ has dimension cm^{-4}.

It is customary to define the dimension of a field in terms of inverse centimeters (or, equivalently, grams). If $[\phi]$ represents the dimension of the field in inverse centimeters, then the dimensions of the fields in d space–time dimensions can be easily computed by analyzing the free action, which must be dimensionless:

$$
\begin{aligned}
[\phi] &= \frac{d-2}{2} \\
[\psi] &= \frac{d-1}{2}
\end{aligned}
\tag{7.1}
$$

The simplest example of a nonrenormalizable theory is one that has a coupling constant with negative dimension, like ϕ^5 theory in four dimensions. To keep the action dimensionless, the coupling constant g must have dimension -1. Now let us analyze the behavior of an N-point function. If we insert a $g\phi^5$ vertex into the N-point function, this increases the number of g's by one, decreasing the dimension of the graph. This must be compensated by an increase in the overall power of k by 1, which increases the dimension of the graph, such that the total dimension of the graph remains the same. Inserting a vertex into the N-point graph has thus made it more divergent by one factor of k. By inserting an arbitrary number of vertices into the N point function, we can arbitrarily increase the overall power of k and hence make the graph arbitrarily divergent. The same remarks apply for ϕ^n for $n > 4$ in four dimensions.

(Since the presence or absence of dimensional coupling constants depends so crucially on the dimension of space–time, we will find that in different space–times the set of renormalizable and nonrenormalizable theories are quite different.)

Some examples of non-renormalizable theories include:

1. *Nonpolynomial actions*

   These actions have an infinite number of terms in them, and typically look like $\sum_{n=3}^{\infty} \phi^n$. They necessarily have coupling constants with negative dimension, and hence are not renormalizable.

2. *Gravity*

   Quantum gravity has a coupling constant κ with negative dimension. ($\kappa^2 \sim G_N$, where G_N is Newton's constant, which has dimension -2.) This means that we cannot perform the standard renormalization program. Also, a power expansion in the coupling constant yields a nonpolynomial theory. Thus, quantum gravity is not renormalizable and is infinitely ambiguous.

3. *Supergravity*

   By the same arguments, supergravity is also nonrenormalizable. Even though it possesses highly nontrivial Ward identities that kill large classes of divergences, the gauge group is not large enough to kill all the divergences.

4. *Four-fermion interactions*

   These actions, like the original Fermi action or the Nambu–Jona–Lasinio action, contain terms like $(\bar{\psi}\psi)^2$. By power counting, we know that ψ has dimensions $\mathrm{cm}^{-3/2}$, so the four-fermion action has dimension 6. This requires a coupling constant with dimension -2, so the theory is nonrenormalizable.

5. $\bar{\psi}\sigma^{\mu\nu}\psi F_{\mu\nu}$

   This coupling, which seems to be perfectly well defined and gauge invariant, is not renormalizable because it has dimension 5, so its coupling constant has dimension -1.

6. *Massive vector theory with non-Abelian group*

   A propagator like:

$$\delta_{ab} \frac{g_{\mu\nu} - \frac{k_\mu k_\nu}{M^2}}{k^2 - M^2 + i\epsilon} \tag{7.2}$$

goes like $O(1)$ for large k, and hence does not damp fast enough to give us a renormalizable theory. So the theory fails to be renormalizable by power counting arguments.

7. *Theories with anomalies*
   Ward–Takahashi identities are required to prove the renormalizability of gauge theories. However, sometimes a classical symmetry of an action does not survive the rigorous process of quantization, and the symmetry is broken. We say that there are anomalies in the theory that destroy renormalizability. Anomalies will be studied in greater length in Chapter 12.

7.2.2 Renormalizable Theories

The renormalizable theories only form a tiny subset of possible quantum field theories. They have only a finite number of counterterms. They also have no dimensional coupling constants; so the dimension of each term in the Lagrangian is cm^{-4}.

Some well-known renormalizable theories include:

1. ϕ^4
   This is the simplest renormalizable theory one can write in four dimensions. Because ϕ has dimension 1, this interaction has dimension 4, and hence the coupling constant is dimensionless. This theory is a prototype of much more complicated actions. (However, it should be pointed out that this theory, when summed to all orders, is probably a free theory.)

2. *Yukawa theory*
   The Yukawa theory has a coupling between fermions and scalars given by:

$$\mathscr{L} = g\bar{\psi}\tau^a\psi\phi^a \tag{7.3}$$

where τ^a is the generator of some Lie group and g is dimensionless.

3. *Massive vector Abelian theory*
   Although this theory has a propagator similar to the massive vector non-Abelian gauge theory, the troublesome $k_\mu k_\nu$ term drops out of any Feynman graph by $U(1)$ gauge invariance. (This term cannot be dropped in a gauge theory with a non-Abelian group.)

4. *QED*
   There are no dimensional coupling constants, and, by power counting arguments, we need only a finite number of counterterms.

5. *Massless non-Abelian gauge theory*
   By power counting, this theory is renormalizable. A more detailed proof of the renormalizability of this theory will be shown in Chapter 13.

6. *Spontaneously broken non-Abelian gauge theory*

Although massive non-Abelian gauge theories are, in general, nonrenormalizable, there is one important exception. If the gauge symmetry of a massless non-Abelian theory is spontaneously broken, then renormalizability is not destroyed. The fact that renormalizability persists even after the gauge group is spontaneously broken helped to spark the gauge revolution.

Since the renormalizability of a theory is dependent on the dimension of space–time, we also have the following renormalizable theories:

7. $\bar{\psi}\psi\phi^2$

This is renormalizable in three dimensions.

8. ϕ^3

This is renormalizable in six dimensions.

9. ϕ^5, ϕ^6

These are renormalizable in three dimensions.

10. $(\bar{\psi}\psi)^2$

This is renormalizable in two dimensions. Although this interaction is non-renormalizable in four dimensions, in two dimensions it requires no dimensional coupling constant.

Finally, we can write down the complete set of interactions for spin 0, $\frac{1}{2}$, and 1 fields that are potentially renormalizable. In four dimensions, they are given symbolically by (if we omit isospin and Lorentz indices):

$$\phi^4, \quad \bar{\psi}\psi\phi, \quad (A^2)^2, \quad \bar{\psi}\slashed{A}\psi, \quad \phi^\dagger\partial_\mu\phi A^\mu, \quad \phi^\dagger\phi A^2 \qquad (7.4)$$

(The Yang–Mills theory has an additional ∂A^3 interaction, which we will discuss separately.)

7.2.3 Super-renormalizable Theories

Super-renormalizable theories converge so rapidly that there are only a finite number of graphs that diverge in the entire perturbation theory. The degree of divergence actually goes down as we add more internal loops. The simplest super-renormalizable theories have coupling constants with positive dimension, such as ϕ^3 in four dimensions. Repeating the argument used earlier, this means that increasing the order g of an N-point function must necessarily decrease the number of momenta k appearing in the integrand, such that the overall dimension of the graph remains the same. Thus, as the order of the graph increases, sooner or later the graph becomes convergent. Thus, there are only a finite number of divergent graphs in the theory.

Some examples of super-renormalizable theories include:

1. ϕ^3

   This is super-renormalizable in four dimensions (but the theory is sick because it is not bounded from below and the vacuum is not stable).

2. ϕ^4

   This is super-renormalizable in three dimensions because it has three superficially divergent graphs, which contribute to the two-point function.

3. $P(\phi)$

   ϕ has zero dimension in two dimensions; so we can have an arbitrary polynomial in the action yet still maintain renormalizability. The interaction $P(\phi)$ produces only a finite number of divergences, all of them due to self-contractions of the lines within the various vertices.

4. $P(\phi)\bar{\psi}\psi$

   This is also super-renormalizable in two dimensions.

7.2.4 Finite Theories

Although Dirac was one of the creators of quantum field theory, he was dissatisfied with the renormalization approach, considering it artificial and contrived. Dirac, in his later years, sought a theory in which renormalization was not necessary at all. Dirac's verdict about renormalization theory was, "This is just not sensible mathematics. Sensible mathematics involves neglecting a quantity when it turns out to be small—not neglecting it because it is infinitely great and you do not want it!"

Instead, Dirac believed that a new theory was needed in which renormalizations were inherently unnecessary. Until recently, it was thought that Dirac's program was a dead end, that renormalization was inherent within any quantum field theory. However, because of the introduction of supersymmetry, we have two possible types of theories that are finite to any order in perturbation theory:

1. *Super Yang–Mills theory*

   Supersymmetry gives us new constraints among the renormalization constants Z that are not found in ordinary quantum theories. In fact, for the $SO(4)$ super Yang–Mills theory, one can show that these constraints are enough to guarantee that $Z = 1$ for all renormalization constants to all orders in perturbation theory. In addition, the $SO(2)$ super Yang–Mills theory, coupled to certain classes of supersymmetric matter, is also finite to all orders in perturbation theory. Although these super Yang–Mills theories are uninteresting from the point of view of phenomenology, the fact that supersymmetry is powerful enough to render certain classes of quantum field theories finite to all orders is reason enough to study them.

2. *Superstrings*

Supersymmetry also allows us to construct actions much more powerful than the super Yang-Mills theory, such as the superstring theory. Superstring theory has two important properties. First, it is finite to all orders in perturbation theory and is free of all anomalies. Second, it contains quantum gravity, as well as all known forces found in nature, as a subset. The fact that superstring theory is the only candidate for a finite theory of quantum gravity is remarkable. (Because there is no experimental evidence at all to support the existence of supersymmetry, we will discuss these supersymmetric theories later in Chapters 20 and 21.)

7.3   Overview of Renormalization in ϕ^4 Theory

Since renormalization theory is rather intricate, we will begin our discussion by giving a broad overview of the renormalization program for the simplest quantum field theory in four dimensions, the ϕ^4 theory, and then for QED. We will stress only the highlights of how renormalization is carried out. After sketching the overall renormalization program for these two theories, we will then present the details, such as the regularization program and the induction argument.

Our goal is to present the arguments that show that (1) the degree of divergence D of any Feynman graph in ϕ^4 theory is dependent only on the number of external lines, and that (2) these divergent classes can be absorbed into the physical parameters.

Given any graph for ϕ^4, we can analyze its divergent structure by power counting as follows:

$$
\begin{aligned}
E &= \text{number of external legs} \\
I &= \text{number of internal lines} \\
V &= \text{number of vertices} \\
L &= \text{number of loops} \qquad\qquad (7.5)
\end{aligned}
$$

The degree of divergence of an arbitrary Feynman graph is easily computed. Each internal propagator contributes $1/p^2$, while each loop contributes d^4p, so the degree of divergence is given by:

$$
D = 4L - 2I \qquad\qquad (7.6)
$$

Now we use some simple counting arguments about graphs to reduce this expression. Each vertex has four lines connecting to it. Each of these lines, in turn, either ends on an external leg, or on one end of an internal leg, which has two ends. Thus, we must have:

$$4V = 2I + E \tag{7.7}$$

Also, the loop number L can be calculated by analyzing the independent momenta in any Feynman graph. The number of independent momenta is equal to the number of internal lines I minus the constraints coming from momentum conservation. There are V such momentum constraints, minus the overall momentum conservation from the entire graph. Since the number of independent momenta in a Feynman graph is also equal to the number of loop momenta, we have, therefore:

$$L = I - V + 1 \tag{7.8}$$

Inserting these graphical rules into our expression for the divergence of a Feynman graph, we now have:

$$D = 4 - E \tag{7.9}$$

This means that the degree of divergence of any graph in four dimensions is strictly dependent on the number of external lines, which is a necessary condition for renormalizability. The degree of divergence is hence independent of the number of internal loops in the graph.

This also a gratifying result because it means that only the two-point and the four-point graphs are divergent. This, in turn, gives us the renormalization of the two physical quantities in the theory: the mass and the coupling constant. Thus, by using only power counting arguments, in principle we can renormalize the entire theory with only two redefinitions corresponding to two physical parameters. [In d dimensions, however, the degree of divergence is $D = d + (1 - d/2)E + (d - 4)V$. Because D increases with the number of internal vertices, there are problems with renormalizing the theory in higher dimensions.]

Next, we want to sketch the two methods by which renormalization is carried out: multiplicative renormalization and counterterms. For ϕ^4 theory, we first begin with an elementary discussion of multiplicative renormalization.

We start with an action defined with "unrenormalized" or "bare" coupling constants and masses:

$$\mathcal{L}_0 = \frac{1}{2} \partial_\mu \phi_0 \partial^\mu \phi_0 - \frac{1}{2} m_0^2 \phi_0^2 - \frac{\lambda_0}{4!} \phi_0^4 \tag{7.10}$$

(We remind the reader that m_0 and λ_0 are formally infinite, but are not measurable.) We define $\Sigma(p^2)$ to be the sum of all *proper* (or one-particle irreducible) two-point graphs. (A graph is called improper or one-particle reducible if we can, by cutting

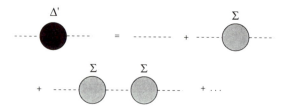

Figure 7.1. The complete propagator Δ' is the sum of an infinite chain of one-particle irreducible graphs Σ. Δ' itself is improper.

just one line, break the graph into two distinct parts. A proper diagram cannot be split into two parts by cutting one line.) For example, the first-loop contribution to $\Sigma(p^2)$ is given by the following:

$$-i\Sigma(p^2) = -i\frac{\lambda_0}{2}\int\frac{d^4p}{(2\pi)^4}\frac{1}{p^2 - m_0^2 + i\epsilon} + \cdots \tag{7.11}$$

which is obtained by taking a four-point vertex and joining two legs together into a loop. It is a proper graph because if we cut the loop, it becomes a four-point vertex and hence does not split apart into two distinct pieces.

Now let $\Delta'(p)$ represent the sum over all possible two-point graphs. We call it the complete or full propagator. $\Delta'(p)$ is obviously the sum of two parts, a proper and improper part. By definition, the proper part cannot be split any further by cutting an intermediate line, but the improper part can. If we split the improper part in half, then each piece, in turn, is the sum of a proper and improper part. Then the smaller improper part can be further split into a proper and improper part. By successively cutting all improper pieces into smaller pieces, we can eliminate all improper parts and write the complete propagator entirely in terms of proper parts. This successive cutting process obviously creates a sequence of proper parts strung together along a string, as in Figure 7.1. $\Delta'(p)$ itself is improper, since it can be split into two pieces by cutting one line.

We can iterate the complete self-energy graph $\Delta'(p^2)$ an infinite number of times with respect to Σ, so that we can formally sum the series, using the fact that:

$$\sum_{n=0}^{\infty}x^n = \frac{1}{1 - x} \tag{7.12}$$

The sum equals:

$$i\Delta'(p) = i\Delta_F(p) + i\Delta_F(p)[-i\Sigma(p^2)]i\Delta_F(p)\cdots$$

$$= i\Delta_F(p)\left(\frac{1}{1 + i\Sigma(p^2)i\Delta_F(p)}\right)$$

$$= \frac{i}{p^2 - m_0^2 - \Sigma(p^2) + i\epsilon} \tag{7.13}$$

Now power expand this self-energy correction $\Sigma(p^2)$ around $p^2 = m^2$, where m is, at this point, finite but arbitrary:

$$\Sigma(p^2) = \Sigma(m^2) + (p^2 - m^2)\Sigma'(m^2) + \tilde{\Sigma}(p^2) \tag{7.14}$$

where $\tilde{\Sigma}(p^2) \sim O(p^2 - m^2)^2$, and where $\Sigma(m^2)$ and $\Sigma'(m^2)$ are divergent. The net effect of summing this infinite series of graphs is that the complete propagator is now modified to:

$$
\begin{aligned}
i\Delta'(p) &= \frac{i}{p^2 - m_0^2 - \Sigma(m^2) - (p^2 - m^2)\Sigma'(m^2) - \tilde{\Sigma}(p^2) + i\epsilon} \\
&= \frac{i}{p^2 - [m_0^2 + \Sigma(m^2)] - (p^2 - m^2)\Sigma'(m^2) - \tilde{\Sigma}(p^2) + i\epsilon} \\
&= \frac{i}{[1 - \Sigma'(m^2)](p^2 - m^2) - \tilde{\Sigma}(p^2) + i\epsilon} \tag{7.15}
\end{aligned}
$$

At this point, m_0 is infinite but arbitrary. Since $\Sigma(m^2)$ is also divergent, we will define m_0 and m such that m_0 cancels against the divergent part coming from $\Sigma(m^2)$, giving us the finite piece m. We will choose:

$$m_0^2 + \Sigma(m^2) = m^2 \tag{7.16}$$

[There is a certain arbitrariness in how two infinite terms cancel, because an infinite term plus a finite term is still infinite. $\Sigma(m^2)$ may have a finite piece in addition to an infinite piece, so the value of m^2 is, at this point, arbitrary. We will comment on this important ambiguity later.]

With this choice, we have:

$$i\Delta'(p) = \frac{iZ_\phi}{p^2 - m^2 - \tilde{\tilde{\Sigma}}(p^2) + i\epsilon} \tag{7.17}$$

where we have made the following redefinitions:

$$
\begin{aligned}
m^2 &= m_0^2 + \Sigma(m^2) = m_0^2 + \delta m^2 \\
Z_\phi &= \frac{1}{1 - \Sigma'(m^2)} \\
\tilde{\tilde{\Sigma}}(p^2) &= \tilde{\Sigma}(p^2)[1 - \tilde{\Sigma}(m^2)]^{-1} = Z_\phi \tilde{\Sigma}(p^2) \tag{7.18}
\end{aligned}
$$

Since the divergence of the propagator is contained within Z_ϕ, we can extract this divergent piece. From the unrenormalized propagator Δ', we will remove this divergence and hence obtain the *renormalized propagator* $\tilde{\Delta}'(p)$:

$$\Delta'(p) = Z_\phi \tilde{\Delta}'(p) \tag{7.19}$$

This simple example demonstrates some of the main features of renormalization theory. First, the pole structure of the bare propagator has changed by summing up all graphs. Although we started with a bare propagator $\Delta_F(p)$ with a simple pole at the (infinite) bare mass m_0, the effect of summing all possible graphs is to shift the bare mass to the renormalized or "dressed" mass m. The p dependence of the complete propagator Δ' has changed in a nontrivial fashion. It no longer consists of just a simple pole. However, the complete propagator Δ' still has a pole at the shifted mass squared $p^2 = m^2$ because $\tilde{\Sigma}(m^2) = 0$,

Second, the divergent unrenormalized propagator $\Delta'(p)$ has been converted to the convergent, renormalized propagator $\tilde{\Delta}'(p)$ by a multiplicative rescaling by Z_ϕ. This is crucial in our discussion of renormalization, because it means that we can extract the divergence of self-energy graphs by a simple rescaling (which will eventually be absorbed by redefining the physical coupling constants and wave functions of the theory).

We could also have written this in the language of proper vertices. Let $\Gamma^{(n)}$ represent the one-particle irreducible n-point vertex function, with the n propagators on the external lines removed. (The proper vertex will be defined more rigorously in Chapter 8.) In the free theory, the proper vertex for the two-point function is defined as the inverse of the propagator:

$$i\Gamma_0^{(2)}(p) = p^2 - m_0^2 \tag{7.20}$$

Once we sum to all orders in perturbation theory, then the proper vertex function becomes infinite. If we divide out by this infinite factor Z_ϕ, then we can write the renormalized vertex as:

$$i\Gamma^{(2)}(0) = -m^2 \tag{7.21}$$

where we have taken $p = 0$.

Now consider the effect of renormalization on the coupling constant λ_0. Let $\Gamma^{(4)}$ represent the four point proper vertex, summed over all possible graphs, with propagators on the external lines removed. To lowest order, this four-point graph equals:

$$i\Gamma_0^{(4)}(p_i) = \lambda_0 \tag{7.22}$$

The one-loop correction to this is given by:

$$i\Gamma_0^{(4)} = \lambda_0 - \frac{i}{2}\lambda_0^2 \int \frac{d^4 p}{[(p-q)^2 - m_0^2 + i\epsilon][p^2 - m_0^2 + i\epsilon]} \tag{7.23}$$

Although we cannot evaluate $\Gamma^{(4)}$ to all orders in perturbation theory, we know that it is Lorentz invariant and hence can be written in terms of the three Mandelstam variables:

$$s = (p_1 + p_2)^2; \quad t = (p_1 + p_3)^2; \quad u = (p_1 + p_4)^2 \tag{7.24}$$

Therefore:

$$i\Gamma_0^{(4)}(p_i) = \lambda_0 + f(s) + f(t) + f(u) \tag{7.25}$$

for some divergent function f. For the value $p = 0$, let Z_λ^{-1} be this overall infinite factor contained within f. If we divide out the infinite factor from the vertex function, then we have:

$$i\Gamma^{(4)}(0) = \lambda \tag{7.26}$$

where λ is the physical, renormalized coupling constant. In fact, we can take Eqs. (7.21) and (7.26) to be the *definition* of the mass and coupling constant, measured at the point $p = 0$. Because of the ambiguity in manipulating finite and infinite quantities, this definition is not unique. We could have defined the physical mass and coupling constant at some arbitrary momentum scale μ as well as $p = 0$. In other words, these quantities are actually functions of μ. (Of course, when we perform an experiment and actually measure the physical masses and coupling constants, we do so at a fixed momentum, so there is no ambiguity. However, if these experiments could be performed at different momenta, then the effective physical constants may change.)

For example, we could have defined the vertex function at the point $p^2 = m^2$ and $s = u = t = 4m^2/3$. In general, we can define the masses and coupling constants at a different momentum $p = \mu$ via:

$$i\Gamma^{(2)}(\mu) = p^2 - m^2(\mu)$$
$$i\Gamma^{(4)}(\mu) = \lambda(\mu) \tag{7.27}$$

This point μ is called the *renormalization point* or the *subtraction point*. The ambiguity introduced by this subtraction point μ will appear repeatedly throughout this book, and will be studied in more detail in Chapter 14.

Now that we have isolated the divergent multiplicative quantities, we can easily pull them out of any divergent Feynman diagram by redefining the coupling constants and masses. To see how this is done, let us split the Z_ϕ factor occurring with every renormalized propagator in a Feynman diagram into two pieces $(\sqrt{Z_\phi})^2$. In this diagram, move each factor of $\sqrt{Z_\phi}$ into the nearest vertex function. Since each vertex function has four legs, it means that the renormalized vertex function will receive the contribution of four of these factors, or $\sqrt{Z_\phi}^4 = Z_\phi^2$. Since the renormalization of the vertex function contributes an additional factor of Z_λ^{-1}, then the λ_0 sitting in front of the vertex function picks up a factor of Z_ϕ^2/Z_λ. But this means that the original bare coupling constant λ_0 is now modified by this multiplicative renormalization as follows:

$$\lambda_0 \rightarrow \lambda = \lambda_0 Z_\lambda^{-1} Z_\phi^2 \tag{7.28}$$

which we define to be renormalized coupling constant.

In this way, we can move all factors of Z_ϕ into the various vertex functions, renormalizing the coupling constant, except for the propagators that are connected to the external legs. We have a left over factor of $\sqrt{Z_\phi}$ for each external leg. This last factor can be eliminated by wave function renormalization. (As we saw in Chapter 5, it was necessary to include a wave function renormalization factor $Z^{-1/2}$ in the definition of "in" and "out" states.) Since the wave function is not a measurable quantity, we can always eliminate the last factors of Z_ϕ by renormalizing the wave function.

In summary, we first began with the ϕ^4 theory defined totally in terms of the unrenormalized, bare coupling constants and masses, m_0^2 and λ_0, which are formally infinite. Then, by summing over infinite classes of diagrams, we found that the modified propagator had a pole at the shifted renormalized mass: $p^2 = m^2$. Furthermore, the propagator and vertex function were multiplied in front by infinite renormalization constants Z_ϕ and Z_λ^{-1}, respectively. By splitting up the Z_ϕ at each propagator into two pieces, we could redistribute these constants such that the new coupling constant became $\lambda_0 Z_\phi^2/Z_\lambda$. The reshuffling of renormalization constants can be summarized as follows:

$$
\begin{aligned}
\phi &= Z_\phi^{-1/2} \phi_0 \\
\lambda &= Z_\lambda^{-1} Z_\phi^2 \lambda_0 \\
m^2 &= m_0^2 + \delta m^2
\end{aligned}
\tag{7.29}
$$

where $\delta m^2 = \Sigma(m^2)$.

This can also be generalized to the n-point function as well. If we take an unrenormalized vertex $\Gamma_0^{(n)}(p_i, \lambda_0, m_0)$, we can begin summing the graphs within the unrenormalized vertex to convert it to the renormalized one. In doing so, we pick up extra multiplicative factors of Z_ϕ and Z_λ, which allow us to renormalize the coupling constant and shift the bare mass. There are n external legs within $\Gamma_0^{(n)}$ that have no propagators, and hence do not contribute their share of the $Z_\phi^{1/2}$ factors. This leaves us with an overall factor of $Z_\phi^{-n/2}$. Then we are left with:

$$\Gamma_0^{(n)}(p_i, \lambda_0, m_0) = Z_\phi^{-n/2} \Gamma^{(n)}(p_i, \lambda, m, \mu) \qquad (7.30)$$

Although renormalization of ϕ^4 seems straightforward, unfortunately, there are two technical questions that remain unanswered.

First, as we mentioned earlier, a more rigorous proof must grapple with the problem of overlapping divergences, which is the primary source of complication in any renormalization program. Salam[4] was perhaps the first to fully appreciate the importance of these overlapping divergences. Because of these divergences, the final steps needed to renormalize ϕ^4 are, in some sense, more difficult than the renormalization of QED. As a result, the problem of overlapping divergences for ϕ^4 theory will be discussed at length later in this chapter. We will then solve the problem of overlapping divergences in Chapter 13, when we study the BPHZ renormalization method. We will then complete the renormalizability of ϕ^4 theory in Chapter 14, where we develop the theory of the renormalization group.

Besides the multiplicative renormalization method, there is yet another way to perform renormalization, and this is to proceed *backwards*, that is, start with the usual ϕ^4 action defined with the physical coupling constants and masses, which, of course, are finite. Then, as we calculate Feynman diagrams to each order, we find the usual divergences. The key point is that we can cancel these divergences by adding counterterms to the original action (which are proportional to terms in the original action). This second method is called the counterterm method.

These counterterms contribute new terms to the Feynman series that cancel the original divergences to that order. At the next order, we then find new divergences, so we add new counterterms (again, proportional to terms in the original action), which cancel the divergences to that order. The final action is then the original renormalized action plus an infinite sequence of counterterms, to all orders. Because all the counterterms are proportional to terms in the original action, we wind up with the unrenormalized action defined in terms of unrenormalized parameters (which was the starting point of the previous procedure).

To see the close link between the multiplicative renormalization approach and the counterterm approach, let us start with the renormalized action:

$$\mathscr{L} = \frac{1}{2}[(\partial_\mu \phi)^2 - m^2 \phi^2] - \frac{\lambda}{4!} \phi^4 \qquad (7.31)$$

where λ and m are the renormalized, finite quantities. We then find divergences with this action, so we add counterterms to the action, such that these counterterms cancel the divergences that appear to that order. Since the infinities that arise are similar to the infinities we encountered with the multiplicative renormalization program, we find that the coefficients of the counterterms can be summed to yield the quantities Z_ϕ, Z_λ, and δm^2 found earlier.

To see this, let us add the counterterm $\Delta\mathscr{L}$ to the Lagrangian:

$$\mathscr{L} \rightarrow \mathscr{L} + \Delta\mathscr{L} \tag{7.32}$$

where the counterterm $\mathscr{L}$ must cancel the divergences coming from the two-point and four-point graphs, which are contained within $\Sigma(p^2)$ and $i\Gamma^{(4)}(p^2)$. Using the subtraction point $\mu = 0$, we find that the counterterm that cancels these divergences can be written as (to lowest loop order):

$$\Delta\mathscr{L} = \frac{\Sigma(0)}{2}\phi^2 + \frac{\Sigma'(0)}{2}(\partial_\mu\phi)^2 + \frac{i\Gamma^{(4)}(0)}{4!}\phi^4 \tag{7.33}$$

Now make the definitions:

$$
\begin{aligned}
\Sigma'(0) &= Z_\phi - 1 \\
\Sigma(0) &= -(Z_\phi - 1)m^2 + \delta m^2 \\
\Gamma^{(4)}(0) &= -i\lambda(1 - Z_\lambda)
\end{aligned}
\tag{7.34}
$$

which are equivalent (to lowest loop order) to the definitions in Eq. (7.18).

With these definitions, we now have:

$$\Delta\mathscr{L} = \frac{1}{2}[Z_\phi - 1][(\partial_\mu\phi)^2 - m^2\phi^2] + \frac{\delta m^2}{2}Z_\phi\phi^2 - \frac{\lambda(Z_\lambda - 1)}{4!}\phi^4 \tag{7.35}$$

If we add $\mathscr{L}$ and $\Delta\mathscr{L}$ together, we find that we retrieve the original unrenormalized action $\mathscr{L}_0$:

$$\mathscr{L}_0 = \mathscr{L} + \Delta\mathscr{L} \tag{7.36}$$

This intuitively shows the equivalence of the multiplicative renormalization and the counterterm methods. Thus, it is a matter of taste which method we use. In practice, however, this second method of adding counterterms is perhaps more widely used. The point is, however, that both the multiplicative approach and the counterterm approach are equivalent. The fact that the counterterms were proportional to terms in the original action made this equivalence possible.

7.4 Overview of Renormalization in QED

Let us now begin an overview of the renormalization of QED. As in any quantum field theory, the first step involves power counting. Once again, our goal is to show that the degree of divergence of any graph is independent of the number of internal loops.

Let us count the superficial degree of divergence of each graph in QED. We define:

$$
\begin{aligned}
L &= \text{number of loops} \\
V &= \text{number of vertices} \\
E_\psi &= \text{number of external electron legs} \\
I_\psi &= \text{number of internal electron legs} \\
E_A &= \text{number of external photon legs} \\
I_A &= \text{number of internal photon legs} \qquad (7.37)
\end{aligned}
$$

Then the superficial degree of divergence is:

$$D = 4L - 2I_A - I_\psi \qquad (7.38)$$

We can rewrite this equation so that it is only a function of the external legs of the graph, no matter how many internal legs or loops it may have. Each vertex, for example, connects to one end of an internal electron leg. For external electron legs, only one end connects onto a vertex. Thus:

$$V = I_\psi + \frac{1}{2}E_\psi \qquad (7.39)$$

Likewise, each vertex connects to one end of an internal photon line, unless it is external. Thus:

$$V = 2I_A + E_A \qquad (7.40)$$

Also, we know that the total number of independent momenta is equal to L, which in turn equals the total number of internal lines in the graph minus the number of vertices (since we have momentum conservation at each vertex) plus one (since we also have overall momentum conservation). Thus:

$$L = I_\psi + I_A - V + 1 \qquad (7.41)$$

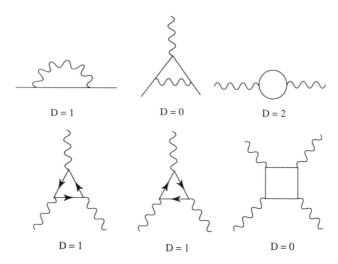

$D = 1$ $D = 0$ $D = 2$

$D = 1$ $D = 1$ $D = 0$

Figure 7.2. The list of divergent classes of graphs in QED. Only the first three graphs, however, are truly divergent if we use gauge invariance and Furry's theorem.

Putting everything together, we find the final formula:

$$D = 4 - \frac{3E_\psi}{2} - E_A \tag{7.42}$$

This is very fortunate, because once again it shows that the total divergence of any Feynman graph, no matter how complicated, is only dependent on the number of external electron and photon legs. In fact, we find that the only graphs that diverge are the following (Fig. 7.2):

1. Electron self-energy graph ($D = 1$).

2. Electron–photon vertex graph ($D = 0$).

3. Photon vacuum polarization graph ($D = 2$) (by gauge invariance, one can reduce the divergence of this graph by two).

4. Three-photon graph ($D = 1$)—these graphs cancel because the internal electron line travels in both clockwise and counterclockwise direction. (If the internal fermion line is reversed, then this can be viewed as reversing the charge of the electron at the vertex. Since there are an odd number of vertices, the overall sign of the graph flips.) Then the two graphs cancel against each other by Furry's theorem.

5. Four-photon graph ($D = 0$)—this graph is actually convergent if we use gauge invariance.

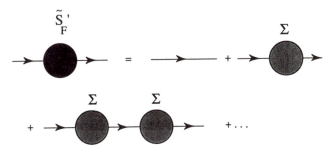

Figure 7.3. The complete propagator S'_F is the sum over one-particle irreducible graphs Σ arranged along a chain.

Only the first three classes of graphs are actually divergent. Fortunately, this is also the set of divergent graphs that can be absorbed into the redefinition of physical parameters, the coupling constant, and the electron mass. (The photon mass, we shall see, is not renormalized because of gauge invariance.) The one-loop graphs that we want particularly to analyze are therefore the electron self-energy correction $\Sigma(\not p)$, the photon propagator $D_{\mu\nu}$, and the vertex correction Γ_μ:

$$-i\Sigma(\not p) \;=\; (-ie)^2 \int \frac{d^4k}{(2\pi)^4}\gamma^\mu\frac{i}{\not p-\not k-m}\frac{-ig_{\mu\nu}}{k^2}\gamma^\nu \tag{7.43}$$

$$i\Pi^{\mu\nu}(k) \;=\; -(-ie)^2 \int \frac{d^4p}{(2\pi)^4}\mathrm{Tr}\left(\gamma^\mu\frac{i}{\not p-m}\gamma^\nu\frac{i}{\not p-\not k-m}\right)$$

$$iD_{\mu\nu}(k) \;=\; -i\frac{g_{\mu\nu}}{k^2}+\left(\frac{-ig_{\mu\alpha}}{k^2}\right)i\Pi^{\alpha\beta}(k)\left(\frac{-ig_{\beta\nu}}{k^2}\right)$$

$$-ie\Lambda_\mu(p,q) \;=\; (-ie)^3 \int \frac{d^4k}{(2\pi)^4}\frac{-ig_{\rho\sigma}}{(k+p)^2}\gamma^\rho\frac{i}{\not k-\not q-m}\gamma_\mu\frac{i}{\not k-m}\gamma^\sigma \tag{7.44}$$

To begin renormalization theory, we will, as in ϕ^4 theory, sum all possible graphs appearing in the complete propagator (Fig. 7.3):

$$iS'_F(\not p) \;=\; iS_F(\not p)+iS_F(\not p)[-i\Sigma(\not p)]iS_F(\not p)+\cdots$$

$$=\; \frac{i}{\not p-m_0-\Sigma(\not p)+i\epsilon} \tag{7.45}$$

Then we make a Taylor expansion of the mass correction $\Sigma(\not{p})$ around $\not{p} = m$, where m is the finite renormalized mass, which is arbitrary:

$$\Sigma(\not{p}) = \Sigma(m) + (\not{p} - m)\Sigma'(m) + \tilde{\Sigma}(\not{p}) \tag{7.46}$$

where $\tilde{\Sigma}(\not{p}) \sim O(\not{p} - m)^2$ and vanishes for $\not{p} = m$. Since m_0 is divergent and arbitrary, we will choose m_0 and m so that m_0 cancels the divergence coming from $\Sigma(m)$. We will choose:

$$m_0 + \Sigma(m) = m \tag{7.47}$$

(Again, there is a certain ambiguity in how this cancellation takes place.)

Inserting the value of $\Sigma(\not{p})$ into the renormalized electron propagator, we now can rearrange terms, just as in the ϕ^4 case, to find:

$$
\begin{aligned}
i S_F'(\not{p}) &= \frac{i}{\not{p} - m_0 - \Sigma(m) - (\not{p} - m)\Sigma'(m) - \tilde{\Sigma}(\not{p})} \\
&= \frac{i}{[1 - \Sigma'(m)](\not{p} - m) - \tilde{\Sigma}(\not{p}) + i\epsilon} \\
&= \frac{i Z_2}{\not{p} - m - \tilde{\tilde{\Sigma}}(\not{p}) + i\epsilon}
\end{aligned}
\tag{7.48}
$$

where we have defined:

$$
\begin{aligned}
m &= m_0 + \Sigma(m) \\
Z_2 &= \frac{1}{1 - \Sigma'(m)} \\
\tilde{\tilde{\Sigma}}(\not{p}) &= \tilde{\Sigma}(\not{p})[1 - \Sigma'(m)]^{-1} = Z_2 \tilde{\Sigma}(\not{p})
\end{aligned}
\tag{7.49}
$$

Since the divergence of the complete propagator S_F' is contained within Z_2, we can remove this term and define the renormalized propagator $\tilde{S}_F'$:

$$S_F'(\not{p}) = Z_2 \tilde{S}_F'(\not{p}) \tag{7.50}$$

(Throughout this chapter, we will label the divergent, complete propagators like S_F' with the prime, and the finite, renormalized propagators like $\tilde{S}_F'$ with a tilde and a prime.) As before, by summing this subclass of diagrams for the electron self-energy, we have been able to redefine the mass of the electron, such that the physical mass m is actually finite, and also show that the electron propagator is multiplicatively renormalized by the factor Z_2:

$$\psi = \sqrt{Z_2}\psi_0 \tag{7.51}$$

Next, we analyze the photon propagator $D_{\mu\nu}$ in the same way, summing over infinite classes of one-particle irreducible graphs:

$$D'_{\mu\nu} = D_{\mu\nu} - D_{\mu\alpha}\Pi^{\alpha\beta}D_{\beta\nu} + \cdots \tag{7.52}$$

However, a naive application of the previous methods would generate a renormalized mass for the photon, which would be disastrous, as we insist that the physical photon be massless. This is the first example of how gauge invariance helps preserve certain properties of the theory to all orders in perturbation theory. The photon propagator must be gauge invariant, meaning that $k_\mu \Pi^{\mu\nu} = 0$. This constraint, in turn, allows us to make the following decomposition of the second-rank tensor into a scalar quantity $\Pi(k^2)$:

$$\Pi^{\mu\nu}(k^2) = (k^\mu k^\nu - g^{\mu\nu}k^2)\Pi(k^2) \tag{7.53}$$

As before, we now make the infinite summation of corrections to the photon propagator:

$$D'_{\mu\nu} = \frac{1}{k^2[1 + \Pi(k^2)]}\left(-g_{\mu\nu} - \frac{k_\mu k_\nu}{k^2}\Pi(k^2)\right) \tag{7.54}$$

We then power expand $\Pi(k^2)$ around $k^2 = 0$:

$$\Pi(k^2) = \Pi(0) + \tilde{\Pi}(k^2) \tag{7.55}$$

Inserting this back into the propagator, we find:

$$D'_{\mu\nu} = -\frac{g_{\mu\nu}Z_3}{k^2[1 + \tilde{\Pi}(k^2)]} + \text{gauge terms} \tag{7.56}$$

where the last term is proportional to $k_\mu k_\nu$ and hence will be dropped. We have also defined:

$$Z_3 = \frac{1}{1 + \Pi(0)}$$

$$\tilde{\Pi}(k^2) = \tilde{\Pi}(k^2)[1 + \Pi(0)]^{-1} = Z_3\tilde{\Pi}(k^2) \tag{7.57}$$

Then the finite, renormalized propagator $\tilde{D}'$ can be defined by extracting out Z_3:

$$D'_{\mu\nu} = Z_3\tilde{D}'_{\mu\nu} \tag{7.58}$$

As before, this allows us to make the following wave function renormalization:

$$A_\mu = \sqrt{Z_3} A_{\mu 0} \tag{7.59}$$

Finally, we wish to study the effect of renormalization on the electron–photon vertex function. We first define the vertex function:

$$\Gamma_\mu(p, p') = \gamma_\mu + \Lambda_\mu(p, p') \tag{7.60}$$

After summing all graphs, we find that the vertex graph is infinite, and that we can parametrize this divergence by introducing a third infinite quantity, Z_1, such that:

$$\Gamma_\mu(p, p') = \frac{1}{Z_1} \tilde{\Gamma}_\mu(p, p') \tag{7.61}$$

or:

$$\gamma_\mu + \Lambda_\mu(p, p') = \frac{1}{Z_1} \left[\gamma_\mu + \tilde{\Lambda}_\mu(p, p') \right] \tag{7.62}$$

As in the case of the electron mass, which was renormalized by the infinite quantity δm, we will also use Z_1 to renormalize the coupling constant. The renormalized coupling constant e, as shown above, receives a multiplicative correction $1/Z_1$ from the divergent part of the vertex graph. However, we also know that we have to renormalize the fermion and photon lines. Since there are two fermion lines and one photon line attached to each vertex function, we must multiply the coupling constant by another factor $\sqrt{Z_2}^2 \sqrt{Z_3}$.

The renormalized coupling constant is therefore:

$$e = \frac{Z_2 \sqrt{Z_3}}{Z_1} e_0 \tag{7.63}$$

(Later, using what are called Ward–Takahashi identities, we will show that some of these renormalization constants are equal, i.e. $Z_1 = Z_2$; so that the condition on the renormalized coupling constant reduces to: $e = \sqrt{Z_3} e_0$.)

As we mentioned earlier, instead of using multiplicative renormalization, we could have alternatively used the counterterm approach. This means starting with the theory defined with the physical parameters and then computing divergent self-energy and vertex corrections. Then we add counterterms to the action which, order by order, cancel these divergences. If we then add the renormalized action to the counterterms, we will reproduce the bare action.

If we start with the action defined with renormalized parameters:

$$\mathcal{L} = -\frac{1}{4}(F_{\mu\nu})^2 + \bar{\psi}(i\not{\partial} - m)\psi - e\bar{\psi}\not{A}\psi \tag{7.64}$$

then the theory produces divergent amplitudes, which can be cancelled order by order by adding counterterms to the action. Since the infinities that we encounter are identical to the ones we found with the multiplicative renormalization program, it is not surprising that the coefficients of these counterterms can be written in terms of the same Z's. A careful analysis yields:

$$\mathcal{L} \rightarrow \mathcal{L} + \Delta\mathcal{L} \tag{7.65}$$

where:

$$\Delta\mathcal{L} = -\frac{1}{4}(Z_3 - 1)(F_{\mu\nu})^2 + (Z_2 - 1)\bar{\psi}(i\not{\partial} - m)\psi + Z_2\delta m\bar{\psi}\psi - e(Z_1 - 1)\bar{\psi}\not{A}\psi \tag{7.66}$$

Adding these two terms together, we find:

$$\mathcal{L}_0 = -\frac{1}{4}Z_3(F_{\mu\nu})^2 + Z_2\bar{\psi}(i\not{\partial} - m)\psi - Z_1 e\bar{\psi}\not{A}\psi \tag{7.67}$$

If we change variables to the unrenormalized quantities:

$$
\begin{aligned}
\psi_0 &= \sqrt{Z_2}\,\psi \\
A_{\mu 0} &= \sqrt{Z_3}\,A_\mu \\
e_0 &= Z_1 Z_2^{-1} Z_3^{-1/2} e \\
m_0 &= m - \delta m
\end{aligned}
\tag{7.68}
$$

then our action becomes the unrenormalized one:

$$\mathcal{L} + \Delta\mathcal{L} = \mathcal{L}_0 \tag{7.69}$$

Finally, we would like to clarify the arbitrariness introduced into the theory by the subtraction point μ, the point at which we define the masses and coupling constants in Eq. (7.27). This ambiguity arose because of the way that finite parts were handled when canceling infinities. For example, we recall that the one-loop calculation of the photon vacuum polarization graph in Section 6.6, using a cutoff Λ, yielded:

$$\Pi(k^2) = \frac{\alpha_0}{3\pi} \log \frac{\Lambda^2}{k^2} \tag{7.70}$$

In order to extract out the renormalization constant Z_3, we must split $\Pi(k^2)$ into two parts, a constant part and a momentum-dependent part, as in Eq. (7.55). However, *there are an infinite number of ways that we can perform this split.* We could equally well have split the momentum-dependent and momentum-independent parts as:

$$\Pi(k^2) = \frac{\alpha_0}{3\pi} \left(\log \frac{\Lambda^2}{\mu^2} + \log \frac{\mu^2}{k^2} \right) = \Pi(\Lambda, \mu) + \tilde{\Pi}(k^2, \Lambda, \mu) \qquad (7.71)$$

where μ is arbitrary. Of course, we have done nothing. We have added and subtracted the same term, $\log \mu^2$. However, by adding finite terms to infinite quantities, we have conceptually made a significant change by altering the nature of the split. With this new split, the photon propagator in Eq. (7.56) can now be rewritten as:

$$D'_{\mu\nu} = \frac{-g_{\mu\nu} Z_3(\Lambda, \mu)}{k^2 \left[1 + \tilde{\Pi}(k^2, \Lambda, \mu) \right]} \qquad (7.72)$$

It is essential to notice that Z_3 now has an explicit dependence on μ:

$$Z_3(\Lambda, \mu) = \frac{1}{1 + \frac{\alpha_0}{3\pi} \log \frac{\Lambda^2}{\mu^2}}$$

$$\tilde{\Pi}(k^2, \Lambda, \mu) = Z_3(\Lambda, \mu) \tilde{\Pi}(k^2, \Lambda, \mu) \qquad (7.73)$$

However, since the renormalized coupling constant α is a function of Z_3, we can isolate the dependence of α on μ. To do this, we note that the unrenormalized α_0 is, by definition, independent of μ. Now choose two different renormalization points, μ_1 and μ_2. Since α can be written in terms of μ_1 or μ_2, we have:

$$\alpha_0 = \frac{\alpha(\mu_1)}{Z_3(\Lambda, \mu_1)} = \frac{\alpha(\mu_2)}{Z_3(\Lambda, \mu_2)} \qquad (7.74)$$

where we have set $Z_1 = Z_2$. Substituting in the value of $Z_3(\Lambda, \mu)$, we have:

$$\frac{1}{\alpha(\mu_1)} = \frac{1}{\alpha(\mu_2)} - \frac{2}{3\pi} \log \frac{\mu_1}{\mu_2} \qquad (7.75)$$

To lowest order, we have therefore derived a nontrivial relation, following from the *renormalization group* analysis,[5,6] which expresses how the effective coupling constants depend on the point μ, where we define the masses and coupling constants via Eq. (7.27). These renormalization group equations will prove crucial in our discussion of the asymptotic behavior of gauge theory in Chapter 14, where we give a more precise derivation of these relations. They will prove essential in demonstrating that QCD is the leading theory of the strong interactions.

7.5 Types of Regularization

This completes a brief sketch of how the renormalization program works for ϕ^4 and QED. Now comes the difficult part of filling in the essential details of the renormalization program. We will now concentrate on the regularization schemes, certain complications such as the Ward–Takahashi identity and overlapping graphs, and then we will present the proof that QED is renormalizable.

Over the decades, a wide variety of regularization schemes have been developed, each with their own distinct advantages and disadvantages. Each regularization scheme necessarily breaks some feature of the original action:

1. *Pauli–Villars regularization*
   Until recently, this was one of the most widely used regularization scheme. We cutoff the integrals by assuming the existence of a fictitious particle of mass M. The propagator becomes modified by:

$$\frac{1}{p^2 - m^2} - \frac{1}{p^2 - M^2} = \frac{m^2 - M^2}{(p^2 - m^2)(p^2 - M^2)} \tag{7.76}$$

   The relative minus sign in the propagator means that the new particle is a ghost; that is, it has negative norm. This means that we have explicitly broken the unitarity of the theory. The propagator now behaves as $1/p^4$, which is usually enough to render all graphs finite. Then, we take the limit as $M^2 \to \infty$ so that the unphysical fermion decouples from the theory. The advantage of the Pauli–Villars technique is that it preserves local gauge invariance in QED; hence the Ward identities are preserved (although they are broken for higher groups). There have been a large number of variations proposed to the Pauli–Villars technique, such as higher covariant derivatives in gauge theory and higher R^2 terms in quantum gravity.

2. *Dimensional regularization*
   This is perhaps the most versatile and simplest of the recent regularizations. Dimensional regularization involves generalizing the action to arbitrary dimension d, where there are regions in complex d space in which the Feynman integrals are all finite. Then, as we analytically continue d to four, the Feynman graphs pick up poles in d space, allowing us to absorb the divergences of the theory into the physical parameters. Dimensional regularization obviously preserves all properties of the theory that are independent of the dimension of space-time, such as the Ward–Takahashi identities.

3. *Lattice regularization*

This is the most widely used regularization scheme in QCD for nonperturbative calculations. Here, we assume that space–time is actually a set of discrete points arranged in some form of hypercubical array. The lattice spacing then serves as the cutoff for the space–time integrals. For QCD, the lattice is gauge invariant, but Lorentz invariance is manifestly broken. The great advantage of this approach is that, with Monte Carlo techniques, one can extract qualitative and some even some quantitative information from QCD. One disadvantage with this approach is that it is defined in Euclidean space; so we are at present limited to calculating only the static properties of QCD; the lattice has difficulty describing Minkowski space quantities, such as scattering amplitudes.

In this chapter, we will mainly stress the dimensional regularization method.[7–17] We will now show explicitly, to lowest order only, how dimensional regularization can regulate the divergences of the theory so that we are only manipulating finite quantities (albeit in an unphysical dimension). Then later we will show how to put this all together and renormalize field theory to all orders in perturbation theory.

Our starting point is the action for scalar mesons and for QED in d dimensions:

$$\mathscr{L} = \frac{1}{2}(\partial_\mu \phi_0)^2 - \frac{m_0^2}{2}\phi_0^2 - \frac{\mu^{4-d}}{4!}\lambda_0\phi_0^4$$

$$\mathscr{L} = \bar{\psi}_0(i\not{\partial} - m_0)\psi_0 - e_0\mu^{2-d/2}A_0^\mu\bar{\psi}_0\gamma_\mu\psi_0 - \frac{1}{4}F_{0\mu\nu}^2 \qquad (7.77)$$

where μ is an arbitrary parameter with the dimension of mass. It is necessary to insert this dimensional parameter because the dimension of the fermion field is $[\psi] = (d-1)/2$ while the dimension of the boson field is $[A_\mu] = (d/2) - 1$. (Our final result must be independent of the choice of μ. In this formalism, μ takes the place of the subtraction point introduced earlier in the renormalization-group equations.)

Generalizing space–time to d dimensions, we are interested in evaluating the integral in Eq. (7.11):

$$-i\Sigma(p^2) = -i\frac{\lambda_0\mu^{4-d}}{2}\int \frac{d^d p}{(2\pi)^d}\frac{1}{p^2 - m_0^2 + i\epsilon} \qquad (7.78)$$

Our goal is to find a region in complex d space where this integral is finite, and then analytically continue to $d = 4$, where we expect to pick up poles in divergent quantities. Then the renormalization scheme consists of absorbing all these poles into the coupling constants and masses of the theory.

To begin the dimensional regularization process, we first must express the integration over $d^d p$ for arbitrary dimension. To calculate the volume integral over

d space, we remind ourselves that the change from Cartesian to polar coordinates in two dimensions is given by:

$$
\begin{aligned}
x_1 &= r \cos \theta_1 \\
x_2 &= r \sin \theta_1
\end{aligned}
\tag{7.79}
$$

In three dimensions, we use the transformation to spherical coordinates given by:

$$
\begin{aligned}
x_1 &= r \cos \theta_1 \\
x_2 &= r \sin \theta_1 \cos \theta_2 \\
x_3 &= r \sin \theta_1 \sin \theta_2
\end{aligned}
\tag{7.80}
$$

Given these two examples, it is not hard to write down the transformation from Cartesian to d-dimensional spherical coordinates, which spans all of d space. (We will make a change of variables in the time parameter, converting real time to imaginary time. This is called a Wick rotation, which takes us from Minkowski space to Euclidean space, where we can perform all d-dimensional integrals all at once. We then Wick rotate back to Minkowski coordinates at the end of the calculation. Alternatively, we could have done the calculation completely in Minkowski space, where the dx^0 integration is handled differently from the other integrations.) The d-dimensional transformation is given by:

$$
\begin{aligned}
x_1 &= r \cos \theta_1 \\
x_2 &= r \sin \theta_1 \cos \theta_2 \\
x_3 &= r \sin \theta_1 \sin \theta_2 \cos \theta_3 \\
x_4 &= r \sin \theta_1 \sin \theta_2 \sin \theta_3 \cos \theta_4 \\
&\vdots \\
x_k &= r \left(\prod_{i=1}^{k-1} \sin \theta_i \right) \cos \theta_k \\
&\vdots \\
x_d &= r \prod_{i=1}^{d-1} \sin \theta_i
\end{aligned}
\tag{7.81}
$$

By induction, one can prove that the Jacobian from Cartesian to spherical coordinates is given by:

$$J = \det\left(\frac{\partial(x_1, x_2, \cdots, x_d)}{\partial(r, \theta_1, \theta_2, \cdots, \theta_{d-1})}\right)$$

$$= r^{d-1}\prod_{i=1}^{d-1}\sin^{i-1}\theta_i \tag{7.82}$$

The volume element in d space is therefore given by:

$$\prod_{i=1}^{d}dx_i = J dr \prod_{i=1}^{d-1}d\theta_i$$

$$= r^{d-1}dr d\Omega_{d-1}$$

$$= r^{d-1}dr \prod_{i=1}^{d-1}\sin^{i-1}\theta_i d\theta_i \tag{7.83}$$

To evaluate this integral, we use the fact that:

$$\int_0^{\pi}\sin^m\theta d\theta = \frac{\sqrt{\pi}\,\Gamma\left(\frac{1}{2}(m+1)\right)}{\Gamma\left(\frac{1}{2}(m+2)\right)} \tag{7.84}$$

Then in d dimensions, we have:

$$\int d^d p = \int_0^{\infty}r^{d-1}dr \int_0^{2\pi}d\theta_1 \int_0^{\pi}\sin\theta_2\, d\theta_2 \int_0^{\pi}\sin^2\theta_3\, d\theta_3 \cdots$$

$$\times \int_0^{\pi}\sin^{d-2}\theta_{d-1}\, d\theta_{d-1}$$

$$= \frac{2\pi^{d/2}}{\Gamma(d/2)}\int_0^{\infty}r^{d-1}\,dr \tag{7.85}$$

The integration that we want to perform can therefore be written in the form:

$$I = \int\frac{d^d p}{(p^2 + 2p\cdot q - m^2)^{\alpha}} = \int d\Omega_{n-1}\int dr \frac{r^{n-1}}{(r^2 - q^2 - m^2)^{\alpha}} \tag{7.86}$$

(In the last expression, we have changed coordinates $p \to p + q$ in order to eliminate the cross term between p and q. We are also working in Euclidean space and will Wick rotate back to Minkowski space later.) The integral over the solid angle is given by:

$$\int d\Omega_{d-1} = \frac{2\pi^{d/2}}{\Gamma(d/2)} \tag{7.87}$$

We are left, therefore, with an integral over r. This integral over r is of the form of an Euler Beta function:

$$B(x, y) = \frac{\Gamma(x)\Gamma(y)}{\Gamma(x + y)} = 2 \int_0^\infty dt \, t^{2x-1}(1 + t^2)^{-x-y} \tag{7.88}$$

With this formula, we can prove:

$$\int_0^\infty \frac{r^\beta dr}{(r^2 + C^2)^\alpha} = \frac{\Gamma\left(\frac{1}{2}(1 + \beta)\right)\Gamma\left(\alpha - \frac{1}{2}(1 + \beta)\right)}{2(C^2)^{\alpha-(1+\beta)/2}\Gamma(\alpha)} \tag{7.89}$$

We are left with the integral:

$$I = \left(\frac{2\pi^{d/2}}{\Gamma(d/2)}\right)\left(\frac{\Gamma(d/2)\Gamma(\alpha - d/2)}{2(-q^2 - m^2)^{\alpha-d/2}\Gamma(\alpha)}\right) \tag{7.90}$$

The final result is therefore:

$$\int \frac{d^d p}{(p^2 + 2p \cdot q - m^2)^\alpha} = \frac{i\pi^{d/2}\Gamma(\alpha - d/2)}{\Gamma(\alpha)(-q^2 - m^2)^{\alpha-d/2}} \tag{7.91}$$

(The extra i appears when we rotate back from Euclidean space to Minkowski space.)

Given these formulas, let us first analyze the one-loop correction to the ϕ^4 theory, which is given by Σ, and then generalize this discussion for the Dirac theory. The one-loop correction in ϕ^4 theory in Eq. (7.78) is given by:

$$-i\Sigma(p^2) = \frac{-i\lambda_0 m_0^2 \Gamma\left(1 - (d/2)\right)}{32\pi^2}\left(\frac{4\pi\mu^2}{-m_0^2}\right)^{2-d/2}$$

$$\rightarrow \frac{i\lambda_0 m_0^2}{16\pi^2}\left(\frac{1}{4 - d}\right) + \cdots \tag{7.92}$$

where we have extracted the pole term in the Γ function as $d \rightarrow 4$. These poles in d space, as we pointed out earlier, can all be absorbed into the renormalization of physical coupling constants and masses. This is our final result for the one-loop, two-point divergence to this order.

To generalize this discussion to the Dirac propagator, we must perform the dimensional integral with two Feynman propagators, so there are two terms in the denominator. It is easier if we combine the two factors in the denominator together; so we will once again use Feynman's parameter trick. First, let us calculate the electron self-energy correction, which can be written in d-dimensional space as:

$$
\begin{aligned}
\Sigma(p) &= -ie_0^2 \mu^{4-d} \int \frac{d^d k}{(2\pi)^d} \gamma_\mu \left(\frac{\not{p} - \not{k} + m_0}{[(p-k)^2 - m_0^2]k^2} \right) \gamma^\mu \\[2mm]
&= -ie_0^2 \mu^\epsilon \int_0^1 dx \int \frac{d^d k}{(2\pi)^d} \gamma_\mu \frac{\not{p} - \not{k} + m_0}{[(p-k)^2 x - m_0^2 x + k^2(1-x)]^2} \gamma^\mu \\[2mm]
&= -ie_0^2 \mu^\epsilon \int_0^1 dx \, \gamma_\mu (\not{p} - \not{p}x - \not{q} + m_0) \gamma^\mu \\[2mm]
&\quad \times \int \frac{d^d q}{(2\pi)^d} \frac{1}{[q^2 - m_0^2 x + p^2 x(1-x)]^2} \\[2mm]
&= \mu^\epsilon e_0^2 \frac{\Gamma(2 - d/2)}{(4\pi)^{d/2}} \int_0^1 dx \, \gamma_\mu [\not{p}(1-x) + m_0] \gamma^\mu \\[2mm]
&\quad \times [m_0^2 x - p^2 x(1-x)]^{d/2 - 2} \tag{7.93}
\end{aligned}
$$

where we have made a sequence of steps: (1) we used the Feynman parameter trick; (2) we made the substitution $q = k - px$; (3) integrated over q, dropping terms linear in q. Finally, we look for the pole in $\Gamma(2 - d/2)$ as $d \to 4$. Let us perform the integration over x:

$$
\begin{aligned}
\Sigma(p) &= -\frac{e_0^2}{16\pi^2} \Gamma(\epsilon/2) \int_0^1 dx \, \{2\not{p}(1-x) - 4m_0 - \epsilon[\not{p}(1-x) + m_0]\} \\[2mm]
&\quad \times \left(\frac{m_0^2 x - p^2 x(1-x)}{4\pi \mu^2} \right)^{-\epsilon/2} \\[2mm]
&= \frac{e_0^2}{8\pi^2 \epsilon} (-\not{p} + 4m_0) + \frac{e_0^2}{16\pi^2} \Big[\not{p}(1+\gamma) - 2m_0(1+2\gamma) \\[2mm]
&\quad + 2 \int_0^1 dx \, [\not{p}(1-x) - 2m_0] \log \left(\frac{m_0^2 x - p^2 x(1-x)}{4\pi \mu^2} \right) \Big] \\[2mm]
&= \frac{e_0^2}{8\pi^2 \epsilon} (-\not{p} + 4m_0) + \text{finite} \tag{7.94}
\end{aligned}
$$

(where γ is the Euler–Mascheroni constant).

Comparing this with the expansion for $\Sigma(p)$ in Eq. (7.49), we now have a result for:

$$Z_2 = 1 - \frac{e_0^2}{8\pi^2\epsilon} \tag{7.95}$$

Next, we evaluate the vacuum polarization contribution to the photon propagator. Once again, we will use the Feynman parameter trick:

$$
\begin{aligned}
\Pi_{\mu\nu} &= ie_0^2\mu^{4-d} \int \frac{d^d p}{(2\pi)^d} \left(\frac{\text{Tr}\left[\gamma_\mu(\not{p}+m_0)\gamma_\nu(\not{p}-\not{k}+m_0)\right]}{(p^2-m_0^2)[(p^2-k^2)^2-m_0^2]} \right) \\[2mm]
&= ie_0^2\mu^\epsilon \int_0^1 dx \int \frac{d^d q}{(2\pi)^d} \\[2mm]
&\quad \times \left(\frac{\text{Tr}\left\{\gamma_\mu(\not{q}+\not{k}x+m_0)\gamma_\nu[\not{q}-\not{k}(1-x)+m_0]\right\}}{[q^2-m_0^2+k^2x(1-x)]^2} \right) \\[2mm]
&= ie_0^2\mu^\epsilon f(d) \int_0^1 dx \int \frac{d^d q}{(2\pi)^d} \left(\frac{2q_\mu q_\nu}{[q^2-m_0^2+k^2x(1-x)]^2} \right. \\[2mm]
&\quad \left. - \frac{2x(1-x)(k_\mu k_\nu - g_{\mu\nu}k^2)}{[q^2-m_0^2+k^2x(1-x)]^2} - \frac{g_{\mu\nu}}{[q^2-m_0^2+k^2x(1-x)]} \right)
\end{aligned} \tag{7.96}
$$

where $q \rightarrow q - kx$ and $f(d) = \text{Tr } I$. The first and third terms on the right-hand side of the equation cancel, leaving us with only a logarithmic divergence. (Gauge invariance has thus reduced a quadratically divergent graph to only a logarithmically divergent one.) Extracting the pole as $d \rightarrow 4$, we find:

$$
\begin{aligned}
\Pi_{\mu\nu} &= \frac{e_0^2}{2\pi^2}(k_\mu k_\nu - g_{\mu\nu}k^2) \\[2mm]
&\quad \times \left[\frac{1}{3\epsilon} - \frac{\gamma}{6} - \int_0^1 dx\, x(1-x)\log\left(\frac{m_0^2 - k^2x(1-x)}{4\pi\mu^2}\right) + \cdots \right] \\[2mm]
&= \frac{e_0^2}{6\pi^2\epsilon}(k_\mu k_\nu - g_{\mu\nu}k^2) + \cdots
\end{aligned} \tag{7.97}
$$

Using Eq. (7.57), this means that we can write:

$$Z_3 = 1 - \frac{e_0^2}{6\pi^2\epsilon} \tag{7.98}$$

Finally, we must write down the dimensionally regularized vertex correction:

$$\Lambda_\mu(p, q, p') = -i(e\mu^{2-d/2})^2 \int \frac{d^d k}{(2\pi)^d} \frac{\gamma_\nu(p' - k + m_0)\gamma_\mu(p - k + m_0)\gamma^\nu}{k^2[(p - k)^2 - m_0^2][(p' - k)^2 - m_0^2]}$$

$$= \frac{2ie_0^2\mu^\epsilon}{(2\pi)^d} \int_0^1 dx \int_0^{1-x} dy \int d^d k$$

$$\times \frac{\gamma_\nu(p' - k + m_0)\gamma_\mu(p - k + m_0)\gamma^\nu}{[k^2 - m_0^2(x + y) - 2k(px + p'y) + p^2 x + p'^2 y]^3}$$

$$= \frac{2ie_0^2\mu^\epsilon}{(2\pi)^d} \int_0^1 dx \int_0^{1-x} dy \int d^d k$$

$$\times \frac{\gamma_\nu[p'(1 - y) - px - k + m_0]\gamma_\mu[p(1 - x) - p'y - k + m_0]\gamma^\nu}{[k^2 - m_0^2(x + y) + p^2 x(1 - x) + p'^2 y(1 - y) - 2p \cdot p'xy]^3}$$

$$(7.99)$$

where we have made the substitution $k \to k - px - p'y$.

This expression, in turn, contains a divergent and a convergent part. The convergent part, as before, gives us the contribution to the anomalous magnetic moment of the electron.

From Eq. (7.61), the divergent part can be isolated as:

$$\Lambda_\mu^{(1)} = \frac{e_0^2\mu^\epsilon}{2(4\pi)^{d/2}}\Gamma(2 - d/2) \int_0^1 dx \int_0^{1-x} dy$$

$$\times \frac{\gamma_\nu\gamma_\rho\gamma_\mu\gamma^\rho\gamma^\nu}{[m_0^2(x + y) + p^2 x(1 - x) + p'^2 y(1 - y) + 2p \cdot p'xy]^{2-d/2}}$$

$$= \frac{e_0^2}{8\pi^2\epsilon}\gamma_\mu + \cdots \qquad (7.100)$$

This, in turn, means that we can write:

$$Z_1 = 1 - \frac{e_0^2}{8\pi^2\epsilon} \qquad (7.101)$$

Notice that $Z_1 = Z_2$, as we expected from the Ward–Takahashi identity.

In summary, we have, to lowest order, the correction to the electron self-energy, the photon propagator, and the vertex function:

$$\Sigma(p) = \frac{e_0^2}{8\pi^2\epsilon}(-p + 4m_0) + \text{finite}$$

$$\Pi_{\mu\nu} = \frac{e_0^2}{6\pi^2\epsilon}(k_\mu k_\nu - k^2 g_{\mu\nu}) + \text{finite}$$

$$\Lambda_\mu^{(1)} = \frac{e_0^2}{8\pi^2\epsilon}\gamma_\mu + \text{finite} \tag{7.102}$$

These, in turn, give us the expression for the renormalization constants:

$$Z_1 = Z_2 = 1 - \frac{e^2}{8\pi^2\epsilon}$$

$$Z_3 = 1 - \frac{e^2}{6\pi^2\epsilon}$$

$$e_0 = \frac{Z_1}{Z_2\sqrt{Z_3}}e \tag{7.103}$$

Comparing this result with the Pauli–Villars regularization method as in Eq. (6.98), we find that, to lowest order, the divergences are identical if we make the following correspondence between the pole $1/\epsilon$ and the cutoff Λ:

$$\frac{1}{\epsilon} \leftrightarrow \frac{1}{2}\log\frac{\Lambda^2}{m^2} \tag{7.104}$$

Thus, to this order, our results are independent of the regularization procedure.

7.6 Ward–Takahashi Identities

When we generalize our discussion to all orders in perturbation theory, our work will be vastly simplified by a set of *Ward–Takahashi* identities,[18,19] which reduce the number of independent renormalization constants Z_i. When we generalize our discussion to include gauge theories in later chapters, we will see that Ward–Takahashi identities are an essential ingredient in proving that a theory is renormalizable.

Specifically, we will use the fact that:

$$\Lambda_\mu(p, p) = -\frac{\partial}{\partial p_\mu}\Sigma(\not p) \tag{7.105}$$

where we have written the vertex correction in terms of the electron self-energy correction.

To prove this and more complicated identities arising from gauge theories, the most convenient formalism is the path integral formalism. However, for our purposes, we can prove the simplest Ward–Takahashi identities from graphical methods. To prove this identity, we observe that:

$$-\frac{\partial}{\partial p_\mu}\frac{1}{\not p - m + i\epsilon} = \frac{1}{\not p - m + i\epsilon}\gamma_\mu\frac{1}{\not p - m + i\epsilon} \tag{7.106}$$

(To prove this matrix equation, we set up a difference equation and then take the limit for small δp_μ.) Now let us use the above identity on an arbitrary graph that appears within $\Sigma(\not p)$. In general, we can write $\Sigma(\not p)$ as follows to bring out its p dependence:

$$\Sigma(\not p) \sim \cdots \frac{1}{(\not p + \not q_i) - m} \cdots \tag{7.107}$$

where the ellipses may represent very complicated integrals that are not important to our discussion. Also, there may be a large number of propagators with p dependence within $\Sigma(\not p)$, which we suppress for the moment. Now take the derivative with respect to p_μ, and we find:

$$-\frac{\partial}{\partial p_\mu} \Sigma(\not p) = \sum_i \cdots \frac{1}{\not p + \not q_i - m} \gamma_\mu \frac{1}{\not p + \not q_i - m} \cdots \tag{7.108}$$

where we must differentiate over all propagators that have a p dependence. The term on the right is therefore the sum over all terms that have electron propagators containing p dependence.

The right-hand side of the previous equation is now precisely the graph corresponding to a vertex correction connected to an external photon line with zero momentum. This in turn means that the right-hand side equals $\Gamma_\mu(p, p)$, as promised. The purpose of this rather simple exercise is to prove a relation between renormalization constants. Now let us insert all this into the Ward–Takahashi identity. Differentiating the full $\Sigma(\not p)$ to all orders, we find:

$$
\begin{aligned}
-\frac{\partial}{\partial p_\mu} \Sigma(\not p) &= -\frac{\partial}{\partial p_\mu} [\Sigma(m) + (\not p - m)\Sigma'(m) + \tilde{\Sigma}(\not p)] \\
&= -\gamma_\mu \Sigma'(m) - \partial_\mu \tilde{\Sigma}(\not p) \\
&= \Lambda_\mu(p, p) \\
&= \gamma_\mu(Z_1^{-1} - 1) + Z_1^{-1}\tilde{\Lambda}_\mu(p, p) \tag{7.109}
\end{aligned}
$$

where we have used Eq. (7.61).

Comparing terms proportional to the Dirac matrix, we see that the term $(Z_1^{-1} - 1)\gamma_\mu$ must be proportional to $-\Sigma'(m)$, which we calculated earlier to be equal to $-1 + (1/Z_2)$ in Eq. (7.49).

Comparing these two coefficients of γ_μ, we easily find:

$$Z_1 = Z_2 \tag{7.110}$$

which is our desired result. For a complete proof of renormalization, however, we need a more powerful version of the Ward–Takahashi identity, rather than the one defined for zero-momentum external photons. We will use the version:

$$(p' - p)_\mu \Gamma^\mu(p', p) = \left[S'^{-1}_F(p') - S'^{-1}_F(p) \right] \tag{7.111}$$

This is easily shown to lowest order, where $S_F^{-1} = \not{p} - m$. Then the Ward–Takahashi identity reduces to the simple expression $(p' - p)_\mu \gamma^\mu = (\not{p}' - m - \not{p} + m)$. We can also show that this expression, in the limit that $p \to p'$, reduces to the previous version of the Ward–Takahashi identity.

This can also be rewritten in terms of the one-particle irreducible graph Σ. Because $S'^{-1}_F = \not{p} - m - \Sigma(\not{p})$, we have:

$$(p' - p)_\mu \Lambda^\mu = -\left[\Sigma(p') - \Sigma(p) \right] \tag{7.112}$$

The proof of this identity is also performed by graphical methods. We use the fact that:

$$\frac{1}{\not{p} + \not{q} - m} \not{q} \frac{1}{\not{p} - m} = \frac{1}{\not{p} - m} - \frac{1}{\not{p} + \not{q} - m} \tag{7.113}$$

which uses the identity:

$$\frac{1}{A}(A - B)\frac{1}{B} = \frac{1}{B} - \frac{1}{A} \tag{7.114}$$

To prove the general form of the Ward–Takahashi identity, we simply use this identity everywhere along a fermion line or along a fermion loop. This procedure is schematically represented in Figure 7.4.

Consider a specific member of $\Sigma(p)_{\alpha\beta}$ and follow the fermion line that connects the index α with β. Then schematically, we have:

$$\Sigma(p_0) = \sum \int \cdots \int \cdots \left(\prod_{i=0}^{n} \not{a}_i \frac{1}{\not{p}_i - m} \right) \not{a}_n \tag{7.115}$$

where $p_n = p_0$, where we have omitted all momentum integrations, and where a_i denote the momentum-dependent photon lines that are connected to other fermion loops (which are temporarily dropped). The matrices are ordered sequentially. Also, the total momenta contained within the various a_i must sum to zero.

Let us do the same for Λ_μ. If we follow the fermion line connecting the two fermion ends of Λ_μ, we find:

$$\Lambda_\mu(p', p) = \sum \int \cdots \int \cdots \sum_{r=0}^{n-1} \left(\prod_{i=0}^{r} \not{a}_i \frac{1}{\not{p}_i - m} \right) \gamma_\mu \frac{1}{\not{p}_r + \not{q} - m}$$

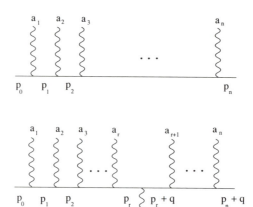

Figure 7.4. The Ward–Takahashi identity is proved by examining the insertion of a photon line along a fermion line contained within Σ and showing that we retrieve Λ_μ.

$$\times \left(\prod_{s=r+1}^{n-1} \rlap{/}{d}_s \frac{1}{\rlap{/}{p}_s + \rlap{/}{q} - m} \right) \rlap{/}{d}_n \qquad (7.116)$$

The γ_μ is inserted at the rth point along the fermion line, and we sum over all possible values of r. In other words, we sum over all possible insertion points of γ_μ along the fermion line. The essential point in the proof is the observation that there is a graphical similarity between the vertex correction and the self-energy correction. Graphically speaking, if we take all the graphs within Σ and attach an extra photon line along all possible fermion propagators, then the resulting graphs are identical to the graphs within Λ_μ.

To show this rigorously, let us now contract Λ_μ with q^μ and use the identity shown previously. Then each term in the rth sum in Λ_μ splits into two parts. We now have $2r$ possible terms, which cancel pairwise because of the minus sign. The only terms that do not cancel are the first and last terms:

$$
\begin{aligned}
q^\mu \Lambda_\mu &= \sum \int \cdots \int \cdots \sum_{r=0}^{n-1} \left(\prod_{i=0}^{r-1} \rlap{/}{d}_i \frac{1}{\rlap{/}{p}_i - m} \right) \rlap{/}{d}_r \left(\frac{1}{\rlap{/}{p}_r - m} - \frac{1}{\rlap{/}{p}_r + \rlap{/}{q} - m} \right) \\
&\quad \times \left(\prod_{s=r+1}^{n-1} \rlap{/}{d}_s \frac{1}{\rlap{/}{p}_s + \rlap{/}{q} - m} \right) \rlap{/}{d}_n \\
&= \Sigma(p_0) - \Sigma(p_0 + q) \qquad (7.117)
\end{aligned}
$$

where $p_0 = p$ and $p' = p_0 + q$.

The last step in the proof is to notice that we could have attached the photon leg of momentum q along any closed fermion loop as well. However, then the pairwise cancellation is exact, and the contribution of all closed loops is zero. This completes the proof of the Ward–Takahashi identity to all orders.

7.7 Overlapping Divergences

As we mentioned earlier, the renormalization program either cuts off the divergences with counterterms or absorbs them by multiplicative renormalization. In order to study how this is done systematically, we can use the method of *skeletons*. Draw a box around each divergent electron and photon self-energy graph and each vertex graph. Then we can replace the self-energy insertions with a line and the vertex insertions with a point. In this way, we obtain a *reduced graph*. Then we repeat this process, drawing boxes around the self-energy and vertex insertions in the reduced graph, and reduce it once more. Eventually, we obtain a graph that can no longer be reduced; that is, it is *irreducible*. The final graph after all these reductions is called a skeleton. An irreducible graph is one that is its own skeleton.

The advantage of introducing this concept is that the renormalization program is reduced to canceling the divergences within each box. For example, consider the complete vertex function Γ_μ summed to all orders. If we make the reduction of Γ_μ, we wind up with a sum over skeleton graphs, such that each line in the skeleton corresponds to the proper self-energy graph S'_F and $D'_{\mu\nu}$ and each vertex is the proper vertex. This can be summarized as:

$$\Gamma_\mu(p, p') = \gamma_\mu + \Lambda_\mu^S(S', D', \Gamma, e_0^2, p, p') \tag{7.118}$$

In this way, all complete vertex functions can be written entirely in terms of skeleton graphs over proper self-energy graphs and proper vertices.

This process of drawing boxes around the self-energy insertions and vertex insertions is unambiguous as long as the boxes are disjoint or nested; that is, the smaller boxes lie wholly within the larger one. Then the skeleton graph is unique. However, the skeleton reduction that we have described has an ambiguity that infests self-energy diagrams. For example, in Figure 7.5, we see examples of *overlapping diagrams* where the boxes overlap and the skeleton is not unique. One can show that these overlapping divergences occur for vertex insertions within self-energy parts. Unless these overlapping divergences appearing in self-energy graphs are handled correctly, we will overcount the number of graphs. These overlapping divergences are the only real difficulty in proving the renormalization of ϕ^4 and QED.

Figure 7.5. Examples of overlapping divergences found in QED.

There are several solutions to this delicate problem of overlapping divergences. The cleanest and most powerful is the BPHZ program, which we will present in Chapter 13. For QED, the most direct solution was originally given by Ward, who used the fact that, although $\Sigma(p\!\!\!/)$ has overlapping divergences, the vertex function Λ_μ does not. Therefore, by using the Ward–Takahashi identity:

$$\frac{\partial \Sigma(p)}{\partial p^\mu} = -\Lambda_\mu \qquad (7.119)$$

we can reduce all calculations to functions in which overlapping divergences are absent.

Mathematically, taking the derivative with respect to p^μ is identical to inserting a zero photon line at every electron propagator, as we saw in Eq. (7.106). This means that every time an electron propagator appears in Σ, we replace it with two electron propagators sandwiching a zero momentum photon insertion matrix γ^μ. (However, we stress that there are still some serious, unresolved questions concerning overlapping divergences that emerge in 14th order diagrams, which we will discuss later.)

To see how this actually works, let us take the derivative of an overlapping divergence appearing in Σ. In Figure 7.6, we see the effect that $\partial/\partial p_\mu$ has on the overlapping divergence.

A single overlapping divergence has now split up into three pieces, each of which has a well-defined skeleton decomposition. Therefore, the Ward–Takahashi identity has helped to reduce the overlapping divergences within Σ to the more

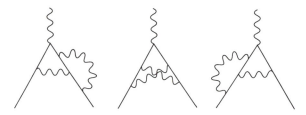

Figure 7.6. The effect that taking the partial derivative with respect to p_μ has on the overlapping divergence.

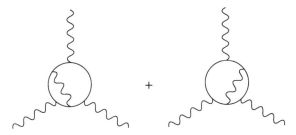

Figure 7.7. Taking the derivative of the photon self-energy graph with respect to the momentum yields these diagrams.

manageable problem of calculating the skeleton decomposition of the vertex Λ_μ. As long as we reduce all electron self-energy ambiguities to vertex insertions via the Ward–Takahashi identity, the overlapping divergence problem is apparently solved.

For photon self-energy graphs, there is also an overlapping divergence problem, which is also treated in much the same way. Although there is no Ward–Takahashi identity for the photon self-energy graph, we can solve the problem by introducing a new function W_μ, which is defined by:

$$W_\mu(k) \equiv \frac{\partial}{\partial k^\mu} \left[i D'(k^2)\right]^{-1} \tag{7.120}$$

where:

$$D'(k^2) = -\frac{1}{k^2 \left(1 - e_0^2 \Pi(k^2)\right)} \tag{7.121}$$

More explicitly, we can write:

$$
\begin{aligned}
W_\mu(k) &= 2i k_\mu + i k_\mu T(k^2) \\
T(k^2) &= \frac{e_0^2}{k^2} k^\mu \frac{\partial}{\partial k^\mu} \left[k^2 \Pi(k^2)\right]
\end{aligned}
\tag{7.122}
$$

The whole point of introducing this function $W_\mu(k)$ is that it has no overlapping divergences. Once again, the net effect of the partial derivative is to convert the photon self-energy graph into a vertex graph for T, which has no overlapping divergences. To see how this works, in Figure 7.7 we have performed the differentiation of an overlapping photon self-energy graph and created a vertex graph that has no ambiguous skeleton decomposition.

In summary, as long as we properly replace all electron and photon self-energy graphs with their vertex counterparts, there is usually no need to worry about overlapping divergences.

7.8   Renormalization of QED

Over the years, a large number of renormalization programs have been developed, with various degrees of rigor, and each with their own advantages and disadvantages. Since the reader will encounter one or more of these renormalization programs in his or her research, it is important that the reader be familiar with a few of these approaches. In this book, we will present three renormalization proofs:

1. The original Dyson/Ward proof, (which apparently breaks down at the fourteenth order, due to ovelapping divergences).

2. The BPHZ proof.

3. Proof based on the Callan–Symanzik equations.

We will first present the Dyson/Ward proof, which uses the auxiliary function W_μ to handle the overlapping divergence problem. In a later chapter, we will renormalize quantum field theory using more modern and sophisticated techniques, such as the renormalization group and the BPHZ program, which are more general than the proof that we will present here.

The Dyson/Ward proof of renormalization can be described in four steps:

1. First write down the complete set of coupled equations containing all the divergences of the theory for all self-energies and vertices. Everything is written in terms of skeleton graphs defined over the (divergent) complete propagators S_F' and $D_{\mu\nu}'$.

2. Subtract off the infinite divergences only in the vertex functions T and Λ_μ, which are free of overlapping divergences. Then define the new renormalized self-energy functions $\tilde{S}_F'$ and $\tilde{D}_{\mu\nu}'$ in terms of these subtracted vertex parts. This will enable us to write down an equivalent set of coupled equations for the finite set of self-energy functions free of overlapping divergences.

3. Rewrite the subtraction process as a multiplicative rescaling of the vertex and self-energy parts.

4. Absorb all multiplicative renormalizations into the coupling constant, masses, and wave functions.

7.8.1   Step One

The proof begins by writing down the expression for the unrenormalized self-energy graphs and vertices, summed to all orders, in terms of their skeletons. It will prove useful to summarize the notation that we will use:

$$
\begin{aligned}
\Sigma(p) &\rightarrow \text{proper } e \text{ self-energy graph} \\
\Pi_{\mu\nu} &\rightarrow \text{proper } \gamma \text{ self-energy graph} \\
\Gamma_{\mu} &\rightarrow \text{proper vertex (unrenormalized)} \\
S'_F &\rightarrow \text{complete } e \text{ self-energy graph} \\
D'_{\mu\nu} &\rightarrow \text{complete } \gamma \text{ self-energy graph} \\
\tilde{S}'_F &\rightarrow \text{renormalized } e \text{ self-energy graph} = Z_1^{-1} S'_F \\
\tilde{D}'_{\mu\nu} &\rightarrow \text{renormalized } \gamma \text{ self-energy graph} = Z_3^{-1} D'_{\mu\nu} \\
\tilde{\Gamma}_{\mu} &\rightarrow \text{renormalized vertex} = Z_1 \Gamma_{\mu}
\end{aligned}
\tag{7.123}
$$

Then the unrenormalized self-energy graphs and vertices satisfy:

$$
\begin{aligned}
\Gamma_{\mu}(p, p') &= \gamma_{\mu} + \Lambda_{\mu}^{S}(S', D', \Gamma, e_0^2, p, p') \\
W_{\mu}(k^2) &= 2ik_{\mu} + ik_{\mu}\Gamma^{S}(S', D', \Gamma, W, e_0^2, k^2) \\
S'(p)^{-1} &= S'(p_0)^{-1} + (p - p_0)^{\mu}\Gamma_{\mu}(p, p_0) \\
D'(k^2)^{-1} &= \int_0^1 dx \, k^{\mu} W_{\mu}(xk)
\end{aligned}
\tag{7.124}
$$

where the superscript S denotes skeleton graphs, the prime denotes the complete propagator, which is the sum over divergent one-particle irreducible graphs.

The first two equations are really definitions, telling us that the vertex graphs Γ_{μ} and W_{μ} can be written in terms of skeleton graphs. (Because they have no overlapping divergences, this can always be done.) The third equation is the Ward–Takahashi identity, and the last equation is an integrated version of the definition of W_{μ}. Since these equations are all divergent, we must make the transition to the renormalized quantities.

7.8.2 *Step Two*

To find the convergent set of equations, we want to perform a subtraction only on those functions that have no overlapping divergences; that is, we perform the subtraction on T and Λ. In this way, we remove the possibility of overcounting. The subtracted functions are denoted by a tilde:

$$
\begin{aligned}
\tilde{\Lambda}_{\mu}(p, p') &\equiv \Lambda_{\mu}^{S}(p, p') - \Lambda_{\mu}^{S}(p_0, p_0)\big|_{p_0=\mu} \\
\tilde{T}(k^2) &\equiv T^{S}(k^2) - T^{S}(\mu^2)
\end{aligned}
\tag{7.125}
$$

Only one subtraction is necessary since the diagrams are only logarithmically divergent. (As before, we stress that there is an infinite degree of freedom in choosing μ, the subtraction point. Whatever value of μ we choose, we demand that the physics be independent of the choice.)

We note that $\Lambda_\mu^S(p_0, p_0)$ can be further reduced. Since it is defined for $p_0 = \mu$, it can only be a function of γ_μ. Thus, we can also write:

$$\Lambda_\mu^S(p_0, p_0)\big|_{p_0=\mu} = L\gamma_\mu \tag{7.126}$$

where L is divergent.

The important point is that we have only performed the subtractions on the quantities that have no overlapping divergences and no overcounting ambiguities, that is, the vertex functions T and Λ. This, in turn, allows us to *define* the finite self-energy parts (with a tilde) via the following equations:

$$\tilde{\Gamma}_\mu(p, p') \equiv \gamma_\mu + \tilde{\Lambda}_\mu(\tilde{S}, \tilde{D}, \tilde{\Gamma}, e^2, p, p')$$

$$\tilde{W}_\mu(k) \equiv 2ik_\mu + ik_\mu \tilde{T}(\tilde{S}, \tilde{D}, \tilde{\Gamma}, \tilde{W}, e^2, k^2)$$

$$\tilde{S}(p)^{-1} \equiv \tilde{S}(p_0)^{-1} + (p - p_0)^\mu \tilde{\Gamma}_\mu(p, p_0)$$

$$\tilde{D}(k^2) \equiv \int_0^1 dx \, k^\mu W_\mu(xk) \tag{7.127}$$

The advantage of these definitions is that everything is now defined in terms of $\tilde{T}$ and $\tilde{\Lambda}$, which have no overlapping divergences. However, the quantities $\tilde{\Gamma}_\mu$, $\tilde{W}_\mu$ were simply defined by the previous equations. We still have no indication that these subtracted functions have any relationship to the actual renormalized self-energy and vertex parts.

7.8.3   Step Three

Now comes the important step. Up to now, we have made many rather arbitrary definitions that, as yet, have no physical content. We must now show that these quantities with the tilde are, indeed, the renormalized quantities that we want.

To see how this emerges, let us focus on the vertex graph. In terms of the subtraction, we can write:

$$\begin{aligned} \tilde{\Gamma}_\mu &\equiv \gamma_\mu + \tilde{\Lambda}_\mu \\ &= \gamma_\mu + \Lambda_\mu^S - L\gamma_\mu \\ &= (1 - L)\left(\gamma_\mu + \frac{1}{1 - L}\Lambda_\mu^S\right) \end{aligned}$$

$$= Z_1 \left(\gamma_\mu + \frac{1}{Z_1} \Lambda_\mu^S \right) \tag{7.128}$$

where we have defined:

$$Z_1 = 1 - L \tag{7.129}$$

Although we have factored out the renormalization constant Z_1 from this equation, the right-hand side is still not in the correct form. We want the right-hand side to be written in terms of the unrenormalized quantities, not the renormalized ones.

To find the scaling properties, it is useful to write the vertex functions symbolically in terms of propagators S and D and vertices γ_μ. Symbolically, by deleting integrals, traces, etc., the vertices can be written as products of propagators and vertices:

$$\Lambda_{2n}^S \sim e_0^{2n} S^{2n} (D)^n (\gamma_\mu)^{2n+1}$$

$$T_{2n}^S \sim e_0^{2n} S^{2n+\sigma} (D)^{n-\sigma} (\gamma_\mu)^{2n+\sigma} (2ik_\mu)^{1-\sigma} \tag{7.130}$$

where σ is an integer that increases by one for every differentiation of an electron line.

Given this symbolic decomposition, we want to study their behavior under a scaling given by:

$$\gamma_\mu \rightarrow a\gamma_\mu$$

$$D_{\mu\nu} \rightarrow b D_{\mu\nu}$$

$$S \rightarrow a^{-1} S \tag{7.131}$$

(where a and b are proportional to renormalization constants).

Then it is easy to show that the vertex functions scale as:

$$a\Lambda_\mu^S(S, D, \gamma_\mu, e_0^2, p, p') = \Lambda_\mu^S(a^{-1}S, bD, a\gamma_\mu, b^{-1}e_0^2, p, p)$$

$$b^{-1}T^S(S, D, \gamma_\mu, 2ik_\mu, e_0^2, k^2) = T^S(a^{-1}S, bD, a\Gamma_\mu, b^{-1}2ik_\mu, b^{-1}e_0^2, k^2) \tag{7.132}$$

With these rescaling relations, we can now absorb the factor Z_1 back into the renormalized vertex function Λ_μ^S and convert it into an unrenormalized one. Let $a = Z_1^{-1}$ and $b = Z_3$. We then have:

$$\tilde{\Gamma}_\mu = Z_1 \left(\gamma_\mu + \frac{1}{Z_1} \Gamma_\mu^S(\tilde{S}, \tilde{D}, \tilde{\Gamma}, e^2, p, p') \right)$$

$$
\begin{aligned}
&= \; Z_1 \left[\gamma_\mu + \Lambda_\mu^S(Z_1 \tilde{S}, Z_3 \tilde{D}, Z_1^{-1} \tilde{\Gamma}, Z_3^{-1} e^2, p, p') \right] \\
&= \; Z_1 \left[\gamma_\mu + \Lambda_\mu^S(S', D', \Gamma, e_0^2, p, p') \right] \\
&= \; Z_1 \Gamma_\mu
\end{aligned}
\tag{7.133}
$$

where we have used $Z_1 = Z_2$. This is the result that we wanted. We have now shown that the subtracted quantity $\tilde{\Gamma}_\mu$, after a rescaling by Z_1 and Z_3, can be written multiplicatively in terms of the unrenormalized quantity Γ_μ. This justified the original definition of $\tilde{\Gamma}_\mu$ that we introduced earlier.

7.8.4   Step Four

Now that we have renormalized the vertex, the rest is now easy. The vertex $\tilde{W}_\mu$ can now be renormalized in the same way. With $a = Z_1^{-1}$ and $b = Z_3$, we have:

$$
\begin{aligned}
\tilde{W}_\mu(k) &= \; 2ik_\mu + ik_\mu \left(T^S(k^2) - T^S(\mu^2) \right) \\
&= \; \left(1 - \frac{1}{2} T^S(\mu^2) \right) \left(2ik_\mu + \frac{1}{1 - \frac{1}{2} T^S(\mu^2)} ik_\mu T^S(k^2) \right) \\
&= \; Z_3 \left(2ik_\mu + \frac{1}{Z_3} ik_\mu T^S(k^2) \right) \\
&= \; Z_3 \left[2ik_\mu + ik_\mu T^S(Z_2 \tilde{S}, Z_3 \tilde{D}, Z^{-1} \tilde{\Gamma}_\mu, Z^{-1} \tilde{W}, Z_3^{-1} e^2, k^2) \right] \\
&= \; Z_3 \left[2ik_\mu + ik_\mu T^S(S', D', \Gamma, W, e_0^2, k^2) \right]
\end{aligned}
\tag{7.134}
$$

Thus, we have the other renormalized relation:

$$
\tilde{W}_\mu = Z_3 W_\mu
\tag{7.135}
$$

From the renormalization of these vertex functions, it is easy to renormalize the self-energy terms as well, since everything is multiplicative. We now have the following relations that show the link between renormalized and unrenormalized quantities:

$$
\begin{aligned}
\Gamma_\mu &= Z_1^{-1} \tilde{\Gamma}_\mu & W_\mu &= Z_3^{-1} \tilde{W}_\mu \\
S' &= Z_2 \tilde{S} & D'_{\mu\nu} &= Z_3 D_{\mu\nu} \\
e_0^2 &= Z_3^{-1} e^2 & Z_1 &= Z_2
\end{aligned}
\tag{7.136}
$$

In summary, we have proved that a subtraction of the divergences can be re-expressed in terms of a multiplicative rescaling of the vertex and self-energy

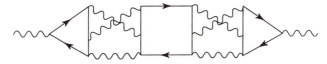

Figure 7.8. This fourteenth-order photon self-energy Feynman diagram was shown by Yang and Mills to suffer from a serious overlapping divergence problem, because of ambiguities introduced when we take the momentum derivative. This graph apparently invalidates the original Dyson/Ward renormalization proof.

parts. This means that all divergences are multiplicative and can be absorbed into a renormalization of the coupling constants, masses, and wave functions.

Although this was thought to be the first complete proof of the renormalizability of QED, this proof may be questioned in terms of its rigor. For example, it was shown by Yang and Mills[20,22] that there is an overlapping ambiguity at the fourteenth order in QED, thereby ruining this proof. For the electron self-energy, the Ward–Takahashi identity solves the problem of overlapping divergences, but for the photon self-energy, they showed that the operation of taking a momentum derivative is ambiguous at that level, thereby invalidating the proof (Fig. 7.8). (They also showed how it might be possible to remedy this problem, but did not complete this step.)

Second, another criticism of this proof is that we necessarily had to manipulate functions that were sums of an infinite number of graphs. Although each graph may be finite, the sum certainly is not, because QED certainly diverges when we sum over all orders; that is, it is an asymptotic theory, not a convergent one. More specifically, what we want is a theory based on an induction process, such that at any finite order, all functions are manifestly finite. We need a functional equation that allows one to calculate all self-energy and vertex parts at the $n + 1$st level when we are given these functions at the nth level.

Since there are many functional equations that link the nth-order functions to the $n + 1$st-order functions, there are also many inductive schemes that can renormalize field theory.[23,24] One of the most useful of these schemes is the inductive process using the BPHZ and renormalization group equations (which will be discussed in further detail in Chapters 13 and 14).

In summary, we have seen that renormalization theory gives us a solution to the ultraviolet divergence problem in quantum field theory. The renormalization program proceeded in several steps. First, by power counting, we isolated the divergences of all graphs, which must be a simple function of the number of external lines. Second, we regulated these divergences via cutoff or dimensional regularization. Third, we showed that these divergences can be canceled either by adding counterterms to the action, or by absorbing them into multiplicative renormalizations of the physical parameters. Finally, we showed that all divergences can be absorbed in this way.

The weakness of our proof, however, is that it handles overlapping divergences in an awkward (possibly incorrect) way, and that it is not general enough to handle different kinds of field theories. It remains to be seen if the overlapping divergence problem can be truly solved in this formalism. In Chapters 13 and 14, we will present the BPHZ and renormalization group proofs of renormalization, which are not plagued by overlapping divergences and are much more versatile.

This completes our discussion of QED. Next, in Part II we will discuss the Standard Model.

7.9   Exercises

1. Show that $Z_1 = Z_2$ to one-loop order for the electromagnetic field coupled to a triplet of π mesons, where Z_1 is the vertex renormalization constant, and Z_2 is for the π self-energy. (Hint: use the Ward–Takahashi identity.)

2. Do a power counting analysis of ϕ^p in q dimensions; isolate the graphs that are divergent. Confirm the statements made in the text concerning the renormalizability or super-renormalizability of the theory in various dimensions.

3. Draw all possible graphs necessary to prove the Ward–Takahashi identity for QED to order α^3.

4. Do a power counting of the massive Yukawa theory, with interaction term $\bar{\psi}\psi\phi$. Isolate all divergent graphs. Show that all divergences can be, in principle, moved into the physical parameters of the system. Outline the renormalization program.

5. Analyze the renormalization properties of a derivative coupling theory:

$$g\bar{\psi}\gamma^\mu\psi\partial_\mu\phi \tag{7.137}$$

   Is the S matrix equal to one? Is the theory trivial? Consider making a field redefinition on the ψ and $\bar{\psi}$. Calculate the self-energy correction to the fermion propagator and verify your conjecture.

6. Consider the one-loop electron self-energy diagram in QED. Let this electron also interact with an external scalar field via a derivative coupling, as in the previous problem. To first order in g, attach this derivative coupling term in all possible places along each electron propagator. There are three such graphs. Show that the sum of these three graphs is zero.

7. Repeat the previous problem, except consider an electron line with all possible photon lines attached to it to all orders in e. To first order in g, attach the

derivative coupling interaction along all electron propagators and show that this also sums to zero.

8. Consider the one-loop correction to the propagator in a massive ϕ^3 theory. Calculate the one-loop correction using both the Pauli–Villars method and the dimensional regularization method in order to find a relationship between ϵ and Λ/m.

9. Prove Eq. (7.91). Fill in the missing steps in its derivation that were omitted in the text.

10. Fill in the missing steps in Eqs. (7.93), (7.94), (7.96), and (7.97).

11. Let $K_{\alpha\beta\gamma\delta}$ equal the complete four-electron Green's function, where the Greek letters label the Dirac spinor indices. From this, we can construct what are called the Schwinger–Dyson equations for the electron vertex, with electron momenta p and p':

$$
\begin{aligned}
\tilde{\Gamma}_\mu(p', p)_{\gamma\delta} \;=\; & Z_1(\gamma_\mu)_{\gamma\delta} - \int \frac{d^4q}{(2\pi)^4} \big[\tilde{S}'_F(p'+q)\tilde{\Gamma}_\mu(p'+q, p+q) \\
& \times \tilde{S}'_F(p+q)\big]_{\beta\alpha} \tilde{K}_{\alpha\beta,\delta\gamma}(p+q, p'+q, q) \qquad (7.138)
\end{aligned}
$$

and for the photon propagator:

$$
\Pi_{\mu\nu}(q) = iZ_1 \int \frac{d^4k}{(2\pi)^4} \mathrm{Tr}\big[\gamma_\mu \tilde{S}'_F(k)\tilde{\Gamma}_\nu(k, k+q)\tilde{S}'_F(k+q)\big] \qquad (7.139)
$$

Graphically, write down what these recursion relations look like. Then show that they are graphically correct to two-loop order.

12. Show that K does not suffer from overlapping divergences, which means that the Schwinger–Dyson equations (instead of the Ward–Takahashi identities) may be used to renormalize QED to all orders.[23,24]

Part II

Gauge Theory
and the Standard Model

Chapter 8
Path Integrals

> *One feels as Cavalieri must have felt calculating the volume of a pyramid before the invention of calculus.*
>
> —R. Feynman

8.1 Postulates of Quantum Mechanics

Previously, we outlined how to quantize field theories with various spins using the canonical quantization approach. However, for increasingly complex systems, such as gauge theory, quantum gravity, and superstring theory, canonical quantization proves to be a very clumsy formalism since manifest Lorentz invariance is broken. Instead, we will explore a new method in this chapter.

Perhaps the most powerful quantization method is the *path integral* approach, which was developed by Feynman,[1,2] based on an idea of Dirac.[3] The path integral method is versatile enough to handle a variety of different types of gauge theories. The path integral approach has many advantages over the other techniques:

1. The path integral formalism yields a simple, covariant quantization of complicated systems with constraints, such as gauge theories. While calculations with the canonical approach are often prohibitively tedious, the path integral approach yields the results rather simply, vastly reducing the amount of work.

2. The path integral formalism allows one to go easily back and forth between the other formalisms, such as the canonical or the various covariant approaches. In the path integral approach, these various formalisms are nothing but different choices of gauge.

3. The path integral formalism is based intuitively on the fundamental principles of quantum mechanics. Quantization prescriptions, which may seem rather arbitrary in the operator formalism, have a simple physical interpretation in the path integral formalism.

4. The path integral formalism can be used to calculate nonperturbative as well as perturbative results.

5. The path integral formalism is based on c-number fields, rather than q-number operators. Hence, the formalism is much easier to manipulate.

6. At present, there are a few complex systems with constraints that can only be quantized in the path integral formalism.

7. Renormalization theory is much easier to express in terms of path integrals.

Our discussion of the path integral formalism begins with two deceptively simple principles:

8.1.1 Postulate I

The probability $P(b, a)$ of a particle moving from point a to point b is the square of the absolute value of a complex number, the transition function $K(b, a)$:

$$P(b, a) = |K(b, a)|^2 \tag{8.1}$$

8.1.2 Postulate II

The transition function $K(b, a)$ is given by the sum of a phase factor $e^{iS/\hbar}$, where S is the action, taken over all possible paths from a to b:

$$K(b, a) = \sum_{\text{paths}} k e^{iS/\hbar} \tag{8.2}$$

where the constant k can be determined by:

$$K(c, a) = \sum_{\text{paths}} K(c, b) K(b, a) \tag{8.3}$$

where we sum over all intermediate points b connecting a and c.

These postulates incorporate the essence of the celebrated double slit experiment, where a beam of electrons passes through a barrier with two small holes. A screen is placed behind the barrier to detect the presence of the electrons. As a point particle, an electron cannot, of course, go through both holes simultaneously. Classically, therefore, we expect that the electrons will go through one slit or the other, leaving two distinct marks on the screen just behind the two holes.

However, experiments show that the pattern created on the screen by repeated passages of the electrons through these holes is an interference pattern, associated with wave-like, not particle-like, behavior. Classically, we are therefore left with a paradox. A point particle cannot go through both holes at once, yet the passage of a large number of electrons successively going past the barrier clearly leaves an interference pattern, with minima and maxima, as if the electron somehow went through both holes.

In the path integral approach, as in quantum mechanics, this puzzle can be resolved. The postulates of the path integral approach and quantum mechanics do not allow us to calculate the precise motion of a single point particle. They only allow us to calculate probability amplitudes. The probability that an electron will go from the source past the slits to the screen is given by summing over *all* possible paths. These probabilities, in turn, may have wave-like behavior, even if the electron itself is a point particle.

The sum over paths reproduces the interference pattern that is experimentally seen on the screen. Thus, the path integral approach incorporates the philosophy behind the double-slit experiment, which, in turn, embodies the essence of the quantum principle.

As in quantum mechanics, we make the transition to classical mechanics by taking the limit $\hbar \to 0$. For large values of S, the exponential of $iS/\hbar$ undergoes large fluctuations, and hence cancels out to zero. Hence, the contribution of the paths that maximize the action S do not contribute much to the sum over paths:

$$\delta S \gg \hbar : \quad \sum_{\text{paths}} e^{iS/\hbar} \sim 0 \tag{8.4}$$

In the classical limit, the paths that dominate the sum are the ones where $\delta S/\hbar$ is as small as possible. However, the path for which δS is minimized is just the classical path:

$$\delta S = 0 \quad \to \quad \text{classical mechanics} \tag{8.5}$$

Thus, we recover classical mechanics in the limit as $\hbar \to 0$. The picture that emerges from the path integral approach is therefore intuitively identical to the principles of quantum mechanics. To calculate the probability that a particle at point a goes to a point b, one must sum over all possible paths connecting these two points, including the classical one. The path preferred by classical mechanics is the one that minimizes the action for $\delta S \ll \hbar$ (Fig. 8.1).

Although the path integral method gives us an elegant formalism in which to reformulate all of quantum field theory, one should also point out the potential drawbacks of the formalism. One problem is that the path integral is not well defined in Minkowski space. In this chapter, we will assume that all path integrals are computed with the Euclidean metric. Then, the functional integral is taken over e^{-S}, which has much better convergence properties than integrals over e^{iS}. At the end of the calculation, we assume that we can analytically continue back

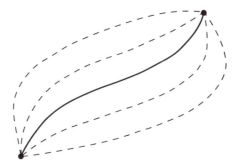

Figure 8.1. The path integral sums over all possible paths connecting two points, including the one favored by classical mechanics. In this way, the path integral sums over quantum corrections to classical mechanics.

to Minkowski space. (The question of whether this analytic continuation from Minkowski space to Euclidean space and back again is rigorously defined is a highly nontrivial question. This is a delicate matter, the subject of a field called axiomatic field theory, which is beyond the scope of this book.)

Another problem is that the transition between c numbers and operators becomes illdefined when the Hamiltonian has ordering problems. The path integral over a system with the Hamiltonian of the form $p^2 f(q)$, for example, becomes ambiguous when making the transition to the operator language, since p and q do not commute. For systems more complex than the harmonic oscillator, the integrals may not be Gaussian, and ordering problems may creep into the path integral. For complicated systems, one must often use "point splitting" methods, that is, separating two fields by a small infinitesimal amount in space-time in order to regularize the integrals. Unfortunately, a detailed elaboration of these delicate points is also beyond the scope of this book.

With these problems in mind, let us now compute with the path integral approach. We first divide up a path by discretizing space–time. Let us divide up each path in three-space into N points (Fig. 8.2). Then the "sum over all paths" can be transformed into a functional integral:

$$\sum_{\text{paths}} = \lim_{N \to \infty} \prod_{i=1}^{3} \prod_{n=1}^{N} \delta x_n^i \to \int Dx \qquad (8.6)$$

The integral $\int Dx$ is not an ordinary integral. It is actually an infinite product of integrals, taken over all possible $dx(t)$. Whenever we use the differential symbol D, we should remember that it is actually an infinite product of differentials taken

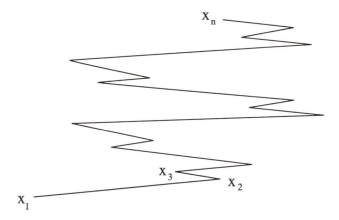

Figure 8.2. To calculate with path integrals, we break the path into a discrete number of intermediate points, and then integrate over the position of these intermediate points.

over all points. In this functional language, the transition function becomes:

$$K(b, a) = k \int_a^b Dx \, e^{iS/\hbar} \qquad (8.7)$$

where k can be determined as follows:

$$K(c, a) = \int K(c, b) K(b, a) \, Dx_b \qquad (8.8)$$

where we integrate over all possible intermediate points x_b which link points a and c.

To give this approach some substance, let us begin with the simplest of all possible classical systems, the free nonrelativistic point particle in the first quantized formalism. Our discussion begins with the classical action:

$$S = \int dt \, \frac{1}{2} m \dot{x}_i^2 \qquad (8.9)$$

Let us now discretize the paths. We take dt to be a small interval ϵ and then discretize the Lagrangian:

$$dt \quad \rightarrow \quad \epsilon$$

$$\frac{1}{2} m \dot{x}_i^2 dt \quad \rightarrow \quad \frac{1}{2} m (x_n - x_{n+1})_i^2 \epsilon^{-1} \qquad (8.10)$$

The transition function $K(b, a)$ can be written as the path integral over e^{iS}:

$$
\begin{aligned}
K(b, a) \;=\; & \lim_{\epsilon \to 0} \int \int \cdots \int dx_2 \, dx_3 \cdots dx_{N-1} \\
& \times \; k \exp\left(\frac{im}{2\epsilon} \sum_{n=1}^{N-1} (x_n - x_{n+1})^2 \right)
\end{aligned}
\tag{8.11}
$$

Unfortunately, one of the drawbacks of the path integral formalism is that embarrassingly few functional integrals can actually be performed. However, we will find that the simplest Gaussian path integral is also the one most frequently found for free systems. Specifically, we will repeatedly use the Gaussian integral:

$$
\int_{-\infty}^{\infty} dx \, x^{2n} e^{-r^2 x^2} = \frac{\Gamma(n + \frac{1}{2})}{r^{2n+1}}
\tag{8.12}
$$

To evaluate the expression for $K(b, a)$, we now perform one of these Gaussian integrations:

$$
\begin{aligned}
& \int_{-\infty}^{\infty} dx_2 \, \exp\left[-a(x_1 - x_2)^2 - a(x_2 - x_3)^2 \right] \\
& = \sqrt{\frac{\pi}{2a}} \exp\left[-\frac{1}{2} a(x_1 - x_3)^2 \right]
\end{aligned}
\tag{8.13}
$$

The key point is that the Gaussian integral over x_2 has left us with another Gaussian integral over the remaining variables. This process can be repeated an arbitrarily large number of times: Each time we perform a Gaussian integral on an intermediate point, we find a Gaussian integral among the remaining variables. After repeated integrations, we find:

$$
\begin{aligned}
& \int_{-\infty}^{\infty} dx_2 \cdots dx_{N-1} \, \exp\left[-a(x_1 - x_2)^2 - \cdots - a(x_{N-1} - x_N)^2 \right] \\
& = \sqrt{\frac{\pi^{N-2}}{(N-1)a^{N-2}}} \exp\left[-\frac{a}{N-1}(x_1 - x_N)^2 \right]
\end{aligned}
\tag{8.14}
$$

This, in turn, allows us to calculate the constant k:

$$
k = \left(\frac{2\pi i \epsilon}{m} \right)^{-(1/2)N}
\tag{8.15}
$$

If we take the limit as the number of intermediate points goes to infinity, then we are left with the final result for the transition function:

$$K(b, a) = \left| \frac{m}{2\pi(t_b - t_a)} \right|^{1/2} \exp \frac{(1/2)im(x_b - x_a)^2}{t_b - t_a} \tag{8.16}$$

This is a pleasant result. This is exactly the Green's function we derived in Eq. (3.58) for nonrelativistic quantum mechanics. Beginning with only the postulates of the path integral approach and the simplest possible classical action, we have derived the Green's function found in quantum mechanics that propagates Schrödinger waves. It obeys the equation:

$$-\frac{1}{2m} \frac{\partial^2}{\partial x_b^2} K(b, a) = i \frac{\partial}{\partial t_b} K(b, a) \tag{8.17}$$

Our first exercise in the path integral formalism gave us encouraging results. Now let us tackle more general and more difficult problems, such as (1) the transition between the Lagrangian and the Hamiltonian approaches and (2) the transition from c-number expressions to operator q-number expressions. In the usual canonical approach, these two transitions appear rather ad hoc and counterintuitive.

In the path integral approach, the transition between the Lagrangian and Hamiltonian systems is easily performed by adding an infinite sequence of Gaussian integrations for the momentum p_i. For each infinitesimal integration, we use the fact that:

$$\int_{-\infty}^{\infty} dp \, e^{iap^2 + ibp} = \sqrt{\frac{i\pi}{a}} e^{-ib^2/4a} \tag{8.18}$$

which can be proved by completing the square. If we let $a = -1/2m$ and $b = \dot{x}$ and integrate over an infinite number of these momenta, then we have:

$$\begin{aligned} K(b, a) &= \int_{x_a}^{x_b} Dx \exp i \int_{t_a}^{t_b} dt \left[\frac{1}{2} m(\dot{x}_i)^2 - V(x) \right] \\ &= \int_{x_a}^{x_b} Dx \, Dp \exp i \int_{t_a}^{t_b} dt \left(p\dot{x}_i - \frac{p_i^2}{2m} - V(x) \right) \end{aligned} \tag{8.19}$$

The Lagrangian appears on the first line, but the Hamiltonian, defined by $H(p, x) = p^2/2m + V(x)$, appears on the second line. By performing the functional integral over Dp, we can go back and forth between the Lagrangian and Hamiltonian formalisms. In the path integral formalism, the relationship between the Lagrangian and the Hamiltonian formalism is no mystery, but simply the byproduct of performing an additional functional integration over momentum.

(For clarity, because normalization factors, such as $1/2\pi$, appear repeatedly throughout our discussion, we have absorbed them into the definition of Dx and Dy. We will henceforth drop these trivial normalization factors, since they can can always be explicitly written out later.) Thus, in the path integral formalism, the difference between the two formalisms only lies in a Gaussian integration over momentum. The path integral formalism allows us to go between these formalisms with ease:

$$L = \frac{1}{2}m(\dot{x}_i)^2 - V(x) \leftrightarrow H = \frac{p_i^2}{2m} + V(x) \tag{8.20}$$

So far, everything has been defined in terms of c-number expressions. Operators, which are the basis of the canonical approach, do not enter into the picture at all. Now, let us make the second transition, this time from the path integral formalism to the operator formalism, to show that the operator formalism that we have patiently developed in Chapters 3 and 4 is nothing but a specific representation of the path integral.

We recall that in the canonical formalism, the starting point was the canonical equal-time commutation relation between fields $\phi(x)$ and their conjugates $\pi(x)$. Only later could we calculate the propagators and finally the S matrix. In the path integral formalism, the sequence is roughly the reverse. We begin with the S matrix as the starting point, and we later derive the operator formalism as a consequence.

To see how operators naturally emerge in a formalism defined entirely without operators, let us write the transition function between point x_1 at time t_1 to point x_N at time t_N in the Heisenberg representation. We will carefully divide the path into N intermediate points. In this formalism, the transition probability of a particle at point x_1 and time t_1 going to x_N and time t_N is given by the matrix element between eigenstates $|x, t\rangle$

The Heisenberg representation, we recall, is based on a complete set of position eigenstates $|x\rangle$ of the position operator $\hat{x}$, which is now treated as an operator with eigenvalue x:

$$\hat{x}|x\rangle = x|x\rangle \tag{8.21}$$

We also introduce eigenstates of the momentum operator p:

$$1 = \int |x\rangle\, dx\, \langle x|$$

$$1 = \int |p\rangle\, dp\, \langle p| \tag{8.22}$$

such that they are normalized as follows:

$$\langle x|y \rangle \;=\; \delta(x-y)$$

$$\langle p|x \rangle \;=\; \frac{e^{ipx}}{\sqrt{2\pi}}; \quad \langle x|p \rangle = \frac{e^{-ipx}}{\sqrt{2\pi}} \tag{8.23}$$

To check the consistency of this normalization, we perform the following manipulations:

$$
\begin{aligned}
\langle x|y \rangle &= \langle x|p \rangle \int dp \langle p|y \rangle \\
&= \int dp \frac{e^{-ipx}}{\sqrt{2\pi}} \frac{e^{ipy}}{\sqrt{2\pi}} \\
&= \int \frac{dp}{2\pi} e^{-ip(x-y)} \\
&= \delta(x-y) \tag{8.24}
\end{aligned}
$$

Our normalizations are thus consistent.

Our task is now to rewrite the functional integration over p at an intermediate point along the path in terms of an operator expression defined in the Heisenberg picture. We will use the fact that the transition element between two neighboring points can be written as:

$$\langle x_2|e^{-iH(t_2-t_1)}|x_1 \rangle = \langle x_2, t_2|x_1, t_1 \rangle \tag{8.25}$$

Let us take a specific value of $\dot{x} \sim (x_1 - x_2)\delta t$ and dp that appears within the functional integral and carefully rewrite the integral over dp and its integrand as follows:

$$
\begin{aligned}
\int \frac{dp}{2\pi} e^{i(p\dot{x} - H(x,p))\delta t} &= \int \frac{dp}{2\pi} e^{-iH(x,p)\delta t} e^{ip(x_1-x_2)} \\
&= e^{-iH(x,\partial_x)\delta t} e^{-ix_2 p} \int \frac{dp}{\sqrt{2\pi}} e^{ipx_1} \\
&= e^{-iH(x,\partial_x)\delta t} \langle x_2|p \rangle \int dp \langle p|x_1 \rangle \\
&= e^{-iH(x,\partial_x)\delta t} \langle x_2|x_1 \rangle \\
&= \langle x_2|e^{-iH(x,\partial_x)\delta t}|x_1 \rangle \\
&= \langle x_2, t_2|x_1, t_1 \rangle \tag{8.26}
\end{aligned}
$$

We have now made the transition between a Lagrangian defined in terms of x and $\dot{x}$ and a Hamiltonian defined in terms of x and its derivative ∂_x. The transition was made possible because the derivative of the exponential brings down a p:

$$\partial_x e^{ipx} = ip e^{ipx}$$

$$e^{-iH(x,\partial_x)\delta t} e^{ipx} = e^{-iH(x,p)\delta t} e^{ipx} \tag{8.27}$$

In the path integral formalism, this is the origin of the transition between c numbers and q-number operators; that is, the insertion of intermediate states defined in p space allows us to replace the p variable with a ∂_x operator. Thus, we have made the transition between:

$$H(x, p) \quad \longleftrightarrow \quad H(x, \partial_x)$$

$$p \quad \longleftrightarrow \quad -i \frac{\delta}{\delta x} \tag{8.28}$$

In summary, we have now shown that the path integral formalism can express the propagator $K(b, a)$ in three different ways, in the Lagrangian or Hamiltonian formalism, or in the operator formalism in the Heisenberg picture. This can be summarized by the following identity:

$$
\begin{aligned}
K(N, 1) &= \langle x_N, t_N | x_1 t_1 \rangle \\
&= \langle x_N, t_N | x_{N-1}, t_{N-1} \rangle \int dx_{N-1} \langle x_{N-1}, t_{N-1} | \\
&\quad \dots | x_2, t_2 \rangle \int dx_2 \langle x_2, t_2 | x_1, t_1 \rangle \\
&= \int Dx \, \exp \left(i \int_{t_1}^{t_N} dt \, L(x, \dot{x}) \right) \\
&= \int Dx \, Dp \, \exp \left(i \int_{t_1}^{t_N} dt \, (p\dot{x} - H(x, p)) \right) \tag{8.29}
\end{aligned}
$$

Finally, let us reanalyze, from the point of view of path integrals, how the time ordering operator T enters into the propagator. Let us analyze the matrix element of an initial state $|x_i, t_i\rangle$ with a final state $\langle x_n, t_n |$, with the operators $x_j(t_j)$ and $x_k(t_k)$ sandwiched between them. We will assume that $t_j > t_k$. As before, we will take time slices and insert a series of complete intermediate states between the states at each slice:

$$\langle x_n, t_n | x(t_j) x(t_k) | x_1, t_1 \rangle = \langle x_n, t_n | x_{n-1}, t_{n-1} \rangle \int dx(t_{n-1}) \langle x_{n-1}, t_{n-1} |$$

$$\cdots |x_{j+1}, t_{j+1}\rangle \int dx(t_{j+1}) \langle x_{j+1}, t_{j+1} | x_j, t_j \rangle x(t_j)$$

$$\cdots |x_{k+1}, t_{k+1}\rangle \int dx(t_{k+1}) \langle x_{k+1}, t_{k+1} | x_k, t_k \rangle x(t_k)$$

$$\cdots |x_2, t_2\rangle \int dx(t_2) \langle x_2, t_2 | x_1, t_1 \rangle \tag{8.30}$$

Taking the limit as the number of time slices goes to infinity, we have:

$$\langle x_n, t_n | x(t_j) x(t_k) | x_1, t_1 \rangle = \int Dx \, Dp \; x(t_j) x(t_k)$$

$$\times \exp\left(i \int_{t_1}^{t_n} (p\dot{x} - H(p, x)) \, dt \right)$$

$$= \int Dx \; x(t_j) x(t_k) \exp i \left(\int_{t_1}^{t_n} L(x, \dot{x}) \, dt \right)$$

$$\tag{8.31}$$

Now, let us reverse the order of the times, such that $t_j < t_k$. In this case, the previous formula must be modified because we can no longer take time slices. Thus, whenever $t_j < t_k$, we cannot make the transition from operators to path integrals unless we reverse the ordering of the operators. In order for this formalism to make sense, we will always reverse the order of the operators, such that the later times always appear to the left, so that we can proceed with taking time slices. To enforce this condition, we must use the time ordered product in this case:

$$\langle x_n, t_n | T \left[x(t_j) x(t_k) \right] | x_1, t_1 \rangle$$

$$= \int Dx \, Dp \; x(t_j) x(t_k) \exp\left(i \int_{t_1}^{t_n} dt \, (p\dot{x} - H(p, x)) \right) \tag{8.32}$$

For a large number of insertions, we have obviously:

$$\langle x_n, t_n | T \left[x(t_j) x(t_k) \cdots x(t_m) \right] | x_1, t_1 \rangle$$

$$= \int Dx \, Dp \; x(t_j) x(t_k) \cdots x(t_m) \exp\left(i \int_{t_1}^{t_n} [p \cdot x - H(p, x)] \right) \tag{8.33}$$

We emphasize that the left-hand side consists of operators, so the ordering of the times is important. However, on the right-hand side we have a c-number expression, where the ordering of the $x(t_i)$ makes no difference. The correspondence between operators and these c-number expressions in the path integral only holds when we can make time slices, that is, when the operators are time ordered. From the path integral point of view, this is the origin of the time ordering in the matrix elements.

8.2   Derivation of the Schrödinger Equation

In a first quantized formalism, where the action is a function of x^i and not fields, the path integral formalism gives us an added bonus: It gives us a derivation of the Schrödinger equation. Usually, introductory courses in quantum mechanics begin by postulating the Schrödinger wave equation. Certain conventions, such as the quantization of x and p, seem rather arbitrary. Only later emerges the probabilistic interpretation. Here, we reverse this order: we begin with the probabilistic postulates of quantum mechanics and derive the Schrödinger wave equation as a consequence, thus giving a new physical interpretation to that equation.

In the path integral approach, the evolution of a state is given by the transition function $K(b, a)$. From a classical point of view, this can be viewed as the analogue of Huygen's principle, where the evolution of a wave can be determined by assuming that each point along a wave front emits a new wave front. The integration over all these infinitesimal wave fronts then gives us the overall evolution of the wave front. Mathematically, this is given by:

$$\psi(x_j, t_j) = \int_{-\infty}^{\infty} K(x_j, t_j; x_i, t_i)\psi(x_i, t_i)\, dx_i \tag{8.34}$$

Earlier, we derived, assuming only the Lagrangian $(1/2)mv_i^2$, an expression for the nonrelativistic transition function. Now let us calculate how wave fronts move with this transition function. The time evolution, from t to $t + \delta t$, is given by:

$$\psi(x, t + \epsilon) = \int_{-\infty}^{\infty} A^{-1} \exp\left(\frac{im(x - y)^2}{2\epsilon}\right)\psi(y, t)\, dy \tag{8.35}$$

where:

$$A = \left(\frac{2\pi i \epsilon}{m}\right)^{1/2} \tag{8.36}$$

To perform this integration, let dy be replaced by $d\eta$, where $\eta = y - x$:

$$\psi(x, t + \epsilon) = \int_{-\infty}^{\infty} A^{-1} e^{im\eta^2/2\epsilon}\psi(x + \eta, t)\, d\eta \tag{8.37}$$

Now Taylor expand the left-hand side in terms of t, and the right-hand side in terms of η:

$$\psi(x,t) + \epsilon \frac{\partial \psi}{\partial t} = \int_{-\infty}^{\infty} A^{-1} e^{im\eta^2/2\epsilon}$$

$$\times \left(\psi(x,t) + \eta \frac{\partial \psi}{\partial x} + \frac{1}{2}\eta^2 \frac{\partial^2 \psi}{\partial x^2} + \cdots \right) d\eta \quad (8.38)$$

The integration over $d\eta$ is easily performed. The integration over the linear term in η vanishes because it is linear, and the integration over the higher terms vanish in the limit $\epsilon \to 0$. This gives us:

$$i\frac{\partial \psi}{\partial t} = -\frac{1}{2m}\frac{\partial^2 \psi}{\partial x^2} \quad (8.39)$$

This is the Schrödinger wave equation, as desired. It is straightforward to insert a potential into the path integral, in which case we derive the Schrödinger wave equation in a potential, which is the traditional starting point for quantum mechanics.

8.3 From First to Second Quantization

So far, we have only investigated the path integral formalism in the first quantized formalism, reproducing known results. The reader may complain that the path integral formalism is an elaborate, powerful machinery that has only rederived simple results. However, when we make the transition to the second quantized formalism and eventually to gauge theory, we will find that the path integral approach is the preferred formalism for quantum field theory. We saw earlier that the integration over all intermediate points along a path was enforced by inserting the number "1" at each intermediate point:

$$1 = |x_i, t_i\rangle \int dx_i \langle x_i, t_i| \quad (8.40)$$

The transition to field theory is made by introducing yet another expression for the number "1," this time based on an integration over an infinite number of degrees of freedom. We will use the familiar Gaussian integration:

$$\delta_{ij} = \frac{2}{(\sqrt{\pi})^n} \int \prod_{l=1}^{n} dx_l \, (x_i \, x_j) \exp\left(-\sum_{k=1}^{n} (x_k)^2 \right) \quad (8.41)$$

Now let us replace the variable x_i with a function $\psi(x)$, which is temporarily viewed as a discretized number ψ_x, where x is now seen as an infinite discrete index.

The transition from finite degrees of freedom to infinite degrees of freedom is then made by inserting the following expression for "1" into the path integral:

$$\delta_{x,y} \sim \int D\psi \, D\psi^* \, \psi_x^* \psi_y \exp - \left(\sum_z \psi_z^* \psi_z \right) \tag{8.42}$$

where:

$$D\psi = \prod_z d\psi_z \tag{8.43}$$

Written in terms of functions $\psi(x)$ rather than discretized variables ψ_z, we now have:

$$\delta(x - y) = \int D\psi \, D\psi^* \, \psi^*(x)\psi(y) \exp - \left(\int Dx |\psi(x)|^2 \right) \tag{8.44}$$

These expressions can also be rewritten in terms of bra and ket vectors as follows:

$$\psi(x) \quad = \quad \langle x|\psi \rangle$$
$$\psi^*(x) \quad = \quad \langle \psi|x \rangle \tag{8.45}$$

This allows us to write:

$$\begin{aligned}
\delta(x - y) \quad &= \quad \int D^2\psi \, \psi^*(x)\psi(y) \exp - \left(\int Dx \, \psi^*(z)\psi(z) \right) \\
&= \quad \langle x|\psi \rangle \int D^2\psi \, \exp - \left(\int \langle \psi|z \rangle \, Dz \, \langle z|\psi \rangle \right) \langle \psi|y \rangle \\
&= \quad \langle x|1|y \rangle \tag{8.46}
\end{aligned}$$

Written in this language, the number "1" now becomes:

$$1 = |\psi \rangle \int d^2\psi \, e^{-\langle \psi|\psi \rangle} \langle \psi| \tag{8.47}$$

We can now repeat all the steps used in making the transition from the Lagrangian approach to the operator approach in the Heisenberg picture by inserting this new expression for the number "1" into the path integral. When we do this, we then have an expression for the transition function written entirely in terms of $\psi(x)$. A straightforward insertion of this new set of intermediate states yields:

$$
\begin{aligned}
K(b, a) &= \int_{x_a}^{x_b} Dx \, e^{i \int dt \, \frac{1}{2} m \dot{x}_i^2} \\
&= \int D\psi \, D\psi^* \, \psi(x_a)^* \psi(x_b) e^{i \int dt \, \psi^* (i\partial_t - H)\psi}
\end{aligned}
\tag{8.48}
$$

where the Lagrangian is equal to:

$$
L = \psi^* \left(i \frac{\partial}{\partial t} - H \right) \psi
\tag{8.49}
$$

At this point, we have now derived a second quantized version of the non-relativistic Schrödinger equation. This may seem odd, since usually quantum field theory is associated with the merger of relativity and quantum mechanics. But quantum field theory can be viewed independently from relativity; that is, the essence of quantum field theory is that it has an infinite number of quantum degrees of freedom. In this sense, the path integral formalism can accomodate a nonrelativistic Schrödinger field theory.

Next, we would like to compute the familiar expressions found in Chapter 3 and 4 in terms of the path integral approach. First, we define the average $\langle O \rangle$ of the expression O by inserting it into the integral:

$$
\langle O \rangle \equiv N \int DX \exp \left(- \sum_{i,j=1}^{n} \frac{1}{2} x_i D_{ij} x_j \right) O
\tag{8.50}
$$

Our goal is to find an expression for $\langle O \rangle$, and later make the transition to an infinite number of degrees of freedom ($n \to \infty$). To find an expression for this average, we will find it convenient to introduce an intermediate stage in the calculation. We define the generating functional as follows:

$$
I(D, J) \equiv \int \prod_{i=1}^{n} dx_i \exp \left(- \sum_{i,j=1}^{n} \frac{1}{2} x_i D_{ij} x_j + \sum_{i=1}^{n} J_i x_i \right)
\tag{8.51}
$$

where we fix N by setting $I(D, 0) = N^{-1}$.

To find an expression for $I(D, J)$, we first make a similarity transformation $x_i' = S_{ij} x_j$, such that S diagonalizes the matrix D. We are then left with the eigenvalues of the matrix D in the integral. The integration separates into a product of independent integrations over x_i'. We then perform each integration separately, giving us the square root of the eigenvalues of D matrix. The product of the eigenvalues, however, is equal to the determinant of the D matrix, giving us the final result:

$$I(D, J) = (2\pi)^{n/2}(\det D_{ij})^{-1/2} \exp\left(\sum_{i,k} \frac{1}{2} J_i(D^{-1})_{ik} J_k\right) \tag{8.52}$$

Our goal is to evaluate the average of an arbitrary product $x_1 x_2 \cdots x_n$:

$$\langle x_1 x_2 \cdots x_n \rangle \equiv N \int \prod_{i=1}^{n} dx_i \, x_1 x_2 \cdots x_n \exp\left(-\sum_{i,j} \frac{1}{2} x_i D_{ij} x_j\right) \tag{8.53}$$

where the normalization constant N can be fixed via:

$$I(D, 0) = N^{-1} = (2\pi)^{n/2} \left(\det D_{ij}\right)^{-1/2} \tag{8.54}$$

We can also take repeated derivatives of $I(D, J)$ with respect to J, and then set J equal to zero. Each time we take the derivative $\partial/\partial J_i$, we bring down a factor of x_i into the integral:

$$\begin{aligned}
\langle x_1 \cdots x_n \rangle &= \prod_{i=1}^{n} \frac{\partial}{\partial J_i} I(D, J)\Big|_{J=0} \\
&= \sum_{\text{pairings}} D_{k_1 k_2}^{-1} \cdots D_{k_{n-1} k_n}^{-1}
\end{aligned} \tag{8.55}$$

This expression, for an odd number of x's, vanishes. However, for two x's, we have:

$$\langle x_i x_j \rangle = (D^{-1})_{ij} \tag{8.56}$$

For four x's, we have:

$$\begin{aligned}
\langle x_i x_j x_k x_l \rangle &= \Big[(D^{-1})_{ij}(D^{-1})_{kl} \\
&\quad + (D^{-1})_{ik}(D^{-1})_{jl} + (D^{-1})_{il}(D^{-1})_{jk}\Big]
\end{aligned} \tag{8.57}$$

Not surprisingly, if we analyze the way in which these indices are paired off, we see Wick's theorem beginning to emerge. The point here is that Wick's theorem, which was based on arguments concerning normal ordering of operators, is now emerging from entirely c-number integrations.

The transition to quantum field theory, as before, is now made by making the transition from a finite number of variables x_i to an infinite number of variables

$\phi(x)$. As before, we are interested in evaluating the transition probability between a field at point x and a field at point y:

$$\int D\phi\, \phi(x)\phi(y) \exp i \left(\int d^4x\, L(\phi) \right) \tag{8.58}$$

To evaluate this integral, we will find it convenient, as before, to introduce the generating functional:

$$Z(J) = N \int D\phi\, e^{i \int d^4x[L(x)+J(x)\phi(x)]} \tag{8.59}$$

where:

$$D\phi \equiv \prod_x d\phi(x)$$

$$N^{-1} \equiv \int D\phi\, e^{i \int d^4x\, L(x)} \tag{8.60}$$

To perform this integration for a Klein–Gordon field, we will repeat the steps we used for the simpler theory based on finite number of degrees of freedom. We first make a shift of variables:

$$\phi(x) \rightarrow \phi(x) + \phi(x)_{\text{cl}} \tag{8.61}$$

where ϕ_{cl} satisfies the Klein–Gordon equation with a source term. We recall that the Feynman propagator is defined via:

$$(\partial_\mu \partial^\mu + m^2)_x \Delta_F(x - y) = -\delta^4(x - y) \tag{8.62}$$

A classical solution can then be defined via:

$$\phi_{\text{cl}} = - \int \Delta_F(x - y) J(y)\, d^4y \tag{8.63}$$

which satisfies:

$$(\partial_\mu \partial^\mu + m^2)\phi_{\text{cl}} = J(x) \tag{8.64}$$

We can now perform the integral by performing a Gaussian integration:

$$Z(J) = \exp\left(-\frac{i}{2} \int d^4x\, d^4y\, J(x)\Delta_F(x - y)J(y) \right) \tag{8.65}$$

where:

$$N^{-1} = \left[\det\left(\partial_\mu^2 + m^2\right)\right]^{1/2} = \int D\phi \, \exp i \left(\int d^4x \, L(\phi)\right) \qquad (8.66)$$

Using the fact that:

$$\frac{\delta}{\delta J(x)} J(y) = \delta^4(x - y) \qquad (8.67)$$

we find that the transition function is given by:

$$-i\Delta_F(x - y) = \frac{\delta}{\delta J(x)} \frac{\delta}{\delta J(y)} Z(J)\Big|_{J=0} \qquad (8.68)$$

The average of several fields taken at points x_i is now given as follows:

$$\begin{aligned}
\Delta(x_1, x_2, \cdots, x_n) &= i^n \langle 0|T\phi(x_1)\phi(x_2)\cdots\phi(x_n)|0\rangle \\
&= \frac{\delta^n Z(J)}{\delta J(x_1)\delta J(x_2)\cdots\delta J(x_n)}\bigg|_{J=0} \qquad (8.69)
\end{aligned}$$

By explicit differentiation, we can take the derivatives for four fields and find:

$$\begin{aligned}
-\left(\prod_{i=1}^{4} \frac{\delta}{\delta J(x_i)}\right) Z(J)\Big|_{J=0} &= \Delta_F(x_1 - x_2)\Delta_F(x_3 - x_4) \\
&+ \Delta_F(x_1 - x_3)\Delta_F(x_2 - x_4) + \Delta_F(x_1 - x_4)\Delta_F(x_2 - x_3) \quad (8.70)
\end{aligned}$$

In this way, we have derived Wick's theorem starting with purely c-number expressions.

$Z(J)$ can also be written as a power expansion in J. If we power expand the generating functional, then we have:

$$Z(J) = \sum_{n=0}^{\infty} \frac{1}{n!} \int \cdots \int d^4x_1 \cdots d^4x_n J(x_1)\cdots J(x_n)Z^{(n)}(x_1 \cdots x_n) \qquad (8.71)$$

where:

$$\begin{aligned}
Z^{(n)}(x_1, \cdots, x_n) &= \frac{\delta^n}{\delta J(x_1)\cdots J(x_n)} Z(J)\Big|_{J=0} \\
&= i^n \langle 0|T\phi(x_1)\cdots\phi(x_n)|0\rangle \qquad (8.72)
\end{aligned}$$

In this way, the path integral method can derive all the expressions found earlier in the canonical formalism.

8.4   Generator of Connected Graphs

In analyzing complicated Feynman diagrams, we must distinguish between two types of graphs: connected and disconnected graphs. A graph is called disconnected when it can be separated into two or more distinct pieces without cutting any line.

The generating functional $Z(J)$ that we have been analyzing generates all types of Feynman graphs, both connected and disconnected. However, when we apply the formalism of path integrals to a variety of physical problems, including renormalization theory, it is often desirable to introduce a new functional that generates just the connected graphs. The path integral formalism is versatile enough to give us this new generating functional, which is denoted $W(J)$. We define this generator as follows:

$$Z(J) = e^{iW(J)}$$

$$W(J) = -i \log Z(J) \tag{8.73}$$

If we take repeated derivatives of $W(J)$ to calculate the relationship between Z and W, we find:

$$\frac{\delta^2 W}{\delta J(x_1)\,\delta J(x_2)} = \frac{i}{Z^2}\frac{\delta Z}{\delta J(x_1)}\frac{\delta Z}{\delta J(x_2)} - \frac{i}{Z}\frac{\delta^2 Z}{\delta J(x_1)\,\delta J(x_2)} \tag{8.74}$$

and:

$$\frac{\delta^4 W}{\delta J(x_1)\,\delta J(x_2)\,\delta J(x_3)\,\delta J(x_4)} = \left(\frac{i}{Z^2}\frac{\delta^2 Z}{\delta J(x_1)\,\delta J(x_2)}\frac{\delta^2 Z}{\delta J(x_3)\,\delta J(x_4)} + \text{perm.}\right)$$

$$- \frac{i}{Z}\frac{\delta^4 Z}{\delta J(x_1)\,\delta J(x_2)\,\delta J(x_3)\,\delta J(x_4)} \tag{8.75}$$

To analyze the content of these equations, let us power expand $W(J)$ in powers of J:

$$W(J) = \sum_{n=0}^{\infty}\frac{1}{n!}\int dx_1\cdots dx_n\, J(x_1)\cdots J(x_n) W^{(n)}(x_1,\cdots,x_n) \tag{8.76}$$

Taking $J = 0$, we arrive at:

$$i W^{(2)}(x_1, x_2) = Z^{(2)}(x_1, x_2) \tag{8.77}$$

This is not surprising, since the propagator is connected. The expansion, however, becomes nontrivial when we consider expanding out to fourth order, where disconnected graphs enter into the functional:

$$W^{(4)}(x_1, x_2, x_3, x_4) = i\left[Z^{(2)}(x_1, x_2)Z^{(2)}(x_3, x_4) + \text{perm.}\right]$$
$$- iZ^{(4)}(x_1, x_2, x_3, x_4) \tag{8.78}$$

This equation can be checked to show that W generates only connected graphs. For example, in ϕ^4 theory to order λ, we can show that this works as indicated. In this case:

$$Z^{(2)}(x_1, x_2) = -i\Delta_F(x_1 - x_2) + \frac{\lambda}{2}\int d^4z\, \Delta_F(x_1 - z)\Delta_F(z - x_2)\Delta_F(z, z) \tag{8.79}$$

while:

$$Z^{(4)}(x_1, x_2, x_3, x_4) = -[\Delta_F(x_1 - x_2)\Delta_F(x_3 - x_4) + 2\,\text{terms}]$$
$$- \frac{i\lambda}{2}\left(\int d^4z\Delta_F(x_1 - z)\Delta_F(z, z)\Delta_F(z - x_2)\Delta_F(x_3 - x_4) + 5\,\text{terms}\right)$$
$$- \frac{i\lambda}{4!}\left(\int d^4z\Delta_F(x_1 - z)\Delta_F(x_2 - z)\right.$$
$$\left. \times \Delta_F(x_3 - z)\Delta_F(x_4 - z) + 23\,\text{terms}\right) \tag{8.80}$$

Inserting these factors back into the identity for $W^{(4)}(x_1, x_2, x_3, x_4)$, we find that the disconnected pieces cancel, and the only term which survives is the connected piece, which forms the topology of a cross.

Next, we would like to find the generating functional for proper vertices Γ, which is essential in a discussion of renormalization theory. Proper vertices (or one-particle irreducible vertices), we recall, appear when we consider renormalizing coupling constants. We define $\Gamma(\phi)$, the generator of proper vertices, via a Legendre transformation as:

$$\Gamma(\phi) = W(J) - \int d^4x\, J(x)\phi(x) \tag{8.81}$$

(From now on, we will use the symbol ϕ to represent the c-number field.)

The fields ϕ and J have a nontrivial relationship between them. They are not independent of each other. In fact, by taking repeated derivatives, we can establish the relationship between them. Let us take the partial derivative of the previous equations with respect to J (keeping ϕ fixed) and with respect to ϕ, keeping J

fixed. Then by differentiating both sides, we have:

$$\frac{\delta W(J)}{\delta J(x)} = \phi(x); \qquad \frac{\delta \Gamma(\phi)}{\delta \phi(x)} = -J(x) \tag{8.82}$$

Let us take repeated differentials of the above equations. Differentiating by J and by ϕ, we find:

$$G(x, y) = -\frac{\delta^2 W}{\delta J(x)\, \delta J(y)} = -\frac{\delta \phi(x)}{\delta J(y)}$$

$$\Gamma(x, y) = \frac{\delta^2 \Gamma}{\delta \phi(x)\, \delta \phi(y)} = -\frac{\delta J(x)}{\delta \phi(y)} \tag{8.83}$$

If we treat $\Gamma(x, y)$ and $G(x, y)$ as matrices with continuous space–time indices, then they are inverses of each other, as can be seen as follows:

$$\int d^4y\, G(x, y)\Gamma(y, z) = -\int d^4y\, \frac{\delta^2 W}{\delta J(x)\, \delta J(y)} \frac{\delta^2 \Gamma}{\delta \phi(y)\, \delta \phi(z)}$$

$$= \int d^4y\, \frac{\delta \phi(x)}{\delta J(y)} \frac{\delta J(y)}{\delta \phi(z)}$$

$$= \frac{\delta \phi(x)}{\delta \phi(z)}$$

$$= \delta^4(x - z) \tag{8.84}$$

We would now like to establish a relationship between third derivatives of the functionals. If we differentiate the previous equation by $J(u)$, we find that it vanishes. Thus, we find the relationship:

$$\int d^4y\, \frac{\delta^3 W}{\delta J(x)\, \delta J(u)\, \delta J(y)} \frac{\delta^2 \Gamma}{\delta \phi(y)\, \delta \phi(z)}$$

$$= \int d^4y\, \frac{\delta^2 W}{\delta J(x)\, \delta J(y)} \int d^4y'\, G(u, y') \frac{\delta^3 \Gamma}{\delta \phi(y)\, \delta \phi(z)\, \delta \phi(y')} \tag{8.85}$$

where we have the fact that:

$$\frac{\delta}{\delta J(u)} = \int d^4y'\, \frac{\delta \phi(y')}{\delta J(u)} \frac{\delta}{\delta \phi(y')} = -\int d^4y'\, G(u, y') \frac{\delta}{\delta \phi(y')} \tag{8.86}$$

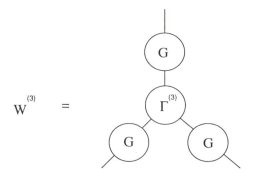

Figure 8.3. Graphic representation of the relationship between $W(J)$ and $\Gamma(\phi)$ out to third order.

We can simplify this equation a bit by inverting the matrix $\Gamma(y, z)$ that appears. Then we find:

$$
\frac{\delta^3 W}{\delta J(x)\, \delta J(y)\, \delta J(z)} = -\int d^4x'\, d^4y'\, d^4z'
$$

$$
\times \quad G(x, x')G(y, y')G(z, z')\frac{\delta^3 \Gamma}{\delta\phi(x')\, \delta\phi(y')\, \delta\phi(z')}
$$

$$(8.87)$$

Graphically, this is represented in Figure 8.3. In this way, we can derive relationships between the generating functionals. However, taking repeated derivatives becomes quite involved when we increase the number of legs. There is yet another, perhaps more direct way in which to see the relationship between the various generating functions by taking power expansions. As before, we can power expand as follows:

$$
\Gamma(\phi) = \sum_{n=0}^{\infty} \frac{1}{n!} \int dx_1 \cdots dx_n\, \phi(x_1) \cdots \phi(x_n)\Gamma^{(n)}(x_1, \cdots x_n) \qquad (8.88)
$$

We want a way to compare $\Gamma^{(n)}$ and $W^{(n)}$. To solve this problem, we would like to power expand ϕ in terms of J. By Taylor's theorem, we know that a power expansion of ϕ can be expressed in terms of powers of J, with coefficients given by the nth derivative of ϕ with respect to J. However, we already know from Eq. (8.83) that the derivative of ϕ with respect to J is given by the second derivative of W with respect to J. Thus, from Eq. (8.82), we find:

$$
\phi(x) = \int d^4x_1\, W^{(2)}(x, x_1)J(x_1) + \frac{1}{2}\int d^4x_1\, d^4x_2\, W^{(3)}(x, x_1, x_2)J(x_1)J(x_2) + \cdots
$$

$$(8.89)$$

We also know the converse, that we can power expand J in terms of ϕ. Likewise, we know from Taylor's theorem that the coefficient of each term is given by the nth derivative of J with respect to ϕ. Thus, Eqs. (8.82) and (8.87) give us:

$$
\begin{aligned}
J(x) \quad = \quad & - \int d^4 x_1 \left[W^{(2)} \right]^{-1} (x, x_1) \phi(x_1) \\
& - \frac{1}{2!} \int d^4 y_1 d^4 y_2 d^4 y_3 d^4 x_1 d^4 x_2 \left[W^{(2)} \right]^{-1} (x, y_3) \left[W^{(2)} \right]^{-1} (x_1, y_1) \\
& \times \left[W^{(2)} \right]^{-1} (x_2, y_2) W^{(3)}(y_1, y_2, y_3) \phi(x_1) \phi(x_2) + \cdots
\end{aligned}
\tag{8.90}
$$

Let us introduce what is called the "amputated" functional:

$$
\tilde{W}^{(n)}(x_1, \cdots, x_n) = \int \prod_{i=1}^{n} d^4 y_i \, G^{-1}(x_i, y_i) W^{(n)}(y_1, \cdots y_n)
\tag{8.91}
$$

This has a simple meaning. Since the connected piece $W^{(n)}$ always has propagators connected to each external leg, this means that $\tilde{W}^{(n)}$ is just the connected part minus these external legs. That is the reason why it is often called the "amputated" Green's functions for connected graphs.

Now that we have solved for $\tilde{W}^{(n)}$, we can now solve for the relationship between $\tilde{W}$ and Γ. Equating terms with the same power of ϕ, we find:

$$
\begin{aligned}
W^{(2)}(x_1, x_2) \quad &= \quad -\left[\Gamma^{(2)}(x_1, x_2) \right]^{-1} \\
\tilde{W}^{(3)}(x_1, x_2, x_3) \quad &= \quad \Gamma^{(3)}(x_1, x_2, x_3) \\
\tilde{W}^{(4)}(x_1, x_2, x_3, x_4) \quad &= \quad \Gamma^{(4)}(x_1, x_2, x_3, x_4) \\
& \quad + \int d^4 y d^4 z \, \Gamma^{(3)}(x_1, x_2, y) W^{(2)}(y, z) \\
& \quad \times \Gamma^{(3)}(z, x_3, x_4) + 2 \text{ terms}
\end{aligned}
\tag{8.92}
$$

Thus, to any desired level of expansion, we can find the relationship between the proper vertices and the amputated Green's function for connected graphs. The advantage of this path integral approach is that our results are independent of perturbation theory. Without having to use complicated graphical techniques, we can rapidly prove nontrivial relations between different types of vertices and propagators. This will prove useful in renormalization theory.

8.5 Loop Expansion

Up to now, we have only explored the power expansion of the S matrix in powers of the coupling constant. However, in quantum field theory it is often convenient to expand in a different power series, one based on the number of loops in a Feynman diagram. In this section, we will show that the loop expansion, in turn, corresponds to a power expansion in $\hbar$, Planck's constant. (To show this, we must reinsert all $\hbar$ factors that were eliminated when we originally set $\hbar = 1$.)

The expansion in loop number or $\hbar$ has important implications. For example, in Chapter 6 we learned that the Feynman tree diagrams for various scattering amplitudes simply reproduced the results of the classical theory. A complicated tree diagram may appear with a large number of coupling constants, but it still only corresponds to the classical theory. To see the true effects of quantization, we have to go beyond the tree diagrams and study loops. A power expansion in the loop number or $\hbar$, rather than the coupling constant, therefore measures the deviation of the quantum theory from the classical theory.

Similarly, loop effects are also important when discussing radiative corrections to a quantum field theory. If a Klein–Gordon theory, for example, has an interacting potential $V(\phi)$, then radiative corrections will modify this potential. These radiative corrections, in turn, are calculated in the loop expansion. These loop corrections are important because they shift the minimum of $V(\phi)$ and hence change the vacua of the theory. This expansion will prove useful when calculating radiative corrections to the effective potential in Chapter 10.

In this section, we will show that the path integral gives a very convenient way in which to power expand a theory in the loop order. Let us now rewrite our previous expressions for the generating functional, explicitly putting back all factors of $\hbar$ that we previously omitted. We know from dimensional arguments and the definition of the path integral that the action appears in the functional as $S/\hbar$, so the generating functional $Z(J)$ can be written as:

$$Z(J) = \int D\phi \exp\left(\frac{i}{\hbar}\int [\mathscr{L} + \hbar J(x)\phi(x)]\, d^4x\right) \tag{8.93}$$

Since we are interested in the relation between $\hbar$ and the standard perturbation theory, let us divide the Lagrangian into the free and interacting parts, $\mathscr{L} = \mathscr{L}_0 + \mathscr{L}_I$. We can extract $\mathscr{L}_I$ from the path integral by converting it into an operator:

$$Z(J) = \exp\left[\frac{i}{\hbar}\int d^4x\, \mathscr{L}_I\left(-\frac{\delta}{\delta J}\right)\right] Z_0(J) \tag{8.94}$$

where $Z_0(J)$ is just a function of the free Lagrangian $\mathcal{L}_0$. If we evaluate the derivatives, then the exponential in front of Z_0 simply reproduces $\mathcal{L}_I(\phi)$ in the exponential, giving us back the original expression for $Z(J)$.

The advantage of writing the functional in this fashion is that we can now explicitly perform the functional integral over ϕ in the free partition functions $Z_0(J)$, leaving us with the standard expression:

$$Z_0(J) = \exp\left(-\frac{i\hbar}{2} \int d^4x \, d^4y \, J(x)\Delta_F(x - y)J(y)\right) \qquad (8.95)$$

Now we can begin the counting of $\hbar$ in a typical Feynman graph by inserting Z_0 back into the expression for $Z(J)$. For any Feynman diagram, each propagator, from the previous expression for Z_0, is multiplied by $\hbar$. However, each vertex, because it appears in the combination $\mathcal{L}_I/\hbar$, is multiplied by a factor of $\hbar^{-1}$.

For an arbitrary Feynman graph, the total counting of $\hbar$ is given by $\hbar$ raised to the power of $P - V$, that is, the number of propagators minus the number of vertices. However, we know from our discussion of renormalization theory in the previous chapter that:

$$L = P - V + 1 \qquad (8.96)$$

where L is the number of loops in a Feynman diagram. For any Feynman diagram, we therefore pick up an overall factor of:

$$\hbar^{P-V} = \hbar^{L-1} \qquad (8.97)$$

It is now easy to see that a power expansion in $\hbar$ is also a power expansion in the loop number. In Chapter 10, we will use this formalism of loop expansions to calculate the radiative corrections to several quantum field theories, showing that the loop expansion is powerful enough to shift the minimum of the potential $V(\phi)$ via radiative corrections. This will prove essential in isolating the true vacuum of a theory with a broken symmetry.

8.6 Integration over Grassmann Variables

Matter, of course, is not just bosonic. To incorporate fermions in the path integral formalism, we must define how to integrate over anticommuting variables. To do this, we must use the *Grassmann variables*, which are a set of anticommuting numbers satisfying $\theta_i\theta_j = -\theta_j\theta_i$. Integration over Grassmann variables is problematic, since $\theta^2 = 0$ for an anticommuting number, and hence the entire foundation of calculus seems to collapse. However, a clever choice allows us to generalize the path integral formalism to fermions.

One of the features that we would like to incorporate in an integration over Grassmann variables is the fact that the integral over all space is translationally invariant:

$$\int_{-\infty}^{\infty} dx\, \phi(x) = \int_{-\infty}^{\infty} dx\, \phi(x+c) \tag{8.98}$$

Let us try to incorporate this feature in an integration over a Grassmann variable:

$$\int d\theta\, \phi(\theta) = \int d\theta\, \phi(\theta+c) \tag{8.99}$$

An arbitrary function $\phi(\theta)$ can be easily decomposed in a Taylor expansion, which terminates after only one term, since higher terms are zero:

$$\phi(\theta) = a + b\theta \tag{8.100}$$

Let us now define:

$$\begin{aligned} I_0 &= \int d\theta \\ I_1 &= \int d\theta\, \theta \end{aligned} \tag{8.101}$$

Inserting this power expansion into the integral, we find:

$$\int d\theta\, \phi(\theta) = aI_0 + bI_1 = (a + bc)I_0 + bI_1 \tag{8.102}$$

In order to maintain this identity, we choose the following normalizations:

$$\begin{aligned} I_0 &= 0 \\ I_1 &= 1 \end{aligned} \tag{8.103}$$

This, in turn, forces us to make the following unorthodox definitions:

$$\int d\theta = 0; \quad \int d\theta\, \theta = 1 \tag{8.104}$$

This is a novel choice of definitions, for it means that the integral over a Grassmann variable equals the derivative:

$$\int d\theta = \frac{\partial}{\partial \theta} \tag{8.105}$$

Now, we generalize this discussion to a nontrivial case of many Grassmann variables. With N such variables, we wish to perform the integration:

$$I(A) = \int \prod_{i=1}^{N} d\theta_i d\bar{\theta}_i \, \exp\left(\sum_{i,j=1}^{N} \bar{\theta}_i A_{ij} \theta_j\right) \tag{8.106}$$

where θ_i and $\bar{\theta}_i$ are two distinct sets of Grassmann variables. To evaluate this integral, we simply power expand the exponential. Because the Grassmann integral of a constant is zero, the only term that survives the integration is the Nth expansion of the exponential:

$$I(A) = \int \prod_{i=1}^{N} d\theta_i \, d\bar{\theta}_i \, \frac{1}{N!} \left(\sum_{i,j=1}^{N} \bar{\theta}_i A_{ij} \theta_j\right)^N \tag{8.107}$$

Most of the terms in the integral of the Nth expansion are zero. The only terms that survive are given by:

$$\begin{aligned}
I(A) &= \int \left(\prod_{i=1}^{N} d\theta_i \, d\bar{\theta}_i \, \bar{\theta}_i \theta_i\right) \left(\sum_{\text{perm}} \epsilon^{i_1 i_2 \cdots i_N} A_{1i_1} A_{2i_2} A_{3i_3} \cdots A_{Ni_N}\right) \\
&= \det A \tag{8.108}
\end{aligned}$$

An essential point is that the determinant appears in the numerator, rather than the denominator. This will have some significant implications later when we discuss ghosts and the Faddeev–Popov quantization program.

Now we make the transition from θ_i to the fermionic field $\psi(x)$. We introduce two sources η and $\bar{\eta}$ and define the generating functional:

$$Z(\eta, \bar{\eta}) = N \int D\psi \, D\bar{\psi} \, e^{i \int d^4 x [\mathscr{L}(x) + \bar{\eta}\psi + \bar{\psi}\eta]} \tag{8.109}$$

where:

$$\mathscr{L} = \bar{\psi}(i\gamma^\mu \partial_\mu - m)\psi \tag{8.110}$$

As before, we can perform the functional integral by shifting variables:

$$\psi(x) \rightarrow \psi(x) + \psi_{\text{cl}}(x) \tag{8.111}$$

where:

$$\psi_{cl}(x) = - \int S_F(x - y)\eta(y) \, d^4y \qquad (8.112)$$

Performing the integral, we find:

$$Z(\eta, \bar{\eta}) = \exp \int d^4x \, d^4y \left(-i\bar{\eta}(x) S_F(x - y)\eta(y) \right) \qquad (8.113)$$

where:

$$N = \det (i \not{\partial} - m) \qquad (8.114)$$

The averages over the field variables can now be found by successively differentiating with respect to the η and $\bar{\eta}$ fields and then setting them equal to zero:

$$
\begin{aligned}
-i S_F(x - y) &= \frac{\delta}{\delta\bar{\eta}(x)} \frac{\delta}{\delta\eta(y)} Z(\eta) \Big|_{\eta=0, \bar{\eta}=0} \\
&= N \int D\psi \, D\bar{\psi} \, \bar{\psi}(y)\psi(x) e^{i \int d^4x \, L(x)}
\end{aligned} \qquad (8.115)
$$

Successive integrations over the source fields gives us:

$$
\begin{aligned}
\prod_{i=1}^{n} \frac{\delta}{\delta\eta(x_i)} \prod_{j=1}^{n} \frac{\delta}{\delta\bar{\eta}(y_j)} Z(\eta, \bar{\eta}) \Big|_{\eta=\bar{\eta}=0} \\
= (i)^{2n} \langle 0|T \bar{\psi}(x_1) \cdots \bar{\psi}(x_n)\psi(y_1) \cdots \psi(y_n)|0 \rangle
\end{aligned} \qquad (8.116)
$$

By performing the functional integration with respect to η and $\bar{\eta}$, we once again retrieve Wick's expansion for fermionic fields.

8.7 Schwinger–Dyson Equations

The functional technique allows us to formulate QED based on the Schwinger–Dyson integral equation. This equation, when power expanded, yields the standard perturbation theory. But since we do not necessarily have to power expand these equations, these integral equations also apply to bound-state and nonperturbative problems. The Schwinger–Dyson equation is based on the deceptively simple observation that the integral of a derivative is zero:

$$\int D\phi \frac{\delta}{\delta\phi} \equiv 0 \tag{8.117}$$

Although this statement appears trivial from the point of view of functional analysis, it yields highly nontrivial relations among generating functionals in quantum field theory.

In particular, let us act on the generating functional $Z(J)$ for a scalar field theory:

$$0 = \int D\phi \frac{\delta}{\delta\phi} \exp\left(iS(\phi) + i\int d^4x J\phi\right) \tag{8.118}$$

The functional derivative of the source term simply pulls down a factor of J. We simply get:

$$0 = \int D\phi \left[iS'(\phi) + iJ\right] \exp\left(iS + i\int d^4x \, J(x)\phi(x)\right) \tag{8.119}$$

This can be rewritten as:

$$\left[S'\left(-i\frac{\delta}{\delta J}\right) + J\right] Z(J) = 0 \tag{8.120}$$

This is the Schwinger–Dyson relation, which is independent of perturbation theory. At this point, we can take any number of derivatives of this equation with respect to the fields and obtain a large number of integral equations involving various Green's functions. Or, we can power expand this equation and reproduce the known perturbation theory.

For QED, the generalization of Eq. (8.120) reads:

$$\left[\frac{\delta}{\delta A_\mu} S\left(-i\frac{\delta}{\delta J_\nu}, -i\frac{\delta}{\delta\bar\eta}, i\frac{\delta}{\delta\eta}\right) + J_\mu\right] Z(J, \eta, \bar\eta) = 0 \tag{8.121}$$

Our strategy will be to convert this expression for $Z(J)$ into an expression for $W(J)$ and to an expression for $\Gamma(\phi)$. Then we will take a derivative with respect to A_μ and set all sources to zero. We begin by using the fact that:

$$\frac{\delta S}{\delta A_\mu} = \left[\partial^2 g_{\mu\nu} - (1 - \alpha^{-1})\partial_\mu\partial_\nu\right] A^\nu - e\bar\psi\gamma_\mu\psi \tag{8.122}$$

We can rewrite the previous equation as an equation on $W(J)$:

$$J_\mu + \left[\partial^2 g_{\mu\nu} - (1 - \alpha^{-1})\partial_\mu\partial_\nu\right] \frac{\delta W}{\delta J_\nu}$$

$$-e\frac{\delta W}{\delta \eta}\gamma_\mu \frac{\delta W}{\delta \bar{\eta}} - e\frac{\delta}{\delta \eta}\left(\gamma_\mu \frac{\delta W}{\delta \bar{\eta}}\right) = 0 \tag{8.123}$$

Now let us convert this expression for $W(J)$ into an expression for $\Gamma(\phi)$, defined by the Legendre transformation:

$$W(J, \eta, \bar{\eta}) = \Gamma(A, \psi, \bar{\psi}) + \int d^4x \left(J_\mu A^\mu + \bar{\psi}\eta + \bar{\eta}\psi\right) \tag{8.124}$$

We must make the substitution:

$$A_\mu = \frac{\delta W}{\delta J^\mu}; \qquad \psi = \frac{\delta W}{\delta \bar{\eta}}$$

$$J_\mu = -\frac{\delta \Gamma}{\delta A^\mu}; \qquad \eta = -\frac{\delta \Gamma}{\delta \bar{\psi}} \tag{8.125}$$

Then the Schwinger–Dyson equation can be written as:

$$\frac{\delta \Gamma}{\delta A^\mu(x)}\bigg|_{\psi=\bar{\psi}=0} = \left[\partial^2 g_{\mu\nu} - (1 - \alpha^{-1})\partial_\mu \partial_\nu\right]A^\nu(x)$$

$$- ie\text{Tr}\left[\gamma_\mu \left(\frac{\delta^2 \Gamma}{\delta \bar{\psi} \delta \psi}\right)^{-1}(x, x)\right] \tag{8.126}$$

where the last term on the right is proportional to the electron propagator, and we have used the fact that:

$$- \delta_{\alpha\beta}\delta^4(x - y) = \int d^4z \frac{\delta^2 W}{\delta \eta_\alpha(x)\delta \bar{\eta}_\gamma(z)} \frac{\delta^2 \Gamma}{\delta \psi_\gamma(z)\delta \bar{\psi}_\beta(y)}\bigg|_{\eta=\bar{\eta}=\psi=\bar{\psi}=0} \tag{8.127}$$

Our last step is to take the derivative with respect to A^ν. Then the term on the right is related to the photon propagator. The final expression becomes:

$$\frac{\delta^2 \Gamma}{\delta A^\mu(x)\delta A^\nu(y)}\bigg|_{A=\psi=\bar{\psi}=0} = \left[\partial^2 g_{\mu\nu} - (1 - \alpha^{-1})\partial_\mu \partial_\nu\right]\delta^4(x - y)$$

$$+ ie^2 \int d^4u\, d^4v\, \text{Tr}\left[\gamma_\mu S_F'(x, u)\Lambda_\nu(y, u, v)S_F'(v, x)\right] \tag{8.128}$$

where we have defined the vertex function as:

$$\frac{\delta^3 \Gamma}{\delta A^\mu(x)\, \delta\bar{\psi}(y)\, \delta\psi(z)}\bigg|_{A=\psi=\bar{\psi}=0} = e\Lambda_\mu(x, y, z) \tag{8.129}$$

We have also used the formula for taking the derivative of an inverse matrix:

$$\frac{\delta}{\delta A_\mu} \mathcal{M}^{-1} = -\mathcal{M}^{-1} \frac{\delta}{\delta A_\mu} \mathcal{M} \mathcal{M}^{-1} \tag{8.130}$$

where $\mathcal{M} = \delta^2 \Gamma / \delta \bar{\psi} \, \delta \psi$ and:

$$
\begin{aligned}
S'_F(x, y) = \mathcal{M}^{-1} &= \left(\frac{\delta^2 \Gamma}{\delta \bar{\psi} \, \delta \psi} \right)^{-1} \\
&= \int \frac{d^4 p}{(2\pi)^4} \frac{e^{-ip \cdot (x-y)}}{\not{p} - m - \Sigma(p)} \tag{8.131}
\end{aligned}
$$

These functional relations, in turn, are identical to the Schwinger–Dyson equations introduced in Exercise (7.11).

In summary, we have seen that the path integral method of Feynman is not only elegant and powerful, it is also very close to the original spirit of quantum mechanics. The formalism is so versatile that we can reproduce the canonical formalism discussed earlier, as well as quantize increasingly complicated theories, such as Yang–Mills theory and quantum gravity. The path integral formalism, in fact, has become the dominant formalism for high-energy physics. In Chapter 9, we will see the power of the path integral approach when we quantize the Yang–Mills theory.

8.8 Exercises

1. Using path integrals for a free Dirac theory, calculate the expectation value of the product of six fermionic fields in terms of propagators. Show that the resulting expression is equivalent to the decomposition given by Wick's theorem.

2. Prove Eq. (8.14).

3. Prove that the disconnected pieces in Eq. (8.78) cancel to second order in λ. Sketch how the cancellation works at third order.

4. Let θ_i be a Grassmann column vector. Let us make the transformation of variables: $\phi_i = M_{ij} \theta_j$. We define:

$$\int d\phi_1 \, d\phi_2 \cdots d\phi_n \ (\phi_1 \phi_2 \cdots \phi_n) = \int d\theta_1 \, d\theta_2 \cdots d\theta_n \ (\theta_1 \theta_2 \cdots \theta_n) \tag{8.132}$$

Prove that this implies:

$$d\phi_1 \, d\phi_2 \cdots d\phi_n = (\det M)^{-1} d\theta_1 \, d\theta_2 \cdots d\theta_n \qquad (8.133)$$

which is the opposite of the usual rule for differentials.

5. Derive the Schrödinger equation for an electron in the presence of a potential $V(x)$ using path integrals.

6. For invertible, square matrices A, B, C, D, prove:

$$\begin{pmatrix} A & C \\ D & B \end{pmatrix} = \begin{pmatrix} A - CB^{-1}D & CB^{-1} \\ 0 & 1 \end{pmatrix} \begin{pmatrix} 1 & 0 \\ D & B \end{pmatrix}$$

$$= \begin{pmatrix} A & 0 \\ D & 1 \end{pmatrix} \begin{pmatrix} 1 & A^{-1}C \\ 0 & B - DA^{-1}C \end{pmatrix}$$

7. For matrices A, B, prove:

$$\log(AB) = \log A + \log B + \frac{1}{2}[\log A, \log B] + \cdots \qquad (8.134)$$

8. Prove:

$$\det (1 + M) = 1 + \text{Tr} \, M + \frac{1}{2}\left[(\text{Tr}M)^2 - \text{Tr}(M^2)\right] + \cdots \qquad (8.135)$$

9. For matrices A and B, prove, by power expansion, that:

$$e^A e^B = \exp\left(A + B + \frac{1}{2}[A, B] + \frac{1}{12}[A, [A, B]] + \frac{1}{12}[B, [B, A]] + \cdots\right) \qquad (8.136)$$

10. For matrices A and B, prove that:

$$e^A e^B = \exp[A, B] \, e^B e^A$$

$$e^{A+B} = \exp\left(-\frac{1}{2}[A, B]\right) e^A e^B \qquad (8.137)$$

Under what conditions are these identities valid?

11. For matrices A and B, prove:

$$(e^A e^B)^n = \exp\left(\frac{n(n+1)}{2}[A, B]\right) e^{nA} e^{nB} \qquad (8.138)$$

Are there any restrictions on this formula?

12. Prove:

$$
\int d\theta_n \cdots d\theta_2 \, d\theta_1 \exp \frac{1}{2} \sum_{i,j=1}^{n} \theta_i D_{ij} \theta_j = \sqrt{\det D} \tag{8.139}
$$

for the case $n = 3$ by an explicit calculation.

13. Prove the previous relation for arbitrary n. Prove it in two ways: first, by diagonalizing the D matrix and then performing the integration over the eigenvalues of D; second, by power expanding the expression and using the known identities for the antisymmetric ϵ tensor.

14. In Eq. (8.92), we established a relationship between the $\tilde{W}^{(4)}$ and $\Gamma^{(4)}$. Find the relationship between the fifth orders, then graphically illustrate what it means.

15. For ϕ^4 theory, show that $\Gamma^{(4)}$ is actually one-particle irreducible. Expand it only to fourth order in the coupling.

Chapter 9
Gauge Theory

We did not know how to make the theory fit experiment. It was our judgment, however, that the beauty of the idea alone merited attention.
                                                                    —C. N. Yang

9.1 Local Symmetry

An important revolution in quantum field theory took place in 1971, when the Yang–Mills theory was shown to be renormalizable, even after symmetry breaking, and therefore was a suitable candidate for a theory of particle interactions. The theoretical landscape in particle physics rapidly changed; a series of important papers emerged in which the weak and strong interactions quickly yielded their secrets. This revolution was remarkable, given the fact that in the relative confusion of the 1950s and 1960s, it appeared as if quantum field theory was an unsuitable framework for particle interactions.

Historically, gauge theory had a long but confused past. Although the Yang–Mills equation had been discovered in 1938 by O. Klein[1] (who was studying Kaluza–Klein theories), it was promptly forgotten during World War II. It was resurrected independently by Yang and Mills[2] in 1954 (and also by Shaw[3] and Utiyama[4]), but it was unsuitable for particle interactions because it only described massless vector particles. The discovery by 't Hooft[5] that the theory could be made massive while preserving renormalizability sparked the current gauge revolution.

Previously, we studied theories which were symmetric under *global* symmetries, so the group parameter ϵ was a constant. But the essence of Maxwell and Yang–Mills theory is that they are invariant under a symmetry that changes at every space–time point; that is, they are *locally* invariant. This simple principle of local gauge invariance, as we shall see, imposes highly nontrivial and nonlinear constraints on quantum field theory.

We begin with the generators of some Lie algebra:

$$[\tau^a, \tau^b] = i f^{abc} \tau^c \tag{9.1}$$

Let the fermion field ψ_i transform in some representation of $SU(N)$, not necessarily the fundamental representation. It transforms as:

$$\psi_i(x) \rightarrow \Omega_{ij}(x)\psi_j(x) \tag{9.2}$$

where Ω_{ij} is an element of $SU(N)$.

The essential point is that the group element Ω is now a function of space–time; that is, it changes at every point in the universe. It can be parametrized as:

$$\Omega_{ij}(x) = \left(e^{-i\theta^a(x)\tau^a}\right)_{ij} \tag{9.3}$$

where the parameters $\theta^a(x)$ are local variables, and where τ^a is defined in whatever representation we are analyzing.

The problem with this construction is that derivatives of the fermion field are not covariant under this transformation. A naive transformation of the derivatives of these fields picks up terms like $\partial_\mu \Omega$. In order to cancel this unwanted term, we would like to introduce a new derivative operator D_μ that is truly covariant under the group. To construct such an operator D_μ, let us introduce a new field, called the *connection* A_μ:

$$D_\mu \equiv \partial_\mu - ig A_\mu \tag{9.4}$$

where:

$$A_\mu(x) \equiv A_\mu^a(x)\tau^a \tag{9.5}$$

The essential point of this construction is that the covariant derivative of the ψ field is gauge covariant:

$$
\begin{aligned}
(D_\mu \psi)' &= \partial_\mu \psi' - ig A_\mu' \psi' \\
&= \Omega \partial_\mu \psi + (\partial_\mu \Omega)\psi - ig A_\mu' \Omega \psi \\
&= \Omega D_\mu \psi
\end{aligned}
\tag{9.6}
$$

The troublesome term $\partial_\mu \Omega$ is precisely cancelled by the variation of the A_μ term if we set:

$$A'_\mu(x) = -\frac{i}{g}[\partial_\mu \Omega(x)]\Omega^{-1}(x) + \Omega(x)A_\mu(x)\Omega^{-1}(x) \tag{9.7}$$

Infinitesimally, this becomes:

$$\begin{cases} \delta A_\mu^a = -\frac{1}{g}\partial_\mu \theta^a + f^{abc}\theta^b A_\mu^c \\ \delta\psi = -ig\theta^a \tau^a \psi \end{cases} \tag{9.8}$$

[If we reduce $SU(N)$ down to the group $U(1)$, then we recover the field transformations for QED.]

It is also possible to construct the invariant action for the connection field itself. Since D_μ is covariant, then the commutator of two covariant derivatives is also covariant. We define the commutator as follows:

$$\begin{aligned} F_{\mu\nu} &= \frac{i}{g}[D_\mu, D_\nu] \\ &= \partial_\mu A_\nu - \partial_\nu A_\mu - ig[A_\mu, A_\nu] \\ &= \left(\partial_\mu A_\nu^a - \partial_\nu A_\mu^a + g f^{abc} A_\mu^b A_\nu^c\right)\tau^a \end{aligned} \tag{9.9}$$

Because D_μ is genuinely covariant, this means that the $F_{\mu\nu}^a$ tensor is also covariant:

$$F_{\mu\nu} \rightarrow \Omega F_{\mu\nu}\Omega^{-1} \tag{9.10}$$

We can now construct an invariant action out of this tensor. We want an action that only has two derivatives (since actions with three or higher derivatives are not unitary, i.e., they have ghosts). The simplest invariant is given by the trace of the commutator. This is invariant because:

$$\text{Tr}\left(\Omega F_{\mu\nu}\Omega^{-1}\Omega F^{\mu\nu}\Omega^{-1}\right) = \text{Tr}\left(F_{\mu\nu}F^{\mu\nu}\right) \tag{9.11}$$

The unique action with only two derivatives is therefore given by:

$$S = \int d^4x \left(-\frac{1}{2}\text{Tr}\, F_{\mu\nu}F^{\mu\nu}\right) = \int d^4x \left(-\frac{1}{4}F_{\mu\nu}^a F^{a\mu\nu}\right) \tag{9.12}$$

This is the action for the Yang–Mills theory, which is the starting point for all discussions of gauge theory.

The field tensor $F_{\mu\nu}$, we should point out, also obeys the Bianchi identities. We know, by the Jacobi identity, that certain multiple commutators vanish identically. Therefore, we have:

$$\left[D_\mu, [D_\nu, D_\rho]\right] + \left[D_\nu, [D_\rho, D_\mu]\right] + \left[D_\rho, [D_\mu, D_\nu]\right] \equiv 0 \tag{9.13}$$

This is easily checked by explicitly writing out the terms in the commutators. Written in terms of the field tensor, this becomes:

$$[D_\mu, F_{\nu\rho}] + [D_\nu, F_{\rho\mu}] + [D_\rho, F_{\mu\nu}] = 0 \tag{9.14}$$

(It is important to stress that these are exact identities. They are not equations of motion, nor are they new constraints on the field tensor.)

Lastly, since $\bar\psi \to \bar\psi\Omega^\dagger$ and $D_\mu\psi \to \Omega D_\mu\psi$, it is easy to show that the invariant fermion action coupled to the gauge field is given by:

$$S = \int d^4x\, \bar\psi(i\,\slashed{D} - m)\psi \tag{9.15}$$

9.2 Faddeev–Popov Gauge Fixing

The real power of the path integral approach is that we have the freedom to choose whatever gauge we desire. This is impossible in the canonical approach, where the gauge has already been fixed. However, in the path integral approach, because gauge fixing is performed by inserting certain delta functions into the path integral, we can change the gauge by simply replacing these factors. This formalism was introduced by Faddeev and Popov.[6]

Historically, however, before the Faddeev–Popov method, the quantization of Yang–Mills theory was not clear for many years. In 1962, Feynman[7] showed that the theory suffered from a strange kind of disease: The naive quantization of the theory was not unitary. In order to cancel the nonunitary terms from the theory, Feynman was led to postulate the existence of a term that did not emerge from the standard quantization procedure. Today, we know this ghost, first revealed by Feynman using unitarity arguments, as the *Faddeev–Popov ghost*.

To begin, we first stress that the path integral of a theory like Maxwell's theory of electromagnetism is, in principle, undefined because of the gauge degree of freedom. Because the path integral DA and the action S are both gauge invariant, it means that functionally integrating over DA will eventually overcount the degrees of freedom of the theory. Because the Maxwell theory is invariant under the gauge transformation:

$$A_\mu^\Omega = A_\mu + \partial_\mu\Omega \tag{9.16}$$

this means that functionally integrating over both A_μ^Ω and A_μ will overcount the integrand repeatedly. In fact:

$$\int DA_\mu e^{iS} = \infty \tag{9.17}$$

If we begin with one field configuration A_μ and consider all possible A_μ^Ω, then we are sweeping out an "orbit" in functional space. The problem is that, as Ω changes along the orbit, we repeat ourselves an infinite number of times in the path integral. Our problem is therefore to "slice" the orbit once so that we do not have this infinite overcounting.

This is the origin of the gauge-fixing problem. To solve this problem, one is tempted to insert factors like:

$$\delta(\partial_\mu A^\mu); \quad \delta(\nabla \cdot \mathbf{A}) \tag{9.18}$$

into the path integral, forcing it to respect the gauge choice $\partial_\mu A^\mu = 0$ or $\nabla \cdot \mathbf{A} = 0$.

More generally, we would like to fix the gauge with an arbitrary function of the fields:

$$\delta\left(F(A_\mu)\right) \tag{9.19}$$

which would fix the gauge to be $F(A_\mu) = 0$.

The source of the problem is that inserting a delta function into the functional integration DA changes the measure of the integration. For example, we know that the delta function changes when we make seemingly trivial changes in it. For example, if we have a function $f(x)$ that has a zero at $x = a$, we recall that:

$$\delta\left(f(x)\right) = \frac{\delta(x - a)}{|f'(x)|} \tag{9.20}$$

The choice $\delta\left(f(x)\right)$ differs from the choice $\delta(x - a)$ by a term $f'(x)$. Thus, there is an ambiguity in the measure of integration. To solve this ambiguity, we insert the number "1" into the path integral, which we know has the correct measure:

$$1 = \Delta_{FP} \int D\Omega \, \delta\left(F(A_\mu^\Omega)\right) \tag{9.21}$$

where Δ_{FP} is the Faddeev–Popov determinant, which guarantees the correct measure, and $D\Omega = \prod_x d\Omega(x)$ is the invariant group measure. It satisfies the invariance property:

$$D\Omega = D(\Omega'\Omega) \tag{9.22}$$

if Ω' is a fixed element of $SU(N)$. Since $\Omega = 1 - i\theta^a \tau^a + \cdots$, the group measure (for small θ) is equal to:

$$D\Omega = \prod_{a,x} d\theta^a(x) \qquad (9.23)$$

(More details concerning the invariant group measure will be presented in Chapter 15.)

Inserting the number "1" into the functional, we now have:

$$\int DA_\mu \left(\Delta_{FP} \int D\Omega \, \delta \left(F(A_\mu^\Omega) \right) \right) e^{i \int d^4x \, \mathcal{L}(A)} \qquad (9.24)$$

Our task is now to calculate an explicit result for the Faddeev–Popov determinant. To do this, we first notice that it is gauge invariant:

$$\Delta_{FP}(A_\mu) = \Delta_{FP}(A_\mu^\Omega) \qquad (9.25)$$

This is because it is equal to the integration over all gauge Ω factors, and hence is independent of the gauge. However, it will be instructive to see this more explicitly.

Let us change the gauge, replacing A_μ in the Faddeev–Popov determinant with A_μ^Ω. Then the definition becomes:

$$\begin{aligned}
\Delta_{FP}^{-1}(A_\mu^{\Omega'}) &= \int D\Omega' \, \delta \left[F(A_\mu^{\Omega'\Omega}) \right] \\
&= \int D[\Omega'\Omega] \delta \left(F(A_\mu^{\Omega'\Omega}) \right) \\
&= \int D\Omega'' \delta \left(F(A_\mu^{\Omega''}) \right) \\
&= \Delta_{FP}^{-1}(A_\mu)
\end{aligned} \qquad (9.26)$$

Thus, it is gauge invariant. Now we come to the crucial part of the calculation. Let us make a gauge transformation on the entire path integral, so that $A_\mu \rightarrow A_\mu^{\Omega^{-1}}$. The measure DA, the Faddeev–Popov determinant, and the action S are all gauge invariant. The factor that is not gauge invariant is $F(A_\mu^\Omega)$, which changes into $F(A_\mu)$. The integral, after the gauge transformation, now becomes:

$$\left(\int D\Omega \right) \int DA_\mu \, \Delta_{FP} \, \delta \left(F(A_\mu) \right) e^{i \int d^4x \, \mathcal{L}(A)} \qquad (9.27)$$

We have now accomplished the following:

1. We have explicitly isolated the infinite part of the matrix element, which is given by:

$$\int D\Omega = \infty \qquad (9.28)$$

By simply dividing out by the $\int D\Omega$, we remove the infinite overcounting.

2. The gauge choice is now enforced by $\delta\left(F(A_\mu)\right)$.

3. The essential point is that the factor Δ_{FP} gives the correct measure in the path integral. This is the factor that was missing for so many years in previous attempts to quantize gauge theories.

The problem of gauge fixing is therefore reduced to finding an explicit expression for Δ_{FP}. From Eq. (9.21), we know that the Faddeev–Popov determinant is written in terms of $\delta(F(A_\mu^\Omega))$, which in turn can be re-expressed via Eq. (9.20) as:

$$\delta\left[F(A_\mu^\Omega)\right] = \delta(\Omega - \Omega_0) \det \left| \frac{\delta F(A_\mu^\Omega)(x)}{\delta\Omega'(x')} \right|^{-1} \qquad (9.29)$$

where the determinant is a functional generalization of the factor $|f'(x)|$.

Thus, the Faddeev–Popov term can be written as:

$$\Delta_{FP} = \det \left| \frac{\delta F(A_\mu^\Omega)(x)}{\delta\Omega'(x')} \right| \qquad (9.30)$$

for $F(A_\mu) = 0$. To perform explicit calculations with this determinant, it is convenient to rewrite this expression in a form where we can extract new Feynman rules. To solve for the determinant, let us power expand the factor $F(A_\mu^\Omega)$ for a small group parameter θ:

$$F\left(A_\mu^\Omega(x)\right) = F(A_\mu(x)) + \int d^4y \, M(x, y)\theta(y) + \cdots \qquad (9.31)$$

In this approximation, only the matrix M survives when we take the derivative. The Faddeev–Popov term can be converted into the determinant of the matrix M, which in turn can be converted into a Gaussian integral over two fields c and $c^\dagger$:

$$\Delta_{FP} = \det M = \int Dc \, Dc^\dagger \exp i \left(\int d^4x \, d^4y \, c^\dagger(x) M(x, y) c(y) \right) \qquad (9.32)$$

The determinant $\det M$ appears in the numerator of the path integral, rather than the denominator. This means that when we exponentiate the Faddeev–Popov term into the Lagrangian, we must integrate over *Grassmann variables*, rather than bosonic variables, as we discussed in Section 8.6. Thus, we find that c and $c^\dagger$ are actually "ghost" fields, that is, scalar fields obeying Fermi–Dirac statistics. This is the origin of the celebrated Faddeev–Popov ghosts.

Alternatively, we could have rewritten the determinant as follows:

$$\Delta_{FP} = \det M = \exp\left(\mathrm{Tr}\log M\right) \tag{9.33}$$

where the determinant is understood to be taken over discretized x and y and any isospin indices. (To prove that the determinant of a matrix M can be written as the exponential of the trace of the logarithm of M, simply diagonalize the M matrix and re-express the formula in terms of the eigenvalues of M. Then this identity is trivial.) This term can be written in more familiar language if we write the matrix as $M = 1 + L$ and expand as follows:

$$
\begin{aligned}
\det(1 + L) &= \exp \mathrm{Tr}\left[\log(1 + L)\right] \\
&= \exp\left(\sum_{n=1}^{\infty} \frac{(-1)^{n-1}}{n} \mathrm{Tr}\, L^n\right)
\end{aligned}
\tag{9.34}
$$

The trace over L, we shall see shortly, can now be interpreted as closed loops in the Feynman expansion of the perturbation theory.

To gain some familiarity with this Faddeev–Popov term, let us compute the Faddeev–Popov determinant for the simplest gauge theory, Maxwell's theory. Earlier, in the canonical and covariant approaches, we fixed the gauge without even thinking about complications due to the functional measure. We will find that we were lucky: The Faddeev–Popov measure is trivial for the gauges found in Maxwell theory, but highly nontrivial for non-Abelian gauge theory. To see this, let us choose the gauge:

$$F(A_\mu) = \partial^\mu A_\mu = 0 \tag{9.35}$$

The variation of this gauge fixing gives us:

$$F(A_\mu^\Omega) = \partial^\mu A_\mu + \partial^\mu \partial_\mu \theta \tag{9.36}$$

Then the M matrix becomes:

$$M(x, y) = \left[\partial_\mu \partial^\mu\right]_{x,y} \tag{9.37}$$

Written in terms of ghost variables c and $c^\dagger$, this means we must add the following term to the action:

$$\int d^4x\, d^4y\, c^\dagger(x)\partial^\mu \partial_\mu c(y) \tag{9.38}$$

where the integration over $c^\dagger$ and c must now be viewed as being Grassmannian; that is, these fields are scalar Grassmann ghosts. Fortunately, this determinant decouples from the rest of the theory. The determinant of the Laplacian does not couple to any of the fermions or gauge fields in QED, and hence gives only an uninteresting multiplicative factor in front of the S matrix, which we can remove at will.

For the Coulomb gauge, we find a similar argument, except that the gauge variation gives us a factor:

$$\det \nabla^2 \tag{9.39}$$

which also decouples from the theory. Thus, from the path integral point of view, we were fortunate that we could take these gauges in quantizing QED without suffering any problems. We are not so fortunate, however, for gauge theories, where the Faddeev–Popov term gives us highly nontrivial corrections to the naive theory. We begin by choosing a gauge for the theory, such as:

$$\partial^\mu A_\mu^a = 0 \tag{9.40}$$

If we place this into the path integral, then we must at the same time insert the Faddeev–Popov measure:

$$\Delta_{FP} = \det\,(M)_{x,y;a,b} \tag{9.41}$$

where x and y are discretized space–time variables, and a, b are isospin variables. The matrix M is easily found:

$$
\begin{aligned}
M_{ab}(x-y) &= \frac{\delta}{\delta\Omega^a(x)} A_\mu^{b,\Omega}(y) = \frac{\delta}{\delta\theta^a(x)}\left(\partial^\mu \frac{-1}{g}(\partial_\mu\theta^b + gf^{bcd}A_\mu^c\theta^d)\right)(y) \\
&= \frac{1}{g}\left(-\delta^{ab}\partial^\mu\partial_\mu + gf^{abc}A_\mu^c\right)_{x,y}\delta^4(x-y)
\end{aligned} \tag{9.42}
$$

If we rescale and exponentiate this into the functional integral, we find an additional term in the action given by:

$$\int d^4x\, c^{\dagger a}\left(\delta^{ab}\partial^\mu\partial_\mu - gf^{abc}A_\mu^c\right)c^b \tag{9.43}$$

The $c(x)$ are the celebrated Faddeev–Popov ghosts. These ghosts have peculiar properties:

1. They are scalar fields under the Lorentz group, but have anticommuting statistics, violating the usual spin-statistics theorem. They are therefore ghosts. However, we do not care, since their job is to cancel the ghosts coming from the quantization of the A_μ^a field.

2. These ghosts couple only to the gauge field. They do not appear in the external states of the theory. Therefore, they cannot appear in tree diagrams at all; they only make their presence in the loop diagrams, where we have an internal loop of circulating ghosts.

3. These ghosts are an artifact of quantization. They decouple from the physical spectrum of states.

9.3 Feynman Rules for Gauge Theory

Let us now put all the various ingredients together and write down the Feynman rules for the theory. The action plus the gauge-fixing contribution becomes:

$$\mathcal{L} + \mathcal{L}_{FP} = -\frac{1}{4} F_{\mu\nu}^a F^{a\mu\nu} - \frac{1}{2\alpha}(\partial_\mu A^{a\mu})^2 \tag{9.44}$$

while the ghost contribution becomes:

$$\mathcal{L}_g = \int d^4x d^4y \sum_{ab} c_a^\dagger(x)[M(x, y)]_{ab} c_b(y) \tag{9.45}$$

where:

$$[M(x, y)]_{ab} = -\frac{1}{g} \partial^\mu \left(\delta^{ab}\partial_\mu - g\epsilon^{abc} A_\mu^c\right)\delta^4(x - y) \tag{9.46}$$

To extract the Feynman rules from this action, we will decompose the action into a free part and interacting part:

$$\mathcal{L}_0 = -\frac{1}{4}(\partial_\mu A_\nu^a - \partial_\nu A_\mu^a)^2 - \frac{1}{2\alpha}(\partial^\mu A_\mu^a)^2 + c_a^\dagger \partial^2 c_a \tag{9.47}$$

and:

$$\mathcal{L}_I = -\frac{1}{2} g(\partial_\mu A_\nu^a - \partial_\nu A_\mu^a)\epsilon^{abc} A^{b\mu} A^{c\nu}$$

$$+ \quad \frac{1}{4} g^2 \epsilon^{abc} \epsilon^{ade} A_\mu^b A_\nu^c A^{d\mu} A^{e\nu}$$

$$- \quad ig c^{a\dagger} \epsilon^{abc} \partial^\mu A_\mu^c c^b \qquad\qquad (9.48)$$

From this, we can read off the Feynman rules for Yang–Mills theory:

1. The gauge meson propagator is given by:

$$i\Delta_{\mu\nu}^{ab}(k) = -i\delta^{ab}\left(g_{\mu\nu} - (1-\alpha)\frac{k_\mu k_\nu}{k^2}\right)\frac{1}{k^2 + i\epsilon} \qquad (9.49)$$

2. The (directed) ghost propagator is given by:

$$i\Delta^{ab}(k) = -i\delta^{ab}\frac{1}{k^2 + i\epsilon} \qquad (9.50)$$

3. The three-gauge meson vertex function for mesons with momenta and quantum numbers (k_1, μ, a), (k_2, ν, b), and (k_3, λ, c) is given by:

$$i\Gamma_{\mu\nu\lambda}^{abc} = ig\epsilon^{abc}\Big[(k_1 - k_2)_\lambda g_{\mu\nu}$$

$$+ \quad (k_2 - k_3)_\mu g_{\nu\lambda} + (k_3 - k_1)_\nu g_{\mu\lambda}\Big] \qquad (9.51)$$

with $\sum_{i=1}^3 k_i = 0$.

4. The four-gauge vertex is given by:

$$i\Gamma_{\mu\nu\lambda\rho}^{abcd} = ig^2\Big[\epsilon^{abe}\epsilon^{cde}\left(g_{\mu\lambda}g_{\nu\rho} - g_{\nu\lambda}g_{\mu\rho}\right)$$

$$+ \quad \epsilon^{ace}\epsilon^{bde}\left(g_{\mu\nu}g_{\lambda\rho} - g_{\lambda\nu}g_{\mu\rho}\right)$$

$$+ \quad \epsilon^{ade}\epsilon^{cbe}\left(g_{\mu\lambda}g_{\rho\nu} - g_{\rho\lambda}g_{\mu\nu}\right)\Big] \qquad (9.52)$$

with $\sum_{i=1}^4 k_i = 0$

5. The two-ghost/gauge meson vertex function is given by:

$$i\Gamma_\mu^{abc} = g\epsilon^{abc} k_\mu \qquad (9.53)$$

where k_μ is the incoming ghost's momentum.

Now let us derive the Feynman rules for the gauge field coupled to a fermion and a scalar field. We assume that the coupling to a scalar field is given by the matrix R^a, taken in whatever representation we desire. The coupling is given by:

$$
\begin{aligned}
\mathscr{L} = {} & \bar{\psi}\left[i\gamma^\mu(\partial_\mu - igA_\mu^a\tau^a) - m\right]\psi \\
& + \left[(\partial^\mu - igA^{a\mu}R^a)\phi\right]^\dagger\left[(\partial_\mu - igA_\mu^aR^a)\phi\right] - m^2\phi^\dagger\phi - \frac{\lambda}{4}(\phi^\dagger\phi)^2
\end{aligned}
$$

$$(9.54)$$

Then the Feynman rules for this theory are given by:

1. The fermion propagator is given by:

$$
iS_F^{ij}(p)_{\alpha\beta} = \left(\frac{i\delta^{ij}}{\not{p} - m + i\epsilon}\right)_{\alpha\beta} \tag{9.55}
$$

2. The scalar boson propagator is given by:

$$
i\Delta_F^{lm}(p) = \frac{i\delta^{lm}}{p^2 - m^2 + i\epsilon} \tag{9.56}
$$

3. The fermion–gauge-meson coupling is given by:

$$
ig(\gamma_\mu)_{\alpha\beta}\left(\tau^a\right)_{ij} \tag{9.57}
$$

4. The boson–gauge-meson coupling is given by:

$$
ig(p + p')_\mu R_{lm}^a \tag{9.58}
$$

5. The two-boson–gauge-meson coupling is given by:

$$
-ig^2 g_{\mu\nu}\{R^a, R^b\} \tag{9.59}
$$

6. The four-scalar couplings are given by $-i\lambda$.

9.4   Coulomb Gauge

To begin a discussion of gauge theory in the Coulomb gauge, it is first instructive to rewrite the original action as follows:

$$\mathcal{L} = \frac{1}{4} F_{\mu\nu}^a F^{a\mu\nu} - \frac{1}{2} F^{a,\mu\nu} \left(\partial_\mu A_\nu^a - \partial_\nu A_\mu^a + g f^{abc} A_\mu^b A_\nu^c \right) \tag{9.60}$$

where it is essential to notice that $F_{\mu\nu}^a$ is an *independent field*, totally unrelated to the vector field A_μ^a. (However, by eliminating the $F_{\mu\nu}$ field by its equations of motion, we can show the equivalence with the standard Yang–Mills action.) This new version of the action is invariant under:

$$\begin{aligned}
\delta A_\mu^a &= -\frac{1}{g} \partial_\mu \theta^a + f^{abc} \theta^b A_\mu^c \\
\delta F_{\mu\nu}^a &= f^{abc} \theta^b F_{\mu\nu}^c
\end{aligned} \tag{9.61}$$

Now let us take the variation of the action with respect to $F_{\mu\nu}^a$ as well as with respect to A_μ^a. The two equations of motion yield:

$$\begin{aligned}
F_{\mu\nu}^a &= \partial_\mu A_\nu^a - \partial_\nu A_\mu^a + g f^{abc} A_\mu^b A_\nu^c \\
0 &= \partial_\mu F^{a\mu\nu} + g f^{abc} A^{b,\mu} F_{\mu\nu}^c
\end{aligned} \tag{9.62}$$

This new action will prove to be useful when quantizing the theory in the Coulomb gauge. First, we will eliminate F_{ij}^a in terms of the A_i^a fields, keeping F_{0i}^a an independent field. Written in terms of these variables, we find that the Lagrangian becomes:

$$\mathcal{L} = -\frac{1}{4} F_{ij}^a(A) F^{aij}(A) + \frac{1}{2} (F_{0i}^a)^2 - F^{a,0i} \left(\partial_i A_0^a - \partial_0 A_i^a + g f^{abc} A_0^b A_i^c \right) \tag{9.63}$$

Let us define:

$$\begin{aligned}
E_i^a &= F_{0i}^a \\
B_i^a &= \frac{1}{2} \epsilon_{ijk} F^{a,jk}(A)
\end{aligned} \tag{9.64}$$

In terms of these fields, we now have:

$$\mathcal{L} = E_i^a \dot{A}_i^a - \mathcal{H} \tag{9.65}$$

where:

$$\mathcal{H} = \frac{1}{2}\left[(E_i^a)^2 + (B_i^a)^2\right] + A_0^a\left(\partial_i E_i^a + g f^{abc} E_i^b A_i^c\right) \qquad (9.66)$$

Written in this form, the similarity to the quantization of Maxwell's theory in the Coulomb gauge is apparent. Of particular importance is the Lagrange multiplier A_0^a, which multiplies the covariant divergence of E_i^a, which is the gauge generalization of Gauss's Law found earlier for QED in Eq. (4.54). The equation of motion for A_0^a yields:

$$D_i E_i^a = \partial_i E_i^a + g f^{abc} E_i^b A_i^c = 0 \qquad (9.67)$$

which indicates that not all conjugate momenta are independent, a situation common to all gauge theories.

There are two ways in which to solve this important constraint. First, as in the Gupta–Bleuler approach, we can apply this constraint directly onto the state vectors of the theory:

$$D_i E_i^a |\Psi\rangle = 0 \qquad (9.68)$$

However, it is easy to show that $D_i E_i^a$, acting on a field A_i^a, generates the standard gauge transformation; that is:

$$\left[\int d^3x\, \Lambda^a D_i E_i^a(x),\, A_j^b(y)\right]_{x_0=y_0} = -i\left(\partial_j \Lambda^b(y) + g f^{bcd} \Lambda^c A_j^d(y)\right) \qquad (9.69)$$

Therefore the constraint equation means that the state vectors $|\Psi\rangle$ of the theory must be *singlets* under the gauge group. This is a rather surprising result, because it implies that free vector mesons A_i^a are not part of the physical spectrum. This is consistent, however, with our understanding of the quark model, where nonperturbative calculations indicate that the only allowed states are singlets under the "color" gauge group $SU(3)$. The allowed singlets under the color group include quark–antiquark and three-quark bound states, which are the only ones seen experimentally. This will be important when we discuss the phenomenon of confinement in Chapters 11 and 15.

The second approach in solving this constraint is to assume, for the moment, that perturbation theory is valid and simply drop the higher-order terms. Then the constraint equation reduces to the statement that the E_i^a field is transverse. To see this, we will decompose the E_i^a field into transverse and longitudinal modes:

$$E_i^a = E_i^{aT} + E_i^{aL} \qquad (9.70)$$

where:

$$\nabla_i E_i^{aT} = 0 \tag{9.71}$$

An explicit decomposition of any field E_i^a into transverse and longitudinal parts can be given as:

$$E_i^a = \left(\delta_{ij} - \nabla_i \frac{1}{\nabla^2} \nabla_j \right) E_j^a + \nabla_i \frac{1}{\nabla^2} \nabla_j E_j^a \tag{9.72}$$

We will find it convenient to factor out the longitudinal mode explicitly by introducing the field f^a:

$$E_i^{aL} = -\nabla_i f^a \tag{9.73}$$

Now insert these definitions back into the Gauss's Law constraint, which now becomes:

$$
\begin{aligned}
&\nabla_i E_i^a + g f^{abc} E_i^b A_i^c \\
&= \quad -\nabla_i^2 f_i^a - g f^{abc} \nabla_i f^b A_i^c + g f^{abc} E_i^{bT} A_i^c = 0 \tag{9.74}
\end{aligned}
$$

For the case of Maxwell's field, this constraint was trivial to solve, since the cross term was not present. For the Yang–Mills theory, this cross term gives us a nontrivial complication.

Fortunately, we can now solve for f^a by inverting this equation as a perturbation series. Let us rewrite it as:

$$D_{ab} f^b = f^{abc} E_i^{bT} A_i^c \tag{9.75}$$

where:

$$D_{ab} = \nabla^2 \delta_{ab} - g f^{abc} A_i^c \nabla_i \tag{9.76}$$

To solve this equation, we must introduce the Green's function:

$$\left(\nabla^2 \delta^{ab} - g f^{abd} A_i^d \nabla_i \right) \Delta^{bc}(x, y; A) = \delta^{ac} \delta^3(\mathbf{x} - \mathbf{y}) \tag{9.77}$$

Let us assume that we can invert this equation. (This statement, as we shall shortly see, is actually incorrect.) Inverting this equation, we find the solution for f^a:

$$f^a(x) = g \int d^3 y \, \Delta^{ab}(x, y; A) f^{bcd} A_k^c(y) E_k^d(y) \tag{9.78}$$

where:

$$\Delta^{ab}(x, y; A) = \frac{\delta^{ab}}{4\pi |\mathbf{x} - \mathbf{y}|} + g \int d^3z \frac{1}{4\pi |\mathbf{x} - \mathbf{z}|} f^{acb} A_k^c \nabla_k \frac{1}{4\pi |\mathbf{x} - \mathbf{y}|} + \cdots \quad (9.79)$$

Let us now insert this expression back into the Hamiltonian, where we have explicitly solved for the Gauss's Law constraint. The Hamiltonian becomes:

$$H = \frac{1}{2} \int d^3x \left[(E_i^{aT})^2 + (B_i^a)^2 + (\nabla_i f^a)^2 \right] \quad (9.80)$$

and the generating functional becomes:

$$Z(J) = \int DE_i^a \, DA_i^a \, \delta \left(\nabla_i E_i^a \right) \delta \left(\nabla_i A_i^a \right)$$
$$\times \exp \left(i \int d^4x \left[E_k^a \dot{A}_k^a - \frac{1}{2}(E_k^a)^2 - \frac{1}{2}(B_k^a)^2 - \frac{1}{2}(\nabla_k f^a)^2 - A_k^a J_k^a \right] \right)$$
$$(9.81)$$

From this, we can read off the Feynman rules for the Yang–Mills theory in the Coulomb gauge. We have written the path integral entirely in terms of the physical, transverse states. The theory is hence unitary, since all longitudinal modes with negative norm have been explicitly removed.

There is, however, another form for the Coulomb path integral that is more covariant looking and whose Feynman rules are easier to work with. Both forms of the Coulomb path integral, of course, yield the same S matrix. If we start with the original action in Eq. (9.60) written in terms of the auxiliary field $F_{\mu\nu}$, then the path integral can be written as:

$$Z(J) = \int DA_\mu^a \, DF_{\mu\nu}^a \, \delta \left(\nabla_i A_i^a \right) \Delta_{FP}(A) \exp \left(i S(A, F) + i \int d^4x \, J^\mu A_\mu \right)$$
$$(9.82)$$

where the Faddeev–Popov factor for the Coulomb gauge $\nabla_i A_i^a = 0$ can be written as:

$$\Delta_{FP}(A) = \det \left(\frac{\delta \nabla_i A_i^\Omega(x)}{\delta \Omega(y)} \right)_{x,y} = \det \left[M^{ab}(x, y) \right] \quad (9.83)$$

where:

$$M^{ab}(x, y) = D^{ab}(A)\delta^4(x - y) \quad (9.84)$$

To calculate the Feynman rules from this action, we simply note that we can write:

$$M^{ab} = \nabla^2(\delta^{ab} + L^{ab})$$ (9.85)

where:

$$L^{ab} = g f^{abc} \frac{1}{\nabla^2} A_i^c \nabla_i$$ (9.86)

Now we use the expression:

$$
\begin{aligned}
\det M \quad &= \quad \exp\left(\operatorname{Tr}\,\log M\right) = \left[\det \nabla^2\right](\det(1+L)) \\
&\sim \quad \exp\left[\operatorname{Tr}\,\log(1+L)\right] \\
&= \quad \exp\sum_{n=1}^{\infty} \frac{(-1)^{n-1}}{n} \\
&\quad \times \int d^4x_1\, d^4x_2 \cdots d^4x_n \,\operatorname{Tr} L(x_1, x_2)L(x_2, x_3)\cdots L(x_n, x_1)
\end{aligned}
$$
(9.87)

where we have thrown away a factor of $\det \nabla^2$, which contributes closed loops that do not couple to anything.

In summary, we have shown that there are two equivalent ways of writing the Coulomb gauge. The first, with all redundant ghost modes removed, is explicitly unitary (but difficult to calculate with). The second form, although not manifestly unitary, has a covariant form. The only correction to this covariant form is the determinant factor, which we can see from the previous expression consists of nothing but closed loops coupled to the A_μ field. Thus, from a calculational point of view, the only correction to the Feynman rules is to insert closed loops into the theory connected to gauge fields.

9.5 The Gribov Ambiguity

There is, however, one tricky point that we have glossed over. As we mentioned earlier, the operator $D^{ab}(A)$ in Eq. (9.76) is actually not invertible, and hence the Coulomb gauge does not, in fact, completely eliminate the gauge degree of freedom of the theory.

This unexpected result is due to the fact that the Coulomb gauge $\nabla_i A_i^a = 0$ does not uniquely fix the gauge; that is, it is possible to find another A_i', which is gauge equivalent to A_i, that satisfies:

$$\nabla_i A_i' = 0 \tag{9.88}$$

To see this, we write A_i' out in more detail:

$$A_i' = \Omega A_i \Omega^{-1} - \frac{i}{g}(\partial_i \Omega)\Omega^{-1} \tag{9.89}$$

The important point is that A_i' contains a term proportional to $1/g$. Perturbation theory, as a power expansion in g, will never pick up this factor. As an example, take the group $SU(2)$, and take A_i to be gauge equivalent to the number 0:

$$A_i = -\frac{i}{g}(\partial_i \Omega)\Omega^{-1} \tag{9.90}$$

with $\nabla_i A_i^a = 0$. If the Coulomb gauge were a good gauge, then the only solution of this equation should be $A_i = 0$. However, this is not so. For example, we can parametrize the previous gauge using radial coordinates:

$$\Omega = \cos\omega(r)/2 + i\boldsymbol{\sigma} \cdot \mathbf{n} \sin\omega(r)/2 \tag{9.91}$$

where $n^i n^i = 1$ and $n^i = x^i/r$.

Then the Coulomb gauge condition becomes:

$$\frac{d^2\omega}{dt^2} + \frac{d\omega}{dt} - \sin 2\omega = 0; \quad t = \log r \tag{9.92}$$

which is the equation of a damped pendulum in a constant gravitational field. If $\omega = 0$, then $A_i = 0$, and this is the solution we desire. The problem, however, is that there are obviously many other solutions to this equation other than $\omega = 0$, so the uniqueness of the Coulomb gauge-fixing procedure is violated. For a nonsingular solution, we want $\omega = 0, 2\pi, 4\pi, \dots$ at $t = -\infty$. But then the pendulum can either fall clockwise or counterclockwise many times and then eventually wind up at a position of stable equilibrium at $\omega = -\pi$. For these nontrivial solutions at $t \to \infty$, this means that we have the asymptotic condition:

$$\Omega \to \pm i \frac{\sigma^i x^i}{r} \tag{9.93}$$

Because the Coulomb gauge does not uniquely fix the gauge, we will, in general, find an infinite sequence of identical copies, each related by a gauge transformation, each satisfying the Coulomb constraint. We will call these *Gribov copies.*[8] Another way of saying this is that $M^{ab}(x, y; A)$ is a matrix that has a zero eigenvalue. Not only do we have nonzero eigenvalues λ_n:

$$M^{ab}\psi_n^b(x) = \lambda_n \psi_n^a \tag{9.94}$$

but we also have eigenfunctions with zero eigenvalue:

$$M^{ab}\phi_m^b(x) = 0 \tag{9.95}$$

Since the determinant of M can be written as the product of its eigenvalues:

$$\det M^{ab} = \prod_n \lambda_n \prod_m 0 = 0 \tag{9.96}$$

the determinant itself is zero and hence M is noninvertible. Thus, M^{ab} cannot be inverted. As a consequence, the formulas we have given for the Coulomb gauge are actually slightly incorrect. The presence of these zero eigenvalue functions $\phi_n(x)$ spoils the inversion process, and hence spoils the Coulomb gauge fixing. Thus the canonical quantization program for gauge theory, based on quantizing the physical fields, does not exist, technically speaking.

Although we cannot fully fix the gauge with the choice $\nabla_i A_i^a = 0$, we can still salvage our calculation in the Coulomb gauge. In fact, we can show, to any order in perturbation theory, that these zero eigenvalues do not affect our perturbative results. The reason we can ignore these states with zero eigenvalues is that they do not couple to the physical Hilbert space. For example, we can show:

$$\langle \nabla^2 f^a | \phi_n^a \rangle = 0 \tag{9.97}$$

To see how this helps us, let us construct a modified propagator:

$$D^{ab}(A)\tilde{G}^{bc}(x, y; A) = \delta^{ac}\delta^3(\mathbf{x} - \mathbf{y}) - \sum_n \phi_n^a(x)\phi_n^{c*}(y) \tag{9.98}$$

With these zero modes explicitly subtracted out, we can define the inverse operator $\tilde{G}^{ab}$ without any problems. The perturbation theory with $\tilde{G}^{ab}$ can be used because:

$$\tilde{G}\nabla^2 f = \Delta\nabla^2 f \tag{9.99}$$

because Δ [in Eq. (9.77)] and $\tilde{G}$ only differ by the zero eigenvalues ϕ_n^a, which vanish when contracted onto $\nabla^2 f$.

The moral of this exercise is that, although the Coulomb gauge is riddled with Gribov copies, we can safely ignore them as long as one stays within perturbation theory. These Gribov copies carry an additional $1/g$ dependence beyond the first copy, and hence cannot be picked up in perturbation theory. Once one leaves perturbation theory to discuss nonperturbative phenomena, such as confinement, then presumably these Gribov copies become important.

The fact that these Gribov copies can be ignored in perturbation theory is fortunate, because the Faddeev–Popov quantization procedure becomes difficult to work with in the presence of these copies. To see how the Faddeev–Popov program is modified, let us enumerate the Gribov copies with an index n. Then are an infinite number of solutions to the equation:

$$\nabla_i A_i^{a(n)} = 0 \tag{9.100}$$

The Faddeev–Popov determinant presumably becomes modified as follows:

$$\sum_n \Delta_{FP}^{-1}[A_\mu^{a(n)}] = \int D\Omega \delta\left(\nabla_i A_i^\Omega\right) \tag{9.101}$$

Then the generating functional becomes:

$$\begin{aligned}
Z(J) &= \int DA_\mu \left(\sum_n \Delta_{FP}^{-1}(A_\mu^{a(n)})\right)^{-1} \delta\left(\nabla_i A_i^{a(n)}\right) \\
&\quad \times \exp\left(iS + i \int d^4x \, J^\mu A_\mu\right)
\end{aligned} \tag{9.102}$$

It is quite difficult to extract Feynman rules from a path integral as complicated as this. Thus, the standard ghost/loop interpretation of the Faddeev–Popov factor is now lost, and we have difficulty setting up the Feynman rules. However, as we have stressed, in perturbation theory the part analytic at $g = 0$ is only a function of the first Gribov copy, and so we can throw away all the infinite contributions from the ambiguity as long as we stick to perturbation theory.

9.6 Equivalence of the Coulomb and Landau Gauge

In general, the Green's functions of a gauge theory are dependent on the gauge. However, because the Green's functions are not directly measurable, this does not cause any harm. But the S matrix, because it is, by definition, measurable, should be independent of the gauge.

We will now present a functional proof that the S matrix is independent of the gauge. We will first establish how the generator of the Green's functions changes when we go from the Coulomb gauge to the Landau gauge, and then show that these modifications vanish when we go on-shell. We stress that our proof easily generalizes to the arbitrary case, proving that the S matrix for gauge theory is gauge independent. This will prove useful in the next chapter, where we discuss gauge theory in different gauges. In one gauge, the "unitary" gauge, the theory is unitary but not manifestly renormalizable. In the other, the "renormalizable" gauge, the theory is renormalizable but not manifestly unitary. Because the S matrix is gauge independent, this will show that the theory is both unitary and renormalizable.

Our starting point is the generator of the Green's functions in the Coulomb gauge:

$$Z_C(J) = \int DA_\mu \, \Delta_C[A_\mu] \prod_x \delta(\nabla_i A_i(x)) \exp\left(iS[A_\mu] + i \int d^4x \, J^\mu A_\mu\right)$$
(9.103)

where $\Delta_C[A_\mu]$ is the Faddeev–Popov measure coming from the Coulomb gauge. Using functional methods, we will now change the gauge to the Landau gauge. We will use the fact that the number "1" can be written, as usual, as:

$$1 = \Delta_L[A_\mu] \int D\Omega(x) \prod_x \delta\left(\partial^\mu A_\mu^\Omega(x)\right)$$
(9.104)

where this expression is written in the Landau gauge.

We will now insert the number "1" (written in the Landau gauge) into the expression for the generator of Green's functions (written in the Coulomb gauge). This insertion does not change anything, so the generator becomes:

$$Z_C(J) = \int DA_\mu \left(\int D\Omega(x) \prod_x \delta\left(\partial^\mu A_\mu^\Omega(x)\right) \Delta_L[A_\mu]\right) \Delta_C[A_\mu]$$

$$\times \prod_x \delta(\nabla_i A_i) \exp\left(iS(A_\mu) + i \int d^4x \, J^\mu A_\mu\right)$$
(9.105)

The insertion is contained within the first set of parentheses. Next, in order to remove the Ω in the previous expression, we will make an inverse gauge transformation on A_μ:

$$A_\mu \to A_\mu^{\Omega^{-1}}$$
(9.106)

We will now use the fact that $S(A_\mu)$, the Faddeev–Popov determinant, and the measure DA_μ are all gauge independent. The term $\delta\left(\partial^\mu A_\mu^\Omega\right)$ loses its dependence on Ω, as desired. Thus, all terms involving the Landau gauge lose their dependence on Ω. On the other hand, we must carefully keep track of those terms in the Coulomb gauge that are affected by this transformation. Then the expression for $Z_C(J)$ becomes:

$$Z_C(J) = \int DA_\mu \, \Delta_L(A_\mu) \prod_x \delta \left(\partial^\mu A_\mu \right) \exp \left[iS(A_\mu) \right]$$

$$\times \left(\Delta_C(A_\mu) \int D\Omega(x) \prod_x \delta \left(\nabla_i A_i^{\Omega^{-1}} \right) \right) \exp \left(i \int d^4x \, J^\mu A_\mu^{\Omega^{-1}} \right)$$

(9.107)

Almost all the dependence on the Coulomb gauge is now concentrated within the first set of large parentheses. Our next step is to show that the factor appearing within these large parentheses can be set equal to one, thereby eliminating almost all trace of the Coulomb gauge. To analyze the term within these parentheses, we first observe that the integration over $D\Omega$ contains the delta function, which forces us to pick out a particular value for Ω^{-1}, which we will call Ω_0. This specific value of Ω_0, because of the delta function constraint, must satisfy:

$$\nabla_i A_i^{\Omega_0} \equiv 0$$

(9.108)

In other words, we can rewrite the delta function term as:

$$\int D\Omega \prod_x \delta \left(\nabla_i A_i^{\Omega^{-1}} \right) = \Delta_0 \int D\Omega \delta \left(\Omega^{-1} - \Omega_0 \right)$$

(9.109)

where Δ_0 and Ω_0 are the terms that we must calculate. In general, this task is not an easy one, because Ω_0 is a function of A_i itself. However, we can determine Ω_0 because it must, by construction, satisfy:

$$\nabla_i A_i^{\Omega_0} = \nabla_i \left\{ \Omega_0 \left[A_i - ig^{-1}\Omega_0^{-1}\nabla_i\Omega_0 \right] \Omega_0^{-1} \right\} = 0$$

(9.110)

This expression cannot be solved exactly for $A_i^{\Omega_0}$. However, we can always power expand this expression to find the perturbative solution, which is given by:

$$A_i^{\Omega_0} = \left(\delta_{ij} - \nabla_i \frac{1}{\nabla^2} \nabla_j \right) A_j + O(A^2)$$

(9.111)

Thus, to any order of accuracy, we can solve for $A_\mu^{\Omega_0}$ and Ω_0. The advantage of introducing this expression for Ω_0 is that we can now show that the object in the first set of large parentheses in Eq. (9.107) equals one:

$$\Delta_C(A_\mu) \int D\Omega \prod_x \delta\left(\nabla_i A_i^{\Omega^{-1}}\right) = \Delta_C(A_\mu) \det \left(\frac{\delta\left(\nabla_i A_i^{\Omega^{-1}}(x)\right)}{\delta\Omega(y)}\right)_{x,y}$$

$$= \Delta_C(A_\mu) \Delta_C^{-1}[A_\mu^{\Omega_0}] \qquad (9.112)$$

where we have used the fact that Δ_{FP} is gauge invariant. Thus, the term in the large parentheses can be set equal to one, and the expression for $Z_C(J)$ can be written as:

$$Z_C(J) = \int D A_\mu \, \Delta_L(A_\mu) \prod_x \delta\left(\partial^\mu A_\mu\right) \exp\left(iS(A_\mu) + i \int d^4x \, J^\mu A_\mu^{\Omega_0}\right)$$

$$(9.113)$$

The important point here is that we have lost all dependence on the Coulomb gauge, except for the term $J^\mu A_\mu^{\Omega_0}$. This means that the Green's functions, as expected, are all gauge dependent. The next task is to show that the S matrix is independent of the gauge choice, even if the Green's functions are gauge dependent. To do this, we must extract out the dependence on Coulomb gauge parameters from the Landau gauge parameters. The source J_μ in the functional, because it originally came from the Coulomb gauge, satisfies:

$$J_0 = \nabla_i J_i = 0 \qquad (9.114)$$

We may therefore write:

$$\int d^4x \, J^\mu A_\mu^{\Omega_0} = \int d^4x \, J^\mu F_\mu(A) \qquad (9.115)$$

where F is defined, to lowest order, as:

$$F_\mu(A) = A_\mu(x) + O(A_\mu^2) \qquad (9.116)$$

Now let us extract out all dependence on the Coulomb gauge out of the generator of the Green's function:

$$Z_C(J) = \exp\left[i \int d^4x J^\mu F_\mu\left(-i\frac{\delta}{\delta j_\nu}\right)\right] Z_L(j)\big|_{j=0} \qquad (9.117)$$

We now have a functional expression relating the generating functional of the Landau gauge to the generating functional of the Coulomb gauge. Next, we must investigate the dependence of the S matrix on the gauge-dependent parameters.

In order to compare our results for the S matrix, we must go on-shell; that is, i.e., we must set the $p^2 \to 0$ on the external legs of any graph. Then, if we look at the contribution of the F term to any Feynman graph in this on-shell limit, the only terms that survive are self-energy corrections (i.e., radiative corrections to the propagator). In Chapter 7, we showed that the net effect of self-energy corrections is to give us a multiplicative renormalization of the overall diagram. This renormalization, in turn, can be absorbed in the overall renormalization of the S matrix that must always be performed when calculating radiative corrections. Thus, the S matrices calculated in the Coulomb or Landau gauges only differ by a multiplicative constant, given by the renormalization of the self-energy parts, which in turn can be absorbed into the overall renormalization of the S matrix. We therefore find that the two formalisms give the same on-shell S matrix, as desired.

In summary, we have seen how the path integral method gives us a convenient formalism in which to quantize gauge theory. The only complications are the Faddeev–Popov ghosts, which arise from the functional measure of integration. The power of the path integral method is that we can rapidly move from one gauge choice to another in gauge theory. This is crucial in order to show that gauge theory is both unitary and renormalizable.

In the next chapter, we will construct a realistic theory of the weak interactions from gauge theory. The essential ingredient will be spontaneous symmetry breaking, which allows us to have massive vector mesons without spoiling renormalization.

9.7 Exercises

1. Choose the gauge $A_1^a = A_2^a$. Is this a legitimate gauge? If so, construct the Faddeev–Popov ghost term.

2. Quantize the theory in the gauge:

$$\partial_\mu A^\mu + c A_\mu A^\mu = 0 \qquad (9.118)$$

   Calculate the Faddeev–Popov ghosts.

3. Repeat the analysis for the gauge:

$$\partial_\mu A_0 \partial^\mu A_0 = 0 \qquad (9.119)$$

4. Calculate the propagator in the gauge $\eta_\mu A^\mu = 0$, where η_μ is a constant and normalized to $\eta_\mu^2 = 1$.

5. Analyze whether the axial gauge $A_3 = 0$ suffers from a Gribov ambiguity or not. (Hint: see whether the gauge constraint is invertible or not.) Do other gauges suffer from a Gribov ambiguity? Discuss the delicate points involved with this problem.

6. A spin-3/2 field ψ_μ has both vector and spinor indices (we will suppress the spinor index). Its action is given by:

$$\mathcal{L} = \epsilon^{\mu\nu\rho\sigma} \bar{\psi}_\mu \gamma_5 \gamma_\nu \partial_\rho \psi_\sigma \tag{9.120}$$

Show that it is invariant under a local gauge symmetry $\delta\psi_\mu = \partial_\mu\alpha$, where α is a spinor. Break the gauge by adding $-\frac{1}{2}(\bar{\psi} \cdot \gamma) \not{\partial} (\gamma \cdot \psi)$ to the action. Show that the propagator equals $\gamma_\mu \not{k} \gamma_\nu / k^2$.

7. If the spin-3/2 field also carries an isospin index, can it be coupled to the Yang–Mills field in a gauge covariant fashion? Justify your answer.

8. Consider an $SU(N)$ Yang–Mills field coupled locally to a multiplet of massive mesons in the adjoint representation of $SU(N)$. Write down the Feynman rules for the scalar–vector interaction vertices.

9. From a canonical point of view, the purpose of the Faddeev–Popov ghost is to cancel the ghost modes coming from the Yang–Mills propagator. In the Landau gauge, prove that this cancellation occurs at the one loop level for vector meson scattering.

10. Prove Eq. (9.34).

11. Prove that Eq. (9.78) solves Eq. (9.75).

12. By repeating the same steps used to show the equivalence between the Coulomb and Landau gauge, show the equivalence between *any* two gauges allowed by the Faddeev–Popov formalism.

Chapter 10
The Weinberg–Salam Model

If my view is correct, the universe may have a kind of domain structure.
In one part of the universe, you may have one preferred direction of the
axis; in another part, the direction of the axis may be different.

—Y. Nambu

10.1 Broken Symmetry in Nature

In nature, a variety of beautiful and elegant symmetries surrounds us. However, there are also many examples of symmetries in nature that are broken. Rather than put explicit symmetry-breaking terms into the Hamiltonian by hand, which seems artificial and unappealing, we would like to break these symmetries in a way such that the equations retain their symmetry.

Nature seems to realize this by exploiting the mechanism of *spontaneous symmetry breaking*; that is, the Hamiltonian is invariant under some symmetry, but the symmetry is broken because the vacuum state of the Hamiltonian is not invariant. The simplest examples come from solid-state physics, where the phenomenon of spontaneous symmetry breaking is quite common. Consider a ferromagnet, such that the atoms possess a spin σ_i. Although the Hamiltonian does not select out any particular direction in space, the ground state of the theory, however, can consist of atoms whose spins are all aligned in the same direction. Thus, rotational symmetry can be broken by the vacuum state, even when the Hamiltonian remains fully symmetric. To restore the symmetry, we have to heat the ferromagnet to a high temperature T, where the atoms once again become randomly aligned.

In addition, spontaneous symmetry breakdown may also be associated with the creation of massive vector fields. In the theory of superconductivity, for example, spontaneous symmetry breaking occurs at extremely low temperatures, giving us the Meissner effect, in which magnetic flux lines are expelled from the interior of a superconductor. However, the magnetic field penetrates slightly into

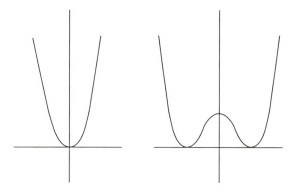

Figure 10.1. The first potential corresponds to a unique vacuum, or a Klein-Gordon field with a positive mass. The second potential, which exhibits spontaneous symmetry breaking, corresponds to a scalar field with a tachyon mass.

the medium, so there is a finite-range electromagnetic field, which corresponds to a "massive photon." We conclude that spontaneous symmetry breaking can, under certain circumstances, give mass to a massless vector field.

Similarly, spontaneous symmetry breaking gives us a solution to the problem that faced physicists trying to write down a theory of the weak interactions in 1950s and 1960s. One candidate was the massive vector meson theory. However, the massive vector meson theory was known to be nonrenormalizable by simple power counting arguments. Spontaneous symmetry breaking, however, solves this problem. It preserves the renormalizability of the original gauge theory even after symmetry breaking, giving us a renormalizable theory of massive vector mesons.

To illustrate spontaneous symmetry breaking, let us begin our discussion with a scalar field with a ϕ^4 interaction which has the symmetry $\phi \rightarrow -\phi$:

$$\mathscr{L} = \frac{1}{2}\partial_\mu \phi \partial^\mu \phi - \frac{1}{2}m^2 \phi^2 - \frac{\lambda}{4!}\phi^4 \tag{10.1}$$

If m^2 is negative, then m is imaginary; that is, we have tachyons in the theory. However, quantum mechanically, we can reinterpret this theory to mean that we have simply expanded around the wrong vacuum.

In Figure 10.1, we see two potentials, one described by $m^2\phi^2/2 + \lambda\phi^4/4!$ with positive m^2, which gives us a unique vacuum, and another with negative m^2, which corresponds to a tachyon mass.

For the second potential, a particle would rather not sit at the usual vacuum $\phi = 0$. Instead, it prefers to move down the potential to a lower-energy state given by the bottom of one of the wells. Thus, we prefer to power expand around the new minimum $\phi = v$.

(Classically, a rough analogy can be made with a vertical, stationary rod that is hanging from the ceiling. Normally, the rod's lowest-energy state is given by $\theta = 0$. If we displace the rod by an angle θ, then the potential resembles the first figure in Fig. 10.1. However, if the rod is sent spinning around the vertical axis, then it will reach a new equilibrium at some fixed angle $\theta = \theta_0$, and its effective potential will resemble the second figure in Fig. 10.1. For the spinning rod, $\theta = 0$ is no longer the lowest energy state.)

For the scalar particle we discussed before, if m^2 is positive, then there is only one minimum at the usual vacuum configuration $\phi = 0$ and the potential is still symmetric; that is, it has the symmetry $\phi \rightarrow -\phi$.

However, if m^2 is negative, then quantum mechanically the vacuum expectation value of the ϕ field no longer vanishes because we have chosen the wrong vacuum. For the tachyonic potential, we can easily find the location of the new minimum:

$$\phi_0 = v = \pm\sqrt{-6m^2/\lambda} \tag{10.2}$$

Normally, we demand that the vacuum expectation value of the scalar field vanishes. However, if the naive vacuum is the incorrect one, then we find instead:

$$\langle 0|\phi|0 \rangle = v \tag{10.3}$$

Clearly, our troubles have emerged because we have power expanded around the wrong vacuum. To correct this situation, we must shift the value of the ϕ field as follows:

$$\tilde{\phi} = \phi - v \tag{10.4}$$

In terms of the shifted field $\tilde{\phi}$, we now have broken the original symmetry $\phi \rightarrow -\phi$ and:

$$\langle 0|\tilde{\phi}|0 \rangle = 0 \tag{10.5}$$

We also have a new action given by:

$$\frac{1}{2}\partial_\mu\tilde{\phi}\partial^\mu\tilde{\phi} + m^2\tilde{\phi}^2 - \frac{1}{6}\lambda v\tilde{\phi}^3 - \frac{1}{4!}\lambda\tilde{\phi}^4 \tag{10.6}$$

Because m^2 is negative, we have an ordinary scalar particle with a positive mass squared given by $-2m^2$. The original symmetry between ϕ and $-\phi$ has now been spontaneously broken because the field has been shifted and there is a new vacuum.

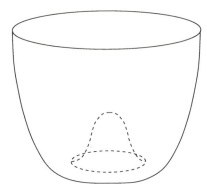

Figure 10.2. The new solution to the minimum of the potential corresponds to a ring of solutions.

Let us now examine a less trivial example given by the global $O(N)$ scalar theory, where the field ϕ^i transforms as an N-vector:

$$\mathcal{L} = \frac{1}{2}\partial_\mu \phi^i \partial^\mu \phi^{i\mu} - \frac{1}{2}m^2 \phi^i \phi^i - \frac{\lambda}{4!}(\phi^i \phi^i)^2 \tag{10.7}$$

Again, if the mass term has the wrong sign, (i.e., if m^2 is negative), then there is a new vacuum solution given by:

$$\phi^i \phi^i = v^2 = -6m^2/\lambda \tag{10.8}$$

Contrary to the previous potential that we examined, which had two minima, this theory has an infinite number of vacua. In Figure 10.2, we can see that there is actually a degenerate ring of solutions sitting in the bottom of the potential well.

Any solution of this equation yields a new vacuum. Now let us break this degeneracy by singling out a specific direction in isospin space, the last entry in the column vector:

$$\langle \phi \rangle_0 = \begin{pmatrix} 0 \\ 0 \\ 0 \\ \vdots \\ v \end{pmatrix} = (v_i) \tag{10.9}$$

where $v_i = 0$ if $i = 1, \cdots, N-1$ and $v_N = v$.

This new vacuum is still invariant under the group that leaves the last entry $v_N = v$ invariant, that is, the subgroup $O(N-1)$. We are thus breaking the original symmetry group $O(N)$ down to $O(N-1)$ with this choice of vacuum.

Let τ^i be the generators of $O(N)$. These generators, in turn, can be divided into two groups: the generators of $O(N-1)$, and everything else. Let $\tilde{\tau}^j$ equal these remaining generators. The simplest parametrization of this new vacuum is to let the last component of the ϕ^i field develop a nonzero vacuum expectation value: $\phi_N \rightarrow \phi_N + v$, keeping the other ϕ_i the same. However, we will find it convenient to introduce a slightly more complicated parametrization of the N fields within the vector ϕ^i. We will choose instead:

$$\phi = e^{i\xi_j\tilde{\tau}_j/v} \begin{pmatrix} 0 \\ 0 \\ 0 \\ \vdots \\ v + \sigma(x) \end{pmatrix} \tag{10.10}$$

where we have replaced the original N fields ϕ^i with a new set, given by $N-1$ fields ξ_i and by σ.

To see the reason for this particular parametrization, we recall that $O(N)$ has $(1/2)N(N-1)$ generators. The number of generators in $O(N)$, after we have subtracted out the generators of $O(N-1)$, is:

$$\frac{1}{2}N(N-1) - \frac{1}{2}(N-1)(N-2) = N - 1 \tag{10.11}$$

Thus, there are $N-1$ generators $\tilde{\tau}^i$ that are not generators of the $O(N-1)$ subgroup.

A particularly useful parametrization for the generators of $O(N)$ is given by a series of delta functions:

$$\begin{aligned} (\tau_{ij})_{kl} &= -i(\delta_{ik}\delta_{jl} - \delta_{il}\delta_{jk}) \\ (\tilde{\tau}_{iN})_{kl} &= -i(\delta_{ik}\delta_{Nl} - \delta_{il}\delta_{Nk}) \end{aligned} \tag{10.12}$$

To lowest order, we have $\phi_i = \xi_i$ for $i < N$ and $\phi_N = v + \sigma$, and the action becomes:

$$\begin{aligned} \mathcal{L} &= \frac{1}{2}\left(\partial_\mu\sigma\partial^\mu\sigma + \partial_\mu\xi_i\partial^\mu\xi_i\right) \\ &\quad - \frac{1}{2}m^2(v+\sigma)^2 - \frac{1}{4!}\lambda(v+\sigma)^4 + \text{higher terms} \end{aligned} \tag{10.13}$$

The $N-1$ ξ_i fields have become massless, and σ is now massive but no longer a tachyon. The action is still invariant under a residual symmetry, $O(N-1)$.

We emphasize that the number of massless bosons ξ_i is equal to the number of broken generators $\tilde{\tau}^i$. We call these massless bosons, which signal the spontaneous breakdown of the theory, *Nambu–Goldstone bosons*.[1–3] These bosons will play a special role in constructing realistic gauge theories of the weak interactions.

In summary, we have:

$$O(N) \rightarrow O(N-1) \;\Rightarrow\; N-1 \text{ Nambu–Goldstone bosons} \qquad (10.14)$$

that is, the breaking of $O(N)$ symmetry down to $O(N-1)$ symmetry leaves us with $N-1$ Nambu–Goldstone bosons ξ_i, or one boson for each of the broken generators $\tilde{\tau}^i$.

10.2   The Higgs Mechanism

We now have a rule that the number of massless Nambu–Goldstone bosons equals the number of broken generators of the theory. Let us discuss a more general action to see if this result still holds:

$$\frac{1}{2}\partial_\mu \phi \cdot \partial^\mu \phi - V(\phi) \qquad (10.15)$$

where ϕ transforms under some representation of a group G, which has N generators.

Let us say that there is a nontrivial vacuum, which we can find by calculating the minimum of the potential:

$$\left. \frac{\delta V}{\delta \phi_i} \right|_{\phi = v} = 0 \qquad (10.16)$$

Let $\langle \phi_i \rangle_0 = v_i$ be a solution of this equation that minimizes the potential. We find that this new vacuum is still invariant under a subgroup of G, called H, which has M generators. This means that there are M generators $\bar{L}^a_{ij}$ that leave v_i unchanged; that is, they satisfy:

$$\bar{L}^a_{ij} v_j = 0 \qquad (10.17)$$

There are also $N-M$ generators $\tilde{L}^a$ for which $\tilde{L}^a_{ij} v_j \neq 0$. In other words, the generators L^a_{ij} are of two types: $\bar{L}^a_{ij}$, which generate the subgroup H, and $\tilde{L}^a_{ij}$, which are all the remaining generators. The first set of generators annihilate on v_i, by construction, while the second set does not.

Next, we wish to show that expanding around this new vacuum will create $N-M$ massless boson fields. To begin, we define the variation of the scalar field as:

$$\delta\phi = -i\theta^a L^a \phi \qquad (10.18)$$

The potential $V(\phi)$ is invariant under this transformation, so we have:

$$\frac{\delta V}{\delta\phi_i}\delta\phi_i = -i\frac{\delta V}{\delta\phi_i}\theta^a L^a_{ij}\phi_j = 0 \qquad (10.19)$$

Since the parameters θ^a are arbitrary, we have N equations:

$$\frac{\delta V}{\delta\phi_i}L^a_{ij}\phi_j = 0 \qquad (10.20)$$

Differentiating this, we arrive at:

$$\frac{\delta^2 V}{\delta\phi_i\delta\phi_k}L^a_{ij}\phi_j + \frac{\delta V}{\delta\phi_i}L^a_{ik} = 0 \qquad (10.21)$$

Substituting the value of ϕ at the minimum into the previous equation, we find:

$$\frac{\delta^2 V}{\delta\phi_i\delta\phi_k}\bigg|_{\phi=v}L^a_{ij}v_j = 0 \qquad (10.22)$$

If we Taylor expand the potential around the new minimum $\phi_i = v_i$, then the mass matrix can be defined as:

$$V(\phi) = -\frac{1}{2}(M^2)_{ij}(\phi - v)_i(\phi - v)_j + \cdots \qquad (10.23)$$

Now let us insert the previous equation for the mass matrix into Eq. (10.22). This gives us:

$$(M^2)_{ij}L^a_{jk}v_k = 0 \qquad (10.24)$$

This equation is trivially satisfied if L^a is a generator of the subgroup H, since $\bar{L}^a_{ij}v_j = 0$. The situation is more interesting, however, when the generator is one of the $N - M$ generators $\tilde{L}^a_{ij}$ of G that are outside H.

For these $N - M$ generators $\tilde{L}^a_{ij}$, Eq. (10.24) is an eigenvalue equation for the matrix M^2. It states that for each of the $N - M$ generators $\tilde{L}^a_{ij}$, there is a zero eigenvalue of the M^2 matrix. Since the eigenvalues of the M^2 matrix give us the mass spectrum of the fields, we have $N - M$ massless bosons in the theory.

Thus, after symmetry breaking there are $N - M$ massless Nambu–Goldstone bosons, one for each broken generator.

In summary:

$$G \to H \; \Rightarrow \; N - M \text{ Nambu–Goldstone bosons} \qquad (10.25)$$

A surprising feature of this method, however, arises when we apply the Nambu–Goldstone theorem to gauge theories. In this case, will find that these Nambu–Goldstone bosons are "eaten up" by the gauge particles, converting massless Yang–Mills vector particles into massive ones. This is called the *Higgs–Kibble mechanism*.[4–6]

Furthermore (and this is the key point), the Yang–Mills theory remains renormalizable even after the Higgs mechanism has generated massive vector particles. In other words, this is the long-sought-after mechanism that can render a massive vector theory renormalizable. It is not an exaggeration to say that this discovery, by 't Hooft, changed the landscape of theoretical particle physics.

To see how the Higgs mechanism works, let us begin with a theory of complex scalar particles coupled to Maxwell's theory. The action is:

$$\mathscr{L} = D_\mu \phi^* D^\mu \phi - m^2 \phi^* \phi - \lambda (\phi^* \phi)^2 - \frac{1}{4} F_{\mu\nu} F^{\mu\nu} \qquad (10.26)$$

The coupled system, as before, is invariant under:

$$\begin{aligned}
\phi &\to e^{-i\theta(x)} \phi \\
\phi^* &\to e^{+i\theta(x)} \phi^* \\
A_\mu &\to A_\mu - \frac{1}{e} \partial_\mu \theta(x)
\end{aligned} \qquad (10.27)$$

The action is invariant under $U(1) = SO(2)$.

Now we break this symmetry; the new vacuum is given by:

$$\langle \phi \rangle_0 = v/\sqrt{2} \qquad (10.28)$$

We will find it convenient to parametrize the complex field ϕ by introducing two fields σ and ξ:

$$\begin{aligned}
\phi &= e^{i\xi/v}(v + \sigma)/\sqrt{2} \\
&= (v + \sigma + i\xi + \cdots)/\sqrt{2}
\end{aligned} \qquad (10.29)$$

We can now make a gauge rotation on both ϕ and A_μ:

$$\begin{aligned}
\phi &\to \phi' = e^{-i\xi/v} \phi = (v + \sigma)/\sqrt{2} \\
A_\mu &\to A'_\mu = A_\mu - \frac{1}{ev} \partial_\mu \xi
\end{aligned} \qquad (10.30)$$

(We have gauge rotated the ξ field away in the definition of ϕ' and A'_μ.)

The final action now reads:

$$\mathcal{L} = -\frac{1}{4} F'_{\mu\nu} F^{\mu\nu\prime} + \frac{1}{2} \partial_\mu \sigma \partial^\mu \sigma + \frac{1}{2} e^2 v^2 A'_\mu A^{\mu\prime}$$
$$+ \frac{1}{2} e^2 (A'_\mu)^2 \sigma (2v + \sigma) - \frac{1}{2} \sigma^2 (3\lambda v^2 + m^2) - \lambda v \sigma^3 - \frac{1}{4} \lambda \sigma^4$$

$$(10.31)$$

The A'_μ field has become massive. The key point is that the ξ has disappeared completely. In other words, it has been "gauged away," or "eaten up" by the vector field.

The Nambu–Goldstone boson, corresponding to the breaking of $U(1)$ invariance, has disappeared and reappeared in a new guise, as the massive component of the massless vector field. Since $U(1)$ is now broken, there is no longer any gauge symmetry left to prevent the Yang–Mills field from acquiring a mass, and hence it acquires a massive mode (at the expense of the ξ mode, which vanishes). Thus, the new field A'_μ has three components (while a massless vector field has two helicities or transverse polarizations), with one of these fields corresponding to the old ξ field. To see that the A'_μ field has gobbled up the ξ Nambu–Goldstone boson, we note that its definition was: $A'_\mu = A_\mu - (1/ev)\partial_\mu \xi$, which clearly shows that the ξ field has been incorporated into the A'_μ field.

Let us now discuss two slightly more difficult examples of the Higgs mechanism with non-Abelian gauge fields. Then we will see that only some of the gauge fields become massive. First, let us discuss a $SU(2)$ gauge theory coupled to an isovector triplet of scalar fields. We will break this down to $U(1)$, so that only one of the three gauge fields remains massless, while the other two acquire a mass by eating up the Nambu–Goldstone bosons.

The action is:

$$\mathcal{L} = \frac{1}{2} D_\mu \phi^i D^\mu \phi^i - V(\phi^i \phi^i) - \frac{1}{4} F^a_{\mu\nu} F^{a\mu\nu} \qquad (10.32)$$

As before, we will choose $V(\phi^i \phi^i)$ such that there is a tachyon in the theory, meaning that we have chosen the wrong vacuum. When we shift to a new vacuum, we find that it is degenerate. We will break this degeneracy by shifting the third component of the isovector:

$$\langle \phi \rangle_0 = \begin{pmatrix} 0 \\ 0 \\ v \end{pmatrix}; \quad \phi = e^{i(\xi_1 L^1 + \xi_2 L^2)/v} \begin{pmatrix} 0 \\ 0 \\ v + \sigma \end{pmatrix} \qquad (10.33)$$

where L^i are the generators of $SU(2)$, ξ_i are the Nambu–Goldstone bosons, and we choose the third isospin generator L^3 to be the generator of the unbroken $U(1)$ symmetry.

Next, we make a local gauge transformation on the fields, such that we remove the ξ fields appearing in the previous equation for ϕ:

$$\phi \quad \rightarrow \quad \phi' = \Omega\phi$$

$$\mathbf{L} \cdot \mathbf{A}_\mu \quad \rightarrow \quad \mathbf{L} \cdot \mathbf{A}'_\mu = \Omega\mathbf{L} \cdot \mathbf{A}_\mu \Omega^{-1} - \frac{i}{g}(\partial_\mu \Omega)\Omega^{-1} \qquad (10.34)$$

where:

$$\Omega = e^{-i(\xi_1 L^1 + \xi_2 L^2)/v} \qquad (10.35)$$

Let us expand the action around the new vacuum. The vector fields now have the following action:

$$-\frac{1}{4}F'_{\mu\nu}F^{\mu\nu\prime} + \frac{1}{2}g^2 v^2(A^1_\mu A^{1\mu} + A^2_\mu A^{2\mu}) \qquad (10.36)$$

Two of the gauge fields have acquired a mass, but A^3_μ, corresponding to a residual $U(1)$ symmetry, is still massless. Again, the number of Nambu–Goldstone bosons equals the number of broken generators.

The other terms in the action are:

$$\frac{1}{2}\partial_\mu\sigma\partial^\mu\sigma - V(v+\sigma) + \text{higher terms} \qquad (10.37)$$

The $\xi_{1,2}$ fields, as expected, have disappeared completely by being absorbed into the gauge fields.

Finally, we must now show that the Higgs mechanism works for an arbitrary gauge theory. The important feature that we want to demonstrate is that the number of gauge fields that become massive is equal to the number of broken generators of the gauge group.

We start with a gauge group G, which has N generators, and hence N gauge fields A^a_μ. We also have the real scalar field ϕ transforming under some n-dimensional representation of the group G. We start with the action:

$$\mathcal{L} = -\frac{1}{4}F^a_{\mu\nu}F^{\mu\nu a} + \frac{1}{2}(\partial_\mu - igL^a A^a_\mu)\phi \, (\partial^\mu - igL^b A^{\mu b})\phi - V(\phi) \qquad (10.38)$$

where L^a is an $n \times n$ representation of the group G, that has N generators.

Let us choose $V(\phi)$ so that symmetry breaking occurs. Let $\phi = v$ be a matrix equation defining the minimum of the potential. We want a vacuum that is invariant under an M-dimensional subgroup H of G. The generators of H, because they leave the vacuum invariant, satisfy $L^a v = 0$.

We now parametrize the scalar field as:

$$\phi = \exp\left(i \sum_{i=1}^{N-M} \xi_i \tilde{L}^i / v\right)(v + \sigma) \tag{10.39}$$

where we sum over the $N - M$ generators that do not correspond to the subgroup H and do not annihilate v. For these generators $\tilde{L}^i v \neq 0$. Also, ξ_i are the Nambu–Goldstone bosons.

The trick is now to choose a gauge transformation which swallows up the ξ fields appearing in the above definition. As before, we choose:

$$\phi \rightarrow \phi' = \exp\left(-i \sum_{i=1}^{N-M} \xi_i \tilde{L}^i / v\right)\phi = \Omega\phi \tag{10.40}$$

The Ω in front of ϕ precisely cancels against the Ω^{-1} appearing in front of the parametrization of ϕ, so the ξ_i fields disappear.

Inserting this parametrization into the action, we can collect the terms responsible for the vector meson masses, which is given by:

$$\frac{1}{2} A^i_\mu (M^2)^{ij} A^{j\mu} = \frac{1}{2} \left(g\tilde{L}^i v \big| g\tilde{L}^j v\right) A^i_\mu A^{j\mu} \tag{10.41}$$

where the brackets represent the matrix contraction or scalar product between two vectors.

Thus, the masses of the gauge fields are given by the eigenvalues of the following matrix:

$$(M^2)^{ij} = g^2 \left(v \big| \tilde{L}^i \tilde{L}^j v\right) \tag{10.42}$$

There are $N - M$ non-zero eigenvalues to this equation, and hence there are $N - M$ massive vector fields (which have absorbed the remnants of the $N - M$ Nambu–Goldstone bosons).

In retrospect, the Higgs mechanism has a rather simple interpretation. We know that a gauge theory locally invariant under a group G has no mass term in the action for the vector field in perturbation theory. However, if the group G breaks down to a subgroup H via symmetry breaking, then we know that the gauge fields corresponding to the H subgroup must still remain massless. However, the gauge fields corresponding to the broken generators of G are now free to become massive.

Let us now present a more formal, model-independent proof of the Nambu–Goldstone theorem. We begin with the observation that spontaneous symmetry breaking occurs because the vacuum is not invariant under a certain symmetry, although the action is. In other words, beginning with a symmetry and its conserved current J_μ, we can construct its conserved charge Q such that:

$$Q = \int d^3x\, J^0; \quad Q|0\rangle \neq 0 \tag{10.43}$$

The essential point is that Q no longer annihilates the vacuum in a broken theory; that is, the vacuum state $|0\rangle$ is not the true vacuum, so it is not annihilated by Q. Current conservation means that the following commutator vanishes:

$$\begin{aligned}
0 &= \int d^3x \left[\partial_\mu J^\mu(x), \phi(0)\right] \\
&= \partial_0 \int d^3x \left[J^0(x), \phi(0)\right] + \int_S d\mathbf{S} \cdot [\mathbf{J}(x), \phi(0)]
\end{aligned} \tag{10.44}$$

for some scalar boson field $\phi(0)$. Then we make the assumption that for large enough surfaces S, we can ignore the term on the right-hand side of the equation. Hence:

$$\frac{d}{dt}\, [Q(t), \phi(0)] = 0 \tag{10.45}$$

Then:

$$\langle 0| \, [Q(t), \phi(0)] \, |0\rangle = C \neq 0 \tag{10.46}$$

where C is a nonzero constant. Now insert a complete set of intermediate states inside the commutator. After making a translation, we can write this expression as:

$$\begin{aligned}
\sum_n (2\pi)^3 \delta^3(\mathbf{p}_n) &\Big[\langle 0|J^0(0)|n\rangle\langle n|\phi(0)|0\rangle e^{-iE_n t} \\
&- \langle 0|\phi(0)|n\rangle\langle n|J^0(0)|0\rangle e^{iE_n t} \Big] = C \neq 0
\end{aligned} \tag{10.47}$$

In general, unless $E_n = 0$, the positive and negative frequency parts cannot possibly cancel, and hence the contributions to the sum cannot give us a constant. In other words, it is impossible to satisfy the previous equation unless the mass of the intermediate states vanishes. (Furthermore, these massless states *must* exist, so that the sum adds up to a non zero value.) Thus, there must be a massless Nambu–Goldstone state in the theory, with the property that:

$$\langle n|\phi(0)|0\rangle \neq 0, \quad \langle 0|J^0(0)|n\rangle \neq 0 \tag{10.48}$$

This completes the proof.

10.3   Weak Interactions

Now that we have investigated the possibility of spontaneous symmetry breaking, we can apply this theory to the weak interactions, where it has enjoyed great success.

To appreciate this breakthrough, we must realize that, from an historical point of view, important progress in the field theory of weak interactions was relatively stagnant for many decades, from the original Fermi action[7] of the 1930s, to the overthrow of parity, to the advent of gauge theories in the 1970s.

Fermi originally tried to explain the decay of the neutron into a proton, an electron, and an anti-neutrino:

$$n \rightarrow p + e + \bar{\nu} \qquad (10.49)$$

by postulating the phenomenological Lagrangian:

$$
\begin{aligned}
\mathscr{L} &= \sum_A \frac{G_F}{\sqrt{2}} (\bar{\Psi}_p \Gamma_A \Psi_n)(\bar{\psi}_e \Gamma^A \psi_\nu) + \text{h.c.} \\
&= \sum_A \frac{G_F}{\sqrt{2}} J^\dagger_{A,\text{had}} J^A_{\text{lept}} + \text{h.c.} \qquad (10.50)
\end{aligned}
$$

(Experimentally, $G_F = (1.16639 \pm 0.00002) \times 10^{-5} \text{ GeV}^{-2}$.) This action, from the very start, was known to suffer from a series of diseases. First, the Γ_A matrices could in principle consist of all possible combinations of the 16 Dirac matrices. The lack of precise experimental data for years prevented a decisive determination of the action. It took many decades finally to resolve that the correct combination should be $V - A$,[8-9] rather than scalar or tensor combinations.

The Fermi action also suffered from a fatal theoretical disease: It was non-renormalizable. Four-fermion interactions must be accompanied by a dimension-ful coupling constant (since the dimension of a spinor is $3/2$). The Fermi constant thus has the dimensions -2, and hence the theory was nonrenormalizable. Finally, the theory violated unitarity. If we calculate the high energy behavior of the differential cross section of any weak process, we can write (purely by dimensional arguments):

$$\frac{d\sigma}{d\Omega} \rightarrow \frac{G_F^2}{4\pi} s \qquad (10.51)$$

(The differential cross section is a pure s wave because the four-fermion interaction takes place at a single point.)

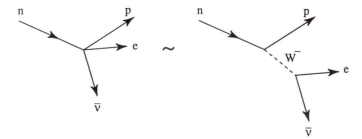

Figure 10.3. If four fermions interact via a massive vector meson, then the exchange of the vector meson, for a large mass, mimics the original Fermi four-fermion interaction.

On the other hand, we know from unitarity that the S matrix for s-wave scattering must obey the law:

$$\frac{d\sigma}{d\Omega} \rightarrow \frac{1}{s} \tag{10.52}$$

These two results are obviously in contradiction for high energies. The discrepancy between these two results becomes serious at around $\sqrt{s} \sim 500\,\mathrm{GeV}$.

To extend the Fermi action, physicists tried to emulate the success of the Yukawa theory for the strong interactions. If the Yukawa meson could mediate the strong interactions, then perhaps a massive vector meson could mediate the weak interactions. The obvious proposal was to treat the Fermi action as a byproduct of vector meson exchange (Fig. 10.3):

$$\mathscr{L} = g_W^2 (\bar{\Psi}_p \gamma^\mu \Psi_n) \frac{g_{\mu\nu} - k_\mu k_\nu / M_W^2}{k^2 - M_W^2 + i\epsilon} (\bar{\Psi}_e \gamma^\nu \psi_\nu) \tag{10.53}$$

This, in turn, gave us a rough determination of the mass of the vector meson:

$$\frac{G_F}{\sqrt{2}} \sim \frac{g_W^2}{M_W^2} \tag{10.54}$$

A rough calculation put the vector meson mass at around 50-100 GeV.

The advantage of the vector meson approach was that it smoothed out the singular behavior of the original Fermi action. However, the massive vector meson theory was still nonrenormalizable. The propagator behaved as $k_\mu k_\nu / M^2 k^2$ in the ultraviolet region; so the higher loop graphs all diverged.

Meanwhile, the experimental confusion in the weak interactions continued with the discovery of three identical sets of lepton pairs, corresponding to the electron, muon, and τ particles and their corresponding neutrino partners:

$$\begin{pmatrix} \nu_e \\ e \end{pmatrix}; \quad \begin{pmatrix} \nu_\mu \\ \mu \end{pmatrix}; \quad \begin{pmatrix} \nu_\tau \\ \tau \end{pmatrix} \tag{10.55}$$

To this day, no one knows why there are three identical copies of lepton families, as well as quark families. The masses of the e, μ, and τ leptons in MeV are .51099906(15), 105.658389(34), and 1784. Their neutrino masses have upper limits given by 10 eV, .27MeV, and 35MeV.

10.4 Weinberg–Salam Model

The Weinberg–Salam model,[10,11] one of the most successful quantum theories besides the original QED, is a curious amalgam of the weak and electromagnetic interactions. Strictly speaking, it is not a "unified field theory" of the weak and electromagnetic interactions, since we must introduce two distinct coupling constants g and g' for the $SU(2)$ and $U(1)$ interactions. Nonetheless, it represents the one of the most important extensions of QED in the past quarter century.

We begin by discussing the $SU(2)$ sector. Observationally, we must incorporate a neutral, left-handed Weyl neutrino along with a Dirac electron, which can be considered to be the sum of left-handed and right-handed Weyl spinors. The left–handed fermions form an isodoublet, consisting of the Weyl neutrino and electron:

$$L \equiv \begin{pmatrix} \nu_e \\ e \end{pmatrix}_L \tag{10.56}$$

while the right-handed sector consists of an isosinglet, the right-handed electron:

$$R \equiv (e)_R \tag{10.57}$$

This curious feature, that the electron is split into two parts, with the left- and right-handed sectors transforming differently, is a consequence of the fact that the weak interactions violate parity and are mediated by $V - A$ interactions.

These two lepton sectors transform under $SU(2)$ in different ways:

$$\begin{aligned} L &\rightarrow & e^{(i/2)\boldsymbol{\theta}\cdot\boldsymbol{\sigma}}L \\ R &\rightarrow & R \end{aligned} \tag{10.58}$$

Now let us examine the transformation of these fields under:

$$L \quad \longrightarrow \quad e^{(i/2)\beta} L$$
$$R \quad \longrightarrow \quad e^{i\beta} R \tag{10.59}$$

R and L transform slightly differently under the $U(1)$ transformation.

The charge Q of the pair (ν_e, e) is $(0, -1)$, which is almost equal to the eigenvalue of the isospin operator T^3. In fact, the correct formula for the charge is:

$$Q = T^3 + \frac{Y}{2} \tag{10.60}$$

where $T^3 = \pm 1/2$ and $Y = -1$ for the left doublets, and $T^3 = 0$ and $Y = -2$ for the right singlets. Thus, we need both the $SU(2)$ and $U(1)$ sectors to get the charge correctly.

The final action consists of three parts. $\mathcal{L}_1$ is the gauge part; $\mathcal{L}_2$ is the fermionic part; and $\mathcal{L}_3$ is the scalar Higgs sector:

$$\mathcal{L} = \mathcal{L}_1 + \mathcal{L}_2 + \mathcal{L}_3 \tag{10.61}$$

where:

$$
\begin{aligned}
\mathcal{L}_1 &= -\frac{1}{4} W^a_{\mu\nu} W^{a\mu\nu} - \frac{1}{4} F_{\mu\nu} F^{\mu\nu} \\[6pt]
\mathcal{L}_2 &= i\bar{R}\gamma^\mu D_\mu R + i\bar{L}\gamma^\mu D_\mu L \\[6pt]
\mathcal{L}_3 &= D_\mu \phi^\dagger D^\mu \phi - m^2 \phi^\dagger \phi - \lambda(\phi^\dagger \phi)^2 \\
&\quad + G_e(\bar{L}\phi R + \bar{R}\phi^\dagger L)
\end{aligned}
\tag{10.62}
$$

where:

$$
\begin{aligned}
W^a_{\mu\nu} &= \partial_\mu W^a_\nu - \partial_\nu W^a_\mu + gf^{abc} W^b_\mu W^c_\nu \\[4pt]
F_{\mu\nu} &= \partial_\mu B_\nu - \partial_\nu B_\mu \\[4pt]
D_\mu R &= (\partial_\mu + ig' B_\mu) R \\[4pt]
D_\mu L &= \left[\partial_\mu + (i/2)g' B_\mu - (i/2)g\sigma_i W^i_\mu \right] L \\[4pt]
D_\mu \phi &= \left[\partial_\mu - (i/2)g\sigma_i W^i_\mu - (i/2)g' B_\mu \right] \phi
\end{aligned}
\tag{10.63}
$$

The scalar multiplet is a complex isodoublet given by:

$$\phi = \begin{pmatrix} \phi^+ \\ \phi^0 \end{pmatrix} \tag{10.64}$$

where the doublet has charge $(1, 0)$, which can be given by $Q = T^3 + (1/2)Y$, such that $T^3 = \pm 1/2$ and $Y = 1$.

Symmetry breaking is induced by:

$$\langle \phi \rangle = \begin{pmatrix} 0 \\ v/\sqrt{2} \end{pmatrix} \tag{10.65}$$

After symmetry breaking, the fields W_μ^a and B_μ recombine and reemerge as the physical photon field A_μ, a neutral massive vector particle Z_μ, and a charged doublet of massive vector particles $W_\mu^\pm$:

$$
\begin{aligned}
Z_\mu &= \frac{g W_\mu^3 + g' B_\mu}{(g^2 + g'^2)^{1/2}} \equiv \cos\theta_W W_\mu^3 + \sin\theta_W B_\mu \\[2mm]
A_\mu &= \frac{-g' W_\mu^3 + g B_\mu}{(g^2 + g'^2)^{1/2}} \equiv -\sin\theta_W W_\mu^3 + \cos\theta_W B_\mu \\[2mm]
W_\mu^\pm &= \frac{1}{\sqrt{2}}(W_\mu^1 \pm i W_\mu^2)
\end{aligned}
\tag{10.66}
$$

where the Weinberg angle θ_W is defined via:

$$
\begin{aligned}
\cos\theta_W &= \frac{g}{(g^2 + g'^2)^{1/2}} \\[2mm]
\tan\theta_W &= \frac{g'}{g}
\end{aligned}
\tag{10.67}
$$

By examining the mass sector, we can read off the masses of the resulting vector particles:

$$
\begin{aligned}
M_{W_1}^2 &= M_{W_2}^2 \\[2mm]
M_Z^2 &= \frac{M_W^2}{\cos^2\theta_W} \\[2mm]
M_A &= 0
\end{aligned}
\tag{10.68}
$$

Finally, the electric charge emerges as:

$$e = g\sin\theta_W \tag{10.69}$$

Experimentally, the predictions of the Weinberg–Salam model have been tested to about one part in 10^3 or 10^4. At present, we have the following values for the parameters of the Weinberg–Salam model[12]:

$$\sin^2 \theta_W = 0.2325 \pm 0.008$$

$$M_Z = 91.173 \pm 0.020 \, \text{GeV}/c^2$$

$$M_W = 80.22 \pm 0.26 \, \text{GeV}/c^2 \tag{10.70}$$

The Weinberg-Salam model has been one of the outstanding successes of field theory, gradually rivalling the predictive power of QED. The rest of this chapter will be devoted to studying the many consequences of the model.

10.5 Lepton Decay

Let us now use the Weinberg–Salam model to do some simple calculations, such as the decay of the muon or the τ lepton to lowest order. Although the Born term resembles the calculation that one might perform with the old massive vector meson theory, the Weinberg–Salam model allows us to calculate quantum corrections to the massive vector meson theory that we can then compare with experiment.

We are interested in purely leptonic decays, such as:

$$\mu^- \rightarrow e^- + \bar{\nu}_e + \nu_\mu \tag{10.71}$$

More generally, we can have:

$$l_a(p_1, s_a) \rightarrow \bar{\nu}_b(p_2) + \nu_a(p_3) + l_b(p_4, s_b) \tag{10.72}$$

where $p_1 = p_2 + p_3 + p_4$ and a and b represent lepton generations. Thus, (l_a, ν_a) and (l_b, ν_b) form two lepton generations.

A straightforward use of the Feynman rules for the Weinberg–Salam model yields, to lowest order:

$$\mathcal{M} = \frac{-ig^2}{8} \left(\bar{u}_3 \gamma_\mu (1 - \gamma_5) u_1 \right) \left(\frac{-g^{\mu\nu} + q^\mu q^\nu / M_W^2}{q^2 - M_W^2} \right) [\bar{u}_4 \gamma_\nu (1 - \gamma_5) v_2] \tag{10.73}$$

The decay rate $d\omega$ is given by:

$$d\omega = (2\pi)^4 \frac{\delta^4(P_i - P_f)|\mathcal{M}|^2 d^3 p_2 d^3 p_3 d^3 p_4}{(2\pi)^9 2E_1 2E_2 2E_3 2E_4} \tag{10.74}$$

We can write $\mathcal{M}$ as follows:

$$|\mathcal{M}|^2 = \frac{g^4}{64 M_W^4} L_{\mu\nu} M^{\mu\nu} \tag{10.75}$$

where:

$$
\begin{aligned}
L_{\mu\nu} &= \mathrm{Tr}\left[(u_3\bar{u}_3)\gamma_\mu(1-\gamma_5)(u_1\bar{u}_1)\gamma_\nu(1-\gamma_5)\right] \\
&= \frac{1}{2}\mathrm{Tr}\left[\slashed{p}_3\gamma_\mu(1-\gamma_5)(\slashed{p}_1+m_a)(1+\gamma^5\slashed{s}_a)\gamma_\nu(1-\gamma_5)\right] \tag{10.76}
\end{aligned}
$$

where we have used the Gordon identity and have taken M_W to be larger than the momenta q^2 and the mass term $m_a m_b$.

Using the standard identities, we can write:

$$
\begin{aligned}
L_{\mu\nu} &= \mathrm{Tr}\left[\slashed{p}_3\gamma_\mu(\slashed{p}_1+m_a\gamma^5\slashed{s}_a)\gamma_\nu(1-\gamma_5)\right] \\
&= p_3^\alpha(p_1-m_a s_a)^\beta \mathrm{Tr}\left[\gamma_\alpha\gamma_\mu\gamma_\beta\gamma_\nu(1-\gamma_5)\right] \tag{10.77}
\end{aligned}
$$

Similarly, $M^{\mu\nu}$ has almost the same structure, so it can be written:

$$M^{\mu\nu} = (p_4 - m_b s_b)_\alpha p_{2\beta} \,\mathrm{Tr}\left[\gamma^\alpha\gamma^\mu\gamma^\beta\gamma^\nu(1-\gamma_5)\right] \tag{10.78}$$

so the final result for $\mathcal{M}$ is:

$$|\mathcal{M}|^2 = \frac{g^4}{M_W^4}\left[p_3\cdot(p_4-m_b s_b)p_2\cdot(p_1-m_a s_a)\right] \tag{10.79}$$

Inserting this expression back into the decay rate, we find:

$$d\omega = \frac{g^4(p_4-m_b s_b)^\mu(p_1-m_a s_a)^\nu}{16(2\pi)^5 M_W^4 E_1 E_4} d^3 p_4\, I_{\mu\nu} \tag{10.80}$$

where:

$$
\begin{aligned}
I_{\mu\nu} &= \int \frac{d^3 p_2\, d^3 p_3\, p_{3\mu} p_{2\nu} \delta^4(p-p_2-p_3)}{E_2 E_3} \\
&= \frac{\pi}{6}(g_{\mu\nu}p^2 + 2 p_\mu p_\nu) \tag{10.81}
\end{aligned}
$$

where $p = p_1 - p_4$.

To simplify this calculation, we will take the rest frame of the decaying lepton l_a. Then we have the following rest frame decomposition:

$$
\begin{aligned}
p_1 &= (m_a, \mathbf{0}) \\
s_a &= (0, \mathbf{s_a}) \\
p_4 &= (E_4, \mathbf{p_4}) \\
s_b &= [\mathbf{p_4} \cdot \mathbf{s_b}/m_b, \, \mathbf{s_b} + (\mathbf{p_4} \cdot \mathbf{s_b})\mathbf{p_4}/m_b(E_4 + m_b)]
\end{aligned}
\tag{10.82}
$$

Then the differential decay rate becomes:

$$
\begin{aligned}
d\omega = {} & \frac{g^4 p_4 \, dE_4 \, d\Omega}{192(2\pi)^4 M_W^4 m_a} \Bigg\{ (m_a^2 + m_b^2 - 2m_a E_4) \Big[m_a(E_4 - \mathbf{p_4} \cdot \mathbf{s_b}) \\
& + m_a \left(\mathbf{p_4} - m_b \mathbf{s_b} - \frac{(\mathbf{p_4} \cdot \mathbf{s_b})\mathbf{p_4}}{E_4 + m_b} \right) \cdot \mathbf{s_a} \Big] + 2(m_a^2 - m_a E_4 - m_a \mathbf{p_4} \cdot \mathbf{s_a}) \\
& \times \left[(m_a - E_4)(E_4 - \mathbf{p_4} \cdot \mathbf{s_b}) + \mathbf{p_4} \cdot \left(\mathbf{p_4} - m_b \mathbf{s_b} - \frac{(\mathbf{p_4} \cdot \mathbf{s_b})\mathbf{p_4}}{E_4 + m_b} \right) \right] \Bigg\}
\end{aligned}
\tag{10.83}
$$

This formidable expression can be simplified if we let $m_b \sim 0$. Let $\mathbf{n}$ be a unit vector pointing along $\mathbf{p_4}$, such that $\cos\theta = \mathbf{s_a} \cdot \mathbf{n}$. Let $x = E_4/E_4^{\max}$, where $E_4^{\max}$ is the maximum allowed value of E_4, or $m_a/2$. Then we have:

$$
d\omega = \frac{g^4 m_a^5}{32 M_W^2 (192\pi)^3} n(x) [1 + \alpha(x)\cos\theta] [1 - \mathbf{n} \cdot \mathbf{s_b}] \frac{dx \, d\cos\theta \, d\phi}{8\pi}
\tag{10.84}
$$

where:

$$
n(x) = 2x^2(3 - 2x); \qquad \alpha(x) = (1 - 2x)/(3 - 2x)
\tag{10.85}
$$

Now sum over the spin states of the bth particle, average over the spin states $\mathbf{s_a}$, and integrate over $d\Omega$. We then find:

$$
\frac{d\omega}{dx} = \frac{g^4}{32 M_W^4} \frac{m_a^5 n(x)}{192\pi^3}
\tag{10.86}
$$

Integrating over x, we find:

$$
\Gamma = \frac{g^4 m_a^5}{32 M_W^4 (192)\pi^3} = \frac{G_F^2 m_a^5}{192\pi^3}
\tag{10.87}
$$

We performed this calculation for small m_b. However, the calculation can also be performed by adding in the corrections for small $\epsilon = m_b/m_a$. Power expanding in this variable, we find:

$$\Gamma = \frac{G_F^2 m_a^5}{192\pi^3} F(\epsilon); \quad F(\epsilon) = 1 - 8\epsilon^2 - 24\epsilon^4 \log \epsilon + 8\epsilon^6 - \epsilon^8 + \cdots \tag{10.88}$$

We should point out that this result can be generalized to allow for couplings other than V and A, which allows for a test of the accuracy of the electroweak theory. We could start with the transition probability:

$$\mathcal{M} = \sum_i (\bar{u}_4 \Gamma_i u_1) \left[\bar{u}_3 \Gamma_i (g_i + g_i' \gamma_5) v_2 \right] \tag{10.89}$$

where we include all possible 16 Dirac matrices in the transition element, not just $V - A$.

For $m_b = 0$, the calculation of the decay rate is long but very straightforward, and yields[13]:

$$d\omega = \frac{A}{4} \frac{m_a^5 x^2 \, dx}{192\pi^3} \frac{d\Omega}{4\pi}$$

$$\times \left\{ 6(1-x) + 4\rho(\frac{4}{3}x - 1) - \xi \cos\theta \left[2(1-x) + 4\delta(\frac{4}{3}x - 1) \right] \right\} \tag{10.90}$$

where we have:

$$
\begin{aligned}
a &= |g_S|^2 + |g_S'|^2 + |g_P|^2 + |g_P'|^2 \\
a' &= 2\text{Re}\,(g_S g_P'^* + g_P g_S'^*) \\
b &= |g_V|^2 + |g_V'|^2 + |g_A|^2 + |g_A'|^2 \\
b' &= -2\text{Re}\,(g_V g_A'^* + g_A g_V'^*) \\
c &= |g_T|^2 + |g_T'|^2 \\
c' &= 2\text{Re}\,(g_T g_T'^*)
\end{aligned}
\tag{10.91}
$$

The Michel parameters are defined as:

$$
\begin{aligned}
A\rho &= 3b + 6c \\
A\xi &= -3a' - 4b' + 14c' \\
\delta &= (3b' - 6c')/(3a' + 4b' - 14c')
\end{aligned}
\tag{10.92}
$$

where $A = a + 4b + 6c$.

When radiative corrections are included in the calculation, we find, for muon decay:[14]

$$\Gamma = \frac{G_\mu m_\mu^5}{192\pi^3}\left[1 - \frac{8m_e^2}{m_\mu^2}\right]\left[1 + \frac{\alpha}{2\pi}\left(\frac{25}{4} - \pi^2\right) + \frac{3}{5}\frac{m_\mu^2}{M_W^2}\right] \qquad (10.93)$$

These and other theoretical calculations have shown good agreement with experiment.

10.6 R_ξ Gauge

Now that we have described the Weinberg–Salam model, we will study, in this and the next section, how to quantize it. We know that massive gauge theories are not renormalizable by a simple analysis of the ultraviolet behavior of their propagators. [These terms diverge as $O(1)$ as k_μ becomes large, which spoils renormalizability.] Unlike the situation in QED, we cannot appeal to the Ward identities to eliminate the troublesome term in the propagator: $k_\mu k_\nu/[M^2(k^2 - m^2)]$.

This makes us wonder how spontaneously broken gauge theories can preserve both unitarity and renormalizability, which seems totally contradictory. On the surface, it seems impossible to preserve both features, which was one reason why massive gauge theories were rejected as a model for the weak interactions.

To see how spontaneously broken gauge theories can be both unitary and renormalizable at the same time, we will use the R_ξ gauge,[15] which has the advantage that it interpolates between two sets of propagators. We will then specialize this to the 't Hooft gauge when we consider the case of the Weinberg–Salam model.

To obtain the R_ξ gauge, we will insert a new term into the action. Let us impose the gauge:

$$F(A_\mu) = a(x) \qquad (10.94)$$

on our theory, where $a(x)$ is an arbitrary, real field. We can insert the following term into the path integral:

$$\int Da\, e^{-\frac{i}{2a}\int d^4x\, a^2(x)} \qquad (10.95)$$

The path integral, with the new gauge constraint, now becomes:

$$\int DA_\mu^a \, Da \, \Delta_{FP} \delta \left(F(A_\mu) - a \right) \exp\left[i \int d^4x \left(-\frac{1}{2}a^2 + \mathscr{L}(x) \right) \right] \quad (10.96)$$

By performing the path integral over $a(x)$, we find that the Lagrangian is altered by:

$$\mathscr{L} \to \mathscr{L} - \frac{1}{2\alpha} F^T F \quad (10.97)$$

In particular, we will insert the gauge fixing term:

$$F(A_\mu) = \sqrt{\xi} \left(\partial^\mu A_\mu^a \right) \quad (10.98)$$

With this new gauge-fixing term, the action becomes (with $\alpha = 1/\xi$):

$$\begin{aligned} \mathscr{L} &= -\frac{1}{4} F_{\mu\nu} F^{\mu\nu} - \frac{1}{2\alpha} (\partial_\mu A^\mu)^2 \\ &= \frac{1}{2} A^\mu \left[g_{\mu\nu} \partial^2 + \left(\alpha^{-1} - 1 \right) \partial_\mu \partial_\nu \right] A^\nu \end{aligned} \quad (10.99)$$

Inverting this expression and solving for the propagator, we find:

$$D_{\mu\nu} = -\frac{1}{k^2} \left(g_{\mu\nu} - (1 - \alpha) \frac{k_\mu k_\nu}{k^2} \right) \quad (10.100)$$

Although the Green's functions for the theory are all gauge dependent (i.e., dependent on the parameter $\alpha = 1/\xi$), we will find that the S matrix elements are all independent of α, which is now seen to be an unphysical artifact of the gauge-fixing procedure.

For various values of α, we have various gauges [see Eq. (4.44)]:

$$\begin{cases} \alpha = 1 : & \text{Feynman gauge} \\ \alpha = 0 : & \text{Landau gauge} \end{cases} \quad (10.101)$$

Now let us discuss the massive case, which describes spontaneously broken gauge theories. When the gauge meson develops a mass, the action becomes:

$$\mathscr{L} - \frac{1}{2\alpha} F^T F = \frac{1}{2} A_\mu \left[g^{\mu\nu} \partial^2 + \partial^\mu \partial^\nu (\alpha^{-1} - 1) + g^{\mu\nu} m^2 \right] A_\nu \quad (10.102)$$

To find the Green's function, we need to solve:

$$\left[(\partial^2 + m^2) g^{\mu\nu} - \partial^\mu \partial^\nu (1 - \alpha^{-1}) \right] \Delta(x - y; \alpha)_{\nu\rho} = -\delta_\rho^\mu \delta^4(x - y) \quad (10.103)$$

Inverting this expression, we find that the propagator is given by:

$$\Delta_{\mu\nu}(k, \alpha) = -\left(g_{\mu\nu} - (1 - \alpha) k_\mu k_\nu \frac{1}{k^2 - \alpha m^2 + i\epsilon}\right) \frac{1}{k^2 - m^2 + i\epsilon} \quad (10.104)$$

(The propagator has a pole at $k^2 = \alpha m^2$, which represents a fictitious particle. This pole is cancelled in the S matrix by other contributions, which preserves unitarity.)

In the limit $\alpha \to 0$, we find:

$$\lim_{\alpha \to 0} \Delta(k, \alpha)_{\mu\nu} = -\frac{g_{\mu\nu} - k_\mu k_\nu/k^2}{k^2 - m^2 + i\epsilon} \quad (10.105)$$

In the ultraviolet limit, this propagator is much better behaved than the usual massive vector propagator; it goes as $O(1/k^2)$, which gives us good power counting behavior in the Feynman graphs. The price we pay for such a propagator, however, is that the theory is not manifestly unitary. If we take the diagonal elements of the propagator, they alternate in sign, an indication that there are longitudinal ghosts.

In the limit $\alpha \to \infty$, however, we have the propagator:

$$\lim_{\alpha \to \infty} \Delta(k, \alpha)_{\mu\nu} = -\frac{g_{\mu\nu} - k_\mu k_\nu/m^2}{k^2 - m^2 + i\epsilon} \quad (10.106)$$

This propagator, by contrast, has very bad convergence properties. For large k, it behaves like a constant, which is disastrous from a power counting point of view. The advantage of this limit, however, from the S matrix point of view, is that it is unitary. The $0, 0$ component of the propagator, taken in the rest frame, vanishes.

We now have the strange situation where for $\alpha = 0$, the theory appears renormalizable but not manifestly unitary, but for $\alpha = \infty$, the theory appears unitary, but not manifestly renormalizable.

In summary:

$$\begin{cases} \alpha \to 0: & \text{Renormalizable, not manifestly unitary} \\ \alpha \to \infty: & \text{Unitary, not manifestly renormalizable} \end{cases} \quad (10.107)$$

Although the Green's functions are ξ dependent, we know that the S matrix must be independent of ξ, which is a gauge artifact; therefore the theory is both unitary and renormalizable. Although this argument is not totally rigorous, the R_ξ gauge, because it smoothly interpolates between two gauges, intuitively shows how unitarity and renormalizability are complementary, not contradictory. To see how to apply this to the Weinberg–Salam model, we now define the 't Hooft gauge.

10.7   't Hooft Gauge

One of the most convenient gauges to quantize the Weinberg–Salam model are the 't Hooft gauges,[16] which reveal the close link between unitarity and renormalizability. We begin by analyzing a $O(3)$ gauge theory coupled to a triplet of Higgs bosons:

$$\mathscr{L} = -\frac{1}{4} F_{\mu\nu}^a F^{a\mu\nu} + \frac{1}{2}(\partial_\mu \phi_i + g\epsilon_{ijk} A_\mu^j \phi^k)^2 - V(\phi^i \phi^i) \qquad (10.108)$$

where $(\tau^i)_{jk} = -i\epsilon_{ijk}$ form the adjoint representation of $O(3)$.

As before, we will parametrize the Higgs bosons via:

$$\phi = \exp\left(-\frac{i}{v}(\xi_1\tau_1 + \xi_2\tau_2)\right)\begin{pmatrix} 0 \\ 0 \\ v + \eta \end{pmatrix}$$

$$= \begin{pmatrix} \xi_2 \\ -\xi_1 \\ v + \eta \end{pmatrix} + \text{higher terms} \qquad (10.109)$$

We are replacing the original triplet of Higgs bosons ϕ^1, ϕ^2, ϕ^3 with the set ξ_1, ξ_2, η.

Now let us substitute this new parametrization of the Higgs sector into the original action. After a bit of algebra, we find that the action can be written as the sum of four pieces. The first piece is the usual gauge action plus scalar fields, with a massive η field:

$$\mathscr{L}_1 = -\frac{1}{4}(F_{\mu\nu}^a)^2 + \frac{1}{2}\left[(\partial_\mu\xi_1)^2 + (\partial_\mu\xi_2)^2 + (\partial_\mu\eta)^2\right] + M_s^2\eta^2 \qquad (10.110)$$

The second term involves a cross-term between the A_μ and ξ_i field that we want to eliminate:

$$\mathscr{L}_2 = M\left(A_\mu^1 \partial^\mu \xi_1 + A_\mu^2 \partial^\mu \xi_2\right) \qquad (10.111)$$

The third term consists of the interactions between the gauge field and the scalar fields:

$$\mathscr{L}_3 = \left(\frac{1}{2}M^2 + g^2 v\eta + \frac{1}{2}g^2\eta^2\right)\left(A_\mu^1 A^{1\mu} + A_\mu^2 A^{2\mu}\right) + g\eta(A_\mu^1 \partial^\mu\xi_1 + A_\mu^2 \partial^\mu\xi_2) \qquad (10.112)$$

The last term contains the scalar self-interactions:

$$\mathscr{L}_4 = -\frac{1}{4}M_s^2 v^2 + \frac{3M_s^2}{2v}\eta^3 + \frac{M_s^2}{4v^2}\eta^4 + \frac{M_s^2}{v}\xi^2\eta + \frac{M_s^2}{2v^2}\xi^2\eta^2 + \frac{M_s^2}{4v^2}\xi^4 \qquad (10.113)$$

where $\xi^2 = \xi_1^2 + \xi_2^2$.

Our goal is to choose a gauge in which the second term $\mathscr{L}_2$ with the gauge–scalar coupling vanishes. To kill this term, we choose the gauge:

$$\mathscr{L}_{GF} = -\frac{1}{2\alpha}\left(\partial_\mu A^{a\mu} - \alpha M \xi_a\right)^2 \qquad (10.114)$$

where $M = gv$. The cross term coming from the gauge fixing kills $\mathscr{L}_3$ after a partial integration.

Now let us write down the propagators for the various fields (disregarding the ghosts):

$$A_\mu^{1,2} \quad \rightarrow \quad -i\frac{g_{\mu\nu} - (1-\alpha)k_\mu k_\nu/(k^2 - \alpha M^2)}{k^2 - M^2 + i\epsilon}$$

$$A_\mu^3 \quad \rightarrow \quad i\frac{g_{\mu\nu}}{k^2 + i\epsilon}$$

$$\xi_{1,2} \quad \rightarrow \quad \frac{i}{k^2 - \alpha M^2}$$

$$\eta \quad \rightarrow \quad \frac{i}{k^2 + 2M^2 + i\epsilon} \qquad (10.115)$$

In the limit $\alpha \rightarrow \infty$, we have a unitary theory, but one in which manifest renormalizability is lost. For this choice, called the unitary gauge, the spurious pole for the $\xi_{1,2}$ field at αM^2 disappears from the theory, and we are only left with the physical fields propagating in the theory.

In the limit $\alpha \rightarrow 0$, the theory is renormalizable by power counting, but not manifestly unitary. For intermediate values of α, the poles in the propagator of the gauge fields $A_\mu^{1,2}$ at $k^2 = \alpha M^2$ cancel with the poles in the propagators of the $\xi_{1,2}$ field. Thus, it is no contradiction to have a theory which is both unitary and renormalizable.

The point of this exercise is to quantize the Weinberg–Salam model. The Higgs sector is given by:

$$\left|\partial_\mu \phi - \frac{ig'}{2}B_\mu\phi - \frac{ig}{2}\tau^i W_\mu^i\phi\right|^2 - m^2\phi^\dagger\phi - \lambda(\phi^\dagger\phi)^2 \qquad (10.116)$$

We will choose a parametrization that exchanges the complex doublet of Higgs mesons ϕ_1, ϕ_2, which contains four separate fields, with the four real fields

$\xi_1, \xi_2, \xi_3, \eta$:

$$\phi = \frac{1}{\sqrt{2}}\left(1 - \frac{i}{v}\tau^i\xi^i\right)\begin{pmatrix} 0 \\ v+\eta \end{pmatrix} + \cdots$$

$$= \frac{1}{\sqrt{2}}\begin{pmatrix} -i\xi_1/v - \xi_2/v \\ 1 - i\xi_3/v \end{pmatrix}(v+\eta) + \cdots \tag{10.117}$$

Expanding out the action generates a large number of terms. However, the term in which we are interested is the cross term between the gauge and scalar part, which is given by:

$$- M_B B_\mu \partial^\mu \xi_3 + M_W W^i_\mu \partial^\mu \xi_i + \frac{1}{2}\partial_\mu \xi_i \partial^\mu \xi_i \tag{10.118}$$

where we sum over $i = 1, 2$ and $M_B = g'v/2$ and $M_W = gv/2$.

We now choose the 't Hooft gauge so that this cross term is cancelled:

$$\mathcal{L}_{GF} = -\frac{1}{2\alpha}\left(\partial_\mu W^{a\mu} - \alpha M_W \xi_a\right)^2 - \frac{1}{2\alpha}\left(\partial_\mu B^\mu + \alpha M_B \xi_3\right)^2 \tag{10.119}$$

where we sum over $a = 1, 2$.

The cross terms cancel, and we are left with the mass terms:

$$-\frac{1}{2}\alpha M_W^2(\xi_1^2 + \xi_2^2) - \frac{1}{2}\alpha M_Z^2 \xi_3^2 \tag{10.120}$$

This can be rewritten in terms of the physical fields:

$$\begin{aligned} W^3_\mu &= -\sin\theta_W A_\mu + \cos\theta_W Z_\mu \\ B_\mu &= \cos\theta_W A_\mu + \sin\theta_W Z_\mu \\ W^\pm_\mu &= \frac{1}{\sqrt{2}}(W^1_\mu \pm W^2_\mu) \end{aligned} \tag{10.121}$$

Now we can write down the new action, and from it extract the Feynman rules for the propagator. The relevant terms are:

$$\begin{aligned} \mathcal{L} = &-\frac{1}{4}(\partial_\mu W^i_\nu - \partial_\nu W^i_\mu)^2 - \frac{1}{4}(\partial_\mu B_\nu - \partial_\nu B_\mu)^2 \\ &-\frac{1}{2\alpha}\left(\partial^\mu W^i_\mu - \alpha M_W \xi^i\right)^2 - \frac{1}{2\alpha}(\partial_\mu A^\mu)^2 \\ &-\frac{1}{2\alpha}(\partial_\mu Z^\mu + \alpha M_Z \xi^3)^2 \end{aligned}$$

$$+ \frac{1}{2}M_W^2(W_\mu^1 W^{1\mu} + W_\mu^2 W^{2\mu}) + \frac{1}{2}M_Z^2 Z_\mu Z^\mu + \cdots \quad (10.122)$$

The propagators can now be read off from the action:

$$W_\mu^\pm \quad \rightarrow \quad -i\frac{g_{\mu\nu} - (1-\alpha)k_\mu k_\nu/(k^2 - \alpha M_W^2)}{k^2 - M_W^2 + i\epsilon}$$

$$A_\mu \quad \rightarrow \quad \frac{ig_{\mu\nu}}{k^2 + i\epsilon}$$

$$Z_\mu \quad \rightarrow \quad -i\frac{g_{\mu\nu} - (1-\alpha)k_\mu k_\nu/(k^2 - \alpha M_Z^2)}{k^2 - M_Z^2 + i\epsilon}$$

$$\xi_{1,2} \quad \rightarrow \quad \frac{i}{k^2 - \alpha M_W^2}$$

$$\xi_3 \quad \rightarrow \quad \frac{i}{k^2 - \alpha M_W^2} \qquad\qquad (10.123)$$

Again, we have a theory in which unitarity is manifest for $\alpha \rightarrow \infty$ but manifest renormalizability is lost. The fictitious poles vanish, and we are left with a theory defined only in terms of the physical fields. For $\alpha \rightarrow 0$, the theory is renormalizable, but not manifestly unitary. In general, the fictitious poles coming from the propagators of the $\xi_{1,2,3}$ fields cancel against the poles coming from the propagators of the Z_μ and $W_\mu^\pm$ fields, giving us a theory that is both unitary and renormalizable.

10.8   Coleman–Weinberg Mechanism

Although spontaneous symmetry breaking lies at the heart of the Weinberg–Salam model, one of its weaknesses is the arbitrariness of the Higgs potential. This is a serious criticism, since many of the physical parameters depend crucially on the precise form of the Higgs potential.

In principle, one would like to derive the Higgs potential from more fundamental principles, with as few arbitrary parameters as possible.

One interesting approach is the Coleman–Weinberg method,[17] where the Higgs potential is induced by radiative corrections, rather than being inserted by hand. In this approach, we sum over higher-loop graphs to induce an effective potential, which may then produce spontaneous symmetry breaking.

Ideally, one would like to start with a theory that is massless from the very beginning and then induce the mass corrections appearing in the action by radiative corrections. This is called *dimensional transmutation*, where a dimensionless

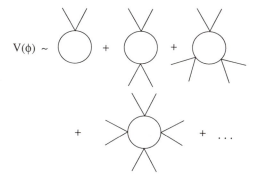

Figure 10.4. The sum over one loop graphs with an arbitrary number of ϕ external lines generates an effective potential, which in turn can induce, under certain circumstances, spontaneous symmetry breaking.

theory trades one of its dimensionless coupling constants for a dimensionful one. This, in turn, gives us the hope of deriving all masses from first principles.

To illustrate this procedure, let us first study the simplest possible example, the ϕ^4 theory. Although this theory is too simple to give us a reliable mechanism to break gauge symmetries, this example reveals the basic principles.

Let us begin with the usual ϕ^4 theory, with the action:

$$L = \frac{1}{2}(\partial_\mu \phi)^2 - \frac{1}{2}m^2\phi^2 - \frac{\lambda}{4!}\phi^4 \tag{10.124}$$

(We will eventually take the limit as $m \to 0$ at the end of the calculation.) Our task is to sum over an infinite series of one-loop graphs with an arbitrary number of ϕ^4 vertices attached to it (Fig. 10.4).

After this sum is performed, the net effect of this series is to generate a new, effective action where the potential is nonpolynomial. For example, let us use Feynman's rules to give us the contribution to the single-loop potential. Each single-loop diagram is given by an integral over the internal momentum. Feynman's rules give us the contribution to the single-loop graph with n vertex insertions as:

$$i\frac{(2n)!}{2^n 2n}\int \frac{d^4k}{(2\pi)^4}\left((-i\lambda)\frac{i}{k^2 - m^2 + i\epsilon}\right)^n \tag{10.125}$$

where the symmetry factor must be inserted because there are $(2n)!$ ways to distribute $2n$ particles among the external lines. (We have taken the external lines to have zero momentum. This will be justified later.)

Now we would like to write down an effective action that generates this series at the tree level. To obtain this new, effective potential, we simply multiply the term with n insertions by ϕ^{2n}, where ϕ from now on will represent the classical value of the field. We must also correct for symmetry factors and then sum. The effective potential now contains the term:

$$i \int \frac{d^4k}{(2\pi)^4} \sum_{n=1}^{\infty} \frac{1}{2n} \left(\frac{(\lambda/2)\phi^2}{k^2 - m^2 + i\epsilon} \right)^n \tag{10.126}$$

This series, fortunately, can be easily summed by using the Taylor expansion for the function $\log(1 + x)$. Let us sum the series (which yields a divergent integral) and then perform the integration by putting in an explicit cutoff Λ:

$$
\begin{aligned}
V_{\text{eff}} &= \frac{1}{2}m^2\phi^2 + \frac{\lambda}{4!}\phi^4 + \frac{i}{2} \int \frac{d^4k}{(2\pi)^4} \log\left[1 + \frac{\lambda\phi^2/2}{k^2 - m^2 + i\epsilon} \right] \\
&= \frac{1}{2}m^2\phi^2 + \frac{\lambda}{4!}\phi^4 + \frac{\Lambda^2}{32\pi^2} \left(m^2 + \lambda\phi^2/2 \right) \\
&\quad + \frac{1}{64\pi^2} \left(m^2 + \frac{\lambda\phi^2}{2} \right)^2 \left[\log\left(\frac{m^2 + \lambda\phi^2/2 + i\epsilon}{\Lambda^2} \right) - \frac{1}{2} \right]
\end{aligned} \tag{10.127}
$$

where we have used the summation:

$$\frac{1}{k^2 - m^2} + \frac{1}{k^2 - m^2} \frac{1}{2}\lambda\phi^2 \frac{1}{k^2 - m^2} + \cdots = \frac{1}{k^2 - (m^2 + \lambda\phi^2/2)} \tag{10.128}$$

Since the original theory was renormalizable, this means that, with the addition of counterterms into the action, we can absorb all cutoff-dependent infinities into a renormalization of the parameters of the theory.

We add the following counterterms to the action:

$$V_{\text{eff}} \rightarrow V_{\text{eff}} + \frac{A}{2}\phi^2 + \frac{B}{4!}\phi^4 \tag{10.129}$$

and then solve for parameters A, B by making the following definitions of the renormalized parameters:

$$
\begin{aligned}
m^2 &= \left. \frac{d^2 V_{\text{eff}}}{d\phi^2} \right|_{\phi=M} \\
\lambda &= \left. \frac{d^4 V_{\text{eff}}}{d\phi^4} \right|_{\phi=M}
\end{aligned} \tag{10.130}
$$

(We have taken the condition that the classical field $\phi = M$ in order to avoid infrared divergences as we take the limit $m \rightarrow 0$.)

Now let us insert this effective potential (with the counterterms added) and then use these conditions to determine A and B. For small m, the calculation is straightforward and yields:

$$
\begin{aligned}
V_{\text{eff,r}} &= \frac{1}{2}m^2\phi^2 + \frac{\lambda}{4!}\phi^4 \\
&\quad + \frac{1}{64\pi^2}\left[\left(m^2 + \frac{\lambda}{2}\phi^2\right)^2 \log\left(\frac{m^2 + \lambda\phi^2/2}{m^2}\right)\right. \\
&\quad \left. - \frac{1}{2}\lambda m^2\phi^2 - \frac{25}{24}\lambda^2\phi^4 + \frac{1}{4}\lambda^2\phi^4 \log\left(\frac{2m^2}{\lambda M^2}\right)\right]
\end{aligned}
\tag{10.131}
$$

We now perform our last step, by taking the limit as $m \to 0$:

$$
V_{\text{eff,r}} = \frac{1}{4!}\lambda\phi^4 + \frac{\lambda^2\phi^4}{256\pi^2}\left(\log\frac{\phi^2}{M^2} - \frac{25}{6}\right)
\tag{10.132}
$$

This is our final result. On the surface, it appears that we have accomplished our goal: We have traded a potential (with no mass terms) with a new potential that has a new minimum away from the origin (where a new mass scale has been introduced by radiative corrections).

However, this example has been too simple. The new minimum lies too far from the origin, beyond the reliability of the one-loop potential:

$$
\lambda \log\frac{\langle\phi\rangle^2}{M^2} = -\frac{32}{3}\pi^2 + O(\lambda)
\tag{10.133}
$$

The term on the left is greater than one, so that the loop contribution is larger than that of the tree contribution; so we are outside the region where the single-loop approximation is reliable. We have gained some insight into the use of radiative corrections to drive the minimum of the potential away from the origin, but our example has been too simple, with only one coupling constant.

Next, we couple charged scalars to QED, where we now have enough coupling constants to make the Coleman–Weinberg mechanism work. We start with a new action, with two coupling constants e and λ:

$$
\mathcal{L} = -\frac{1}{4}F_{\mu\nu}^2 + \frac{1}{2}\left|(\partial_\mu - ieA_\mu)\phi\right|^2 - \frac{\lambda}{4!}(\phi^*\phi)^2
\tag{10.134}
$$

We choose the Landau gauge, where the propagator becomes:

$$
i\Delta_{\mu\nu}(k) = -i\frac{g_{\mu\nu} - k_\mu k_\nu/k^2}{k^2 + i\epsilon}
\tag{10.135}
$$

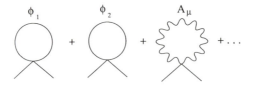

Figure 10.5. These are the only diagrams which contribute to the effective potential in the problem.

We choose this gauge because $k^\mu \Delta_{\mu\nu} = 0$. In this gauge, the only diagrams that contribute to the effective potential are given in Figure 10.5.

The graph in Figure 10.6 does not contribute to the action because k_μ does not couple to the propagator in the Landau gauge. There are thus only three types of diagrams that have to be computed.

We perform the calculation in the same way as before:

1. We sum over each set of diagrams separately in the one-loop approximation by using the power expansion for $\log(1 + x)$.

2. We use a cutoff to render the integrals finite.

3. We introduce new parameters into the theory via counterterms.

4. We calculate the value of these new parameters at the classical value of $\phi = M$.

All steps are exactly the same as before; the only difference comes from the value of the coupling constant contribution of each of the three diagrams.

The result is given by:

$$V_{\text{eff,r}} = \frac{1}{4!}\lambda\phi^4 + \frac{\phi^4}{64\pi^2}\left(\log\frac{\phi^2}{M^2} - \frac{25}{6}\right)C$$

$$C = (\lambda/2)^2 + (\lambda/2)^2(1/9) + 3e^4 \tag{10.136}$$

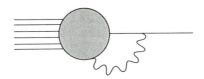

Figure 10.6. This graphs does not contribute because the momentum vector does not couple to the propagator.

[This result is almost identical to Eq. (10.132), except that the last factor of $3e^4$ comes from the trace of the Landau propagator, and the extra $1/3$ ratio between the $\phi_1^2 \phi_2^2$ contribution and the ϕ_1^4 contribution comes from different Wick expansion coefficients.]

We now make the assumption that λ is of the order of e^4 (which we will show is self-consistent). Choosing the new minimum to be $\langle \phi \rangle = M$, we find the potential to be:

$$V_{\text{eff,r}} = \frac{1}{4!}\lambda\phi^4 + \frac{3e^4}{64\pi^4}\phi^4\left(\log\frac{\phi^2}{\langle\phi\rangle^2} - \frac{25}{6}\right) \tag{10.137}$$

We know that the slope of the potential at the minimum is related to the mass, which we set to zero:

$$V'(\langle\phi\rangle) = \left(\frac{\lambda}{6} - \frac{11e^4}{16\pi^2}\right)\langle\phi\rangle^3 = 0 \tag{10.138}$$

Solving, we find:

$$\lambda = \frac{33}{8\pi^2}e^4 \tag{10.139}$$

Thus, our assumption $\lambda \sim e^4$ is self-consistent. Moreover, we find a nontrivial constraint between two previously arbitrary coupling constants. We have traded the two dimensionless coupling constants e and λ for a dimensionful parameter $\langle\phi\rangle$ and a dimensionless parameter e. As we said earlier, this is an example of dimensional transmutation. At first, this might seem strange, because the original theory had no mass parameters at all, and yet a new mass parameter seems to have mysteriously entered into our theory.

The origin of this new mass comes from renormalization theory. Even in scale-invariant theories with no mass parameters, renormalization theory introduces a mass parameter because we must perform the subtraction of divergent diagrams at some mass scale μ. Changes in M simply involve a change in the definition of the coupling constant. (This forms the basis of renormalization group theory, which will be discussed in further detail in Chapter 14.)

Finally, with this new value of λ, we can now write:

$$V_{\text{eff}} = \frac{3e^4}{64\pi^2}\phi^4 \log\left(\frac{\phi^2}{\langle\phi\rangle^2} - \frac{1}{2}\right) \tag{10.140}$$

This is our final result, which shows that there is indeed a new minimum of the potential away from the origin, as claimed. We also mention that the generalization to gauge theory proceeds as expected, with little change. The only complication is that new graphs are generated from the interaction Lagrangian that contains the coupling:

$$\phi^\dagger (\mathbf{A}_\mu \cdot \mathbf{A}^\mu)\phi \tag{10.141}$$

As we said earlier, to illustrate this method we have made a number of assumptions; that is, the effective potential in Eq. (10.125) is defined at zero external momenta. Let us now justify this assumption from a more general point of view. To generalize our calculation, we will use the path integral method of effective actions. We recall that $Z(J)$ is the generator of Green's functions, and that $W(J) = e^{iZ}$ is the generator of connected graphs. In Section 8.4, we showed that a Legendre transformation produces the effective potential:

$$\Gamma(\phi) = W(J) - \int d^4x \, J(x)\phi(x) \tag{10.142}$$

We can power expand Γ as:

$$\Gamma(\phi) = \sum_{n=1}^{\infty} \frac{1}{n!} \int d^4x_1 \ldots d^4x_n \, \Gamma^n(x_1, \cdots, x_n)\phi(x_1) \cdots \phi(x_n) \tag{10.143}$$

Each of the $\Gamma^n(x_1, \ldots, x_n)$ is the sum over all one-particle irreducible Feynman graphs. What we are interested in, however, is the effective potential V_{eff}, which is defined by taking the position space expansion of Γ:

$$\Gamma(\phi) = \int d^4x \left[-V(\phi)_{\text{eff}} + \frac{1}{2}(\partial_\mu\phi)^2 Z(\phi) + \cdots \right] \tag{10.144}$$

where the term without any derivatives is defined to be the effective potential. To calculate a manageable expression for the effective potential, we will take the Fourier transform of Γ :

$$\Gamma^n(x_1, \ldots, x_n) = \int \frac{d^4k_1}{(2\pi)^4} \cdots \frac{d^4k_n}{(2\pi)^4}(2\pi)^4\delta^4(k_1 + \cdots + k_n)$$
$$\times e^{i(k_1 \cdot x_1 + \cdots k_n \cdot x_n)}\Gamma^n(k_1, k_2, \cdots, k_n) \tag{10.145}$$

Now let us insert this equation for $\Gamma(\phi)$ as a power expansion in ϕ. Inserting one into the other, we find:

$$\Gamma(\phi) = \sum_n \frac{1}{n!} \int d^4x_1 \ldots d^4x_n \int \frac{d^4k_1}{(2\pi)^4} \cdots \frac{d^4k_n}{(2\pi)^4}$$
$$\times e^{i(k_1 \cdot x_1 + \cdots + k_n \cdot x_n)} \left[\Gamma^n(0, 0, \cdots, 0)\phi(x_1)\phi(x_2) \cdots \phi(x_n) + \cdots \right]$$

$$= \int d^4x \sum_n \frac{1}{n!} \left[\Gamma^n(0, 0, \ldots, 0)\phi(x)^n + \cdots \right] \tag{10.146}$$

In the last step, we have power expanded $\Gamma^n(k_1, \ldots, k_n)$ and have taken only the lowest term. The higher-order terms contribute higher order derivatives of the fields, in which we are not interested.

Now comes the key step, comparing this equation with the power expansion of $\Gamma(\phi)$ in terms of ϕ. Comparing only the lowest-order term (which contains no derivatives of the field ϕ), we can now extract the effective potential:

$$V(\phi)_{\text{eff}} = -\sum_n \frac{1}{n!}\Gamma^n(0, 0, \ldots, 0)\phi^n(x) \tag{10.147}$$

This is the desired expression. It simply says that the effect of summing over the loop expansion produces a series of Feynman diagrams with zero momenta $\Gamma^n(0, 0, \ldots, 0)$, such that they act as the effective potential for a new action. In this way, we can justify all the steps that we made earlier from more intuitive arguments.

The ultimate use of the Coleman–Weinberg method, however, remains unclear, especially since our accelerators have not been able to pin down the Higgs particle and its interactions, other than the fact that its mass must be greater than 90 GeV. At the very least, we must use the Coleman–Weinberg mechanism to calculate radiative corrections to standard spontaneously broken theories to show that radiative corrections do not spoil the breakdown of symmetry. In other words, the Coleman–Weinberg mechanism can erase minima as well as create them in the potential. In this way, we find that the mechanism gives us bounds on the hypothetical mass of the Higgs particle.

For the Weinberg–Salam model, a very straightforward summing of radiative corrections coming from the scalar, fermion, and vector meson loops gives the correction to the potential:

$$V(\phi) = C\phi^4 \log(\phi^2/M^2) \tag{10.148}$$

where:

$$C = \frac{1}{16\pi v^4} \left(3\sum_V m_V^4 + m_\phi^4 - 4\sum_f m_f^4 \right) \tag{10.149}$$

where V represents the sum over Z and W bosons. The value of v can be determined by solving for the minimum of the potential:

$$\frac{\partial V}{\partial \phi}\bigg|_{\phi=v/\sqrt{2}} = 0 \tag{10.150}$$

The mass of the Higgs field then becomes:

$$m_\phi^2 = \frac{1}{2}\frac{\partial^2 V}{\partial \phi^2}\Big|_{\phi=v/\sqrt{2}} = 2v^2\left\{\lambda + C\left[\log\left(\frac{v^2}{2M^2}\right) + \frac{3}{2}\right]\right\} \tag{10.151}$$

Putting everything together, we find:

$$m_\phi > Cv^2 = \frac{1}{16\pi^2 v^4}\sum_V 3m_V^4 = \frac{3\alpha^2(2 + \sec^4\theta_W)}{16\sqrt{2}G_F\sin^4\theta_W} \tag{10.152}$$

For the Weinberg–Salam model, we must have:

$$m_\phi > 7.9\,\text{GeV} \tag{10.153}$$

or else the radiative corrections will overwhelm the theory and destabilize the vacuum. This bound is easily met.

Alternatively, one might postulate that the Higgs mechanism is driven entirely by radiative corrections. In this interesting case, we find:

$$m_\phi \sim 11\,\text{GeV} \tag{10.154}$$

(which is experimentally ruled out).

In closing, we should also mention that a broken symmetry may be restored under certain conditions. If we consider a ferromagnet, for example, we know that the Hamiltonian does not select out any preferred direction, but the vacuum state may consist of atoms that are all aligned. However, if we heat the magnet sufficiently, the spins become more disordered until a phase transition occurs. At even higher temperatures, the spin alignment is completely lost, and randomness is restored.

Likewise, in a quantum field theory a spontaneously broken symmetry may also be restored if we place the system in a hot enough environment. This is called *symmetry restoration*. Although the temperature necessary to restore a broken symmetry is extraordinarily high, this is not an academic question. It may have great physical implications if we consider the temperatures found originally near the Big Bang.

Perhaps, at the instant of the Big Bang, a unified theory of all known quantum forces possessed a symmetry large enough to include the strong, weak, and electromagnetic interactions and possibly even gravity. As the temperature of the universe rapidly cooled, the original symmetry broke down in several stages. The gravitational interactions first broke off from the particle interactions, then the GUT symmetry broke down into the $SU(3) \otimes SU(2) \otimes U(1)$ symmetry of the Standard Model, then this group broke down into $SU(3) \otimes U(1)$.

If this general picture is correct, then the study of symmetry restoration gives us a useful tool by which to probe the universe at early cosmological times. This is discussed in more detail in the exercises.

In summary, we have seen how spontaneous symmetry breaking is perhaps the most elegant way in which symmetries can be broken. We retain all the symmetries of the theory in the Lagrangian, but the symmetry is broken via the vacuum state. In particular, spontaneous symmetry breaking allows us to generate a mass for the Yang–Mills theory without spoiling renormalizability. This was the crucial step in creating the Weinberg–Salam model, which successfully unites the electromagnetic interactions with the weak interactions.

In the next chapter, we will discuss how the Yang–Mills theory also forms the basis of the strong interactions, giving us the possibility of splicing all quantum interactions into a single Standard Model.

10.9 Exercises

1. Write down the Lagrangian for a model of Higgs mesons with local $O(N)$ symmetry, broken down to $O(M)$ symmetry, for $N > M$, with the Higgs transforming in the vector representation.

2. Do the same for a Lagrangian of Higgs with local $SU(N)$ symmetry broken down to $SU(M)$ symmetry $(N > M)$, with the Higgs transforming in the fundamental representation.

3. Derive the Feynman rules in the 't Hooft gauge for Exercises 1 and 2.

4. Calculate the Coleman–Weinberg potential for self-interacting $O(N)$ mesons in the vector representation.

5. Consider the two-dimensional Gross–Neveu model, with N massless fermions ψ^a. The action is given by:

$$\bar{\psi}^a i \not{\partial} \psi^a + \frac{g_0}{N} (\bar{\psi}^a \psi^a)^2 \tag{10.155}$$

(This contains a four-fermion interaction, which is nonrenormalizable in four dimensions.) By power counting, is this theory renormalizable in two dimensions? Why? Write down all possible divergent graphs. Show that the action is invariant under a discrete transformation:

$$\psi^a \rightarrow \gamma^5 \psi^a, \quad \bar{\psi}^a \rightarrow -\bar{\psi}^a \gamma^5 \tag{10.156}$$

6. Show that the Gross–Neveu Lagrangian can be rewritten as:

$$\bar{\psi}^a i \not\partial \psi^a - \frac{N}{2g_0}\sigma^2 + \sigma \bar{\psi}^a \psi^a \tag{10.157}$$

where σ is a scalar field.

7. Examine the one-loop graphs in this theory with external σ legs and an internal fermion loop. Calculate the effective potential $V(\sigma)$ for the σ field by summing over one-loop graphs. Show that it equals:

$$-iV = -i\frac{N}{2g_0}\sigma^2 - \sum_{n=1}^{\infty} \frac{N}{2n}\,\mathrm{Tr}\int \frac{d^2p}{(2\pi)^2}\left(\frac{-\not p \sigma}{p^2 + i\epsilon}\right)^{2n} \tag{10.158}$$

8. Show that the potential can be written as:

$$\begin{aligned} V &= N\left[\frac{\sigma^2}{2g_0} - \int \frac{d^2p}{(2\pi)^2}\log\left(1 + \frac{\sigma^2}{p^2}\right)\right] \\ &= N\left[\frac{\sigma^2}{2g_0} + \frac{1}{4\pi}\sigma^2\left(\log\frac{\sigma^2}{\Lambda} - 1\right)\right] \end{aligned} \tag{10.159}$$

where we have taken a Euclidean integral and cut it off at momentum Λ.

9. Define a renormalization mass M, defined by:

$$\frac{1}{g} \equiv N^{-1}\frac{d^2V}{d\sigma^2}\bigg|_M \tag{10.160}$$

Solve for g, and write the potential as:

$$V = N\left[\frac{\sigma^2}{2g} + \frac{1}{4\pi}\sigma^2\left(\log\frac{\sigma^2}{M^2} - 3\right)\right] \tag{10.161}$$

Show that there is a minimum to this potential at a negative value, less than $V(0)$. Show that the theory has spontaneous symmetry breakdown, and that dimensional transmutation has occurred. Which parameter has been exchanged for which parameter?

10. A four-dimensional precursor to the Gross–Neveu model is the Nambu–Jona - Lasinio model,[18] with an interaction Lagrangian given by:

$$g\left[(\bar{\psi}\psi)^2 - (\bar{\psi}\gamma_5\psi)^2\right] \tag{10.162}$$

Show that it is invariant under a global continuous transformation given by a chiral $U(1) \otimes U(1)$. Is the theory renormalizable? Perform the same analysis as before. Add in an extra σ field, and perform the integration over ψ; that is, sum the fermion bubble graphs. Show that chiral symmetry breaking occurs dynamically; that is, the fundamental action has no scalars, so the symmetry breaking occurs via a pseudoscalar bound state of the fermions.

11. Add in a Maxwell term to make the Nambu–Jona-Lasinio action locally chirally invariant. Show that the massless particles are removed in this case.

12. Although we have mainly discussed symmetry breaking, consider a model where a broken symmetry may be *restored* if we heat the system sufficiently. Consider a theory defined with potential $V = \frac{1}{2}m^2\phi^2 + (\lambda/4!)\phi^4$ for $m^2 < 0$ with a Euclidean metric. Consider the finite-temperature Green's function:

$$G_\beta(x_1, x_2, \cdots x_N) = \frac{\mathrm{Tr}\, e^{-\beta H} T \phi(x_1)\phi(x_2)\cdots \phi(x_N)}{\mathrm{Tr}\, e^{-\beta H}} \tag{10.163}$$

$\beta = 1/kT$, where T is the temperature and k is the Boltzmann constant. Show that the one-loop correction to the potential is given by:

$$V_1 = \frac{1}{2\beta} \sum_n \int \frac{d^3k}{(2\pi)^3} \log\left(-\frac{4\pi^2 n^2}{\beta^2} - k^2 - M^2\right) \tag{10.164}$$

where $M^2 = m^2 + \frac{1}{2}\lambda\varphi^2$ and where the theory is *periodic* in time and therefore has integral Fourier moments labelled by n.

13. The sum in the previous problem diverges, so use the trick:

$$v(E) \equiv \sum_n \log\left(\frac{4\pi^2 n^2}{\beta^2} + E^2\right)$$

$$\frac{\partial v(E)}{\partial E} = \sum_n \frac{2E}{4\pi^2 n^2/\beta^2 + E^2} \tag{10.165}$$

Using the fact that:

$$\sum_{n=1}^\infty \frac{y}{y^2 + n^2} = -\frac{1}{2y} + \frac{1}{2}\pi \coth \pi y \tag{10.166}$$

show that:

$$\frac{\partial v(E)}{\partial E} = 2\beta\left(\frac{1}{2} + \frac{1}{e^{\beta E} - 1}\right)$$

$$v(E) = 2\beta\left[(E/2) + \beta^{-1} \log(1 - e^{-\beta E})\right] + \ldots \tag{10.167}$$

Then show that this implies:

$$V_1 = \int \frac{d^3k}{(2\pi)^3} \left[(E_M/2) + \beta^{-1} \log(1 - e^{-\beta E_M}) \right] \tag{10.168}$$

where $E_M^2 = \mathbf{k}^2 + M^2$.

14. Show that, for high temperature (small β), the expression for the potential is:

$$V_1 \sim -\frac{\pi^2}{90\beta^4} + \frac{M^2}{24\beta^2} - \frac{1}{12\pi} \frac{M^3}{\beta}$$

$$- \frac{1}{64\pi^2} M^4 \log M^2 \beta^2 + \frac{c}{64\pi^2} M^4 + O(M^6\beta^2) \tag{10.169}$$

where $c \sim 5.41$. Take the second derivative of V_1 with respect to φ. From this, show that the symmetry is restored when:

$$\frac{1}{\beta_c^2} = -\frac{24m^2}{\lambda} \tag{10.170}$$

Calculate the order of magnitude of this temperature for the Weinberg–Salam model. At what temperature is the $SU(2) \otimes U(1)$ symmetry restored? Can these temperatures be found on the earth, in a star, or in the early universe?

15. Prove Eqs. (10.136) and (10.149).

16. For superconductors, assume that there is an attractive force between electrons that forms Cooper pairs. Assume that this many-body system can be described by the Ginzburg–Landau action, which couples a ϕ field to Maxwell's theory:

$$\mathcal{L} = -\frac{1}{4}F_{\mu\nu}^2 + D_\mu \phi D^\mu \phi^* - m^2|\phi|^2 - \lambda|\phi|^4 \tag{10.171}$$

For small enough temperature, spontaneous symmetry breaking occurs at the minimum $|\phi|^2 = -m^2/2\lambda > 0$. Construct the conserved current j_μ. For the static case, calculate the vector current $\mathbf{j}$. Assume that ϕ varies slowly over the medium, then show that this implies London's equation (i.e., $\mathbf{j} = -k^2\mathbf{A}$), where $k^2 = -em^2/2\lambda$.

17. By Ohm's law, we have $\mathbf{E} = R\mathbf{j}$. For the previous problem, using London's equation, show that this means that the resistance is zero. Now take the curl of Ampere's equation. Show that this implies:

$$\nabla^2\mathbf{B} = k^2\mathbf{B} \tag{10.172}$$

so that (in one dimension) $B_z \sim e^{-kx}$. This means that magnetic fields are expelled in a superconductor (Meissner effect), with penetration depth characterized by $1/k$. This implies that spontaneous symmetry breaking has made the Maxwell field massive, with mass k^2.

Chapter 11
The Standard Model

This was a great time to be a high-energy theorist, the period of the famous triumph of quantum field theory. And what a triumph it was, in the old sense of the word: a glorious victory parade, full of wonderful things brought back from far places to make the spectator gasp with awe and laugh with joy.

—S. Coleman

11.1 The Quark Model

The Standard Model, based on the gauge group $SU(3) \otimes SU(2) \otimes U(1)$, is one of the great successes of the gauge revolution. At present, the Standard Model can apparently describe all known fundamental forces (excluding gravity).

The Standard Model is certainly not the final theory of particle interactions. It was created by crudely splicing the electroweak theory and the theory of quantum chromodynamics (QCD). It cannot explain the origin of the quark masses or the various coupling constants. The theory is rather unwieldy and inelegant. However, at present, it seems to be able to explain an enormous body of experimental data. Not only is it renormalizable, it can explain a vast number of results from all areas of particle physics, such as neutrino scattering experiments, hadronic sum rules, weak decays, current algebras, etc. In fact, there is no piece of experimental data that violates the Standard Model.

In this chapter, we will discuss the Standard Model by first reviewing the experimental situation in the 1960s with regard to the quark model. Then we will present compelling evidence, from a wide variety of quarters, that the strong interactions can be described by QCD. Then we will marry QCD to the Weinberg–Salam model to produce the Standard Model.

Although the Standard Model makes the situation seem so clear today, back in the 1960s the experimental situation with the strong interactions was totally confused, with hundreds of "elementary particles" pouring out of our particle accelerators. J. Robert Oppenheimer, in exasperation, said that the Nobel Prize should be given to the physicist who *did not* discover a new particle. Although the Yukawa theory of strong interactions was fully renormalizable, the coupling constant of the strong interactions was large, and hence perturbation theory was unreliable:

$$\frac{g_{\pi N}^2}{4\pi} \sim 14 \tag{11.1}$$

One important observation was that the existence of resonances usually indicated the presence of bound states, so Sakata[1] in the 1950s postulated that the hadrons could be considered to be composite states built out of p, n, and Λ particles. Then Ikeda, Ohnuki, and Ogawa[2] in 1959 made the suggestion that this triplet of particles transformed in the fundamental representation **3** of $SU(3)$. They correctly said that the mesons could be built out bound states of **3** and $\bar{\mathbf{3}}$:

$$\mathbf{3} \otimes \bar{\mathbf{3}} = \mathbf{8} \oplus \mathbf{1} \tag{11.2}$$

However, several of their assignments were incorrect.

In 1961, the correct $SU(3)$ assignments were finally found by Gell-Mann[3,4] and Ne'eman,[5] who postulated that the baryons and mesons could be arranged in what they called the Eightfold Way. Then Gell-Mann[6] and Zweig[7] proposed that these $SU(3)$ assignments could be generated if one postulated the existence of new constituents, called "quarks," which transformed as a triplet **3**. Since all representations of $SU(N)$ can be generated by taking multiple products of the fundamental representation, in this way we could generate all higher representations beginning with the quarks.

The quarks belonged to the fundamental representation of $SU(3)$:

$$\mathbf{3} = q_i = \begin{pmatrix} u \\ d \\ s \end{pmatrix} \tag{11.3}$$

where the quarks were called the "up," "down," and "strange" quark, for historical reasons. The u and the d quarks formed a standard $SU(2)$ isodoublet, but the addition of the third quark was necessary because it was observed in the 1950s that a new quantum number in addition to isospin was conserved by hadronic processes, called "strangeness." This new quantum number could be explained in terms of $SU(3)$, which is a rank 2 Lie group. Its representations are therefore labeled by two numbers, the third component of isospin T_3 and also a new quantum number Y, called "hypercharge."

The new quantum number of strangeness and hypercharge could be related to each other via the Gell-Mann–Nishijima[8,9] formula:

$$Q = T_3 + \frac{Y}{2} = \begin{pmatrix} \frac{2}{3} & 0 & 0 \\ 0 & -\frac{1}{3} & 0 \\ 0 & 0 & -\frac{1}{3} \end{pmatrix} \tag{11.4}$$

where $Y = B + S$. B is the baryon number, S is the strangeness number, T_3 is the third component of isospin, and Q is the charge.

To fit the known spectrum, the mesons were postulated to be composites of a quark and an antiquark, while the baryons were postulated to be composites of three quarks. Thus, we expect to see the mesons and baryons arranged according to the following tensor product decomposition:

$$\text{Meson} = \mathbf{3} \otimes \bar{\mathbf{3}} = \mathbf{8} \oplus \mathbf{1}$$

$$\text{Baryon} = \mathbf{3} \otimes \mathbf{3} \otimes \mathbf{3} = \mathbf{10} \oplus \mathbf{8} \oplus \mathbf{8} \oplus \mathbf{1} \tag{11.5}$$

The theory predicted that the mesons should be arranged in terms of octets and singlets, while baryons should be in octets as well as decuplets. The fact that this simple picture could arrange the known mesons and baryons in such an elegant picture was remarkable.

In order to reproduce the known charges of the mesons and baryons, it was necessary to give the quarks fractional charges:

$$Q_u = \frac{2}{3}e; \quad Q_d = -\frac{1}{3}e; \quad Q_s = -\frac{1}{3}e \tag{11.6}$$

Since three of them were required to make up a single baryon, this meant that each of them had baryon number $\frac{1}{3}$. We summarize the quantum numbers of the quarks in the following chart:

$$
\begin{array}{c|c|c|c|c|c|c}
 & Q & T & T_3 & Y & S & B \\
\hline
u & \frac{2}{3} & \frac{1}{2} & \frac{1}{2} & \frac{1}{3} & 0 & \frac{1}{3} \\
d & -\frac{1}{3} & \frac{1}{2} & -\frac{1}{2} & \frac{1}{3} & 0 & \frac{1}{3} \\
s & -\frac{1}{3} & 0 & 0 & -\frac{2}{3} & -1 & \frac{1}{3} \\
\end{array}
\tag{11.7}
$$

For the meson spectrum, we can get a rough classification by considering the bound states generated by a $q\bar{q}$ pair. The bound states arrange themselves roughly in the following angular momentum series (similar to the familiar series found in spectroscopy using the notation $^{2S+1}L_J$):

$$
{}^1S_0,\ {}^3S_1,\ {}^1P_1,\ {}^3P_0,\ {}^3P_1,\ {}^3P_2,\ {}^1D_2, \ldots
\tag{11.8}
$$

The fit between experiment and the predicted bound states of the quark model was exceptionally good. For example, the octet containing the π meson and K meson corresponds to the $^1S_0$ bound state, while the K^* multiplet is part of a $^3S_1$ bound state:

$$
\begin{array}{c|c|c}
q_i\bar{q}_j & J = 0 & J = 1 \\
\hline
|u\bar{d}\rangle & \pi^+(140) & \rho^+(770) \\
2^{-1/2}|d\bar{d} - u\bar{u}\rangle & \pi^0(135) & \rho^0(770) \\
|u\bar{d}\rangle & \pi^-(140) & \rho^-(770) \\
2^{-1/2}|d\bar{d} + u\bar{u}\rangle & \eta(549) & \omega(783) \\
|u\bar{s}\rangle & K^+(494) & K^{*+}(892) \\
|d\bar{s}\rangle & K^0(498) & K^{*0}(892) \\
|\bar{u}s\rangle & K^-(494) & K^{*-}(892) \\
|\bar{d}s\rangle & \bar{K}^0(498) & \bar{K}^{*0}(892) \\
|s\bar{s}\rangle & \eta'(958) & \phi(1020) \\
\end{array}
\tag{11.9}
$$

Similarly, we can also analyze the baryons. The familiar proton and neutron belong to the octet, while the Δ resonance (found in pion–nucleon scattering) belongs to the decuplet:

$q_i q_j q_k$	$J = 1/2$	$J = 3/2$
$\lvert uuu \rangle$		$\Delta^{++}(1230)$
$\lvert uud \rangle$	$p(938)$	$\Delta^{+}(1231)$
$\lvert udd \rangle$	$n(940)$	$\Delta^{0}(1232)$
$\lvert ddd \rangle$		$\Delta^{-}(1234)$
$\lvert uus \rangle$	$\Sigma^{+}(1189)$	$\Sigma^{+}(1383)$
$2^{-1/2}\lvert (ud+du)s \rangle$	$\Sigma^{0}(1192)$	$\Sigma^{0}(1384)$
$\lvert dds \rangle$	$\Sigma^{-}(1197)$	$\Sigma^{-}(1387)$
$2^{-1/2}\lvert (ud-du)s \rangle$	$\Lambda(1116)$	
$\lvert uss \rangle$	$\Xi^{0}(1315)$	$\Xi^{0}(1532)$
$\lvert dss \rangle$	$\Xi^{-}(1321)$	$\Xi^{-}(1535)$
$\lvert sss \rangle$		$\Omega^{-}(1672)$

$$(11.10)$$

To see how the bound states are constructed, it is sometimes useful to rearrange the meson and baryon matrices according to their quark wave functions. Let us define:

$$
M = q \otimes \bar{q} = \begin{pmatrix} u\bar{u} & u\bar{d} & u\bar{s} \\ d\bar{u} & d\bar{d} & d\bar{s} \\ s\bar{u} & s\bar{d} & s\bar{s} \end{pmatrix}
$$

$$
= \begin{pmatrix} (2u\bar{u} - d\bar{d} - s\bar{s})/3 & u\bar{d} & u\bar{s} \\ d\bar{u} & (2d\bar{d} - u\bar{u} - s\bar{s})/3 & d\bar{s} \\ s\bar{u} & s\bar{d} & (2s\bar{s} - u\bar{u} - d\bar{d})/3 \end{pmatrix}
$$

$$
+ \; (1/3)\mathbf{1}(u\bar{u} + d\bar{d} + s\bar{s}) \tag{11.11}
$$

In terms of the familiar pseudoscalar mesons, we have the following arrangement of the meson matrix:

$$M = \begin{pmatrix} \frac{1}{\sqrt{2}}\pi^0 + \frac{1}{\sqrt{6}}\eta & \pi^+ & K^+ \\ \pi^- & -\frac{1}{\sqrt{2}}\pi^0 + \frac{1}{\sqrt{6}}\eta & K^0 \\ K^- & \bar{K}^0 & -\frac{2}{\sqrt{6}}\eta \end{pmatrix} \tag{11.12}$$

Likewise, the baryon matrix can be arranged as:

$$B = q \otimes q \otimes q = \begin{pmatrix} \frac{1}{\sqrt{2}}\Sigma^0 + \frac{1}{\sqrt{6}}\Lambda^0 & \Sigma^+ & p \\ \Sigma^- & -\frac{1}{\sqrt{2}}\Sigma^0 + \frac{1}{\sqrt{6}}\Lambda^0 & n \\ \Xi^- & \Xi^0 & -\frac{2}{\sqrt{6}}\Lambda^0 \end{pmatrix} \tag{11.13}$$

To actually perform any calculations with $SU(3)$, we need, of course, an explicit representation of the generators of the algebra. We will choose the standard Gell-Mann representation of $SU(3)$ generators in terms of 3×3 Hermitian matrices:

$$\lambda_1 = \begin{pmatrix} 0 & 1 & 0 \\ 1 & 0 & 0 \\ 0 & 0 & 0 \end{pmatrix}, \quad \lambda_2 = \begin{pmatrix} 0 & -i & 0 \\ i & 0 & 0 \\ 0 & 0 & 0 \end{pmatrix}, \quad \lambda_3 = \begin{pmatrix} 1 & 0 & 0 \\ 0 & -1 & 0 \\ 0 & 0 & 0 \end{pmatrix}$$

$$\lambda_4 = \begin{pmatrix} 0 & 0 & 1 \\ 0 & 0 & 0 \\ 1 & 0 & 0 \end{pmatrix}, \quad \lambda_5 = \begin{pmatrix} 0 & 0 & -i \\ 0 & 0 & 0 \\ i & 0 & 0 \end{pmatrix}, \quad \lambda_6 = \begin{pmatrix} 0 & 0 & 0 \\ 0 & 0 & 1 \\ 0 & 1 & 0 \end{pmatrix}$$

$$\lambda_7 = \begin{pmatrix} 0 & 0 & 0 \\ 0 & 0 & -i \\ 0 & i & 0 \end{pmatrix}, \quad \lambda_8 = \frac{1}{\sqrt{3}}\begin{pmatrix} 1 & 0 & 0 \\ 0 & 1 & 0 \\ 0 & 0 & -2 \end{pmatrix} \tag{11.14}$$

where:

$$\mathrm{Tr}\,(\lambda_i \lambda_j) = 2\delta_{ij}$$
$$\left[\frac{\lambda_i}{2}, \frac{\lambda_j}{2}\right] = if_{ijk}\frac{\lambda_k}{2} \tag{11.15}$$

The structure constants are given by:

$$f_{123} = 1; \; f_{147} = -f_{156} = f_{246} = f_{257} = f_{345} = -f_{367} = \frac{1}{2}$$

$$f_{458} = f_{678} = \frac{\sqrt{3}}{2} \tag{11.16}$$

From these commutation relations, we can work out the representations of the group (see Appendix). Because $SU(3)$ is a rank 2 Lie group, we can chart the various representations of the group in a two-dimensional space, plotting the eigenvalues of λ_3 of the $SU(2)$ subgroup against λ_8, which is proportional to the hypercharge. Thus, on a two-dimensional graph (isospin plotted against hypercharge) we can pictorially represent the triplet, antitriplet, octet, decuplet, etc. (Fig. 11.1)

Although hadronic masses are not exactly $SU(3)$ invariant (i.e., the masses of particles within a multiplet vary slightly), it is reasonable to assume that the terms that break $SU(3)$ symmetry should themselves transform covariantly under $SU(3)$. We assume, for example, that the mass term in the Hamiltonian includes a term that breaks the symmetry transforms as λ_8, as hypercharge. We assume the mass term has the form:

$$\bar{\psi}(a + b\lambda_8)\psi + \ldots \tag{11.17}$$

This, in turn, gives us nontrivial relations between the masses of the various particles within a multiplet called the Gell-Mann–Okubo[10] mass relation, which provided experimental verification of the theory. It gives us the mass relation:

$$m_N + m_\Xi = \frac{1}{2}(m_\Sigma + 3m_\Lambda) \tag{11.18}$$

which agrees well with experiment. The left-hand side equals 2.25 GeV experimentally, while the right-hand side equals 2.23 GeV.

For the spin-3/2$^+$ decuplet, we also find the equal-spacing rule:

$$m_\Omega - m_{\Xi^*} = m_{\Xi^*} - m_{\Sigma^*} = m_{\Sigma^*} - m_{N^*} \tag{11.19}$$

The experimental mass differences are 139, 149, and 152 MeV, respectively.

Historically, the prediction of the Ω^- mass from this formula gave a boost to the wide acceptance of $SU(3)$ symmetry.

Because of the success of the $SU(3)$ quark model, attempts were made to generalize this to larger groups. One attempt, merging $SU(3)$ with the $SU(2)$ of spin to create $SU(6)$,[11–13] tried to mix an internal symmetry with a space symmetry. This was possible because $SU(2) \otimes SU(3) \subset SU(6)$. $SU(6)$ had some success in predicting the magnetic moments of baryons. [However, attempts to generalize $SU(6)$ to the relativistic case floundered because of the Coleman–Mandula theorem.]

Attempts were also made to generalize $SU(3)$ to $SU(4)$[14] and beyond by adding more quarks. This approach received experimental vindication with the discovery

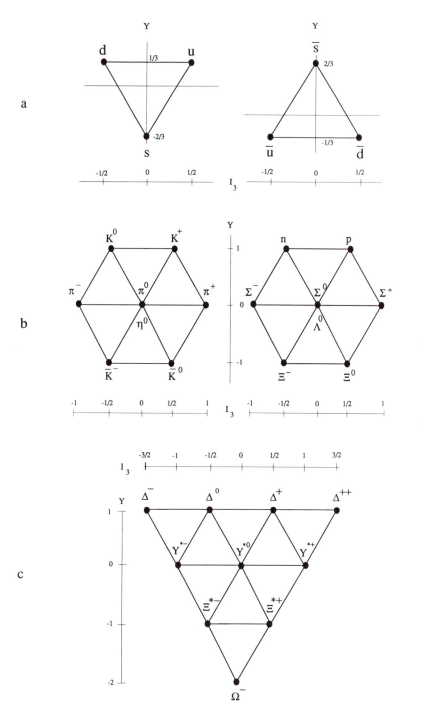

Figure 11.1. When we plot isospin against hypercharge, we can represent the triplet, antitriplet, octet, decuplet, and higher multiplets in simple geometrical patterns.

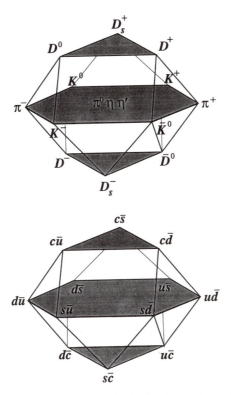

Figure 11.2. With charm plotted on the vertical axis, the quark model gives an excellent fit to the charmed meson multiplet.

of the charmed quark c in 1974 and the bottom quark b in 1977. [However, because the masses of the charmed and bottom quark are so large, global $SU(4)$ and $SU(5)$ are less reliable than $SU(2)$ and $SU(3)$ experimentally.]

For the charmed quark system, the new $q\bar{q}$ and qqq states are given the following group-theoretical assignments:

$$4 \otimes \bar{4} = 15 \oplus 1$$
$$4 \otimes 4 \otimes 4 = \bar{4} \oplus 20 \oplus 20 \oplus 20 \qquad (11.20)$$

The charmed quark bound states are given the following names for the 0^- and 1^- multiplets (Fig. 11.2):

$q_i \bar{q}_j$	$J = 0$	$J = 1$
$c\bar{c}$	$\eta_c(2980)$	$J/\Psi(3097)$
$c\bar{d}$	$D^+(1869)$	$(D^*)^+(2010)$
$c\bar{u}$	$D^0(1865)$	$(D^*)^0(2007)$
$\bar{c}u$	$\bar{D}^0(1865)$	$(\bar{D}^*)^0(2007)$
$\bar{c}d$	$D^-(1869)$	$(D^*)^-(2010)$
$c\bar{s}$	$D_s^+(1970)$	$D_s^{*+}(2109)$
$\bar{c}s$	$D_s^-(1970)$	$D_s^{*-}(2109)$

$$(11.21)$$

We can also write down an explicit form for the generators of the $SU(4)$ global symmetry. In fact, it is possible to find a simple iterative algorithm to write down the generators of $SU(N)$ almost by inspection. We first notice that we can write the first three λ_i matrices as follows:

$$\lambda_a = \begin{pmatrix} \sigma_a & 0 \\ 0 & 0 \end{pmatrix}; \quad a = 1, 2, 3 \tag{11.22}$$

where this symbolically means that the Pauli spin matrices are placed in the upper left-hand corner for $a = 1, 2, 3$.

Next, λ_{4-7} obey a simple pattern. Along the right column and bottom row, we insert the numbers 1 and 1 (as well as $-i$ and i) symmetrically in all possible slots. Finally, the generator λ_8 has the unit 2×2 matrix in the upper left-hand corner and we choose the last number along the diagonal to make it traceless.

From this algorithm, we can easily write down the generators of $SU(N)$ if we know the generators of $SU(N - 1)$. For example, we can now write down the generators of $SU(4)$ almost by inspection. To see this, we place the generators of $SU(3)$ in the upper left-hand corner for $a = 1 - 8$:

$$\lambda_a = \begin{pmatrix} \lambda_a & 0 \\ 0 & 0 \end{pmatrix}; \quad a = 1, \ldots, 8 \tag{11.23}$$

For the generators λ_{9-14}, we place the pairs of numbers 1, 1 and $-i$, i symmetrically in the right column and bottom row, while for the last generator λ_{15} we put the unit 3×3 matrix in the upper left-hand side and make it traceless:

$$\lambda_9 = \begin{pmatrix} 0 & 0 & 0 & 1 \\ 0 & 0 & 0 & 0 \\ 0 & 0 & 0 & 0 \\ 1 & 0 & 0 & 0 \end{pmatrix}; \quad \lambda_{10} = \begin{pmatrix} 0 & 0 & 0 & -i \\ 0 & 0 & 0 & 0 \\ 0 & 0 & 0 & 0 \\ i & 0 & 0 & 0 \end{pmatrix}$$

$$\lambda_{11} = \begin{pmatrix} 0 & 0 & 0 & 0 \\ 0 & 0 & 0 & 1 \\ 0 & 0 & 0 & 0 \\ 0 & 1 & 0 & 0 \end{pmatrix}; \quad \lambda_{12} = \begin{pmatrix} 0 & 0 & 0 & 0 \\ 0 & 0 & 0 & -i \\ 0 & 0 & 0 & 0 \\ 0 & i & 0 & 0 \end{pmatrix}$$

$$\lambda_{13} = \begin{pmatrix} 0 & 0 & 0 & 0 \\ 0 & 0 & 0 & 0 \\ 0 & 0 & 0 & 1 \\ 0 & 0 & 1 & 0 \end{pmatrix}; \quad \lambda_{14} = \begin{pmatrix} 0 & 0 & 0 & 0 \\ 0 & 0 & 0 & 0 \\ 0 & 0 & 0 & -i \\ 0 & 0 & i & 0 \end{pmatrix}$$

$$\lambda_{15} = \frac{1}{\sqrt{6}} \begin{pmatrix} 1 & 0 & 0 & 0 \\ 0 & 1 & 0 & 0 \\ 0 & 0 & 1 & 0 \\ 0 & 0 & 0 & -3 \end{pmatrix} \tag{11.24}$$

Similarly, the generators of $SU(5)$ can be constructed in this way. Although less is known about the quark spectrum for mesons containing the bottom quark, all states discovered so far obey the quark model predictions. The lowest lying states include:

$q_i \bar{q}_j$	$J = 0$	$J = 1$
$b\bar{b}$	η_b	$\Upsilon(9460)$
$u\bar{b}$	$B^+(5278)$	$(B^*)^+(5324)$
$d\bar{b}$	$B^0(5278)$	$(B^*)^0(5324)$
$\bar{u}b$	$\bar{B}^0(5278)$	$(\bar{B}^*)^0(5324)$
$\bar{d}b$	$B^-(5278)$	$(B^*)^-(5324)$

(The η_b has not been firmly established, and the charges of the B^* are not yet confirmed.)

Today, the original three quarks have been expanded to six quarks: the up, down, strange, charmed, bottom, and top quark. All but the last have been discovered. The global symmetry group $SU(N)$ for N quarks is now called the "flavor" symmetry.

The rough values for the constituent quark masses in GeV are given by:

$$
\begin{aligned}
m_u &= 0.35; & m_c &= 1.5 \\
m_d &= 0.35; & m_b &= 5 \\
m_s &= 0.5; & m_t &> 91
\end{aligned}
\tag{11.25}
$$

These quarks, in turn, can be arranged in three identical families or generations, each having the same quantum numbers: (u, d), (c, s), and (t, b). (The reason why nature should prefer three identical generations of quarks and leptons is one of the great mysteries of subatomic physics.)

Historically, although the quark model had great success in bringing order out of the chaos of the hundreds of resonances found in scattering experiments, it also raised a host of other problems. In fact, each year, even as the successes of the quark model began to pile up, the questions raised by the quark model also began to proliferate. For example, why were the quarks not observed experimentally? Were they real, or were they just a useful mathematical device? And what was the binding force that held the quarks together? For example, some believed that the glue that held the quarks together might be a vector meson; however, to be renormalizable, it had to be massless. But this was impossible, because if it was massless, then it should generate a long-range force, like gravity and electromagnetism, rather than being a short-range force like the strong force.

11.2   QCD

After years of confusion, the theory that has emerged to give us the best under-standing of the strong interactions is called QCD, which has the Lagrangian:

$$
\mathcal{L} = -\frac{1}{4} F_{\mu\nu}^a F^{a\mu\nu} + \sum_{i=1}^{6} \bar{\psi}_i (i\,\slashed{D} - m_i)\psi_i
\tag{11.26}
$$

where the Yang–Mills field is massless and carries the $SU(3)$ "color" force [not to be confused with the global $SU(3)$ flavor symmetry introduced earlier]. Unlike the electroweak theory, where the gauge group is broken and the Z and W become massive, the color group is unbroken and the gluons remain massless.

The quarks have *two* indices. The i index is taken over the flavors, which labels the up, down, strange, charm, top, and bottom quarks. The flavor index is *not* gauged; it represents a global symmetry. However, the quarks also carry the important local color $SU(3)$ index (which is suppressed here). In other words, quarks come in six flavors and three colors, but only the color index participates in the local gauge symmetry. From the point of view of QCD, the flavor index, which dominated most of the phenomenology of the 1960s and 1970s, is now relegated to a relatively minor role compared to the color force, which binds the quarks together.

In fact, from the perspective of QCD, we can see the origin of the early phenomenological successes of global $SU(3)$. In the limit of equal quark masses, the QCD action possesses an additional symmetry, global $SU(N)$ symmetry for N flavors of quarks. For the u and d quarks, this is a very good approximation; so $SU(2)$ global symmetry is experimentally seen in the hadron spectrum. The s quark mass, although larger, is still relatively close to the u and d mass when compared to the baryon mass; so we expect flavor $SU(3)$ symmetry to be a relatively good one. However, the masses of the c and b are much larger; so we expect $SU(4)$ or $SU(5)$ flavor symmetry breaking to be quite large. The higher flavor symmetry groups are hence less useful phenomenologically.

Although quarks have never been seen in the laboratory, there is now an overwhelming body of data supporting the claim that QCD is the leading theory of the strong interactions. This large body of theoretical and experimental results and data, accumulated slowly and painfully over the past several decades, can be summarized in the following sections.

11.2.1 Spin-Statistics Problem

According to the spin-statistics theorem, a fermion must necessarily be totally antisymmetric with respect to the interchange of the quantum numbers of its constituents. One long-standing problem, however, was that certain baryon states, such as the **10** representation, which includes the Δ^{++} resonance, were purely symmetric under this interchange, violating the spin-statistics theorem.

For example, the wave function for this resonance is naively given by:

$$\Psi_\Delta = \psi_{SU(3)}\psi_{\text{orbital}}\psi_{\text{spin}} \tag{11.27}$$

This wave function is symmetric under the interchange of any two quarks, which is typical of bosonic, not fermionic, states. To see this, notice that the $SU(3)$ flavor part of the wave function is symmetric, since it is composed of three u quarks, all pointing in the same direction in isospin space, as in (11.10). Also, since the spin of this resonance is 3/2, all three quark spins are pointing spatially in the same direction, so the spin wave function is also symmetric. Finally, the interchange of the quarks in the orbital part yields a factor $(-1)^L$, which is one because $L = 0$ for this resonance. Thus, under an interchange of any two quarks, the wave function picks up a factor of $(+1)(+1)(+1) = 1$, so the overall wave function is symmetric under the interchange of the quarks, which therefore violates the spin-statistics theorem for fermions.

Since physicists were reluctant to abandon the spin-statistics theorem, one resolution of this problem was to postulate the existence of yet another isospin symmetry, a new "color" symmetry, so that the final state could be fully antisymmetric. This was the original motivation behind the Han–Nambu[15] quark model, a precursor of QCD.

11.2.2   Pair Annihilation

The simplest way experimentally to determine the nature of this mysterious new color symmetry is to perform electron–antielectron collision experiments. Pair annihilation creates an off-shell photon, which then decays into various possible combinations. We are interested in the process:

$$e^+ + e^- \rightarrow \gamma \rightarrow \bar{q} + q \rightarrow \text{hadrons} \tag{11.28}$$

This process is highly sensitive to the number of quarks and their charges that appear in the calculation. Using Feynman's rules, we find that the cross section must be proportional to the number of quark colors times the sum of the squares of the quark charges. In practice, it is convenient to divide by the leptonic contribution to the cross section:

$$e^+ + e^- \rightarrow \mu^+ + \mu^- \tag{11.29}$$

By taking the ratio of these two cross sections, we should therefore find the pure contribution of the quark color sector. In particular, we find:

$$R \equiv \frac{\sigma(e^+ e^- \rightarrow \text{hadrons})}{\sigma(e^+ e^- \rightarrow \mu^+ \mu^-)} = N_c \sum_{i=1} Q_i^2 \tag{11.30}$$

where N_c is the number of quark colors and Q_i is the charge of each quark.

For low energies, when we excite just the u, d, s quarks, we expect R to equal $3(4 + 1 + 1)/9 = 2$. When we hit the threshold for creating charm–anticharm intermediate states, then this ratio rises to over 4. If we include the u, d, s, c, b quarks, then we have $R = 11/3$. Experimentally, this agrees rather well with experiment, assuming that there are three colors (Fig. 11.3).

11.2.3   Jets

High-energy scattering experiments should be able to knock individual gluons and quarks out of the nucleus. Although they quickly reform into bound states and hence cannot be isolated, they should make a characteristic multiprong event

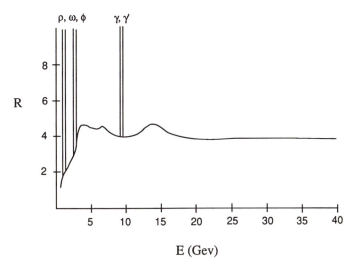

Figure 11.3. The plot of R against energy, which agrees with the value of R found in QCD.

in the scattering apparatus. These multiprong events, as predicted, have been produced in high-energy scattering experiments.

For example, two-prong jets have a characteristic distribution dependence given by $(1 + \cos^2 \theta)$, where θ is the angle between the beam and the jets. This is consistent with the process $e^- + e^+ \to q + \bar{q}$ for spin $\frac{1}{2}$ quarks, as expected. This important topic will be discussed in the next section.

11.2.4 Absence of Exotics

Although the quark model had great success in fitting the known hadrons into $3 \otimes \bar{3}$ and $3 \otimes 3 \otimes 3$ bound states, it was at a loss to explain why exotic states, such as $3 \otimes 3$, etc., should not be formed as well. Because the original quark model gave no indication of what the binding force was, this question could never be answered within the context of the old quark model.

QCD, however, gives a simple reason why these exotic states are absent. We learned earlier in Section 9.4 that the states of the unbroken Yang–Mills theory are singlets under the gauge group. We notice immediately that $\bar{q}q$ and qqq states are invariant under the color group because they are contracted by constant invariant tensors, like the delta function and the structure constant f_{ijk}. Low-lying exotic states, because they are nonsinglets under the color group, are either absent or will decay into the usual bound states. (Later, when we discuss lattice gauge theory, we will see that QCD can, in principle, give us numerical results to back up this heuristic result.)

11.2.5   Pion Decay

The Feynman diagram for the decay $\pi \rightarrow 2\gamma$ consists of an internal quark triangle loop, with the pion and the gamma rays attached to the three corners of the triangle. Thus, this decay rate is proportional to the sum over all the quarks that occur in this internal triangle loop. By comparing the experimental decay rate of the pion into two gamma rays, we can therefore calculate the number of quark colors. The experimental evidence supports the presence of 3.01 ± 0.08 colors.

11.2.6   Asymptotic Freedom

Historically, it was the discovery of asymptotic freedom that elevated QCD into the leading theory of the strong interactions. Deep inelastic experiments, such as $e + p \rightarrow e +$ anything, showed that the cross sections exhibited scale invariance at high energies; that is, the form factors lost their dependence on certain mass parameters at high energies. This scale invariance could be interpreted to mean that the quark constituents acted as if they were free particles at extremely high energies.

QCD offers a simple explanation of this scale invariance. Using the renormalization group, which will be discussed at length in Chapter 14, one could show that the coupling constant became smaller at high energies, which could explain the reason why the quarks behaved as if they were free. Asymptotic freedom gave a simple reason why the naive quark model, which described complex scattering experiments with free quarks fields, had such phenomenological success.

11.2.7   Confinement

The renormalization group also showed the converse, that the coupling constant should become large at low energies, suggesting that the quarks were permanently bound inside a hadron. This gave perhaps the most convincing theoretical justification that the quarks should be permanently "confined" inside the bound states. Although a rigorous proof that the Yang–Mills theory confines the quarks and gluons has still not been found, the renormalization group approach gives us a compelling theoretical argument that the coupling constant is large enough at small energies to confine the quarks and gluons. If correct, this approach also explains why the massless gluon field does not result in a long-range force, like gravity or the electromagnetic force. Although the range of a massless field is formally infinite, the gluon field apparently "condenses" into a stringlike glue that binds the quarks together at the ends. [In Chapter 15, we will present some compelling (but not rigorous) numerical justification for this picture using lattice gauge theory.]

Thus, with one theory, we are able to interpret two divergent facts, that quarks appear to be confined at low energies but act as if they are free particles at high energy.

The phenomenological success of the quark model, of course, is not exclusively confined to the strong interactions. Quarks also participate in the weak interactions, and have greatly clarified the origins of certain phenomenological models proposed in the 1960s (such as current algebras). By studying the weak currents generated by the quarks, one can find a simple quark model explanation for a number of phenomenologically important results from weak interaction physics, such as are given in the following sections.

11.2.8 Chiral Symmetry

In the limit that the quarks have vanishingly small mass, the QCD action possesses yet another global flavor symmetry, chiral symmetry. For N flavors, the QCD action for massless quarks is invariant under chiral $SU(N) \otimes SU(N)$, generated by:

$$q \rightarrow e^{i\lambda^a \theta^a} q; \quad q \rightarrow e^{i\gamma_5 \lambda^a \theta^a} q \qquad (11.31)$$

[Actually, the QCD action has the additional chiral symmetry $U(1) \otimes U(1)$ if we drop the λ^a in the previous expression. The first $U(1)$ symmetry gives the usual baryon number conservation. The second chiral $U(1)$ symmetry will be studied in more detail in the next chapter.]

Since the u, d, and s quarks have relatively small mass when compared with the scale of the strong interactions, then QCD has a global chiral symmetry in this approximation:

$$m_u \sim m_d \sim m_s \sim 0 \rightarrow SU(3) \otimes SU(3) \qquad (11.32)$$

The $SU(3) \otimes SU(3)$ chiral symmetry of QCD, in turn, allows us to compute a large number of relations and sum rules between different physical processes.

Since chiral symmetry is broken in nature, for small quark masses we can, in fact, view the π meson as the Goldstone boson for broken chiral symmetry. The fact that the π meson has an exceptionally light mass is a good indicator of the validity of chiral symmetry as an approximate symmetry. A large number of successful sum rules for the hadronic weak current, as we shall see, can be written down as a byproduct of the smallness of the pion mass.

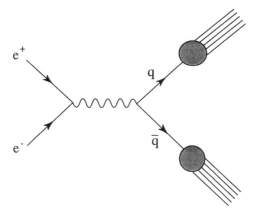

Figure 11.4. Typical Feynman diagrams describing the collision of electrons and positrons, creating jets.

11.2.9   No Anomalies

The theory of leptons given by the Weinberg–Salam model in Chapter 10 is actually fatally flawed by the presence of something called "anomalies," which will be discussed in more detail in the next chapter. These anomalies sometimes arise when a classical symmetry of an action does not survive the process of quantization. In particular, there are certain divergent fermionic triangle graphs that can potentially destroy the Ward–Takahashi identities and hence ruin renormalizability. However, when the quarks are inserted into the Weinberg–Salam model, they also produce anomalies, but of the opposite sign. In fact, the charge assignments of the quarks and leptons in the Standard Model are precisely the ones that cancel the anomaly. The vanishing of the lepton anomalies against the quark anomalies in the Weinberg–Salam model can be seen as one more theoretical justification of the Standard Model.

11.3   Jets

Now that we have completed a broad overview of the theoretical and experimental successes of the Standard Model, let us now focus in detail on some of the specifics. One of the most graphic reasons for supporting QCD is the existence of "jets" in electron–positron collisions. In Figure 11.4, we see some of the typical Feynman diagrams that arise when electrons collide with positrons.

The momentum transfer is so large that the quarks, antiquarks, and gluons in the final states are scattered in different directions. They later regroup into standard hadrons to form a jet-like structure, which has been found experimentally.

We will analyze the process with the following labeling of momenta:

$$e^-(q) + e^+(q') \to q(p) + \bar{q}(p') + g(k) \tag{11.33}$$

with $Q = q + q' = p + p' + k$ and $S = Q^2$.

For the two-jet event (where we drop the final gluon), we can factorize the transition matrix into a leptonic and a hadronic part:

$$|\mathcal{M}|^2 = \frac{1}{S^2} l_{\mu\nu} H^{\mu\nu} \tag{11.34}$$

where:

$$
\begin{aligned}
l_{\mu\nu} &= \frac{1}{4} \sum_{\text{spins}} \langle 0|J_\mu|e^+e^-\rangle \langle 0|J_\nu|e^+e^-\rangle^* \\
&= \frac{1}{4} \text{Tr} \left(\gamma_\mu \frac{\slashed{q}}{2E} \gamma_\nu \frac{\slashed{q}'}{2E} \right) \\
&= \frac{e^2}{4E^2} \left(q_\mu q_\nu' + q_\nu q_\mu' - g_{\mu\nu} q \cdot q' \right)
\end{aligned} \tag{11.35}
$$

The hadronic part, to lowest order, can also be calculated in the same way:

$$H_{\mu\nu} = \frac{e_f^2}{E^2} \left(p_\mu p_\nu' + p_\nu p_\mu' - g_{\mu\nu} p \cdot p' \right) \tag{11.36}$$

where e_f is the electric charge of the quark flavor index f. Then we have:

$$|\mathcal{M}|^2 = \frac{e^2 e_f^2}{4E^4} (1 + \cos^2 \theta) \tag{11.37}$$

where θ is the center-of-mass scattering angle, so the differential and total cross sections for the two-jet process are given by:

$$
\begin{aligned}
\left(\frac{d\sigma}{d\Omega} \right)_0 &= \frac{\alpha^2}{4S} (1 + \cos^2 \theta) \sum_{f,c} \left(\frac{e_f}{e} \right)^2 \\
\sigma_0 &= \frac{4\pi\alpha^2}{3S} \sum_{f,c} \left(\frac{e_f}{e} \right)^2
\end{aligned} \tag{11.38}
$$

This $(1 + \cos^2 \theta)$ dependence on the angle θ has actually been seen experimentally in two-jet processes, strengthening our belief in spin-$\frac{1}{2}$ quarks.

For the three-jet event the calculation is a bit more difficult because of the kinematics. The only complication is the hadronic part:

$$
\begin{aligned}
H_{\mu\nu} &= \sum_{s,s',c,f,t} \langle q\bar{q}g|J_\mu|0\rangle \langle q\bar{q}g|J_\nu|0\rangle^* \\
&= \frac{1}{8p_0 p_0' k_0} \sum_{f,t} e_f^2 g^2 \, \mathrm{Tr}\left[\not{p}\left(\not{v}\frac{1}{p\cdot k}(\not{p}+\not{k})\gamma_\mu \right.\right. \\
&\quad\left. - \gamma_\mu \frac{1}{p'\cdot k}(\not{p}+\not{k})\not{v}\right) \\
&\quad\left.\times \not{p}'\left(\gamma_\nu \frac{1}{p\cdot k}(\not{p}+\not{k})\not{v}- \not{v}\frac{1}{p'\cdot k}(\not{p}'+\not{k})\gamma_\nu \right)\right] \quad (11.39)
\end{aligned}
$$

where t is the gluon polarization. This then becomes:

$$
\begin{aligned}
H_{\mu\nu} &= \frac{2}{p_0 p_0' k_0} \sum_f e_f^2 g^2 \\
&\quad\times \left(\frac{p\cdot p'}{(p\cdot k)(p'\cdot k)} \left(2[p,p']_{\mu\nu} + [k,p+p']_{\mu\nu}\right) \right. \\
&\quad + \frac{1}{p\cdot k}\left([p',p+k]_{\mu\nu} - [p,p]_{\mu\nu}\right) \\
&\quad\left. + \frac{1}{p'\cdot k}\left([p,p'+k]_{\mu\nu} - [p',p']_{\mu\nu}\right)\right) \quad (11.40)
\end{aligned}
$$

where we have defined $[p,q]_{\mu\nu} \equiv p_\mu q_\nu + q_\mu p_\nu - g_{\mu\nu} p\cdot q$. Then the transition element:

$$
|\mathcal{M}|^2 = \left(\frac{1}{4E^2}\right)^2 l_{\mu\nu} H^{\mu\nu} \quad (11.41)
$$

becomes:

$$
\begin{aligned}
|\mathcal{M}|^2 &= \frac{1}{16E^6 p_0 p_0' k_0} \sum_f \left(ee_f g\right)^2 \\
&\quad\times \left(\frac{p\cdot p'}{(p\cdot k)(p'\cdot k)}\left[2(p\cdot q)(p'\cdot q') + 2(p\cdot q')(p'\cdot q)\right.\right. \\
&\quad\left. - 2(k\cdot q)(k\cdot q') + (Q\cdot q)(Q\cdot k)\right] \\
&\quad\left. + \frac{1}{p\cdot k}\left[(Q\cdot q)(Q\cdot p') - 2(p\cdot q)(p\cdot q') - 2(p'\cdot q)(p'\cdot q')\right]\right.
\end{aligned}
$$

$$+ \frac{1}{p' \cdot k}\left[(Q \cdot q)(Q \cdot p) - 2(p \cdot q)(p \cdot q') - 2(p' \cdot q)(p' \cdot q')\right]\right)$$

$$(11.42)$$

In practice, it is difficult to tell which jet emerged from which constituent. Therefore, let us number the jets. Let $p_{i\mu}$ represent the momentum of the ith jet. We define the variables:

$$x_i \equiv p_{i0}/E \qquad (11.43)$$

where $p_3 = k$. Let us also choose the center-of-mass frame, so that: $q = (E, \mathbf{q})$ and $q' = (E, -\mathbf{q})$.

Then the kinematics gives us:

$$x_1 + x_2 + x_3 = 2 \qquad (11.44)$$

The differential cross section is then given by:

$$d\sigma = \frac{\pi}{(8\pi^3)^2} \int |\mathcal{M}|^2 d^3p \, d^3p' \, d^3k \, \delta^4\left(p + p' + k - Q\right) \qquad (11.45)$$

Inserting the value of $\mathcal{M}$ into the cross section, we get:

$$d\sigma = \frac{2\alpha^2}{3E^2} \frac{g^2}{4\pi} dx_1 \, dx_2 \sum_f \left(\frac{e_f}{e}\right)^2$$

$$\times \left(\frac{x_1^2 + x_2^2}{(1 - x_1)(1 - x_2)} + \text{perm}\right) \qquad (11.46)$$

The experimental data is sometimes analyzed in terms of a quantity called "thrust," which can be defined as:

$$T = \frac{2\left(\sum_i' p_{\|i}\right)_{\max}}{\sum_i |p_i|} \qquad (11.47)$$

where the prime in the numerator means we sum over all particles in just one hemisphere. The thrust is a good variable because it varies from $T = \frac{1}{2}$ (isotropy) to $T = 1$ (perfect jet behavior). In Figure 11.5, we see some typical three-jet events found at PETRA compared to the theoretical prediction. The agreement with the experimental data is excellent.

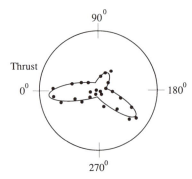

Figure 11.5. Experimentally, jets are found in collision processes, confirming the prediction of QCD (solid line).

11.4   Current Algebra

In the 1960s, before the rise of gauge theories and the Weinberg–Salam model, the original four-fermion Fermi action was a useful phenomenological guide that could account for some of the qualitative features of the weak interactions.

The hadrons participate in the weak interactions. For example, the hadrons are mostly unstable and decay via the weak interactions. The β decay of the neutron is the classic example of the weak interaction of the hadrons. Mimicking the success of the simple Fermi action, attempts were made to describe the hadronic weak interactions by postulating that the action was the product of two currents. The phenomenology of the weak interactions in the 1960s was dominated by something called *current algebra*,[16,17] which postulated the commutation relations of the currents among themselves. By taking their matrix elements, one could therefore derive sum rules that linked different physical processes. Although there was no understanding why the effective action should be the product of two currents, or why they should obey a chiral algebra, these current algebra relations agreed rather well with experiment. From the perspective of the Standard Model, however, we can see the simple origin of the current algebras.

To be specific, let us analyze the weak interactions of the quarks. We will insert quarks into the Weinberg–Salam model, arranging them into standard $SU(2)$ doublets and $U(1)$ singlets, which gives us the Glashow–Weinberg–Salam model. We now pair off the three generations of quarks u, d, c, s, t, and b with the three generations of leptons:

$$\begin{pmatrix} \nu_e \\ e \end{pmatrix}_L \begin{pmatrix} u \\ d \end{pmatrix}_L ; \quad \begin{pmatrix} \nu_\mu \\ \mu \end{pmatrix}_L \begin{pmatrix} c \\ s \end{pmatrix}_L ; \quad \begin{pmatrix} \nu_\tau \\ \tau \end{pmatrix}_L \begin{pmatrix} t \\ b \end{pmatrix}_L \qquad (11.48)$$

Notice that we are suppressing the color index on the quarks. From the point of view of the weak interactions, the color force is not important.

As with the electron, the quarks also have right-handed $SU(2)$ singlets: $u_R, d_R, c_R, s_R, t_R, b_R$, which are necessary when we construct mass eigenstates for the quarks. (We will describe the question of mixing between these various generations shortly.)

The Lagrangian for the Standard Model then consists of three parts:

$$\mathscr{L}_{\text{Standard Model}} = \mathscr{L}_{\text{WS}}(\text{lept.}) + \mathscr{L}_{\text{WS}}(\text{had.}) + \mathscr{L}_{\text{QCD}} \qquad (11.49)$$

where W-S stands for the Weinberg–Salam model, and lept. and had. stand for the leptons and quarks that are inserted into the Weinberg–Salam model with the correct $SU(2) \otimes U(1)$ assignments. (We assume that both the leptons and hadrons couple to the same Higgs field in the usual way. We also ignore quark mixing here.)

In order to get the correct quantum numbers, such as the charge, we must choose the following covariant derivatives for the left-handed and right-handed quarks in $\mathscr{L}_{\text{WS}}(\text{had.})$:

$$D_\mu \begin{pmatrix} u \\ d \end{pmatrix}_L = \left[\partial_\mu - i \left(\frac{g}{2} \right) \sigma^i W^i_\mu - i \left(\frac{g'}{6} \right) B_\mu \right] \begin{pmatrix} u \\ d \end{pmatrix}_L \qquad (11.50)$$

and:

$$\begin{aligned} D_\mu u_R &= \left[\partial_\mu - i(2g'/3)B_\mu \right] u_R \\ D_\mu d_R &= \left[\partial_\mu + i(g'/3)B_\mu \right] d_R \end{aligned} \qquad (11.51)$$

Since the quarks have different charges than the leptons, the coefficients appearing in the covariant derivatives are different from Eq. (10.63), in order to reproduce the correct coupling of the photon A_μ to the quarks.

From this form of the Standard Model action, several important conclusions can be drawn. First, the gluons from QCD only interact with the quarks, not the leptons. Thus, discrete symmetries like parity are conserved for the strong interactions. Second, the chiral $SU(N) \otimes SU(N)$ symmetry, which is respected by the QCD action in the limit of vanishing quark masses, is violated by the weak interactions. Third, quarks interact with the leptons via the exchange of W and Z vector mesons. Since the action couples two fermions and one vector meson together, the exchange of a W or Z meson couples four fermionic fields together. As in the previous chapter, one can therefore write down a phenomenological action involving four fermions, similar to the original Fermi action. In the limit of large vector meson mass, the Standard Model action can therefore be written as the sum of four-fermion terms that link quarks and leptons together. The important observation is that the charged W and neutral Z mesons couple to the currents, so the effective four-fermion action is the sum of the products of two currents:

$$\mathscr{L}_{\text{eff}} = \frac{G_F}{\sqrt{2}} \sum J_\mu^\dagger J^\mu \tag{11.52}$$

where we sum over charged and neutral currents, and where the charged weak current is the sum of a leptonic and a hadronic part:

$$J^\mu = J_{\text{lept}}^\mu + J_{\text{had}}^\mu \tag{11.53}$$

The leptonic part is given by:

$$J_{\text{lept}}^\mu = \bar{\psi}_e \gamma^\mu (1 - \gamma_5) \psi_{\nu_e} + \bar{\psi}_\mu \gamma^\mu (1 - \gamma_5) \psi_{\nu_\mu} \tag{11.54}$$

and the hadronic part is given by a vector and axial-vector piece:

$$J_{\text{had}}^\mu = V^\mu - A^\mu \tag{11.55}$$

where the vector and axial vector currents are made out of quark fields. (There are corresponding expressions for the neutral current, which is mediated by Z meson exchange and is diagonal in the lepton fields.) If we neglect strangeness, then the charged hadronic weak current can be written as:

$$
\begin{aligned}
J_{\text{had}}^\mu &= \bar{d}\gamma^\mu(1 - \gamma_5)u = (\bar{u}\ \bar{d}\ \bar{s})\gamma_\mu(1 - \gamma_5)\begin{pmatrix} 0 & 0 & 0 \\ 1 & 0 & 0 \\ 0 & 0 & 0 \end{pmatrix}\begin{pmatrix} u \\ d \\ s \end{pmatrix} \\
&= \bar{q}\lambda^{1-i2}\gamma_\mu(1 - \gamma_5)q/2 \tag{11.56}
\end{aligned}
$$

In other words, it transforms as the $1 \pm i2$ component of a $SU(2)$ triplet.

As before, the effective coupling constant G_F, because it is generated by the exchange of W and Z mesons, is related to the vector meson coupling constant; that is, $G_F/\sqrt{2} \sim g^2/M_W^2$.

Because the weak current is the sum of a leptonic and hadronic part, and because the effective action is a product of two such currents, the effective weak action can be broken up into three pieces, generating hadronic–hadronic, hadronic–leptonic, and leptonic–leptonic interactions:

$$\mathscr{L}_{\text{eff}} = \frac{G_F}{\sqrt{2}} \sum \{J_{\text{lept}} + J_{\text{had}}\}_\mu^\dagger \{J_{\text{lept}} + J_{\text{had}}\}^\mu \tag{11.57}$$

where we sum over both charged and neutral currents. The Standard Model effective action can therefore describe, in principle, all possible interactions between subatomic particles.

In the previous chapter, we discussed the leptonic–leptonic weak interactions. In this section, we will discuss the hadronic–hadronic weak interactions induced by the hadronic part of the weak current. Then in a later section, we will discuss the hadronic–leptonic interactions, which can mediate the weak decay of hadrons.

The current algebra relations can also be easily deduced from the Standard Model. The vector and axial vector currents generated by chiral $SU(N) \otimes SU(N)$ symmetry of the quark model are:

$$V_\mu^a = \bar{q} \gamma_\mu \lambda^a q / 2; \quad A_\mu^a = \bar{q} \gamma_\mu \gamma_5 \lambda^a q / 2 \tag{11.58}$$

where λ^a are the generators of $SU(N)$, and q transforms in the fundamental representation of $SU(N)$. For the free quark model, it is then easy to show that these $SU(N) \otimes SU(N)$ currents generate a closed algebra, which forms the basis of current algebra.

To see this more explicitly, we note that our action is invariant under:

$$\delta \phi_i = -i \epsilon^a (\tau^a)_{ij} \phi_j \tag{11.59}$$

(for generality, ϕ_i can be either fermionic or bosonic).

From our previous discussion, we know that, for every symmetry of the action, we have a conserved charge:

$$Q^a(t) = \int J_0^a(x) \, d^3 x \tag{11.60}$$

where the current is:

$$J_\mu^a = -i \frac{\delta \mathcal{L}}{\delta \partial^\mu \phi_i} (\tau^a)_{ij} \phi_j \tag{11.61}$$

We know that the conjugate field is given by:

$$\pi_i(x) = \frac{\delta \mathcal{L}}{\delta \partial^0 \phi_i} \tag{11.62}$$

with canonical commutation (anticommutation) relations:

$$[\pi_i(\mathbf{x}, t), \phi_j(\mathbf{y}, t)] = -i \delta^3(\mathbf{x} - \mathbf{y}) \delta_{ij} \tag{11.63}$$

Assuming that the current is composed out of free fields, we can take the commutator of two charges:

$$
\begin{aligned}
[Q^a(t), Q^b(t)] &= -\int d^3x\, d^3y\, \left[\pi_i(\mathbf{x}, t)(\tau^a)_{ij}\phi_j(\mathbf{x}, t), \pi_k(\mathbf{y}, t)(\tau^b)_{kl}\phi_l(\mathbf{y}, t)\right] \\
&= -\int d^3x\, d^3y\, \Big\{\pi_i(\mathbf{x}, t)(\tau^a)_{ij}\left[\phi_j(\mathbf{x}, t), \pi_k(\mathbf{y}, t)\right](\tau^b)_{kl}\phi_l(\mathbf{y}, t) \\
&\qquad + \pi_k(\mathbf{y}, t)(\tau^b)_{kl}\left[\pi_i(\mathbf{x}, t), \phi_l(\mathbf{y}, t)\right](\tau^a)_{ij}\phi_j(\mathbf{x}, t)\Big\} \\
&= -\int d^3x\, \left\{\pi_k(\mathbf{x}, t)i[\tau^a, \tau^b]_{kj}\phi_j(\mathbf{x}, t)\right\}
\end{aligned}
\tag{11.64}
$$

Finally, we find:

$$
[Q^a(t), Q^b(t)] = if^{abc}Q^c(t) \tag{11.65}
$$

where the derivation works equally well for commutators with bosons and anti-commutators with fermions.

Let us assume that we have, in addition to the usual symmetry, an axial symmetry, generated by:

$$
Q^{5a}(t) = \int J_0^{5a}(x)\, d^3x \tag{11.66}
$$

Then it is easy to show that the vector and axial vector currents generate the algebra:

$$
\begin{aligned}
{[Q^a(t), Q^b(t)]} &= if^{abc}Q^c(t) \\
{[Q^a(t), Q^{5b}(t)]} &= if^{abc}Q^{5c}(t) \\
{[Q^{5a}(t), Q^{5b}(t)]} &= if^{abc}Q^c(t)
\end{aligned}
\tag{11.67}
$$

For a unitary group, this generates the group $SU(N) \otimes SU(N)$. To see this, we redefine:

$$
\begin{aligned}
Q_L^a &= \frac{1}{2}(Q^a - Q^{5a}) \\
Q_R^a &= \frac{1}{2}(Q^a + Q^{5a})
\end{aligned}
\tag{11.68}
$$

These generators have the commutation relations:

$$
[Q_L^a(t), Q_L^b(t)] = if^{abc}Q_L^c(t)
$$

$$[Q_R^a(t), Q_R^b(t)] = if^{abc} Q_R^c(t)$$

$$[Q_L^a(t), Q_R^a(t)] = 0 \tag{11.69}$$

We can also show that the unintegrated currents satisfy:

$$[Q^a(t), J_\mu^b(\mathbf{x}, t)] = if^{abc} J_\mu^c(\mathbf{x}, t) \tag{11.70}$$

as well as:

$$[J_0^a(\mathbf{x}, t), J_0^b(\mathbf{y}, t)] = if^{abc} J_0^c(\mathbf{x}, t)\delta^3(\mathbf{x} - \mathbf{y}) \tag{11.71}$$

For our purposes, we are interested in the current algebra that generates $SU(N) \otimes SU(N)$ from Eq. (11.55):

$$[V_0^a(\mathbf{x}, t), V_0^b(\mathbf{y}, t)] = if_{abc} V_0^c(\mathbf{x}, t)\delta^3(\mathbf{x} - \mathbf{y})$$

$$[V_0^a(\mathbf{x}, t), A_0^b(\mathbf{y}, t)] = if_{abc} A_0^c(\mathbf{x}, t)\delta^3(\mathbf{x} - \mathbf{y})$$

$$[A_0^a(\mathbf{x}, t), A_0^b(\mathbf{y}, t)] = if_{abc} V_0^c(\mathbf{x}, t)\delta^3(\mathbf{x} - \mathbf{y}) \tag{11.72}$$

We should be careful to state, however, that the current algebra for the other components of J_μ do not form such a simple algebra. For example, one can show that the commutator between a current J_0 and J_i does not close properly:

$$[J_0^a(\mathbf{x}, t), J_i^b(\mathbf{y}, t)] = if^{abc} J_i^c(\mathbf{x}, t)\delta^3(\mathbf{x} - \mathbf{y})$$

$$+ S_{ij}^{ab} \frac{\partial}{\partial y_j}\delta^3(\mathbf{x} - \mathbf{y}) \tag{11.73}$$

where the last term is called a *Schwinger term*. Very general arguments show that such a term must necessarily exist in the algebra. In our discussion of current algebra, we must be aware of the presence of these Schwinger terms.

11.5 PCAC and the Adler–Weisberger Relation

By analyzing the properties of these weak currents, we can derive a large body of relations between different physical processes, which agree remarkably well with experiment. In this section, we will study the Conserved Vector Current (CVC) hypothesis, the Partially Conserved Axial Current (PCAC) hypothesis, and the Adler–Weisberger relation.

Although these relations were originally derived from the effective action and current algebra, we can see that they are all rather simple consequences of the Standard Model.

11.5.1   CVC

To see the origin of the CVC hypothesis,[18] notice that the muon decay constant G_μ is remarkably similar to the coefficient C_V appearing in the strong current of the old Fermi model for beta decay:

$$J_\mu^{\text{had.}} = \frac{1}{\sqrt{2}} \left(C_V \bar{\psi}_p \gamma_\mu \psi_n + C_A \bar{\psi}_p \gamma_\mu \gamma_5 \psi_n \right) \tag{11.74}$$

In fact, the coupling constants for muon decay and neutron decay differ by only 2.2%:

$$\frac{G_\mu - G_n}{G_\mu} = 2.2\% \tag{11.75}$$

To explain this, we assume that the strong electromagnetic current, which transforms like J_{had}^3 under $SU(2)$, must be part of the same $SU(2)$ multiplet as the strangeness-preserving hadronic weak currents J_{had}^{1+i2} and J_{had}^{1-i2}. The CVC hypothesis simply says that J_{had}^{1+i2} is conserved:

$$\text{CVC}: \quad \partial_\mu J_{\text{had}}^{\mu,1+i2} = 0 \tag{11.76}$$

just like the strong electromagnetic current, which can now be written as:

$$J_{\text{em}}^\mu = V_3^\mu + \frac{1}{\sqrt{3}} V_8^\mu \tag{11.77}$$

Since the electromagnetic current and the hadronic weak current $J_{\text{had}}^{1\pm i2}$ now transform as part of the same $SU(2)$ multiplet, there should be relations among the couplings for this current. This simple observation has had experimental success, for example, in explaining the beta decay rate of pions.

From the point of view of the quark model, CVC has a simple interpretation. The CVC relations can be derived by writing down the quark representation of the various currents:

$$
\begin{aligned}
J_{\text{em}}^\mu &= \frac{2}{3}\bar{u}\gamma^\mu u - \frac{1}{3}\bar{d}\gamma^\mu d = \frac{1}{6}\bar{q}\gamma^\mu I q + \bar{q}\gamma^\mu \frac{\lambda^3}{2} q \\
V_{\text{had}}^\mu &= \bar{d}\gamma^\mu u + \cdots = \bar{q}\gamma^\mu \lambda^{1-i2} q + \cdots
\end{aligned}
\tag{11.78}
$$

(where the ellipses represent the strangeness-changing part of the hadronic weak current, which we will discuss shortly). Written in terms of the quark fields, then it is obvious that the electromagnetic current and the strangeness-preserving hadronic weak current are part of the same $SU(2)$ multiplet. Since $SU(2)$ is a reasonably good symmetry of QCD, we expect CVC to hold (but be broken by electromagnetic interactions and the fact that $m_u \neq m_d$).

11.5.2 PCAC

Now let us assume, because of approximate chiral symmetry, that a modified current conservation rule applies for the axial current A_μ^a as well. The approximate $SU(2) \otimes SU(2)$ chiral symmetry of QCD allows us to write down new relations based on the PCAC (partially conserved axial current) hypothesis.[19-21] This states that the divergence of the axial current is exact in the limit of $SU(2) \otimes SU(2)$ symmetry, but is broken by the quark masses. Phenomenologically, this means that the conservation of the axial current is broken because of the small pion mass (which is now viewed as a Nambu–Goldstone boson).

To see how PCAC provides nontrivial relations between scattering amplitudes, let us construct the matrix element of the axial current A_μ^a between the vacuum and a pion state $|\pi^b\rangle$. This matrix element can be coupled to the matrix element of the leptonic weak current, so that it governs the decay of the pion into an electron and neutrino. By Lorentz symmetry, this matrix element can only be proportional to the momentum of the pion:

$$\langle 0|A_\mu^a(0)|\pi^b(p)\rangle = i f_\pi \delta^{ab} p_\mu \tag{11.79}$$

where f_π is the pion decay constant, and experimentally it is equal to 93 MeV.

We normalize the pion state by:

$$\langle 0|\phi^a(0)|\pi^b(p)\rangle = \delta^{ab} \tag{11.80}$$

Now let us take the divergence of this equation:

$$\langle 0|\partial^\mu A_\mu^a(0)|\pi^b(p)\rangle = \delta^{ab} m_\pi^2 f_\pi = f_\pi m_\pi^2 \langle 0|\phi^a(0)|\pi^b(p)\rangle \tag{11.81}$$

where we have integrated by parts. It is therefore reasonable to set, as an effective relation:

$$\partial^\mu A_\mu^a = f_\pi m_\pi^2 \phi^a \tag{11.82}$$

which is the PCAC hypothesis. In the limit of vanishing pion mass, we have exact chiral symmetry, and hence these equations give us an exactly conserved axial vector current in this limit. (From the point of view of the quark model, one can explicitly take the derivative of the axial current and one finds that it is not conserved. The right-hand side of the equation is proportional to $\bar{q}\lambda^a\gamma_5 q$, which has the quantum numbers of the π meson. When matrix elements of this relation are taken, one can, with a few assumptions, show that the PCAC relation holds.)

One of the great successes of PCAC was the derivation of the Goldberger–Treiman relation.[22] The origin of this new relation was simple. The PCAC hypothesis links the axial current to the pion field. By taking different matrix elements of the PCAC relation, we can derive different relations between different physical processes. For example, if we take the matrix element of the PCAC equation between neutron and proton wave functions, then we can establish a relationship between the pion–nucleon coupling constant $g_{\pi NN}$ and the decay constant of the pion f_π. Thus, PCAC is able to link two unrelated physical processes, pion–nucleon scattering and pion decay.

To see how this happens, let us take the matrix element of the axial current between neutron and proton states. The only axial vectors that we have at our disposal are $q_\mu\gamma_5$ and $\gamma_\mu\gamma_5$. Thus, the matrix element must have the following form:

$$\langle p(k')|A_\mu^{1+i2}|n(k)\rangle = \bar{u}_p(k')\left[\gamma_\mu\gamma_5 g_A(q^2) + q_\mu\gamma_5 h_A(q^2)\right]u_n(k) \qquad (11.83)$$

Now take the matrix element of the pion between the proton and neutron states:

$$\langle p(k')|\phi^+|n(k)\rangle = \frac{\sqrt{2}}{-q^2 + m_\pi^2} g_{\pi NN}(q^2) i\bar{u}_p(k')\gamma_5 u_n(k) \qquad (11.84)$$

which is dominated by the pion pole term and where the coupling constant $g_{\pi NN}(q^2)$ is related to the physical coupling constant for pion–nucleon scattering by:

$$
\begin{aligned}
g_{\pi NN} &= g_{\pi NN}(m_\pi^2) \\
\frac{g_{\pi NN}^2}{4\pi} &\sim 14.6
\end{aligned}
\qquad (11.85)
$$

Now let us put everything together. Let us take the divergence of the matrix element of the PCAC equation. This easily gives us:

$$\frac{2f_\pi m_\pi^2}{-q^2 + m_\pi^2} g_{\pi NN}(q^2) = 2M g_A(q^2) + q^2 h_A(q^2) \qquad (11.86)$$

Let us now set $q^2 = 0$, and make the crucial assumption that $g_{\pi NN}(q^2)$ does not vary much between $q^2 = 0$ and the pion mass squared. (This assumption is based on the fact that the pion mass is small, and that the analytic behavior of the function is relatively smooth in q^2 space.)

When we make this assumption, we find the celebrated Goldberger–Treiman relation, which agrees well with experiment:

$$\frac{f_\pi g_{\pi NN}}{M} = g_A(0) \tag{11.87}$$

Experimentally, $g_A(0) \sim 1.22$, while $f_\pi g_{\pi NN}/m_N \sim 1.34$.

This relation, we saw, depended crucially on the mass of the pion being small; that is, the masses of the u and d quarks are relatively small on a hadronic scale. However, the mass of the strange quark is larger than the others, and hence we expect the approximation to be less reliable for the K meson.

11.5.3 Adler–Weisberger Relation

Finally, one of the most important relations that one can derive from the current algebra is the celebrated Adler–Weisberger[23,24] sum rule, which relates the integral of pion–nucleon cross sections to known form factors.

Our goal is to write the pion–nucleon scattering amplitude in terms of the scattering of nucleons and axial currents using current algebra and PCAC. Therefore, we wish to study the scattering of nucleon N and the axial current:

$$N(p_1) + A_\nu^b(q_1) \to N(p_2) + A_\mu^a(q_2) \tag{11.88}$$

where $p_1 + q_1 = p_2 + q_2$, and where we will eventually set $q_1 = q_2$. The matrix element for this process is given by:

$$T_{\mu\nu}^{ab} = \int d^4x\, e^{iq\cdot x} \langle N(p_2)|T A_\mu^a(x) A_\nu^b(0)|N(p_1)\rangle \tag{11.89}$$

Let us contract this amplitude with $q^\mu q^\nu$. By integration by parts, we can convert this into a divergence. Each time the derivative hits a θ function within the time-ordered product, it creates a delta function. Then it is not hard to show:

$$
\begin{aligned}
q_1^\mu q_2^\nu T_{\mu\nu}^{ab} &= i \int d^4x\, e^{iq\cdot x} \Big\{ \langle N(p_2)|\partial^\mu A_\mu^a(x)\partial^\nu A_\nu^b(0)|N(p_1)\rangle \\
&\quad - iq_1^\mu \langle N(p_2)|\delta(x_0)[A_0^b, A_\mu^a(x)]|N(p_1)\rangle \\
&\quad + \langle N(p_2)|\delta(x_0)[A_0^a(x), \partial^\nu A_\nu(0)]|N(p_1)\rangle \Big\}
\end{aligned} \tag{11.90}
$$

In general, almost nothing is known about the left-hand side of this equation. However, the three terms on the right-hand side of this equation can, by PCAC and current algebra, be reduced into known quantities.

For example, the first term on the right-hand side of this equation can be related to the pion–nucleon scattering amplitude via PCAC. Using the LSZ and PCAC relations, we can rewrite the usual pion–nucleon amplitude as:

$$
\begin{aligned}
T_{\pi N}^{ab} &= i \int d^4x \, e^{iq_1 \cdot x}(q_1^2 - m_\pi^2)(q_2^2 - m_\pi^2) \\
&\quad \times \langle N(p_2)|T\phi^a(x)\phi^b(0)|N(p_1)\rangle \\
&= i(q_1^2 - m_\pi^2)(q_2^2 - m_\pi^2)m_\pi^{-4} f_\pi^{-2} \int d^4x \, e^{iq_1 \cdot x} \\
&\quad \times \langle N(p_2)|T[\partial^\mu A_\mu^a(x)\partial^\lambda A_\lambda^b(0)]|N(p_1)\rangle
\end{aligned}
\tag{11.91}
$$

The second term on the right-hand side of Eq. (11.90) can also be reduced if we use the following current algebra relation:

$$
\delta(x_0)[A_0^b(0), A_\mu^c(x)] = -i\delta(x_0)\epsilon^{abc} V_\mu^c(x)
\tag{11.92}
$$

(There is a potential Schwinger term in this commutator, but one can show that it cancels out.)

Because the iso-vector current appears on the right-hand side, its matrix element is proportional to τ^c, so we can now reduce this expression down to:

$$
\begin{aligned}
-iq^\mu \int d^4x \, e^{iq \cdot x} &\langle N(p)|\delta(x_0)[A_0^b(0), A_\mu^a(x)]|N(p)\rangle \\
&= \epsilon^{abc} q^\mu \bar{u}(p)\gamma_\mu \tau^c u(p)/2 \\
&= 2p \cdot q\epsilon^{abc}\tau^c/2 = -iv[\tau^a, \tau^b]/2
\end{aligned}
\tag{11.93}
$$

where $v = p \cdot q$.

The third term on the right-hand side of Eq. (11.90), which we will call σ_N^{ab}, can also be simplified. It is easy to show that this term is symmetric in a and b. (To prove this, we drop the term with the integral over the spatial derivative $\partial^i A_i^b$, since we assume that the fields vanish sufficiently rapidly at infinity. Then we are left with a commutator between A_0^a and $\partial^0 A_0^b$. By integrating by parts, we can move ∂^0 to the other current. Then by reinserting the spatial derivatives, we find that $\sigma_N^{ab} = \sigma_N^{ba}$.)

Putting everything together, we can write Eq. (11.90) as:

$$
q^\mu q^\nu T_{\mu\nu}^{ab} = -i(q^2 - m_\pi^2)m_\pi^4 f_\pi^2 T_{\pi N}^{ab} + iv[\tau^a, \tau^b]/2 - i\sigma_N^{ab}
\tag{11.94}
$$

Next, we wish to reduce the left-hand side of this equation. In the low-energy limit $q \to 0$, the term that dominates this expression consists of the one-nucleon pole term, since the Feynman propagator of the one-nucleon pole term diverges. Using Feynman's rules, we can add the two one-nucleon pole diagrams that contribute to the pion–nuclear amplitude in this limit.

The poles coming from the two graphs are proportional to $1/[(p \pm q)^2 - M^2] = 1/(\pm 2v + q^2)$, where $v = p \cdot q$. To calculate the residue of these two pole terms, we use the fact that the matrix element of A_μ^a between two nucleon states is given by Eq. (11.83).

Using the Gordon identity, we find that the sum of the two one-nucleon exchange graphs becomes:

$$q^\mu q^v T_{\mu v}^{ab} \sim 2i g_A^2 \left([\tau^a, \tau^b]v - \delta^{ab}q^2\right)(v^2 - M^2 q^2)/(q^4 - 4v^2) + \cdots$$
$$\sim i g_A^2 v[\tau^a, \tau^b] + \cdots \tag{11.95}$$

where $q^2 \ll v = p \cdot q$ for small q.

To simplify matters, we will be interested only in the amplitude that is anti-symmetric in a and b (so we can drop σ_N^{ab}). We will make the following isotopic decomposition of T^{ab}:

$$T_{\pi N}^{ab} = T^+ \delta^{ab} + i\epsilon^{abc} \tau^c T^- \tag{11.96}$$

We are only interested in T^-.

We can now put all the pieces together in Eq. (11.90). Taking the limit as $q^2 \to 0$ in the expression for $T_{\pi N}^{ab}(q^2, v)$, Eq. (11.90) becomes:

$$\lim_{v \to 0} v^{-1} T^-(v, 0) = (1 - g_A^2)/f_\pi^2 \tag{11.97}$$

This is our primary result. However, left in this fashion, this sum rule is rather useless. In order to make comparisons with experimental data, we must rewrite this in terms of measurable cross sections. Since T^- is analytic and odd under $v \to -v$, $v^{-1}T^-$ satisfies an unsubtracted dispersion relation:

$$\frac{T^-(v, 0)}{v} = \frac{2}{\pi} \int_{v_0}^\infty \frac{\operatorname{Im} T^-(v', 0)\, dv'}{v'^2 - v^2} \tag{11.98}$$

Putting $v = 0$, we arrive at:

$$\frac{1}{g_A^2} = 1 + \frac{2M^2}{\pi g_{\pi NN}^2} \int_{v_0}^\infty \frac{dv}{v^2} \operatorname{Im} T^-(v, 0) \tag{11.99}$$

where we have taken advantage of the Goldberger–Treiman relation.

Using the optical theorem, we can write:

$$\text{Im}\, T^-(\nu, m_\pi^2) = \nu[\sigma^{\pi^- p}(\nu) - \sigma^{\pi^+ p}(\nu)] \tag{11.100}$$

Inserting this back into the dispersion relation, we find the Adler–Weisberger relation:

$$\frac{1}{g_A^2} = 1 + \frac{2M^2}{\pi g_{\pi NN}^2} \int_{\nu_0}^{\infty} \frac{d\nu}{\nu} \left[\sigma^{\pi^- p} - \sigma^{\pi^+ p} \right] \tag{11.101}$$

Putting the experimental values into the relation, we can solve for g_A, which yields 1.24, in good agreement with the experimental value of 1.259.

11.6 Mixing Angle and Decay Processes

In the previous section, we saw how the hadronic–hadronic part of the effective Standard Model action in Eq. (11.57) gave us a wealth of weak interaction relations that agreed well with experiment. In this section, we will examine the hadronic–leptonic part of the effective Standard Model action. Specifically, we will study the decays of hadrons via the weak interactions, which is mediated by the hadronic–leptonic effective action. However, since there is a vast number of decays, we will not catalog them. Although the Standard Model gives us the ability, in principle, to calculate them all, we will only use the Standard Model to make certain qualitative observations concerning these decays.

In the 1960s, many of these important decays were carefully experimentally studied, although there was no comprehensive, underlying explanation for their behavior. From the perspective of the Standard Model, many of the mysteries of these decays can be easily explained. In particular, we will be interested in the decays of the K mesons, which are bound states of the strange quark with the u or d quarks. The addition of the strange quark to our discussion, however, brings in an important complication: the quark mixing angles. Since the weak interactions do not respect chiral $SU(3) \otimes SU(3)$ symmetry, there is no unique way in which to insert the strange quark into the Weinberg–Salam model. In principle, since the d and s quarks have the same charges, and since the weak interactions do not respect global $SU(3)$ symmetry, there is nothing to prevent the d and s quarks from mixing within the same $SU(2)$ doublet. We can parametrize this ambiguity by taking the following $SU(2)$ doublet:

$$\begin{pmatrix} u \\ d \cos \theta_C + s \sin \theta_C \end{pmatrix}_L \tag{11.102}$$

where θ_C is called the Cabibbo angle.[25] The Standard Model does not explain the origin of this mixing. The Cabibbo angle, however, allows us to parametrize our ignorance. Like the quark masses, the Weinberg angle, etc., it is one of the many undetermined parameters within the Standard Model (which indicates that the Standard Model is only a first approximation to the correct theory of subatomic particles). Experimentally, we find:

$$\sin\theta_C = 0.231 \pm 0.003 \qquad (11.103)$$

so that the Cabibbo angle is equal to $\theta_C \sim 15°$.

In the Weinberg–Salam model, the W and Z mesons then couple to the strange current, given by $\bar{u}\gamma_\mu(1 - \gamma_5)s$ times the sine of the Cabibbo angle. If we write this in terms of its $SU(3)$ content, this strange current transforms as the $4 + i5$ component. Thus, the vector and axial charged vector hadronic currents can be written as:

$$
\begin{aligned}
V_\mu &= \cos\theta_C V_\mu^{1+i2} + \sin\theta_C V_\mu^{4+i5} \\
A_\mu &= \cos\theta_C A_\mu^{1+i2} + \sin\theta_C A_\mu^{4+i5}
\end{aligned}
\qquad (11.104)
$$

If we write these charged currents in terms of their quark content, we find:

$$J_\mu = V_\mu - A_\mu = \cos\theta_C \left(\bar{u}\gamma_\mu d - \bar{u}\gamma_\mu\gamma_5 d\right) + \sin\theta_C \left(\bar{u}\gamma_\mu s - \bar{u}\gamma_\mu\gamma_5 s\right) \quad (11.105)$$

Since the Cabibbo angle is experimentally found to be relatively small, $\sin\theta_C$ is suppressed relative to $\cos\theta_C$. The effective action, written in terms of this quark representation, automatically reproduces the fact that the $1+i2$ reactions are larger than the $4 + i5$ reactions. This is because the $u - s$ quark current (transforming like the $4 + i5$ current) is suppressed by a factor of $\tan\theta_C$ relative to the $u - d$ quark current (transforming like the $1 + i2$ current).

Now that we have parametrized the strange hadronic current, there is a wide variety of decay processes that can be described by the Standard Model. It will be helpful to divide these decays into several classes.

11.6.1 Purely Leptonic Decays

These decays involve no hadronic particles at all. For example, the decay of the muon is a purely leptonic decay process.

11.6.2   Semileptonic Decays

Semileptonic decays are those that involve both hadrons and leptons. For example, the decay of hadrons into leptons is a typical example. These decays, in turn, are divided further into two types, $\Delta S = 0$ and $\Delta S \neq 0$.

For $\Delta S = 0$, the beta decay of the neutron is one of the most important examples. Other $\Delta S = 0$ semileptonic decays include the following hyperon and pion decays:

$$\begin{aligned}
\Sigma^- &\rightarrow \Lambda + e + \bar{\nu} \\
\pi^- &\rightarrow \pi^0 + e + \bar{\nu}
\end{aligned} \qquad (11.106)$$

Strangeness-changing decays include: $\Lambda \rightarrow p + e + \bar{\nu}$.

Some other examples of semileptonic decays are given by π_{l3} and K_{l3} decays:

$$\begin{aligned}
\pi^+ &\rightarrow \pi^0 + l^+ + \nu_l \\
K^+ &\rightarrow \pi^0 + l^+ + \nu_l
\end{aligned} \qquad (11.107)$$

where $l = e, \mu$.

In an obvious notation, the decays of the K meson are sometimes designated K_{e3}, $K_{\mu3}$, K_{e4}, and $K_{\mu4}$.

11.6.3   Nonleptonic Decays

These decays involve the decay of hadrons into other hadrons. These decays can also be divided into two classes, $\Delta S = 0$ or $\Delta S \neq 0$.

Some of the hyperon nonleptonic decays are:

$$\begin{aligned}
\Sigma^+ &\rightarrow p + \pi^0 \\
\Lambda &\rightarrow n + \pi^0 \\
\Omega^- &\rightarrow \Lambda + \pi^-
\end{aligned} \qquad (11.108)$$

Nonleptonic K decays include:

$$\begin{aligned}
K^0 &\rightarrow \pi^+ + \pi^- \\
K^+ &\rightarrow \pi^+ + \pi^0 + \pi^0
\end{aligned} \qquad (11.109)$$

Although the Standard Model gives us the ability to calculate these decays from Feynman's rules, we will only make a few brief qualitative observations concerning these decays, from the point of view of the Standard Model.

First, the Standard Model forbids a large number of decays that cannot be described by the current–current effective action. This is totally consistent with the experimental data. Second, the various rules that have been accumulated over the years (such as the $\Delta I = \frac{1}{2}$ rule) can be explained qualitatively by analyzing the isospin nature of the currents of the Standard Model. Third, we find that certain decays are suppressed relative to others because the Cabibbo angle is small.

As an example, consider the decays:

$$
\begin{aligned}
\pi^+ &\rightarrow \pi^0 + e^+ + \nu_e \\
K^+ &\rightarrow \pi^0 + e^+ + \nu_e
\end{aligned}
\tag{11.110}
$$

The first transition from π^+ to π^0 is mediated by a current transforming as $1 + i2$, while the transition from K^+ to π^0 is mediated by a $4 + i5$ current. Thus, we expect that the coupling constants for these decays to be related to each other via the Cabibbo angle, which agrees with experiment.

11.7   GIM Mechanism and Kobayashi–Maskawa Matrix

Experimentally, there is very strong experimental evidence that strangeness changing neutral currents are suppressed. For example, we have the experimental results:

$$
\begin{aligned}
\frac{\Gamma(K_L^0 \rightarrow \mu^+ \mu^-)}{\Gamma(K_L^0 \rightarrow \text{all})} &\sim 10^{-8} \\[2mm]
\frac{\Gamma(K^\pm \rightarrow \pi^\pm \nu \bar{\nu})}{\Gamma(K^\pm \rightarrow \text{all})} &< 0.6 \times 10^{-6}
\end{aligned}
\tag{11.111}
$$

The numerator is sensitive to the existence of a weak current that couples to the strange quark s, is electrically neutral, and changes the strangeness number. The fact that these processes are extremely rare indicates that such currents should be absent in our action, at least to lowest order.

One of the triumphs of this simple picture is the success of the GIM (Glashow–Iliopoulos–Maiani) mechanism,[26] which uses a fourth, charmed quark to cancel such currents at the tree level.

So far, the model that we been describing allows strangeness-changing neutral currents, which arise when we analyze the following part of the neutral current: $J_\mu^0 = \bar{d}' \gamma_\mu (1 - \gamma_5) d'$. If we expand out this current, we find the piece:

$$
\bar{s} \gamma_\mu (1 - \gamma_5) d \sin \theta_C \cos \theta_C
\tag{11.112}
$$

which introduces strangeness-changing processes through the $s - d$ coupling. We want to cancel this term.

To explain the absence of such a current, we will use the fourth charmed quark, which will give us a global $SU(4)$ flavor symmetry. The hadronic weak current can now be represented in terms of four quarks as:

$$J_\mu^a = \bar{q}\gamma_\mu(1 - \gamma_5)\lambda^a q/2 \tag{11.113}$$

where q consists of two $SU(2)$ doublets:

$$\begin{pmatrix} u \\ d' \end{pmatrix}; \quad \begin{pmatrix} c \\ s' \end{pmatrix} \tag{11.114}$$

where:

$$\begin{aligned} d' &= d \cos\theta_C + s \sin\theta_C \\ s' &= -d \sin\theta_C + s \cos\theta_C \end{aligned} \tag{11.115}$$

The key observation is that the existence of the charmed fourth quark allows us to express the neutral current, which is diagonal in the fermion fields, as the sum of two terms:

$$\bar{d}'\gamma_\mu(1 - \gamma_5)d' + \bar{s}'\gamma_\mu(1 - \gamma_5)s' \tag{11.116}$$

Because of the presence of s' in the neutral current, there is an additional piece to the strangeness changing neutral current given by:

$$- \bar{s}\gamma_\mu(1 - \gamma_5)d \sin\theta_C \cos\theta_C \tag{11.117}$$

If we add the two pieces in Eqs. (11.117) and (11.112) together, we find an exact cancellation, meaning that (at the tree level) there are no strangeness-changing neutral currents. The mixing of the various quarks therefore gives us new physically interesting results.

(Another way of saying this is that the total neutral current has the quark content: $\bar{d}'d' + \bar{s}'s' = \bar{d}d + \bar{s}s$ if we drop the Dirac matrices. This combination is invariant even after rotating the quarks by the Cabibbo angle θ_C. The neutral, strangeness-changing current vanishes because there is no term proportional to $\bar{s}d$ in this combination.)

We can now appreciate the importance of the angle θ_C, which not only serves to suppress some reactions that occur with $\sin\theta_C$, but also eliminates strangeness-changing neutral currents via charm. Given the importance of mixing between generations, let us now try to analyze the question of mixing systematically between three generations of quarks and leptons. The Cabibbo angle, for example, was the unique way in which to mix two generations of quark flavors. It is possible, with a few simple arguments, to write down the complete set of these mixing angles for three generations.

To construct the most general charged weak current, we first notice that the u, c, t all have charge $\frac{2}{3}$, while d, s, b all have charge $-\frac{1}{3}$. Since the weak interactions do not respect global flavor symmetry, there is nothing to prevent mixing within these groups of quarks of the same charge. We thus have the freedom to rearrange the charge $\frac{2}{3}$ and the charge $-\frac{1}{3}$ quarks into two multiplets called U and D, respectively:

$$U = \begin{pmatrix} u \\ c \\ t \end{pmatrix}; \quad D = \begin{pmatrix} d \\ s \\ b \end{pmatrix} \tag{11.118}$$

where the space in which we are working is labeled by the generation or family. Since we have three families, we can mix the three families of quarks within U and D. The most arbitrary mixing between them can be parametrized by:

$$U'' = M_U U; \quad D'' = M_D D \tag{11.119}$$

where M_U and M_D are 3×3 unitary matrices.

Then the charged weak current can be written as:

$$
\begin{aligned}
J_\mu &= \bar{U}'' \gamma_\mu (1 - \gamma_5) D'' \\
&= \bar{U} \gamma_\mu (1 - \gamma_5) M D
\end{aligned}
\tag{11.120}
$$

where we have defined the Kobayashi–Maskawa matrix[27] as:

$$M \equiv M_U^\dagger M_D \tag{11.121}$$

The unitary matrix M is a $N_f \times N_f$ matrix for N_f families. This matrix has, in general, N_f^2 real parameters. However, since there are $2N_f$ quarks, $2N_f - 1$ of these parameters can be absorbed into the quark wave functions. We are then left with $(N_f - 1)^2$ real mixing angles that cannot be absorbed by any field redefinition.

For $N_f = 2$, we have one mixing angle, which is the just the original Cabibbo angle. In that case, the charged weak current is:

$$J_\mu = (\bar{u}\ \bar{c})\gamma_\mu(1 - \gamma_5) \begin{pmatrix} \cos\theta_C & \sin\theta_C \\ -\sin\theta_C & \cos\theta_C \end{pmatrix} \begin{pmatrix} d \\ s \end{pmatrix} \tag{11.122}$$

However, for $N_f = 3$, we have the possibility of four mixing angles. Traditionally, these four mixing angles are parametrized with three angles θ_i, $i = 1, 2, 3$ and one phase δ. Then the matrix M is usually written as:

$$
M = \begin{pmatrix} 1 & 0 & 0 \\ 0 & C_2 & S_2 \\ 0 & -S_2 & C_2 \end{pmatrix} \times \begin{pmatrix} C_1 & S_1 & 0 \\ -S_1 & C_1 & 0 \\ 0 & 0 & 1 \end{pmatrix}
$$

$$
\times \begin{pmatrix} 1 & 0 & 0 \\ 0 & 1 & 0 \\ 0 & 0 & e^{i\delta} \end{pmatrix} \times \begin{pmatrix} 1 & 0 & 0 \\ 0 & C_3 & S_3 \\ 0 & -S_3 & C_3 \end{pmatrix} \qquad (11.123)
$$

where $C_i = \cos \theta_i$ and $S_i = \sin \theta_i$.

Written out explicitly, this is:

$$
M = \begin{pmatrix} C_1 & S_1 C_3 & S_1 S_3 \\ -S_1 C_2 & C_1 C_2 C_3 - S_2 S_3 e^{i\delta} & C_1 C_2 S_3 + S_2 C_3 e^{i\delta} \\ S_1 S_2 & -C_1 S_2 C_3 - C_2 S_3 e^{i\delta} & -C_1 S_2 S_3 + C_2 C_3 e^{i\delta} \end{pmatrix} \qquad (11.124)
$$

Experimentally, the three mixing angles θ_i are either smaller than or comparable to the Cabibbo angle. In the limit $\theta_2 = \theta_3 = 0$, then θ_1 reduces to the Cabibbo angle. Also, one of the advantages of the KM formalism is that it gives us a convenient way of parametrizing CP violation, which is found experimentally in K meson decays.[128] The angle δ gives us a complex M matrix, thereby violating CP invariance. (CP invariance demands that $M^* = M$.)

In summary, the Standard Model is created by splicing QCD with the Weinberg–Salam model. It allows us to unify all known experimental data concerning particle interactions via the gauge group $SU(3) \otimes SU(2) \otimes U(1)$. The gauge fields of color $SU(3)$ are responsible for binding the quarks together, while the gauge fields of $SU(2) \otimes U(1)$ mediate the electromagnetic and weak interactions. Altogether, there are quite a few free parameters in the theory: three coupling constants for the groups in $SU(3) \otimes SU(2) \otimes U(1)$, two parameters in the Higgs sector (the Higgs mass and the Higgs vacuum expectation value v), $N_f^2 + 1$ quark parameters [$2N_f$ quark masses for N_f families and $(N_f - 1)^2$ KM mixing angles and phases], an equal number $N_f^2 + 1$ of lepton parameters (for massive neutrinos), and the angle θ_{QCD} (coming from instanton contributions). For a Standard Model with massive neutrinos, we thus have $2(N_f^2 + 1) + 6$ free parameters. For three families or generations, that makes 26 free parameters. (For massless neutrinos and no leptonic mixing angles, we have 19 free parameters.) With so many free parameters, the Standard Model should be viewed as the first approximation to the true theory of subatomic particles.

In the next chapter, we will discuss quantum anomalies that arise in any naive attempt to quantize chiral fermions. The marriage between quarks and leptons in the Standard Model is not a trivial one, because the anomalies of the leptons in the Weinberg–Salam model cancel precisely the anomalies coming from the quarks.

11.8   Exercises

1. Calculate the tensor product reduction of $3 \otimes 8$ and $6 \otimes 6$ for $SU(3)$ using Young tableaux. Identify the dimension of each of the Young tableaux in the decomposition.

2. For $SU(4)$, calculate the decomposition of:

$$\mathbf{15} \otimes \mathbf{15} \otimes \mathbf{15} \tag{11.125}$$

3. If we adopt the $SU(3)$ particle assignments in Eqs. (11.9) and (11.10), show that the meson and baryon matrices are given by Eqs. (11.12) and (11.13).

4. Using Feynman's rules, prove Eq. (11.30).

5. Why must a Schwinger term exist in Eq. (11.73)? Hint: assume that the Schwinger term is missing, and then prove:

$$
\begin{aligned}
0 &= \langle 0 | [J_0(\mathbf{x}, t), \partial_0 J_0(\mathbf{y}, t)] | 0 \rangle \\
&= i \sum_n \left(e^{i \mathbf{p}_n \cdot (\mathbf{x} - \mathbf{y})} + e^{-i \mathbf{p}_n \cdot (\mathbf{x} - \mathbf{y})} \right) \\
&\quad \times E_n |\langle 0 | J_0(0) | n \rangle|^2
\end{aligned}
\tag{11.126}
$$

Then prove that this relation shows that $J_0 = 0$, and so, by contradiction, there must be a Schwinger term.

6. Let the electromagnetic current J_{em}^μ be sandwiched between two proton states. By invariance arguments, the most general matrix element is given by:

$$
\begin{aligned}
\langle p', s' | J_{\text{em}}^\mu(x) | p, s \rangle &= e^{i(p' - p) \cdot x} \bar{u}(p', s') \Big(F_1(q^2) \gamma^\mu \\
&\quad + i \frac{F_2(q^2)}{2M} \sigma^{\mu\nu} q_\nu + F_3(q^2) q^\mu \Big) u(p, s)
\end{aligned}
\tag{11.127}
$$

and q_μ is the difference in momenta. Using the fact that the electromagnetic current is conserved, prove that $F_3(q^2) = 0$. Using time-reversal invariance, prove that F_1 and F_2 are both real.

7. Write down an explicit matrix representation of the generators of $SU(5)$ and $SU(6)$ using the algorithm mentioned in the book.

8. Can the bottom quark and top quark be added and still have no flavor changing neutral currents? Examine the KM matrix.

9. Prove Eq. (11.90).

10. In the Standard Model with two generations, we could have mixed the u and c quarks together as well as the d and s quarks. This would give us two Cabibbo angles. How do we reconcile this with a single Cabibbo angle?

11. Since the experimental evidence points to three quark colors, why cannot the color gauge group be $SO(3)$ instead of $SU(3)$? $SO(3)$ would apparently satisfy many of the experimental tests. (Hint: analyze if the triplet representations are real or complex for the antiquarks.)

12. Show explicitly how the $N_f \times N_f$ K–M matrix, with N_f^2 real parameters, can be reduced to a matrix with only $(N_f - 1)^2$ unknowns after a re-definition of the quark wave functions.

13. Let the matrix element of the strangeness-preserving vector hadronic weak current between neutron and proton states be:

$$\langle p | J_{\text{had}}^{1+i2,\mu} | n \rangle = \bar{u}' \left(f_1(q^2) + i \frac{f_2(q^2)}{2M} \sigma^{\mu\nu} q_\nu + f_3(q^2) q^\mu \right) \lambda^{1+i2} u \quad (11.128)$$

Using the CVC hypothesis, prove:

$$
\begin{aligned}
f_1(q^2) &= F_1^p(q^2) - F_1^n(q^2) \to 1 \text{ (as } q^2 \to 0) \\
f_2(q^2) &= F_2^p(q^2) - F_2^n(q^2) \to \mu_p - \mu_n \text{ (as } q^2 \to 0) \\
f_3(q^2) &= 0 \qquad\qquad\qquad\qquad\qquad\qquad (11.129)
\end{aligned}
$$

where F_i are defined in Problem 6.

14. For the massive quark model, prove to lowest order that the divergence of the axial current can be written as:

$$\partial_\mu A^{a\mu} = 2im\bar{q}\lambda^a \gamma_5 q \quad (11.130)$$

(where we integrate by parts). What assumptions are necessary to convert this into the PCAC relationship?

15. Prove that the strangeness-preserving charged hadronic weak current $J_{\text{had}}^{\mu,1+i2}$ and its conjugate are responsible for inducing the following weak reactions:

$$\pi^\pm \rightarrow \text{vacuum state}$$
$$\pi^\pm \rightarrow \pi^0$$
$$n \rightarrow p$$
$$\Sigma^\pm \rightarrow \Lambda^0 \tag{11.131}$$

where we omit the effect of the leptonic weak current.

16. Similarly, prove that the strangeness-changing charged hadronic weak current $J_{\text{had}}^{\mu,4+i5}$ and its conjugate are responsible for inducing the following reactions:

$$K^\pm \rightarrow \text{vacuum state}$$
$$K^\pm \rightarrow \pi^0$$
$$\Sigma^\pm \rightarrow n$$
$$\Lambda^0 \rightarrow p \tag{11.132}$$

where we omit the effect of the leptonic weak current.

17. Show that the covariant derivatives for the quarks in Eqs. (11.50) and (11.51) yield the correct charge assignments of the quarks after symmetry breaking.

Chapter 12
Ward Identities, BRST, and Anomalies

12.1 Ward–Takahashi Identity

In the case of QED, we found that the Ward–Takahashi identities[1,2] were a powerful way in which to prove important relations between renormalization constants. In this chapter, we will examine these identities from the path integral point of view, and show how they can be generalized to gauge theories with very little extra effort. We also explore perhaps the most convenient way in which to summarize the information contained within the Ward–Takahashi identities, which is the BRST approach. And finally, we will show that these Ward–Takahashi identities actually break down in certain circumstances due to anomalies.

The origin of the WT identities lies in the gauge invariance of the generating functional of QED, which is given by:

$$Z(J_\mu, \bar{\eta}, \eta) = \int DA_\mu\, D\bar{\psi}\, D\psi\, \exp i \int d^4x \left[\mathscr{L}(\psi, A_\mu) + A_\mu J^\mu + \bar{\eta}\psi + \bar{\psi}\eta \right] \tag{12.1}$$

If we make a field redefinition of ψ and A_μ (i.e., make an arbitrary, field-dependent redefinition of the fields), we know that the generating functional $Z(J_\mu, \bar{\eta}, \eta)$ remains the same. This is because changing variables in an integral never affects its value.

We can consider a gauge transformation to be a specific type of field redefinition. Thus, the generating functional is trivially invariant under a gauge transformation. This will allow us to derive nontrivial identities on the generating functional.

We know that the action and the measure are all gauge invariant. The generating functional is also invariant; so the only terms that are not gauge invariant are the gauge-fixing term and the coupling to the sources. The non-gauge-invariant terms, we see, must all vanish because of the overall invariance of the generating functional under a field redefinition.

We begin with the gauge-fixed action:

$$\mathcal{L} = -\frac{1}{4}F_{\mu\nu}^2 + \bar{\psi}(i\,\not{D} - m)\psi - \frac{1}{2\alpha}(\partial \cdot A)^2 \tag{12.2}$$

and the following gauge transformations on the fields:

$$\delta A_\mu \;=\; \partial_\mu \Lambda$$

$$\delta \psi \;=\; -ie\Lambda\psi$$

$$\delta \bar{\psi} \;=\; ie\Lambda\bar{\psi} \tag{12.3}$$

The variation of the generating functional is given as:

$$Z + \delta Z \;=\; \int DA_\mu\, D\bar{\psi}\, D\psi \exp i \int d^4x \left(\mathcal{L}(A_\mu, \psi, \bar{\psi}) + A_\mu J^\mu + \bar{\eta}\psi + \bar{\psi}\eta \right.$$

$$\left. + \partial_\mu \Lambda J^\mu - ie\Lambda(\bar{\eta}\psi - \bar{\psi}\eta) - \frac{\Lambda}{\alpha}\partial^2(\partial \cdot A) \right)$$

$$\delta Z \;=\; \int DA_\mu\, D\bar{\psi}\, D\psi \int \Lambda \left(-\partial_\mu J^\mu + ie(\bar{\eta}\psi - \bar{\psi}\eta) - \frac{1}{\alpha}\partial^2(\partial \cdot A) \right) d^4x$$

$$\times \exp i \int d^4x \left[\mathcal{L}(A_\mu, \bar{\psi}, \psi) + A_\mu J^\mu + \bar{\eta}\psi + \bar{\psi}\eta \right] \tag{12.4}$$

To lowest order in Λ, this can be written as the functional version of the WT identity:

$$i\Lambda \left[-\partial_\mu J^\mu - e\left(\bar{\eta}\frac{\delta}{\delta\bar{\eta}} - \eta\frac{\delta}{\delta\eta} \right) + \frac{i}{\alpha}\partial^2\partial^\mu \frac{\delta}{\delta J^\mu} \right] Z(J_\mu, \bar{\eta}, \eta) = 0 \tag{12.5}$$

where we have made the substitutions:

$$A_\mu \to -i\frac{\delta}{\delta J^\mu}; \quad \psi \to -i\frac{\delta}{\delta\bar{\eta}}; \quad \bar{\psi} \to -i\frac{\delta}{\delta\eta} \tag{12.6}$$

We now have established the nontrivial WT constraint on the generating functional. This identity, however, is not yet written in a form recognizable from our previous discussion of the WT identities. To derive the first consequence of the identity, let us differentiate the expression with respect to $J_\nu(y)$ and then set $J = \eta = \bar{\eta} = 0$. The terms that survive are:

$$\frac{i}{\alpha} \partial^2 \partial^\mu \frac{\delta^2 Z}{\delta J^\mu(x) \delta J_\nu(y)} = \partial^\nu \delta(x - y) \tag{12.7}$$

Now let us take the Fourier transform of the expression. In momentum space, we have:

$$-\frac{1}{\alpha} k^2 k^\mu \Delta_{\mu\nu}(k) = k_\nu \tag{12.8}$$

where $\Delta_{\mu\nu}$ is the connected part of the interacting photon propagator. The general solution to this equation is easy to find in terms of a longitudinal and a transverse part:

$$\Delta_{\mu\nu}(k) = -\alpha k_\mu k_\nu/k^2 + (g_{\mu\nu} - k_\mu k_\nu/k^2) f(k^2) \tag{12.9}$$

for some function $f(k^2)$. This means that the longitudinal part is unchanged by higher-order corrections, as expected.

However, to make contact with the Ward–Takahashi identity found earlier in QED, it is necessary to convert the constraint into one for the proper vertices. We recall first that $Z = e^{iW}$, so we can write the identity as a constraint on W:

$$\left[ie \left(\eta \frac{\delta}{\delta\eta} - \bar{\eta} \frac{\delta}{\delta\bar{\eta}} \right) - \frac{1}{\alpha} \partial^2 \partial^\mu \frac{\delta}{\delta J^\mu} \right] W(J_\mu, \bar{\eta}, \eta) = \partial^\mu J_\mu \tag{12.10}$$

The next step is to convert the identity to a constraint over Γ, the generator of proper vertices. As usual, we now make a Legendre transformation on the fields:

$$\Gamma(A_\mu, \bar{\psi}, \psi) \equiv W(J_\mu, \bar{\eta}, \eta) - \int d^4x \left(A_\mu J^\mu + \bar{\eta}\psi + \bar{\psi}\eta \right) \tag{12.11}$$

Then we can make the following substitutions:

$$\frac{\delta W}{\delta J^\mu} = A_\mu; \qquad \frac{\delta\Gamma}{\delta A^\mu} = -J^\mu$$

$$\frac{\delta W}{\delta\bar{\eta}} = \psi; \qquad \frac{\delta\Gamma}{\delta\psi} = -\bar{\eta}$$

$$\frac{\delta W}{\delta\eta} = \bar{\psi}; \qquad \frac{\delta\Gamma}{\delta\bar{\psi}} = -\eta \tag{12.12}$$

Making these substitutions, we then have the formula for the proper vertices:

$$\partial_\mu \frac{\delta \Gamma}{\delta A_\mu} - ie\left(\psi \frac{\delta \Gamma}{\delta \psi} - \bar{\psi} \frac{\delta \Gamma}{\delta \bar{\psi}}\right) - \alpha^{-1}\partial^2 \partial^\mu A_\mu = 0 \qquad (12.13)$$

This is the form for the Ward–Takahashi identity that yields the various identities found in QED. Notice that by repeated differentiation of this identity, we can derive more and more complicated versions of the WT identity.

As an example, let us take the simplest case, where we differentiate the above equation by $\bar{\psi}(x)$ and $\psi(y)$ and set $\bar{\psi} = \psi = A_\mu = 0$ at the end of the calculation. After differentiation, we find:

$$\left[-\partial^\mu_z \frac{\delta^3}{\delta\bar{\psi}(x)\delta\psi(y)\delta A^\mu(z)} - ie\left(\delta(x-z)\frac{\delta^2}{\delta\bar{\psi}(x)\delta\psi(y)}\right.\right.$$
$$\left.\left. - \delta(y-z)\frac{\delta^2}{\delta\bar{\psi}(x)\delta\psi(y)}\right)\right]\Gamma(0,0,0) = 0 \qquad (12.14)$$

where $\Gamma(0,0,0)$ equals $\Gamma(A_\mu, \bar{\psi}, \psi)$ where all fields have been set to zero.

The Fourier transform of Γ, in turn, can be related to both the electron propagator S_F as well as the proper vertex function Γ_μ via:

$$\int d^4x \int d^4y \, e^{ip'x-ipy} \frac{\delta^2\Gamma(0,0,0)}{\delta\bar{\psi}(x)\delta\psi(y)}$$
$$= (2\pi)^4\delta(p'-p)i\,S_F(p)'^{-1} \qquad (12.15)$$

(where the prime indicates that we are taking the interacting electron propagator) and:

$$\int d^4x\, d^4y\, d^4z\, e^{i(p'x-py-qz)} \frac{\delta^3\Gamma(0,0,0)}{\delta\bar{\psi}(x)\,\delta\psi(y)\,\delta A^\mu(z)}$$
$$= ie(2\pi)^4\delta(p'-p-q)\Gamma_\mu(p,p',q) \qquad (12.16)$$

Taking the Fourier transform of the Ward–Takahashi identity, we have:

$$q^\mu\Gamma_\mu(p,q,p+q) = S_F'^{-1}(p+q) - S_F'^{-1}(p) \qquad (12.17)$$

which is the more familiar form of the identity found in Chapter 7.

If we take the limit as $q_\mu \to 0$, we find:

$$\Gamma_\mu(p,0,p) = \frac{\partial S_F'^{-1}}{\partial p^\mu} \qquad (12.18)$$

12.2 Slavnov–Taylor Identities

As in the case of ordinary QED, we may derive a set of identities on the generating functionals of gauge theory. However, the corresponding identities, called the Slavnov–Taylor identities,[3,4], are much more involved. Later, we will use what is called the BRST formalism in order to simplify the complications found in the Slavnov–Taylor identities.

The Slavnov–Taylor identities are complicated because of two factors: the nonlinear nature of the gauge transformation, and also the presence of the Faddeev–Popov ghosts.

The generating functional (for just the gauge field) can be written as:

$$
Z(J_\mu) = N \int DA_\mu \Delta_{FP} \exp i \int d^4x \left(-\frac{1}{4}(F^a_{\mu\nu})^2 - \frac{1}{2\alpha} F^T F + J^a_\mu A^{a\mu} \right)
$$

$$(12.19)$$

As we saw earlier, Δ_{FP} can be written as the determinant:

$$
\Delta_{FP} = \det (M)_{x,y;ab}
$$

$$(12.20)$$

where the M matrix is defined in terms of the gauge-fixing function $F^a(A_\mu)$:

$$
\delta F^a(A_\mu(x)) = \int d^4y \, M(x, y)^{ab} \Lambda^b(y)
$$

$$(12.21)$$

As before, we know that $Z(J^a_\mu)$ is invariant under a field redefinition. Since a gauge transformation is also a field redefinition, it means that the generating functional is gauge invariant. Thus, $\delta Z = 0$ under this transformation. This means:

$$
\begin{aligned}
0 &= \delta Z(J^a_\mu) = \int DA_\mu \, \Delta_{FP} \exp i \int d^4x \left[\mathscr{L}(A) + J^a_\mu A^{a\mu} \right] \\
&\times \left(-\frac{1}{\alpha} F^a(x) \int d^4y M^{ab}(x, y) \Lambda^b(y) + J_\mu D^\mu \Lambda \right)
\end{aligned}
$$

$$(12.22)$$

For small Λ, we can bring the last term in the exponential into the integrand. Finally, we can choose $\Lambda \to M^{-1}\Lambda$. With this substitution, we now have the new identity:

$$
\left[\frac{1}{\alpha} F^c \left(-i \frac{\delta}{\delta J_\mu} \right) - \int d^4y \, J^a_\mu D^{ab\mu} M^{-1}_{bc} \left(x, y; \frac{\delta}{\delta J_\mu} \right) \right] Z = 0 \qquad (12.23)
$$

This is a rather complicated nonlinear identity, and in general it is quite difficult to extract simple identities on the proper vertices. Also, this identity is rather awkward to work with from the point of view of renormalization theory.

There are, however, some clever tricks that one can use in rendering this problem tractable. The key to this simpler construction is to use the BRST construction.

12.3   BRST Quantization

After fixing a gauge, we know that a gauge theory is no longer gauge invariant. All the local gauge invariances have been removed from the theory. Thus, it is rather surprising that, even after gauge fixing has been performed, a new symmetry arises involving the Faddeev–Popov ghosts. This new symmetry, however, is a global one, and hence no new degrees of freedom can be eliminated from the gauge-fixed theory.

We recall that the gauge-fixed action is given by:

$$\mathscr{L} = -\frac{1}{4}(F^a_{\mu\nu})^2 - \frac{1}{2\alpha}(\partial \cdot A)^2 - \bar{\eta}^a \partial^\mu D_\mu \eta^a \tag{12.24}$$

The original action, of course, was invariant under:

$$\delta A^a_\mu = \frac{1}{g}\partial_\mu \Lambda^a + f^{abc} A^b_\mu \Lambda^c \tag{12.25}$$

Now make the replacement:

$$\Lambda^a = -\eta^a \lambda \tag{12.26}$$

where η^a and λ are both Grassmann variables and λ is constant. Then the gauge-fixed action is invariant under a new global symmetry[5]:

$$\begin{aligned}
\delta A^a_\mu &= -\frac{1}{g}(D_\mu \eta^a)\lambda \\
\delta \eta^a &= -\frac{1}{2} f^{abc} \eta^b \eta^c \lambda \\
\delta \bar{\eta}^a &= -\frac{1}{\alpha g}(\partial^\mu A^a_\mu)\lambda
\end{aligned} \tag{12.27}$$

To prove the invariance of the action, we note that:

$$\delta \mathscr{L}_{GF} = \frac{1}{\alpha g} \left(\partial^\mu A_\mu^a \right) \left(\partial^\nu D_\nu \eta^a \right) \lambda \tag{12.28}$$

and:

$$\delta \mathscr{L}_{FP} = - \left(\delta \bar{\eta}^a \right) \partial^\mu D_\mu \eta^a - \bar{\eta}^a \partial^\mu \delta \left(D_\mu \eta^a \right) \tag{12.29}$$

Adding these together, we find:

$$\delta \mathscr{L} = -\bar{\eta}^a \partial^\mu \left(\delta D_\mu \eta^a \right) \tag{12.30}$$

We can use also prove that:

$$\delta (D_\mu \eta^a) = \delta (f^{abc} \eta^b \eta^c) = 0 \tag{12.31}$$

where we have used the Jacobi identity on the structure constants:

$$f^{abk} f^{kde} + f^{adk} f^{keb} + f^{aek} f^{kbd} = 0 \tag{12.32}$$

Thus, the action is BRST invariant.

Since the original variation was nilpotent, one can also show that:

$$\delta^2_{BRST} = 0 \tag{12.33}$$

Using Noether's method, we can also construct the current that corresponds to the BRST variation. Using the Noether prescription, we find:

$$\begin{aligned} J_\mu &= \sum_j \frac{\delta \mathscr{L}}{\delta \partial_\mu \phi_j} \frac{\delta_{BRST} \phi_j}{\delta \lambda} \\ &= \left(-F_{\mu\nu}^a D^\nu \eta^a - \frac{g}{2} \partial_\mu \bar{\eta}^a f^{abc} \eta^b \eta^c \right) \end{aligned} \tag{12.34}$$

From the Noether current, we can also construct the BRST charge:

$$Q_{BRST} = \int J_0 \, d^3 x \tag{12.35}$$

which satisfies the nilpotency condition:

$$Q^2_{BRST} = 0 \tag{12.36}$$

In general, the states of the theory are constructed from all possible monomials that one can construct out of η^a and $\bar{\eta}^a$. Thus, the Fock space has increased enormously with the presence of these ghosts and antighosts. However, one can show that the physical state condition (similar to the Gupta–Bleuler condition) is given by[6]:

$$Q_{BRST}|\Psi\rangle = 0 \qquad (12.37)$$

The states that satisfy this condition are the physical states of the system. We now have a compact and elegant statement of the physical state condition.

12.4 Anomalies

Because of the subtle manipulations that must be performed on potentially divergent quantities when we renormalize a theory, there may be unexpected surprises. One of these is the existence of Adler–Bardeen–Jackiw (ABJ) anomalies.[7,8]

An anomaly is the failure of a classical symmetry to survive the process of quantization and regularization. For example, in a chiral gauge theory, we naively expect axial currents to be conserved. However, we will find that actions that are classically chiral symmetric can develop anomalies that spoil the conservation of the axial current.

If we start with a gauge theory that naively is invariant under axial gauge symmetry:

$$\psi \rightarrow e^{i\epsilon(x)\gamma_5}\psi \qquad (12.38)$$

then we can define:

$$
\begin{aligned}
V_\mu(x) &= \bar{\psi}(x)\gamma_\mu\psi(x) \\
A_\mu(x) &= \bar{\psi}(x)\gamma_\mu\gamma_5(x) \\
P(x) &= \bar{\psi}(x)\gamma_5\psi(x)
\end{aligned}
\qquad (12.39)
$$

Using the naive equations of motion, we can easily show:

$$
\begin{aligned}
\partial^\mu V_\mu &= 0 \\
\partial^\mu A_\mu &= 2im\,P(x)
\end{aligned}
\qquad (12.40)
$$

The last equation vanishes in the limit of zero mass, that is, when chiral symmetry is restored. It appears as if we have an exact conservation of both the vector current and the axial current in the zero mass limit. However, we will see that this current conservation is anomalous, that the divergence of the axial current is not equal to zero, even in the zero mass limit.

Specifically, we will examine the "triangle graph," which consists of an internal fermion loop connected to two vector fields and to one axial vector field. This is appropriately called the V-V-A triangle graph.

If we perform power counting on this graph, we find that the integration over d^4k gives us four powers of momentum in the numerator, but the fermionic propagators only give us three powers of momentum in the denominator. Thus, the graph should diverge linearly.

The origin of this anomaly is rather subtle. In performing the integration over the loop variable, we will cancel certain graphs by performing a shift of the integration variable. Normally, one expects that integrals like this vanish:

$$\int_{-\infty}^{\infty} dx \, [f(x+a) - f(x)] = 0 \qquad (12.41)$$

because, by shifting $x + a \to x$, we get an exact cancellation. However, we have tacitly made certain unjustified assumptions.

To see how this integral may not vanish, let us power expand it:

$$\int_{-\infty}^{\infty} dx \left(af'(x) + \frac{a^2}{2} f''(x) + \cdots \right)$$

$$= a[f(\infty) - f(-\infty)] + \frac{a^2}{2}[f'(\infty) - f'(-\infty)] + \cdots \qquad (12.42)$$

If the integral of f converges, then there is no problem in setting the above equal to zero. However, if the integral diverges linearly, then Eq. (12.41) need not vanish. In fact, it can equal $a[f(\infty) - f(-\infty)]$.

We can also generalize this ambiguity to arbitrary (Euclidean) dimensions. Let us define the function:

$$\Delta(a) \equiv \int d^N x \, [f(x+a) - f(x)]$$

$$= \int d^N x \left[a^\mu \partial_\mu f(x) + \frac{1}{2} (a^\mu \partial_\mu)^2 f(x) + \cdots \right]$$

$$= a^\mu \frac{X_\mu}{R} f(R) S_N(R) \qquad (12.43)$$

In performing the integral over the volume element $d^N x$, we used Gauss's theorem to drop all but the first term in the expansion. In the last line, the volume integral reduces to a surface integral over a large hypersphere with radius R, surface area

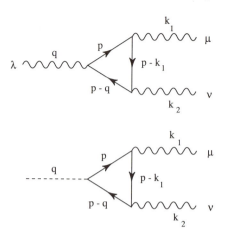

Figure 12.1. The V-V-A and V-V-P triangle graphs, which give rise to the anomaly.

$S_N(R)$, labeled by the vector X_μ. For the case of four dimensions, we can perform the integral, and take the limit as the hypersphere's radius R expands to infinity. The result is:

$$\Delta(a) = \lim_{R \to \infty} (2i\pi^2) a^\mu X_\mu R^2 f(R) \qquad (12.44)$$

Now that we see the inherent ambiguity in shifting the integration variable in a linearly divergent integral, let us apply this knowledge to gauge theory. To begin, let us examine the following two matrix elements, corresponding to the V-V-A and V-V-P triangle (Fig. 12.1):

$$T_{\mu\nu\rho}(k_1, k_2, q) = i \int d^4x_1 \, d^4x_2 \langle 0|T\left[V_\mu(x_1)V_\nu(x_2)A_\rho(0)\right]|0\rangle e^{ik_1 \cdot x_1 + ik_2 \cdot x_2}$$

$$T_{\mu\nu}(k_1, k_2, q) = i \int d^4x_1 \, d^4x_2 \langle 0|T\left[V_\mu(x_1)V_\nu(x_2)P(0)\right]|0\rangle e^{ik_1 \cdot x_1 + ik_2 \cdot x_2}$$

$$(12.45)$$

Our next step is to differentiate the previous expressions. This will pull down a factor of q^ρ, but there are complications when we take the derivative of a time-ordered product, which contains theta functions in the time variable. In particular, we have:

$$\partial_\mu \theta(x^0 - y^0) = \delta_{\mu 0}\delta(x^0 - y^0) \qquad (12.46)$$

This means that:

$$\partial_x^\mu T\left(V_\mu(x)O(y)\right) = T\left(\partial^\mu V_\mu(x)O(y)\right) + [V_0(x), O(y)]\delta(x^0 - y^0) \quad (12.47)$$

for an arbitrary operator $O(y)$.

Taking different derivatives of the matrix element, we easily derive:

$$k_1^\mu T_{\mu\nu\rho} = 0$$

$$k_2^\nu T_{\mu\nu\rho} = 0$$

$$q^\rho T_{\mu\nu\rho} = 2m T_{\mu\nu} \quad (12.48)$$

Written out explicitly using Feynman rules, we find that the matrix elements can be written as:

$$\begin{aligned}
T_{\mu\nu\rho}(a) &= -\int \frac{d^4p}{(2\pi)^4} \,\mathrm{Tr}\left[\frac{i}{\not p + \not q - m}\gamma_\rho\gamma_5\frac{i}{\not p + \not q - \not q - m}\right.\\
&\quad \times \left. \gamma_\nu\frac{i}{\not p + \not q - \not k_1 - m}\gamma_\mu\right]\\
&\quad + \left\{k_1 \leftrightarrow k_2;\ \mu \leftrightarrow \nu\right\} \quad (12.49)
\end{aligned}$$

It is important to notice that we have explicitly made a shift $p \to p + a$ in performing the integral where $a = \alpha k_1 + (\alpha - \beta)k_2$, where α and β are arbitrary. Normally, for convergent integrals, $T_{\mu\nu\rho}(a)$ is independent of a by shifting the region of integration. However, we now see that $T_{\mu\nu\rho}(a)$ is linearly divergent, and hence inherently ambiguous. We can, of course, explicitly calculate the value of $T_{\mu\nu\rho}(a) - T_{\mu\nu\rho}(0)$ using the formula derived earlier.

Dropping the cross term for the moment, we find:

$$T_{\mu\nu\rho}(a) - T_{\mu\nu\rho}(0)$$

$$\begin{aligned}
&= -\int \frac{d^4p}{(2\pi)^4} a^\lambda \frac{\partial}{\partial p_\lambda} \mathrm{Tr}\left(\frac{1}{\not p - m}\gamma_\rho\gamma_5\frac{1}{\not p - \not q - m}\gamma_\nu\frac{1}{\not p - \not k_1 - m}\gamma_\mu\right) + \cdots\\
&= -i\frac{2\pi^2 a^\lambda}{(2\pi)^4} \lim_{p\to\infty} p^2 p_\lambda \mathrm{Tr}\left(\gamma_\alpha\gamma_\rho\gamma_5\gamma_\beta\gamma_\nu\gamma_\delta\gamma_\mu\right)p^\alpha p^\beta p^\delta / p^6 + \cdots\\
&= \frac{2i\pi^2 a_\lambda}{(2\pi)^4} \lim_{p\to\infty} \frac{p^\lambda p^\delta}{p^2} 4i\epsilon_{\mu\nu\rho\delta} + \left\{k_1 \leftrightarrow k_2; \mu \leftrightarrow \nu\right\}\\
&= \epsilon_{\sigma\mu\nu\rho}a^\sigma / 8\pi^2 + \left\{k_1 \leftrightarrow k_2; \mu \leftrightarrow \nu\right\} \quad (12.50)
\end{aligned}$$

This then gives us:

$$T_{\mu\nu\rho}(a) - T_{\mu\nu\rho}(0) = -\frac{\beta}{8\pi^2}\epsilon_{\mu\nu\rho\sigma}(k_1 - k_2)^\sigma \tag{12.51}$$

where β is a constant that is not yet determined and is inherently ambiguous.

Finally, the anomalous Ward–Takahashi identity can be written as:

$$q^\rho T_{\mu\nu\rho} = 2m T_{\mu\nu} - \frac{1-\beta}{4\pi^2}\epsilon_{\mu\nu\sigma\rho}k_1^\sigma k_2^\rho \tag{12.52}$$

This equation, which expresses the divergence of the axial current, implies that axial current conservation is anomalous.

At this point, the value of β is arbitrary. β, in turn, can be calculated by the fact that we would like to preserve the vector current conservation. Thus, we demand that $k_1^\mu T_{\mu\nu\rho} = k_2^\nu T_{\mu\nu\rho} = 0$, even though they, too, are anomalous. However, demanding that the vector current be exactly conserved serves to fix the ambiguity in β.

To fix the value of β, we now calculate the anomaly coming from the vector current and then set it equal to zero.

We must thus calculate:

$$\begin{aligned}
k_1^\mu T_{\mu\nu\rho}(0) &= (-1)\int \frac{d^4p}{(2\pi)^4}\left[\text{Tr}\left(\frac{1}{\not{p}-m}\gamma_\rho\gamma_5\frac{1}{\not{p}-\not{q}-m}\gamma_\nu\frac{1}{\not{p}-\not{k}_1-m}\not{k}_1\right)\right.\\
&\quad + \left.\text{Tr}\left(\frac{1}{\not{p}-m}\gamma_\rho\gamma_5\frac{1}{\not{p}-\not{q}-m}\not{k}_1\frac{1}{\not{p}-\not{k}_2-m}\gamma_\nu\right)\right]
\end{aligned} \tag{12.53}$$

Using the identity:

$$\begin{aligned}
\not{k}_1 &= (\not{p}-m) - [(\not{p}-\not{k}_1)-m]\\
&= [(\not{p}-\not{k}_2)-m] - [(\not{p}-\not{q})-m]
\end{aligned} \tag{12.54}$$

we can write the expression as the difference between two shifted integrands, which in turn allows us to write everything in terms of a limit on a hypersphere:

$$\begin{aligned}
k_1^\mu T_{\mu\nu\rho} &= (-1)\int \frac{d^4p}{(2\pi)^4}\text{Tr}\left(\gamma_\rho\gamma_5\frac{1}{(\not{p}-\not{q})-m}\gamma_\nu\frac{1}{(\not{p}-\not{k}_1)-m}\right)\\
&\quad - \text{Tr}\left(\gamma_\rho\gamma_5\frac{1}{(\not{p}-\not{k}_2)-m}\gamma_\nu\frac{1}{\not{p}-m}\right)\\
&= \frac{k_1^\sigma}{(2\pi)^4}\int d^4p\,\frac{\partial}{\partial p_\sigma}\left(\frac{\text{Tr}\,[\gamma_\rho\gamma_5(\not{p}-\not{k}_2+m)\gamma_\nu(\not{p}+m)]}{[(p-k_2)^2-m^2](p^2-m^2)}\right)
\end{aligned}$$

$$= \frac{2i\pi^2 k_1^\sigma}{(2\pi)^4} \lim_{p\to\infty} \frac{p_\sigma}{p^2} \mathrm{Tr}\,(\gamma_5 \gamma_\rho \gamma_\alpha \gamma_\nu \gamma_\beta) k_2^\alpha p^\beta$$

$$= -\frac{1}{8\pi^2} \epsilon_{\rho\sigma\nu\tau} k_1^\tau k_2^\sigma \tag{12.55}$$

This means that:

$$k_1^\mu T_{\mu\nu\rho}(\beta) = \frac{1+\beta}{8\pi^2} \epsilon_{\nu\rho\sigma\tau} k_1^\sigma k_2^\tau \tag{12.56}$$

The key point is now this: For arbitrary values of β, it is impossible to keep both vector current conservation and axial vector conservation. We will keep the vector current conserved and push the anomaly entirely onto the axial current conservation.

With the choice $\beta = -1$, we find:

$$q^\rho T_{\mu\nu\rho} = 2m T_{\mu\nu} - \frac{1}{2\pi^2} \epsilon_{\mu\nu\sigma\tau} k_1^\sigma k_2^\tau \tag{12.57}$$

Written in x space, the ABJ anomaly can be summarized as follows:

$$\partial^\mu A_\mu = 2im P(x) + \frac{1}{8\pi^2} \tilde{F}_{\mu\nu} F^{\mu\nu} \tag{12.58}$$

where $\tilde{F}_{\mu\nu} = \frac{1}{2} \epsilon_{\mu\nu\alpha\beta} F^{\alpha\beta}$.

12.5 Non-Abelian Anomalies

For the non-Abelian case,[9] we must study the following anomalous V-V-A graph:

$$T_{\mu\nu\lambda}^{abc}(k_1, k_2, q) = i \int d^4 x_1\, d^4 x_2 \langle 0|T\left[V_\mu^a(x_1)V_\nu^b(x_2)A_\lambda^c(0)\right]|0\rangle e^{ik_1\cdot x_1 + ik_2\cdot x_2} \tag{12.59}$$

The anomalous Ward–Takahashi identity becomes:

$$q^\rho T_{\mu\nu\rho}^{abc} = 2m T_{\mu\nu}^{abc} - \frac{1}{2\pi^2} \epsilon_{\mu\nu\rho\sigma} k_1^\rho k_2^\sigma D^{abc} + \cdots \tag{12.60}$$

where:

$$D^{abc} = \frac{1}{2} \mathrm{Tr}\left(\{\tau^a, \tau^b\}\tau^c\right) \tag{12.61}$$

(The anticommutator in the trace comes from the fact that we must add two triangle diagrams together to produce the anomaly. The only difference between these two diagrams is that the a and b lines are interchanged; so this explains the anticommutator.)

The origin of the axial anomaly, however, has much deeper significance at the quantum level, persisting for every possible regularization scheme. The method of Pauli–Villars regularization, for example, violates chiral symmetry, because we have explicitly added a massive fermion into the theory. Thus, a theory that is classically chiral invariant does not necessarily maintain chiral invariance if we use the Pauli–Villars regularization method.

This anomaly persists even if we use other regularization schemes. For example, in the dimensional regularization scheme, there is no higher-dimensional counterpart of γ_5, so we expect that dimensional regularization will also spoil chiral symmetry.

In Chapter 15, we will discuss yet another regularization scheme, putting space–time on a discrete lattice. In contrast to the previous regularization schemes, lattice regularization does preserve chiral invariance at every step of the transition from the classical theory to the quantum theory. Putting fermions on a lattice does not spoil chiral symmetry at all. Then, the theory is chirally invariant even as we perform the quantization program. (However, there is still a catch to this, as we shall see.)

12.6 QCD and Pion Decay into Gamma Rays

One of the earliest discoveries in this area was the realization that these anomalies may actually solve the $\pi \rightarrow 2\gamma$ puzzle. Historically, it was noticed that π meson decay into two photons was not occurring with the expected rate. However, by correcting for the presence of an anomaly, we can obtain the experimentally observed decay rate. (The presence of the anomaly does not necessarily spoil renormalization, because here there is no Ward–Takahashi identity that is destroyed, since there is no local conserved axial current.)

The Feynman graph that mediates pion decay into two photons is a triangle graph, in which the two photons couple to an internal fermion loop via two currents J_μ. The pion also couples to this internal fermion loop, but, because of PCAC, the pion couples via the axial hadronic current. Thus, we have the classic V-V-A triangle, which we know is anomalous.

The decay of a pion of momentum p into two photons of momenta k_1 and k_2 is denoted by:

$$\pi(p) \rightarrow \gamma(k_1) + \gamma(k_2) \tag{12.62}$$

and governed by the following matrix element:

$$\langle \gamma(k_1, \epsilon_1), \gamma(k_2, \epsilon_2) | \pi^0(p) \rangle \quad = \quad i(2\pi)^4 \Gamma_{\mu\nu}(p, k_1, k_2)$$

$$\times \quad \delta^4(p - k_1 - k_2) \epsilon_1^\mu(k_1) \epsilon_2^\nu(k_2) \quad (12.63)$$

The tensor $\Gamma_{\mu\nu}$, in turn, is given by:

$$\Gamma_{\mu\nu}(p, k_1, k_2) = e^2 \int d^4x \, d^4y \, \langle 0 | T \left[J_\mu(x) J_\nu(y) \right] | \pi^0 \rangle e^{ik_1 x + ik_2 y} \qquad (12.64)$$

where J_μ is the electromagnetic current.

By Lorentz invariance, we know that the only tensors that we can use to construct this matrix element are k_1, k_2, and $\epsilon_{\mu\nu\lambda\rho}$. Since the pion is a pseudoscalar particle, we must choose:

$$\Gamma_{\mu\nu}(p, k_1, k_2) = i \epsilon_{\mu\nu\sigma\rho} k_1^\sigma k_2^\rho \Gamma(p^2) \qquad (12.65)$$

Next, we can use LSZ methods to reduce out the pion field appearing in the state vector $|\pi(p)\rangle$. Then we use PCAC, given by $\partial^\lambda A_\lambda^a = f_\pi m_\pi^2 \pi^a$, to replace the pion field with the divergence of the axial current. Then our tensor becomes:

$$\Gamma_{\mu\nu}(p, k_1, k_2) \quad = \quad \frac{i e^2 (q^2 - m_\pi^2)}{f_\pi m_\pi^2} \int d^4x \, d^4y \, e^{ik_2 \cdot y - ip \cdot x}$$

$$\times \quad \langle 0 | T \left[\partial^\lambda A_\lambda^3(x) J_\nu(y) J_\mu(0) \right] | 0 \rangle \qquad (12.66)$$

(Because we are analyzing the π^0 field, we must use the third component of isospin A_μ^3 in the PCAC relations.)

Our goal is to derive a low-energy theorem on this matrix element. To do this, let us define a new matrix element, which will prove useful in our discussion:

$$\Gamma_{\mu\nu\lambda}(p, k_1, k_2) = \int d^4x \, d^4y \, e^{ik_2 \cdot y - ip \cdot x} \langle 0 | T \left[A_\lambda^3(x) J_\nu(y) J_\mu(0) \right] | 0 \rangle \qquad (12.67)$$

The trick is to find a relationship between our $\Gamma_{\mu\nu}$ and the new $\Gamma_{\mu\nu\lambda}$ that we have just written.

Next, we will hit this tensor with p^λ. Contracting the left-hand side with p^λ is equivalent to taking the derivative with respect to x of the right-hand side. However, taking the x derivative of the right-hand side will pick up the derivative of a θ function appearing in the time-ordered product, which in turn will yield delta functions. Performing the derivative, we find:

$$
\begin{aligned}
p^\lambda T_{\mu\nu\lambda}(p, k_1, k_2) \;=\; & -i \int d^4x \, d^4y \, e^{ik_2 \cdot y - ip \cdot x} \\
& \times \Big(\langle 0 | T \left[\partial^\lambda A_\lambda^3(x) J_\nu(y) J_\mu(0) \right] | 0 \rangle \\
& + \langle 0 | T \left\{ \delta(x_0 - y_0) \left[A_0^3(x), J_\nu(y) \right] J_\mu(0) \right\} | 0 \rangle \\
& + \langle 0 | T \left\{ \delta(x_0) [A_0^3(x), J_\mu(0)] J_\nu(y) \right\} | 0 \rangle \Big) \qquad (12.68)
\end{aligned}
$$

In the limit as $p \to 0$, the left-hand side of the equation vanishes. Also, the two commutators on the right-hand side of the equation also vanish, using the current algebra relations. Thus, all terms have vanished except $\Gamma_{\mu\nu}$, which therefore must also vanish. This means that the entire equation has collapsed, showing that π can never decay to two photons in this limit, which violates the experimental data. This problem can be resolved by noting that the Feynman graph that dominates this process to lowest order is the V-V-A triangle graph, which we know is anomalous.

Inserting the anomaly back into the previous relationship, we therefore have:

$$
\lim_{p \to 0} \Gamma_{\mu\nu}(p, k_1, k_2) = \frac{ie^2 D}{2\pi^2 f_\pi} \epsilon_{\mu\nu\sigma\rho} k_1^\sigma k_2^\rho \qquad (12.69)
$$

where the value of D depends on the fermions moving within the triangle graph.

Comparing this with our previous Lorentz decomposition of this tensor, we therefore have:

$$
\Gamma(0) = \frac{e^2 D}{2\pi^2 f_\pi} \qquad (12.70)
$$

Now let us calculate D. To lowest order, we can assume that the naive quark model is correct. Using free-field representations of the currents, we find that the electromagnetic current is related to the charge Q matrix by the following:

$$
J_\mu(x) = \bar{q}(x) \gamma_\mu Q q(x) \qquad (12.71)
$$

where:

$$
Q = \text{diag}\,(2/3, -1/3, -1/3) \qquad (12.72)
$$

and that the axial current is given by:

$$
A_\mu^3(x) = \bar{q}(x) \gamma_\mu \gamma_5 \frac{\lambda^3}{2} q(x) \qquad (12.73)
$$

where:

$$\lambda_3 = \text{diag}\,(1, -1, 0) \tag{12.74}$$

Inserting these expressions back into the value for D, we find:

$$D = \frac{N}{2}\text{Tr}\,\left(\{Q, Q\}\frac{\lambda_3}{2}\right) = \frac{N}{6} \tag{12.75}$$

where N is the number of colored quarks. Assuming $N = 3$, this gives us a value of:

$$\Gamma(0) = 0.037\,m_\pi^{-1} \tag{12.76}$$

This is to be compared with the experimental value:

$$\Gamma(m_\pi^2) = 0.0375\,m_\pi^{-1} \tag{12.77}$$

if we have three colors of quarks.

Yet another check on the Standard Model is the fact that the anomaly contribution of the leptonic and hadronic sectors of the Weinberg–Salam model just cancel each other. The leptonic sector of the Weinberg–Salam model, by itself, is not renormalizable because of the chiral anomaly. However, the true anomaly is the sum of the anomalies coming from the leptonic and hadronic sectors of the Glashow–Weinberg–Salam model, and these cancel perfectly, giving us confidence once again of the correctness of the Standard Model.

To see how this works, let us calculate the anomaly contribution from the leptonic sector of the Weinberg–Salam model. In particular, the calculation simplifies if we just calculate the anomaly coming from the coupling of the Z_0 with W^+ and W^-. (This WWZ triangle graph appears, for example, in neutrino–neutrino scattering, where a triangle graph is exchanged between the two neutrinos.)

Since right-handed fermions do not couple to the W vector meson, we are only interested in the left-hand anomaly:

$$\text{Tr}\,\tau^a\{\tau^b, \tau^c\}_L \tag{12.78}$$

For the W mesons, the isospin coupling is easy to find, since they couple to fermions via $\tau_\pm$:

$$\tau^b \sim \tau_+; \quad \tau^c \sim \tau_- \tag{12.79}$$

To find the contribution from the Z vertex is a bit more complicated, but it can be read off the Lagrangian using Eqs. (10.62) and (10.66). The Z gives an isospin coupling of:

$$\tau_c \sim \sec\theta_W(\tau_3 + \sin^2\theta_W Q) \tag{12.80}$$

Now let us insert everything into the anomaly:

$$\text{Anomaly} \sim \text{Tr}\left[(\tau_3 + \sin^2\theta_W Q)\{\tau_+, \tau_-\}\right] \tag{12.81}$$

All terms in the trace vanish, except the one containing the charge Q, so we have:

$$\text{Anomaly} \sim \sum_i Q_L^i \tag{12.82}$$

In other words, the sum of the left-handed changes must sum to zero. However, it is easy to see that the sum of the electron and neutrino charge does not vanish, and hence the Weinberg–Salam model, for the leptons, is not renormalizable. In other words, the leptonic sector by itself is not self-consistent.

In the Standard Model, however, we add the contribution of both the leptonic and the hadronic sector. The right-handed quarks do not couple to the W meson, so we only have to sum the contributions of the charges of the left-handed quarks. The sum of the two sectors is given by:

$$\text{Anomaly} \sim Q(e) + Q(\nu) + 3\left[Q(u) + Q(d)\right] = -1 + 0 + 3\left(\frac{2}{3} - \frac{1}{3}\right) = 0 \tag{12.83}$$

Thus, for one generation of quarks and leptons, we have an exact cancellation. This result is also welcomed, because it helps to explain the rough symmetry in the number of leptons and quarks that have been discovered over the years. Every time a new lepton was discovered, a new quark would be discovered soon afterwards, and vice versa. From this point of view, we need an exact balancing between the lepton and quark sector to give us a renormalizable, anomaly-free theory. (However, this still does not explain why leptons and quarks come in three distinct generations.)

12.7 Fujikawa's Method

There is another method of obtaining the anomaly that is much simpler and more conceptually intuitive using path integrals.[10] We notice that the anomaly arises because of a failure of the regularization scheme to accommodate the axial current conservation. Thus, we might expect the failure of the symmetry to take place at a more fundamental level, such as the quantum measure.

Under the chiral gauge transformation:

$$\psi \rightarrow e^{i\theta\gamma_5}\psi$$

$$\bar{\psi} \rightarrow \bar{\psi}e^{i\theta\gamma_5} \tag{12.84}$$

we wish to calculate the change both in the action as well as in the functional measure.

The action transforms as:

$$\int d^4x \; \bar{\psi} i \; \not{D}\psi \rightarrow \int d^4x \; \bar{\psi} i \; \not{D}\psi - \int d^4x \; \theta(x)\partial_\mu J_5^\mu \tag{12.85}$$

where the axial current is given by:

$$J_5^\mu = \bar{\psi}\gamma^\mu\gamma_5\psi \tag{12.86}$$

The measure transforms as:

$$D\psi \; D\bar{\psi} \rightarrow \det\left(e^{i\theta(x)\gamma_5}\right) D\psi \; D\bar{\psi} \tag{12.87}$$

Normally, we discard the determinant because it appears to be a constant. However, closer analysis of this term shows that it is actually divergent, and hence requires regularization. This process of regularizing the determinant, in turn, will generate the anomaly. To determine the value of the determinant carefully, let us introduce a complete set of eigenfunctions ϕ_n of the operator $\not{D}$:

$$\not{D}\phi_n(x) = \lambda_n\phi_n(x) \tag{12.88}$$

We assume that the eigenvalues λ_n are all discrete, although this is not necessary. We will normalize these eigenfunctions as follows:

$$\int d^4x \; \phi_n^\dagger(x)\phi_m(x) = \delta_{nm} \tag{12.89}$$

Then the Dirac spinor can be decomposed in terms of this complete set of eigenfunctions:

$$\psi(x) = \sum_n a_n\phi_n(x); \quad \bar{\psi} = \sum_n \phi_n^\dagger(x)\bar{b}_n \tag{12.90}$$

The functional measure can be rewritten as differentials over da_n and $d\bar{b}_n$:

$$D\psi \; D\bar{\psi} \rightarrow \prod_n da_n \prod_m d\bar{b}_m \tag{12.91}$$

We are now in a position to determine the determinant in the functional measure. The transformation of the field variables is now written as:

$$\psi'(x) = e^{i\theta\gamma_5}\psi \rightarrow \sum_n a'_n\phi_n = e^{i\theta\gamma_5}\sum_m a_m\phi_m \tag{12.92}$$

Let us multiply both sides by $\phi_n^\dagger$ and integrate over x. Then we find:

$$a'_n = \sum_m C_{nm}a_m$$

$$C_{nm} \equiv \int d^4x\ \phi_n^\dagger(x)e^{i\theta(x)\gamma_5}\phi_m(x) \tag{12.93}$$

Thus, the change in functional measure is given by:

$$\prod_m da'_m = \det(C_{nm})^{-1}\prod_n da_n \tag{12.94}$$

If the determinant of $e^{i\theta\gamma_5}$ were equal to one, then the functional measure would be invariant under a chiral transformation. However, a careful analysis shows that this determinant is not equal to one and, in fact, is potentially divergent. (The determinant occurs with exponent minus one because we are dealing with Grassmann variables, not ordinary c numbers.)

For small $\theta(x)$, we can make some approximations and rewrite the determinant factor as:

$$
\begin{aligned}
\det(C_{nm})^{-1} &= \det\left(\delta_{nm} + i\int\theta(x)\phi_n^\dagger(x)\gamma_5\phi_m(x)\,dx\right)^{-1}\\
&= \exp\left(-i\sum_n\int dx\,\theta(x)\phi_m^\dagger(x)\gamma_5\phi_m(x)\right)\\
&= \exp\left(-i\int dx\,\theta(x)A(x)\right)
\end{aligned} \tag{12.95}
$$

where we have defined:

$$A(x) \equiv \sum_n \phi_n^\dagger(x)\gamma_5\phi_n \tag{12.96}$$

Since the $D\bar{\psi}$ yields the same determinant, we find that the overall measure transforms as:

$$D\psi\,D\bar{\psi} \rightarrow \exp\left(-2i\int d^4x\,\theta(x)A(x)\right)D\psi\,D\bar{\psi} \tag{12.97}$$

Written in this fashion, the determinant in the functional measure is actually divergent, and hence must be regularized. This process of regularization, in turn, will generate the anomaly, since the axial current conservation cannot be maintained by any regularization scheme.

To regularize this sum, we will find it convenient to introduce the convergent factor $\exp -(\lambda_n/M)^2$ and take the limit as $M \to \infty$. Inserting this converging factor into the sum, we have:

$$
\begin{aligned}
A(x) &= \lim_{M\to\infty} \sum_n \phi_n^\dagger(x)\gamma_5 e^{-(\lambda_n/M)^2} \phi_n(x) \\
&= \lim_{M\to\infty} \sum_n \phi_n^\dagger(x)\gamma_5 e^{-(\not{D}/M)^2} \phi_n(x)
\end{aligned}
\tag{12.98}
$$

where we have replaced λ_n with $\not{D}$.

Since we are taking the trace with respect to ϕ_n, we are free to change the basis of the trace. Using Eq. (8.23), we can change the basis to $|k\rangle$ eigenstates instead, as follows:

$$
\begin{aligned}
\phi_n(x) &= \langle x|n\rangle = \langle x|k\rangle \int d^4k\, \langle k|n\rangle \\
&= e^{-ik\cdot x} \int \frac{d^4k}{(2\pi)^2} \langle k|n\rangle
\end{aligned}
\tag{12.99}
$$

Then the trace of a arbitrary matrix $\mathcal{M}$ can be expressed as:

$$
\begin{aligned}
\mathrm{Tr}\,\mathcal{M}(x) &= \sum_n \phi_n^\dagger(x)\mathcal{M}(x)\phi_n(x) \\
&= \sum_n \langle n|x\rangle \mathcal{M}(x)\langle x|n\rangle \\
&= \sum_n \langle n|k\rangle \int d^4k \langle k|x\rangle \mathcal{M}(x)\langle x|k'\rangle \int d^4k' \, \langle k'|n\rangle \\
&= \int \frac{d^4k}{(2\pi)^4} e^{ik\cdot x} \mathcal{M}(x) e^{-ik\cdot x}
\end{aligned}
\tag{12.100}
$$

where we have removed the sum over n because $1 = \sum_n |n\rangle\langle n|$.

The trace over γ_5 can now be written as:

$$
A(x) = \lim_{M\to\infty} \mathrm{Tr} \int \frac{d^4k}{(2\pi)^4} e^{ik\cdot x} \gamma_5 e^{-(\not{D}/M)^2} e^{-ik\cdot x}
\tag{12.101}
$$

Next, we must decompose $(\not{D})^2$. Because D_μ is an operator, we must be careful in handling this expression. This factor can be decomposed into an odd piece proportional to $[\gamma^\mu, \gamma^\nu]$ and an even piece proportional to $\{\gamma^\mu, \gamma^\nu\} = 2g^{\mu\nu}$.

The odd piece, in turn, is proportional to $[D_\mu, D_\nu]$, which gives us $F_{\mu\nu}$. Putting everything together, we now have:

$$
\begin{aligned}
A(x) &= \lim_{M \to \infty} \text{Tr} \int \frac{d^4 k}{(2\pi)^4} \gamma_5 \exp \\
&\quad \times \left(-\frac{1}{2M^2} \left\{ \left[ik_\mu + D_\mu(x) \right]^2 + [\gamma^\mu, \gamma^\nu] F_{\mu\nu}(x) \right\} \right) \\
&= \lim_{M \to \infty} \text{Tr}\, \gamma_5 \left([\gamma^\mu, \gamma^\nu] F_{\mu\nu} \right)^2 \left(\frac{1}{2M^2} \right)^2 \frac{1}{2!} \int \frac{d^4 k}{(2\pi)^4} e^{-k^2/M^2} \\
&= -\frac{1}{16\pi^2} \text{Tr}\, F^{\mu\nu} \tilde{F}_{\mu\nu}
\end{aligned}
\tag{12.102}
$$

In conclusion, we find that the trace of γ_5 can be written as:

$$
A(x) = \text{Tr}\, (\gamma_5) = -\frac{1}{16\pi^2} \text{Tr}\, F^{\mu\nu} \tilde{F}_{\mu\nu}
\tag{12.103}
$$

Now let us put the total variation of the action and the measure together. From Eqs. (12.85), (12.95), and (12.103), we find that:

$$
\begin{aligned}
D\psi\, D\bar{\psi}\, e^{i \int d^4 x \mathcal{L}(x)} &\to \exp \left(i \int d^4 x \left[\mathcal{L}(x) - \theta(x) \partial_\mu J_5^\mu \right] \right) \\
&\quad \times \exp \left(-2i \int d^4 x \frac{-\theta(x)}{16\pi^2} \text{Tr}\, F^{\mu\nu} \tilde{F}_{\mu\nu} \right) D\psi\, D\bar{\psi}
\end{aligned}
\tag{12.104}
$$

This functional is invariant if we choose:

$$
\partial_\mu J_5^\mu = \frac{1}{8\pi^2} \text{Tr} \left(F^{\mu\nu} \tilde{F}_{\mu\nu} \right)
\tag{12.105}
$$

which is the same result that we found before in Eq. (12.58).

In summary, we have seen that the path integral method allows us to generalize the Ward–Takahashi identities found earlier for QED. These identities arise because the generating functional $Z(J)$ is gauge invariant. When applied to gauge theory, these identities become the Slavnov–Taylor identities and the BRST identities. The BRST symmetry arises because there is a residual (global) symmetry left over after the gauge symmetry is broken and Faddeev–Popov ghosts are allowed into the action.

These identities are crucial for renormalization. However, they can be violated by chiral anomalies, which must therefore be cancelled. In the Standard Model, the anomalies from leptons in the Weinberg–Salam model cancel against the anomalies coming from the quarks, giving us a renormalizable, anomaly-free theory.

12.8 Exercises

1. For gauge theory, prove that the functional measure for the various fields is invariant under a BRST transformation.

2. Calculate explicitly the anomaly contribution of $SO(3)$ and show that it vanishes.

3. Discuss the generalization of the chiral anomaly in higher dimensions, such as $d = 6, 8, 10$. What kinds of graphs are divergent? Using the Fujikawa method, calculate what the anomalous term to current conservation might look like. In 10 dimensions, show that the hexagon graph is anomalous.

4. Fill in the missing steps leading up to Eqs. (12.28), (12.29), and (12.31).

5. A representation λ_a is called *real* if there exists a unitary matrix U such that:

$$\lambda_a = -U\lambda_a^* U^\dagger \qquad (12.106)$$

Show that the anomaly cancels for a real representation.

6. For the antisymmetric representation of $SO(N)$ defined by M_{ab}, the anomaly is proportional to $\mathrm{Tr}\,(\{M_{ab}, M_{cd}\}M_{ef})$. Show that an invariant tensor cannot be constructed out of Kronecker delta functions and antisymmetric ϵ tensors with the proper symmetry/antisymmetry properties of the anomaly (except for $N = 6$). Therefore, the anomaly vanishes for all $SO(N)$ except for $SO(6)$, where we have the invariant tensor ϵ_{abcdef}.

7. Consider a Maxwell field locally coupled to a charged triplet meson field. Construct the Ward–Takahashi identity for this theory.

8. Calculate the Ward–Takahashi identity in a theory of spin 3/2 particles coupled to the Maxwell field, where the action contains $\epsilon^{\mu\nu\rho\sigma}\psi_\mu\gamma_5 D_\nu\gamma_\rho\psi_\sigma$. (Note: this action is actually inconsistent.)

9. Prove that the Z meson contributes the isospin factor given in Eq. (12.80) to the anomaly.

10. Prove that the condition $Q_{BRST}|\Psi\rangle = 0$ eliminates not only the ghost states within $|\Psi\rangle$, but also the longitudinal mode of the gauge field, leaving only the

transverse, physical states. (Work only the lowest order in the ghost expansion of $|\Psi\rangle$.) Show that this condition reduces back to Gauss's Law.

11. Fill in the missing steps in Eq. (12.50).

12. Fill in the missing steps in Eq. (12.55).

Chapter 13
BPHZ Renormalization
of Gauge Theories

> Veltman: *I do not care what or how, but what we must have is at least one*
> *renormalizable theory with massive charged vector bosons, and whether*
> *that looks like Nature is of no concern, those are details that will be fixed*
> *later by some model freak...*
> 't Hooft: *I can do that.*
> Veltman: *What do you say?*
> 't Hooft: *I can do that.*

13.1   Counterterms in Gauge Theory

The renormalization of spontaneously broken gauge theories, proved by 't Hooft, using powerful techniques developed by Veltman, Faddeev, Popov, Higgs, and others, opened the floodgates for acceptable quantum field theories of massive vector mesons, which were previously thought to be nonrenormalizable.

In Chapter 7, we presented the proof of the renormalizability of QED based on the original Dyson–Ward multiplicative renormalization scheme. Although a number of proofs of the renormalization of non-Abelian gauge theories have been proposed, we present two such proofs that are quite general and can be applied to a wide variety of quantum field theories, including those that do not have gauge symmetries. We will present the proof based on the BPHZ method and, in the next chapter, a proof based on the renormalization group.

The renormalization program for gauge theories proceeds much the same as for ϕ^4 and QED; that is,

1. First, by power counting arguments, we must isolate the superficially divergent diagrams, show that their degree of divergence depends only on the number of external lines, and that there are only a finite number of classes of these divergent diagrams.

2. We must regularize the divergent diagrams in order to perform manipulations on them.

3. We must show that we can absorb the divergences into the physical parameters of the system, either by extracting out multiplicative renormalization constants, or by subtracting off counterterms. Slavnov–Taylor or BRST identities are needed to show that gauge invariance is maintained and that the renormalized coupling constants have the correct value.

4. We then must prove, via an induction argument, that the theory is renormalizable to all orders.

Of course, we must also check that the renormalization program does not spoil the original physical properties of the theory, such as unitarity. For gauge theories, for example, the proof that the renormalized theory is unitary is actually nontrivial.

We begin this program by power counting to determine the superficial degree of divergence of the Feynman diagrams. We define:

$$
\begin{aligned}
L &= \text{number of loops} \\
E_\psi &= \text{number of external fermion legs} \\
E_A &= \text{number of external vector lines} \\
I_\psi &= \text{number of internal fermion lines} \\
I_G &= \text{number of internal ghost lines} \\
V_A^3 &= \text{number of three-vector vertices} \\
V_A^4 &= \text{number of four-vector vertices} \\
V_G &= \text{number of ghost-vector vertices} \\
V_\psi &= \text{number of fermion-vector vertices}
\end{aligned}
\tag{13.1}
$$

By now familiar arguments, we can show that the superficial degree of divergence of any Feynman diagram is equal to:

$$
D = 4L - 2I_A - I_\psi - 2I_G + V_G^3 + V_G
\tag{13.2}
$$

In addition, we have various identities among these numbers that eliminate all internal lines and vertices from D. As in QED, we now observe that two fermion lines connect with one vector meson line in a vertex. Thus, we have, as before:

$$
V_\psi = I_\psi + \frac{1}{2} E_\psi
\tag{13.3}
$$

Each ghost propagator connects onto one end of a ghost vertex, so that:

$$V_G = I_G \tag{13.4}$$

We also have the constraint that there are no external ghost lines:

$$E_A + 2I_A = 4V_A^4 + 3V_A^3 + V_G + V_\psi \tag{13.5}$$

Finally, we can count the number of loop variables in the theory. Each internal leg I_A, I_ψ, I_G is associated with a momentum. However, there are restrictions on these momenta. Each vertex V_A^3, V_A^4, V_ψ, V_G contributes a delta function constraint that enforces conservation of momentum at that point. We also have the overall momentum conservation of the entire diagram. Thus:

$$L = I_A + I_\psi + I_G - V_A^3 - V_A^4 - V_G + 1 \tag{13.6}$$

Putting everything together, our final result is that the superficial degree of divergence is:

$$D = 4 - E_A - \frac{3}{2} E_\psi \tag{13.7}$$

which is the same as for QED, as in Eq. (7.42).

This means that gauge theory is, in principle, renormalizable. The degree of divergence is a function only of the number of external lines on any Feynman graph, and it decreases for higher point functions. Furthermore, it is easily checked that, as in QED, the classes of diagrams that diverge correspond to the renormalized quantities of the theory. Thus, by renormalizing these physical parameters, we can absorb all the divergences of the theory into these parameters.

Next, we try to isolate the possibly divergent graphs in Figure 13.1. To be concrete, let us begin with the effective action defined in terms of the finite, physical parameters g and m (in Euclidean metric):

$$\mathscr{L} = \frac{1}{4}(F_{\mu\nu}^a)^2 + \frac{1}{2\alpha} \partial \cdot A^a \partial \cdot A^a + \bar{\psi}(\not{D} + im)\psi \tag{13.8}$$

By power counting, we can easily categorize which classes of diagrams are divergent. To this effective action, we can then add the counterterms. It is therefore just a matter of counting to show that the counterterms we must add to the action have the form:

$$\Delta \mathscr{L}_{\text{gauge}} = \frac{1}{4}(Z_3 - 1)(\partial_\mu A_\nu^a - \partial_\nu A_\mu^a)^2 - (Z_4 - 1)g\mu^{\epsilon/2} f^{abc} A_\mu^b A_\nu^c \partial^\mu A^{a\nu}$$

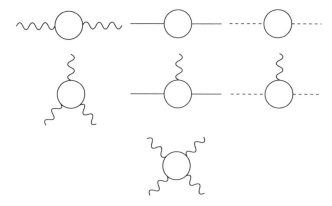

Figure 13.1. The set of diagrams in gauge theory which are potentially divergent.

$$+ \quad \frac{1}{4}(Z_5 - 1)g^2\mu^\epsilon f^{abc} f^{ade} A_\mu^b A_\nu^c A^{d\mu} A^{e\nu} + \frac{1}{2\alpha}(Z_\alpha - 1)\partial \cdot A^a \partial \cdot A^a$$

$$(13.9)$$

[We will use dimensional regularization, so we will find it convenient to perform all integrations by working in the Euclidean metric and hence some of our signs will be reversed due to the metric; that is, $(\gamma_\mu)^2 = -4$.]

Also:

$$\Delta\mathscr{L}_{\text{fermion}} = (Z_2 - 1)\bar\psi \not\partial\psi + im(Z_m - 1)\bar\psi\psi + i(Z_1 - 1)g\mu^{\epsilon/2} A_\mu^a \bar\psi\tau^a\gamma^\mu\psi \quad (13.10)$$

and:

$$\Delta\mathscr{L}_{\text{ghost}} = i(Z_6 - 1)\partial_\mu\eta^{*a}\partial_\mu\eta^a - \frac{i}{2}(Z_7 - 1)g\mu^{\epsilon/2} f^{abc} A_\mu^c\eta^{*a} \overset{\leftrightarrow}{\partial^\mu} \eta^b$$

$$- \frac{i}{2}(Z_8 - 1)g\mu^{\epsilon/2} f^{abc}\eta^{*a}\eta^b\partial_\mu A^{c\mu} \quad (13.11)$$

where $\epsilon = 4 - d$. Each counterterm was chosen to kill off a divergence among the Feynman diagrams generated by our action.

If we add the two pieces $\mathscr{L}$ and $\Delta\mathscr{L}$ together, we arrive at the action defined in terms of the bare, infinite quantities:

$$\mathscr{L} + \Delta\mathscr{L} = \frac{1}{4}(\partial_\mu A_\nu^a - \partial_\nu A_\mu^a)_0^2 - g_0' f^{abc} A_{\mu0}^b A_{\nu0}^c \partial^\mu A_0^{a\nu}$$

$$+ \frac{1}{4}g_0''^2 f^{abc} f^{ade} A_{\mu0}^b A_{\nu0}^c A_0^{d\mu} A_0^{e\nu} + \frac{1}{2\alpha_0}\partial \cdot A_0^a\partial \cdot A_0^a$$

$$+ i\partial_\mu\eta_0^{*a}\partial^\mu\eta_0^a - \frac{i}{2}g_0''' f^{abc} A_{\mu0}^c\eta_0^{*a} \overset{\leftrightarrow}{\partial}_\mu \eta_0^b - \frac{i}{2}g_0'''' f^{abc}\eta_0^{*a}\eta_0^b\partial \cdot A_0^c$$

$$+ \quad \bar{\psi}_0 \not{\partial} \psi_0 + i g_0 A^a_{\mu 0} \bar{\psi}_0 \gamma^\mu \tau^a \psi_0 + i m_0 \bar{\psi}_0 \psi_0 \qquad (13.12)$$

Let us compare the equation on the left-hand side, which is defined in terms of the Z's, with the right-hand side, which is defined in terms of the g_0's. Setting the two sides equal, we find the relation between the multiplicative renormalization constants and the counterterms:

$$
\begin{aligned}
g_0 &= g\mu^{\epsilon/2} Z_1/Z_2\sqrt{Z_3}; & \psi_0 &= \sqrt{Z_2}\psi \\
g_0' &= g\mu^{\epsilon/2} Z_4/Z_3^{3/2}; & A_{\mu 0} &= \sqrt{Z_3} A_\mu \\
g_0'' &= g\mu^{\epsilon/2} \sqrt{Z_5}/Z_3; & \eta_0 &= \sqrt{Z_6}\eta \qquad (13.13)\\
g_0''' &= g\mu^{\epsilon/2} Z_7/\sqrt{Z_3}Z_6; & m_0 &= mZ_m/Z_2 \\
g_0'''' &= g\mu^{\epsilon/2} Z_8/\sqrt{Z_3}Z_6;
\end{aligned}
$$

In principle, the various coupling constants do not have to be equal. In the original bare action, these coupling constants were, of course, all identical, but after renormalization there is no guarantee that these coupling constants will remain equal. In other words, there is the possibility that renormalization will destroy gauge invariance. If they are not equal, then gauge invariance is broken. Gauge invariance, therefore, demands that the various coupling constants be identical. This is where we need the Slavnov–Taylor identities, to guarantee that we can maintain gauge invariance during renormalization.

The Slavnov–Taylor identities (the gauge generalization of the Ward–Takahashi identities) preserve gauge invariance and hence keep all the coupling constants equal:

$$g_0 = g_0' = g_0'' = g_0''' = g_0'''' \qquad (13.14)$$

Setting the coupling constants to be equal, we arrive at:

$$\frac{Z_1}{Z_2} = \frac{Z_4}{Z_3} = \frac{\sqrt{Z_5}}{\sqrt{Z_3}} = \frac{Z_7}{Z_6} = \frac{Z_8}{Z_6} \qquad (13.15)$$

These identities are the gauge counterparts of the relation $Z_1 = Z_2$ found in ordinary QED.

To prove that a theory is renormalizable, it is necessary (but not sufficient) to show that, by power counting, we can cancel all potential divergences by adding counterterms into the action, which in turn gives us a simple renormalization of the physical parameters. To complete the proof, we must show that we can write a recursion relation that proves that all diagrams are finite to all orders in perturbation theory. This recursion relation, in turn, must be able to handle overlapping divergences.

To begin this inductive procedure, let us now show, to lowest order, that we can explicitly eliminate all divergences via this renormalization procedure. We will use the dimensional regularization approach, which is perhaps one of the most convenient regularization approaches for gauge theories since it respects the Ward–Takahashi identities. (The Pauli–Villars method, by contrast, violates gauge invariance for non-Abelian theories. To apply it to gauge theories, one must make a nontrivial generalization of this method involving higher derivatives.)

13.2 Dimensional Regularization of Gauge Theory

The task of demonstrating that all divergences at the first loop can be absorbed into a renormalization is simplified by repeating some of the calculations that we found in QED, except that we must include more diagrams with additional isospin indices. We will only analyze the fermion self-energy graph, the vertex correction, and the vector meson self-energy graph. The other divergences can be analyzed in a straightforward fashion.

For example, the fermion self-energy diagram is identical to the QED electron self-energy diagram, except that we must add in the isospin indices:

$$\Sigma(\not{p}) = \tau^a \tau^a \Sigma_{\text{QED}}(\not{p}) \tag{13.16}$$

(We work in the Feynman gauge.) To calculate this, we must be more specific about the structure of the Lie algebra. In general, for Lie algebra generators τ^a which are $d_f \times d_f$ matrices in some R representation of the algebra, we have:

$$\text{Tr}\, \tau^a \tau^b = C_R \delta^{ab} \tag{13.17}$$

where C_R is called the Dynkin index of the representation R of the algebra.

$\tau^a \tau^a$ is a Casimir operator of the Lie algebra; that is, it commutes with all members of the Lie algebra. It can be chosen to be proportional to a $d_f \times d_f$ unit matrix times δ^{ab}. To calculate the coefficient, we contract over a:

$$\sum_a \text{Tr}\, \left(\tau^a \tau^a\right) = N C_R \tag{13.18}$$

where N is the number of generators in the algebra.

Thus, we have:

$$\tau^a \tau^a = \frac{N}{d_f} C_f \tag{13.19}$$

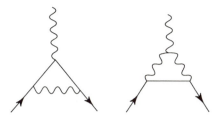

Figure 13.2. The vertex correction for gauge theory has an additional graph not found in QED because of the three-boson interaction.

summed over a. This then gives us:

$$\Sigma(p) = -i\frac{NC_f}{d_f}\frac{g_0^2}{8\pi^2\epsilon}(p + 4m) \tag{13.20}$$

[Notice that the sign appearing in this equation differs from Eq. (7.94) because of our choice of Euclidean metric. Also, for $SU(M)$, we have $N = M^2 - 1$ and $d_f = M$. For the fundamental representation, we have $C_f = \frac{1}{2}$.] Likewise, the vertex correction graph resembles the vertex correction graph found in QED, except that there is an extra graph coming from the three-boson graph (Fig. 13.2). The first vertex correction graph is directly related to the QED result:

$$\Gamma_\mu^{(1)} = \tau^b\tau^a\tau^b\Gamma_{\mu\text{QED}} \tag{13.21}$$

We use the fact that:

$$\begin{aligned}
\tau^b\tau^a\tau^b &= [\tau^b, \tau^a]\tau^b + \tau^a\tau^b\tau^b \\
&= if^{bac}\tau^c\tau^b + \frac{N}{d_f}C_f\tau^a \\
&= \frac{1}{2}f^{bac}f^{dbc}\tau^d + \frac{N}{d_f}C_f\tau^a \\
&= \frac{1}{2}C_{ad}\tau^a + \frac{N}{d_f}C_f\tau^a \tag{13.22}
\end{aligned}$$

where C_{ad} is the Dynkin index in the adjoint representation of the group (the same representation as the generators) and equals M for $SU(M)$.

Our final result for the first vertex correction graph is then:

$$\Gamma_\mu^{(1)}(p, q) = -ig_0\mu^{\epsilon/2}\tau^a\gamma_\mu\left(\frac{1}{2}C_{ad} + C_f\frac{N}{d_f}\right)\frac{g_0^2}{8\pi^2\epsilon} + \cdots \tag{13.23}$$

Next, we must calculate the vertex correction piece coming from the three-boson graph that does not appear in QED:

$$\Gamma_\mu^{(2)} = -g_0^3 \mu^{3\epsilon/2} f^{abc} \tau^b \tau^c \int \frac{d^d k}{(2\pi)^d} \gamma^\sigma \frac{1}{\not{k} - m} \gamma^\nu$$

$$\times \frac{(k+p)_\sigma g_{\nu\mu} + (q - 2p - k)_\nu g_{\mu\sigma} + (p + q - 2k)_\mu g_{\sigma\nu}}{(p-k)^2 (k-q)^2} \quad (13.24)$$

We now contract over the isospin indices:

$$f^{abc} \tau^b \tau^c = \frac{i}{2} C_{ad} \tau^a \quad (13.25)$$

and introduce Feynman parameters and integrate to zero any terms that are purely linear in momenta:

$$\Gamma_\mu^{(2)} = -i g_0^2 \mu^{3\epsilon/2} C_{ad} \tau^a \int_0^1 dx \int_0^{1-x} dy \int \frac{d^d k}{(2\pi)^d}$$

$$\times \frac{2 k_\mu \gamma_\nu \not{k} \gamma^\nu}{[k^2 + m^2(1 - x - y) + q^2 x + p^2 y - (qx - py)^2]^3}$$

$$= -i g_0^3 \mu^{\epsilon/2} \tau^a \frac{C_{ad}}{16\pi^2} \int_0^1 dx \int_0^{1-x} dy \left[\gamma_\mu (1 - \epsilon/2) \Gamma(\epsilon/2) \right.$$

$$\times \left. \left(\frac{4\pi \mu^2}{m^2(1 - x - y) + q^2 x + p^2 y - (qx - py)^2} \right)^{\epsilon/2} \right] + \cdots$$

$$= -i g_0 \mu^{\epsilon/2} \gamma_\mu \tau^a \frac{g_0^2 C_{ad}}{16\pi \epsilon} + \cdots \quad (13.26)$$

The sum of the two contributions to the vertex correction gives us:

$$\Gamma_\mu = -i g_0 \mu^{\epsilon/2} \gamma_\mu \tau^a \frac{g_0^2}{8\pi^2 \epsilon} \left(C_{ad} + C_f \frac{N}{d_f} \right) \quad (13.27)$$

$$Z_1 = 1 - \frac{g_0^2}{8\pi^2 \epsilon} \left(C_{ad} + C_f \frac{N}{d_f} \right) + \cdots \quad (13.28)$$

Last, we would like to calculate the vacuum polarization graph for the gauge field. There are, unfortunately, four graphs that must be computed (Fig. 13.3), only one of which can be read off from our QED calculations.

Figure 13.3. Of the four graphs contributing to the meson self-energy graph, only one has a counterpart in QED.

That contribution to the vacuum polarization contains internal fermion lines. It is given by:

$$
\begin{aligned}
\Pi_{\mu\nu}^{(f)ab} &= \operatorname{Tr}(\tau^a\tau^b)\Pi_{\mu\nu}^{\text{QED}} \\[2mm]
&= -C_f\delta^{ab}\,\frac{g_0^2}{16\pi^2\epsilon}\,\frac{2}{3}\left(g_{\mu\nu}p^2 - p_\mu p_\nu\right)
\end{aligned}
\tag{13.29}
$$

By a straightforward application of Feynman's rules, we can also calculate the contribution in which gauge mesons circulate in the interior loop. We merely contract over two gauge meson vertices:

$$
\Pi_{\mu\nu}^{(1)ab}(p) = -\frac{1}{2}g_0^2\mu^\epsilon f^{acd}f^{bdc}\int \frac{d^dk}{(2\pi)^d}\frac{V_{\mu\nu}}{k^2(p+k)^2}
\tag{13.30}
$$

where:

$$
\begin{aligned}
V_{\mu\nu} &= \left[(2k+p)_\mu g_{\rho\sigma} - (k+2p)_\sigma g_{\mu\rho} + (p-k)_\rho g_{\mu\sigma}\right] \\
&\quad\times \left[(2k+p)_\nu g^{\sigma\rho} - (2p+k)^\sigma \delta_\nu^\rho + (p-k)^\rho \delta_\nu^\sigma\right] \\
&= (4d-6)k_\mu k_\nu + (2d-3)(k_\mu p_\nu + k_\nu p_\mu) + (d-6)p_\mu p_\nu \\
&\quad + \left[(p-k)^2 + (2p+k)^2\right]g_{\mu\nu}
\end{aligned}
\tag{13.31}
$$

Let us now introduce Feynman parameters into the calculation:

$$\Pi_{\mu\nu}^{(1)ab}(p) = -\frac{1}{2}g_0^2\mu^\epsilon f^{acd} f^{bdc} \int_0^1 dx \int \frac{d^d k}{(2\pi)^d} \frac{1}{[k^2 + p^2 x(1-x)]^2}$$

$$\times \left((4d-6)k_\mu k_\nu + [(4d-6)x(x-1)+d-6]p_\mu p_\nu \right.$$

$$+ \left. \{2k^2 + p^2[2x(x-1)+5]\}g_{\mu\nu}\right)$$

$$= -\frac{g_0^2\mu^\epsilon}{2(4\pi)^{d/2}} f^{acd} f^{bcd} \int_0^1 dx \left(\frac{(3d-3)g_{\mu\nu}\Gamma(1-d/2)}{[p^2 x(1-x)]^{1-d/2}}\right.$$

$$+ \frac{\Gamma(2-d/2)}{[p^2 x(1-x)]^{2-d/2}} \left\{g_{\mu\nu}p^2[5-2x(1-x)]\right.$$

$$+ \left. p_\mu p_\nu[d-6-(4d-6)x(1-x)]\right\}\right)$$

$$= \frac{g_0^2}{16\pi^2\epsilon} f^{acd} f^{bcd}\left(\frac{19}{6}g_{\mu\nu}p^2 - \frac{11}{3}p_\mu p_\nu\right) + \cdots \qquad (13.32)$$

where we only keep the pole term and drop finite parts, and where we eliminate momentum integration over terms linear in the momentum. (We note that the finite parts to this integral contain infrared divergences.)

Now we must also calculate the contribution to $\Pi_{\mu\nu}^{(2)ab}$ coming from the ghost loop. We find:

$$\Pi_{\mu\nu}^{(2)ab}(p) = g_0^2\mu^\epsilon f^{dca} f^{cdb} \int \frac{d^d k}{(2\pi)^d} \frac{(k+p)_\mu k_\nu}{k^2(k+p)^2}$$

$$= -g_0^2\mu^\epsilon f^{acd} f^{bcd} \int_0^1 dx \int \frac{d^d k}{(2\pi)^d} \frac{(k-px)_\nu[k+p(1-x)]_\mu}{[k^2 + p^2 x(1-x)]^2}$$

$$= -\frac{g_0^2\mu^\epsilon}{(4\pi)^{d/2}} f^{acd} f^{bcd} \int_0^1 \frac{dx}{2}\left(\frac{g_{\mu\nu}\Gamma(1-d/2)}{[p^2 x(1-x)]^{1-d/2}}\right.$$

$$\left. -2\frac{p_\mu p_\nu x(1-x)\Gamma(2-d/2)}{[p^2 x(1-x)]^{2-d/2}}\right)$$

$$= \frac{g_0^2}{16\pi^2\epsilon} f^{acd} f^{bcd}\left(\frac{1}{6}g_{\mu\nu}p^2 + \frac{1}{3}p_{\mu\nu}\right) + \cdots \qquad (13.33)$$

where we drop all finite parts.

There is also the zero-momentum loop diagram (and also two tadpole graphs) that do not contribute anything at all. We know that they will give us mass corrections that do not have any momentum dependence, being proportional to $\delta^d(0)$. However, we know that, by gauge invariance, the mass of the gluon is zero even after renormalization. Therefore, we can drop these potential mass corrections from our calculation.

We summarize our final results for some of the renormalization constants:

$$
\begin{aligned}
Z_1 &= 1 - \frac{g_0^2}{8\pi^2\epsilon}\left(C_{ad} + C_f\frac{N}{d_f}\right) + \cdots \\
Z_2 &= 1 - \left(\frac{g_0^2}{8\pi^2}\frac{NC_f}{d_f\epsilon}\right) + \cdots \\
Z_3 &= 1 + \frac{g_0^2}{8\pi^2\epsilon}\left(\frac{5}{3}C_{ad} - \frac{4}{3}C_f\right) + \cdots
\end{aligned}
\tag{13.34}
$$

which is consistent with Eq. (7.103).

This now completes the first step in the induction process. Now, we must tackle the most difficult part of the program, which is to write down the recursion relations and show they are actually satisfied.

13.3 BPHZ Renormalization

The multiplicative renormalization procedures that we developed for QED are quite awkward when applied to gauge theories, since we have many more inter-action vertices and fields. We now present a different renormalization scheme, the BPHZ renormalization program,[1-3] which is one of the most powerful and versatile of the various renormalization programs. Although it has a reputation of being a formidable, difficult formalism, the essential features of this approach are easy to summarize.

There are several important reasons for analyzing the BPHZ renormalization prescription:

1. The BPHZ approach easily handles overlapping divergences, which are diffi-cult to manipulate in other formalisms. In fact, overlapping divergences are the chief complication in any renormalization program.

2. It is independent of the regularization prescription, and hence may be used to show that renormalization theory is independent of the regularization scheme. Since we use a subtraction on the integrand of the Feynman integral, we never need to make any explicit mention of a regularization scheme. There is no

need to discuss the details of Feynman graph divergences. All we need to know is that a prescription exists to render a graph convergent.

3. Although the Dyson renormalization program outlined earlier is ideally suited for multiplicative renormalization, the BPHZ formalism is more closely related to the counterterm method.

In the BPHZ formalism, we assume that the usual power-counting analysis has been performed, leaving us with the final induction step. We begin by first showing that we can, via a Taylor expansion at zero external momentum, eliminate the divergent quantities of any graph by a subtraction. This is called the Bogoliubov R operation. In our discussion of the BPHZ technique, we will derive an explicit expression for the subtractions. We will then show that this method of subtractions can be rewritten in terms of counterterms added to the action.

In this section, we will first try to outline the intuitive ideas behind the BPHZ program, in order to stress the simplicity of its basic ideas, and then later we will be more precise in our definitions. (We omit detailed proofs.)

We begin by defining the *superficial degree of divergence* of a graph as the degree of divergences given by power counting. We define a *renormalization part* as a proper (1PI) diagram that is superficially divergent. Let Γ be a particular Feynman graph to which we associate a Feynman integral:

$$
F_\Gamma = \lim_{\epsilon \to 0^+} \int dk_1 \cdots dk_\Gamma \, I_\Gamma
$$

$$
I_\Gamma = \prod_{a,b} \Delta_F(x_a - x_b) \prod_c V_c \tag{13.35}
$$

where the integrand consists of a certain number of propagators and vertices. This graph, in general, is divergent as $\epsilon \to 0^+$. (Our results, however, will be independent of any particular regularization scheme.)

We will now define the *finite part* of this graph, denoted by J_Γ:

$$
J_\Gamma = \lim_{\epsilon \to 0^+} \int dk_1 \cdots dk_\Gamma R_\Gamma \tag{13.36}
$$

The goal of the BPHZ renormalization scheme is to find a prescription or a set of rules by which we can extract R_Γ from any I_Γ. We define a graph to be *primitively divergent* if it (1) is 1PI (one-particle irreducible), (2) is superficially divergent, (3) becomes convergent if any line is broken up. For these primitively divergent graphs, let us introduce an operator t^Γ that has the ability to extract out the divergent part of a graph via a Taylor expansion at zero momentum: Then:

$$
J_\Gamma = \int dk_1 \cdots dk_\Gamma (1 - t^\Gamma) I_\Gamma \tag{13.37}
$$

To define this operator, we define a Taylor expansion about the point where all external momenta are set equal to zero. We define:

$$t^\Gamma I_\Gamma(p_i) \equiv I_\Gamma(0) + \cdots + \frac{1}{D(\Gamma)!} \sum_{j_1 \cdots j_D = 1}^{E-1} (p_{j_1})_{\lambda_1} (p_{j_2})_{\lambda_2} \cdots (p_{j_D})_{\lambda_D}$$

$$\times \frac{\partial^D I_\Gamma}{(\partial p_{j_1})_{\lambda_1} (\partial p_{j_2})_{\lambda_2} \cdots (\partial p_{j_D})_{\lambda_D}} \qquad (13.38)$$

where $E - 1$ is the number of external lines and D is the superficial degree of divergence. The operator $(1 - t^\Gamma)$ has a simple interpretation: It just subtracts the divergent part of an integral at zero momentum, with the number of subtractions determined by the superficial degree of divergence D.

The more general case, however, is much more complicated than this because a graph Γ may have divergent subgraphs γ_i. In fact, a graph Γ may be superficially convergent but may contain divergent subgraphs. The bulk of our work is to find a way in which to catalog and then subtract each of these divergent subgraphs. Because of the large number of definitions we must make, we will first intuitively sketch the outline of the BPHZ program, without regard to rigor, in order to display the essence of the technique. In the next section, we will be more precise in our definitions.

Let us define $\bar{R}_\Gamma$ as the integrand of a graph with all subgraph divergences subtracted out. The only divergence left is therefore the overall divergence of the entire graph. Once we subtract out this overall divergence, then we are left with all divergences subtracted, so we have the renormalized integrand R_Γ:

$$R_\Gamma = \bar{R}_\Gamma - t^\Gamma \bar{R}_\Gamma \qquad (13.39)$$

There are two equivalent approaches to finding the solution for R_Γ. Historically, the first approach was pioneered by Bogoliubov and Parasiuk[1] and Hepp,[2] who wrote down a recursion relation for R_Γ in terms of lower-order graphs. In the second approach, Zimmerman[3] wrote down the explicit solution of these recursion relations for R_Γ.

To understand both approaches, we first recall that divergent subgraphs γ_i can be one of three possible types. If we draw boxes around each subgraph, then these boxes are either

1. Disjoint (the boxes are separated, with no common region).

2. Nested (one box appears entirely within another).

3. Overlapping (the boxes share some common lines).

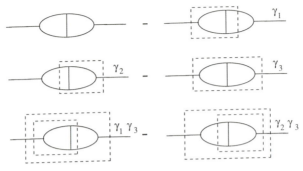

Figure 13.4. The 6 ways we can draw boxes in the BPHZ approach, as shown here for the two-loop graph, avoid the overlapping divergence problem that is a major obstacle in other renormalization methods.

One can, of course, construct R_Γ by simply subtracting off all possible subdivergences within I_Γ. In Zimmerman's approach, however, one omits the overlapping divergences among the subdivergences. The subtractions are taken only over nested and disjoint graphs. To see this, let U be any particular set of boxes. Let $\mathcal{F}$ be the total set of all possible combinations of boxes. For example, in Figures 13.4 and 13.5, we show how to draw boxes around the various subgraphs for a two-loop and three-loop diagram, such that we ignore all overlapping subgraphs.

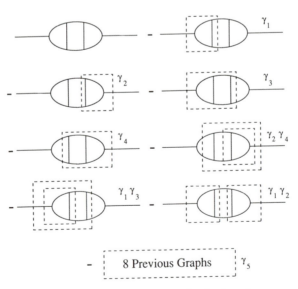

Figure 13.5. The 16 ways that boxes can be drawn for the three-loop case avoid overlapping divergences.

There are 6 ways in which to draw these boxes for the two-loop diagram (dropping overlapping combinations):

$$\mathscr{F} = \left\{ \emptyset, \; \gamma_1, \; \gamma_2, \; \gamma_3, \; \gamma_3\gamma_1, \; \gamma_3\gamma_2 \right\} \tag{13.40}$$

It is essential to notice that we have omitted the overlapping cases: $\{\gamma_2\gamma_1\}$, and $\{\gamma_3\gamma_2\gamma_1\}$. Symbolically, we may therefore write R_Γ as the usual Feynman integral minus the divergences associated with each of these subgraphs:

$$R_\Gamma = \left[1 - t^{\gamma_1} - t^{\gamma_2} - t^{\gamma_3} + (-t^{\gamma_3})(-t^{\gamma_1}) + (-t^{\gamma_3})(-t^{\gamma_2}) \right] I_\Gamma \tag{13.41}$$

The generalization to the three-loop case is straightforward. The decomposition into boxes is given by:

$$\begin{aligned} \mathscr{F} = \big\{ & \emptyset, \; \gamma_1, \; \gamma_2, \; \gamma_3, \; \gamma_4, \; \gamma_5, \\ & \gamma_2\gamma_1, \; \gamma_3\gamma_1, \; \gamma_4\gamma_2, \; \gamma_5\gamma_2, \; \gamma_5\gamma_1, \; \gamma_5\gamma_3, \\ & \gamma_5\gamma_4, \; \gamma_5\gamma_1\gamma_2, \; \gamma_5\gamma_3\gamma_1, \; \gamma_5\gamma_4\gamma_3 \big\} \end{aligned} \tag{13.42}$$

Then R_Γ is given by:

$$\begin{aligned} R_\Gamma = \Big[& 1 - t^{\gamma_1} - t^{\gamma_2} - t^{\gamma_3} - t^{\gamma_4} - t^{\gamma_5} + (-t^{\gamma_4})(-t^{\gamma_2}) \\ + \; & (-t^{\gamma_1})(-t^{\gamma_2}) + (-t^{\gamma_3})(-t^{\gamma_1}) + (-t^{\gamma_5})(-t^{\gamma_1}) \\ + \; & (-t^{\gamma_5})(-t^{\gamma_2}) + (-t^{\gamma_5})(-t^{\gamma_3}) + (-t^{\gamma_5})(-t^{\gamma_4}) \\ + \; & (-t^{\gamma_5})(-t^{\gamma_2})(-t^{\gamma_1}) + (-t^{\gamma_5})(-t^{\gamma_3})(-t^{\gamma_1}) + (-t^{\gamma_5})(-t^{\gamma_4})(-t^{\gamma_2}) \Big] I_\Gamma \end{aligned} \tag{13.43}$$

What is remarkable is that this subtraction process works even if we simply drop the troublesome overlapping divergences. These terms, we recall, invalidated the naive multiplicative renormalization scheme of Dyson/Ward for QED, which broke down at the 14th-order level. So it is rather surprising that we can simply drop them in the BPHZ counterterm approach.

To see why the overlapping divergences can be dropped in this approach, consider the two-loop case shown in Figure 13.4. A direct calculation of the double-loop graph shows that it contains divergences proportional to $1/\epsilon^2$, which can be cancelled, as well as $\log p^2/\epsilon$. This second type of divergence is the celebrated overlapping divergence and cannot, at first glance, be cancelled by adding any counterterm to the action. A term like $\log \partial_\mu^2$ is required to cancel this diagram, and such a term does not appear in the action and hence cannot be absorbed into any renormalization of a physical parameter. The method of counterterms seems, at first glance, to fail.

Miraculously, however, such terms can, indeed, be cancelled if we take a closer look at the method of counterterms. A single-loop counterterm for ϕ^3 can, of course, cancel the subdivergence in γ_1. When applied to the double-loop graph, this counterterm gives us $1/\epsilon$ multiplied by the loop integration of the rest of the graph, which produces $\log p^2$. The product of the two gives us $\log p^2/\epsilon$, which is the term needed to cancel the overlapping divergence. Thus, the single-loop counterterms, when applied to the double-loop graph, give us a product that can cancel the overlapping divergence $\log p^2/\epsilon$.

In BPHZ language, this cancellation is written as: $(1 - t^{\gamma_3})t^{\gamma_1}t^{\gamma_2} = 0$.

Thus, overlapping divergences, which are difficult to handle in the multiplicative renormalization scheme, can be cancelled by carefully iterating lower-order counterterms for higher-order graphs. This demonstrates the superiority of the counterterm method over the multiplicative renormalization.

In the same manner, one can show that all overlapping divergences drop out to all orders, although we will not present the proof. BPHZ showed that this cancellation can be generalized for an arbitrary number of t^{γ_i} even if the γ_i are overlapping.

Although this result is gratifying, there is still one last step that we must complete. Zimmerman's solution, although explicit, still has one serious disadvantage. It contains nested graphs, which cannot be cancelled by the counterterm method. This is because counterterms in the action only cancel against disjoint graphs, never against nested graphs. (A simple application of Wick's theorem and Feynman's rules for the counterterms shows that nested subdivergences are never generated.)

To make contact with the counterterm method, we will now use an equivalent method pioneered by BPH, which is equivalent to the Zimmerman solution. It is possible to absorb all unwanted nested graphs into purely disjoint graphs (which can be cancelled against counterterms) if we write down recursion relations for lower-order subgraphs.

For example, in Figure 13.5, we notice that a nested graph arises from γ_3 and γ_1. This nested divergence can be absorbed by introducing a new subtraction operator $\bar{R}_{\gamma_3}$ which operates on subdivergences: $\bar{R}_{\gamma_3} I_\Gamma = I_\Gamma + (-t)^{\gamma_1} I_\Gamma$, where $\bar{R}_{\gamma_3}$ is an operator that subtracts out the divergences contained within the subgraph γ_3, which is due to the subgraph γ_1. Therefore, the nested graph can be absorbed by introducing this subtraction operator for subgraphs:

$$(-t)^{\gamma_3}\bar{R}_{\gamma_3} I_\Gamma = (-t)^{\gamma_3} I_\Gamma + (-t)^{\gamma_3}(-t)^{\gamma_1} I_\Gamma \qquad (13.44)$$

The last term is the nested graph, which has now been absorbed into the operator $\bar{R}_{\gamma_3}$. This method is quite general: All nested graphs can be absorbed into the subtraction operator of some subgraph. BPHZ proved that this process allows one to express R_Γ by iterating R_γ for lower-order disjoint subgraphs and dropping all nested ones.

Let now summarize both the Zimmerman and the BPH formalism. Let γ be a divergent subgraph. Let $\mathscr{D}$ be the set of all possible combinations of just disjoint subgraphs and $\mathscr{D} \cup \mathscr{N}$ be the set of both disjoint and nested graphs. Then the formulas of Zimmerman and BPHZ, respectively, can be written symbolically as:

$$\bar{R}_\Gamma I_\Gamma = \sum_{\mathscr{D}\cup\mathscr{N}} \left(\prod_{\gamma\in\mathscr{D}\cup\mathscr{N}} (-t)^\gamma \right) I_\Gamma$$

$$= \sum_{\mathscr{D}} \left(\prod_{\gamma\in\mathscr{D}} (-t)^\gamma \bar{R}_\gamma \right) I_\Gamma \qquad (13.45)$$

For example, for Figure 13.5, the set $\mathscr{D}$ is given by just the disjoint set $\{\emptyset, \gamma_1, \gamma_2, \gamma_3, \gamma_4, \gamma_1\gamma_2\}$. By expanding out all the terms in the BPH recursion relation on the second line, we recover Zimmerman's formula on the first line.

The advantage of the BPH recursion relation is that we sum solely over divergent disjoint graphs, which in turn can be cancelled against the counterterms appearing in the action. The recursion relation is then the last step in demonstrating that the BPHZ method guarantees that counterterms in the action can cancel against all potential divergences of field theory.

13.4 Forests and Skeletons

So far, our discussion has tried to emphasize the intuitive nature of this BPHZ approach, which is a specific prescription by which to subtract out all possible divergent subgraphs. This intuitive discussion, however, will now be repeated and strengthened by making a few rigorous definitions. Specifically, these definitions will allow us to show the equivalence of BPH's recursion formula and Zimmerman's explicit solution.

Let γ be a subgraph within a graph Γ. Two graphs are mutually disjoint if they have no lines or vertices in common:

$$\gamma_1 \cap \gamma_2 = \emptyset \qquad (13.46)$$

Now define $\{\gamma_1, \gamma_2, \ldots, \gamma_n\}$ to be a set of mutually disjoint connected subdiagrams of the same graph Γ. Then we define the reduction operation:

$$\Gamma \rightarrow \frac{\Gamma}{\{\gamma_1, \gamma_2, \ldots, \gamma_n\}} \tag{13.47}$$

which contracts each subgraph γ_i down to a point.

We say that two subgraphs *overlap* if they share some lines and vertices. More precisely, they overlap if none of the following holds:

$$\gamma_1 \cap \gamma_2 = \emptyset; \quad \gamma_1 \subset \gamma_2; \quad \gamma_2 \subset \gamma_1 \tag{13.48}$$

Both overlapping and nested graphs are omitted in Eq. (13.47).

Now we come to the definition of a *forest* (which includes nested graphs). A forest U of Γ is a hierarchy of subdiagrams such that:

1. The elements of U are all renormalization parts.

2. Any two elements of U are nonoverlapping.

(Loosely speaking, as we saw before, a forest U is a set of subgraphs that can be either nested *or* disjoint, but not overlapping. Each subgraph is superficially divergent. For example, there are 16 forests in Figure 13.5.)

A forest is called full if it contains Γ itself. And it is called normal if it does not. A forest is called empty if it contains only the null set.

To define this subtraction scheme, we introduce the *Bogoliubov R operation*. Then BPH proved that $\bar{R}_\Gamma$ can be expressed recursively as:

$$\bar{R}_\Gamma = I_\Gamma + \sum_{\gamma_1, \cdots \gamma_c} I_{\Gamma/\{\gamma_1, \cdots, \gamma_c\}} \prod_{\tau=1}^{c} O_{\gamma_\tau} \tag{13.49}$$

where we define:

$$O_\gamma = -t^\gamma \bar{R}_\gamma \tag{13.50}$$

Then R_Γ can now be defined as follows:

$$R_\Gamma = \bar{R}_\Gamma \text{ if } \Gamma = \text{renorm. part}$$
$$R_\Gamma = (1 - t^\Gamma)\bar{R}_\Gamma \text{ if } \Gamma \neq \text{renorm. part} \tag{13.51}$$

Notice that this definition of the R operation is recursive and that we only subtract disjoint graphs. $\bar{R}_\gamma$ is always defined in terms of $\bar{R}_\gamma$ of lower order.

The remarkable thing about this procedure is that it is equivalent to Zimmerman's forest formula:

$$R_\Gamma = \sum_{\text{all } U} \prod_{\lambda \in U} (-t^\lambda) I_\Gamma \tag{13.52}$$

where we now subtract *both* disjoint and nested graphs, and where the product over λ are ordered, such that t^λ is to the left of t^σ if $\lambda \supset \sigma$.

We now sketch the proof that Zimmerman's Eq. (13.52) satisfies the recursive BPH definition of R_Γ in Eq. (13.49). We can always find the unique set of biggest disjoint subgraphs $M_1, M_2, \ldots, M_n$ of any forest U. Each biggest subgraph M_i contained within a forest U may have smaller nested subgraphs contained within it. To construct this unique set, we take any two nonoverlapping subgraphs γ_i and γ_j within the forest U. Then we must have one of the three possibilities:

$$\gamma_i \subset \gamma_j$$
$$\gamma_j \subset \gamma_i$$
$$\gamma_i \cap \gamma_j = \emptyset \tag{13.53}$$

For the first possibility, we remove γ_i as a candidate for a biggest subgraph. For the second possibility, we remove γ_j from consideration. For the last possibility, we leave both in. By successively eliminating the various subgraphs in this way, we are left with only the biggest subgraphs $\{M_i\}$, which are disjoint and unique.

The forest U is then the union of full forests, one for each M_i. We can therefore rewrite Zimmerman's forest formula in Eq. (13.52) as:

$$\bar{R}_\Gamma = I_\Gamma + \sum_{M_1,\ldots,M_n} \left[\prod_{i=1}^{n} (-t)^{M_i} \left(\sum_{U_1 \in \mathcal{F}(M_1)} \cdots \right. \right.$$

$$\left. \left. \times \sum_{U_n \in \mathcal{F}(M_n)} \prod_{\gamma_1 \in U_1} (-t)^{\gamma_1} \cdots \prod_{\gamma_n \in U_n} (-t)^{\gamma_n} \right) I_\Gamma \right] \tag{13.54}$$

For example, consider Figure 13.5. The set of disjoint biggest subgraphs is $\{M_i\} = \{\emptyset, \gamma_1, \gamma_2, \gamma_3, \gamma_4, \gamma_1\gamma_2\}$. Then the terms farthest to the right contain the nested combinations $\{\gamma_3\gamma_1, \gamma_4\gamma_2\}$. In this way, Eq. (13.54) separates the forests into two sets: the disjoint set $\{M_i\}$ and the nested set.

The point of this construction is that we have rewritten the forest formula so that all nested sequences of graphs appear within the parenthesis. This allows us to regroup these nested formulas into the form $(-t)^{M_i} \bar{R}_{M_i} I_\Gamma$. Since $\{M_i\}$ is the unique set of disjoint biggest subgraphs within any forest, we have now converted sequences of nested subgraphs into a recursion relation involving only these biggest subgraphs. The nested graphs in Figure 13.5 have not disappeared; they have simply been hidden within the M_i.

With this regrouping of graphs, Eq. (13.54) has now been converted to the expression $\sum \prod_i (-t)^{M_i} \bar{R}_{M_i} I_\Gamma$. But written in this way, we recover the BPH formula of Eq. (13.49), based entirely on disjoint graphs M_i. This completes the sketch that Zimmerman's forest formula in Eq. (13.52) (based on nested and disjoint graphs) can be reexpressed in terms of the BPH recursion formula in Eq. (13.49) (based only on disjoint graphs M_i). This demonstrates the equivalence of the two formulas given earlier in Eq. (13.45).

Now that we have rendered all graphs finite, the last step in the proof of BPHZ renormalization is to show that this subtraction technique can be accomplished by adding counterterms into the action. This is easy, since the subtraction process on disjoint graphs that we have outlined is equivalent to the process of adding counterterms into the action. Since the counterterms correspond to the set of divergent disjoint graphs, the procedure of subtracting off the divergences is identical to adding counterterms into the action. Since we saw earlier that these counterterms are proportional to the original action, we have now demonstrated that the BPHZ method is equivalent to multiplicative renormalization.

The Yang–Mills theory, because it satisfies all the properties required by BPHZ, is therefore renormalizable. Not only does the Yang–Mills theory satisfy all the requirements coming from power counting, it also satisfies all the properties demanded by the BPHZ recursion method. (Since the BPHZ method makes no mention of gauge invariance, we must also impose the additional constraint of the Slavnov–Taylor identities to keep the renormalized coupling constants for gauge theory equal.)

Finally, it is useful to compare the BPHZ method with the Dyson renormalization program mentioned earlier. In retrospect, there are some key differences between these two approaches. The Dyson renormalization program was based on defining *skeleton graphs* constructed out of renormalized vertices and self-energy graphs, such as $\tilde{S}'_F$. The Dyson approach from the very beginning tried to lump infinite classes of divergences into these renormalized vertices and self-energy graphs. The advantage of doing this is, of course, that one can immediately extract out the multiplicative renormalization constants Z_i. However, the price we paid for grouping the graphs from the very start into renormalized propagators and vertices was that we were plagued with overlapping divergences. Thus, the recursion relations had to be written out entirely in terms of vertices without the overlapping divergences, which often gave us clumsy equations. Another disadvantage of the Dyson approach is that it was not very general. It was constructed explicitly for QED, and hence must be modified in significant ways to handle more general theories.

This, however, is precisely the advantage of the BPHZ method: It is quite general. The BPHZ approach abandons the skeleton method of trying to lump divergent graphs from the very beginning into $\tilde{S}'_F$, etc. The BPHZ approach is based on successively adding counterterms to the action. These counterterms are chosen to subtract out the divergent integrand of any Feynman diagram, without performing any regrouping of diagrams into renormalized vertices and propagators. As a result, *we lose multiplicative renormalization* at each intermediate step. However, the advantage of this is that we are no longer plagued by overlapping divergences. Only at the last step do we recognize that these subtractions give us counterterms that are proportional to terms in the original action, which in turn finally gives us multiplicative renormalization.

(We should also point out some drawbacks of the BPHZ method. Because we subtracted all diagrams at zero momentum, infrared divergences are more difficult to handle in this approach. Also, the method must be modified to handle gauge invariances, since the Slavnov–Taylor identity must be added as an additional constraint.)

13.5 Does Quantum Field Theory Really Exist?

Because of the remarkable experimental success of quantum field theory in describing the interactions of electrons and photons, we might be surprised to find that, strictly speaking, quantum field theory as a perturbation theory may not exist. This is because although we can successfully renormalize the perturbation series, there exists the possibility that the entire perturbation theory diverges. Simple arguments, in fact, show that perturbative quantum field theory may likely diverge at extremely high order. Although the perturbation theory for QED seems to converge rapidly at low orders because $\alpha \sim 1/137$, eventually the Feynman graphs themselves may overwhelm the smallness of the fine structure constant and yield a divergent sum.

For example, Dyson pointed out many years ago that for negative α, QED should be unstable, with unlimited virtual pair production from the vacuum. However, virtual pair production with sufficiently small separation may become real pair production by separating to larger distances. Thus, real pair production from the vacuum could progress unimpeded, and the theory could collapse with an unstable vacuum. Thus, QED may have a zero radius of convergence in α space.

To see how the sum of a perturbation might diverge, let us take the much simpler example of ϕ^4 theory without any kinetic term, and let us replace a functional integral over ϕ with an ordinary integral. Already, at this simple level, we can see how the perturbation theory, although perfectly well behaved at any finite order, diverges at infinite order.

Let us examine the behavior of the following partition function at high order:

$$Z(g) = \frac{1}{\sqrt{2\pi}} \int_{-\infty}^{\infty} e^{-\phi^2/2 - g\phi^4} d\phi \tag{13.55}$$

This function is interesting because the coefficients in the expansion in g equal the number of vacuum diagrams in ϕ^4 theory. Although this integral cannot be performed exactly, we can always power expand this function in powers of g and then try to sum the perturbation theory. A simple power expansion yields:

$$Z(g) = \sum_{n=0}^{\infty} g^n Z_n$$

$$Z_n = \frac{(-1)^n}{\sqrt{2\pi}} \int_{-\infty}^{\infty} d\phi \frac{\phi^{4n}}{n!} e^{-\frac{1}{2}\varphi^2} = (-1)^n 4^n \frac{\Gamma(2n + \frac{1}{2})}{\sqrt{\pi} n!} \tag{13.56}$$

Our goal is to examine the behavior of the perturbation theory for large n. We can use the Stirling approximation formula:

$$n! \sim \sqrt{2\pi n} e^{n \log n - n} \tag{13.57}$$

For large n, the perturbation theory therefore behaves as:

$$Z_n \sim \frac{(-16)^n}{\sqrt{\pi}} e^{(n-1/2) \log n - n} \tag{13.58}$$

Although this simple example is unrealistic, we can already see the nontrivial behavior of the theory in g space. The perturbation theory diverges with large n.

In fact, a more careful analysis shows that the theory, in complex g space, has an essential singularity at $g = 0$. For any negative g, the integral over ϕ blows up and the theory breaks down. The potential is no longer bounded from below and the integral diverges. Thus, there is ample reason to believe that QED may suffer the same fate.

The tremendous experimental accuracy of the theory, however, shows us that QED cannot be simply discarded as a physical theory just because the perturbation theory may not converge. QED has been able to withstand all challenges over the last 6 decades, and, not surprisingly, there is a resolution to this problem.

We can consider QED to be an *asymptotic theory*, that is, a theory that can, for fixed n and α small enough, approach a definite result. For example, in our simple example, we may treat the perturbation series as an asymptotic series:

$$\left| Z(g) - \sum_{i=0}^{n} Z_i g^i \right| < \frac{4^{n+1} \Gamma(2n + \frac{3}{2})}{\sqrt{\pi}(n+1)!} \frac{|g|^{n+1}}{[\cos(\frac{1}{2} \operatorname{Arg} g)]^{2n+3/2}} \tag{13.59}$$

For our purposes, we may consider QED to be an asymptotic theory that will allow us to obtain perfectly convergent results, even though the original perturbation series, in principle, may not exist. One may also approach this problem from another direction. One may be able to generalize the definition of the original divergent function $Z(g)$ even if the perturbation theory was divergent. To do this, we will use the method of the Borel transform, which allows us to extract meaningful information from divergent series. For example, let us begin with a function $G(g)$ whose power expansion in g diverges:

$$G(g) = \sum_{n=1}^{\infty} a_n g^n = \infty \qquad (13.60)$$

Although the original power expansion of $G(g)$ makes no sense, it is possible to define a new power series that has much better convergence properties. To see this, let us divide each coefficient by $n!$ in order to obtain a more convergent series:

$$F(g) = \sum_{n=1}^{\infty} \frac{a_n}{n!} g^n \qquad (13.61)$$

Although the original power expansion diverged, this new function has a radius of convergence given by:

$$\frac{1}{R_1} = \lim_{n \to \infty} \sup \left| \frac{a_n}{n!} \right|^{1/n} \qquad (13.62)$$

With this new function, we can reintroduce the original function $G(g)$ by defining it to be:

$$G(g) \equiv \int_0^{\infty} e^{-t} F(tg) dt \qquad (13.63)$$

This new definition of $G(g)$ reduces to the old one if we perform the integration over dt:

$$
\begin{aligned}
G(g) &= \sum_{n=1}^{\infty} \frac{a_n}{n!} g^n \int_0^{\infty} e^{-t} t^n dt \\
&= \sum_{n=1}^{\infty} a_n g^n \qquad (13.64)
\end{aligned}
$$

Although the original power expansion diverged, the advantage of this new definition of $G(g)$ is that it may have a finite radius of convergence, while the old one was given by:

$$\frac{1}{R_2} = \lim_{n \to \infty} \sup |a_n|^{1/n} \tag{13.65}$$

Thus, if $R_2 > 0$, then $R_1 = \infty$. Similarly, if there is a singularity in the Borel plane for $F(g)$, then $R_2 = 0$.

Now let us use this technique to analyze the Borel transform for the function $Z(g)$. The key to this method is to define a new function $B(t)$ that is constructed from the same coefficients Z_n found in the divergent series except that we divide each term by new factors sufficient to make the series converge. Then we take the inverse Borel transform in order to recover $Z(g)$ from $B(t)$.

For example, we can define:

$$B(t) = \sum_{i=0}^{\infty} \frac{Z_n t^n}{\Gamma(n + \frac{3}{2})} \tag{13.66}$$

Because we have divided each term by the Γ function, the series may now converge in a finite radius in t space.

Now that we have defined a function $B(t)$ that exists, then we define the inverse transform to recover $Z(g)$:

$$Z(g) = \int_0^{\infty} dt\, e^{-t} \sqrt{t}\, B(gt) \tag{13.67}$$

If this process of recovering the function $Z(g)$ from its divergent perturbation series exists, then we say that the theory is Borel summable.

Now let us analyze quantum field theories that might be Borel summable, even if the original perturbation theory diverges. We would like to analyze theories more realistic than this toy model that we have been studying. Our starting point will be the usual N-point Green's function, but defined in Euclidean space:

$$\langle 0 | T \, [\phi(x_1)\phi(x_2) \cdots \phi(x_N)] \, | 0 \rangle$$
$$= \frac{\int D\phi\, e^{-S(\phi)} \phi(x_1)\phi(x_2) \cdots \phi(x_N)}{\int D\phi\, e^{-S(\phi)}} \tag{13.68}$$

where $\phi(x)$ is a generic field for an arbitrary field theory of arbitrary spin. Our task is to take the Borel transform of this function in order to find when the Borel transform diverges.

To analyze this Green's function, we will rewrite the numerator of this function as:

$$N(g) = g \int_0^{\infty} e^{-t} F(gt)\, dt \tag{13.69}$$

where:

$$F(z) = \int D\phi \, \delta \, (z - S(\phi)) \, \phi(x_1)\phi(x_2) \cdots \phi(x_\mathbf{N}) \tag{13.70}$$

(To prove the equivalence of this expression with the original numerator of the Green's function, simply insert the expression for F into N and perform the integration over t, which is trivial because of the delta function.)

We recognize F to be the Borel transform. In order to analyze the singularities of the transform, it is helpful to analyze the singularities of a much simpler expression. We would like to analyze the singularities of the following function:

$$\int du_1 \, du_2 \cdots du_N \quad \delta \, (z - f(u_1, u_2, \ldots, u_\mathbf{N}))$$

$$= \int_\Sigma d\Sigma \, |\nabla f|^{-1} \tag{13.71}$$

where Σ is a hypersurface in $\{u\}$ space and where $f(u) = z$. (To prove this identity, perform the integration over, say, u_1. Then invert this implicit function, and rewrite the expression in a more symmetric fashion.)

This function obviously diverges if there is a point where:

$$|\nabla f|^2 = \sum \left| \frac{\partial f}{\partial u_i} \right|^2 = 0 \tag{13.72}$$

Then the denominator blows up, and the function becomes singular.

Now replace u_i with $\phi(x_i)$ and $f(u)$ with $S(\phi)$. Then, if we perform the functional integral over ϕ, we find that the resulting integral is singular if, for some $S(\phi) = z$, there is a point satisfying the usual equations of motion:

$$\frac{\delta S(\phi)}{\delta \phi(x)} = 0 \tag{13.73}$$

In summary, we have shown that the Borel transform F blows up if there is a solution to the Euclidean equations of motion where the action $S(\phi)$ is finite. These finite-action, Euclidean solutions spoil Borel summability.

Unfortunately, such finite-action solutions to the Euclidean equations of motion actually exist. They are called *instantons*, and represent genuine solutions to the gauge theory with Euclidean metric. Instantons will be discussed in greater depth in Chapter 16, where they will play a key role in our understanding of the stability of the vacuum. Thus the perturbation theory of gauge theory is neither convergent nor is it Borel summable. QCD, for example, has zero radius of convergence. We must, as a consequence, treat it strictly as an asymptotic theory.

This also has practical implications for QCD. There are, as we have noted, an infinite number of possible renormalization schemes. Usually, we say that the sum of the perturbation series is independent of whichever scheme we choose. However, in actual practice, as we have seen, the various subtraction schemes have different convergence properties.

In summary, the divergences of gauge theory are only a bit more complicated than those of QED. In both cases, power counting arguments show that the divergences of a graph are functions of the number of external lines, and these divergences can be absorbed into a renormalization of the physical parameters.

We have also seen that the BPHZ method gives us a powerful method of renormalizing quantum field theories, including gauge theory. The advantage of the BPHZ method is that overlapping divergences, which give rise to severe complications for the Dyson approach, do not have to be treated separately. The BPHZ method also gives us a simple formalism in which to handle counterterms. No explicit regularization is needed.

In the next chapter, we will use renormalization theory to give us perhaps the most important experimental verification of QCD.

13.6 Exercises

1. Draw all the Feynman graphs in gauge theory with fermions necessary to calculate Z_4, Z_5, Z_6, Z_7, and Z_8, to one-loop order.

2. Consider a φ^3 theory. Consider (a) a four-loop diagram with the topology of a ladder with five rungs; (b) a four-loop self-energy graph, consisting of a circle containing three interior parallel vertical lines, with two external lines coming out from the left and right. Break them both down in terms of a skeleton and a forest decomposition.

3. Couple $SU(N)$ Yang–Mills theory to a Yukawa theory of mesons with quartic interactions. By power counting, find all primitively divergent graphs including ghosts. Show which graphs correspond to the renormalization of which physical parameters.

4. For the Yang–Mills theory coupled to Yukawa mesons, write down the counterterms that must be added to the action to renormalize it. Find the relations between the various Z_i that are preserved by the Slavnov–Taylor identity.

5. From Feynman's rules for this same theory, setup the dimensionally regulated integrals necessary to compute the scalar meson self-energy diagram and scalar–scalar–vector meson vertex to lowest order. Do not solve.

6. Consider the Slavnov–Taylor identity to one loop order in gauge theory coupled to fermions. Prove two of the relations appearing in Eq. (13.15) to that order.

7. Beginning with Feynman's rules, fill in the missing steps in Eqs. (13.24) and (13.26).

8. Beginning with Feynman's rules, fill in the missing steps in Eqs. (13.30) and (13.32).

9. Prove Eqs. (13.56) and (13.58).

10. Prove that Eq. (13.43) in Zimmerman's approach can be re-expressed as a recursion relation, as in Eq. (13.49) in BPH's approach.

Chapter 14
QCD and the Renormalization Group

There's a long tradition in theoretical physics, which by no means affected everyone but certainly affected me, that said the strong interactions are too complicated for the human mind.

— S. Weinberg

14.1 Deep Inelastic Scattering

One of the great theoretical breakthroughs in gauge theory was the realization that the renormalization theory of gauge theories may explain many of the curious features found in deep inelastic scattering. In fact, it was the remarkable success of gauge theory in explaining the Stanford Linear Accelerator Center (SLAC) experiments on electron–proton scattering that helped to elevate QCD into the leading theory of the strong interactions. At very high energies, the form factors begin to lose some of their dependence on certain low-energy dimensional parameters for $|q|^2 \geq 2\mathrm{Ge}V^2$. This phenomenon is called *scaling*. For the deep inelastic scattering experiments at SLAC, where a high energy beam of electrons was scattered off a proton target, Bjorken[1] predicted that scaling should occur, (using current algebra, Regge asymptotics, and kinematics).

The deep inelastic scattering amplitude was calculated for the process (Fig. 14.1):

$$e^- + p \rightarrow e^- + \text{anything} \tag{14.1}$$

for large momentum transfers of the electron. This was an ideal experiment to analyze the structure of the proton, since the probe was an off-shell photon, which has a relatively clean interaction with the hadrons.

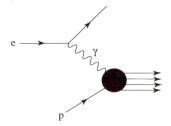

Figure 14.1. Deep inelastic scattering: in electron-proton scattering, an off-shell photon probes the structure of the proton.

The simplest explanation of scaling comes from Feynman's *parton* model, where the proton is assumed to consist of point-like constituents.[2,3] Remarkably, such a simple picture explained many of the qualitative features of the SLAC experiments, including scaling.

There was a puzzle, however. If the proton was a bound state of some mysterious force, then presumably nonperturbative effects were dominant. However, the parton model indicated that, at high energies, the partons (e.g., quarks) could be considered to act like free point-like particles. Apparently, nonperturbative effects could somehow be neglected, and we could assume the quarks were free to roam inside the proton.

This simple experimental picture was then explained through QCD. Using the theory of the *renormalization group*, it could be shown that the renormalized coupling constant varied with the energy scale. At increasingly high energies, the coupling constant of the strong force became smaller and smaller, so that the quarks could be treated as if they were free point-like particles in the asymptotic domain. This effect was called *asymptotic freedom*. A general analysis revealed that non-Abelian gauge theories were the *only* field theories in which asymptotic freedom was exhibited.

The flip side of asymptotic freedom was that, at smaller and smaller energies, the coupling constant became increasingly large. This could, in principle, explain why the quarks were permanently confined within the hadrons.

Let us explain the development of asymptotic freedom by first giving the experimental results at SLAC on scaling, and then continue our discussion of renormalization theory and the renormalization group, leading up to the celebrated result that non-Abelian gauge theories are asymptotically free.

We will close this chapter by showing that the renormalization group equations give us a recursion relation that yields yet another method of renormalizing field theory.

We begin by defining the kinematics of electron–proton deep inelastic scattering. Let the incoming electron have momentum k, and the outgoing electron have momentum k'. Then we define:

$$q = k - k'$$
$$\nu = \frac{p \cdot q}{M}$$
$$x = -\frac{q^2}{2M\nu} \tag{14.2}$$

The SLAC experiments probed the interior of the proton with a photon that was very much off-shell, (i.e., $q^2 \to -\infty$).

In the lab frame, where the proton is at rest, we have the following:

$$p_\mu = (M, 0, 0, 0); \quad k_\mu = (E, \mathbf{k}); \quad k'_\mu = (E', \mathbf{k'}) \tag{14.3}$$

Therefore, in the limit of small electron mass, we have:

$$\nu = E - E'$$
$$q^2 = -4EE' \sin^2(\theta/2) \leq 0 \tag{14.4}$$

where θ is the scattering angle.

We will be interested in the deep inelastic region, which is defined by:

$$\text{Deep inelastic region} = \left\{ \begin{array}{ccc} \nu & \to & \infty \\ -q^2 & \to & \infty \\ x & \to & \text{fixed} \end{array} \right\} \tag{14.5}$$

We can show that $0 \leq x \leq 1$. (This parameter measures how far we are from elastic scattering, which corresponds to the point $x = 1$.)

Using Feynman's rules, let us construct the scattering amplitude of an electron colliding with a proton of polarization σ, emitting some unknown state $|n\rangle$:

$$\mathcal{M}_n = \left[e^2 \bar{u}(k', s') \gamma^\mu u(k, s) \right] \left(\frac{1}{q^2} \right) \left[\langle n | J_\mu(0) | p, \sigma \rangle \right] \tag{14.6}$$

where J_μ is the electromagnetic current, and the matrix element of this current between hadronic states is unknown.

Using the standard rules for constructing differential cross sections, we find that the scattering into the nth final state is given by:

$$d\sigma_n = \frac{d^3k'}{|\mathbf{v}| 2M2E(2\pi)^3 2k'_0} \prod_{i=1}^{n} \frac{d^3 p_i}{(2\pi)^3 2p_{i0}}$$

$$\times \ \frac{1}{4} \sum_{\sigma, s, s'} |\mathcal{M}_n|^2 (2\pi)^4 \delta^4 (p + k - k' - p_n) \tag{14.7}$$

where p_n is the sum of the momenta of the various hadronic final states.

Now let us sum over all the hadronic final states n, and we obtain the inclusive cross section:

$$\frac{d^2\sigma}{d\Omega \, dE'} = \frac{\alpha^2}{q^4} \left(\frac{E'}{E} \right) l^{\mu\nu} W_{\mu\nu} \tag{14.8}$$

where the leptonic tensor $l_{\mu\nu}$ is given by $(\bar{u}\gamma_\mu u)(\bar{u}\gamma_\nu u)$:

$$l_{\mu\nu} = \frac{1}{2}\mathrm{Tr}\left(\not{k}'\gamma_\mu \not{k}\gamma_\nu \right) = 2(k'_\mu k_\nu + k_\mu k'_\nu + \frac{q^2}{2}g_{\mu\nu}) \tag{14.9}$$

The hadronic tensor $W_{\mu\nu}$ is the object we wish to study, since it is basically unknown. It can be vastly simplified, however, by explicitly performing the sum over the unknown final state $|n\rangle$. Using completeness arguments, dependence on $|n\rangle$ disappears:

$$W_{\mu\nu} = \frac{1}{4M} \sum_\sigma \sum_n \int \prod_{i=1}^{n} \left(\frac{d^3 p_i}{(2\pi)^3 2p_{i0}} \right)$$

$$\times \langle p, \sigma | J_\mu(0) | n \rangle \langle n | J_\nu(0) | p, \sigma \rangle (2\pi)^3 \delta^4(p_n - p - q)$$

$$= \frac{1}{4M} \sum_\sigma \int \frac{d^4 x}{(2\pi)} e^{iq \cdot x} \langle p, \sigma | [J_\mu(x), J_\nu(0)] | p, \sigma \rangle \tag{14.10}$$

[In the last step, we have written the product of two currents as a commutator. We have dropped the term $J_\nu(x)J_\mu(0)$ because it occurs with a momentum constraint $p_n = q - p$. In the lab frame, this means that $E_n = M - \nu$, which cannot be satisfied.]

We know from current conservation that $\partial_\mu J^\mu = 0$, or:

$$q^\mu W_{\mu\nu} = W_{\mu\nu} q^\nu = 0 \tag{14.11}$$

Thus, using general invariance arguments, we can re-express $W_{\mu\nu}$ in terms of only two form factors W_1 and W_2:

$$W_{\mu\nu} = -\left(g_{\mu\nu} - \frac{q_\mu q_\nu}{q^2} \right) W_1 + \left(p_\mu - q_\mu \frac{p \cdot q}{q^2} \right) \left(p_\nu - q_\nu \frac{p \cdot q}{q^2} \right) \frac{W_2}{M^2} \tag{14.12}$$

Inserting this expression back into the differential cross section, we find:

$$\frac{d\sigma}{dq^2 d\nu} = \frac{4\pi\alpha^2}{q^4} \frac{E'}{E} \left[W_2 \cos^2\left(\frac{\theta}{2}\right) + 2W_1 \sin^2\left(\frac{\theta}{2}\right) \right] \tag{14.13}$$

Experimentally, it was discovered that, in the deep inelastic region, the dependence on q^2 and ν was replaced by the dependence on $x = -q^2/2M\nu$ alone in the structure functions:

$$MW_1(q^2, \nu) \rightarrow F_1(x)$$
$$\nu W_2(q^2, \nu) \rightarrow F_2(x) \tag{14.14}$$

This relation is called Bjorken scaling.

14.2 Parton Model

The most intuitive explanation of the scaling relations came from the parton model. The parton model simply assumed that the dominant contribution to the hadronic tensor $W_{\mu\nu}$ came from the scattering of point-like constituents within the proton of unknown spin. It was a very naive picture of the proton, but it worked surprisingly well. In fact, it became a central mystery as to why such a naive model worked so well, far beyond its hypothetical range of validity.

The essence of the parton model can be summarized in Figure 14.2, where the dominant contribution to the hadronic tensor comes from the scattering of the off-shell photon with a parton.

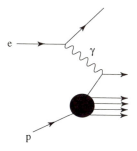

Figure 14.2. The parton model: an off-shell photon scatters off a point-like constituent of the proton. Comparing the resulting sum rules with experiment shows that the parton has spin $\frac{1}{2}$ and most likely corresponds to a quark.

We assume that the parton has negligible transverse momentum with respect to the proton, so the parton momentum is in the same direction as the proton momentum; that is, the parton has momentum ξp_μ, where $0 \leq \xi \leq 1$.

As one might suspect, the secret of the parton model's ability to explain scaling lies in the kinematics of Figure 14.2. To see how scaling emerges from this simple picture, notice that momentum conservation forces us to have:

$$p' = \xi p + q \tag{14.15}$$

Now square both sides of this equation. We arrive at:

$$p'^2 = \xi^2 p^2 + 2M\nu\xi + q^2 \tag{14.16}$$

In the scaling region, where p^2 and p'^2 can be neglected, we have:

$$q^2 + 2M\nu\xi \sim 0 \tag{14.17}$$

In other words:

$$\xi = x \tag{14.18}$$

This is important, because it means that all structure functions will become functions of ξ or x alone. This, of course, is the essence of scaling. Thus, a very simple kinematic picture of partons yields scaling behavior.

The naive parton model tells us more. From Eq. (14.15), one concludes that x is the fraction of the momentum carried by the parton in the nucleon. For a given spin, it allows us to calculate restrictions on W_1 and W_2. By checking these structure constants against experiment, one can therefore determine the spin of the parton. To see how the spin of the parton is determined, we note that, in this approximation, the matrix element $\langle \xi p, \sigma | J_\mu(0) | p', \sigma' \rangle$ is proportional to $\bar{u}(\xi p)\gamma_\mu u(p')$ for spin $\frac{1}{2}$ partons. Thus, the contribution to the hadronic tensor coming from a parton of momentum ξp is given by:

$$K_{\mu\nu}(\xi) = \frac{1}{4M\xi} \sum_{\text{spins}} \left[\bar{u}(\xi p)\gamma_\mu u(p') \right] \left[\bar{u}(p')\gamma_\nu u(\xi p) \right] \frac{\delta(p_0' - \xi p_0 - q_0)}{2p_0'} \tag{14.19}$$

The total hadronic tensor is given by integrating this over all ξ. Let the number of partons of momentum ξp be proportional to some unknown function $f(\xi)$. Then the total hadronic current from all partons is given by:

$$W_{\mu\nu} = \int_0^1 f(\xi) K_{\mu\nu}(\xi) \, d\xi \tag{14.20}$$

Let us now calculate all sums appearing in the scattering amplitude for the partons. The sum over the spinors is easily calculated using the usual rules:

$$\frac{1}{2}\sum_{\text{spins}} \bar{u}(\xi p)\gamma_\mu u(\xi p + q)\bar{u}(\xi p + q)\gamma_\nu u(\xi p)$$

$$= (\xi/2)\,\text{Tr}\left[\not{p}\gamma_\mu(\not{p}\xi + \not{q})\gamma_\nu\right]$$

$$= 4\xi^2 p_\mu p_\nu - 2M\nu\xi g_{\mu\nu} + \cdots \qquad (14.21)$$

Now comes the crucial step. We will rewrite the delta function over momenta to explicitly display the fact that $\xi = x$:

$$\delta(p_0' - \xi p_0 - q_0)/2p_0'$$

$$= \theta(p_0')\delta[p'^2 - (\xi p + q)^2]$$

$$= \theta(\xi p_0 + q_0)\delta(2M\nu\xi + q^2)$$

$$= \theta(\xi p_0 + q_0)\delta(\xi - x)/(2M\nu) \qquad (14.22)$$

It is important to note that we have generated the factor $\delta(\xi - x)$ from kinematic arguments alone. Now insert everything back into the hadronic tensor. The integral over ξ is now trivial to perform, and we arrive at:

$$W_{\mu\nu} = p_\mu p_\nu \left(\frac{xf(x)}{\nu M^2}\right) - g_{\mu\nu}\left(\frac{f(x)}{2M}\right) + \cdots \qquad (14.23)$$

Now let us compare this tensor with Eq. (14.12). We find that we have now derived:

$$MW_1 \quad \rightarrow \quad F_1(x) = \frac{1}{2}f(x)$$

$$\nu W_2 \quad \rightarrow \quad F_2(x) = xf(x) \qquad (14.24)$$

Not only have we established scaling, we have also derived the simple relation:

$$2xF_1(x) = F_2(x) \qquad (14.25)$$

which is the Callan–Gross relation.[4]

The usefulness of the parton model is that we can compare the scaling behavior of $W_{1,2}$ against the various predictions for spin-0 and spin-$\frac{1}{2}$ partons. For example, for spin-0 partons, general invariance arguments show that we have:

$$\langle xp|J_\mu|xp + q\rangle \sim (2xp + q)_\mu \qquad (14.26)$$

Therefore:

$$W_{\mu\nu} \sim (2xp + q)_{\mu}(2xp + q)_{\nu} \tag{14.27}$$

Comparing this with the previous expression for the hadronic tensor in Eq. (14.12), we find:

$$\text{Spin } 0 : \quad F_1(x) = 0 \tag{14.28}$$

Experimentally, the Callan–Gross relation is reasonably satisfied, while the spin-0 parton relation is not. This gives us confidence that the partons are, in fact, just the quarks.

Next, we want to calculate the form factors $F_{1,2}$ in terms of the various quark constituents contained within the nucleon. In the naive quark model, as well as in QCD, the electromagnetic current appearing in the scattering amplitude was given by:

$$J_{\mu} = \frac{2}{3}\bar{u}\gamma_{\mu}u - \frac{1}{3}\bar{d}\gamma_{\mu}d - \frac{1}{3}\bar{s}\gamma_{\mu}s \tag{14.29}$$

since the charges of the quarks are given by $2/3$, $-1/3$, $-1/3$.

Each piece of the electromagnetic current, given by the respective quark fields, contributes to the structure function, which is now the sum of the squares of the various contributions from each quark. Let us now separate out each individual contribution of each quark current to F_1. Since F_1 is written in terms of the square of the current, it can be written as the sum over the square of the quark charge times the individual distribution function:

$$2F_1(x) = \sum_{i=u,d,s} Q_i^2 [q_i(x) + \bar{q}_i(x)] \tag{14.30}$$

where the charge of the quark is given by Q_i and $q_i(x)$ is the distribution function for the ith quark.

Then we have:

$$2F_1^{ep} = \frac{4}{9}(u_p + \bar{u}_p) + \frac{1}{9}(d_p + \bar{d}_p) + \frac{1}{9}(s_p + \bar{s}_p)$$

$$2F_1^{en} = \frac{4}{9}(u_n + \bar{u}_n) + \frac{1}{9}(d_n + \bar{d}_n) + \frac{1}{9}(s_n + \bar{s}_n) \tag{14.31}$$

where we have used the symbol $u_p(x)$, etc. to represent the u-quark contribution to the structure function for $e + p$ scattering. These functions represent the probability of finding a quark-parton with x fraction of longitudinal momentum for the given process. The coefficient appearing before each quark contribution is nothing but the square of the quark charge. Let us assume $SU(2)$ isospin symmetry holds. Then the u_p parton distribution function equals the isospin partner d_n. Isospin invariance gives us the equivalence:

$$
\begin{aligned}
u_p &= d_n \equiv u \\
d_p &= u_n \equiv d \\
s_p &= s_n \equiv s
\end{aligned}
\tag{14.32}
$$

We will now drop the subscript p on the proton quark distribution functions. Then we can write:

$$
\frac{F_1^{ep}(x)}{F_1^{en}(x)} = \frac{4(u + \bar{u}) + (d + \bar{d}) + (s + \bar{s})}{(u + \bar{u}) + 4(d + \bar{d}) + (s + \bar{s})}
\tag{14.33}
$$

Therefore we have the constraint:

$$
\frac{1}{4} \le \frac{F_1^{en}(x)}{F_1^{ep}(x)} \le 4
\tag{14.34}
$$

which agrees with the data.

14.3   Neutrino Sum Rules

Next, we would like to study neutrino–nucleon inclusive reactions:

$$
\nu + N \rightarrow e^- + \text{anything}
\tag{14.35}
$$

which resemble the electron–nucleon inclusive reactions except that we use different currents within the Hamiltonian, and we have more invariant tensors in the decomposition of the transition function.

For neutrino scattering, the hadronic current is given by Eq. (11.105):

$$
J_{\text{had}}^{\mu} = \bar{u}\gamma^{\mu}(1 - \gamma_5)(d\cos\theta_C + s\sin\theta_C)
\tag{14.36}
$$

and the leptonic part is given by:

$$
J_{\text{lept}}^{\mu} = \bar{\nu}\gamma^{\mu}(1 - \gamma_5)e + \bar{\nu}_{\mu}\gamma^{\mu}(1 - \gamma_5)\mu + \cdots
\tag{14.37}
$$

Once again, the cross sections can be expressed in terms of various structure functions W_i:

$$
\begin{aligned}
W_{\mu\nu} &= \frac{1}{4M} \sum_{\text{spins}} \int \frac{d^4x}{2\pi} e^{iq\cdot x} \langle p, s | \left[J_\mu(x), J_\nu^\dagger(0) \right] | p, s \rangle \\
&= -W_1 g_{\mu\nu} + W_2 p_\mu p_\nu / M^2 - i W_3 \epsilon_{\mu\nu\lambda\rho} p^\lambda q^\rho / M^2 \\
&\quad + W_4 q_\mu q_\nu / M^2 + W_5 (p_\mu q_\nu + p_\nu q_\mu)/M^2 \\
&\quad + i W_6 (p_\mu q_\nu - p_\nu q_\mu)/M^2
\end{aligned}
\tag{14.38}
$$

where, because of the nature of weak interactions, we have more possible tensors in the decomposition.

Then the cross section can be written in terms of these structure functions as:

$$
\begin{aligned}
\frac{d^2\sigma^{\nu,\bar\nu}}{d\Omega\, dE'} &= \frac{G_F^2 E'^2}{2\pi^2} \left[2\sin^2\left(\frac{\theta}{2}\right) W_1 + \cos^2\left(\frac{\theta}{2}\right) W_2 \right. \\
&\quad \left. \mp \frac{(E+E')}{M} \sin^2\left(\frac{\theta}{2}\right) W_3 \right]
\end{aligned}
\tag{14.39}
$$

where the $-$ (+) sign corresponds to ν ($\bar\nu$) scattering.

In the Bjorken scaling limit, we find:

$$
\begin{aligned}
M W_1(q^2, \nu) &\rightarrow F_1(x) \\
\nu W_2(q^2, \nu) &\rightarrow F_2(x) \\
\nu W_3(q^2, \nu) &\rightarrow F_3(x)
\end{aligned}
\tag{14.40}
$$

where the neutrino scattering amplitude has one additional structure function W_3.

As before, we can now write down a number of relations for the structure functions W_i using the fact that the scattering process probes the quark structure of the nucleon. By analyzing the quantum numbers of the $\nu + N$ reaction, the hadronic current induces the transitions:

$$
\begin{aligned}
d &\rightarrow u \\
s &\rightarrow c \\
\bar u &\rightarrow \bar d \\
\bar c &\rightarrow \bar s
\end{aligned}
\tag{14.41}
$$

which appear multiplied by the factor $\cos^2 \theta_C$, which we will take to be equal to 1. Similarly, the Cabibbo suppressed transitions:

$$
\begin{aligned}
d &\rightarrow c \\
s &\rightarrow u \\
\bar{u} &\rightarrow \bar{s} \\
\bar{c} &\rightarrow \bar{s}
\end{aligned}
\tag{14.42}
$$

are proportional to $\sin^2 \theta_C$ and will be dropped. For $\bar{\nu} + N$, the favored and unfavored reactions can be found by simply reversing the direction of the arrow.

As before, we can write various sum rules by calculating the contribution of the various quark distribution functions to the structure functions. In $e + N$ scattering, we found earlier that the structure constants were proportional to Q^2 times the quark distribution function, as in Eq. (14.30). For $\nu + N$ scattering, the contribution of the ith quark to F_2 or $x F_3$ is proportional to $g_i^2 q_i(x)$, where q_i is the distribution function of the ith quark, and g_i^2 is either $\cos^2 \theta_C$ or $\sin^2 \theta_C$. We will set $\theta_C \sim 0$ for now. The total contribution of the quarks to the structure constants F_2 and $x F_3$ is then the sum over the various quark contributions:

$$
\begin{aligned}
F_2(x) &= 2x \sum_{i,j} \left[g_i^2 q_i(x) + g_j^2 \bar{q}_j(x) \right] \\
x F_3(x) &= 2x \sum_{i,j} \left[g_i^2 q_i(x) - g_j^2 \bar{q}_j(x) \right]
\end{aligned}
\tag{14.43}
$$

We can read off the quark functions q_i that have a nonzero contribution to this sum by analyzing Eq. (14.41). For example, for $\nu + p$ scattering, Eq. (14.41) shows us that only the d, s, $\bar{u}$, and $\bar{c}$ quark functions contribute with coefficient $\cos^2 \theta_C \sim 1$.

Then the complete list of structure functions, written as sums over various quark probability distribution functions, is:

$$
\begin{aligned}
\nu p &: \quad F_2 = 2x(d + s + \bar{u} + \bar{c}); \quad x F_3 = 2x(d + s - \bar{u} - \bar{c}) \\
\nu n &: \quad F_2 = 2x(u + s + \bar{d} + \bar{c}); \quad x F_3 = 2x(u + s - \bar{d} - \bar{c}) \\
\bar{\nu} p &: \quad F_2 = 2x(u + c + \bar{d} + \bar{s}); \quad x F_3 = 2x(u + c - \bar{d} - \bar{s}) \\
\bar{\nu} n &: \quad F_2 = 2x(d + c + \bar{u} + \bar{s}); \quad x F_3 = 2x(d + c - \bar{u} - \bar{s})
\end{aligned}
\tag{14.44}
$$

For the most part, we will ignore the contribution of the strange and charmed quarks to the proton and neutron scattering function, since the nucleon is primarily made of up and down quarks. Then we have:

$$F_2^{\bar{v}p} - F_2^{vp} = 2x\left[u(x) - \bar{u}(x) - d(x) + \bar{d}(x)\right]$$

$$= 4xT_3(x) \tag{14.45}$$

where T_3 is the isospin density, which integrates to one-half. By integrating this expression, we then arrive at the Adler sum rule[5]:

$$\int_0^1 \frac{dx}{x}\left[F_2^{\bar{v}p}(x) - F_2^{vp}(x)\right] = 4\int_0^1 T_3(x)\,dx = 2 \tag{14.46}$$

We can take the sum of the third structure function:

$$F_3^{vp} + F_3^{vn} = -2\left[u(x) + d(x) - \bar{u}(x) - \bar{d}(x)\right] \tag{14.47}$$

We can therefore write (for zero strangeness):

$$F_3^{vp} + F_3^{vn} = -6B(x) \tag{14.48}$$

Since the proton has baryon number B equal to one, we then find the Gross–Llewellyn Smith sum rule[6]:

$$\int_0^1 dx\,\left[F_3^{vp}(x) + F_3^{vn}(x)\right] = -6 \tag{14.49}$$

The experimental value for this is roughly -6.4 ± 1.2.

Historically, many of these sum rules were derived from a variety of related viewpoints, such as current algebra and the parton model. Although it was gratifying to see the success of these methods, they basically relied on a simplistic, free-field approach to the strong interactions. It was a puzzling question why this naive approach should work so well and at such low energies. Given the complicated nature of the strong interactions, the quark–parton model was working well beyond the range of validity originally postulated for it.

14.4   Product Expansion at the Light-Cone

Yet another way to see that scaling emerges in the high-energy limit is to use Wilson's *operator product expansion*,[7,8] where we can show that the space–time region explored by the deep inelastic experiments is near the light–cone. Again, the mystery is why the free-field approximation should work so well in describing strong interactions.

The scattering amplitude can be written as the matrix element of the commutator of two currents:

$$W_{\mu\nu} = \frac{1}{4M} \sum_s \int \frac{d^4 x}{(2\pi)} e^{iq \cdot x} \langle p, s | [J_\mu(x), J_\nu(0)] | p, s \rangle \qquad (14.50)$$

We will show that, using the operator product expansion, we can rederive the scaling behavior of the form factors found earlier with the parton model.

We will show this in two parts. First, we will show that the deep inelastic experiments probe a region of space–time near the light–cone (i.e., $x_\mu^2 \sim 0$). Second, we will then show that the operator product expansion of the currents near the light–cone give us the desired scaling of the form factors.

To see this, let us explore the high q behavior of the integral, which is dominated by a region where $q \cdot x$ does not oscillate appreciably. (Regions of rapid oscillation cancel each other out.) We expand $q \cdot x$ into its components:

$$q \cdot x = \frac{(q_0 + q_3)}{\sqrt{2}} \frac{(x_0 - x_3)}{\sqrt{2}} + \frac{(q_0 - q_3)}{\sqrt{2}} \frac{(x_0 + x_3)}{\sqrt{2}} - \sum_{i=1}^2 q_i x_i \qquad (14.51)$$

Let us go to the rest frame of the proton, so that:

$$p_\mu = (M, 0, 0, 0); \quad q_\mu = (\nu, 0, 0, \sqrt{\nu^2 - q^2}) \qquad (14.52)$$

In the deep inelastic limit, where $\nu, -q^2 \to \infty$ with $-q^2/2M\nu$ held fixed, we can show that:

$$q_0 + q_3 \sim \nu; \quad q_0 - q_3 \sim q^2/2\nu \qquad (14.53)$$

Since we are interested in the region of space–time where $q \cdot x \sim 1$, we thus have:

$$x_0 - x_3 \sim O(1/\nu); \quad x_0 + x_3 \sim O(1/xM) \qquad (14.54)$$

Therefore:

$$x_0^2 - x_3^2 \sim O(-1/q^2) \qquad (14.55)$$

This means that:

$$x_\mu^2 = x_0^2 - \mathbf{x}^2 \leq x_0^2 - x_3^2 \sim O(-1/q^2) \qquad (14.56)$$

In other words, in the limit $q^2 \to -\infty$, the integral is dominated by:

$$x_\mu^2 \to 0 \tag{14.57}$$

Therefore $W_{\mu\nu}$ is dominated by the region of space–time near the light–cone. Now that we have established the importance of the light–cone, we will now show that the operator product expansion near the light–cone yields scaling behavior.

To see the importance of the operator product expansion, we note that the product of two fields taken at the same point is divergent. Our job is to calculate the short-distance behavior of the product of two currents $J_\mu(x)J_\nu(0)$ and insert this expression back into the integral. For free fields, we have:

$$J_\mu(x) =: \bar{\psi}(x)\gamma_\mu Q\psi(x) : \tag{14.58}$$

where Q is a matrix whose eigenvalues give the charges of the various fermions in the theory.

To calculate the commutator, it will be useful to use Wick's theorem to decompose this product of currents. We will use a simple trick. We will analyze the time-ordered product, which yields propagators that have well-known power expansions. Then we will convert this time-ordered product into a commutator by a change in the singular structure of the fields. We begin by writing:

$$
\begin{aligned}
T\left[J_\mu(x)J_\nu(0)\right] \;=\; & \operatorname{Tr}\left[iS_F(-x)\gamma_\mu iS_F(x)\gamma_\nu Q^2\right] \\
& + \; : \bar{\psi}(x)\gamma_\mu Q iS_F(x)\gamma_\nu Q\psi(0) : \\
& + \; \bar{\psi}(0)\gamma_\nu Q iS_F(-x)\gamma_\mu Q\psi(x) : \\
& + \; \bar{\psi}(x)\gamma_\mu Q\psi(x)\bar{\psi}(0)\gamma_\nu Q\psi(0)
\end{aligned}
\tag{14.59}
$$

The advantage of using the time-ordered expression (rather than the commutator) is that the propagator has an explicit expression in terms of x space variables:

$$
\begin{aligned}
\Delta_F(x) \;=\; & -\frac{1}{4\pi}\delta(x^2) + \frac{m}{8\pi\sqrt{x^2}}\theta(x^2)\left[J_1(m\sqrt{x^2}) - iN_1(m\sqrt{x^2})\right] \\
& - \frac{im}{4\pi^2\sqrt{-x^2}}\theta(-x^2)K_1(m\sqrt{-x^2})
\end{aligned}
\tag{14.60}
$$

where J_n, N_n and K_n are the standard Bessel functions. For our purposes, however, we are only interested in the behavior of this function near the light–cone: $x^2 \sim 0$.

In this approximation, we have the simple result:

$$\Delta_F(x) \sim \frac{i}{4\pi^2(x^2 - i\epsilon)} + O(m^2 x^2) \tag{14.61}$$

Near the light–cone, the Feynman propagator for spin-$\frac{1}{2}$ fields can therefore be written as:

$$S_F(x) = (i\gamma \cdot \partial - m)\Delta_F(x) \sim (i\not{\partial})\left(\frac{i}{4\pi^2(x^2 - i\epsilon)}\right) + \cdots \tag{14.62}$$

We now make the switch from the time-ordered product to the commutator. In space–time, this transition is possible if we make the substitution:

$$\left(\frac{1}{-x^2 - i\epsilon}\right)^n \rightarrow \frac{2\pi i\epsilon(x_0)\delta^{(n-1)}(x^2)}{(n-1)!} \tag{14.63}$$

With this substitution, we can now write, using Wick's theorem:

$$
\begin{aligned}
[J_\mu(x), J_\nu(0)] \quad \sim \quad & i\frac{\text{Tr } Q^2}{\pi^3}\left[\frac{2}{3}g_{\mu\nu}\delta''(x^2)\epsilon(x_0) + \frac{1}{6}\partial_\mu\partial_\nu\left[\delta'(x^2)\epsilon(x_0)\right]\right] \\
& + \left\{S_{\mu\alpha\nu\beta}\left[V^\beta(x,0) - V^\beta(0,x)\right]\right. \\
& + \left. i\epsilon_{\mu\alpha\nu\beta}\left[A^\beta(x,0) - A^\beta(0,x)\right]\right\}\partial^\alpha\left[\delta(x^2)\epsilon(x_0)\right]/(2\pi) \\
& + \; :\bar{\psi}(x)\gamma_\mu Q\psi(x)\bar{\psi}(0)\gamma_\nu Q\psi(0):
\end{aligned} \tag{14.64}
$$

where:

$$
\begin{aligned}
V^\beta(x,y) &= \; :\bar{\psi}(x)\gamma^\beta Q^2\psi(y): \\
A^\beta(x,y) &= \; :\bar{\psi}(x)\gamma^\beta\gamma_5 Q^2\psi(y):
\end{aligned} \tag{14.65}
$$

and where we have used the fact that:

$$
\begin{aligned}
\gamma_\mu\gamma_\nu\gamma_\lambda &= (S_{\mu\nu\lambda\rho} + i\epsilon_{\mu\nu\lambda\rho}\gamma_5)\gamma^\rho \\
S_{\mu\nu\lambda\rho} &\equiv g_{\mu\nu}g_{\lambda\rho} + g_{\mu\rho}g_{\nu\lambda} - g_{\mu\lambda}g_{\nu\rho}
\end{aligned} \tag{14.66}
$$

The first term in the commutator does not contribute, since it is a c-number. The second term involves bilocal currents defined at two distinct space–time points. To evaluate them, we can take a Taylor expansion of the fields:

$$\bar{\psi}(x/2)\psi(-x/2) = \bar{\psi}(0)\left(1 + \overleftarrow{\partial}_{\mu_1}\frac{x^{\mu_1}}{2} + \frac{1}{2!}\overleftarrow{\partial}_{\mu_1}\overleftarrow{\partial}_{\mu_2}\frac{x^{\mu_1}}{2}\frac{x^{\mu_2}}{2} + \cdots\right)$$

$$\times \left(1 - \frac{x^{\nu_1}}{2} \overrightarrow{\partial}_{\nu_1} + \frac{1}{2!} \frac{x^{\nu_1}}{2} \frac{x^{\nu_2}}{2} \overrightarrow{\partial}_{\nu_1} \overrightarrow{\partial}_{\nu_2} \right) \psi(0) + \cdots$$

$$= \sum_n \frac{1}{n!} \frac{x^{\mu_1}}{2} \frac{x^{\mu_2}}{2} \cdots \frac{x^{\mu_n}}{2} \bar\psi(0) \overleftrightarrow{\partial}_{\mu_1} \overleftrightarrow{\partial}_{\mu_2} \cdots \overleftrightarrow{\partial}_{\mu_n} \psi(0) + \cdots$$

$$(14.67)$$

Putting this value back into the commutator of two currents, we find:

$$\left[J_\mu(x/2), J_\nu(-x/2) \right] = \Bigg\{ \sum_{\text{odd } n} \frac{1}{n!} \frac{x^{\mu_1}}{2} \frac{x^{\mu_2}}{2} \cdots \frac{x^{\mu_n}}{2} O^{(n+1)}_{\beta\mu_1\mu_2\cdots\mu_n}(0) S_{\mu\alpha\nu\beta}$$

$$+ \sum_{\text{even } n} \frac{1}{n!} \frac{x^{\mu_1}}{2} \frac{x^{\mu_2}}{2} \cdots \frac{x^{\mu_n}}{2} \bar O^{(n+1)}_{\beta\mu_1\mu_2\cdots\mu_n}(0) i\epsilon_{\mu\alpha\nu\beta} \Bigg\} \frac{\partial^\alpha \left[\delta(x^2)\epsilon(x_0) \right]}{(2\pi)}$$

$$(14.68)$$

where:

$$O^{(n+1)}_{\beta\mu_1\mu_2\cdots\mu_n} = \bar\psi(0) \overleftrightarrow{\partial}_{\mu_1} \overleftrightarrow{\partial}_{\mu_2} \cdots \overleftrightarrow{\partial}_{\mu_n} \gamma^\beta Q^2 \psi(0)$$

$$\bar O^{(n+1)}_{\beta\mu_1\mu_2\cdots\mu_n}(0) = \bar\psi(0) \overleftrightarrow{\partial}_{\mu_1} \overleftrightarrow{\partial}_{\mu_2} \cdots \overleftrightarrow{\partial}_{\mu_n} \gamma^\beta \gamma_5 Q^2 \psi(0)$$

$$(14.69)$$

Now insert this expansion back into the expression for $W_{\mu\nu}$. We are interested in the averaged matrix element of these operators. When we perform the average, the matrix element of $\bar O$ vanishes because $\epsilon_{\mu\alpha\nu\beta}$ is antisymmetric. We only need to define:

$$\frac{1}{2} \sum_s \langle p, s | O^{(n+1)}_{\beta\mu_1\mu_2\cdots\mu_n}(0) | p, s \rangle = A^{(n+1)} p_\beta p_{\mu_1} p_{\mu_2} \cdots p_{\mu_n} + \cdots \qquad (14.70)$$

where $A^{(n+1)}$ are undetermined constants.

Putting everything back into the expression for the deep inelastic scattering amplitude, we now have:

$$W_{\mu\nu}(p, q) \sim \frac{1}{2M} \int \frac{d^4x}{(2\pi)^2} e^{ix\cdot q} \sum_{\text{odd } n}^{\infty} (x \cdot p/2)^n \frac{A^{(n+1)}}{n!} S_{\mu\alpha\nu\beta} p^\beta \partial^\alpha \left[\delta(x^2)\epsilon(x_0) \right]$$

$$(14.71)$$

Now, let us introduce yet another unknown function:

$$\sum_{\text{odd } n} (x \cdot p/2)^n \frac{A^{(n+1)}}{n!} \equiv F(x \cdot p) \qquad (14.72)$$

Thus, our expression for the scattering amplitude has now reduced down to:

$$W_{\mu\nu}(p, q) \sim \frac{i}{2M} S_{\mu\alpha\nu\beta} p^\beta \int \frac{d^4x}{(2\pi)^2} \left[e^{ix \cdot q} F(x \cdot p) \partial^\alpha \delta(x^2) \epsilon(x_0) \right] \quad (14.73)$$

Everything has now been concentrated into this one integration. To perform this integral, it is useful to take the Fourier transform of $F(x \cdot p)$:

$$F(x \cdot p) = \int d\xi \, e^{ix \cdot p\xi} \tilde{F}(\xi) \quad (14.74)$$

Then the expression in the brackets becomes:

$$\int d\xi \, e^{ix \cdot (p+\xi p)} \delta(x^2) \epsilon(x_0) \tilde{F}(\xi) \quad (14.75)$$

We now take the Fourier transform of $\delta(x^2)\epsilon(x_0)$:

$$\int d^4k \, e^{ik \cdot x} \delta(k^2) \epsilon(k_0) = -i(2\pi)\epsilon(x_0)\delta(x^2) \quad (14.76)$$

Putting this back into the expression for the scattering amplitude, we find:

$$W_{\mu\nu} \sim \frac{1}{M} \int d\xi \, \tilde{F}(\xi) \delta(q^2 + 2M\nu\xi) S_{\mu\alpha\nu\beta}(q + \xi p)^\alpha p^\beta$$

$$\sim \tilde{F}(x) \left[-\frac{g_{\mu\nu}}{2M} + \frac{x}{\nu} \frac{p_\mu p_\nu}{M^2} + \cdots \right] \quad (14.77)$$

Thus, we have now shown that scaling occurs, that is, that the form factors are functions of $x = -q^2/2M\nu$. Furthermore, we reproduce Eq. (14.24); that is, we have the scaling behavior of spin-$\frac{1}{2}$ partons:

$$MW_1 \rightarrow F_1(x) = \frac{1}{2}\tilde{F}(x)$$

$$\nu W_2 \rightarrow F_2(x) = x\tilde{F}(x) \quad (14.78)$$

(Actually, this last relation is not surprising, since we have taken a representation of the hadronic current in terms of free spin-$\frac{1}{2}$ quarks. If we had taken a representation of the hadronic current in terms of free fields with different spins, we would have derived different relations among the structure functions $W_{1,2}$.)

In conclusion, any theory of the strong interactions must reproduce two seemingly contradictory experimental results: that the quarks seem to be strongly bound together in the low-energy region, but that they act as if they are free in the high-energy region; that is, they act as partons.

Remarkably, we will now see that the gauge theory of QCD can successfully reproduce both behaviors. We will now develop the theory of the renormalization group, and see that QCD is asymptotically free; that is, the quarks have vanishingly small coupling constant in the high-energy region (they act as if they were free point-like constituents) but they have a large coupling constant in the low-energy region (which binds quarks together into mesons and baryons).

14.5   Renormalization Group

The renormalization group equations[9−12] represent a deceptively simple constraint on the renormalized vertex functions of any renormalizable field theory, yet they yield some of the most nontrivial consequences.

The renormalization group equations are based on the simple observation that the physical theory cannot depend on the subtraction point μ at which we regularized our theory. We recall that the subtraction point μ was introduced purely as a mathematical device to begin the process of renormalization, and that no physical consequences could emerge from it.

This means that if we change the subtraction point μ, other parameters, such as the masses and coupling constants, must also change in order to compensate for this effect. *In order to keep the physics invariant, changing the subtraction point must be offset by changes in the renormalized physical parameters as a function of the energy.*

There are several equivalent ways in which to view this highly nontrivial feature of renormalization theory:

1. If we adopt the formalism of counterterms and subtractions, then there are an infinite number of ways in which to split the unrenormalized action $\mathscr{L}_0$ into the renormalized piece $\mathscr{L}$ and its counterterm $\Delta\mathscr{L}$. This is because there is the ambiguity of how to split $\mathscr{L}_0$ between the renormalized action and the counterterm, as we saw in Chapter 7. Changing the subtraction point μ creates a corresponding change in the value of the renormalized physical parameters, so that there are an infinite number of possible renormalizations [see Eqs. (7.71)–(7.75)]. However, the physical quantities at fixed energy must be independent of how we make the split, and this independence is mathematically expressed in terms of the renormalization group.

2. If we adopt the alternative viewpoint of multiplicative renormalization, then we have a multiplicative relation between the vertex functions of the un-renormalized theory $\Gamma_0^{(n)}$ and the vertex functions of the renormalized theory $\Gamma^{(n)}$. However, the unrenormalized vertex function $\Gamma_0^{(n)}$ is totally indepen-

dent of the subtraction point μ (since subtractions are computed only for the renormalized vertex):

$$\frac{\partial}{\partial \mu} \Gamma_0^{(n)} = 0 \tag{14.79}$$

Thus, to keep the unrenormalized vertex function $\Gamma_0^{(n)}$ independent of μ, it means that there is a nontrivial relation between the renormalized $\Gamma^{(n)}$ and Z, which is expressed mathematically as the renormalization group equations.

3. The group nature of the renormalization group can be seen more abstractly if we let R represent some (unspecified) renormalization scheme. If Γ_0 is an unrenormalized quantity and Γ_R is same quantity renormalized by the scheme R, then:

$$\Gamma_R = Z(R)\Gamma_0 \tag{14.80}$$

where $Z(R)$ represents some renormalization constant under the renormalization scheme R.

Let us now choose a different renormalization scheme R'. Since the unrenormalized quantity Γ_0 was independent of the renormalization scheme, then:

$$\Gamma_{R'} = Z(R')\Gamma_0 \tag{14.81}$$

Then the relationship between these two renormalized quantities is given by:

$$\Gamma_{R'} = Z(R', R)\Gamma_R \tag{14.82}$$

where:

$$Z(R', R) \equiv Z(R')/Z(R) \tag{14.83}$$

Trivially, this satisfies a group multiplication law:

$$Z(R'', R')Z(R', R) = Z(R'', R) \tag{14.84}$$

where the identity element is given by:

$$Z(R, R) = 1 \tag{14.85}$$

Now that we have explained the origin of the renormalization group equations, let us try to find a mathematical expression for these relations. While the unrenormalized vertex functions are independent of μ, the renormalized ones are not. For example, in ϕ^4 theory, we have the following relationship between unrenormalized and renormalized quantities:

$$\Gamma_0^{(n)}(p_i, g_0, m_0) = Z_\phi^{-n/2}\Gamma^{(n)}(p_i, g, m, \mu) \tag{14.86}$$

where μ is the subtraction point, and we assume that we have used some regularization scheme to render all expressions finite for the moment.

Now let us differentiate this via the dimensionless derivative $\mu(d/d\mu)$. We know that the unrenormalized bare quantity is independent of the subtraction point, so that the derivative acting on the unrenormalized quantity must, by construction, be zero:

$$
\begin{aligned}
0 &= \mu\frac{\partial}{\partial\mu}\Gamma_0^{(n)} \\
&= \left(\mu\frac{\partial}{\partial\mu}Z_\phi^{-n/2}\right)\Gamma^{(n)} + Z_\phi^{-n/2}\left(\mu\frac{\partial}{\partial\mu}\Gamma^{(n)}\right)
\end{aligned}
\tag{14.87}
$$

We now use the chain rule. We choose as our independent variables μ, g, and m:

$$\frac{d}{d\mu} = \frac{\partial}{\partial\mu} + \frac{\partial g}{\partial\mu}\frac{\partial}{\partial g} + \frac{\partial m}{\partial\mu}\frac{\partial}{\partial m} \tag{14.88}$$

Let us make the following definitions (where we now take the limit as $\epsilon \to 0$):

$$
\begin{aligned}
\beta(g) &\equiv \mu\frac{\partial g}{\partial\mu} \\
\gamma(g) &\equiv \mu\frac{\partial}{\partial\mu}\log\sqrt{Z_\phi} \\
m\gamma_m(g) &\equiv \mu\frac{\partial m}{\partial\mu}
\end{aligned}
\tag{14.89}
$$

With these definitions, we now have the compact expression:

$$\left(\mu\frac{\partial}{\partial\mu} + \beta(g)\frac{\partial}{\partial g} - n\gamma(g) + m\gamma_m(g)\frac{\partial}{\partial m}\right)\Gamma^{(n)}(p_i, g, m, \mu) = 0 \tag{14.90}$$

These are the renormalization group equations, and they express how the renormalized vertex functions change when we make a change in the subtraction point μ.

(In principle, the parameters like β can also depend on the dimensionless quantity m/μ. Then the renormalization group equations become difficult to solve as a function of two independent variables g and m/μ. However, we can ignore this dependence on m/μ if we adopt the "mass-independent regularization scheme," or the "minimal subtraction scheme," which we will discuss later in Section 14.8. We tacitly assume that we adopt this regularization scheme.)

The importance of the renormalization group equations is that they tell us how the renormalized functions change as we vary the subtraction point μ. We know that no physics can emerge by a change in the subtraction point; so a change in the subtraction point must be compensated by a change in how we define the renormalized coupling constants and renormalized masses. The renormalization group equations perform the book-keeping necessary to keep track of how these other variables change when we change the subtraction point.

From our point of view, the most important parameter is β. Knowledge of β determines the behavior of the coupling constant as a function of the mass scale. (We should also point out that the functions β, etc. are dependent on the regularization scheme that we use. Although the physics remains the same, the exact form that these functions take varies with different regularization schemes.)

We first note that we can solve the expression for the β function. We simply divide by β and multiply by $d\mu$:

$$\frac{d\mu}{\mu} = \frac{dg}{\beta(g)} \tag{14.91}$$

Integrating, we have:

$$\log \frac{\mu}{\mu_0} = \int_{g(\mu_0)}^{g(\mu)} \frac{dg}{\beta(g)} \tag{14.92}$$

where μ_0 is some arbitrary reference point. For the moment, let us assume that, for small g, we Taylor expand β:

$$\beta \sim bg^n + \cdots \tag{14.93}$$

for some coupling constant g and integer n. Then, inserting this value of β into the integral equation, we can perform the integration and arrive at:

$$g^{n-1}(\mu) = \frac{g(\mu_0)^{n-1}}{1 - (n-1)bg(\mu_0)^{n-1} \log (\mu/\mu_0)} \tag{14.94}$$

Our goal, however, is to analyze the behavior of the theory at high energies, so let us make the following scale transformation and derive a slightly different constraint on the vertex functions. If we scale the momenta via:

$$p_i \rightarrow e^t p_i = \lambda p_i \tag{14.95}$$

then, using dimensional arguments, the vertex function behaves as:

$$\Gamma^{(n)}(\lambda p_i, g, \mu) = \mu^D f(\lambda^2 p_i \cdot p_j/\mu^2) \qquad (14.96)$$

where D is the dimension of the vertex function. (This is because Γ is a Lorentz invariant, and hence can only be a function of the various dot products $p_i \cdot p_j$. To create a dimensionless quantity out of this, we must divide this by μ^2. The over all scaling quantity μ^D means that the function has dimension D.)

This, in turn, implies that the vertex function obeys the following equation:

$$\left[\mu \frac{\partial}{\partial \mu} + \frac{\partial}{\partial t} - D \right] \Gamma^{(n)}(\lambda p_i, g, \mu) = 0 \qquad (14.97)$$

where $\lambda = e^t$.

Now let us eliminate the term $\mu(\partial/\partial \mu)\Gamma$ from this equation using Eq. (14.90). Then we find:

$$\left(-\frac{\partial}{\partial t} + \beta(g)\frac{\partial}{\partial g} + [D + n\gamma(g)] \right) \Gamma^{(n)}(\lambda p_i, g) = 0 \qquad (14.98)$$

If β were equal to zero, then the scaling behavior of the vertex function would be given by:

$$\Gamma^{(n)} \rightarrow (\lambda)^{D+n\gamma(g)} \qquad (14.99)$$

which is the scaling behavior of the vertex function with the additional $\gamma(g)$ correction. This the reason why $\gamma(g)$ is called the "anomalous" dimension. The important point is that β and γ measure the deviation from naive scaling.

Fortunately, we can solve this equation. Let us introduce the function $\bar{g}(g, t)$, called the *running coupling constant*, such that:

$$\frac{d\bar{g}(g, t)}{dt} = \beta(\bar{g}) \qquad (14.100)$$

with the boundary condition that $\bar{g}(g, 0) = g$. Then the solution is given by:

$$\Gamma(\lambda p_i, g) = \Gamma(p_i, \bar{g}) \exp\left(\int_g^{\bar{g}} dg' \frac{\tilde{\gamma}(g')}{\beta(g')} \right) \qquad (14.101)$$

where $\tilde{\gamma} = D + n\gamma(g)$. To prove this, substitute it directly into Eq. (14.98).

To analyze the nature of these solutions, let us make a few definitions.

Let a *fixed point* represent a zero of the β function for some value g_F. The origin of the name comes from the fact that if the coupling constant were near this fixed point g_F, it will remain there as we increase μ.

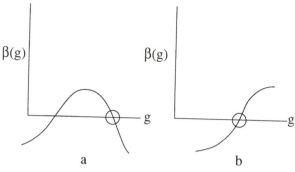

Figure 14.3. In (a), the slope of β is negative, giving us an ultraviolet fixed point. In (b), the slope is positive, giving us an infrared fixed point.

To see this, let us analyze the situation in Figure 14.3 and power expand β around the fixed point:

$$\beta = \mu \frac{\partial}{\partial \mu} g = (g - g_F)\beta'(g_F) + \cdots \tag{14.102}$$

In Figure 14.3(a), the slope of β is negative at the fixed point g_F. Consider what happens, as μ increases, when g is near g_F. If g is less than g_F, and if $\beta' < 0$, then the two signs cancel in the Taylor expansion and $\partial g/\partial \mu$ is positive, so g rises with rising μ. This means that g is driven towards g_F for increasing μ. If, however, g is larger than g_F and $\beta' < 0$, then the derivative $\partial g/\partial \mu$ is negative, so g decreases with increasing μ. Thus, g is driven downwards back toward g_F with increasing μ. In both situations, g is driven towards g_F with increasing μ. We call this an *ultraviolet stable fixed point*.

Now consider the situation in Figure 14.3(b), where the slope of β is positive at g_F. Then g is also driven towards g_F, but for *decreasing* values of μ. If g is less than g_F and $\beta' > 0$, then $\partial g/\partial \mu$ is negative. Thus, for decreasing μ, g will increase in the direction of g_F. Likewise, if g is greater than g_F and $\beta' > 0$, then $\partial g/\partial \mu$ is positive, and hence g will decrease towards g_F if μ decreases.

We can summarize this situation as follows:

$$\begin{cases} \beta'(g_F) < 0: & \text{Ultraviolet stable} \\ \beta'(g_F) > 0: & \text{Infrared stable} \end{cases} \tag{14.103}$$

Let us analyze some theories in this context. We know that, for ϕ^4 theory, the coupling constants are related by:

$$g_0 = g\mu^\epsilon \left(1 + \frac{3g}{16\pi^2\epsilon}\right) + \cdots \tag{14.104}$$

Differentiating, we have:

$$\begin{aligned}
\beta &= \mu\frac{\partial g}{\partial\mu} \\
&= \epsilon g\mu^\epsilon + \frac{3g^2}{16\pi^2}\mu^\epsilon + \cdots \\
&= \frac{3g^2}{16\pi^2} + \cdots \tag{14.105}
\end{aligned}$$

in the limit $\epsilon \to 0$.

The theory is not asymptotically free because of the positive sign of β. In fact, it is easy to integrate the previous equation as a function of μ, and we arrive at:

$$g(\mu) = \frac{g_0(\mu_0)}{1 - (3/16\pi^2)g_0(\mu_0)\log(\mu/\mu_0)} \tag{14.106}$$

Clearly, increasing μ increases g.

Next, let us investigate QED. As before, we know that:

$$\begin{aligned}
e_0 &= e\mu^{\epsilon/2}\frac{Z_1}{Z_2\sqrt{Z_3}} = \frac{e\mu^{\epsilon/2}}{\sqrt{Z_3}} \\
&= e\mu^{\epsilon/2}\left(1 + \frac{e^2}{12\pi^2\epsilon}\right) + \cdots \tag{14.107}
\end{aligned}$$

Differentiating this equation to solve for β, we find:

$$\begin{aligned}
\beta &= \mu\frac{\partial e}{\partial\mu} \\
&= -\frac{\epsilon}{2}e + \frac{e^3}{12\pi^2} + \cdots \\
&= \frac{e^3}{12\pi^2} + \cdots \tag{14.108}
\end{aligned}$$

as $\epsilon \to 0$. As with the ϕ^4 theory, we find that $\beta > 0$, so that the running coupling constant e increases with larger energies. In fact, we can easily integrate this equation, arriving at:

$$e^2(\mu) = \frac{e^2(\mu_0)}{1 - (e^2(\mu_0)/6\pi^2) \log \frac{\mu}{\mu_0}} \tag{14.109}$$

which we derived in Section 7.4, using simpler methods.

The coupling constant increases with μ, and there is the *Landau singularity* at:

$$\mu = \mu_0 \exp\left(\frac{6\pi^2}{e^2(\mu_0)}\right) \tag{14.110}$$

(Although it appears as if the coupling constant blows up at this point, we must realize that the formula breaks down in the approximation we have made, i.e., for small e only.)

14.6 Asymptotic Freedom

One of the theoretical breakthroughs in quantum field theory came when the high-energy scaling behavior found at the SLAC experiments could be explained via non-Abelian gauge theory.

Previously, in Eq. (13.13), we found that the coupling constant renormalization in gauge theory was given multiplicatively by:

$$g_0 = g\mu^{\epsilon/2} \frac{Z_1}{Z_2\sqrt{Z_3}} \tag{14.111}$$

If we put in the values of the Z's, we find:

$$g_0 = g\mu^{\epsilon/2} \left[1 - \frac{g^2}{8\pi^2\epsilon}\left(\frac{11}{6}C_{ad} - \frac{2}{3}C_f\right) \right] + \cdots \tag{14.112}$$

We can then solve for β:

$$\begin{aligned}
\beta &= \mu \frac{\partial g}{\partial \mu} \\
&= \left(g\frac{\partial}{\partial g} - 1\right)\left(\frac{-g^3}{16\pi^2}\right)\left(\frac{11}{6}C_{ad} - \frac{2}{3}C_f\right) + \cdots \\
&= -\frac{g^3}{16\pi^2}\left(\frac{11}{3}C_{ad} - \frac{4}{3}C_f\right) + \cdots
\end{aligned} \tag{14.113}$$

We come to the rather surprising conclusion that the theory is asymptotically free if the following prescription is satisfied:

$$\frac{11}{3}C_{ad} > \frac{4}{3}C_f \tag{14.114}$$

This is the first example of *asymptotic freedom*,[13-15] which was discovered by Gross, Wilczek, Politzer, and independently by 't Hooft. Asymptotic freedom only occurs in the presence of gauge theories. For QCD, we have the group $SU(3)$, so that $C_{ad} = 3$. The final relationship now reads:

$$\beta(g) = -\frac{g^3}{16\pi^2}\left(11 - \frac{2}{3}N_f\right) + \cdots \tag{14.115}$$

where N_f is the number of flavors of fermions in the theory. This is one of the theoretical triumphs of gauge theory, that gauge theory proves to be the most important ingredient in any asymptotically free theory.

The value of the coupling constant can also be integrated explicitly. Performing the integration, we find to the one-loop level:

$$g^2(\mu) = \frac{g^2(\mu_0)}{1 + (g^2(\mu_0)/8\pi^2)\left(\frac{11}{3}C_{ad} - \frac{4}{3}C_f\right)\log\mu/\mu_0} \tag{14.116}$$

Because of the importance of asymptotic freedom, the β function has even been computed to three loops:

$$\beta = -\frac{g^3}{16\pi^2}\left(11 - \frac{2}{3}N_f\right) - \frac{g^5}{(16\pi^2)^2}\left(102 - \frac{38}{3}N_f\right)$$

$$-\frac{g^7}{(16\pi^2)^3}\left(\frac{2857}{2} - \frac{5033}{18}N_f + \frac{325}{54}N_f^2\right) + O(g^9) \tag{14.117}$$

(We should mention that the β function actually vanishes to all orders in perturbation theory for certain forms of super Yang–Mills theories, which are finite to all orders in perturbation theory. This will be discussed in more detail in Chapter 20.)

In summary, asymptotic freedom means that, roughly speaking, at shorter and shorter distances, the coupling constant decreases in size, so that the theory appears to be a free theory. This is the phenomenon of scaling, which is simply interpreted as the quarks acting as if they were free partons in the high-energy realm.

Conversely, at larger and larger distances, the coupling constant increases, so that at a certain point perturbative calculations can no longer be trusted. Large coupling constants, in turn, imply that the quarks bind more tightly together, giving rise to confinement. This is called "infrared slavery," which is the flip side of asymptotic freedom.

Finally, we remark that there is a simple way in which we can describe asymptotic freedom, which only manifests itself in non-Abelian gauge theories. Although this example does not explain asymptotic freedom, it gives us a convenient intuitive model by which to describe it.

In the case of QED, we know that, at large distances, the effective coupling constant α gets smaller. This is because any charged particle is surrounded by a dense cloud of electron–positron virtual pairs that tend to screen the charge of a particle. Thus, the effective coupling constant is reduced by the presence of this screening charge. At smaller distances, and higher energy, a probe can penetrate through this virtual cloud, and hence the QED coupling constant gets larger as we increase the energy of the probe.

Classically, we can think of this in terms of the dielectric constant of the vacuum. If we place a charge in a dielectric, we know that the electric field of the dielectric causes the dipoles within the dielectric medium to line up. The net effect of the dipoles lining up around the charge is to decrease the charge, so the medium has a dielectric constant greater than one.

The situation in QCD, as we have seen, is precisely the opposite. We no longer have an electric charge (since QCD gauge particles have neutral charge), but we have color charges and color coupling constants. This means that, at large distances (low energy) the presence of the cloud of virtual particles creates an antiscreening effect. The net coupling constant get larger at large distances. Contrary to the situation in QED, a probe that comes near a colored particle feels the coupling constant decrease at high energies. Thus, the dielectric constant of the vacuum is less than one for an asymptotically free theory.

14.7 Callan–Symanzik Relation

We now would like to clarify certain points that were ignored earlier. We pointed out after Eq. (14.90) that β is not, strictly speaking, just a function of g. It can also be a function of the dimensionless parameter m/μ, and hence the renormalization-group equations become much more difficult to solve. Therefore our previous derivation of the scaling relations, although correct, was actually incomplete.

There are several ways in which to complete this subtle but important step. The first is to use a slightly different form for the renormalization group equations, called the *Callan–Symanzik* relations,[11,12] which are written as derivatives with respect to the bare masses, rather than the subtraction point μ. In this case, β and γ appear in slightly different form, but are now functions of just one variable, the renormalized coupling constant. Then the renormalization group equations can be rigorously solved, and we find the scaling relations derived earlier.

There is also a second solution, which is to use a different regularization scheme, called the mass-independent minimal subtraction (MS) scheme,[16] where the mass dependence drops out from the very beginning in the definition of β. Then we can ignore the mass dependence of these functions because of the way that we have regularized all divergent integrals.

We will discuss the first solution to this subtle problem using the Callan–Symanzik relations, where the renormalization group equations are derived from a slightly different set of physical assumptions than before. Then later we will discuss the MS scheme.

We begin with the obvious identity that the derivative of a propagator, with respect to the unrenormalized mass squared, simply squares the propagator:

$$\frac{\partial}{\partial m_0^2}\left(\frac{i}{p^2 - m_0^2 + i\epsilon}\right) = \frac{i}{p^2 - m_0^2 + i\epsilon}(-i)\frac{i}{p^2 - m_0^2 + i\epsilon} \tag{14.118}$$

or simply:

$$\frac{\partial}{\partial m_0^2} i\Delta_F = i\Delta_F(-i)i\Delta_F \tag{14.119}$$

Now assume that $i\Delta_F$ occurs in some vertex function $\Gamma_0^{(n)}$ of arbitrary order. Each time a propagator appears, the derivative replaces the propagator with the square of the propagator. From a field theory point of view, the squaring of the propagator (with the same momentum) is equivalent to the insertion of the operator $\phi^2(x)$ in the diagram with zero momentum. [We recall that the addition of the counterterm $\delta m^2 \phi^2$ into the action had the net effect of converting each Δ_F into Δ_F^2. In the same way, the squaring of each propagator can be simulated by the insertion of $\phi^2(x)$, which acts like a counterterm.]

This means that the derivative of an arbitrary vertex function with respect to m_0^2 yields another vertex function where $\phi^2(x)$ with zero momentum has been inserted. In other words:

$$\frac{\partial \Gamma_0^{(n)}(p_i)}{\partial m_0^2} = -i\Gamma_{0,\phi^2}^{(n)}(0; p_i) \tag{14.120}$$

where $\Gamma_{0,\phi^2}^{(n)}$ represents a vertex function with the insertion of this composite operator.

We now make the transition from the unrenormalized vertices to the renormalized ones. This means the introduction of yet another renormalization constant Z_{ϕ^2} to renormalize the insertion of the composite field operator.

As before, the relationship between the renormalized and unrenormalized vertices is given by:

$$\Gamma^{(n)}(p_i, g, m) = Z_\phi^{n/2}\Gamma_0^{(n)}(p_i, g_0, m_0)$$

$$\Gamma_{\phi^2}^{(n)}(p, p_i, g, m) = Z_{\phi^2}^{-1}Z_\phi^{n/2}\Gamma_{0,\phi^2}^{(n)}(p, p_i, g_0, m_0) \qquad (14.121)$$

Now we use the chain rule to write:

$$\frac{\partial}{\partial m_0^2} = \frac{\partial m^2}{\partial m_0^2}\frac{\partial}{\partial m^2} + \frac{\partial g}{\partial m_0^2}\frac{\partial}{\partial g} \qquad (14.122)$$

As before, we now apply this operator on the unrenormalized vertices, which then picks up derivatives of the renormalization constants Z_ϕ and Z_{ϕ^2}.

Putting everything together, and dividing by $\partial m^2/\partial m_0^2$, we find:

$$\left(m\frac{\partial}{\partial m} + \beta\frac{\partial}{\partial g} - n\gamma\right)\Gamma^{(n)}(p_i, g, m) = -im^2\alpha\Gamma_{\phi^2}^{(n)}(0, p_i, g, m) \qquad (14.123)$$

where:

$$\beta = 2m^2\frac{\partial g}{\partial m_0^2}\left(\frac{\partial m^2}{\partial m_0^2}\right)^{-1}$$

$$\gamma = m^2\frac{\partial \log Z_\phi}{\partial m_0^2}\left(\frac{\partial m^2}{\partial m_0^2}\right)^{-1}$$

$$\alpha = \frac{\partial Z_{\phi^2}}{\partial m_0^2}\left(\frac{\partial m^2}{\partial m_0^2}\right) \qquad (14.124)$$

Although these equations look suspiciously like the previous renormalization-group equations in Eq. (14.90), there are many subtle but crucial differences. First, the definitions of the parameters, like β, are different from the usual ones. Most important, the previous renormalization group equations were written as derivatives with respect to the subtraction point μ, while the new ones are written with respect to the unrenormalized mass m_0. Second, it can be shown that these functions are strictly functions of just one variable, the coupling constant. Hence, they can be solved using the methods outlined earlier. Third, there is an inhomogeneous term on the right-hand side of this equation, while the previous renormalization group equation did not have this term.

Next, we would like to show how to eliminate the inhomogeneous term appearing in the Callan–Symanzik relation that does not appear in the original formulation of the renormalization group equations. To eliminate this term, we will appeal to Weinberg's theorem. In the version that we need, this theorem tells us that if we scale the external momenta as $p_i \rightarrow \lambda p_i$ in the deep Euclidean region, the one-particle-irreducible Green's functions $\Gamma^{(n)}(p_i)$ grow as λ^{4-n} times lower polynomials in $\log \lambda$, while the Green's function $\Gamma^{(n)}_{\phi^2}$ grows only as λ^{2-n} times similar polynomials. We note that this divergence is just what one might expect from naive dimensional grounds, and is also the superficial degree of divergence of the graph.

Mathematically, Weinberg's theorem tells us that:

$$\Gamma^{(n)}(\lambda p_i, g, m) \rightarrow \lambda^{4-n} \left(\sum_{n=1}^{\infty} a_n (\log \lambda)^{b_n} g^{n-1} \right) \tag{14.125}$$

for some constants a_n, b_n. (In principle, the logarithms can sum to a nontrivial expression, giving us the possibility that the entire expression scales as $\lambda^{4-n-\gamma}$, where γ is just the familiar anomalous dimension.)

For our purposes, the important point is that scaling takes us into the deep Euclidean region, where $\Gamma^{(n)}$ is much larger than $\Gamma^{(n)}_{\phi^2}$, so that we can drop the latter term.

The Callan–Symanzik equations then become homogeneous, like the previous equations given earlier, in this limit. Then the equations can be solved, much like Eq. (14.101).

14.8   Minimal Subtraction

Finally we remark that it seems remarkable that the renormalization group equations work at all, that is, that we can extract information concerning the higher-order behavior of the coupling constants knowing only the one-loop results.

For example, if we power expand the coupling constant g, we find an infinite series of logarithms. The renormalization group equations, on the basis of just the one-loop results, are able to reproduce the leading logarithmic behavior of the entire function, without having to compute any higher-loop Feynman diagrams.

To see the origin of this rather mysterious but important result, it is perhaps instructive to use what is called the *minimal subtraction scheme*.[16]. The MS scheme defines the renormalized coupling constants strictly in terms of their poles using dimensional regularization. Since we have the freedom to choose where we separate the infinite part from the finite part, we will define the subtraction scheme so that we only take the poles in ϵ, so the counterterms have no finite parts.

The MS scheme has a further advantage because it is a mass-independent regularization scheme. The Z's depend on μ *only* through the renormalized coupling constants. By dimensional arguments, hence the MS scheme produces the functions β, etc., which are independent of the renormalized mass. We mentioned earlier that our original derivation of the renormalization group equations in Eq. (14.90) ignored the fact that β could, in principle, be a function of both g and m/μ, which made solving the renormalization group equations difficult. We ignored the dependence on m/μ because there exists a regularization scheme, the MS, in which β appears strictly independent of the renormalized mass m.

To begin our discussion of the MS, we define our unrenormalized variables as follows:

$$
\begin{aligned}
g_0 &= \mu^\epsilon \left(g + \sum_{n=1}^\infty \frac{g_n(g)}{\epsilon^n} \right) \\
m_0 &= \mu \left(1 + \sum_{n=1}^\infty \frac{m_n(g)}{\epsilon^n} \right) \\
\phi_0 &= \phi \left(1 + \sum_{n=1}^\infty \frac{\phi_n(g)}{\epsilon^n} \right)
\end{aligned}
\tag{14.126}
$$

The coefficients in the expansion are independent of m/μ, without any finite parts at all. We also assume that $\mu(\partial g/\partial \mu)$ is a smooth analytic function of ϵ, so that we can expand:

$$
\mu \frac{\partial g}{\partial \mu} = \sum_{n=0}^\infty d_n \epsilon^n
\tag{14.127}
$$

Now the key physical input is this: the bare quantities are all, by definition, independent of the subtraction point. Thus, we can differentiate them and set them to zero. Thus:

$$
\mu^{-\epsilon} \frac{\partial g_0}{\partial \mu} = 0
\tag{14.128}
$$

or:

$$
\epsilon g + g_1 + \mu \frac{\partial g}{\partial \mu} + \sum_{n=1}^\infty \epsilon^{-n} \left[\frac{\partial g_n}{\partial g} \mu \frac{\partial g}{\partial \mu} + g_{n+1} \right] = 0
\tag{14.129}
$$

This is a set of nontrivial, highly coupled equations linking the various terms in the MS scheme. To solve them, let us insert the power expansion of $\mu(dg/d\mu)$ into the previous equation and sort out powers of ϵ:

$$
\epsilon(g+d_1) + \left(g_1 + d_0 + d_1 \frac{dg_1}{dg} \right) + \sum_{n=1}^\infty \frac{1}{\epsilon^n} \left(g_{n+1} + d_0 \frac{dg_r}{dg} + d_1 \frac{dg_{r+1}}{dg} \right) = 0
\tag{14.130}
$$

Since each order of ϵ must vanish separately, we now have the equations:

$$d_1 = -g; \quad g_1 + d_1 g_1' = -d_0; \quad \left(1 + d_1 \frac{d}{dg}\right) g_{n+1} = -d_0 g_n' \qquad (14.131)$$

as well as:

$$\left(1 - g \frac{d}{dg}\right) [g_{n+1} - g_1] = g_n' \qquad (14.132)$$

This is an important recursion relation, because it shows that the residues of the higher-order pole terms can, in principle, be determined from a knowledge of just the simple pole term.

We can repeat the same steps for the other parameters. Since the other unrenormalized parameters are also independent of the subtraction point μ, we know that $\partial m_0/\partial \mu = 0$ and $\partial \phi_0/\partial \mu = 0$. We can therefore derive the recursion relation for the other residues:

$$
\begin{aligned}
g m_{n+1}' &= m_n g m_1' - m_n' \left(1 - g \frac{d}{dg}\right) g_1 = 0 \\
g \phi_{n+1}' &= \phi_n g \phi_1' - \phi_n' \left(1 - g \frac{d}{dg}\right) g_1
\end{aligned}
\qquad (14.133)
$$

In terms of the original renormalization group parameters, we also have:

$$
\begin{aligned}
\beta(g) &= -g_1 + g a_1' \\
\gamma(g) &= g c_1' \\
\gamma_m(g) &= g b_1'
\end{aligned}
\qquad (14.134)
$$

We can draw several interesting conclusions from this simple exercise. It is possible to construct a self-consistent renormalization scheme based entirely on dropping all finite parts in the counterterms. Thus, the counterterms are chosen to cancel just the poles, nothing more. Then the mass dependence within the renormalization group parameters disappears, and our previous assumption about dropping the m/μ dependence is justified. (We should point out there exists a modified MS procedure, called $\overline{MS}$, which is used extensively in the literature. In the $\overline{MS}$ scheme, we eliminate the poles along with certain finite transcendental constants.) Furthermore, the higher-order terms (in principle) can be determined from lower-order terms by a recursion relation. The nth-order coefficients are all determined from the $(n-1)$st-order terms. In other words, the renormalization group equations tell us that the knowledge of the lowest-order terms will automatically determine much of the higher-order behavior, without actually having to compute all higher-order graphs. Thus, it is now no mystery why the renormalization group equations only need, as input, the lowest-order one-loop results, yet manage to determine much of the higher-order behavior without having to perform multiloop calculations. The mathematical essence of the renormalization group equations is that they are a recursion relation that "bootstraps" all the higher coefficients from a knowledge of just the lower ones.

14.9 Scale Violations

The actual experiments done on deep inelastic scattering not only give us the scaling behavior, they also give us the deviation to exact scaling, which we would now like to calculate using renormalization group methods. In particular, we will write down the renormalization group equations for the structure functions found in lepton–nucleon scattering.

In the language of the operator product expansion given earlier, we can write the behavior of two operators near the light–cone:

$$A(x)B(0) \sim \sum_{i,n} C_i^n(x^2) x_{\mu_1} x_{\mu_2} \cdots x_{\mu_n} O_i^{\mu_1 \mu_2 \cdots \mu_n}(x) \qquad (14.135)$$

where we no longer assume that we are dealing with free quark states. The singularity found earlier for the free quark model can be included in the C_i^n function. The summation is performed over the spin n of the operators, and also the type i, which is not yet specified.

We know that the dimension $d_A + d_B$ of the left-hand side must be equivalent to the dimension of the right-hand side, which is given by $-n + d_{O_i}$. By scaling arguments, we then have the behavior of the coefficients near the light–cone:

$$C_i^n(x^2) \sim \left(x^2\right)^{-(d_A+d_B+n-d_{O_i})/2} \qquad (14.136)$$

Examining the power expansion, we see it is in general dominated by the operator with minimum *twist*, which is defined by:

$$\text{Twist} = \tau = d_{O_i} - n \qquad (14.137)$$

that is, the twist is equal to the dimension of the operator minus the spin. For example, simple operators of $\tau = 1$ are given by:

$$\phi; \ \partial_\mu \phi; \ \partial_\mu \partial_\nu \phi \qquad (14.138)$$

In deep inelastic scattering, the relevant operators are composed of quark fields ψ and gluons A_μ^a, so the minimum twist operators have $\tau = 2$ and their product expansion is given by:

$$
iT\left[J_\mu(x)J_\nu(0)\right] = \sum_{n,i}\left[-g_{\mu\nu}x_{\mu_1}x_{\mu_2}\cdots x_{\mu_n}i^n C_{1,i}^n(x^2,g,\mu)\right.
$$
$$
\left. + g_{\mu\mu_1}g_{\nu\mu_2}x_{\mu_3}\cdots x_{\mu_n}i^{n-2}C_{2,i}^n(x^2,g,\mu)\right]O_i^{\mu_1\mu_2\cdots\mu_n}(0)
$$
$$(14.139)$$

As before, by Lorentz invariance we can write the matrix element of O_i as a function of the momentum p:

$$
\langle p|O_i^{\mu_1\mu_2\cdots\mu_n}|p\rangle = A_i^n\left(p^{\mu_1}p^{\mu_2}\cdots p^{\mu_n} + \text{trace terms}\right) \qquad (14.140)
$$

where A_i^n are undetermined constants, and where the trace terms arise because the operator is traceless and symmetric.

In this form, the traces are a bit unmanageable. But we will use a trick. Each p_{μ_i} is contracted onto a x_{μ_i}, which in turn can be converted into $\partial/\partial q_{\mu_i}$ when we take the Fourier transform of the expression. Then we can use the identity:

$$
\frac{\partial}{\partial q_{\mu_1}}\frac{\partial}{\partial q_{\mu_2}}\cdots\frac{\partial}{\partial q_{\mu_n}} = 2^n q^{\mu_1}q^{\mu_2}\cdots q^{\mu_n}\left(\frac{\partial}{\partial q^2}\right)^n + \text{trace} \qquad (14.141)
$$

With this substitution, we can absorb all the trace terms into a single differential.

The goal of this process is to be able to determine the nature of the structure functions of lepton–nucleon deep inelastic scattering. We are interested in the tensor:

$$
T_{\mu\nu} = \frac{i}{2M}\int\frac{d^4x}{(2\pi)}e^{ip\cdot y}\langle p|T\left(J_\mu(x)J_\nu(0)\right)|p\rangle
$$
$$
= \left(-g_{\mu\nu}+\frac{q_\mu q_\nu}{q^2}\right)T_1 + \frac{1}{M^2}\left(p_\mu - \frac{p\cdot q}{q^2}q_\mu\right)\left(p_\nu - \frac{p\cdot q}{q^2}q_\nu\right)T_2
$$
$$(14.142)$$

The relation of T_i to the previous structure functions found earlier is given by:

$$
W_{1,2} = \frac{1}{\pi}\text{Im }T_{1,2} \qquad (14.143)
$$

Inserting Eq. (14.139) into Eq. (14.142), we find:

$$
T_{\mu\nu} = \frac{1}{2M} \sum_{n,i} \left[-g_{\mu\nu} \left(\frac{2p \cdot q}{-q^2} \right)^n \tilde{C}_{1,i}^n(q^2, g, \mu) \right.
$$
$$
\left. + p_\mu p_\nu (2p \cdot q)^{n-2}(-q^2)^{1-n} \tilde{C}_{2,i}^n(q^2, g, \mu) \right] A_i^n \qquad (14.144)
$$

where $Q^2 = -q^2$ and:

$$
\tilde{C}_{1,i}^n(Q^2, g, \mu) = (Q^2)^n \left(\frac{\partial}{\partial q^2} \right)^n \int d^4x \, e^{iq \cdot x} C_{1,i}^n(x^2, g, \mu)
$$
$$
\tilde{C}_{2,i}^n(Q^2, g, \mu) = (Q^2)^{n-1} \left(\frac{\partial}{\partial q^2} \right)^{n-2} \int d^4x \, e^{iq \cdot x} C_{2,i}^n(x^2, g, \mu)
$$
$$
(14.145)
$$

Comparing this expression for $T_{\mu\nu}$ with the original definition, we easily find:

$$
T_1(x, Q^2) = \frac{1}{2M} \sum_{n,i} x^{-n} \tilde{C}_{1,i}^n(Q^2, g, \mu) A_i^n
$$
$$
T_2(x, Q^2) = \frac{1}{2M} \sum_{n,i} x^{-n+1} \tilde{C}_{2,i}^n(Q^2, g, \mu) A_i^n \qquad (14.146)
$$

Taking the moments of $T_{1,2}$ with respect to powers of x, we then find:

$$
\int_0^1 dx \, x^{n-2} F_2(x, Q^2) \sim \frac{1}{8} \sum_i \tilde{C}_{1,i}^n(Q^2, g, \mu) A_i^n
$$
$$
\int_0^1 dx \, x^{n-1} F_1(x, Q^2) \sim \frac{1}{4} \sum_i \tilde{C}_{2,i}^n(Q^2, g, \mu) A_i^n \qquad (14.147)
$$

Up to now, we have not used the power of the renormalization group. Since these form factors are physical, measurable quantities, we now impose the fact that they must obey renormalization group equations of the form:

$$
\left[\left(\mu \frac{\partial}{\partial \mu} + \beta(g) \frac{\partial}{\partial g} \right) \delta_{jk} - \gamma_{kj}^n \right] \tilde{C}_{\alpha,j}^n(q^2, g, \mu) = 0 \qquad (14.148)
$$

The solution to this equation is easy to find:

$$
\tilde{C}_{\alpha,i}^n(Q^2/\mu^2, g) \sim \sum_i \tilde{C}_{\alpha,j}^n(1, \bar{g}(t)) \exp \left\{ -\int_0^t dt' \, \gamma_{ji}^n \left(\bar{g}(t') \right) \right\} \qquad (14.149)
$$

where $t = (1/2)\log(Q^2/\mu^2)$. We now use the fact that, to lowest order, we have:

$$
\begin{aligned}
\beta &= -bg^3 + O(g^4) \\
\gamma_{ij}^n &= d_{ij}^n g^2 + O(g^3)
\end{aligned}
\tag{14.150}
$$

(The d_{ij}^n, in fact, are exactly computable using lowest order perturbation theory.)

Then the Wilson coefficients obey:

$$
\tilde{C}_{\alpha,i}^n(Q^2/\mu^2, g) \sim \sum_j \tilde{C}_{\alpha,j}^n(1, 0) \left[\log\left(Q^2/\mu^2\right)\right]^{-d_{ij}^n/2b}
\tag{14.151}
$$

Reinserting these equations back into the expression for the integral of the moments, we find:

$$
\begin{aligned}
M_2^n(Q^2) &\equiv \int_0^1 dx\, x^{n-2} F_2(x, Q^2/\mu^2) \\
&\sim \frac{1}{8} \sum_i \tilde{C}_{2,j}^n(1, 0) A_i^n \left[\log\left(Q^2/\mu^2\right)\right]^{-d_{ji}^n/2b} \\
M_1^n(Q^2) &\equiv \int_0^1 dx\, x^{n-1} F_1(x, Q^2/\mu^2) \\
&\sim \frac{1}{4} \sum_i \tilde{C}_{1,j}^n(1, 0) A_i^n \left[\log\left(Q^2/\mu^2\right)\right]^{-d_{ji}^n/2b}
\end{aligned}
\tag{14.152}
$$

This is our final result. With a few modest assumptions and the renormalization group equations, we can compute the logarithmic corrections to Bjorken scaling. The point is that F_1 is now a function of *both* x and Q, but the momentum dependence is given by logs, which gives a weak violation of Bjorken scaling, as observed experimentally.

14.10 Renormalization Group Proof

We began our discussion of renormalization theory in Chapter 7 with ϕ^4 theory, but did not complete it because of the problem of overlapping divergences; for example, there was no unique skeleton expansion of certain graphs, giving us the headache of the overcounting of graphs. Although the ϕ^4 renormalization program was simpler than the one for QED, the final step could not be completed because the skeleton reduction was not unique. For QED, however, the renormalization program, although more difficult, could be completed because the

Ward–Takahashi identity allowed us to write the vertex graphs (which do not have any overlapping divergences) as the derivative of the self-energy graphs (which have overlapping divergences). Then we could write all recursion relations strictly in terms of graphs that have no overlapping divergences. The essential idea was that taking derivatives of self-energy graphs, which have no skeleton reduction, creates insertions of zero momentum photons. The insertions of these photons, we saw, converts a self-energy graph (without a skeleton reduction) into a vertex graph (with a skeleton reduction).

Now, the renormalization group equations can be viewed as the "Ward" identity for scale invariance. They will allow us to complete the renormalization of ϕ^4 theory. The key, once again, is that taking the derivative of a self-energy graph creates insertions of ϕ^2 that give us vertex graphs that have a skeleton reduction.

We remarked earlier than almost any functional recursion relation, linking the $(r + 1)$st order term to the rth term, can be used as a basis to prove the renormalizability of field theory if they have no overlapping divergences. We recall that the Callan–Symanzik relations were derived by taking the derivative of a vertex function with respect to the mass. This squared the propagator, which could then be interpreted as the insertion of the operator $\phi^2 = \theta$ into the theory. By expanding out the derivative with respect to the unrenormalized mass, we found:

$$\left(\mu\frac{\partial}{\partial\mu} + \beta(g)\frac{\partial}{\partial g} + n\gamma(g)\right)\Gamma^{(n)}(p; g, \mu) = -i\mu^2\alpha(g)\Gamma^{(n)}_\theta(0, p, g, \mu)$$

$$\left(\mu\frac{\partial}{\partial\mu} + \beta(g)\frac{\partial}{\partial g} + n\gamma(g)\right)\Gamma^{(n)}_\theta(q, p, g, \mu) = -i\mu^2\alpha(g)\Gamma^{(n)}_{\theta\theta}(0, q, p, g, \mu)$$

$$(14.153)$$

where the second equation arises by taking the second derivative with respect to the mass squared.

Our approach will now be to treat the Callan–Symanzik relations as the equivalent of the Ward–Takahashi identity, giving us functional recursion relations that will allow us to complete the induction step.[17,18] We will discuss the ϕ^4 theory, but the method is quite general. We can renormalize QED and non-Abelian gauge theories without too much difficulty.

We recall that, for ϕ^4 theory, the vertices $\Gamma^{(n)}$ for $n > 4$ all have a degree of divergence less than zero. Furthermore, they have a skeleton expansion. In this case, this means that they do not have subgraphs that have positive degree of divergence (i.e., there are no nontrivial insertions of $\Gamma^{(2)}$ and $\Gamma^{(4)}$). Since both the overall divergence is negative and the divergence of all subgraphs is also negative, then the graph itself is convergent. If we can understand the behavior of $\Gamma^{(2)}$ and $\Gamma^{(4)}$, then we can determine the behavior of all the $\Gamma^{(n)}$ for $n > 4$ by a skeleton expansion. We will thus concentrate on these two types of vertices.

We will thus assume that $\left[\Gamma^{(4)}\right]_{(r+1)}$, $\left[\Gamma^{(2)}\right]_{(r)}$, and $\left[\Gamma^{(2)}_\theta\right]_{(r)}$ are all finite quantities, where r denotes the order of the perturbation expansion. We also assume that $[\beta]_{(r+1)}$ and $[\gamma]_{(r)}$ are all finite as well.

Our task is then to show that renormalization-group equations allow us to complete the induction step, that is, to calculate $\left[\Gamma^{(4)}\right]_{r+2}$, $\left[\Gamma^{(2)}\right]_{r+1}$, and $\left[\Gamma^{(2)}_\theta\right]_{r+1}$ in terms of the known quantities at a lower order. A close look at the renormalization group equations shows that they are ideally suited for such a task.

The calculation is then carried out in three steps:

1. Calculate $\left[\Gamma^{(4)}\right]_{r+2}$ in terms of the finite quantities.

2. Calculate $\left[\Gamma^{(2)}_\theta\right]_{r+1}$.

3. Calculate $\left[\Gamma^{(2)}\right]_{r+1}$.

14.10.1 Step One

The first calculation is rather easy, since we can write the renormalization group equations in the following fashion:

$$\mu \frac{\partial}{\partial \mu} \left[\Gamma^{(4)}\right]_{r+2} = -i\mu^2 \left[\Gamma^{(4)}_\theta\right]_{r+2} - \left[\left(\beta(g)\frac{\partial}{\partial g} + 4\gamma\right)\Gamma^{(4)}\right]_{r+2} \qquad (14.154)$$

The first term on the right-hand side has a skeleton expansion. This means that, at most, it contains $\left[\Gamma^{(2)}\right]_r$, $\left[\Gamma^{(2)}_\theta\right]_r$, and $\left[\Gamma^{(4)}\right]_{r+1}$, and hence, by construction, it is finite. (One might suspect that the overall integration over these finite pieces might contribute a divergence, but since the superficial divergence is -1, there is no problem.)

The last term in the previous expression causes some problems, since we have the term $[\beta]_{r+2}$ and $[\gamma]_{r+1}$ multiplying the lowest-order term in $\Gamma^{(4)}$ (which is $-ig$). Therefore one might worry about the term $[\beta]_{r+1} + 4g[\gamma]_{r+1}$ that appears on the right-hand side.

However, we now use one more bit of information to show that this last remaining term is finite. If we take the zero momentum renormalization group equations, we find that they reduce to:

$$[\beta + 4\gamma g]_{r+2} = \mu^2 \left[\Gamma^{(4)}_\theta(0)\right]_{r+2} \qquad (14.155)$$

But we already showed that this was finite (since it it had a skeleton decomposition given by finite quantities).

In summary, all terms on the right-hand side have been shown to be finite. The last step is trivial, which is to integrate both sides of the equation in order to calculate $\left[\Gamma^{(4)}\right]_{r+2}$.

We can write the renormalization group equation as:

$$\mu \frac{\partial}{\partial \mu} \left[\Gamma^{(4)}\left(p/\mu, \Lambda/\mu, g\right)\right]_{r+2} = \left[\Phi\left(p/\mu, \Lambda/\mu; g\right)\right]_{r+2} \qquad (14.156)$$

where we have now included explicitly the argument of the vertex functions, and have written the right hand side as simply Φ. Integrating, we have:

$$\left[\Gamma^{(4)}\left(p/\mu, \Lambda/\mu; g\right)\right]_{r+2} = -ig - \int_0^1 \frac{d\alpha}{\alpha} \left[\Phi\left(\alpha p/\mu, \alpha\Lambda/\mu; g\right)\right]_{r+2} \qquad (14.157)$$

Thus, we have shown that all terms on the right-hand side of the renormalization-group equations are finite, so therefore $\left[\Gamma^{(4)}\right]_{r+2}$ is also finite.

14.10.2   Step Two

In the second step, we will rewrite the second renormalization group equation as:

$$\mu \frac{\partial}{\partial \mu} \left[\Gamma_\theta^{(2)}\right]_{r+1} = -i\mu^2 \left[\alpha \Gamma_{\theta\theta}^{(2)}\right]_{r+1}$$
$$- \left[\left(\beta(g)\frac{\partial}{\partial g} + 2\gamma + \gamma_\theta\right)\Gamma_\theta^{(2)}\right]_{r+1} \qquad (14.158)$$

This can also be shown to be finite by repeating the same steps given earlier. We first remark that $\Gamma_{\theta\theta}^{(2)}$ is finite since it has a skeleton decomposition and can be calculated in terms of finite, lower-order parts. The only troublesome term is the one on the right, which appears in the combination $[2\gamma + \gamma_\theta]_{r+1}$, which does not appear to be finite. However, as before, we simply take the zero momentum limit of this equation. Then this precise combination $[2\gamma + \gamma_\theta]$ appears in the low-momentum limit of $\Gamma_{\theta\theta}$, which we just showed to be finite. Finally, we then integrate the entire equation in μ to arrive at an expression for $\left[\Gamma_\theta^{(2)}\right]_{r+1}$.

14.10.3   Step Three

Finally, to compute the remaining function $\left[\Gamma^{(2)}\right]_{r+1}$, we write the first renormalization group equation as:

$$\mu \frac{\partial}{\partial \mu} \left[\Gamma^{(2)}\right]_{r+1} = -i\mu^2 \left[\alpha \Gamma_\theta^{(2)}\right]_{r+1}$$

$$- \left[\left(\beta(g)\frac{\partial}{\partial g} + 2\gamma \right) \Gamma^{(2)} \right]_{r+1} \tag{14.159}$$

We also invoke similar arguments to show this is finite. First, we know that $\left[\Gamma^{(2)}_{\theta} \right]_{r+1}$ is finite. Therefore, the only troublesome term is the one involving $[\beta]_{r+2}$ and $[\gamma]_{r+1}$.

Although this last term does not appear to be finite, we can power expand the previous renormalization group equation around $p^2 = 0$, and we have:

$$2[\gamma]_{r+1} = \mu^2 \left(\alpha \frac{d}{dp^2} \Gamma^{(2)}_{\theta}(p^2) \Big|_{p^2=0} \right)_{r+1} \tag{14.160}$$

This allows us to determine that $[\gamma]_{r+1}$ is finite, since the right-hand side is finite. Also, we can show that $[\beta]_{r+2}$ is finite if we review our discussion of the finiteness of $\left[\Gamma^{(4)} \right]_{r+2}$.

Thus, the entire right-hand side of the renormalization group equation can be shown to be finite, so a simple integration over μ yields $\left[\Gamma^{(2)} \right]_{r+1}$ entirely in terms of finite quantities.

In summary, the SLAC deep inelastic scattering experiments demonstrated the importance of Bjorken scaling. The simplest explanation of scaling comes from the naive quark model using either a parton description or light–cone commutators. However, this did not explain why the naive quark model should work so exceptionally well, why strong interaction corrections could be ignored, or why scaling set in so early. Ultimately, the scaling experiments were explained in terms of gauge theory and the renormalization group. We have seen that the coupling constant can change with the energy via the renormalization group equations. Since β is negative near the fixed point for theories with gauge fields, we can prove that QCD is asymptotically free; that is, at asymptotic energies, the coupling constant goes to zero. Historically, the explanation of the SLAC experiments by asymptotically free gauge theory helped to convince the scientific community of the correctness of QCD, even though quarks have never been seen experimentally.

Although the Standard Model has enjoyed great experimental success, there are still important gaps left unanswered by the theory. In particular, we cannot explain the low-energy spectrum of the hadrons until we understand quark confinement. Furthermore, we cannot understand the origin of the generation problem, the origin of the quark masses, etc. unless we go to theories beyond the Standard Model. In Part III, we will turn to these questions.

14.11 Exercises

1. Show Eq. (14.13).

2. Rewrite the light–cone commutator of two currents using scalars, as in Eq. (11.61), rather than spinors. How will Eq. (14.68) be modified?

3. Prove the mass-independent MS prescription for γ in Eq. (14.134).

4. Solve the Callan–Symazik equation in the deep Euclidean region.

5. Derive the analogue of Eq. (14.44) if we include the b and t quarks.

6. Consider the effect of the scale transformation $x^\mu \rightarrow \lambda x^\mu$ on a massless scalar field. The variation of a scalar field is given by:

$$\delta\phi(x) = (1 + x^\mu \partial_\mu)\phi \tag{14.161}$$

Prove that the variation of the Lagrangian is given by:

$$\delta\mathscr{L} = \partial_\mu(x^\mu \mathscr{L}) \tag{14.162}$$

and that the Noether current is given by:

$$J^\mu = x^\nu T^\mu_\nu + \frac{1}{2}\partial^\mu(\phi^2) \tag{14.163}$$

If the scalar particle has a mass, show that the trace of the energy-momentum tensor is proportional to the mass squared.

7. Consider the generators of an algebra given by: $P^\mu \equiv \gamma^\mu/R$ and $M^{\mu\nu} = (1/2)\sigma^{\mu\nu}$. Show that these generate the algebra $O(4, 1)$, the de Sitter group. In the limit of $R \rightarrow \infty$ (i.e., in the limit that the de Sitter sphere approaches ordinary space–time), prove that these generate the Poincaré algebra. (This is called the Wigner–Inönü contraction.)

8. Now consider the algebra generated by:

$$
\begin{aligned}
P_\mu &= i\partial_\mu \\
K_\mu &= 2x_\mu x^\nu \partial_\nu - x^2 \partial_\mu \\
M_{\mu\nu} &= i(x_\mu \partial_\nu - x_\nu \partial_\mu) \\
D &= x^\mu \partial_\mu
\end{aligned}
\tag{14.164}
$$

Prove that they have the commutation relations:

$$[P_\mu, P_\nu] \quad = \quad 0$$

$$[K_\mu, P_\nu] \quad = \quad -2(ig_{\mu\nu}D + M_{\mu\nu})$$

$$[K_\mu, D] \quad = \quad -K_\mu$$

$$[P_\mu, D] \quad = \quad P_\mu \qquad\qquad (14.165)$$

Complete all the other commutators. What transformations do these generators induce on x_μ space? Show that D corresponds to a scale transformation,
and that K_μ corresponds to an operator that is the product of an inversion,
translation, and another inversion. (An inversion is given by $x_\mu \rightarrow x_\mu/x^2$.)

9. Show that this algebra generates $O(4, 2)$.

10. Prove that:

$$O(4, 2) = SU(2, 2) \qquad\qquad (14.166)$$

which is called the *conformal group*.

11. Discuss how to use the renormalization group to prove the renormalizability
of QED. Set up the basic equations, discuss how the recursion relations might
work, but do not solve.

12. Complete the missing steps needed to prove Eqs. (14.131) and (14.132).

13. Prove Eq. (14.60).

14. Prove Eq. (14.61).

15. Prove Eq. (14.64).

16. Prove Eq. (14.68).

17. In the *background field method*, we expand A_μ around a classical background
field B_μ that satisfies the equations of motion:

$$A_\mu = B_\mu + \hat{A}_\mu \qquad\qquad (14.167)$$

where $\hat{A}_\mu$ represents the quantum fluctuations. Let us define their transformation properties as:

$$\delta B_\mu \quad = \quad \partial_\mu \Lambda + g[B_\mu, \Lambda]$$

$$\delta \hat{A}_\mu \quad = \quad g[\hat{A}_\mu, \Lambda] \qquad\qquad (14.168)$$

Choose the gauge fixing term F to be:

$$F = \partial_\mu \hat{A}^\mu + g[B_\mu, \hat{A}^\mu] = D(B)_\mu \hat{A}^\mu \qquad (14.169)$$

Prove that the gauge-fixing term violates gauge invariance, but preserves the new invariance as defined. Prove that the Faddeev–Popov determinant is also invariant. Thus, the gauge-fixed perturbation theory is still invariant under the new gauge. Counterterms are also invariant, which vastly simplifies calculations.

Part III

Nonperturbative Methods and Unification

Chapter 15
Lattice Gauge Theory

Unfortunately, the color gauge theory will remain in limbo unless we learn how to solve it and in particular get the spectrum out of it. So, in particular, I wish to emphasize how one might solve the color gauge theory to get a spectrum . . .

—K. Wilson

15.1 The Wilson Lattice

Although QCD is the leading candidate for a theory of the strong interactions, the embarrassing fact is that perturbation theory fails to reproduce many of the essential low-energy features of the hadron world, such as the spectrum of low-lying hadron states. Perturbation theory seems to be effective only in the asymptotic region, where we can use the arguments of renormalization group theory to make a comparison between theory and experimental data.

Nonperturbative methods, however, have proved to be notoriously difficult in quantum field theory. However, one of the most elegant and powerful nonperturbative methods is Wilson's *lattice gauge theory*,[1] where one may put QCD on a computer and, in principle, calculate the basic features of the low-energy strong-interaction spectrum. In fact, the only apparent limitation facing lattice gauge theory is the available computational power.[2]

Monte Carlo methods,[3,4] in particular, have given us rough qualitative agreement between experiment and theory, giving us the hope that, with a steady increase in computer power, we might be able to reduce the discrepancy between theory and experiment.

We should also point out that we must pay a price for putting QCD on the lattice. First, because the metric is Euclidean, it means that present calculations with lattice gauge theory are limited to the static properties of QCD. Although lattice gauge theory may be good for confinement and perhaps the low-energy spectrum of states, it has difficulty calculating scattering amplitudes, which are defined in Minkowski space. (One can, in principle, make an analytic continuation from Euclidean to Minkowski space, but then we need to have much greater computational power than what is currently available.) Second, lattice gauge theory explicitly breaks continuous rotational and translational invariance, since space–time is discretized. All that is left is symmetry under discrete rotations of the lattice. (Presumably, we can recover continuous symmetries when we let the lattice spacing go to zero.) Third, we are limited by the available computational power. Lattice sizes are thus unrealistically small, on the order of a fermi, so that important effects that enter at larger distances are cutoff. However, since computer power is increasing exponentially, there is hope that we will one day soon extract a realistic spectrum from lattice gauge theory.

Let us begin by defining the simplest lattice in four dimensions, a *Euclidean* hypercubical lattice with equal lattice spacing a in the x, y, z, and t direction. If we take the limit as $a \to 0$, then our action should reduce to the usual Yang–Mills action.

Between two neighboring sites of the lattice, we define a "string bit" or "link," which is a member of $SU(3)$ and is denoted by $U(n, n + \hat{\mu})$. This string bit connects the nth point with the $n + \hat{\mu}$ point, where $\hat{\mu}$ defines a direction in the μth lattice direction.

We define this string bit or link to be unitary:

$$U(n, n + \hat{\mu})^\dagger = U(n, n + \hat{\mu})^{-1} = U(n + \hat{\mu}, n) \tag{15.1}$$

Taking the inverse of a link therefore reverses its orientation. Since a unitary matrix can be written as the exponential of an imaginary matrix, we can write:

$$U(n, n + \hat{\mu}) = \exp iag \frac{\lambda^a}{2} A_\mu^a(n) \tag{15.2}$$

where g is the coupling constant, λ^a the generator of $SU(N)$, and $A_\mu^a(n)$ is the gauge field.

We define a *plaquette* as a square face of the lattice with dimensions $a \times a$ (Fig. 15.1). Our action is equal to tracing Us around each of the squares of the lattice:

$$
\begin{aligned}
S &= -\frac{1}{2g^2} \sum_p \mathrm{Tr}\, U_p \\
U_p &= U(n, n + \hat{\mu})U(n + \hat{\mu}, n + \hat{\mu} + \hat{v})U(n + \hat{\mu} + \hat{v}, n + \hat{v})U(n + \hat{v}, n)
\end{aligned}
\tag{15.3}
$$

where we symbolically sum over all plaquettes p in the four-dimensional lattice, and where with each p we associate the point $(n, \hat{\mu}, \hat{v})$. The essential point is

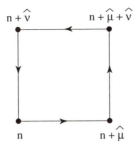

Figure 15.1. A plaquette on a Wilson hypercubical lattice. The action is defined as the sum of traces over these plaquettes.

that this formulation is both gauge invariant and also reproduces the Yang–Mills theory in the continuum limit.

A gauge transformation is defined as:

$$U(n, n + \hat{\mu}) \rightarrow \Omega(n)U(n, n + \hat{\mu})\Omega(n + \hat{\mu})^{-1} \tag{15.4}$$

Notice that a string bit is defined between two neighboring sites, while the gauge parameter $\Omega(n)$ is defined at a lattice site. Our action is invariant under this transformation, since every $\Omega(n)$ in the transformed action cancels against an $\Omega^{-1}(n)$ in $\mathrm{Tr}\, U_p$.

Finally, we take the continuum limit. To do this, we must use the Baker–Campbell–Hausdorff theorem to combine each of the Us in a plaquette into a single exponential. We use the equation:

$$e^A e^B = e^{A+B+\frac{1}{2}[A,B]+\cdots} \tag{15.5}$$

In general, we have an infinite number of terms appearing in this expansion, corresponding to all possible multiple commutators between A and B. However, because we take the limit as $a \rightarrow 0$, we need only keep the first-order terms in this expansion.

For example, if we keep the lowest-order terms and drop all commutators, we find terms like:

$$\exp iag\frac{\lambda^a}{2}\left[A_\nu(n + \hat{\mu}) - A_\nu(n)\cdots\right]^a \rightarrow \exp ia^2 g(\lambda^a/2)\left[\partial_\mu A_\nu(n)\cdots\right]^a \tag{15.6}$$

Putting everything together, we find that we can write the action as:

$$S = -\frac{1}{2g^2}\sum_p \mathrm{Tr} \exp\left[ia^2 g^2 F_{\mu\nu}(n) + \cdots\right] \tag{15.7}$$

where $A_\mu(n) = A_\mu^a \lambda^a / 2$, and:

$$F_{\mu\nu}(n) = \partial_\mu A_\nu(n) - \partial_\nu A_\mu(n) - ig[A_\mu(n), A_\nu(n)] \qquad (15.8)$$

Taking the limit as $a \to 0$, then we find the continuum result:

$$S = -\frac{1}{2g^2} \sum_p \left(1 - \frac{a^4}{2} g^2 F_{\mu\nu}^a F^{\mu\nu a} + \cdots \right)$$

$$\sim \frac{1}{4} \int d^4x\, F_{\mu\nu}^a F^{\mu\nu a} \qquad (15.9)$$

We thus recover the continuum theory in the $a \to 0$ limit.

15.2   Scalars and Fermions on the Lattice

We have placed gauge particles on the lattice in an elegant fashion, preserving exact gauge invariance on the lattice, even with finite spacing. We now generalize these results to put scalars and fermions on the lattice. In particular, we will find curious complications when fermions are introduced.

To put scalars on the lattice, we must make the substitution:

$$\partial_\mu \phi \to \frac{1}{a}(\phi_{n+\hat\mu} - \phi_n) \qquad (15.10)$$

With this simple substitution, we find that the scalar action becomes:

$$S = \int d^4x \left(\frac{1}{2} \partial_\mu \phi \partial^\mu \phi + \frac{1}{2} m^2 \phi^2 + \frac{\lambda}{4!} \phi^4 \right)$$

$$\to \sum_n \left[\frac{a^2}{2} \sum_{\mu=1}^{4} (\phi_{n+\hat\mu} - \phi_n)^2 + a^4 \left(\frac{m^2}{2} \phi_n^2 + \frac{\lambda}{4!} \phi_n^4 \right) \right] \qquad (15.11)$$

To calculate the propagator of the scalar particle on the lattice, we will find it convenient to go to momentum space. We wish to replace ϕ_n with its Fourier transform $\phi(k)$. We will define:

$$\phi_n = \int \frac{d^4k}{(2\pi)^4} e^{ik \cdot n} \phi(k) \qquad (15.12)$$

We will arbitrarily truncate the integral, since wavelengths smaller than twice the size of the lattice can be discarded. We will take:

$$-\frac{\pi}{a} \leq k_\mu \leq \frac{\pi}{a} \tag{15.13}$$

Now let us insert the Fourier expansion of ϕ_n into the free action of the scalar field on the lattice. The free part can be calculated by taking a double integral over k and k':

$$a^4 \sum_n \int \frac{d^4k}{(2\pi)^4} \int \frac{d^4k'}{(2\pi)^4} e^{i(k+k')\cdot n}(e^{iak_\mu} - 1)(e^{iak'_\mu} - 1)$$

$$= \int \frac{d^4k}{(2\pi)^4}(e^{iak_\mu} - 1)(e^{-iak_\mu} - 1)$$

$$= 4 \int \frac{d^4k}{(2\pi)^4} \sin^2(ak_\mu/2) \tag{15.14}$$

Inserting this back into the free action, we now have:

$$S = \frac{1}{2} \int \frac{d^4k}{(2\pi)^4} \left(\sum_\mu \frac{4}{a^2} \sin^2(ak_\mu/2) + m^2 \right) \phi(-k)\phi(k) \tag{15.15}$$

Not surprisingly, this differs from the usual propagator defined in momentum space. Normally, the Euclidean Klein–Gordon equation has a propagator given by $1/(k^2 + m^2)$. On the lattice, the propagator is generated by taking the inverse of:

$$k^2 + m^2 \rightarrow m^2 + \sum_\mu \frac{4}{a^2} \sin^2(ak_\mu/2) \tag{15.16}$$

In the limit as $a \rightarrow 0$, we find that the two expressions are identical (for small k). Both are parabolic, as shown in Figure 15.2. (For large k, the two expressions differ noticeably. However, large values of k are cut off.)

The relative ease with which we could put scalar particles on the lattice compares with the relative difficulty of placing fermions, especially quarks, on the lattice. A number of problems, both conceptual as well as computational, arise.

As before, we make the substitution:

$$\partial_\mu \psi \rightarrow \frac{1}{a}(\psi_{n+\hat\mu} - \psi_n) \tag{15.17}$$

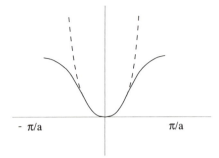

$- \pi/a$ π/a

Figure 15.2. For small k, both the lattice and continuum inverse propagators behave like $k^2 + m^2$. For large k, where the lattice approximation is not reliable, they differ.

With this substitution, our lattice fermionic action becomes:

$$S = \sum_n \left(\frac{a^3}{2} \sum_{\mu=1}^{4} \bar{\psi}_n \gamma^\mu (\psi_{n+\hat{\mu}} - \psi_{n-\hat{\mu}}) + ma^4 \bar{\psi}_n \psi_n \right) \tag{15.18}$$

As before, we take the Fourier transform of the ψ_n field. This gives us the action:

$$S = \int \frac{d^4k}{(2\pi)^4} \bar{\psi}(-k) \left(i \sum_\mu \gamma^\mu \frac{\sin(ak_\mu)}{a} + m \right) \psi(k) \tag{15.19}$$

Therefore, we wish to examine the properties of the expression:

$$\frac{1}{a^2} \sin^2 ak_\mu + m^2 \tag{15.20}$$

Unfortunately, this has bad behavior as we take the continuum limit. In Figure 15.3, this expression contains two equal minimum within the Brillouin zone. One is located at $k = 0$, as before. However, we also have the minimum located at $k = \pm\pi/a$.

Therefore, we have an unphysical doubling problem; that is, the lattice fermion theory does not give us the correct continuum limit. In fact, since we have a doubling for each space–time dimension, we actually have $2^4 = 16$ times too many fermions.

Several solutions have been proposed to cure this problem, none of them without some drawbacks. One convenient solution to the fermion doubling problem is to modify the lattice fermion action by hand, which can cancel the unwanted zero. We can always do this as long as the correct continuum limit is obtained.

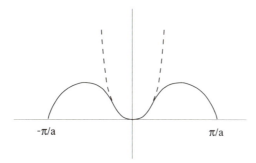

Figure 15.3. Fermion doubling problem: For small k, both the continuum and lattice fermion inverse propagators behave the same; for larger k, the lattice fermion propagator has other minima.

We add to the previous action the following Wilson term:

$$\frac{1}{2a} \bar{\psi}_n \left(\psi_{n+\hat{\mu}} + \psi_{n-\hat{\mu}} - 2\psi_n \right) \tag{15.21}$$

If we now calculate the momentum-space contribution of this term and add it to the previous one, we find:

$$S = \int \frac{d^4 k}{(2\pi)^4} \bar{\psi}(-k) \left[i \sum_{\mu} \gamma^{\mu} \frac{\sin a k_{\mu}}{a} + m - \sum_{\mu} \frac{\cos a k_{\mu} - 1}{a} \right] \psi(k) \tag{15.22}$$

The second term, containing the cosine, preserves the original minimum at $k = 0$ but eliminates the unwanted one.

The existence of this fermion doubling problem is related to the anomaly problem. As we saw previously, the various regulator schemes that we have studied for Feynman integrals violate chiral symmetry. For example, Pauli–Villars regularization adds infinite mass fermions, which violate chiral symmetry. Similarly, dimensional regulation has problems because of the presence of γ_5, which cannot be generalized in d-dimensional space. Thus, chiral symmetry is not respected by these regulator schemes, and hence an anomaly arises.

On the lattice, however, chiral symmetry is exact for massless fermions. Since chiral symmetry is respected by the lattice theory, there can be no anomaly. However, there is a price we pay for the absence of the anomaly, and this is the doubling of the fermion chiralities such that the anomaly cancels. If we calculate the chiralities of the two types of fermions, we find that they are opposite and hence produce no anomaly. The fermion doubling problem is thus deeply connected with the problem of chiral symmetry breaking on the lattice. For example, adding the

Wilson correction term violates chiral symmetry, even for zero mass fermions. Thus, studying chiral symmetry breaking on the lattice is always a bit delicate.

An even more difficult problem, from the point of view of computation, is the problem of quark loops. This is because Grassmann variables cannot be modeled on a computer. We cannot use Monte Carlo methods to minimize an action with Grassmann variables.

However, we can functionally integrate out the fermion contribution entirely, yielding determinant factors. For example, we have:

$$\int D\psi \, D\bar{\psi} \, e^{i \int L(\psi) d^4 x} = \det \left(i\gamma^\mu \partial_\mu - m \right) \tag{15.23}$$

These determinants, in turn, can be modeled on computer, although they are non-local and quite difficult to compute. This is unfortunate, since a computer simulation of QCD necessarily involves quarks. This remains one of the main computational problems facing lattice gauge theory. (However, calculations omitting the fermion loops, called the "quenched approximation," exhibit many of the nonperturbative features we expect in the final theory.)

15.3 Confinement

One of the main reasons for introducing lattice gauge theory is to calculate effects that lie beyond the reach of perturbation theory, such as the confinement of quarks. Although quark confinement has not been rigorously proved within the framework of QCD, we provide compelling reasons for believing that quarks are confined in lattice gauge theory.

In general, if the potential between two quarks is proportional to the distance between them, then the two quarks can never be separated:

$$\text{Confinement}: \quad V(r) \sim \sigma r \tag{15.24}$$

where σ is called the string tension. If we try to separate the quarks by force, then the restoring force of the linear potential between them grows sufficiently rapidly to prevent them from being separated. Furthermore, the string may break, creating a quark–antiquark pair held together by another string. Thus, they can never be separated if they are bound by a linear potential. Similarly, if the quark potential asymptotically becomes a constant or decreases with distance, then the potential is not sufficient to confine the quarks.

We will now see that, in the strong-coupling limit, the lattice gauge theory confines quarks. Let us first set up the Wilson loop:

$$\Phi(C) = P \operatorname{Tr} e^{i \oint_C A_\mu dx^\mu} \tag{15.25}$$

where C represents a closed loop and where P represents the path ordering of the exponential along the loop. (This means that we cannot simply perform the line integral dx^μ along the loop. We must first split up the line integral into an infinite number of infinitesimally small exponentials, and take the product of them ordered sequentially along the loop.) Notice that the Wilson loop is gauge invariant.

We shall primarily be interested in the Wilson loop because it gives us a criterion for confinement. The counterpart of the Wilson loop for the lattice is given by:

$$W(C) = \left\langle \operatorname{Tr} \prod_n U_n \right\rangle \tag{15.26}$$

where we take the product around a discretized loop C. We will be interested in the behavior of $W(C)$ where C is a rectangular loop with width R in one spatial direction and length T in the time direction, in the limit of large T.

Our strategy is to rewrite the path-ordered Wilson loop $W(R, T)$ in terms of the matrix elements of gauge-invariant, two-quark states. The two-quark state at time t is given by:

$$
\begin{aligned}
|\bar{q}(t, 0) q(t, R)\rangle &= \sum_C f(C) \bar{q}(t, 0) P_C \exp\left(ig \int_{(t,0)}^{(t,R)} \frac{\lambda^i}{2} A^i(z)\, dz\right) q(t, R)|0\rangle \\
&\equiv \Gamma(t, R)|0\rangle \tag{15.27}
\end{aligned}
$$

where the quark states are at equal times but are separated by a spatial distance R, where P_C takes the path-ordered exponential along the path C, and $f(C)$ is some function along the path C. The sum over C is taken over all paths that connect the two points. The presence of the path-ordered exponential guarantees that this two-quark state is gauge invariant.

Now construct the overlap function by taking the matrix element of the two-quark state at time $t = 0$ and at some later time $t = T$. After inserting a complete set of intermediate states, and taking the limit $T \to \infty$, we find:

$$
\begin{aligned}
\lim_{T \to \infty} \Omega(T, R) &\equiv \lim_{T \to \infty} \langle 0|q(T, 0)\bar{q}(T, R)|\bar{q}(0, 0)q(0, R)|0\rangle \\
&= \lim_{T \to \infty} \langle 0|\Gamma^\dagger(T, R)\Gamma(0, R)|0\rangle
\end{aligned}
$$

$$= \lim_{T \to \infty} \sum_n |\langle 0|\Gamma^\dagger(0, R)|0\rangle|^2 e^{-E_n T}$$

$$\sim \quad e^{-E_0(R)T} \tag{15.28}$$

(In the $T \to \infty$ limit, only the smallest energy eigenvalue E_0 dominates the right-hand side of the equation.)

Now let us contract the quark wave functions that appear within $\Omega(T, R)$. The quark propagator (in a background of gluon fields) can be approximated as:

$$\langle 0|q^\alpha(t, \mathbf{x})\bar{q}^\beta(t', \mathbf{x})|0\rangle = \exp\left(i \int_t^{t'} A_0(\tau, \mathbf{x})d\tau\right) \langle 0|q^\alpha(t, \mathbf{x})\bar{q}^\beta(t', \mathbf{x})|0\rangle_0 \tag{15.29}$$

where the subscript 0 refers to free quark fields.

Within Ω, we contract the quark field $q(0, 0)$ with $\bar{q}(T, 0)$, and the quark field $q(0, R)$ with $\bar{q}(T, R)$. With this contraction, we find that there are now four contributions to the path-ordered exponential integral. These four contributions complete a path-ordered integral over the sides of a closed rectangle, whose vertices are given by $(0, 0)$, $(0, R)$, $(T, 0)$, and (T, R). Because the exponentials are now taken over a closed loop, $\Omega(T, R)$ is thus proportional to $W(C)$. We have, therefore:

$$\lim_{T \to \infty} \Omega(T, R) \sim \lim_{T \to \infty} W(T, R) \sim e^{-E_0 T} \tag{15.30}$$

If the potential between the quarks grows like the distance of separation R, then the quarks are confined. We therefore have:

$$W(R, T) \to \exp(-\sigma RT) \tag{15.31}$$

Since the area of the Wilson loop is RT, the *area law* for the behavior of the Wilson loop for large T gives us confinement.

If the quark potential goes to a constant m for large distance, then we have:

$$W(R, T) \to \exp(-mT) \tag{15.32}$$

which gives us a *perimeter law* for nonconfining potentials.

15.4 Strong Coupling Approximation

Since the area law gives us a criterion for confinement, our next task is to calculate the functional integration in the path integral for gauge theory to see if QCD gives us a confining theory. We wish to calculate the functional integral of the Wilson loop for small $1/g^2$ (i.e., the strong coupling approximation):

$$\langle W(C) \rangle = \frac{1}{Z} \int DU \, \text{Tr} \, (U_1 U_2 \cdots U_N) \exp \left(-\frac{1}{2g^2} \sum_p \text{Tr} \, U_p \right) \qquad (15.33)$$

where $U_1 U_2 \cdots U_N U_1$ symbolically represents the product of a series of U matrices around some closed path C in the lattice.

In order to perform this integration in the strong-coupling limit, we use the invariant group integration dU introduced in Chapter 9. We recall that if U is an element of a Lie group, then the invariant measure dU obeys the property:

$$d(U'U) = dU \qquad (15.34)$$

for fixed U'. dU is easy to construct for $SU(2)$. For this group, we can reduce the string bit to:

$$
\begin{aligned}
U_{ji} &\sim \exp(i g A_\mu^b \frac{\sigma^b}{2} a) \\
&\sim \mathbf{1} \cos \theta_{ji} + i \boldsymbol{\sigma} \cdot \mathbf{n}_{ji} \sin \theta_{ji} \\
&\sim \mathbf{1} a_0 + i \boldsymbol{\sigma} \cdot \mathbf{a} \qquad (15.35)
\end{aligned}
$$

where:

$$\theta_{ji} = \frac{1}{2} g a |A_\mu^b| \qquad (15.36)$$

and $a_0 + \mathbf{a} \cdot \mathbf{a} = 1$. Then it can be shown that the invariant group measure for $SU(2)$ is given by:

$$dU = \pi^{-2} da_0 \, da_1 \, da_2 \, da_3 \, \delta \left(\sum_{i=0}^{3} a_i^2 - 1 \right) \qquad (15.37)$$

With this explicit representation of the group measure, we can easily prove:

$$
\begin{aligned}
\int dU \, U_{ij} &= 0 \\
\int dU \, U_{i_1 j_1} U_{i_2 j_2} &= \frac{1}{2} \epsilon_{i_1 j_1} \epsilon_{i_2 j_2} \\
\int dU \, U_{ij} (U^{-1})_{kl} &= \frac{1}{2} \delta_{jk} \delta_{il} \qquad (15.38)
\end{aligned}
$$

With these integrals, we can perform the strong coupling expansion of the Wilson loop. First, we reduce the plaquette to:

$$U_P \sim \exp(iga F_{\mu\nu}^b \frac{\sigma^b}{2})$$
$$\sim \mathbf{1}\cos\theta_P + i\boldsymbol{\sigma}\cdot\mathbf{n}_P\sin\theta_P \qquad (15.39)$$

where:

$$\theta_P = \frac{1}{2}ga^2|F_{\mu\nu}^a| \qquad (15.40)$$

The action for the lattice gauge theory then becomes:

$$\frac{1}{g^2}(1 - \frac{1}{2}\operatorname{Tr} U_P) = \frac{1}{g^2}(1 - \cos\theta_P) \qquad (15.41)$$

Then the expectation value of a Wilson loop becomes:

$$\langle W(C)\rangle = Z^{-1}\int DU \operatorname{Tr}(U_1 U_2 \cdots U_N)\exp\frac{1}{g^2}\sum_P(1 - \frac{1}{2}\cos\theta_P) \qquad (15.42)$$

In the strong-coupling limit, we want to power expand this expression in terms of $1/g^2$. If we expand the exponential, to lowest order we have:

$$\langle W(C)\rangle = Z^{-1}\int DU \operatorname{Tr}(U_1 U_2 \cdots U_N)\left(1 - \frac{1}{2g^2}\sum_p \operatorname{Tr} U_p \cdots\right) \qquad (15.43)$$

Because of the identities in Eq. (15.38), the functional integral is zero unless each U within a plaquette is paired off the same U appearing elsewhere in the integral. Unless the pairing takes place, the resulting integral is zero because $\int dU\, U = 0$. The pairing can be performed in two ways: U can pair off with the same U appearing within the Wilson loop, or with the same U appearing within a neighboring plaquette.

This stringent condition sets almost all the terms in the integral to zero; the only nonzero contribution comes from plaquettes that completely fill the two-dimensional space within the Wilson loop. This effect is called "tiling"; that is, the only nonzero contribution comes from the plaquettes arranged like tiles within the loop. Each plaquette borders another plaquette, or borders the Wilson loop. In this way, each U appearing in the integral appears twice, either in neighboring plaquettes or in a plaquette and the Wilson loop.

The functional integral is therefore proportional to the number of integrations that we have performed; that is, it is proportional to A/a^2, where A is the minimal area of a surface that fills up the loop C. The integral can be approximated in this limit, and the strong-coupling expansion gives:

$$\langle W(C) \rangle \sim \exp\left[-\frac{Af(g)}{a^2} \right] \tag{15.44}$$

where $f(g) = \log g^2$ to lowest order.

The crucial point is that this trace goes as the exponential of the area of the enclosed loop. We have, therefore, formally proved that the strong-coupling limit produces a confining theory for the simplest $SU(2)$ gauge theory.

A curious phenomenon occurs, however, when we try to reach the continuum limit in the strong-coupling approximation. In the continuum limit, we need to keep:

$$\frac{f(g)}{a^2} = \text{constant} \tag{15.45}$$

If this condition is not met, then the trace formula becomes singular and meaningless. Therefore, after taking the strong-coupling approximation, we cannot take the continuum limit. Although this seems to be a problem, it is actually a blessing in disguise, because the discussion we have just made applies to QED, which we know is not a confining theory in the weak-coupling regime. Thus, we wish to have a phase transition separating the weak- and strong-coupling regimes for QED.

However, for gauge theory, we do *not* want a phase transition separating these two regimes, because we want a theory of confinement for the quarks. Here we see the crucial role played by non-Abelian gauge theory; QED has a qualitatively different phase structure than non-Abelian gauge theory.

All these results can be generalized to $SU(3)$ and higher. For $SU(3)$, we need the identities:

$$\int dU \ (U)_{m,n} = 0$$

$$\int dU \ (U)_{m,n}(U)^\dagger_{p,q} = \frac{1}{3}\delta_{m,q}\delta_{n,p}$$

$$\int dU \ (U)_{m,n}(U)_{p,q} = 0 \tag{15.46}$$

Then the calculation proceeds as before, giving us the area law and hence a confining theory for $SU(3)$ lattice gauge theory in the strong-coupling limit.

15.5   Monte Carlo Simulations

So far, our results have been qualitative and not quantitative. One of the great advances in lattice gauge theory is the *Monte Carlo method*, where we can use supercomputers to calculate a large number of numerical results for QCD.

A brute force calculation of the path integral, of course, is out of the question. If we have the simplest possible group defined on the lattice, $\mathbf{Z}_2$, with elements ± 1, and if the lattice is $8 \times 8 \times 8 \times 8$ in size, then the sum contains the following number of terms:

$$2^{2^{14}} = 2^{16384} \approx 10^{5460} \tag{15.47}$$

Clearly, this is prohibitive. The Monte Carlo technique, however, evades this problem by making certain approximations to the path integral.

The path integral, in general, sums over an enormous number of configurations that contribute almost nothing to the integral. We wish to throw most of them away, while keeping the ones that tend to minimize the action. The Monte Carlo method gives us a specific algorithm by which to accept only these gauge configurations.

Let Σ_1 be a certain set of initial values for each of the various links for the entire lattice (say, each link equals one). Then the Monte Carlo method generates a sequence of configurations $\Sigma_2, \Sigma_3, \ldots$. When statistical equilibrium is eventually reached, the probability of encountering any specific configuration Σ in the sequence is proportional to $e^{-\Delta S}$. Then the expectation value of any observable O may be approximated as:

$$\langle O \rangle \sim \frac{1}{n} \sum_{i=m+1}^{m+n} O(\Sigma_i) \tag{15.48}$$

where $O(\Sigma_i)$ represents the average of O computed with the set of link variables $\{\Sigma_i\}$, and where the first m steps have brought the system near equilibrium. Notice that we have replaced the original sum over all gauge configurations with this smaller, streamlined sum of configurations $\{\Sigma_i\}$ near equilibrium, which give us the bulk of the nonvanishing contributions to the path integral.

There are several useful algorithms, such as the heat bath and the Metropolis methods, which can generate this sequence of configurations $\{\Sigma_i\}$. We will use the latter. If we make the change from Σ to Σ' (by changing the value of just one link), we can compute the corresponding change in the action:

$$\Delta S = S(\Sigma') - S(\Sigma) \tag{15.49}$$

Now comes the key step: choose a random number r between 0 and 1. If $e^{-\Delta S} > r$, then the change from Σ to Σ' is accepted. If not, it is rejected.

If ΔS is negative, then the change is always accepted since $e^{-\Delta S} > 1$. However, if we *only* accepted negative values of ΔS, then we would always be decreasing the action and hence would tend toward the *classical* equations of motions. This, of course, throws out all quantum mechanical corrections and is to be avoided.

By choosing this random number r, we are allowing positive values of ΔS, so that the action can actually increase as we make the transition from Σ to Σ'. This, in turn, allows for quantum fluctuations around the classical equations of motion.

Now make a change in another link, generating yet another configuration, and test to see if it meets the proper criteria. In this way, we can sweep through the entire lattice, making small changes successively in each link. Once we have swept through the entire lattice, the process is repeated once again. After many sweeps, we gradually reach thermal equilibrium, yielding the set of link variables $\{\Sigma_1\}$. Then the process is repeated once again until, after many sweeps, we obtain the second set of link variables $\{\Sigma_2\}$. Over time, we arrive at a sequence of $\{\Sigma_i\}$, which we then insert into the sum in Eq. (15.48).

The net effect of this algorithm is that the new configuration Σ' is accepted with the conditional probability of $e^{-\Delta S}$. To see this, let $P(\Sigma \rightarrow \Sigma')$ be the probability of making the change from Σ to Σ'. Then this algorithm gives us:

$$P(\Sigma \rightarrow \Sigma') = \begin{cases} 1 & \text{if } S(\Sigma) > S(\Sigma') \\ e^{-\Delta S} & \text{if } S(\Sigma) < S(\Sigma') \end{cases} \tag{15.50}$$

This can also be written as:

$$\frac{P(\Sigma \rightarrow \Sigma')}{P(\Sigma' \rightarrow \Sigma)} = e^{-S(\Sigma')+S(\Sigma)} \tag{15.51}$$

But there is something that still must be checked: Is this algorithm sufficient to force the system into thermal equilibrium?

The advantage of this iteration process is that it does, in fact, automatically tend toward thermal equilibrium. To see this, let us review what we mean by thermal equilibrium. The transition matrix $P(\Sigma \rightarrow \Sigma')$ satisfies the usual properties of stochastic matrices:

$$\sum_{\Sigma'} P(\Sigma \rightarrow \Sigma') = 1$$

$$P(\Sigma \rightarrow \Sigma') \geq 0 \tag{15.52}$$

(The first statement simply says that probability is conserved, i.e., that the sum of probabilities for the transition to all possible configurations is equal to 100%. The second statement says that the probabilities are never negative.)

Next, we want to have the system in thermal equilibrium. We can consider $P(\Sigma \rightarrow \Sigma')$ to be a square matrix, with elements labeled by $\{\Sigma\}$. We demand that the Boltzmann distribution $e^{-S(\Sigma)}$ be an eigenvector of this transition P matrix:

$$\sum_{\{\Sigma\}} e^{-S(\Sigma)} P(\Sigma \rightarrow \Sigma') = e^{-S(\Sigma')} \tag{15.53}$$

(This simply means that, if the system is already in equilibrium, then the transition from Σ to Σ' leaves the system in equilibrium.)

We can also show that these three conditions are consistent with the *detailed balance equation*:

$$\frac{P(\Sigma \rightarrow \Sigma')}{P(\Sigma' \rightarrow \Sigma)} = \frac{e^{-S(\Sigma')}}{e^{-S(\Sigma)}} \tag{15.54}$$

To prove this consistency with the detailed balance equation, we can remove the denominators by cross multiplying and then summing over Σ. This gives us:

$$\sum_{\{\Sigma\}} e^{-S(\Sigma)} P(\Sigma \rightarrow \Sigma') = \sum_{\{\Sigma\}} e^{-S(\Sigma')} P(\Sigma' \rightarrow \Sigma)$$

$$= e^{-S(\Sigma')} \tag{15.55}$$

where we have used Eqs. (15.52) and (15.53).

This shows that the detailed balance equation is a sufficient (but not necessary) condition to prove thermal equilibrium. However, if we compare the detailed balance equation with Eq. (15.51), we find that the Metropolis algorithm satisfies this condition, and hence one can show that the algorithm drives the system to thermal equilibrium, as desired.

Once we have reached equilibrium and have generated a sequence of these configurations, we can calculate many numerical values for physical parameters. The simplest and most convenient is the string tension. By analyzing the behavior of the string tension, we can rapidly get an indication of the existence of a phase transition.

If two quarks are indeed linked together by a thin, condensed glue of gauge fields that behaves like a string, then it should be possible to calculate the tension on that string with these methods.

Let $W(R, T)$ describe the Wilson loop, as before, and define the string tension as:

$$\sigma \equiv \log \left(\frac{W(R, T)W(R - 1, T - 1)}{W(R - 1, T)W(R, T - 1)} \right) \tag{15.56}$$

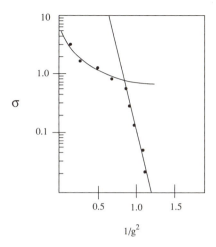

Figure 15.4. Monte Carlo simulation of a $SU(3)$ lattice calculation, with string tension plotted against $1/g^2$. As long as the string tension is nonzero, we have confinement.

Now insert the value of $W(R, T) \sim e^{-E_0 T}$ into this function. We find:

$$\sigma \sim a^2 \frac{d E_0(R)}{d R} \qquad (15.57)$$

which very conveniently gives us the force between the quarks for all values of R.

By plotting the string tension σ versus $1/g^2 N$, we can see in what region the area law is satisfied or violated. In Figure 15.4, for example, we see a typical result from a Monte Carlo calculation for $SU(3)$. We plot the string tension σ on the vertical axis, and $\beta \sim 1/g^2$ on the horizontal axis.

Finally, we remark that since Monte Carlo methods for $SU(3)$ are slow and cumbersome, it is instructive to analyze simpler groups, such as $\mathbf{Z}_n$, defined at each site.

Monte Carlo studies indicate that these systems with $n = 2, 3, 4$ exhibit a two-phase structure. However, for $n \geq 5$, the theory seems to prefer a three-phase structure. One phase corresponds to a confinement phase. Another corresponds to a phase where we have spin waves and free photons. The third phase is peculiar to systems with discrete groups only.

In the limit as $n \to \infty$, the $\mathbf{Z}_n$ models approach $U(1)$ gauge theory. In this limit, one phase of the $\mathbf{Z}_n$ model shrinks to zero, leaving only two phases for large n (and presumably QED).

15.6   Hamiltonian Formulation

Since Lorentz covariance is manifestly broken on the lattice, it is worthwhile to investigate the canonical formulation of lattice gauge theory as a Hamiltonian system, where the time parameter t is continuous.

As before, the continuum Yang–Mills action can be written as:

$$L = p\dot{q} - H(p, q) = E_i^a \dot{A}_i^a - \frac{1}{2} \int d^3x \left((E_i^a)^2 + (F_{ij}^a)^2 \right) + E_0^a G_i^a \qquad (15.58)$$

where G_i^a is the generator of gauge transformations.

We wish to put only the three space directions on the lattice and keep the time component continuous. In the lattice limit, we have the Kogut–Susskind Hamiltonian[5]:

$$H = \sum \frac{g^2}{2a} E_i^a E_i^a - \frac{1}{4ag^2} \sum_P \text{Tr } U_P + \text{h.c.} \qquad (15.59)$$

subject to the constraint:

$$G_i^a |\Psi\rangle = 0 \qquad (15.60)$$

This last constraint (Gauss's Law) forces us to choose only gauge-invariant states for our system.

In this picture, gauge-invariant states include quark–antiquark states:

$$\psi(\mathbf{n}) \left(\prod_{\text{path}} U \right) \psi(\mathbf{n} + \mathbf{R}) \qquad (15.61)$$

as well as glue-balls:

$$\text{Tr} \left(\prod_{\text{closed path}} U \right) \qquad (15.62)$$

This means that we are immediately left with a Hilbert space consisting of strings, without any free gluon states. The advantage of this formalism, therefore, is that, to lowest order, we see only strings. Any approximation we make will be an approximation around string states.

The new commutation relations are given by:

$$[E^a(\mathbf{n}, i), U(\mathbf{m}, j)] = \frac{1}{2} \tau^a U(\mathbf{m}, i) \delta_{ij} \delta_{\mathbf{n}, \mathbf{m}}$$

$$[E^a(\mathbf{n}, i), E^b(\mathbf{m}, j)] = i\epsilon^{abc} E^c(\mathbf{n}, i) \delta_{ij} \delta_{\mathbf{n}, \mathbf{m}} \qquad (15.63)$$

where the lattice site is given by $\mathbf{m}$, and where $E^a(\mathbf{n}, i)$ is the electric field.

Next, we wish to calculate the energy associated with these strings. In the strong coupling limit, we can keep only the E^2 term. For $SU(2)$, we therefore have:

$$(E^a)^2 U|0\rangle = \frac{1}{2}\tau^a \frac{1}{2}\tau^a U|0\rangle = \frac{3}{4}U|0\rangle \tag{15.64}$$

We apply the Hamiltonian on a quark–antiquark state with length R to obtain its energy:

$$H_0|\bar{q}q\rangle = \left(\frac{g^2}{2a}\right)\left(\frac{3}{4}\right)\left(\frac{R}{a}\right)|\bar{q}q\rangle \tag{15.65}$$

Thus, the energy of a string state, to lowest order, is proportional to its length, with a string tension given by $3g^2/8a^2$. The Hamiltonian formulation of lattice gauge theories gives us a quick way in which to see confinement and calculate the string tension.

Using operator techniques, we can calculate the string tension to any arbitrary accuracy. A more precise calculation of the string tension for $SU(3)$ gives us:

$$\sigma = \frac{g^2}{2a^2}\left(\frac{4}{3} - \frac{11}{153}y^2 - \frac{61}{1632}y^3 - 0.041378y^4 - 0.034436y^5 + \cdots\right) \tag{15.66}$$

where $y = 2/g^4$.

15.7 Renormalization Group

We mentioned earlier that the continuum limit $a \to 0$ is a subtle limit requiring renormalization group techniques. This is because the lattice spacing is a regulator on the potentially divergent structure of the theory. To eliminate divergences and take an appropriate continuum limit, the coupling constant g must be taken to depend on a.

For example, let O be a physical observable with dimension d. Since there are no intrinsic dimensional constants in the theory other than the lattice spacing, then by dimensional arguments we can write:

$$O = a^{-d}r(g) \tag{15.67}$$

where r is some function of the coupling constant g. The limit $a \to 0$ is ambiguous. For example, for negative d, $r(g)$ must become singular in order to have a finite result.

A mass m, for example, must obey:

$$m = \frac{1}{a} r(g) \tag{15.68}$$

Demanding that m be independent of a in the continuum limit leads to:

$$\frac{dm}{da} = 0 \tag{15.69}$$

Differentiating, we arrive at:

$$r'(g) = -r(g)/\beta(g)$$

$$\beta(g) = -a \frac{\partial g}{\partial a} \tag{15.70}$$

where β is the usual Callan–Symanzik function found in renormalization-group theory.

We know that, for $SU(N)$ theories, the behavior of β is given by:

$$a \frac{\partial g}{\partial a} = \beta_0 g^3 + \beta_1 g^5 + \cdots \tag{15.71}$$

where:

$$\beta_0 = -\frac{11}{3} \left(\frac{N}{16\pi^2} \right); \quad \beta_1 = \frac{34}{3} \left(\frac{N}{16\pi^2} \right)^2 \tag{15.72}$$

for $SU(N)$ theory. We can then integrate the expression for $r(g)$:

$$r(g) = \exp \left(-\int^g \frac{dg'}{\beta(g')} \right) \tag{15.73}$$

which becomes:

$$r(g) \sim (\beta_0 g^2)^{-\beta_1/2\beta_0^2} \exp \left(-\frac{1}{2\beta_0 g^2} \right) \tag{15.74}$$

In the limit $a \to 0$, we take $g \to 0$, so that $r(g)$ goes to zero in such a way that m is finite and nonzero. In this way, masses can develop even in a theory with no dimensional parameters. This is an example of "dimensional transmutation," where massless theories develop a scale because of renormalization effects.

In summary, lattice gauge theory gives us perhaps the best hope of extracting the low-energy hadron spectrum from QCD. Approximations to lattice gauge theory, such as the strong-coupling approximation, indicate that quarks are confined, as expected. Furthermore, simple computer calculations with Monte Carlo programs show many of the qualitative features associated with nonperturbative phenomena, such as phase transitions.

Some of the important problems facing lattice gauge theory include how to go from the Euclidean metric to the Minkowski metric, how to calculate with quarks on the lattice, and how to increase the computational power of our computers. Although lattice calculations have not yet given us the mass of the proton or other physical parameters of the low-lying hadron states, the qualitative features of the theory are all in agreement with our expectations. The only limitation seems to be the level of our current computer power.

15.8   Exercises

1. Complete all intermediate steps necessary to prove Eq. (15.7).

2. Prove that the Wilson fermion correction gives us the propagator in Eq. (15.22).

3. Let U be an element of $SU(2)$, parametrized as $U = a_0 + i\boldsymbol{\sigma} \cdot \mathbf{a}$. Prove that $\sum_{i=0}^{4} a_i^2 = 1$. Define the measure as follows:

$$dU \equiv \pi^{-2} d^4 a \, \delta(a_i^2 - 1) \tag{15.75}$$

Prove that the measure is invariant by multiplication with another element of $SU(2)$:

$$d(U'U) = dU \tag{15.76}$$

for fixed U'.

4. For $SU(2)$, prove Eq. (15.38).

5. Define the quantity:

$$W(J) \equiv \int dU \, \exp(JU) \tag{15.77}$$

For $SU(N)$, prove that:

$$\int dU \, U_{i_1 j_1} U_{i_2 j_2} \cdots U_{i_N j_N} = \frac{\delta}{\delta J_{j_1 i_1}} \frac{\delta}{\delta J_{j_2 i_2}} \cdots \frac{\delta}{\delta J_{j_N i_N}} W(J) \Big|_{J=0} \tag{15.78}$$

and that:

$$\det \left(\frac{\delta}{\delta J}\right) W(J) = 1 \tag{15.79}$$

[Hint: use the fact that $SU(N)$ matrices have unit determinant.]

6. One can show that an explicit value for $W(J)$ is given by:

$$W(J) = \sum_{i=0}^{\infty} \frac{2! \cdots (N-1)!}{i! \cdots (i+N-1)!} \left(\frac{1}{N!} \epsilon_{i_1 \cdots i_N} \epsilon_{j_1 \cdots j_N} J_{i_1 j_1} J_{i_2 j_2} \cdots J_{i_N j_N}\right)^i \tag{15.80}$$

Using this, prove that:

$$\int dU \, U_{i_1 j_j} \cdots U_{i_N j_N} = \frac{1}{N!} \epsilon_{i_1 \cdots i_N} \epsilon_{j_1 \cdots j_N} \tag{15.81}$$

7. Evaluate:

$$\int dU \, \mathrm{Tr}\,(U^n) \tag{15.82}$$

for the $SU(N)$ matrix U.

8. To construct the lattice version of the Bianchi identities, one must trace over two plaquettes. Construct this trace, and show how to reduce down to the usual Bianchi identity.

9. Consider the $\mathbf{Z}_2$ model in d dimensions, where the spins on the lattice can only equal ± 1. The partition function Z and free energy F are given by:

$$Z = 2^{-Nd} \sum_{\sigma_l = \pm 1} \exp\left(\beta \sum_p \sigma_p\right) = \exp N F(\beta) \tag{15.83}$$

where σ_p is the product of the spins around a plaquette. For large β, the sum over plaquettes can be performed. We find:

$$\begin{aligned}
(\cosh \beta)^{-N^{d(d-1)/2}} Z &= \sum_{\text{closed surfaces}} t^{|S|} \\
&= 1 + (N/6)d(d-1)(d-2)t^6 \\
&\quad + (N/2)d(d-1)(d-2)(2d-5)t^{10} + \cdots \tag{15.84}
\end{aligned}$$

and:

$$\frac{F}{d(d-1)} = \frac{1}{2}\log\cosh\beta + \frac{1}{2}d(d-2)\left[\tanh\beta^6 + (2d-5)\tanh\beta^{10} + \cdots\right]$$
(15.85)

for $t = \tanh\beta$. The first term corresponds to summations over cubes, the second to adjacent cubes, the next to disconnected cubes, etc. Prove these two strong-coupling relations to second order only. Hint: use the fact that:

$$\exp\beta\sigma_p = \cosh\beta(1 + t\sigma_p)$$
(15.86)

10. Prove Eq. (15.64).

11. In order to have the commutation relations in Eq. (15.63), what must the complete lattice Lagrangian look like in terms of independent variables and their conjugates?

12. Draw the graphs necessary for the calculation of Eq. (15.66) to second order. Do not solve.

13. The lattice gauge action makes no mention of gauge fixing, yet all integrals are well defined, without any infinite overcounting. How does the lattice gauge action accomodate gauge fixing?

14. Discuss how lattice gauge theory might be formulated on a noncompact group. Discuss some of the problems.

Chapter 16
Solitons, Monopoles, and Instantons

> *I was observing the motion of a boat which was rapidly drawn along a narrow channel by a pair of horses, when the boat suddenly stopped, [creating] a large solitary elevation, a rounded, smooth and well-defined heap of water .... I followed it on horseback, and overtook it still rolling on at a rate of some eight or nine miles an hour, preserving its original figure ... after a chase of one or two miles, I lost it in the windings of the channel.*
>
> —J. Scott Russel, 1834

16.1 Solitons

Perturbation theory is based upon making power expansions of the path integral around trivial vacua such as $\phi = 0$ or const. However, there are solutions of the classical, nonlinear equations of motion that exhibit particle-like behavior that give us powerful insight into the nonperturbative behavior of these theories. A new quantum power expansion can be developed around each exact solution, allowing us to explore regions that are not accessible by standard perturbation theory. In particular, these solutions give us nonperturbative information about important physical phenomena such as tunneling and bound states.

In this chapter, we will describe three different types of classical solutions that have been intensively studied:

1. *Solitons* are finite-energy, localized solutions to the equations of motion that, after collisions, retain their shape. They were first investigated by J. Scott Russel[1] in 1834. Since then, a large number of different wave equations have been shown to possess soliton solutions, especially in two dimensions.

2. *Monopoles*, or particles with magnetic charge, were first investigated by Dirac. They have been found in gauge theories with spontaneous symmetry breaking and may have cosmological significance.

3. *Instantons* are finite-action solutions to the Euclideanized equations of motion. Their presence signals the possibility of tunneling between degenerate vacua.

One of the surprising features of these solutions is that they can be studied using topological methods.[2] In topology, two geometric surfaces are considered equivalent if they can be smoothly deformed into each other without cutting. For example, a coffee mug, an inner tube, and a doughnut are all topologically equivalent. We will find that certain topological numbers can be assigned to these classical solutions.

Solitons (for solitary waves) exhibit some unusual properties, providing a laboratory in which we can test some of our ideas about bound states. Eventually, the hope is that we can extrapolate some of the qualitative features of solitons to describe more complex bound-state systems, such as the proton. Their main distinguishing feature is that, after they have scattered against each other, they retain their shape (although there is a phase shift after the scattering). They are hence stable against collisions and perturbations.

To exhibit soliton solutions, let us begin with a two-dimensional relativistic Lagrangian:

$$\mathscr{L} = \frac{1}{2}(\dot{\phi})^2 - \frac{1}{2}(\phi')^2 - U(\phi) \tag{16.1}$$

where $U(\phi)$ is some arbitrary potential function. Its classical wave equation is given by the Euler–Lagrange equations:

$$\partial_\mu^2 \phi = \ddot{\phi} - \phi'' = -\frac{\partial U}{\partial \phi} \tag{16.2}$$

The energy is given by:

$$E = \int_{-\infty}^{\infty} dx \left(\frac{1}{2}(\dot{\phi})^2 + \frac{1}{2}(\phi')^2 + U(\phi) \right) \tag{16.3}$$

We can find solutions of the equations of motion by solving them for the static case and then boosting them with a Lorentz transformation. For static solutions, we can set $\dot{\phi} = 0$. If we multiply the static equations of motion by ϕ', we have:

$$\phi'' \phi' = \frac{\partial U(\phi)}{\partial \phi} \phi' \tag{16.4}$$

These equations can be integrated over x, yielding:

$$\frac{1}{2}(\phi')^2 = U \tag{16.5}$$

Taking the square root, this equation can be integrated once again to yield:

$$x - x_0 = \pm \int_{\phi(x_0)}^{\phi(x)} \frac{d\bar{\phi}}{\sqrt{2U(\bar{\phi})}} \tag{16.6}$$

Let us now give some concrete examples of soliton solutions.

16.1.1 Example: ϕ^4

Let us choose the potential:

$$U(\phi) = \frac{1}{4}(\phi^2 - m^2/\lambda)^2 \tag{16.7}$$

The potential has two degenerate minima, given by the values:

$$\phi = \pm m/\sqrt{\lambda} \tag{16.8}$$

This means that soliton solutions, if they exist, must asymptotically tend toward these values as $x \to \pm\infty$; that is:

$$\phi(|x| = \infty) = \pm m/\sqrt{\lambda} \tag{16.9}$$

To solve the system, we can integrate this ϕ^4 theory to yield:

$$x - x_0 = \pm \int_{\phi(x_0)}^{\phi(x)} \frac{d\bar{\phi}}{\sqrt{\lambda/2}(\bar{\phi}^2 - m^2/\lambda)} \tag{16.10}$$

Inverting, we then find:

$$\phi(x) = \pm(m/\sqrt{\lambda}) \tanh\left[(m/\sqrt{2})(x - x_0)\right] \tag{16.11}$$

The $\pm$ sign in front indicates that there are two solitary waves, which are sometimes called the "kink" and "antikink" solitons. This solution approaches the asymptotic solution $\phi = \pm m/\sqrt{\lambda}$ as it should.

The energy density is then given by:

$$E(x) = (m^4/2\lambda) \operatorname{sech}^4\left[m(x - x_0)/\sqrt{2}\right] \tag{16.12}$$

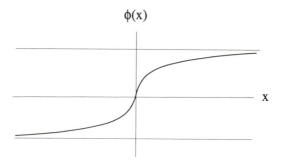

Figure 16.1. A soliton solution in ϕ^4 theory in two dimensions.

The mass of the soliton is given by the integral over the energy density:

$$M = \int_{-\infty}^{\infty} E(x) \, dx = \frac{2\sqrt{2}}{3} \frac{m^3}{\lambda} \qquad (16.13)$$

Because the system is fully relativistic, we can obtain the time-dependent solution by simply boosting the static one. This gives us the soliton moving at velocity u (Fig. 16.1):

$$\phi(x, t) = \frac{m}{\sqrt{\lambda}} \tanh \left[\frac{m}{\sqrt{2}} \left(\frac{(x - x_0) - ut}{\sqrt{1 - u^2}} \right) \right] \qquad (16.14)$$

Then the energy of the soliton is given by:

$$
\begin{aligned}
E &= \int_{-\infty}^{\infty} dx \left(\frac{1}{2}(\dot{\phi})^2 + \frac{1}{2}(\phi')^2 + U(\phi) \right) \\
&= \frac{M}{\sqrt{1 - u^2}} \qquad (16.15)
\end{aligned}
$$

where $M = 2\sqrt{2}m^3/(3\lambda)$.

Perhaps the most interesting feature of the kink and antikink solitons is that they are stable. Because of the way they extend asymptotically to infinity, it takes an infinite amount of energy to change the kink to the constant, vacuum solution. Although there are no Noether currents, we suspect that this stability, in turn, indicates the presence of a conserved current.

In fact, we can define a conserved current as:

$$J^\mu = \frac{\sqrt{\lambda}}{m} \epsilon^{\mu\nu} \partial_\nu \phi \qquad (16.16)$$

which gives us the conserved charge:

$$Q = \frac{1}{2} \int_{-\infty}^{\infty} dx \, J^0 = \frac{\sqrt{\lambda}}{m} [\phi(x = \infty) - \phi(x = -\infty)] \tag{16.17}$$

It is now easy to see that the constant solutions correspond to $Q = 0$ and the kink (antikink) solutions correspond to $Q = +1(-1)$. Since Q is constant in time, kinks with "topological charge" Q can never decay into solutions with different topological charge. In fact, solutions of the equations of motion can be grouped according to their value of Q; that is, they fall into discrete equivalence classes, labeled by Q. Two solutions of the equations of motion, even though they may look quite different, belong to the same equivalence class if they have the same topological charge Q.

The concept of the topological charge (which cannot be derived from Noether's theorem) will surface repeatedly throughout our discussion of solitons and instantons.

16.1.2   *Example: Sine–Gordon Equation*

A more complicated example, the sine-Gordon equation, is given by:

$$\mathcal{L} = \frac{1}{2}\partial_\mu \phi \partial^\mu \phi + (m^4/\lambda) \left[\cos \left(\sqrt{\lambda}\phi/m \right) - 1 \right] \tag{16.18}$$

Its wave equation is given by:

$$\partial_\mu^2 \phi + (m^3/\sqrt{\lambda}) \sin \left[(\sqrt{\lambda}/m)\phi \right] = 0 \tag{16.19}$$

To eliminate some of the unwanted constants, let us make the substitution $x \rightarrow mx$, $t \rightarrow mt$, $\phi \rightarrow (\sqrt{\lambda}/m)\phi$. Then the wave equation simply reads $\partial_\mu^2 \phi + \sin \phi = 0$.

Perhaps the most important way in which to catalog solutions of the sine–Gordon equation is by their topological charge. With the potential rescaled to $U(\phi) = 1 - \cos \phi$, the constant solutions with zero energy are given by:

$$\phi = 2N\pi \tag{16.20}$$

where N is an integer. Therefore, all soliton solutions must, at $x \rightarrow \pm\infty$, tend towards one of these constant values, labeled by an integer N. If the topological charge is defined as:

$$J^\mu = \frac{1}{2\pi}\epsilon^{\mu\nu}\partial_\nu \phi \tag{16.21}$$

then the conserved charge is given by:

$$
\begin{aligned}
Q &= \frac{1}{2\pi} \int_{-\infty}^{\infty} dx \frac{\partial \phi}{\partial x} = \frac{1}{2\pi} [\phi(x = \infty) - \phi(x = -\infty)] \\
&= N_1 - N_2
\end{aligned}
\tag{16.22}
$$

where N_1 and N_2 are the integers that describe the asymptotic value of the field.

Since Q is a constant topological charge, solitons with one value of Q cannot decay into solutions with a differing value of Q; that is, these solutions are stable for topological reasons.

Let us now calculate the value of Q for different soliton solutions. The easiest one is the static case, where we have:

$$
x - x_0 = \pm \int_{\phi(x_0)}^{\phi(x)} \frac{d\phi}{2 \sin(\phi/2)}
\tag{16.23}
$$

Inverting, and then making a Lorentz boost, we now have the solution:

$$
\phi(x) = \pm 4 \tan^{-1} \left[\exp \left(\frac{x - x_0 - ut}{\sqrt{1 - u^2}} \right) \right]
\tag{16.24}
$$

where the $+1$ (-1) sign corresponds to the soliton (antisoliton) solution.

By examining their asymptotic values, we can easily show that the soliton (antisoliton) solution has topological charge $Q = +1(-1)$. Because of the periodicity of the cos function, we can add $2\pi N$ to the soliton solution to generate a new soliton solution with the same value of Q.

More complicated generalized solutions are not difficult to find. For example, the following solution represents the scattering of a soliton off an antisoliton:

$$
\phi_{S-A} = 4 \tan^{-1} \left(\frac{\sinh(ut/\sqrt{1 - u^2})}{u \cosh(x/\sqrt{1 - u^2})} \right)
\tag{16.25}
$$

What is most remarkable about this soliton solution is that the individual soliton and antisoliton waves keep their same shape even after a collision:

$$
\phi_{S-A} \to \phi_S \left(\frac{x + u(t - \Delta/2)}{\sqrt{1 - u^2}} \right) + \phi_A \left(\frac{x - u(t - \Delta/2)}{\sqrt{1 - u^2}} \right)
\tag{16.26}
$$

where $t \to \infty$, where $\phi_S(\phi_A)$ corresponds to the soliton (antisoliton) solution, and where there is a time delay given by:

$$
\Delta = [(1 - u^2)/u] \log u
\tag{16.27}
$$

For $t \to \infty$, the asymptotic solution is the same, except that Δ flips sign. Thus, the only difference between the asymptotic states at negative infinity and the states at positive infinity is that there has been a time delay of Δ. Otherwise, asymptotically it appears as if nothing happened.

We should also mention that the two-soliton solution is given by:

$$\phi = 4 \tan^{-1} \left(\frac{u \sinh x / \sqrt{1 - u^2}}{\cosh ut / \sqrt{1 - u^2}} \right) \tag{16.28}$$

Since this function goes from -2π to 2π as x goes from $-\infty$ to $+\infty$, this two-soliton solution has topological charge $Q = 2$.

Many-soliton solutions can also be found using an ingenious technique called the *Bäcklund transformation*. Given a solution ϕ_0 of these equations, we are able to generate a new solution ϕ_1.

To see this, we write the sine–Gordon equation in terms of light-cone co-ordinates $\sigma = (x + t)/2$ and $\rho = (x - t)/2$. Then the sine–Gordon equation reads:

$$\frac{\partial^2 \phi}{\partial \sigma \, \partial \rho} - \sin \phi = 0 \tag{16.29}$$

We now define the Bäcklund equations as:

$$\frac{1}{2} \frac{\partial}{\partial \sigma} (\phi_1 - \phi_0) = a \sin[\frac{1}{2}(\phi_1 + \phi_0)]$$

$$\frac{1}{2} \frac{\partial}{\partial \sigma} (\phi_1 + \phi_0) = \frac{1}{a} \sin[\frac{1}{2}(\phi_1 - \phi_0)] \tag{16.30}$$

Next, we multiply the first equation by $\partial/\partial\rho$ and use the second equation to arrive at:

$$\frac{1}{2} \frac{\partial^2}{\partial \sigma \, \partial \rho} (\phi_1 - \phi_0) = \cos[\frac{1}{2}(\phi_1 + \phi_0)] \sin[\frac{1}{2}(\phi_1 - \phi_0)]$$

$$= \frac{1}{2} \sin \phi_1 - \frac{1}{2} \sin \phi_0 \tag{16.31}$$

Thus, ϕ_1 is a solution of the sine–Gordon equation if ϕ_0 is. The beauty of this formulation is that we can now solve for ϕ_1 in terms of ϕ_0, thereby allowing us to generate a new solution in terms of the old one. If, for example, we plug in the trivial no-soliton solution $\phi = 0$ into these equations, then we obtain the one-soliton equation found earlier. The equation for ϕ_1 reads:

$$\frac{1}{2} \frac{\partial \phi_1}{\partial \sigma} = \frac{1}{2} a^2 \frac{\partial \phi_1}{\partial \rho} = a \sin(\phi_1/2) \tag{16.32}$$

which is easily integrated back to the one-soliton solution, with $u = (1 - a^2)/(1 + a^2)$.

16.1.3   Example: Nonlinear $O(3)$ Model

Let us start with a triplet of scalar fields ϕ^a with the simple $O(3)$ action:

$$\mathscr{L} = \frac{1}{2}(\partial_\mu \phi^a)(\partial^\mu \phi^a) = \frac{1}{2}\partial_\mu \phi \cdot \partial^\mu \phi \tag{16.33}$$

This is the usual linear $O(3)$ model, except we impose a nontrivial constraint:

$$(\phi_a)^2 = \phi \cdot \phi = 1 \tag{16.34}$$

We can impose this constraint by using a Lagrange multiplier in the action:

$$S \to S + \int d^2x \lambda(\phi \cdot \phi - 1) \tag{16.35}$$

The energy of the system is defined as:

$$E = \frac{1}{2}\int \partial_\mu \phi \cdot \partial_\mu \phi \, d^2x \tag{16.36}$$

(We have reversed the sign of the space derivative term in the Hamiltonian.)

As before, let us analyze the possible soliton solutions according to their topological charge. We must first calculate the constant vacuum solutions, which then fixes the asymptotic value of the solitons. Then we construct the topological charge associated with each asymptotic value of the soliton.

The zero energy vacuum solutions obey $\partial_\mu \phi^a = 0$, so they are just constants pointing in some fixed direction in isotopic space $\phi^a = \phi_0^a$. As before, the soliton solutions at infinity must asymptotically tend to this constant isovector ϕ_0^a.

The field $\phi^a(x)$, by definition, is a function that takes two-dimensional space–time, labeled by t and x, into a vector ϕ^a in $O(3)$ isotopic space. In general, this function therefore defines a map between points in R_2 (the two-dimensional plane) and the space of three real coordinates ϕ^as.

However, as $|x| \to \infty$, this function approaches the same constant value, ϕ_0^a. Therefore x space is actually described by S_2 (a sphere) since the values of the function at infinity are all the same, no matter where we point. In other words, we have replaced the plane with a sphere, where infinity has been transformed into the north pole.

Furthermore, because of the constraint $\sum_{a=1}^{3} \phi^a \phi^a = 1$, the isotopic space is actually a sphere S_2. In conclusion, we find that the function $\phi^a(x)$ is therefore a mapping of S_2 (x, t space) onto S_2 (isotopic ϕ^a space):

$$\pi : S_2 \rightarrow S_2 \tag{16.37}$$

The question is: how many distinct ways can the points of a sphere be smoothly mapped onto the points of another sphere? There is a theorem in topology that says that the topologically distinct ways in which this mapping, called π, can take place is labeled by the integers:

$$\pi_2(S_2) = \mathbf{Z} \tag{16.38}$$

(These mappings actually form a group, since we can "add" these maps by sequentially iterating them.)

To motivate this abstract mathematical result, one can study the simpler example of classifying the number of smooth maps from the circle S_1 onto another circle S_1. Let $\phi(\theta)$ map the circle ($0 \leq \theta \leq 2\pi$) onto the circle given by the function $\phi(0) = \phi(2\pi)$ mod 2π. Construct the charge:

$$Q = \frac{1}{2\pi} \int_0^{2\pi} d\theta \frac{d\phi(\theta)}{d\theta} = \frac{1}{2\pi} [\phi(2\pi) - \phi(0)] \tag{16.39}$$

At first, one might suspect that Q is equal to zero, because $\phi(2\pi) = 0$, or that $Q = 1$, because $\phi(2\pi) = 2\pi$. However, there is also the possibility that $\phi(2\pi) = 2N\pi$, where N is an integer, in which case $Q = N$. In this case, the function $\phi(\theta)$ maps the circle ($0 \leq \theta \leq 2\pi$) onto another circle N times; that is, it repeatedly wraps around the circle an integer number of times. Q is therefore sometimes called the "winding number," and is a topological invariant; that is, it does not change even if we smoothly deform the function $\phi(\theta)$, as long as the boundary conditions remain the same. Thus, Q is sensitive to the overall topological nature of the mapping $\phi(\theta)$, not its specific value. Mathematically, we can say:

$$\pi_1(S_1) = \mathbf{Z} \tag{16.40}$$

Each value of N, in turn, represents an equivalence class of maps. Two functions $\phi(\theta)$ and $\phi'(\theta)$ are members of the same equivalence class or "homotopy class" if they have the same N.

Returning now to the more difficult question of the nonlinear $O(3)$ model, we shall find that the topological charge Q can be defined as:

$$Q = \frac{1}{8\pi} \int \epsilon^{\mu\nu} \phi \cdot (\partial_\mu \phi \times \partial_\nu \phi) \tag{16.41}$$

Our task is now to prove that Q is, in fact, an integer that represents the number of inequivalent smooth maps from S_2 to S_2. Consider a sphere of unit radius, with the surface described by the three-dimensional Cartesian coordinates x_a, such that $\sum x_a^2 = 1$. We can also describe the same sphere with two dimensional coordinates σ_1, σ_2, which can be polar coordinates or any local coordinates we place on the surface of the sphere.

Then one can show that an infinitesimal element of surface area dS_a pointing in the a direction is given by:

$$dS_a = \frac{1}{2}\epsilon^{\mu\nu}\epsilon^{abc}\frac{\partial x_b}{\partial \sigma_\mu}\frac{\partial x_c}{\partial \sigma_\nu}d^2\sigma \qquad (16.42)$$

By a direct calculation, one can show that this expression is independent of the specific choice of two-dimensional coordinates $\{\sigma\}$ one chooses. The surface area of a sphere can then be computed by contracting dS_a onto the unit vector x_a and integrating:

$$4N\pi = \int dS_a x_a \qquad (16.43)$$

The integer N appears because the map $x_a(\sigma_1, \sigma_2)$ may wind around the sphere an integer number of times.

To make contact with the topological charge Q, the crucial step is to make the replacement $x_a \rightarrow \phi_a$. Then Q can be rewritten as:

$$\begin{aligned} Q &= \frac{1}{8\pi}\int \epsilon^{\mu\nu}\epsilon^{abc}\phi^a\,\partial_\mu\phi^b\,\partial_\nu\phi^c \\ &= \frac{1}{4\pi}\int dS_a\,\phi^a = N \end{aligned} \qquad (16.44)$$

The topological charge is therefore equal to the winding number, that is, the number of distinct ways that the points on a sphere S_2 can be mapped smoothly onto another sphere S_2. Each N, in turn, represents a distinct homotopy class of maps.

Q is also important because it appears in the self-dual solutions of the nonlinear sigma model. For example, consider the identity:

$$\int d^2x\left[(\partial_\mu\phi \pm \epsilon_{\mu\nu}\phi \times \partial_\nu\phi)\cdot(\partial_\mu\phi \pm \epsilon_{\mu\rho}\phi \times \partial_\rho\phi)\right] \geq 0 \qquad (16.45)$$

(We are contracting with a Euclidean metric.) This quantity is positive definite because it is the sum of squares of real numbers.

Expanding, we find:

$$\int d^2x \left[(\partial_\mu \phi) \cdot (\partial_\mu \phi) + \epsilon_{\mu\nu}(\phi \times \partial_\nu \phi) \cdot \epsilon_{\mu\sigma}(\phi \times \partial_\sigma \phi) \right]$$

$$\geq \pm 2 \int d^2x \, (\epsilon_{\mu\nu}\phi \cdot \partial_\mu \phi \times \partial_\nu \phi) \qquad (16.46)$$

The two terms on the left are actually the same, since $\epsilon_{\mu\nu}\epsilon_{\rho\sigma} = \delta_{\mu\rho}\delta_{\nu\sigma} + \text{perm.}$ Finally, we arrive at:

$$E \geq 4\pi |Q| \qquad (16.47)$$

The equality $E = 4\pi|Q|$ will be reached for the self-dual solutions:

$$\partial_\mu \phi = \pm \epsilon_{\mu\nu}\phi \times (\partial_\nu \phi) \qquad (16.48)$$

16.2   Monopole Solutions

In addition to these two-dimensional toy models, we have the more complicated monopole solutions of gauge theory. Before we discuss the properties of the gauge monopole, let us review the properties of the Dirac magnetic monopole[3] found in ordinary electrodynamics. The Dirac monopole is based upon a straightforward generalization of the electric monopole. By analogy, the electric field $\mathbf{E}$ of a point electric charge can be generalized to the magnetic field $\mathbf{B}$ of a point magnetic monopole:

$$\mathbf{E} = e\frac{\mathbf{r}}{r^3} \quad \rightarrow \quad \mathbf{B} = g\frac{\mathbf{r}}{r^3} \qquad (16.49)$$

Then Maxwell's equations are generalized to include a nontrivial divergence of the magnetic field:

$$\nabla \cdot \mathbf{E} = 4\pi e\delta^3(r) \quad \rightarrow \quad \nabla \cdot \mathbf{B} = 4\pi g\delta^3(r) \qquad (16.50)$$

If we express these fields in terms of potentials $\mathbf{E} = -\nabla\phi$ and $\mathbf{B} = \nabla \times \mathbf{A}$, then we seem to have a contradiction. Usually, the magnetic field, because it has no sources, can be written in term of the curl of the vector potential. This is because the divergence of a curl is equal to zero; that is, $\nabla \cdot \nabla \times \mathbf{A} = 0$. (This is because $\partial_i\partial_j\epsilon^{ijk} = 0$ since ϵ^{ijk} is antisymmetric.)

However, it is possible to evade this identity if there is a delta function type singularity in the $\mathbf{A}$ field. To see this, let us take a sphere surrounding the point monopole. At the top of the sphere, there is a small circle that is centered around the north pole. The flux of magnetic field through this circle is given by:

$$\int \mathbf{B} \cdot d\mathbf{S} \;=\; \int \nabla \times \mathbf{A} \cdot d\mathbf{S}$$

$$\;=\; \oint \mathbf{A} \cdot d\mathbf{l} \qquad\qquad (16.51)$$

If the circle is infinitesimally small, including only the north pole, then the line integral of the $\mathbf{A}$ field around this infinitesimally small circle is zero. However, if the circle is made successively larger, until it includes the entire sphere, then the surface integral over the $\mathbf{B}$ field is given by $4\pi g$. However, the line integral over the $\mathbf{A}$ field must be zero because the loop has become an infinitesimally small loop surrounding the south pole. To avoid this contradiction, the $\mathbf{A}$ field must be singular along the negative z axis. There must be an unphysical singularity that extends from the origin down to the south pole and beyond. This singularity is called the *Dirac string*.

In addition to the Dirac string, there is yet another curious property of magnetic monopoles. When we apply quantum mechanics to monopoles, we find that the magnetic monopole charge g cannot have arbitrary values; that is, the monopole charge is quantized.

To see this strange effect, notice that a wave function ψ in the presence of a monopole must be single valued when we go around the Dirac string. A plane wave is given by:

$$\psi \sim \exp{(i/\hbar)}(\mathbf{p} \cdot \mathbf{r} - Et) \qquad\qquad (16.52)$$

The wave function, in the presence of a magnetic monopole, can be obtained by making the standard substitution: $\mathbf{p} \to \mathbf{p} - (e/c)\mathbf{A}$. With this substitution, the wave function picks up a new phase factor given by:

$$\exp{\frac{-ie}{c\hbar}}(\mathbf{A} \cdot \mathbf{r}) \qquad\qquad (16.53)$$

In order for the wave function to be single valued when we go around a loop, this factor must be equal to one. The line integral around the Dirac string must therefore be $2\pi n$, where n is an integer. Then we have:

$$2\pi n \;=\; \frac{e}{c\hbar} \oint \mathbf{A} \cdot d\mathbf{l}$$

$$\;=\; \frac{e}{c\hbar} \int \mathbf{B} \cdot d\mathbf{S}$$

$$= \frac{e}{c\hbar} 4\pi g \tag{16.54}$$

Therefore the final quantization condition is given by:

$$e = n\frac{\hbar c}{2g} \tag{16.55}$$

This quantization of the monopole strength is a rather curious result, which shows that the quantum mechanics of magnetic monopoles yields novel features.

We would like to make a final remark about Dirac magnetic monopoles. One can find fault with the previous presentation because of the existence of the singular Dirac string. Although the Dirac string can be moved in any direction and also has no physical consequences, one suspects that there is another formulation of the monopole in which the Dirac string is absent. This new presentation of the magnetic monopole uses the theory of fiber bundles. It has the advantage that the presentation is completely nonsingular and also is formulated in a well-established mathematical formalism.

Let **A** be the vector potential for the previous monopole, in which the Dirac string goes through the south pole. However, there is, of course, another vector potential $\tilde{\mathbf{A}}$ in which the Dirac string runs through the north pole. Our strategy is to split the sphere surrounding the magnetic monopole into two pieces along the equator. For the northern hemisphere, we take the field configuration **A** and simply throw away the Dirac string running through the south pole. In the southern hemisphere we take the field configuration $\tilde{\mathbf{A}}$ (and throw away the Dirac string that runs through the north pole; see Fig. 16.2).

Thus, **A** defines the monopole field in the northern hemisphere, while $\tilde{\mathbf{A}}$ describes the field in the southern hemisphere. Neither **A** nor $\tilde{\mathbf{A}}$ are singular.

However, there is a price we have to pay for this sophisticated construction; that is, we have to piece together these two distinct patches in order to cover the sphere. We will "glue" the two vector potentials along the equator. The final gluing process between these two different field configurations is accomplished by making a gauge transformation between the two configurations along the equator; that is:

$$\mathbf{A} = \tilde{\mathbf{A}} + \nabla\Omega \tag{16.56}$$

Since a gauge transformation cannot affect the physics, we now have a description of the field configuration that covers the entire sphere. To see how this gluing is actually accomplished, let us write down the explicit representation of the vector fields. For **A**, we have:

$$A_x = -g\frac{y}{r(r+z)}$$

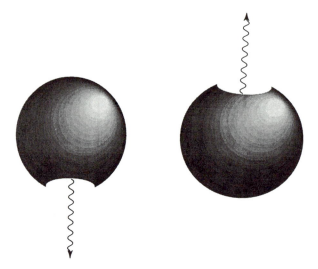

Figure 16.2. A Dirac string. The total magnetic vector potential of a monopole is obtained by splicing the field of the northern hemisphere of the diagram on the left with the field of the southern hemisphere of the one on the right along the equator.

$$A_y = g\frac{x}{r(r+z)}$$

$$A_z = 0 \tag{16.57}$$

Actually, a more convenient description of the monopole field is given in terms of spherical coordinates. Let θ be the polar angle, which is 0 at the north pole and π along the south pole. Let ϕ be the azimuthal angle, which ranges from 0 to 2π. Then the field configuration is given by:

$$A_r = 0$$

$$A_\theta = 0$$

$$A_\phi = \pm g\frac{1 \mp \cos\theta}{r\sin\theta} \tag{16.58}$$

Notice that we have two solutions, given by the sign of $\pm$. The $-$ solution corresponds to $\mathbf{A}$, while the $+$ corresponds to $\tilde{\mathbf{A}}$.

We can now "glue" the two configurations together along the equator by a gauge transformation:

$$\tilde{A}_\phi = A_\phi - \frac{2g}{r\sin\theta} = A_\phi - (i/e)S\nabla_\phi S^{-1} \tag{16.59}$$

where:

$$S = e^{2ige\phi} \tag{16.60}$$

16.3 't Hooft–Polyakov Monopole

The previous discussion of magnetic monopoles, although interesting, was not compelling, because ordinary electrodynamics does not require that monopoles exist. Electrodynamics without monopoles is perfectly consistent. However, in certain gauge theories, we will find that spontaneous symmetry breaking is intimately connected with the existence of monopole solutions. Hence, monopoles must exist for these theories as a consequence of broken gauge symmetry.

It can be shown that pure gauge theory does not, by itself, possess any static nonsingular monopole configurations. However, a more general case, such as gauge theory coupled to scalar fields, does possess monopole solutions.

We begin with the standard gauge action with scalar fields, with the gauge group $O(3)$:

$$\begin{aligned} \mathscr{L} &= -\frac{1}{4}F^a_{\mu\nu}F^{a\mu\nu} + \frac{1}{2}D_\mu\phi^a D^\mu\phi^a \\ &\quad -\frac{1}{2}m^2\phi^a\phi^a - \frac{\lambda}{4!}(\phi^a\phi^a)^2 \end{aligned} \tag{16.61}$$

One can show that there exists a solution with the asymptotic behavior ($r \to \infty$):

$$\begin{aligned} A^a_i &\to -\epsilon_{iab}\frac{r^b}{er^2} \\ A^a_0 &\to 0 \\ \phi^a &\to (-6m^2/\lambda)\frac{r^a}{r} \end{aligned} \tag{16.62}$$

(We have made a nontrivial linkage between three-dimensional physical space and three dimensional isospin space.) One can show from this that ϕ^a is covariantly constant at infinity (i.e., $D_\mu\phi^a = 0$).

This is the 't Hooft–Polyakov monopole.[4,5] To compare this monopole, defined for $O(3)$ symmetry, with the usual Dirac monopole, we will have to define a new Maxwell tensor $F_{\mu\nu}$ that will reduce to the usual one when ϕ^a becomes fixed in isospin space. We define:

$$\begin{aligned} F_{\mu\nu} &\equiv \partial_\mu A_\nu - \partial_\nu A_\mu - \frac{1}{e|\phi|^3}\epsilon_{abc}\phi^a(\partial_\mu\phi^b)(\partial_\nu\phi^c) \\ A_\mu &\equiv \frac{1}{|\phi|}\phi^a A^a_\mu \end{aligned} \tag{16.63}$$

With this definition, we can now calculate the magnetic and electric charge of the monopole. We find that $A_\mu = 0$ and that:

$$F_{0i} = 0, \quad F_{ij} = -\frac{1}{er^3}\epsilon_{ijk}r^k, \quad B_k = \frac{r^k}{er^3} \tag{16.64}$$

With this value of the magnetic field, then, we can show that the total flux through a sphere surrounding the monopole is given by $4\pi/e$. But the total flux of a monopole is $4\pi g$, so the monopole magnetic charge then obeys the constraint:

$$eg = 1 \tag{16.65}$$

which is twice the Dirac case in Eq. (16.55).

To reveal the topological nature of these monopole solutions, we remark that the sole contribution to $F_{\mu\nu}$ comes from the Higgs sector, since $A_\mu = 0$. The magnetic current is given by $K^\mu = \partial_\nu \tilde{F}^{\mu\nu}$ and can be written entirely in terms of Higgs field $\hat{\phi}^a = \phi^a/|\phi|$. A direct calculation shows that the conserved magnetic current equals:

$$K^\mu = -\frac{1}{2e}\epsilon^{\mu\nu\rho\sigma}\epsilon_{abc}\partial_\nu\hat{\phi}^a\partial_\rho\hat{\phi}^b\partial_\sigma\hat{\phi}^c \tag{16.66}$$

Since $\partial_\mu K^\mu = 0$, the conserved magnetic charge can be written as:

$$\begin{aligned}
M &= \frac{1}{4\pi}\int K^0 \, d^3x \\
&= -\frac{1}{8e\pi}\int \epsilon^{ijk}\epsilon_{abc}\partial_i\hat{\phi}^a\partial_j\hat{\phi}^b\partial_k\hat{\phi}^c \, d^3x \\
&= -\frac{1}{8e\pi}\oint_{S_2} \epsilon^{ijk}\epsilon_{abc}\hat{\phi}^a\partial_j\hat{\phi}^b\partial_k\hat{\phi}^c \, dS_i
\end{aligned} \tag{16.67}$$

where we have integrated by parts, so that this volume integral becomes a two-dimensional surface integral taken over S_2 at infinity, which is the boundary of the static field $\hat{\phi}$.

Comparing this with the definition with the winding number in Eq. (16.44), the magnetic charge M is proportional to the winding number that maps the sphere S_2 (in two-dimensional space) onto S_2 (in isotopic space). But we know topologically that:

$$\pi_2(S_2) = \mathbf{Z} \tag{16.68}$$

so we are left with $M = n/e$, where n is the winding number.

Finally, the previous results may be generalized to more complicated, phenomenologically acceptable groups. The key element of this monopole solution was the existence of a function $\hat{\phi}$ that smoothly mapped S_2 onto S_2. If we have a gauge group G that is broken down to the subgroup H, then monopole solutions will exist if there are nontrivial mappings of S_2 onto S_2; that is:

$$\pi_2(G/H) = \mathbf{Z} \tag{16.69}$$

where G/H is called the coset space, (G/H is the set of elements g_i of G such that g_i is equivalent to g_2 if $g_1 = g_2 h$ for some element h in H.) Any gauge theory with this group property may have monopole solutions. For example, this can be satisfied if H has $U(1)$ factors.

For example, the GUT theory based on $SU(5)$ can be shown to have monopole solutions because it has a nontrivial homotopy group. In addition, it can be shown that these monopoles have finite energy and mass given, after symmetry breaking, by roughly $137 M_W$, where M_W is a vector meson mass, so the monopole can be extremely heavy. (Any gauge theory with nontrivial homotopy groups can have monopole solutions, and hence must account for the experimental fact that monopoles have not been conclusively seen. This, in turn, places important limits on the production rates for monopoles in the early universe.)

Finally, we remark that it is possible to develop a complete quantum theory of these classical solutions, for example, a theory in which we can study the quantum scattering of solitons against each other, including loops. The complete quantum theory of solitons, however, is beyond the scope of this book. Instead, we will now turn to another classical solution of field theory, the instantons.

16.4 WKB, Tunneling, and Instantons

One of the oldest nonperturbative methods is the semiclassical or WKB approach used in ordinary quantum mechanics. One of the advantages of the WKB approach is that we can calculate tunneling effects that are beyond the usual perturbative method. To any finite order in perturbation theory, we will never see any of these nontrivial nonperturbative effects. The WKB approach also naturally leads to the concept of *instantons*,[6,7] which have proved to be a powerful tool to probe the nonperturbative regime of gauge theory. In particular, we will show that QCD instantons force us to re-examine the whole question of CP violation.

We begin our discussion of instantons by considering $\hbar$ corrections to the classical limit. To see the relationship between $\hbar$ and the perturbative coupling constant g, let us rescale the ϕ field found in ϕ^4 theory as $\phi \rightarrow \phi/g$. Under this

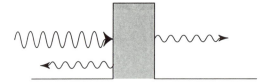

Figure 16.3. Quantum mechanically, a wave can tunnel across a barrier. WKB methods give us the transmission probability.

rescaling, the Lagrangian transforms as:

$$\mathscr{L}(\phi) \rightarrow \frac{1}{g^2}\mathscr{L}'(\phi) \tag{16.70}$$

where the mass also gets rescaled, and where $\mathscr{L}'$ is the action where the coupling constant has been rescaled to unity. Classically, the coupling constant g is not important: if we can solve it classically for any value of g, then we can also solve it for any other value of g. It can always be rescaled to one.

Quantum mechanically, things are a bit different, because we also have the quantity $\hbar$. The factor appearing in the path integral is $S/\hbar$, which can be rescaled as:

$$\frac{S}{\hbar} \rightarrow \frac{S}{g^2\hbar} \tag{16.71}$$

Thus, the weak coupling expansion is identical to an expansion in $\hbar$ in the semi-classical approximation. The essential dimensionless parameter is $g^2\hbar$.

We know from ordinary quantum mechanics that it is possible for a wave to tunnel from one side of a potential well to the other side (Fig. 16.3). The transmission amplitude is given by the WKB result:

$$\Gamma = \exp\left[\left(-\frac{1}{\hbar}\int_{x_1}^{x_2} dx\, [2\sqrt{V-E}]\right)[1 + O(\hbar)]\right] \tag{16.72}$$

The important point is to note that the tunneling amplitude occurs as $\exp(-1/\hbar\cdots)$, and hence tunneling can never be seen to any finite order in $\hbar$. By the previous rescaling argument, this also means that tunneling can never be seen to any finite order in perturbation theory.

The WKB method, as it was originally formulated in nonrelativistic quantum mechanics, consisted of solving the Schrödinger equation separately in different regions. Then, by matching the wave function at the boundary of the potential, we could calculate the leakage through the potential barrier.

To generalize these semiclassical methods and recast them in the language of path integrals, let us first define the partition function of a system with a Hamiltonian H with an Euclidean metric as follows:

$$Z(\beta) = \text{Tr } e^{-\beta H} \tag{16.73}$$

The fact that we have Wick rotated to a Euclidean metric, so the exponential appears with a real argument, is essential to our discussion. Our approach to the WKB method, as applied to path integrals, is to find classical solutions to the Euclidean equations of motion and then to integrate functionally over quantum fluctuations around these classical solutions.

If we trace over the Hilbert space of eigenstates of the Hamiltonian, then the partition function can be written as:

$$Z(\beta) = \sum_{n=0}^{\infty} e^{-\beta E_n} \tag{16.74}$$

We set β to be $1/kT$, where T is the temperature of the system and k is Boltzmann's constant. For our purposes, however, we will interpret β to be the Euclidean time τ.

As $\beta \to \infty$, at large Euclidean times, the right-hand side vanishes, but the state of lowest energy E_0 vanishes slower than the rest. To extract the lowest energy eigenvalue E_0, we therefore take the logarithm of both sides:

$$E_0 = -\frac{1}{\beta} \lim_{\beta \to \infty} \log Z(\beta) \tag{16.75}$$

Thus, the advantage of examining the Euclidean partition function is that we can analyze the ground-state energy of the system. Furthermore, if we calculate the imaginary part of the energy of an unstable state, we can find the decay width, which in turn gives us a derivation of the tunneling rate given earlier.

Let us now write the partition function in terms of path integrals involving a specific potential function $V(x)$:

$$
\begin{aligned}
Z(\beta) &= \text{Tr } e^{-\beta H} \\
&= \int_{x(o)}^{x(\beta)} Dx \, \exp -\frac{1}{g^2} \left(\int_0^\beta dt \, \frac{1}{2}\dot{x}^2 + V(x) \right)
\end{aligned} \tag{16.76}
$$

whre $x(0) = x(\beta)$, and where we have rescaled $x \to x/g$ to extract the coupling constant in front of the action, and where β is treated like a Euclidean time. (Notice that the potential appears with the opposite sign than is usually found in the Minkowski path integral.)

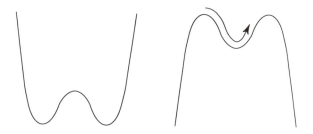

Figure 16.4. To analyze the tunneling between the two minima in the diagram on the left, we must invert the potential and find classical solutions for the inverted potential on the right connecting the two maxima. These are the instantons.

The stationary points of the path integral can be found by using the Euler–Lagrange equations of motion:

$$\ddot{x} = \frac{\partial V}{\partial x} \tag{16.77}$$

Because the sign of the potential is reversed from the usual one, we must now solve the equations of motion in a potential that is upside down.

In Figure 16.4, we see a typical double-well potential V. The quantum-mechanical problem can therefore be solved if we know the classical solutions to the problem with the potential reversed, with Euclidean metric. We know from ordinary quantum mechanics that a state that is concentrated in one part of the well may tunnel into the other. To solve for the tunneling between these two states, we must turn this picture upside down and solve for the motion of a classical body with this new potential. Intuitively, this corresponds to solving the classical problem of a ball rolling down one hill and arriving at the other hill.

The simplest classical solution for the system is just the static one:

$$x(\tau) = \pm a \tag{16.78}$$

where the particle just sits at the top of each potential and remains there. If we insert this solution back into the action S, we find that it corresponds to zero action.

A more interesting case is when the ball rolls down one hill and then up the other, until it stops at the other maximum of the reversed potential. Let the classical solution to this simple problem be given by x_0:

$$x(\tau) = x_0(\tau - \tau_0) \tag{16.79}$$

If we graph what this solution looks like classically, we find Figure 16.5(a). If we then insert this solution into the Lagrangian, we find the graph in Figure 16.5(b),

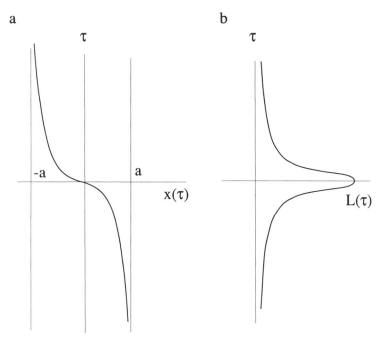

Figure 16.5. In (a), we see a plot of the solution $x(\tau)$. In (b), we see a plot of the Lagrangian evaluated at the classical solution.

which is a rough plot of the function:

$$L(\tau) = \frac{1}{2}\dot{x}_0^2 + V(x_0(\tau)) \tag{16.80}$$

Because this solution creates an almost instantaneous blip in the Lagrangian, we call this finite-action, classical solution to the Euclidean problem an *instanton*. Not surprisingly, we will call the solution that takes us back from the hill to the original one an anti-instanton. Because we are taking the trace in the partition function, we are integrating over all states which start at $x = -a$ and wind up back at $x = -a$. Thus, instantons and anti-instantons occur in pairs in the partition function.

The summation in the partition function, of course, must also sum over multi-instanton solutions as well. Because the instanton and anti-instanton create only a momentary distortion in the Lagrangian, it is a reasonable assumption to replace the sum over the complete multi-instanton solution with the sum over noninteracting instanton and anti-instanton solutions appearing sequentially. Since each instanton and anti-instanton appears only briefly, this approximation is a relatively good one and is called the "dilute gas approximation," after a similar approximation found in statistical mechanics. It treats multi-instanton solutions as if the instantons and anti-instantons were dilute (i.e., their density is low, and they act like free noninteracting gas molecules).

We obviously neglect the overlap between instantons in this approximation, so the contribution to the action by n pairs of instanton–anti-instantons $S^{(2n)}$ is roughly given by the sum of the individual contributions:

$$S^{(2n)} = 2n S_0 \qquad (16.81)$$

Now we would like to calculate the contribution of these instantons to the partition function, with the goal of calculating the ground state and decay rate for this quantum-mechanical problem. We will expand the functional integral around the classical solution for the zero-instanton and the one-instanton case as follows:

$$
\begin{aligned}
x(\tau) &= -a + \xi(\tau - \tau_0) \\
x(\tau) &= x_0(\tau - \tau_0) + \xi(\tau - \tau_0)
\end{aligned}
\qquad (16.82)
$$

where $\xi(\tau)$ represents the quantum fluctuation around the classical solution. $\xi(\tau)$, in turn, can be decomposed into eigenfunctions:

$$\xi(\tau) = \sum_n c_n \xi_n \qquad (16.83)$$

where ξ_n are a complete set of eigenfunctions or normal modes. If we power expand around the classical solution to the action, we find:

$$
\begin{aligned}
V(x) &\rightarrow V(x_0) + \xi^2 V''(x_0) + \cdots \\
S &\rightarrow S_0 + \int d\tau \left(\frac{1}{2} \dot{\xi}^2 + V'' \xi^2 \right) + \cdots
\end{aligned}
\qquad (16.84)
$$

The key assumption we will make is that we can ignore the higher corrections to the potential and the action. This approximation is quite good near the bottom of the well, where the potential is approximated by a quadratic function, but is less reliable away from the minimum.

Let Z_n represent the contribution to the path integral of n instanton–anti-instanton pairs. After we make this approximation, we find:

$$
\begin{aligned}
Z_0 &= \int D\xi \, \exp\left(-\int_0^\beta \frac{1}{2}(\dot{\xi}^2 + \omega \xi^2) \right) d\tau \\
&= \left[\det\left(-\partial_\tau^2 + \omega^2 \right) \right]^{-1/2}
\end{aligned}
\qquad (16.85)
$$

for the zero-instanton contribution, where $\omega = V''(0)$, and:

$$
\begin{aligned}
Z_1 &= e^{-S_0} \int D\xi \, \exp\left(-\int_0^\beta \frac{1}{2}(\dot{\xi}^2 + V''\xi^2)\right) d\tau \\
&= e^{-S_0} \left[\det\left(-\partial_\tau^2 + V''\right)\right]^{-1/2}
\end{aligned}
\tag{16.86}
$$

for the one instanton–anti-instanton contribution. All determinants are evaluated with respect to the eigenfunctions ξ_n.

In the dilute gas approximation, the complete partition function is given by the sum over all the multi-instanton contributions, so:

$$
Z(\beta) = Z_0 + Z_2 + Z_4 \cdots
\tag{16.87}
$$

Our task is now to find an expression for Z_n in terms of Z_0 and Z_1.

In this approximation, the higher Z_n can all be reduced because the functional integral factorizes. The two-instanton contribution, for example, consists of a functional integral over two regions I and II, as in Figure 16.6. The functional integral factorizes as the product of $\prod_\tau dx(\tau) \, e^{-S}$ where τ ranges over regions I and II:

$$
Z_2 = Z_1(I)Z_{\bar{1}}(II)
\tag{16.88}
$$

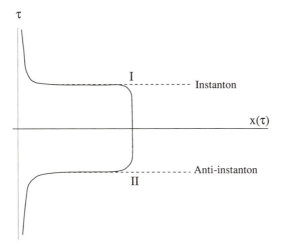

Figure 16.6. The instanton–antiinstanton contribution is shown in this diagram consisting of an integral over regions I and II.

where $Z_n(R)$ is the functional integral over the n-instanton configuration evaluated only in region R:

$$Z_n(R) = \int \cdots \int \prod_{\tau \in R} dx(\tau) \, e^{-\int d\tau \, \mathcal{L}(\tau)} \tag{16.89}$$

Likewise, Z_1 and Z_0 can be factorized as functional integrals over regions I and II:

$$Z_0 = Z_0(I)Z_0(II); \qquad Z_1 = Z_1(I)Z_0(II) \tag{16.90}$$

By multiplying and dividing by Z_0 we now easily have:

$$Z_2 = \frac{1}{2}\frac{Z_1^2}{Z_0} \tag{16.91}$$

(The $\frac{1}{2}$ factor comes from the restriction that the position of the instanton is taken to be larger than the position of the anti-instanton. If we remove this restriction, then we must compensate by dividing by 2.)

By continuing to factorize the functional integral into the product of $Z_0(R)$ and $Z_1(R)$ over different regions, we can then show that:

$$Z_{2n} = \frac{1}{(2n)!} T^{2n} Z_0 \tag{16.92}$$

where $T = Z_1/Z_0$. All Z_{2n} are now expressed in terms of Z_0 and Z_1. If we sum over all the multi-instanton contributions, we find:

$$\begin{aligned} Z(\beta) &= \sum_{n=0}^{\infty} Z_{2n} = Z_0 \sum_{n=0}^{\infty} \frac{1}{(2n)!} T^{2n} \\ &= Z_0 \cosh T \\ &\to e^{-\beta\omega/2 + T} \end{aligned} \tag{16.93}$$

where we took the limit as $\beta \to \infty$ in the last line.

In the limit of large β, the contribution of the vacuum to the partition function gives us the standard harmonic oscillator result:

$$Z_0 = \sum_{n=0}^{\infty} e^{-\beta(n+\frac{1}{2})\omega} \to e^{-\frac{1}{2}\beta\omega} \tag{16.94}$$

In the presence of instantons, however, we expect to find a quantum correction to this:

$$E = \frac{\omega\hbar}{2} + \epsilon \tag{16.95}$$

where:

$$\epsilon = -\lim_{\beta\to\infty} \frac{T}{\beta} \tag{16.96}$$

and:

$$T = \frac{Z_1}{Z_0} = e^{-S_0} \left(\frac{\det(-\partial_\tau^2 + V'')}{\det(-\partial_\tau^2 + \omega^2)} \right)^{-1/2} \tag{16.97}$$

The final answer is therefore:

$$E_0 = \frac{\omega\hbar}{2} + \hbar K e^{-S_0/\hbar} \tag{16.98}$$

where K is the ratio of the two determinants.

Taking the imaginary part, we find that the decay width is:

$$\Gamma = \hbar |K| e^{-S_0/\hbar} \tag{16.99}$$

which is the original WKB result presented earlier. To see this, we note that the classical solution x_{cl} obeys $\frac{1}{2}(\dot{x}_{cl})^2 = V(x_{cl})$ so that:

$$\begin{aligned}
S_0 &= \int_{-\infty}^{\infty} d\tau \left(\frac{1}{2}\dot{x}_{cl}^2 + V \right) = \int_{-\infty}^{\infty} (\dot{x}_{cl})^2 \, d\tau \\
&= \int_{x_1}^{x_2} \dot{x}_{cl} \, dx = \int_{x_1}^{x_2} \sqrt{2V(x)} \, dx
\end{aligned} \tag{16.100}$$

so that the tunneling amplitude is proportional to:

$$\exp\left(-\frac{1}{\hbar} \int_{x_1}^{x_2} \sqrt{2V} \, dx \right) \tag{16.101}$$

as in the WKB result quoted earlier in Eq. (16.72).

[We note that we omitted some subtle details concerning the determination of K in our final expression. In particular, a naive calculation of K actually vanishes because of zero modes, since the determinant in Eq. (16.97) is the product of the eigenvalues, which can be zero. The zero mode is due to the time translation invariance of the system, and hence we must be careful in integrating over all positions of the instanton. For a careful determination of K and how to handle zero modes, the reader is referred to the literature on instantons.]

16.5 Yang–Mills Instantons

The purpose of discussing instantons is to probe the nonperturbative realm of gauge theories. We will see that the theory of finite-action solutions to the Euclidean Yang–Mills theory has profound implications for the nature of QCD. In particular, we will be interested in considering the implications of classical solutions to the Euclidean Yang–Mills equations of motion, which are self-dual.

If we define:

$$\tilde{F}_{\mu\nu} \equiv \frac{1}{2}\epsilon_{\mu\nu\sigma\rho}F_{\sigma\rho} \tag{16.102}$$

then a classical solution is self-dual if:

$$\tilde{F}_{\mu\nu} = F_{\mu\nu} \tag{16.103}$$

Our first task is to calculate the action corresponding to a self-dual solution to the Euclidean Yang–Mills theory. We begin with the simple observation that the sum of squares of a sequence of numbers must necessarily be greater than or equal to zero:

$$\text{Tr}\,(F_{\mu\nu} - \tilde{F}_{\mu\nu})^2 \geq 0 \tag{16.104}$$

Let us now expand the terms in the sum. We use the identity:

$$\epsilon_{\mu\nu\sigma\rho}\epsilon_{\mu\nu\alpha\beta} = 2(\delta_{\rho\beta}\delta_{\sigma\alpha} - \delta_{\rho\alpha}\delta_{\sigma\beta}) \tag{16.105}$$

From this, we can show:

$$\tilde{F}_{\mu\nu}\tilde{F}_{\mu\nu} = F_{\mu\nu}F_{\mu\nu} \tag{16.106}$$

Therefore, our inequality now reads:

$$\text{Tr}\,F_{\mu\nu}F_{\mu\nu} \geq \text{Tr}\,F_{\mu\nu}\tilde{F}_{\mu\nu} \tag{16.107}$$

Our original task, to calculate the action corresponding to a self-dual solution, is now reduced to calculating the integral of $F\tilde{F}$. This is easily accomplished by observing that $F\tilde{F}$ is actually a total derivative:

$$\frac{1}{4} \operatorname{Tr} F_{\mu\nu} \tilde{F}_{\mu\nu} = \partial_\mu K_\mu \tag{16.108}$$

where a direct computation shows that:

$$K_\mu = \epsilon_{\mu\nu\alpha\beta} \operatorname{Tr}\left(\frac{1}{2} A_\nu \partial_\alpha A_\beta - \frac{ig}{3} A_\nu A_\alpha A_\beta\right) \tag{16.109}$$

Because of this, we can integrate over the volume of four-space:

$$\begin{aligned}
\int \operatorname{Tr} F_{\mu\nu} \tilde{F}_{\mu\nu}\, d^4 x &= 4 \int \partial_\mu K_\mu\, d^4 x \\
&= 4 \oint K_\perp\, d^3 \sigma
\end{aligned} \tag{16.110}$$

Normally, in field theory we expect that the integral of a total derivative should vanish. However, the field may vanish slowly enough at infinity so that we can have nonzero values of this integral. In fact, as we shall demonstrate shortly, this integral equals an integer:

$$n = \frac{1}{32\pi^2} \int d^4 x\, F^a_{\mu\nu} \tilde{F}^{a\mu\nu} \tag{16.111}$$

Putting everything together, we now have:

$$\begin{aligned}
S &= \frac{1}{4g^2} \int d^4 x\, \left(F^a_{\mu\nu}\right)^2 \\
&\geq \frac{1}{4g^2} \int F^a_{\mu\nu} \tilde{F}^{a\mu\nu}\, d^4 x \\
&= \frac{8\pi^2 n}{g^2}
\end{aligned} \tag{16.112}$$

and therefore:

$$S \geq S_{\text{self}-\text{dual}} = \frac{8\pi^2 n}{g^2} \tag{16.113}$$

As desired, we have now shown that a Euclidean, self-dual solution, if it exists, has finite action, labeled by n, which will be called the winding number. Inserting this value of the action back into the path integral, the contribution of the self-dual solution to the functional integral is given by:

$$e^{-S} = e^{-8\pi^2/g^2} \tag{16.114}$$

This clearly shows the nonperturbative nature of the instanton. The contribution of the instanton is proportional to $\exp(-1/g^2)$, which can never be approximated to any finite order in perturbation theory. Thus, the instanton contributes nonperturbatively to the gauge theory functional.

We still, however, have not touched upon a few important questions: First, do these self-dual solutions really exist, and, if so, what do they look like, and what possible physical consequence can Euclidean solutions have upon our Minkowski world?

To answer these questions, we start with the one-instanton solution for $SU(2)$ Euclidean Yang–Mills theory. From our previous discussion of instantons, we are led to postulate a form for the gauge field that asymptotically goes to the vacuum solution $A_\mu \to (-i/g)(\partial_\mu\Omega)\Omega^{-1}$. We are led to postulate the form[6]:

$$A_\mu = (-i/g)\frac{x^2}{x^2 + \lambda^2}(\partial_\mu\Omega)\Omega^{-1} \tag{16.115}$$

where λ is an arbitrary parameter and where:

$$\Omega = \frac{x_4 \pm i\sigma_i x_i}{\sqrt{x^2}} \tag{16.116}$$

Since $\Omega^\dagger\Omega = 1$, we have:

$$x^2 = x_4^2 + x_i^2 \tag{16.117}$$

We can also generalize this single-instanton solution to the more general case. Let us define:

$$\begin{aligned}
A_{ia} &= (\epsilon_{aik}\partial_k \mp \delta_{ai}\partial_0)\log f \\
A_a^0 &= \pm\partial_a\log f
\end{aligned} \tag{16.118}$$

where $i, k = 1, 2, 3$ and we have deliberately mixed up space and isospin indices. The condition $F_{\mu\nu} = \pm\tilde{F}_{\mu\nu}$ fixes a constraint on f:

$$f^{-1}\partial_\mu^2 f = 0 \tag{16.119}$$

If we choose $f = 1 + \lambda^2/x^2$, then we recover the previous solution with winding number $n = \pm 1$. However, we can also choose:

$$f^{(n)}(x) = \sum_{i=1}^{n+1} \frac{\lambda_i^2}{(x - x_i)^2} \tag{16.120}$$

which corresponds to a multi-instanton solution, where the various instantons are located at x_i. This solution is parametrized by an arbitrary integer n.

Next, let us explore the asymptotic properties of this instanton. If we take the limit $x_4 \to \pm\infty$, then the self-dual instanton solution reduces to:

$$A_\mu \to (-i/g)\left(\partial_\mu \Omega\right)\Omega^{-1} \tag{16.121}$$

that is, it approaches the vacuum solution at infinity, as desired. In particular, we find that:

$$
\begin{aligned}
x_4 &\to \infty; & A_i &\to i(\Omega_n)^{-1}(\partial_i \Omega_n) \\
x_4 &\to -\infty; & A_i &\to i(\Omega_{n-1})^{-1}(\partial_i \Omega_{n-1})
\end{aligned}
\tag{16.122}
$$

where:

$$\Omega_n = \Omega_1^n; \quad \Omega_1 = \exp\left(-i\pi\,\frac{x_i \sigma_i}{(x^2 + \lambda^2)^{1/2}}\right) \tag{16.123}$$

This is a rather surprising conclusion. It shows that the n-instanton solution, at $x_4 = \pm\infty$, connects two different vacua, which differ by one unit. One vacuum has winding number $n-1$, and the other has winding number n. (This is similar to the instanton solution we found in Eq. (16.79), which connects the two vacua at $x = \pm a$.)

We now can give a mathematical meaning to the index n in Eq. (16.111). Let us specialize our case to the group $SU(2)$. The elements Ω of $SU(2)$, in turn, can be put in correspondence with the points that label a three-dimensional sphere, as in Eq. (16.117). Thus, for each point x_μ on a three-dimensional sphere S_3, we can generate an element Ω of $SU(2)$.

Since, at asymptotic times, the gauge field becomes a pure gauge field:

$$A_\mu \to (-i/g)\left(\partial_\mu \Omega\right)\Omega^{-1} \tag{16.124}$$

then a pure gauge configuration is labeled by a three-dimensional surface, given by a hypersphere S_3.

Let us now insert this value of A_μ into the expression for K_μ:

$$K_\mu = \frac{1}{6g^2}\epsilon_{\mu\nu\alpha\beta}\mathrm{Tr}\left(\Omega^{-1}\partial_\nu\Omega\right)\left(\Omega^{-1}\partial_\alpha\Omega\right)\left(\Omega^{-1}\partial_\beta\Omega\right) \tag{16.125}$$

To make some sense out of this expression, we will parametrize the invariant $SU(2)$ group measure dU (which we introduced in Chapters 9 and 15) as follows:

$$dU = \rho(\sigma_1, \sigma_2, \sigma_3)\,d\sigma_1\,d\sigma_2\,d\sigma_3 \tag{16.126}$$

where U is an element of $SU(2)$ parametrized by some coordinates σ_i. Let U_0 be a fixed element of $SU(2)$, and $U' = U_0 U$. If $\{\sigma\}$ are the coordinates that parametrize U and $\{\sigma'\}$ the coordinates that parametrize U', then the group measure obeys:

$$
\begin{aligned}
dU &= dU' \\
&= \rho(\sigma_1', \sigma_2', \sigma_3')\, d\sigma_1'\, d\sigma_2'\, d\sigma_3'
\end{aligned}
\tag{16.127}
$$

that is, the group measure obeys $dU = d(U_0 U)$ for fixed U_0.

Then there is a theorem from classical group theory that states that the invariant measure $\rho(\sigma_1, \sigma_2, \sigma_3)$ is given by:

$$
\rho(\sigma_1, \sigma_2, \sigma_3) = \epsilon^{ijk}\, \mathrm{Tr}\left(U^{-1}\frac{\partial U}{\partial \sigma_i} U^{-1}\frac{\partial U}{\partial \sigma_j} U^{-1}\frac{\partial U}{\partial \sigma_k} \right)
\tag{16.128}
$$

With this expression, one can check explicitly that:

$$
\rho(\sigma_i) = \rho(\sigma_i')\, \mathrm{Det}\left| \frac{\partial \sigma'}{\partial \sigma} \right|
\tag{16.129}
$$

Then the index n is given by:

$$
\begin{aligned}
n &= \frac{g^2}{4\pi^2}\int \partial_\mu K_\mu\, d^4 x \\
&= \frac{1}{24\pi^2}\oint_{S_3} \epsilon_{\mu\nu\alpha\beta}\, \hat{n}_\mu \mathrm{Tr}\left(\Omega^{-1}\partial_\nu\Omega \right)\left(\Omega^{-1}\partial_\alpha\Omega \right)\left(\Omega^{-1}\partial_\beta\Omega \right) d^3\sigma \\
&= \frac{1}{24\pi^2}\int_G dU
\end{aligned}
\tag{16.130}
$$

In the last line, we have the integral over the invariant volume element in the group space. The surface term in Euclidean space E_4 is taken as $r \to \infty$, where $r = (x_1^2 + x_2^2 + x_3^2 + x_4^2)^{1/2}$. This boundary, of course, is the hypersphere S_3. Thus, the index n gives us the degree of mapping from:

$$
S_3 \to S_3
\tag{16.131}
$$

that is, it gives us the number of topologically distinct ways in which the surface of S_3 can wind around another S_3.

This formalism thus gives us a nontrivial mapping from one S_3 onto another; one S_3 represents the isotopic space of $SU(2)$ denoted by Ω, and the other S_3 represents physical space, the boundary of Euclideanized space, denoted by the boundary of the integral over x-space .

This mapping from $S_3 \rightarrow S_3$ is called $\pi_3(S_3)$. In topology, one can show:

$$\pi_3(S_3) = \mathbf{Z} \tag{16.132}$$

Thus, the mappings of $S_3 \rightarrow S_3$ are characterized by integers; that is, the points of one S_3 can be mapped smoothly to another S_3 such that we wind around S_3 an integer number of times. This now explains the mathematical origin of the index n.

This clearly demonstrates the highly nontrivial nature of the gauge instantons. It reveals the fact that the naive vacuum of Yang–Mills theory is the incorrect one, that there are actually an infinite number of topologically distinct vacua, each labeled by an integer n.

This shows that the vacuum of Yang–Mills theory actually consists of an infinite number of degenerate vacua, so the true vacuum must be a superposition of all of them.

16.6 θ Vacua and the Strong CP Problem

Finally, we comment on the physical interpretation of the theory of instantons[8,9]. In ordinary quantum mechanics, we know that nonperturbative effects, such as tunneling, can be computed using the WKB method. This formalism, in turn, requires finding solutions to the Euclidean equations of motion that connect two classical solutions at $x_4 \rightarrow \pm\infty$. We now see the true significance of instanton solutions: They allow tunnelling between different vacua because they connect these vacua at $x_4 \rightarrow \pm\infty$.

The naive vacuum is thus unstable. The instanton allows tunneling between all possible vacua labeled by winding number n. Thus, the true vacuum must be a superposition of the various vacua $|n\rangle$, each belonging to some different homotopy class.

The effect of a gauge transformation Ω_1 in Eq. (16.122) is to shift the winding number n by one:

$$\Omega_1 : \quad |n\rangle \rightarrow |n + 1\rangle \tag{16.133}$$

Since the effect of Ω_1 on the true vacuum can change it only by an overall phase factor, this fixes the coefficients of the various vacua $|n\rangle$ within the true vacuum. This fixes the coefficients of $|n\rangle$ as follows:

$$|\text{vac}\rangle_\theta = \sum_{n=-\infty}^{\infty} e^{in\theta} |n\rangle \tag{16.134}$$

We can then check that the effect of Ω_1 on the true vacuum is to generate a phase shift:

$$\Omega_1 : \quad |\text{vac}\rangle_\theta \to e^{-i\theta}|\text{vac}\rangle_\theta \tag{16.135}$$

The presence of instantons means that the true vacuum is parametrized by the arbitrary number θ. The effect of this θ dependence can be also expressed by writing down an effective action. To do this, recall that in Chapter 8 we wrote the expectation value $\langle x_2|e^{-iH\delta t}|x_1\rangle$ as a Lagrangian path integral, where the integration over Dx connected the configurations x_1 and x_2. Likewise, we may write the expectation value $\langle m|e^{-iH\delta t}|n\rangle$ as the path integral over DA_μ that connects the mth vacuum with the nth vacuum:

$$\langle m|e^{-iHt}|n\rangle = \int [DA_\mu]_{\nu=m-n} \exp\left(-i\int \mathscr{L}\, d^4x\right) \tag{16.136}$$

where we integrate over all A_μ of the same homotopic class with winding number $\nu = m - n$.

This allows us to write the vacuum-to-vacuum transition as:

$$
\begin{aligned}
_\theta\langle \text{vac}|e^{-iHt}|\text{vac}\rangle_\theta &= \sum_{m,n} e^{im\theta'} e^{-in\theta} \langle m|e^{-iHt}|n\rangle \\
&= \sum_{m,n} e^{-i(n-m)\theta} e^{im(\theta'-\theta)} \int [DA_\mu]_{n-m}\, e^{iS} \\
&= \delta(\theta'-\theta) \sum_n e^{-i\nu\theta} \int [DA_\mu]_\nu \\
&\quad \times \exp\left(-i\int dx^4\, \mathscr{L}\right)
\end{aligned}
\tag{16.137}
$$

where we have obtained the delta function by summing over m, and where $[DA_\mu]$ connects two vacua with different winding numbers. The phase factor $e^{-i\nu\theta}$ can be absorbed into the action. We know that $\nu = (1/16\pi^2)\int d^4x\, \text{Tr}\, F\tilde{F}$, so we can add it to the Lagrangian, giving us an effective Lagrangian:

$$\mathscr{L}_{\text{eff}} = \mathscr{L} + \theta\nu = \mathscr{L} + \frac{\theta}{16\pi^2} \text{Tr}\, F_{\mu\nu}\tilde{F}^{\mu\nu} \tag{16.138}$$

This is a rather surprising result, that the effect of the instantons is to create tunneling between degenerate vacua, which in turn generates an effective action with the additional term $F\tilde{F}$. The presence of this extra term in the action does not alter the theory perturbatively, since it is a total derivative and hence never enters into the perturbation theory. Perturbatively, we therefore never see the effect of this term. However, nonperturbatively, it will have an important effect on the physics.

So far, our discussion has been rather abstract. We will now show that the instanton solution has an immediate impact on QCD. The instanton solves one problem [the $U(1)$ problem] but also raises another (the strong CP problem).

To understand the $U(1)$ problem, let us first catalog all the global symmetries of QCD. In the limit of zero quark masses, QCD for the up and down quarks is invariant under chiral $SU(2) \otimes SU(2)$. This is because:

$$\bar{q} \, \slashed{D} q = \bar{q}_L \, \slashed{D} q_L + \bar{q}_R \, \slashed{D} q_R \qquad (16.139)$$

so the left- and right-handed sectors are separately invariant under $SU(2)_L$ and $SU(2)_R$.

QCD is also invariant under two global $U(1)$ transformations. The first $U(1)$ transformation leads to a conserved current:

$$J_\mu = \sum_a \bar{\psi}_a \gamma_\mu \psi_a \qquad (16.140)$$

which give us baryon number conservation, which is, of course, seen experimentally. However, the second $U(1)$ symmetry is given by the transformation:

$$\psi_a \to e^{i\alpha\gamma_5} \psi_a \qquad (16.141)$$

This leads to the current:

$$J_\mu^5 = \sum_a \bar{\psi}_a \gamma_\mu \gamma_5 \psi_a \qquad (16.142)$$

Although QCD is classically invariant under global $SU(2) \otimes SU(2) \otimes U(1) \otimes U(1)$, quantum corrections to QCD may alter this symmetry in various ways. There are three possibilities:

1. A symmetry may be preserved by quantum corrections, in which case the particle spectrum should manifest this symmetry.

2. The symmetry could be spontaneously broken, in which case there are Nambu–Goldstone bosons.

3. The symmetry may be broken by quantum corrections, in which case the symmetry is not manifested in the particle spectrum and the Nambu–Goldstone boson is absent.

For example, chiral $SU(2)$ symmetry is believed to be spontaneously broken, so there must be Nambu–Goldstone bosons associated with this broken symmetry. The Nambu–Goldstone bosons are the triplet of π mesons.

The axial $U(1)$ symmetry, however, is more problematic. If it is preserved, then all hadrons should be parity doubled. This is not the case, since the π meson has no scalar partner.

The second possibility is that the axial $U(1)$ symmetry is spontaneously broken, in which case there should be a light Nambu–Goldstone boson. However, there is no Nambu–Goldstone boson around the π meson mass. Weinberg has proved a theorem that says that the $U(1)$ Nambu–Goldstone boson should have mass less than $\sqrt{3}m_\pi$. However, there is no such particle. The particles that come closest, the $\eta(549)$ and the $\eta'(985)$, fail to satisfy the Weinberg bound, and $\eta(549)$ is actually part of the pseudoscalar octet.

The $U(1)$ problem, therefore, is to explain the absence of both parity doubling as well as the Nambu–Goldstone boson for this symmetry.

This leaves open the third possibility, that the symmetry is not preserved quantum mechanically. Indeed, one might suspect that the anomaly in the $U(1)$ current makes it impossible to construct conserved currents. There is indeed a triangle anomaly, which breaks the conservation of the axial current. However, this is not enough to solve the $U(1)$ problem. By slightly modifying the calculation of the triangle anomaly presented earlier to accomodate quark flavors, we can calculate the contribution of the anomaly to the current conservation condition:

$$\partial_\mu J_\mu^5 = \frac{N_f g^2}{8\pi^2} \text{Tr}\,(F_{\mu\nu}\tilde{F}_{\mu\nu}) \tag{16.143}$$

where N_f is the number of flavors. However, using the fact that:

$$\text{Tr}\,(F_{\mu\nu}\tilde{F}_{\mu\nu}) = 4\partial_\mu K_\mu \tag{16.144}$$

where:

$$K_\mu = \frac{1}{2}\epsilon_{\mu\nu\rho\sigma}\text{Tr}\,(A^\nu \partial^\rho A^\sigma - \frac{2}{3}ig A^\nu A^\rho A^\sigma) \tag{16.145}$$

we can construct a current that is indeed conserved:

$$\tilde{J}_\mu^5 = J_\mu^5 - \frac{N_f g^2}{2\pi^2} K_\mu \tag{16.146}$$

The modified conserved charge $\tilde{Q}_5$ is given by:

$$\frac{d\tilde{Q}_5}{dt} = \int d^3x\,\partial_0 \tilde{J}^{50} = 0 \tag{16.147}$$

Because of this, it appears that the modified current is still conserved, and that the $U(1)$ problem persists even in the presence of anomalies.

In conclusion, we seem to have exhausted all possibilities for the $U(1)$ problem. However, the solution to the problem was pointed out by 't Hooft,[8] who observed that instantons can render the previous equation incorrect. He pointed out that there is yet another contribution to $\tilde{Q}_5$ that may break the symmetry. If we calculate the change in $\tilde{Q}_5$ between the distant past and the distant future, the presence of instantons can create a nonzero value for $\Delta\tilde{Q}_5$. We observe that:

$$\begin{aligned}
\Delta\tilde{Q}_5 &= \int dt\, \frac{d Q_5}{dt} = \int d^4x\, \partial_\mu J^{5\mu} \\
&= \frac{N_f g^2}{2\pi^2} \int d^4x\, \mathrm{Tr}\,(F_{\mu\nu}\tilde{F}^{\mu\nu})
\end{aligned} \tag{16.148}$$

Usually, $\Delta\tilde{Q}_5$ is equal to zero because the right-hand side is the integral over a pure divergence, which vanishes at infinity. However, in the presence of instantons, the right-hand side does not vanish at all. We know that the instanton has a finite action, so the right-hand side is not zero and $\Delta\tilde{Q}_5$ is not zero. Thus, there is no Nambu–Goldstone boson because the current is not really conserved, and hence the $U(1)$ symmetry was not a good one in the first place.

(An equivalent way of stating this is to notice that the modified current is not gauge invariant. The Green's functions for this modified symmetry may develop poles that naively indicate that there are Nambu–Goldstone bosons in the theory, but these Green's functions are gauge variant, and these poles cancel against other poles. The gauge-invariant amplitudes, which add up both the gauge-variant particle and ghost poles, do not have a net pole, and hence there are no Nambu–Goldstone bosons.)

Instantons, therefore, appear to give us a nice explanation for the fact that the Nambu–Goldstone boson associated with the breaking of axial $U(1)$ symmetry is not experimentally observed. However, instantons solve one problem, only to raise another.

We saw earlier that the instanton contribution to the effective action $\Delta\mathcal{L}$ vanishes perturbatively but may have nontrivial nonperturbative effects. In particular, because of the existence of $\epsilon_{\mu\nu\sigma\rho}$, it indicates that parity is violated by the strong interactions. T is also violated; so there is a violation of CP. This is rather disturbing, because CP is known to be conserved rather well by the strong interactions, as measured by the neutron electric dipole moment, which is known experimentally to obey $d_n \leq 10^{-24}$ e-cm. This serves as an experimental constraint on the parameters of the Standard Model, since we can calculate the perturbative and nonperturbative (instanton) corrections to the neutron dipole moment. The perturbative corrections to the moment can be shown to give a dipole moment much smaller than this, which then gives us a bound on the nonperturbative correction.

This constraint gives us the bound on θ:

$$\theta < 10^{-9} \qquad (16.149)$$

This is the strong CP problem: If instanton effects necessarily contribute an extra parameter to QCD, then why is θ so small?

In principle, if one or more of the quarks had been massless, we could have absorbed this term and preserved CP invariance. For example, if the up quark had been massless, then we could have made the usual chiral transformation on ψ_u, which creates a change in the action given by:

$$\delta S = -i\alpha \int d^4x \, (\partial_\mu J^{5\mu}) \qquad (16.150)$$

so that $\delta S = 2N_f\alpha$. We could thus absorb the θ term by choosing an appropriate α. However, the up quark is massive, so this line of argument is ruled out.

The simplest suggested solution to why θ is so small is to invoke yet another $U(1)$ symmetry, the Peccei–Quinn symmetry,[10] which is preserved by a combined QCD and electroweak theory. The presence of this additional $U(1)$ symmetry would be sufficient to keep $\theta = 0$.

To see how the axion hypothesis works, consider the possibility of CP violation in QCD caused by introducing a complex, nondiagonal mass matrix M for the quarks: $\bar{q}_i M_{ij} q_j$. Classically, M can be diagonalized and made real by making a field redefinition of the quark fields q_i, so CP violation does not appear as a consequence of a complex mass matrix M.

In this field redefinition, we made a chiral transformation on the quark fields to eliminate an overall phase factor. Once quantum corrections are allowed, however, we can no longer eliminate this phase factor with a chiral transformation. Since the functional measure $Dq \, D\bar{q}$ is not invariant under a chiral transformation (see Section 12.7), the chiral anomaly adds a θ term to the measure, given by:

$$Dq \, D\bar{q} \to Dq \, D\bar{q} \exp\left(\text{Arg Det } M \int d^4x \frac{g^2}{16\pi^2} F^a_{\mu\nu} \tilde{F}^{a\mu\nu}\right) \qquad (16.151)$$

Therefore the effective θ is given by:

$$\theta \to \theta + \text{Arg Det } M \qquad (16.152)$$

To eliminate this effective θ term, consider adding a new field σ to the QCD action given by:

$$\mathscr{L}_{\text{axion}} = \bar{\psi}(Me^{-i\sigma})\psi + \frac{1}{2}\partial_\mu\sigma\partial^\mu\sigma \qquad (16.153)$$

where σ is the axion field, which couples to the quark mass term via a phase factor. [The axion arises as a Nambu–Goldstone boson of the new broken $U(1)$ symmetry of the quark and the Higgs sector.]

Now perform another axial $U(1)$ transformation on the quark fields that eliminates the $F\tilde{F}$ term entirely and puts all CP violating terms in the mass matrix. We then find that the mass term in QCD is multiplied by:

$$\exp i\,(\theta + \text{Arg Det } M - \sigma) \tag{16.154}$$

Now make the trivial shift:

$$\sigma \to \sigma + \theta + \text{Arg Det } M \tag{16.155}$$

Since the axion is massless, the kinetic term is invariant under this shift, so the shift is sufficient to absorb all CP violating terms that appear exclusively in the mass matrix.

In this way, the introduction of a massless axion field, to lowest order, can absorb all strong CP violating effects by a shift. (At higher orders, the axion develops a mass, although we can still absorb the CP violating terms.)

Although the axion gives us a way in which the strong CP problem might be solved, experimentally the situation is still unclear. Experimental searches for the axion have been unsuccessful. In fact, the naive axion theory that we have presented can actually be experimentally ruled out. However, it is still possible to revive the axion theory if we assume that it is very light and weakly coupled. Experimentally, this "invisible axion," if it exists, should have a mass between 10^{-6} and 10^{-3} eV. The invisible axion[11] would then be within the bounds of experiments. Phenomenologically, it has been suggested that the axion may solve certain cosmological problems, such as the missing mass problem (i.e., that only 1% to 10% of the mass of the universe is visible, and the remaining mass is invisible, in the form of "dark matter"). However, until the axion is discovered, the strong CP problem is an open question and much of this discussion is speculative.

In summary, we have seen that the theory of solitons, monopoles, and instantons probes an area of quantum field theory that is not accessible by perturbative methods.

Instantons (solitons) are classical finite-action (energy) solutions to the Euclidean (Minkowski) equations of motion which obey special properties. Their existence proves that gauge theories are more sophisticated than previously thought. The existence of instantons, for example, is an indication that tunneling takes place in the theory. Instantons in gauge theory are useful in giving us a solution to the $U(1)$ problem, but they also raise the question of strong CP violation.

16.7   Exercises

1. Show that the N-instanton solution in Eq. (16.118) solves the Yang–Mills equations of motion.

2. Given the one-dimensional Lagrangian:

$$\mathcal{L} = \frac{1}{2}\dot{x}^2 + \frac{1}{2}x^2(1 - \sqrt{g}x)^2 \qquad (16.156)$$

Plot the potential and show that the instanton solution is given by:

$$x(t) = \frac{1}{\sqrt{g}} \frac{1}{(1 + e^{\pm(t-t_0)})} \qquad (16.157)$$

Where on the potential curve does this instanton make tunneling possible?

3. Let us integrate this Lagrangian from $-\beta/2$ to $\beta/2$. Show that the energy and action of this system are finite and are given by:

$$\begin{aligned} E(\beta) &= -2e^{-\beta} + O(e^{-2\beta}) \\ S &= \frac{1}{g}\left(\frac{1}{6} - 2e^{-\beta} + O(e^{-2\beta})\right) \end{aligned} \qquad (16.158)$$

4. Consider the Lagrangian:

$$\mathcal{L} = \frac{1}{2}\dot{x}^2 + g^{-1}\left[1 - \cos(\sqrt{g}x)\right] \qquad (16.159)$$

Again, graph the potential and show that the instanton solution is given by:

$$x(t) = \frac{4}{\sqrt{g}}\tan^{-1}e^{(t-t_0)} \qquad (16.160)$$

Between what states does this instanton make tunneling possible?

5. Show that the action is finite (when integrated as before) and that:

$$S = \frac{8}{g} \qquad (16.161)$$

6. Consider a massless four-dimensional ϕ^4 theory with the action:

$$\frac{1}{2}(\partial_\mu\phi)^2 + \frac{1}{4}g\phi^4 \qquad (16.162)$$

Show that the instanton solution is given by:

$$\phi(x) = \pm \frac{1}{\sqrt{-g}} \frac{2\sqrt{2}\lambda}{1 + \lambda^2(x - x_0)^2} \qquad (16.163)$$

where λ is an arbitrary constant. Show that the action is again finite, with:

$$S = -\frac{8\pi^2}{3g} \qquad (16.164)$$

7. Prove by direct computation that the sine-Gordon equation is solved by the soliton solutions in Eqs. (16.24) and (16.25).

8. Consider the two-dimensional complex scalar theory with Lagrangian:

$$\mathcal{L} = \partial_\mu \sigma^\dagger \partial^\mu \sigma + V(\sigma^\dagger \sigma) \qquad (16.165)$$

Show that if V is given by:

$$V = \frac{\sigma^\dagger \sigma}{1 + \epsilon^2} \left[(1 - \sigma^\dagger \sigma)^2 + \epsilon^2 \right] \qquad (16.166)$$

then a solution is given by:

$$\sigma = \left(\frac{a}{1 + \sqrt{1 - a} \cosh y} \right)^{1/2} e^{-i\omega t} \qquad (16.167)$$

with:

$$
\begin{aligned}
a &= (1 + \epsilon^2)(1 - \omega^2) \\
y &= 2\sqrt{1 - \omega^2}(x - \xi)
\end{aligned} \qquad (16.168)
$$

9. Prove that if the soliton system is translationally invariant, there is a zero mode in Eq. (16.97).

10. For the 't Hooft–Polyakov monopole, prove explicitly that the solution for A_μ^a and ϕ^a in Eq. (16.62) solves the equations of motion of the monopole at large distances.

11. For the nonlinear $O(3)$ model, in Eqs. (16.33) and (16.34), define the variables:

$$
\begin{aligned}
\omega_1 &= 2\phi_1/(1 - \phi_3); \quad \omega_2 = 2\phi_2/(1 - \phi_3) \\
\omega &= \omega_1 + i\omega_2; \quad \phi = \phi_1 + i\phi_2
\end{aligned} \qquad (16.169)
$$

Show that:

$$\partial_1 \omega = [2/(1 - \phi_3)^2](\partial_1 \phi + \phi \overleftrightarrow{\partial}_1 \phi_3) \qquad (16.170)$$

Show that the self-duality condition in Eq. (16.48) becomes:

$$\partial_1 \phi = \mp i(\phi \overleftrightarrow{\partial}_2 \phi_3)$$
$$\partial_2 \phi = \pm i(\phi \overleftrightarrow{\partial}_1 \phi_3) \qquad (16.171)$$

12. Show that this self-duality condition can be rewritten as:

$$\frac{\partial \omega_1}{\partial x_1} = \pm \frac{\partial \omega_2}{\partial x_2}; \quad \frac{\partial \omega_1}{\partial x_2} = \mp \frac{\partial \omega_2}{\partial x_1} \qquad (16.172)$$

This means that the self-duality condition reduces to the Cauchy–Riemann condition. This, in turn, means that *any* analytic function of $z = x_1 + i x_2$ will satisfy the self-duality condition, and hence the equations of motion.

13. Now choose the following analytic function:

$$\omega(z) = [(z - z_0)/\lambda]^n \qquad (16.173)$$

where n is an integer. Show that Q in Eq. (16.41) can be written as:

$$Q = \frac{1}{4\pi} \int d^2 z \, \frac{n^2 |z - z_0|^{2n-1} \lambda^{2n}}{(\lambda^{2n} + \frac{1}{4}|z - z_0|^{2n})^2} \qquad (16.174)$$

Using polar coordinates, perform the integration and show:

$$Q = n \qquad (16.175)$$

as expected.

14. Prove that the invariant measure given in Eq. (16.128) satisfies the property $dU = d(U_0 U)$ if U_0 is a constant.

15. Prove that the measure in Eq. (16.42) is generally covariant under a reparametrization of the coordinates.

16. Another theory with instantons is the CP_N theory. We begin with $N + 1$ complex scalar fields $n_a(x) = \mathbf{n}$. The Euclideanized action is given by:

$$\mathcal{L} = D_\mu \mathbf{n}^* \cdot D_\mu \mathbf{n} \qquad (16.176)$$

where:

$$D_\mu \mathbf{n} = (\partial_\mu + i A_\mu)\mathbf{n} \tag{16.177}$$

We also have the constraint:

$$\mathbf{n}^* \cdot \mathbf{n} = \sum_{a=1}^{N+1} n_a^* n_a = 1 \tag{16.178}$$

Eliminate the A_μ field by its equations of motion. Show that the resulting action is:

$$\mathcal{L} = (\partial_\mu \mathbf{n}^* \cdot \partial_\mu \mathbf{n}) + (\mathbf{n}^* \cdot \partial_\mu \mathbf{n})(\mathbf{n}^* \cdot \partial_\mu \mathbf{n}) \tag{16.179}$$

Show that this action is invariant under:

$$\begin{aligned}
\partial_\mu \mathbf{n} &\longrightarrow (\partial_\mu \mathbf{n} + i\partial_\mu \Lambda \mathbf{n})e^{i\Lambda} \\
\mathbf{n}^* \cdot \partial_\mu \mathbf{n} &\longrightarrow \mathbf{n}^* \cdot \partial_\mu \mathbf{n} + i\partial_\mu \Lambda
\end{aligned} \tag{16.180}$$

17. For the CP_N model, show that the positive-definite quantity:

$$\int d^2x \left(D_\mu \mathbf{n} \pm i\epsilon_{\mu\nu} D_\nu \mathbf{n} \right)^* \cdot \left(D_\mu \mathbf{n} \pm i\epsilon_{\mu\nu} D_\nu \mathbf{n} \right) \geq 0 \tag{16.181}$$

reduces to:

$$2 \int d^2x \left[(D_\mu \mathbf{n})^* \cdot (D_\mu \mathbf{n}) \pm i\epsilon_{\mu\nu}(D_\mu \mathbf{n}^*) \cdot D_\nu \mathbf{n} \right] \geq 0 \tag{16.182}$$

Show that this proves:

$$S \geq 2\pi |Q| \tag{16.183}$$

where we define the topological charge as:

$$\begin{aligned}
Q &= -\frac{1}{2\pi} \int d^2x \, \epsilon_{\mu\nu}\partial_\mu A_\nu \\
&= -\frac{i}{2\pi} \int d^2x \, \epsilon_{\mu\nu}(D_\mu \mathbf{n}^*) \cdot D_\nu \mathbf{n}
\end{aligned} \tag{16.184}$$

18. Prove that Eq. (16.109) solves Eq. (16.108).

Chapter 17
Phase Transitions and Critical Phenomena

17.1 Critical Exponents

Historically, there has been a fair amount of cross pollination between statistical systems and quantum field theory, to the benefit of both disciplines. In the past few decades, many of the successful ideas in quantum field theory actually originated in statistical systems, such as spontaneous symmetry breaking and lattice field theory.

There are several advantages that such statistical systems have over quantum field theory. First, many of them, in lower dimensions, are exactly solvable. Thus, they have served as a theoretical "laboratory" in which to test many of our ideas about much more complicated quantum field theoretical systems. Second, even simple statistical models exhibit nontrivial nonperturbative behavior. While a rigorous nonperturbative treatment of quantum field theory is notoriously difficult, even the simple classical Ising model shows a rich nonperturbative structure.

Thus, statistical systems have helped to enrich our understanding of quantum field theory. Even though they only have a finite number of degrees of freedom, they have served as a surprisingly faithful mirror to the qualitative features of our physical world.

We begin our discussion of statistical mechanics by making a few basic definitions. Whether discussing the properties of a solid, liquid, or gas, we will base our discussion on the classical Boltzmann partition function:

$$Z = \sum_n \exp\left(-\frac{E_n}{kT}\right) \tag{17.1}$$

where E_n represents the energy of the nth state, k represents the Boltzmann constant, and T represents the temperature.

(There is a close relationship between this partition function and the generating functional of quantum field theory:

$$Z = \int D\phi \, \exp i \int L(\phi) \, d^4x \qquad (17.2)$$

The difference, however, is that the field theory generating functional has an imaginary exponent in Minkowski space and is defined over an infinite number of degrees of freedom.)

In statistical mechanics, the fundamental quantity we wish to calculate is called the *free energy*, defined by:

$$F = -kT \, \log Z \qquad (17.3)$$

We say that a statistical model is *exactly solvable* if we can solve for an explicit expression for the free energy.

As in field theory, the statistical average of any observable X is given by:

$$\langle X \rangle = Z^{-1} \sum_n X_n \, \exp\left(-\frac{E_n}{kT}\right) \qquad (17.4)$$

There are only a few models that are exactly solvable (usually in two dimensions).[1,2] Some of them include the Ising model, the ferroelectric six-vertex model, the eight-vertex model, the three-spin model, and the hard hexagon model. There are also classes of solvable models, such as the RSOS (restricted solid-on-solid) models.

One of the earliest successes of these models was their ability to describe the properties of simple ferromagnets. For example, if we know that an atom has a magnetic moment μ, then the energy of the atom in an external magnetic field $\mathbf{H}$ is given by the dot product:

$$E = -\mu \cdot \mathbf{H} \qquad (17.5)$$

For quantum systems, we know that the magnetic moment is proportional to the spin σ_i. Therefore it is customary to add to the action the term:

$$H_0 + H \sum_i \sigma_i \qquad (17.6)$$

For systems with a magnetic field, for example, the *magnetization M* is defined to be the average of the magnetic moment per site:

$$M(H, T) = N^{-1}\langle \sigma_1 + \cdots + \sigma_N \rangle \qquad (17.7)$$

In the limit that $N \to \infty$, we can describe the magnetization as:

$$M(H, T) = -\frac{\partial}{\partial H} F(H, T) \qquad (17.8)$$

because taking the derivative with respect to H simply brings down σ_i into the sum.

The *susceptibility* is then defined as:

$$\chi(H, T) = \frac{\partial M(H, T)}{\partial H} \qquad (17.9)$$

which is related to the second derivative of the free energy.

Similarly, the *specific heat* can be defined as:

$$C = -T \frac{\partial^2 F}{\partial T^2} \qquad (17.10)$$

If there is a collection of spins σ_i arranged in some regular two-dimensional lattice, then we define the *correlation function* g_{ij} between the ith and jth spins as:

$$g_{ij} = \langle \sigma_i \sigma_j \rangle - \langle \sigma_i \rangle \langle \sigma_j \rangle \qquad (17.11)$$

In general, we find that the function g_{ij} will depend on the distance x separating the states, and at large distances, it will behave like some decreasing power of x multiplied by some exponential:

$$g_{ij} \sim x^{-\tau} e^{-x/\xi} \qquad (17.12)$$

where ξ is called the *correlation length*.

Near the critical temperature T_c, we find that these physical parameters, like the magnetization, either vanish or diverge. Intuitively, for example, we know that a magnetized substance begins to lose its magnetic properties as we increase the temperature and the spins become random. At the critical temperature, we find that the magnetic properties of the substance vanish. For example:

$$\chi \sim \begin{cases} |T - T_c|^{-\gamma}, & T > T_c \\ |T - T_c|^{-\gamma'}, & T < T_c \end{cases} \qquad (17.13)$$

where γ is the *critical exponent* that describes the susceptibility slightly above the critical temperature. (We will use primed symbols to represent the critical exponent just below the critical temperature.) This is shown graphically in Figure 17.1 for another quantity, the magnetization.

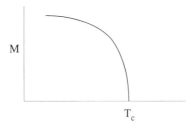

Figure 17.1. The critical exponent β governs the behavior of the magnetization M below the critical temperature.

The magnetization vanishes at T_c, and its behavior is determined by the critical exponent β:

$$M \sim (T_c - T)^\beta \qquad (17.14)$$

Then the critical exponent α characterizes the behavior of the specific heat near the critical temperature:

$$C_v \sim (T - T_c)^{-\alpha} \qquad (17.15)$$

We can describe the behavior of the correlation length near the critical temperature as:

$$\xi \sim (T - T_c)^{-\nu} \qquad (17.16)$$

In what are called second-order transitions, as the temperature approaches the critical temperature, the correlation length goes to infinity. (Because the system at criticality loses its dependence on a length scale for these transitions, the system becomes symmetric under conformal transformations. This means that we can use the constraints coming from conformal invariance to place stringent restrictions on the free energy at criticality. This will prove to be crucial in our discussion of scaling and the renormalization group.)

At the transition point, the system loses all dependence on any fundamental length scale, so the correlation function exhibits a power behavior:

$$g_{ij} \sim x^{-d+2-\eta} \qquad (17.17)$$

Also, the magnetization, at the critical point for weak magnetic fields, obeys the relation:

$$M \sim H^{1/\delta} \qquad (17.18)$$

We can summarize all this with the following simplified chart:

Magnetization	$M = -\partial F/\partial H$	$(T - T_c)^\beta$	
Magnetization	M at $T = T_c$	$M = H^{1/\delta}$	
Susceptibility	$\chi = \partial M/\partial H$	$(T - T_c)^{-\gamma}$	
Specific Heat	$C = -T(\partial^2 F/\partial T^2)$	$(T - T_c)^{-\alpha}$	(17.19)
Corr. Function	$g_{ij} \sim \langle \sigma_i \sigma_j \rangle$	$x^{-\tau} e^{-x/\xi}$	
Corr. Function	g_{ij} at $T = T_c$	$x^{-d+2-\eta}$	
Corr. Length	$-x/\log g(x)$	$(T - T_c)^{-\nu}$	

17.2 The Ising Model

One of the first, and simplest, statistical systems to be analyzed was the Ising model[3] in one dimension. Ising, who proposed and solved the model in 1925, showed that the system had interesting physical properties, with a critical point at $H = T = 0$.

We begin by placing a series of spins σ_i, which can take the values of ± 1, at regular intervals along a line. The energy of the system is given by:

$$E(\sigma) = -J \sum_{j=1}^{N} \sigma_j \sigma_{j+1} - H \sum_{j=1}^{N} \sigma_j \qquad (17.20)$$

where the jth spin only interacts with its nearest neighbors at the $j - 1$ and $j + 1$ sites, and where H is the external magnetic field.

Then the partition function can be written as:

$$Z_N = \sum_{\sigma} \exp \left(K \sum_{j=1}^{N} \sigma_j \sigma_{j+1} + h \sum_{j=1}^{N} \sigma_j \right) \qquad (17.21)$$

where we have rescaled the parameters via $K = J/kT$ and $h = H/kT$.

We will find it convenient to introduce a 2×2 matrix:

$$V(\sigma, \sigma') = \exp \left(K\sigma\sigma' + \frac{h}{2}(\sigma + \sigma') \right) \qquad (17.22)$$

This matrix **V**, which is called the *transfer matrix*, depends on whether the spins are $+1$ or -1; that is:

$$\mathbf{V} = \begin{pmatrix} V(+,+) & V(+,-) \\ V(-,+) & V(-,-) \end{pmatrix} = \begin{pmatrix} e^{K+h} & e^{-K} \\ e^{-K} & e^{K-h} \end{pmatrix} \tag{17.23}$$

Now comes the crucial transformation. We will rewrite the partition function as a sum over a series of matrices:

$$Z_N = \sum_\sigma V(\sigma_1, \sigma_2) V(\sigma_2, \sigma_3) \cdots V(\sigma_{N-1}, \sigma_N) V(\sigma_N, \sigma_1) \tag{17.24}$$

Therefore, the partition function can now be succinctly rewritten as:

$$Z_N = \text{Tr}\,\mathbf{V}^N \tag{17.25}$$

On one hand, we have done nothing. We have merely reshuffled the summation within Z_N by rewriting it as a sum over the 2×2 transfer matrix **V**. On the other hand, we have made an enormous conceptual difference, because we can now diagonalize the transfer matrix in terms of its eigenvalues; that is, there exists a matrix **P** that diagonalizes **V**:

$$\mathbf{V} = \mathbf{P} \begin{pmatrix} \lambda_1 & 0 \\ 0 & \lambda_2 \end{pmatrix} \mathbf{P}^{-1} \tag{17.26}$$

Substituting this into our original expression for the partition function, we now find:

$$Z_N = \text{Tr} \begin{pmatrix} \lambda_1 & 0 \\ 0 & \lambda_2 \end{pmatrix}^N = \lambda_1^N + \lambda_2^N \tag{17.27}$$

Let λ_1 be the larger of the two eigenvalues, which will then dominate the sum in the limit as $N \to \infty$. We then have:

$$\begin{aligned} F(H, T) &= -kT \lim_{N \to \infty} N^{-1} \log Z_N = -kT \log \lambda_1 \\ &= -kT \log \left(e^K \cosh h + \sqrt{e^{2K} \sinh^2 h + e^{-2K}} \right) \end{aligned} \tag{17.28}$$

In addition to having an exact expression for the free energy, we also have an exact expression for the magnetization:

$$M(H, T) = \frac{e^K \sinh h}{\sqrt{e^{2K} \sinh^2 h + e^{-2K}}} \tag{17.29}$$

This is an important result: We have obtained the complete solution for the free energy and magnetization in an exactly solvable statistical mechanical system.

Because we have an exact expression for the transfer matrix, we can now solve for the correlation length and show that it goes to infinity when $H = T = 0$. To do this, we need to calculate the averages $\langle \sigma_i \rangle$ and $\langle \sigma_i \sigma_j \rangle$. We begin by defining the matrix **S** in spin space as:

$$\mathbf{S} = \begin{pmatrix} 1 & 0 \\ 0 & -1 \end{pmatrix} \qquad (17.30)$$

which has elements:

$$S(\sigma, \sigma') = \sigma \delta(\sigma, \sigma') \qquad (17.31)$$

Therefore, the average can be written as:

$$\langle \sigma_1 \sigma_3 \rangle = Z_N^{-1} \sum_\sigma \sigma_1 V(\sigma_1, \sigma_2) V(\sigma_2, \sigma_3) \sigma_3 \cdots = Z_N^{-1} \operatorname{Tr} \mathbf{S} \mathbf{V}^2 \mathbf{S} \mathbf{V}^{N-2} \qquad (17.32)$$

So:

$$
\begin{aligned}
\langle \sigma_i \sigma_j \rangle &= Z_N^{-1} \operatorname{Tr} \mathbf{S} \mathbf{V}^{j-i} \mathbf{S} \mathbf{V}^{N+i-j} \\
\langle \sigma_i \rangle &= Z_N^{-1} \operatorname{Tr} \mathbf{S} \mathbf{V}^N
\end{aligned}
\qquad (17.33)
$$

Now let the matrix **P**, which diagonalizes the transfer matrix, be parametrized by an angle ϕ:

$$\mathbf{P} = \begin{pmatrix} \cos \phi & -\sin \phi \\ \sin \phi & \cos \phi \end{pmatrix} \qquad (17.34)$$

Then we have:

$$
\begin{aligned}
g_{ij} &= \langle \sigma_i \sigma_j \rangle - \langle \sigma_i \rangle \langle \sigma_j \rangle \\
&= \cos^2 \phi + \sin^2 2\phi \left(\lambda_2/\lambda_1 \right)^{j-i} - \cos 2\phi \\
&= \sin^2 2\phi \left(\lambda_2/\lambda_1 \right)^{j-i} \\
&\sim e^{-(j-i)/\xi}
\end{aligned}
\qquad (17.35)
$$

So, we have the desired result:

$$\xi = \left[\log(\lambda_1/\lambda_2) \right]^{-1} \qquad (17.36)$$

At $H = 0$, we have:

$$\lim_{T \to 0^+} \frac{\lambda_2}{\lambda_1} = 1 \qquad (17.37)$$

so ξ tends to ∞ as $H, T \to 0$. Thus, all reference to a length scale has disappeared, as expected, at criticality.

Since the model has been solved exactly, it is now an easy task to calculate the critical exponents for the theory:

$$\text{Ising model :} \quad \begin{cases} 2 - \alpha = \gamma = \nu \\ \beta = 0 \\ \delta = \infty \\ \eta = 1 \end{cases} \qquad (17.38)$$

Another lesson that we have learned from this simple example is that the model was solvable because the partition function could be written in terms of a single matrix, the transfer matrix, which obviously commuted with itself. For more complicated models in two dimensions, we will find more than one transfer matrix, and the essential reason why some of them are exactly solvable is that their transfer matrices commute with each other.

Now that we have some experience using the transfer matrix technique, let us tackle a nontrivial problem, the two-dimensional Ising model, which was first solved exactly by Onsager[4,5] in 1944 for the zero magnetic field case. Its partition function is given by:

$$Z_N = \sum_\sigma \exp \left(K \sum_{(i,j)} \sigma_i \sigma_j + L \sum_{(i,k)} \sigma_i \sigma_k \right) \qquad (17.39)$$

where the (i, j) sum is taken symbolically over the nearest-neighbor horizontal sites on the lattice and the (i, k) sum is taken over the vertical lattice sites.

Now rotate the lattice by 45 degrees so the lattice sites are arranged diagonally, as in Figure 17.2.

Let us perform the sum over these rotated lattice sites first in the horizontal direction over n sites, and then in the vertical direction over m sites. Let W and V represent the partial sums taken in the horizontal direction. W and V alternate as we descend down the lattice in a vertical direction. Then the partition function is the sum of $W V W V W \cdots$ taken in the vertical direction.

To sum the lattice sites horizontally, let $\phi = \{\sigma_1, \sigma_2, \ldots, \sigma_n\}$; that is, ϕ is the set of spins taken along a horizontal direction over n sites. Since each spin can take on two values, ϕ has 2^n possible values. Let $\phi' = \{\sigma_1', \sigma_2', \ldots, \sigma_n'\}$ be the set of horizontal sites just below ϕ.

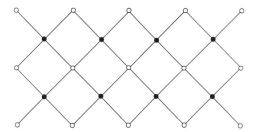

Figure 17.2. In the two-dimensional Ising model, we rotate the lattice by 45 degrees. By summing horizontally across the lattice, we obtain V and W. Then Z_N is the sum over $VWVWVWV\cdots$.

Then we can define W and V as follows:

$$V_{\phi,\phi'} = \exp\left(\sum_{j=1}^{n}(K\sigma_{j+1}\sigma'_j + L\sigma_j\sigma'_j)\right)$$

$$W_{\phi,\phi'} = \exp\left(\sum_{j=1}^{n}(K\sigma_j\sigma'_j + L\sigma_j\sigma'_{j+1})\right) \tag{17.40}$$

where W and V are now $2^n \times 2^n$ matrices. As before, we can perform the sum over the two transfer matrices by summing vertically over the lattice:

$$Z_N = \sum_{\phi_1}\sum_{\phi_2}\cdots\sum_{\phi_m} V_{\phi_1\phi_2} W_{\phi_2\phi_3}\cdots W_{\phi_m\phi_1} \tag{17.41}$$

Written in matrix form, this becomes:

$$Z_N = \mathrm{Tr}(\mathbf{VW})^{m/2} = \sum_{i=1}^{2^n} \Lambda_i^m \tag{17.42}$$

where Λ_i are the eigenvalues. In the thermodynamic limit, as we let the number of points $n, m \to \infty$, the partition function is once again dominated by the largest eigenvalue of the transfer matrix $\mathbf{VW}$:

$$\lim_{n,m\to\infty} Z \sim (\Lambda_{\max})^m \tag{17.43}$$

The one-dimensional and two-dimensional Ising models are therefore closely related to each other, and the calculation of the free energy (which we omit) reduces to calculating the largest eigenvalue of the transfer matrix.

We should also mention that there are a number of models that generalize the behavior of the Ising model and are exactly solvable. More important, there are a number of models that, although they may not be exactly solvable, exhibit critical behavior that can be described by the known conformal field theories. Let us list a few of these models and their properties.

17.2.1 XYZ Heisenberg Model

Closely related to the Ising model is the XYZ Heisenberg model. Here, we replace the spin σ_i with a Pauli matrix. The Hamiltonian is given by:

$$H = -\frac{1}{2} \sum_{\sigma} \left(J_x \sigma_j^x \sigma_{j+1}^x + J_y \sigma_j^y \sigma_{j+1}^y + J_z \sigma_j^z \sigma_{j+1}^z + \cdots \right) \tag{17.44}$$

where the sum is taken both horizontally and vertically over the entire lattice.

We have different models for different values of J_i:

If $J_x = J_y = J_z$, then this is the usual Heisenberg model.

If $J_x = J_y = 0$, then only J_z survives, and hence we obtain the usual Ising model.

If $J_z = 0$, then we have the XY model.

If $J_x = J_y$, then we have the Heisenberg–Ising model.

17.2.2 IRF and Vertex Models

A large number of exactly solvable models can be grouped into two categories, the IRF (interactions around a face) and the vertex models, which differ by the way in which we place spins on a regular lattice.

The IRF model includes the Ising model and many of the other exactly solvable models. If we place four spins a, b, c, and d (which can equal $+1$ or 0) around the four corners of a plaquette, the energy associated with the plaquette will be $\epsilon(a, b, c, d)$; so we define the Boltzmann weight of the plaquette as:

$$w(a, b, c, d) = \exp[-\epsilon(a, b, c, d)/kT] \tag{17.45}$$

For different choices of $\epsilon(a, b, c, d)$, we can represent a wide variety of models. For example, the Ising model can be represented as:

$$\begin{aligned}
\epsilon(a, b, c, d) &= -\frac{1}{2} J[(2a - 1)(2b - 1) + (2c - 1)(2d - 1)] \\
&\quad - \frac{1}{2} J'[(2c - 1)(2b - 1) + (2d - 1)(2a - 1)] \tag{17.46}
\end{aligned}$$

and the eight-vertex model can be written as:

$$\epsilon(a, b, c, d) = -J(2a - 1)(2c - 1) - J'(2b - 1)(2d - 1)$$
$$- J''(2a - 1)(2b - 1)(2c - 1)(2d - 1) \qquad (17.47)$$

for $a, b, c, d = 0, 1$. The partition function for the IRF model is given by:

$$Z = \sum_{\sigma_1} \cdots \sum_{\sigma_N} \prod_{i,j,k,l} w(\sigma_i, \sigma_j, \sigma_k, \sigma_l) \qquad (17.48)$$

The other large class of models is given by the vertex models. For example, in ice, we have the molecules of water held together by electric dipole moments. Let us place water molecules on a square two-dimensional lattice, such that the line segments forming the lattice correspond to the electric fields, represented by arrows.

These arrows have only two directions on any given line segment. If we impose the rule that there are always two arrows pointing out of and two arrows pointing into each vertex, then at any lattice site, there are six different possible orientations of the arrows. Each of these six different orientations will have an energy associated with it, called ϵ_i, for $i = 1, 2, \ldots, 6$. Thus, if ϕ represents the lattice sites along a horizontal line, then we have the six-vertex model:

$$Z = \sum_{\phi_1} \sum_{\phi_2} \cdots \sum_{\phi_M} V(\phi_1, \phi_2) V(\phi_2, \phi_3) \cdots V(\phi_M, \phi_1) = \text{Tr } V^M \qquad (17.49)$$

where:

$$V(\phi, \phi') = \sum \exp \left(-\frac{(m_1 \epsilon_1 + m_2 \epsilon_2 + \cdots + m_6 \epsilon_6)}{kT} \right) \qquad (17.50)$$

The partition function can be totally rewritten in terms of:

$$w(i, j | k, l) \equiv \exp \left[-\epsilon(i, j, k, l)/kT \right] \qquad (17.51)$$

Different values of $\epsilon(i, j, k, l)$ correspond to different models.

17.3 Yang–Baxter Relation

The reason for the exact solvability of these models is that the transfer matrices, which define the partition function and free energy, commute. When expressed mathematically, this relationship becomes the celebrated *Yang–Baxter relation*.[1,6]

In fact, mutually commuting transfer matrices, or equivalently the Yang–Baxter relation, are sufficient conditions for the solvability of any two-dimensional model. (One way to show this is that commuting transfer matrices give us an infinite set of conserved currents, which are sufficient to solve the system exactly. For more precise details and subtleties, the reader is referred to the literature.)

Let us study the Yang–Baxter relation in terms of the vertex models. Let $w(\mu_i, \alpha_i|\beta_i, \mu_{i+1})$ represent the contribution to the sum from the ith site. Each Greek index, in turn, can have values of ± 1. Let us perform the sum horizontally, as before:

$$V_{\alpha,\beta} = \sum_{\mu_1} \cdots \sum_{\mu_N} w(\mu_1, \alpha_1|\beta_1, \mu_2)w(\mu_2, \alpha_2|\beta_2, \mu_3) \cdots w(\mu_N, \alpha_N|\beta_N, \mu_1)$$

$$(17.52)$$

Let V' represent another transfer matrix (with a different Boltzmann weight w'). Let us define the quantity:

$$S(\mu, \nu|\mu', \nu'|\alpha, \beta) \equiv \sum_{\gamma} w(\mu, \alpha|\gamma, \mu')w'(\nu, \gamma|\beta, \nu') \qquad (17.53)$$

Then the matrix product $V V'$ can be represented as:

$$(VV')_{\alpha,\beta} = \sum_{\gamma} V_{\alpha\gamma} V'_{\gamma\beta} = \sum_{\mu_1,\dots,\mu_N} \sum_{\nu_1,\dots,\nu_N} \prod_{i=1}^{N} S(\mu_i, \nu_i|\mu_{i+1}, \nu_{i+1}|\alpha_i, \beta_i) \quad (17.54)$$

We can write this in matrix form by introducing the 4×4 matrix $\mathbf{S}(\alpha, \beta)$, which is a $(\mu, \nu) \times (\mu', \nu')$ matrix whose elements are given by $S(\mu, \nu|\mu', \nu'|\alpha, \beta)$. We can therefore write:

$$\begin{aligned}
(VV')_{\alpha,\beta} &= \operatorname{Tr} \mathbf{S}(\alpha_1, \beta_1)\mathbf{S}(\alpha_2, \beta_2) \cdots \mathbf{S}(\alpha_N, \beta_N) \\
(V'V)_{\alpha,\beta} &= \operatorname{Tr} \mathbf{S}'(\alpha_1, \beta_1)\mathbf{S}'(\alpha_2, \beta_2) \cdots \mathbf{S}'(\alpha_N, \beta_N)
\end{aligned} \qquad (17.55)$$

We now assume that V and V' commute, so that the two previous expressions are identical. This is obviously possible if there exists a 4×4 matrix $\mathbf{M}$ such that:

$$\mathbf{S}(\alpha, \beta) = \mathbf{M}\mathbf{S}'(\alpha, \beta)\mathbf{M}^{-1} \qquad (17.56)$$

Let the matrix $\mathbf{M}$ have elements given by $w''(\mu, \nu|\mu', \nu')$. Let us multiply the previous relation from the right by $\mathbf{M}$, so we have $\mathbf{SM} = \mathbf{MS}'$. Written out explicitly, this matrix equation is:

$$\sum_{\gamma,\mu'',\nu''} w(\mu, \alpha|\gamma, \mu'')w'(\nu, \gamma|\beta, \nu'')w''(\nu'', \mu''|\nu', \mu')$$

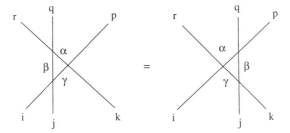

Figure 17.3. The Yang–Baxter relation, shown here pictorially, resembles the topological relations found in knot theory and braid theory, which demonstrates the close relationship between exactly solvable statistical systems and topology.

$$= \sum_{\gamma,\mu'',\nu''} w''(\nu, \mu|\nu'', \mu'')w'(\mu'', \alpha|\gamma, \mu')w(\nu'', \gamma|\beta, \nu') \quad (17.57)$$

If we redefine:

$$w(\mu, \alpha|\mu'', \gamma) = S_{\alpha\gamma}^{\mu\mu''}(u), \qquad w'(\nu, \gamma|\nu'', \beta) = S_{\gamma\beta}^{\nu\nu''}(u+v)$$

$$w''(\nu'', \mu''|\mu', \nu') = S_{\mu''\nu'}^{\nu''\mu'}(v) \qquad (17.58)$$

then we can write the Yang–Baxter relationship in the form:

$$\sum_{\alpha,\beta,\gamma} S_{j\beta}^{i\alpha}(u)S_{k\gamma}^{\alpha p}(u+v)S_{\gamma r}^{\beta q}(v) = \sum_{\alpha,\beta,\gamma} S_{k\gamma}^{j\beta}(v)S_{\gamma r}^{i\alpha}(u+v)S_{\beta q}^{\alpha p}(u) \qquad (17.59)$$

If we graphically represent this relationship, then we find the pattern expressed in Figure 17.3, which pictorially displays the Yang–Baxter relation.

In summary, the reason why many of these two-dimensional models are exactly solvable is because their transfer matrices commute, and the mathematical statement of this fact is the Yang–Baxter relation. The problem of finding exact solutions to these two-dimensional models then reduces to finding solutions to a much simpler problem, the Yang–Baxter equation. Fortunately, a variety of solutions to the Yang–Baxter relation exist. We notice that, as a function of the parameters u, v, the matrices appearing in the Yang–Baxter relation have a vague resemblance to the addition formulas for sines and cosines. By choosing an appropriate ansatz, we can, in fact, reduce the Yang–Baxter relations to the usual trigonometric addition formulas. (More precisely, it can be shown that a large number of solutions to the Yang–Baxter relation can be found using the addition formulas of what are called the "modular functions" ϑ, which are special functions found in solutions to certain periodic boundary value problems.)

Each solution, in turn, corresponds to an exactly solvable statistical mechanics model. Thus, we have now stumbled upon a powerful way in which to catalog the known exactly solvable two-dimensional models and also generate new ones. We will not present the explicit solutions to the Yang–Baxter equation for these models, since they are technically rather involved, so we refer the interested reader to the references.

The Yang–Baxter equation, in turn, is intimately related to several other branches of mathematics, such as *knot theory*, conformal field theory, and *quantum groups*. The topological structure of the Yang–Baxter relation resembles the manipulation of strands of string. Hence, the Yang–Baxter relation can be reduced to the *braid group* relations found in knot theory. Thus, the relationship between knot theory and the Yang–Baxter relation gives us hope that a more or less complete classification of solutions to the Yang–Baxter relation may eventually be found.

17.4   Mean-Field Approximation

Unfortunately, the simplifications that exist in one- and two-dimensional systems that allow us to find exact solutions do not generalize easily to three and four dimensions. The transfer matrix technique, the Yang–Baxter equation, and other techniques devised for one- and two-dimensional systems do not have soluble counterparts for higher dimensional systems. In fact, for years the two dimensional Ising model (with zero magnetic field) was the only exactly soluble two-dimensional system exhibiting a second-order phase transition.

We now must leave the realm of exact solutions and postulate various approximation schemes, with varying degrees of success. We will study approximation schemes that have been proposed over the years, the simplest and most widely used being Landau's *mean-field approximation*.[7]

The essence of the mean-field approximation is that we can substitute the actual field within a substance with an approximate, average field and ignore fluctuations. In practice, the mean-field approximation assumes that the magnetic field felt inside a substance equals the external magnetic field H plus an average field M, which we can calculate by minimizing the action. This assumption, of course, totally ignores the local fluctuations of the magnetic field throughout the substance, but it serves as a rough first approximation.

By assumption, the mean-field approximation assumes that the magnetic field is equal to the external magnetic field H, plus the average field M, plus small corrections:

$$H' = H + aM - bM^3 + \cdots \qquad (17.60)$$

where we ignore fluctuations. By assumption, the M^2 term is missing and b is small.

We also know that the average value of M obeys the Curie law, that is, M is proportional to the magnetic field H' and inversely proportional to the temperature:

$$M = \frac{cH'}{T} \tag{17.61}$$

From these two simple assumptions, we can derive a wide variety of nontrivial, first-order results. Let us now solve for M:

$$M = \frac{cH'}{T} = \frac{c(H + aM - bM^3)}{T} \tag{17.62}$$

Let us define T_c to be ac, and then we have:

$$M(1 - T_c/T + cbM^2/T) = cH/T \tag{17.63}$$

Now set $H = 0$ and determine the behavior of M below the critical temperature. The solution for M becomes:

$$M \sim \sqrt{\frac{T_c - T}{cb}} \tag{17.64}$$

From Eq. (17.13), this implies that $\beta = 1/2$. So the first critical exponent has been determined.

Now let us take the derivative of M with respect to H, and assume that M is small so we can drop higher powers. Then the susceptibility becomes:

$$\chi = \frac{\partial M}{\partial H} \sim \frac{1}{T - T_c} \Rightarrow \gamma = 1 \tag{17.65}$$

where we have used Eq. (17.14). So we have now derived the second critical exponent.

Now set $T = T_c$, so we are sitting at the critical temperature. In this limit, the magnetization M becomes very small. The dependence of M on H can again be calculated from Eq. (17.63), and is therefore given by:

$$M \sim H^{1/3} \tag{17.66}$$

From Eq. (17.18), this therefore gives us $\delta = 3$.

In summary, we have calculated three critical exponents with very little effort by making the key assumption in Eq. (17.60). The mean-field approximation gives us:

$$\text{Mean-field theory}: \quad \begin{cases} \beta = \frac{1}{2} \\ \delta = 3 \\ \gamma = \gamma' = 1 \end{cases} \quad (17.67)$$

Thus, the mean-field approximation gives us a wealth of information about a statistical system with very little physical input. Although the mean-field approximation does not give very reliable results for complicated systems, it does much better than one might first expect for relatively simple systems.

We would like to justify some of these assumptions within the framework of an action and a partition function. We will analyze the mean-field approximation within the context of one of the most widely studied statistical systems, *Ginzburg–Landau* theory.[8]

To define the Ginzburg–Landau model in a way that resembles field theory, it is convenient to introduce the variable $\sigma(x)$ to represent the value of the spin at lattice site x. From a field theoretic point of view, the theory then resembles the ϕ^4 theory, except it has a linear term proportional to the external magnetic field:

$$\frac{H}{T} = \int d^d x \left(r_0 \sigma^2(x) + \frac{u}{4} \sigma^4(x) + c \left[\nabla \sigma(x) \right]^2 - H \cdot \sigma(x) \right) \quad (17.68)$$

where, by the symbol $\int d^d x$, we mean taking the sum over all lattice sites in the limit of small lattice spacing, and $\nabla \sigma$ denotes taking differences along the lattice.

The mean-field approximation, in the context of this field theory, becomes the expansion of the action around a constant solution to the equations of motion, which gives us an average value of the field. In other words, the mean-field theory is based on the Born term of a perturbation theory. The mean-field approximation corresponds to tree diagrams, and the loop diagrams correspond to the fluctuations that we will ignore for the moment.

To first approximation, we find that the solution to the equations of motion is given by a constant:

$$\sigma(x) = \bar{\sigma} \quad (17.69)$$

But this also means, from Eq. (17.7), that the magnetization can be given in terms of the average spin:

$$M \sim \bar{\sigma} \quad (17.70)$$

The gradient term in the action disappears, and then the equations of motion from Eq. (17.68) become:

$$2\bar\sigma[r_0 + (u/2)\bar\sigma^2] - H = 0 \tag{17.71}$$

In studying the various solutions of this equation, Landau observed that the point $r_0 = 0$ marked a qualitative change in the nature of the system for small H, so that a phase transition was evident. So therefore we can make the statement that $r_0 = 0$ must be equivalent to $T = T_c$ for this phase transition:

$$r_0 = t(T - T_c) \tag{17.72}$$

for some constant t. [Eq. (17.72) means that there can be spontaneous symmetry breaking at the critical temperature, since the sign of the mass term changes.]

Inserting this back into the original equation, we find:

$$2M\left(t(T - T_c) + \frac{u}{2}M^2\right) = H \tag{17.73}$$

Now compare this with Eq. (17.60) postulated earlier. We find that there is an exact correspondence, and hence we can derive the critical exponents precisely in the same way as before.

A more detailed examination of the Ginzburg–Landau theory in the mean-field approximation yields the following critical exponents:

$$\text{Ginzburg–Landau model}: \begin{cases} \alpha = \alpha' = 2 - d/2 \\ \beta = \frac{1}{2} \\ \gamma = \gamma' = 1 \\ \delta = 3 \\ \eta = 0 \end{cases} \tag{17.74}$$

The approximation that we made, that the spin configuration that minimizes the action is the constant one, is called the *Gaussian approximation*, since all path integrals to lowest order become Gaussians. It is the particular form that the mean-field approximation takes for the Ginzburg–Landau model.

Historically, when experimental results were not very precise, the mean-field approximation was a valuable theoretical tool that gave good explanations of the experimental situation. However, as the experimental results became more precise over the decades, it became clear that the mean-field approximation gave only a rough fit to the data. Attempts to go beyond the mean-field approximation, however, were met with frustration. New theoretical ideas were necessary to push beyond the mean-field approximation. These new ideas came from scaling and the renormalization group.

17.5   Scaling and the Renormalization Group

Although the mean-field approximation gives us crude but reasonable fits to the data, it is difficult to go beyond the mean-field approximation and derive a perturbation series. Treating the mean-field approximation as the Born term in a perturbation series creates new problems near a second-order phase transition.

In general, at a second-order phase transition the correlation length ξ becomes infinite. At the transition, spins located in different parts of the system have a large effect on each other. This also means that specific features of the model wash out at the phase transition, giving us universality. Since ξ sets the basic scale of the system, at criticality the system usually loses all dependence on length; that is, it becomes scale or conformally invariant.

This means that the behavior of the magnetization, susceptibility, etc. near the transition can be determined by the behavior of ξ. But since $\xi \sim (T - T_c)^{-\nu}$, this means that all critical exponents can be written in terms of more fundamental critical exponents, like ν.

For perturbation theory, however, this causes problems. In d dimensions, the coupling constant g has dimensions. Therefore perturbation theory can be based on the dimensionless quantity:

$$g\,\xi^{4-d} \tag{17.75}$$

However, near a phase transition, we have $\xi \to \infty$, so this clearly diverges if $d < 4$. The coupling constant becomes infinitely strong and the perturbation theory makes no sense. For $d > 4$, the crucial features of the phase transition often disappear, and the approximation becomes useless. For many years, this prevented a perturbative generalization of the mean-field theory near the critical point.

However, it is possible to set up a new perturbation theory that is defined near the phase transition using the renormalization group. The new perturbation theory will be defined in

$$d = 4 - \epsilon \tag{17.76}$$

dimensions. (For example, for three-dimensional systems, $\epsilon = 1$.)

To understand how the ϵ expansion cures the usual problems of ordinary perturbation theory, let us first use a few scaling arguments to derive the relationship between critical exponents. The free energy F has dimension equal to zero, since it is not affected by a scale change. The free energy per unit volume therefore has dimension d. Therefore, by the scaling hypothesis and Eq. (17.16):

$$F \sim \xi^{-d} \sim (T - T_c)^{\nu d} \tag{17.77}$$

From the free energy, we can calculate other physical quantities and their exponents. If we calculate the specific heat C, we find:

$$C = -T \frac{\partial^2 F}{\partial T^2} \sim (T - T_c)^{\nu d - 2} \tag{17.78}$$

It therefore follows from Eq. (17.15) that:

$$\alpha = 2 - \nu d \tag{17.79}$$

From Eq. (17.17), the correlator of two spins has dimension $d - 2 + \eta$. The dimension of spin is therefore half that amount. From Eq. (17.7), we see that the magnetization has dimension $d_M = (d - 2 + \eta)/2$. Therefore we can read off its critical exponent:

$$\begin{aligned} M \quad &\sim \quad \xi^{-(d-2+\eta)/2} \\ &\sim \quad (T - T_c)^{\nu(d-2+\eta)/2} \end{aligned} \tag{17.80}$$

Therefore, from Eq. (17.13), we have:

$$\beta = \frac{1}{2}\nu(d - 2 + \eta) \tag{17.81}$$

We also know that the external field H, because $M = -\partial F/\partial H$, must have dimension equal to:

$$d_H = d - d_M = (d + 2 - \eta)/2 \tag{17.82}$$

Therefore, we also have:

$$M \sim \left(H^{1/d_H}\right)^{d_M} \tag{17.83}$$

in order to make the dimensions match. From Eq. (17.18), this gives us:

$$\delta = d_H/d_M = (d + 2 - \eta)/(d - 2 + \eta) \tag{17.84}$$

In summary, a few simple assumptions about the scaling behavior at criticality give us nontrivial relationships between the various critical exponents:

$$\text{Scaling :} \quad \begin{cases} \alpha = \alpha' = 2 - \nu d \\ \beta = \nu(d - 2 + \eta)/2 \\ \delta = (d + 2 - \eta)/(d - 2 + \eta) \\ \nu = \nu' = \gamma/(2 - \eta) \end{cases} \quad (17.85)$$

To go beyond these simple-minded arguments, we will now use the method of *block spins* in order to calculate, to lowest order in ϵ, the critical exponents.

The block spin method of Kadanoff[9,10] is based on some rather intuitive arguments. We know that, at criticality, the correlation length goes to infinity and many of the features of the model get washed out. At the phase transition, the partition function obeys a new symmetry. At first, this seem strange, since the lattice spacing between spins is equal to a, which is not scale invariant. However, at criticality, the system loses its dependence on a length scale and obeys highly nontrivial scaling properties. This allows us to write down the renormalization group equations for the system, using a prescription slightly different from that used in the previous chapters.

Let us begin with a partition function $Z(a)$ defined on a hypercubical lattice of length L with spacing a. At each lattice site, we have a spin operator σ_i, where $i = 1, \ldots, n$. Now let us decompose this lattice into larger blocks of length b, which is a multiple of a, so $b = sa$. We now perform the spin averaging within each larger block. This averaging within each larger block creates a new average spin σ_i' with a new Hamiltonian. This will create a new partition function $Z'(b)$ that is defined on a new lattice with lattice spacing b, such that the spin operator at the various blocks is defined to be the average spin σ_i'.

Now rescale the new partition function $Z'(b)$ by simply reducing the lattice spacing from b to a. In general, the partition function that we get $Z'(a)$ is not equal to the original $Z(a)$ with which we started. This is because the two operations we have performed are quite distinct. The first operation integrated out the spins within each block to define a new lattice of length b, while the second operation was a trivial rescaling from b back to a. However, at criticality, when we lose all reference to mass scales, these two operations should be roughly inverses of each other.

Let K_s represent an operator that performs the first operation of averaging within each block of size b. The operator K_s transforms the original Hamiltonian $H(\sigma)$ into a new Hamiltonian $H'(\sigma')$ by averaging over the s^d spins in each block (we set $k = 1$):

$$K_s \left[H(\sigma)/T \right] = H'(\sigma')/T \quad (17.86)$$

More precisely, the action of the operator K_s is given by:

$$e^{-H'(\sigma')/T} = \int e^{-H(\sigma)/T} \prod_{i,x} \delta\left(\sigma'_{i,x} - s^{-d} \sum_y^x \sigma_{i,y}\right) \prod_{j,y} d\sigma_{j,y} \qquad (17.87)$$

where $s^{-d} \sum_y \sigma_{i,y}$ represents the average spin over the s^d old blocks, replaced by the new block centered at x. This operation averages over all spins $\sigma_{j,y}$ within each block and replaces them with a function of spin $\sigma'_{i,x}$ defined at the center of each block.

We will find it more convenient to work in momentum space. The spin σ_{ik}, where k represents the momentum, can be represented as:

$$\sigma_{ik} = L^{-d/2} \sum_c \sigma_{i,c} e^{ik \cdot c} \qquad (17.88)$$

$$\sigma_{i,c} = L^{-d/2} \sum_{k < \Lambda} e^{-ik \cdot c} \sigma_{ik} \qquad (17.89)$$

where c is the site of the spin σ_{ic}. The momentum sum only extends as far as a sphere of radius Λ. (Beyond that momentum, we are probing a distance less than the lattice spacing, which is undefined.)

In momentum space, the K_s operation can be written as:

$$e^{-H'(\sigma')/T} = \int e^{-H(\sigma)/T} \prod_{i, \Lambda > k > \Lambda/s} d\sigma_{i,k} \qquad (17.90)$$

We have split the momentum sum over k into two parts. The sum over $\Lambda/s < k < \Lambda$ corresponds, in x space, to the sum over the spins with block size less than b, but larger than a. The sum over $k < \Lambda/s$, which corresponds to taking sums over blocks larger than b, is omitted in the block spin method.

Next, we rescale the size of the lattice from b back to a, which means we also must rescale the following:

$$\sigma(x) \rightarrow \sigma(x) s^{1-d/2}$$

$$x \rightarrow x/s$$

$$\int d^d x \rightarrow s^d \int d^d x \qquad (17.91)$$

where s is the scaling parameter.

The combination of the two operations, K_s and the rescaling shown, gives us the renormalization group operator R_s, which acts on the physical parameters of the original theory, called collectively μ, and creates a new set μ':

$$\mu' = R_s \mu \tag{17.92}$$

where:

$$R_s R_{s'} = R_{ss'} \tag{17.93}$$

Now we apply the method of block spins to the Ginzburg–Landau action:

$$
\begin{aligned}
H(r_0, u, c) &= H_0 + H_I \\
H_0 &= \frac{1}{2} \int d^d x \left[r_0 \sigma^2 + c(\nabla \sigma)^2 \right] \\
H_I &= \frac{1}{2} \int d^d x \frac{u}{4} \sigma^4
\end{aligned}
\tag{17.94}
$$

so that the set of parameters is $\mu = \{r_0, u, c\}$.

The calculation is conceptually simple, but the details are a bit involved, so we will break it up into four steps:

1. First, we perform the block spin integration, which converts H into H':

$$H(r_0, u, c) \rightarrow H'(r_0', u', c') \tag{17.95}$$

   where H' has the same form as H, except that it is defined with parameters r_0', u', c', and spins σ'.

2. Then we go to the critical point, where u is stationary, i.e. $u' = u$. The solution of this gives us the critical value u^*.

3. Near criticality, we solve for r_0^*. This gives us the behavior of r_0' near criticality:

$$(r_0 - r_0^*)' = s^{1/\nu}(r_0 - r_0^*) + \cdots \tag{17.96}$$

   where the ellipsis includes terms of higher order in ϵ. This gives us the value of ν.

4. Last, we will insert this value of ν into the scaling relations in Eq. (17.85), which gives us the remaining critical exponents.

17.5.1 Step One

To begin, we will perform the sum by splitting up σ_i into two pieces:

$$
\begin{aligned}
\sigma_i &= \sigma_i' + \tilde{\sigma}_i \\
\sigma_i' &\equiv L^{-d/2} \sum_{k < \Lambda/s} \sigma_{i,k} e^{ik \cdot x} \\
\tilde{\sigma}_i &\equiv L^{-d/2} \sum_{\Lambda/s < q < \Lambda} \sigma_{i,q} e^{iq \cdot x}
\end{aligned}
\tag{17.97}
$$

where we have split the sum over momentum space into two parts. The purpose of this split is that the summation over the blocks with less than size b is performed by summing over the spins within $\Lambda/s < q < \Lambda$. Thus, we are only interested in the summation over $\tilde{\sigma}_i$, while keeping σ_i' constant.

After performing the averaging over $\tilde{\sigma}_i$, we are left with a new Hamiltonian defined totally with the variable σ_i'. This new Hamiltonian $H'(\sigma_i')$ will have parameters r_0', u', and c' that we want to calculate.

We will only perform the calculation to lowest order, so we will power expand in H_I. After performing the block spin summation over $\tilde{\sigma}$, H changes into H', where:

$$
\begin{aligned}
H' &\equiv \frac{1}{2} \int d^d x \left(r_0'(\sigma')^2 + c'(\nabla \sigma')^2 + \frac{u'}{4} (\sigma')^4 \right) + \cdots \\
&= R_s H = H_0' + \langle H_I \rangle - \frac{1}{2} \langle (H_I - \langle H_I \rangle)^2 \rangle \Big|_{\sigma_k \to s^{1-\eta/2} \sigma_{s,k}} \\
&= H_0' + A + B
\end{aligned}
\tag{17.98}
$$

Our problem, therefore, is to average over $\tilde{\sigma}$, which leaves us with a modified Hamiltonian H' defined in terms of σ', which in turn allows us to compute r_0' and u'.

The key to the calculation is therefore to compute A, B:

$$
\begin{aligned}
A &= \langle H_I \rangle = \frac{u}{8} \int d^d x \, \langle \sigma^4 \rangle \\
B &= -\frac{u^2}{128} \int d^d x \, d^d y \\
&\quad \times \left\langle \left[\sigma^4(x) - \langle \sigma^4(x) \rangle \right] \left[\sigma^4(y) - \langle \sigma^4(y) \rangle \right] \right\rangle
\end{aligned}
\tag{17.99}
$$

A and B, after summing over $\tilde{\sigma}$, are functions of $(\sigma')^2$ and $(\sigma')^4$, which give corrections to r_0 and u.

To perform the averaging over the block, we will use the following relations:

$$\langle \tilde{\sigma}^m \rangle = 0; \text{ m odd} \qquad \langle \tilde{\sigma}_{i,p} \tilde{\sigma}_{j,q} \rangle = \delta_{ij} \delta_{p,-q} \langle \tilde{\sigma}_{i,p} \tilde{\sigma}_{i,-q} \rangle \qquad (17.100)$$

The summation is over a Gaussian, and is hence calculable. From the Hamiltonian in Eq. (17.94), we know that the two-point function is given by the usual propagator, which is $1/(r_0 + cq^2)$. Then the summation over the block spins from $\Lambda/s < q < \Lambda$ can be performed by taking the continuum limit:

$$
\begin{aligned}
\langle \tilde{\sigma}^2 \rangle &= \sum_i \langle \tilde{\sigma}_i^2 \rangle \\
&= L^{-d} \sum_{\Lambda/s < q < \Lambda} \langle \tilde{\sigma}_{i,q} \tilde{\sigma}_{i,-q} \rangle \\
&= n(2\pi)^{-d} \int d^d q \, \frac{1}{r_0 + cq^2} \\
&= n K_d \int_{\Lambda/s}^{\Lambda} dq \, \frac{q^{d-1}}{r_0 + cq^2} \\
&= n_c(1 - s^{-2+\epsilon}) - n K_4 c^{-2} r_0 \log s + O(\epsilon^2) \qquad (17.101)
\end{aligned}
$$

where $d^d q = q^{d-1} dq \, K_d (2\pi)^d$ and:

$$
\begin{aligned}
n_c &= (n/c) K_d \Lambda^{d-2}/(d-2) \\
K_d &= 2^{-d+1} \pi^{-d/2}/\Gamma(d/2) \qquad (17.102)
\end{aligned}
$$

where K_d is the surface area of a unit d dimensional sphere, divided by $(2\pi)^d$, and $d = 4 - \epsilon$.

The point of listing these identities is to find an expression for $\langle \sigma^4 \rangle$, which can now be written as:

$$
\begin{aligned}
\langle \sigma^4 \rangle &= \langle [\sigma' + 2\sigma' \cdot \tilde{\sigma} + \tilde{\sigma}^2]^2 \rangle \\
&= (\sigma')^4 + 2(\sigma')^2 \langle \tilde{\sigma}^2 \rangle + 4 \langle (\tilde{\sigma} \cdot \sigma')^2 \rangle + \langle \tilde{\sigma}^4 \rangle \qquad (17.103)
\end{aligned}
$$

Only the second and third term give a contribution to $(\sigma')^2$.

The third term can be written as:

$$
\begin{aligned}
\langle (\tilde{\sigma} \cdot \sigma')^2 \rangle &= \sum_{i,j} \sigma_i' \sigma_j' \langle \tilde{\sigma}_i \tilde{\sigma}_j \rangle \\
&= \sigma'^2 \langle \tilde{\sigma}^2 \rangle /n \\
&= \sigma'^2 \left[(n_c/n)(1 - s^{2-d}) - K_4 c^{-2} r_0 \log s \right] \qquad (17.104)
\end{aligned}
$$

Now insert Eqs. (17.101) and (17.104) into (17.103) and collect all terms containing $(\sigma')^2$. Inserting these averages back into the expression for A in Eqs. (17.99) and (17.98), we find:

$$
r_0' = s^2 \left[r_0 + (u/c)(n/2 + 1)(\Lambda^2/2)K_d(1 - s^{-2}) + uC\epsilon \right.
$$

$$
\left. - r_0(u/c^2)(n/2 + 1)K_4 \log s + u^2 d \right] + O(\epsilon^3) \tag{17.105}
$$

(where the C and D terms are not important to the final result).

Next, we wish to calculate the term B in Eq. (17.99). This is also straightforward. We define:

$$
\delta_{ij} \tilde{G}(x - y) = \langle \tilde{\sigma}_i(x)\tilde{\sigma}_j(y) \rangle \tag{17.106}
$$

We therefore have:

$$
\tilde{G}(r) = L^{-d} \sum_{\Lambda/s < q < \Lambda} e^{iq \cdot r} q^{-2} c^{-1}
$$

$$
= (2\pi)^{-2} r^{-2} c^{-1} [J_0(\Lambda r/s) - J_0(\Lambda r)] \tag{17.107}
$$

where J_0 is a Bessel function, and:

$$
\int d^4 r \, \tilde{G}^2(r) = \frac{K_4}{c^2} \int_{\Lambda/s}^{\Lambda} q^3 \, dq \, (cq^2)^{-2}
$$

$$
= \frac{K_4 \log s}{c^2} \tag{17.108}
$$

If we expand the terms in B in Eq. (17.97), we find a large number of extraneous terms. After summing over $\tilde{\sigma}$, the only terms that survive are of the form $(\sigma')^4 \tilde{G}^2$ and $(\sigma')^2 \tilde{G}^2$. All summations over the $\tilde{\sigma}$ can be therefore performed over the block spin. When the summation is performed, we find that we have a new Hamiltonian H', which has the same form as the original Ginzburg–Landau Hamiltonian, but is now a function of the spin operator σ_i', with coefficients r_0', u', c'.

Inserting these summations back into Eq. (17.99), we find that the $(\sigma')^4 \tilde{G}^2$ contribution to B gives a correction to u:

$$
u' = s^\epsilon \left[u - (u^2/2c^2)(n + 8)K_4 \log s \right] \tag{17.109}
$$

Now that we have explicit results for r_0' and u', this completes the first step of our calculation.

17.5.2 Step Two

To complete the second step, we go to criticality, where we have $u = u' = u^*$, since u is invariant under the scale transformation. Imposing this restriction, we can now solve for u^* from the previous equation:

$$u^*/c^2 = \frac{2\epsilon}{(n+8)K_4} \tag{17.110}$$

We can also solve for the value of r_0^* at criticality by inserting Eq. (17.110) into Eq. (17.105) and ignoring higher-order terms:

$$
\begin{aligned}
r_0^* &= -(u^*/c)(n/2+1)(\Lambda^2/2)K_4 \\
      &= -\frac{n+2}{n+8}\epsilon(\Lambda^2 c/2)
\end{aligned}
\tag{17.111}
$$

where we have inserted the value of u^*.

17.5.3 Step Three

The third step consists of calculating the value of r_0' as a function of s. Inserting the value of u^* and r_0^* into r_0', we have:

$$(r_0 - r_0^*)' = s^{y_1}(r_0 - r_0^*) + \cdots \tag{17.112}$$

where:

$$y_1 = 2 - \frac{n+2}{n+8}\epsilon = \frac{1}{\nu} \tag{17.113}$$

This fixes the value of ν to be:

$$\nu = \frac{1}{2} + \frac{(n+2)}{4(n+8)}\epsilon \tag{17.114}$$

17.5.4 Step Four

Finally, we now insert this value of ν into the scaling relations in Eq. (17.85) to obtain the rest of the critical exponents. To this order, we have $\eta = 0$; therefore:

$$\text{Block spins :} \quad \begin{cases} \alpha = (4 - n)\epsilon/2(n + 8) \\[4pt] \beta = \frac{1}{2} - 3\epsilon/2(n + 8) \\[4pt] \gamma = 1 + (n + 2)\epsilon/2(n + 8) \\[4pt] \delta = 3 + \epsilon \\[4pt] \eta = 0 \\[4pt] \nu = \frac{1}{2} + (n + 2)\epsilon/4(n + 8) \end{cases} \tag{17.115}$$

17.6 ϵ Expansion

We have seen the importance of the emergence of a new symmetry, scale invariance, at criticality because the system loses all reference to a length scale at the phase transition. This also means, however, that we can use an alternative method of deriving these identities, equivalent to the method of block spins, which is the familiar Callan–Symanzik equations. The usual Callan–Symanzik relations allow us to calculate the critical exponents to arbitrary order in ϵ by calculating loop diagrams.[11]

In familiar field theory language, we start with the action:

$$\mathcal{L} = -\frac{Z_3}{2}\left(\sum_{i=1}^{d}(\partial_i\phi^a)^2 + m^2(\phi^a)^2\right)$$
$$-g\frac{Z_1}{4!}\left[(\phi^a)^2\right]^2 - \frac{Z_3}{2}\delta m^2(\phi^a)^2 \tag{17.116}$$

where we sum over $a = 1, \ldots, N$ and where Z_1 and Z_3 are the usual renormalization constants that correspond to the four-point and two-point functions. Notice that we are taking a Euclidean metric, not a Minkowski metric.

Then we can immediately write down the Callan–Symanzik equations for the s-point function:

$$\left(m\frac{\partial}{\partial m} + \beta(g)\frac{\partial}{\partial g} - \frac{s}{2}\gamma_3\right)\Gamma^{(s)} = \Delta\Gamma^{(s)} \tag{17.117}$$

where we will omit the right-hand side in the asymptotic limit that we are analyzing.

In this asymptotic limit, we can solve the Callan–Symanzik equations in the usual way, and we get:

$$\Gamma^{(s)}(p_i; m/\lambda, g) = \left[\exp\left(-\frac{1}{2} s \int_g^{g(\lambda)} dg' \frac{\gamma_3(g')}{\beta(g')} \right) \right]$$
$$\times \Gamma_{as}^{(s)}(p_i; m, g(\lambda)) \tag{17.118}$$

where $g(\lambda)$ is defined by:

$$\int_g^{g(\lambda)} \frac{dg'}{\beta(g')} = \log \lambda \tag{17.119}$$

However, as we mentioned earlier, this solution is only formal, since the perturbation theory in g near the critical point does not make any sense, since the dimensionless quantity in which we expand is $g\xi^{d-4}$, which blows up for $d < 4$ near criticality. This is the reason why, before renormalization group arguments were developed, the mean field approximation could not be generalized properly near a phase transition.

Thus, we want a revised set of equations defined as a perturbation series in a new, dimensionless quantity called u:

$$u \equiv gm^{d-4} \tag{17.120}$$

In terms of the new dimensionless variable u, the Callan–Symanzik equations are almost identical, except that the independent variable is now given by u. We therefore have:

$$\beta(u) = m \frac{\partial u}{\partial m}\bigg|_{g_0}$$
$$\gamma_3(u) = m \frac{\partial \log Z_3(u)}{\partial m}\bigg|_{g_0} \tag{17.121}$$

and the bare coupling constant g_0 is related to the renormalized one by:

$$g_0 = g \frac{Z_1(u)}{Z_3^2(u)} \tag{17.122}$$

In terms of Z_i, we can write:

$$\beta(u) = -\epsilon \left(\frac{d}{du} \log \frac{u Z_1(u)}{Z_3^2(u)} \right)^{-1}$$
$$\gamma_3(u) = \beta(u) \frac{d \log Z_3(u)}{du} \tag{17.123}$$

To make contact with the usual physical variables, we define the standard renormalized two-point function as the mass squared, and four-point function as the coupling constant at the point $p^2 = 0$:

$$\Gamma^{(2)}(p, -p; m, u)\big|_{p^2=0} = m^2$$

$$\frac{\partial}{\partial p^2}\Gamma^{(2)}(p, -p; m, u)\big|_{p^2=0} = 1$$

$$\Gamma^{(4)}(0, 0, 0; m, u) = g \qquad (17.124)$$

where all vertices are defined as a function of u.

In terms of this new variable, we can extract the scaling property from the solution to the Callan–Symanzik equations. We are interested in rescaling the momenta $p_i \rightarrow \lambda p_i$ in the asymptotic limit $\lambda \rightarrow \infty$. From simple dimensional arguments, we know that naive scaling gives:

$$\Gamma^{(s)}(\lambda p_i; m, u) = \lambda^{d-s(d-2)/2}\Gamma^{(s)}(p_i; m/\lambda, u) \qquad (17.125)$$

So far, everything resembles the ordinary field theoretic discussion. To begin the calculation, let us assume that we are near a fixed point, such that:

$$\beta(u^*) = 0 \qquad (17.126)$$

Then the γ_3 term in Eqs. (17.117) and (17.118) contributes to the asymptotic limit near u^*, such that:

$$\Gamma^{(s)}(\lambda p_i; m, u^*) \sim \lambda^{[d-s(d-2)/2-s\gamma_3(u^*)/2]} \qquad (17.127)$$

However, we know from the definition of the critical exponents that the two-point function scales as:

$$\Gamma^{(s)}(\lambda p_i) \sim \lambda^{d-sd_\phi} \qquad (17.128)$$

where $2d_\phi$ is the anomalous dimension of ϕ. From the chart Eq. (17.19), we can compare this with the two-point correlation function's asymptotic behavior at criticality. Equating Eqs. (17.127) and (17.128) and setting $s = 2$, we have $2d_\phi = d - 2 + \eta$, and thus:

$$\eta = \gamma_3(u^*) \qquad (17.129)$$

Our strategy to calculate the critical exponents is now as follows.

1. First, we calculate the two- and four-point Feynman graphs necessary to evaluate Z_1 and Z_3.

2. We insert these values of Z_1 and Z_3 into Eq. (17.123) and calculate $\beta(u)$ and $\gamma_3(u)$.

3. We calculate the critical point u^* by setting $\beta(u^*) = 0$.

4. Finally, we insert u^* into γ_3, and use the relation $\gamma_3(u^*) = \eta$ to calculate the critical exponent. By similar arguments, we can also calculate γ. With η and γ, we can calculate all the critical exponents via the scaling relations in Eq. (17.85).

The Feynman rules corresponding to our action are easy to calculate. The propagator, for example, is just the usual $1/(q^2 + 1)$. A direct calculation of the renormalization constants for the two- and four-point functions yields:

$$
\begin{aligned}
Z_1 &= 1 + Su\frac{n+8}{6}a \\
&\quad + (Su)^2 \left(\frac{n^2 + 26n + 108}{36}a^2 - \frac{(5n+22)c}{9}\right) + O(u^3) \\
Z_3 &= 1 + \frac{n+2}{18}(Su)^2 b \\
&\quad + \frac{(n+8)(n+2)}{54}(Su)^3(ab - d/2) + O(u^4)
\end{aligned}
\tag{17.130}
$$

where $S = 2\pi^{d/2}/\Gamma(d/2)(2\pi)^d$ and where values of the loop parameters a, b, c, d are given by Feynman's rules:

$$
\begin{aligned}
a &= \frac{1}{S(2\pi)^d} \int \frac{d^d q}{(q^2 + 1)^2} \\
b &= \frac{1}{S^2(2\pi)^{2d}} \frac{d}{dp^2}\bigg|_{p^2=0} \\
&\quad \times \int \frac{d^d q_1\, d^d q_2}{(q_1^2 + 1)(q_2^2 + 1)\left[(p + q_1 + q_2)^2 + 1\right]} \\
c &= \frac{1}{(2\pi)^{2d}S^2} \int \frac{d^d q_1\, d^d q_2}{(q_1^2 + 1)^2(q_2^2 + 1)\left[(q_1 + q_2)^2 + 1\right]} \\
d &= \frac{1}{S^3(2\pi)^{3d}} \frac{d}{dp^2}\bigg|_{p^2=0} \\
&\quad \times \int \frac{d^d q_2}{(p + q_2)^2 + 1} \left(\int \frac{d^d q_1}{(q_1^2 + 1)\left[(q_1 + q_2)^2 + 1\right]}\right)^2
\end{aligned}
\tag{17.131}
$$

These integrals can all be performed using dimensional regularization. We find:

$$a = \epsilon^{-1}(1 - \epsilon/2) + O(\epsilon)$$

$$b = -\frac{1 - \frac{1}{4}\epsilon}{8\epsilon} - I/8 + O(\epsilon)$$

$$c = \frac{1}{2\epsilon^2}\left[1 - \epsilon/2 + O(\epsilon^2)\right]$$

$$d = -\frac{1}{6\epsilon^2}\left[1 - \frac{1}{4}\epsilon + O(\epsilon^2)\right] - I/(4\epsilon) \qquad (17.132)$$

where:

$$I = \int_0^1 dx \left(\frac{1}{1 - x(1 - x)} + \frac{\log x(1 - x)}{[1 - x(1 - x)]^2}\right) \qquad (17.133)$$

We can now calculate β and γ_3 from the definitions of the renormalization-group variables in Eq. (17.123):

$$\beta(u) = -u\left(\epsilon - \frac{n + 8}{6}(1 - \epsilon/2)(Su)\right.$$

$$+ \left.\frac{3n + 14}{12}(Su)^2\right) + O(u^4)$$

$$\gamma_3(u) = -\epsilon\frac{n + 2}{9}(Su)^2\left(b + (Su)\frac{n + 8}{6}(2ab - 3d/2)\right) + O(u^4)$$

$$(17.134)$$

Solving $\beta(u^*) = 0$ at criticality, we find:

$$u^* = \frac{6}{S(n + 8)}\epsilon\left[1 + \epsilon\left(\frac{1}{2} + \frac{3(3n + 14)}{(n + 8)^2}\right)\right] + O(\epsilon^3) \qquad (17.135)$$

If we now plug the value of $u^*(\epsilon)$ into $\gamma_3(u)$ in Eq. (17.123), we then have a power expansion of $\gamma_3 = \eta$ in terms of ϵ, as desired.

In addition to η, we must also evaluate one more critical exponent. In order to determine all the critical exponents, we would also like to solve for the exponent γ, which determines the critical behavior of the magnetic susceptibility. The susceptibility is defined as the second derivative of the free energy, and hence we want to calculate the anomalous dimension of the composite operator $\phi^2(x)$. Although only Z_1 and Z_3 are sufficient to renormalize the original action in the usual way, we will find it convenient to introduce another renormalization constant Z_4, which is the renormalization constant that appears in the vertex of

$\phi^2(x)$ coupled to two other fields:

$$\langle\phi^2(x)\phi(y)\phi(0)\rangle = Z_4(u)\langle\phi^2(x)\phi(y)\phi(0)\rangle_{\text{bare}} \qquad (17.136)$$

We will therefore find it convenient to introduce the vertex function $\Gamma^{(1,s)}$:

$$\Gamma^{(1,s)}(q, p_1, \cdots p_s; m, u) = \int e^{i(q\cdot x + p_1\cdot x_1 + \cdots + p_s\cdot x_s)}dx\, dx_1 \cdots dx_s$$

$$\times \langle\phi^2(x)\phi(x_1)\cdots\phi(x_s)\rangle\Big|_{1\text{PI}} \qquad (17.137)$$

As before, we can show that this new vertex function also satisfies a Callan–Symanzik relation:

$$\left(m\frac{\partial}{\partial m} + \beta(u)\frac{\partial}{\partial u} - (s/2 - 1)\gamma_3(u) - \gamma_4(u)\right)\Gamma^{(1,s)} = \Delta\Gamma^{(1,s)} \qquad (17.138)$$

where:

$$\gamma_4 = m\frac{\partial}{\partial m}\log Z_4\Big|_{g_0} = \beta(u)\frac{d\log Z_4}{du} \qquad (17.139)$$

We can now extract the asymptotic behavior of this vertex function as we rescale the moments $p_i \to p_i\lambda$:

$$\Gamma^{(1,s)}(\lambda q; \lambda p_i) \sim \lambda^{-(s-2)(d-2)/2 - (s/2-1)\gamma_3(u^*) - \gamma_4(u^*)} \qquad (17.140)$$

However, from general asymptotic arguments, we also know that the asymptotic behavior of the vertex function is governed by the anomalous dimensions d_ϕ and d_{ϕ^2}:

$$\Gamma^{(1,s)} \sim \lambda^{-sd_\phi + d_{\phi^2}} \qquad (17.141)$$

So we obtain from Eqs. (17.140) and (17.141), setting $s = 2$:

$$d_{\phi^2} = 2d_\phi - \gamma_4(u^*) \qquad (17.142)$$

From this and the definition of γ in Eq. (17.19), it can be shown that[11]:

$$\gamma = \frac{2 - \eta}{d - d_{\phi^2}} \qquad (17.143)$$

This is the desired relationship between the critical exponent γ and the anomalous dimension of ϕ^2.

Now we repeat the same steps as before. Using Feynman graphs, we can calculate Z_4 in terms of a, b, c, d. We find:

$$Z_4^{-1} = 1 - \frac{n+2}{6}a(Su) + \frac{n+2}{6}(c - a^2)(Su)^2 + O(u^3) \tag{17.144}$$

From this, we can derive an expression for γ_4:

$$\gamma_4(u) = \frac{n+2}{6}\epsilon(Su)\left[-a + (Su)(2c - a^2)\right] + O(u^3) \tag{17.145}$$

To sum up, our strategy, as we mentioned earlier, has been to calculate Z_1 and Z_3, which gives us $\beta(u)$. By solving for $\beta(u^*) = 0$, we can calculate $u^*(\epsilon)$ at the critical point. We insert $u^*(\epsilon)$ into the expression for $\gamma_{3,4}$, that gives us a power expansion for η and γ. Then, by the scaling relations, we can determine all the critical exponents.

This expansion can be carried out to arbitrary accuracy in ϵ. We list some of the critical exponents which have been calculated out to fourth and fifth order[11]:

$$\alpha = -\frac{(n-4)}{2(n+8)}\epsilon - \frac{(n+2)^2}{4(n+8)^3}(n+28)\epsilon^2 - \frac{(n+2)}{8(n+8)^5}\Big[n^4 + 50n^3$$

$$+ 920n^2 + 3472n + 4800 - 192(5n+22)(n+8)T\Big]\epsilon^3 + O(\epsilon^4)$$

$$\beta = \frac{1}{2} - \frac{3}{2(n+8)}\epsilon + \frac{(n+2)(2n+1)}{2(n+8)^3}\epsilon^2 + \frac{(n+2)}{8(n+8)^5}\Big[3n^3 + 128n^2$$

$$+ 488n + 848 - 48(5n+22)(n+8)T\Big]\epsilon^3 + O(\epsilon^4)$$

$$\gamma = 1 + \frac{(n+2)}{2(n+8)}\epsilon + \frac{(n+2)}{4(n+8)^3}(n^2 + 22n + 52)\epsilon^2 + \frac{(n+2)}{8(n+8)^5}\Big[n^4$$

$$6 + 44n^3 + 664n^2 + 2496n + 3104 - 96(5n+22)(n+8)T\Big]\epsilon^3 + O(\epsilon^4)$$

$$\delta = 3 + \epsilon + \frac{1}{2(n+8)^2}(n^2 + 14n + 60)\epsilon^2 + \frac{1}{4(n+8)^4}\Big(n^4 + 30n^3$$

$$+ 276n^2 + 1376n + 3168\Big)\epsilon^3 + \frac{1}{16(n+8)^6}\Big[2n^6 + 96n^5 + 1778n^4$$

$$+ 12760n^3 + 50280n^2 + 147136n + 263040$$

$$+ 768(n+2)(n+8)(5n+22)T\Big]\epsilon^4 + O(\epsilon^5)$$

$$\eta = \frac{(n+2)}{2(n+8)^2}\epsilon^2 + \frac{(n+2)}{8(n+8)^4}(-n^2 + 56n + 272)\epsilon^3$$

$$+ \frac{(n+2)}{32(n+8)^6}\Big[-5n^4 - 230n^3 + 1124n^2 + 17920n + 46144$$

$$768(5n + 22)(n + 8)T\Big]\epsilon^4 + O(\epsilon^5)$$

$$
\begin{aligned}
\nu \;=\; & \frac{1}{2} + \frac{(n+2)}{4(n+8)}\epsilon + \frac{(n+2)}{8(n+8)^3}(n^2 + 23n + 60)\epsilon^2 \\
& + \frac{(n+2)}{32(n+8)^5}\Big[2n^4 + 89n^3 + 1412n^2 + 5904n + 8640 \\
& - 192(5n + 22)(n + 8)T\Big]\epsilon^3 + O(\epsilon^4)
\end{aligned}
$$

$$(17.146)$$

where $T = 0.60103$.

To check the reliability of the methodology, we compare the perturbation calculation at $\epsilon = 1$ with the high-temperature series calculations for the three-dimensional Ising model, using the ϵ expansion and the Landau theory:

Exponent	ϵ Expansion	Ising	Landau
ν	0.626	0.642 ± 0.003	0.5
η	0.037	0.055 ± 0.010	0
γ	1.244	1.250 ± 0.003	1.0
α	0.077	0.125 ± 0.015	0
β	0.340	0.312 ± 0.003	0.5
δ	4.46	5.150 ± 0.02	3

$$(17.147)$$

The ϵ expansion is taken to second order, except for η, which is taken to third order. The agreement is surprisingly good even for $\epsilon = 1$.

(We caution, however, that the ϵ expansion is not convergent but only asymptotic. The convergence properties of the ϵ expansion are not fully understood.)

In summary, in this chapter we have seen how phase transitions can be categorized according to their critical exponents. In two dimensions, a large class of exactly solvable statistical models exist. The reason why they are solvable is because of commuting transfer matrices, or the Yang–Baxter relation. Unfortunately, many of the techniques used to solve these two-dimensional models do not carry over to four dimensions.

The mean-field approximation has been one of the main ways in which to extract qualitative features of more complicated statistical systems. However, trying to treat the mean-field approximation as a Born term to a power expansion fails because, at criticality, the coupling constant $g\xi^{4-d}$ is large. Fortunately, the renormalization group method allows one to expand in $u = gm^{d-4}$. We can then extract meaningful relations by solving for $\beta(u^*) = 0$ near criticality and inserting these relations back into the scaling relations. The results, even for the three-dimensional case ($\epsilon = 1$), are surprisingly good.

17.7 Exercises

1. Show that $\langle \tilde{\sigma}^4 \rangle = (n^2 + 2n)(n_c/n)^2(1 - s^{2-d})^2$ for the Ginzburg–Landau model. (Hint: use Wick's theorem.)

2. Consider the one-dimensional partition function:

$$Z = \prod_m \int_{-\infty}^{\infty} ds_m \exp\left(-\frac{1}{2}bs_m^2 - us_m^4\right) \exp\left(K \sum_n \sum_i s_n s_{n+i}\right) \quad (17.148)$$

where the spins s_n are arranged discretely along a line and they can assume any real value. Take the limit $u \to \infty$ and $b \to -\infty$ with $b = -4u$. Show that this model becomes the familiar Ising model in this limit [if one puts in a factor $(u/\pi)^{1/2} \exp -u$ per spin]. (Hint: show that we recover a Dirac delta function condition on the spin s_n.)

3. In the limit $u \to 0$, this becomes the Gaussian model. Why is it exactly solvable? Rewrite the Hamiltonian totally in terms of the Fourier transform $\sigma_q = \sum_n \exp(-i\mathbf{q} \cdot \mathbf{n})s_n$. Show that the term appearing in the partition function: $\beta H = K \sum_n \sum_i s_n s_{n+i} - \frac{1}{2}b \sum_n s_n^2$ can be rewritten as:

$$
\begin{aligned}
\beta H &= -\frac{1}{2}\int_q \left(K \sum_i |\exp(iq_i) - 1|^2 + (b - 2dK)\right)\sigma_q \sigma_{-q}\frac{d^d q}{(2\pi)^d} \\
&\sim -\frac{1}{2}\int_q (q^2 + r)\sigma_q \sigma_{-q}\frac{d^d q}{(2\pi)^d} \quad\quad\quad (17.149)
\end{aligned}
$$

where K is rescaled to one, and $r = (b - 2dK)/K$ and $|\mathbf{q}| < 1|$.

4. For the Gaussian model, the two-point function is $\Gamma_q = 1/(q^2 + r)$. In x space, we have: $\Gamma(x) = \int_q e^{i\mathbf{q}\cdot\mathbf{x}}\Gamma_q d^d q/(2\pi)^d$. Show that this gives $\Gamma(x) \sim \exp(-\sqrt{r}|x|)$ and hence $\xi \sim 1/\sqrt{r}$. Given the form of r with $K \sim 1/kT$, show that $\nu = 1/2$.

5. Consider the Ising model in d dimensions using mean-field theory. The partition function is given by summing over nearest neighbors i, j:

$$Z = \sum_{\text{spins}} \exp\left(\beta \sum_{ij} s_i s_j\right) \qquad (17.150)$$

The magnetization is given by $M = \langle s_i \rangle$. In the mean-field approximation, assume that all neighboring spins are replaced by their average value M. Then the sum over nearest neighbors picks up the factor $2d\beta M$. Show that the Boltzmann probability for the ith spin to have the value s_i is given by:

$$P(s_i) = \frac{\exp(2d\beta M s_i)}{2\cosh(2d\beta M)} \qquad (17.151)$$

(Hint: perform the sum over $2d$ neighboring sites, and treat the denominator as a normalization factor.)

6. For this Ising system, assume that the average of s_i is also M. Show that this gives us the self-consistency equation for the mean-field approximation:

$$M = \langle s_i \rangle = \tanh(2d\beta M) \qquad (17.152)$$

For small β, the unique solution is $M = 0$. For larger β, at a certain point there are nonzero solutions for M. Show that this phase transition takes place at $\beta > \beta_c = 1/2d$. This crude assumption agrees remarkably well with the correct result, especially for larger d. (Hint: plot the equation for M graphically for various values of β, and show that a phase transition occurs at β_c.)

7. Prove Eq. (17.38).

8. Fill in the steps in Eq. (17.101).

9. Prove Eqs. (17.105) and (17.107).

10. Draw the Feynman graphs that correspond to Eq. (17.131).

11. Prove Eq. (17.132).

12. Nonperturbative information can be extracted from $SU(N)$ gauge theory in the limit that $N \rightarrow \infty$. Write QCD in the fundamental representation, so that the gauge field is written as $A^a_{\mu b}$ for $a, b = 1, 2, 3$, since the adjoint representation can be written as the product of **3** and **3***. Show that the QCD action can be written as follows:

$$L = \frac{N}{g^2}\left[-\frac{1}{4}F^a_{\mu\nu b}F^{\mu\nu b}_a + \bar{\psi}_a\gamma^\mu(i\partial_\mu\delta^a_b + A^a_{\mu b})\psi^b - m\bar{\psi}_a\psi^a\right] \qquad (17.153)$$

where:

$$F_{\mu\nu b}^a = \partial_\mu A_{\nu b}^a - \partial_\nu A_{\mu b}^a + i(A_{\mu c}^a A_{\nu b}^c - A_{\nu c}^a A_{\mu b}^c) \tag{17.154}$$

and $A_{\mu a}^b$ is a Hermitian, traceless matrix, such that $A_{\mu a}^a = 0$ and $A_{\mu a}^{b\dagger} = A_{\mu b}^a$. Show that the gluon and quark propagator become:

$$\langle 0|T\left[A_{\mu b}^a(x)A_{\nu d}^c(y)\right]|0\rangle = \left(\delta_d^a \delta_b^c - \frac{1}{N}\delta_b^a \delta_d^c\right)D_{\mu\nu}(x-y)$$

$$\langle 0|T\left[\psi^a(x)\bar\psi_b(y)\right]|0\rangle = \delta_b^a S_F(x-y) \tag{17.155}$$

13. Consider the large-N limit with $g^2 N$ held fixed. Consider a vacuum Feynman diagram of very large order. It has the shape of a large polyhedron, with F faces, V vertices, and I internal lines. Using Feynman rules, this polyhedron corresponds to a Feynman diagram with I propagators, V vertices, and F traces over internal lines. Show that whenever we trace over a loop (face), we pick up a factor of N, since $\delta_a^a = N$. Show that each gluon vertex contributes a factor of N, and that each internal line I contributes a factor N^{-1}; that is, show that:

$$\begin{aligned} \text{Faces} &\rightarrow N \\ \text{Vertices} &\rightarrow N \\ \text{Lines} &\rightarrow N^{-1} \end{aligned} \tag{17.156}$$

Show that the Feynman diagram for this polyhedron has the overall factor of:

$$N^{F+V-I} = N^\chi \tag{17.157}$$

where χ is called the *Euler characteristic* of a polyhedron. It is a topological invariant.

14. Now show that we can envision the vacuum Feynman graph as a sphere with H handles (holes) and B boundaries, where the surface of the sphere is triangulated by a large number of triangles making up the vertices and propagators of a Feynman diagram. Show that the Euler characteristic becomes:

$$\chi = 2 - 2H - B \tag{17.158}$$

Show that the leading vacuum graphs behave like N^2. They are topologically equivalent to spheres with no handles or boundaries ($H = B = 0$). They have no fermion lines (since fermions punch holes in the sphere, and thereby decrease the Euler number). They correspond to purely *planar diagrams*. At the next order N, show that we have planar surfaces bounded by closed fermion lines; that is, we have a sphere with one boundary $B = 1$. At the next order $O(1)$, we have either a sphere with a hole (i.e., a doughnut), or a sphere with two boundaries (i.e., a disk with an inner and outer boundary). Comment on the physical meaning of the $N \rightarrow \infty$ limit, in terms of the bound states of the theory and the gluon "strings" that form between quarks.

Chapter 18
Grand Unified Theories

We present a series of hypotheses and speculations leading inescapably to the conclusion that $SU(5)$ is the gauge group of the world...

—H. Georgi and S. Glashow

18.1 Unification and Running Coupling Constants

The Standard Model successfully incorporates all the known properties of the strong, weak, and electromagnetic forces. In fact, there is not a single experiment in particle physics that contradicts the results of the Standard Model. Its weakness, however, is that it is ad hoc: It has too many arbitrary parameters (especially quark masses) and absolutely no interaction with the gravitational force. Since the various interactions are simply spliced together, one feels that a more fundamental theory should be possible.

One improvement on the Standard Model are Grand Unified Theories (GUT). GUTs also share many of the weaknesses of the Standard Model (e.g., too many arbitrary parameters, no interaction with gravity). However, they are genuine unified field theories because there is only one gauge group and hence only one coupling constant. Furthermore, they make a prediction that is now the subject of several ongoing experiments: the decay of the proton.

One of the most compelling arguments for the unification of these forces comes from asymptotic freedom. We deduced previously that the β function for Yang–Mills theory can be written as:

$$\beta(g) = -\frac{g^3}{16\pi^2}\left[\frac{11}{3}N - \frac{2}{3}N_f\right] + \cdots \tag{18.1}$$

for a $SU(N)$ gauge theory coupled to N_f fermions transforming as the N-dimensional representation of the group.

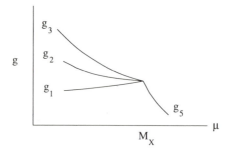

Figure 18.1. The strong, weak, and electromagnetic running coupling constants plotted against the energy. At the GUT scale, all three coupling constants seem to merge into one.

At extremely high energies, the Callan–Symanzik relation shows that the three coupling constants of the strong, weak, and electromagnetic interactions begin to converge, leading us to suspect that all three interactions become part of the same interaction at a very high energy.

For the strong, weak, and electromagnetic interactions, respectively, we have three distinct equations for $\beta(g)$[1]:

$$\beta_3(g_3) = -\frac{g_3^3}{16\pi^2}\left(11 - \frac{2}{3}N_f\right)$$

$$\beta_2(g_2) = -\frac{g_2^3}{16\pi^2}\left(\frac{22}{3} - \frac{2}{3}N_f\right)$$

$$\beta_1(g_1) = N_f\frac{g_1^3}{16\pi^2}\left(\frac{2}{3}N_f\right) \tag{18.2}$$

where we have set the number of Higgs particles to zero. All three equations can be summarized as: $\beta_i(g_i) = b_i g_i^3/16\pi^2$, which then determines the value of b_i.

Let us assume that there is a mass scale, governed by M_X, where all three coupling constants converge:

$$\alpha_1(M_X) = \alpha_2(M_X) = \alpha_3(M_X) \tag{18.3}$$

where $\alpha_i = g_i^2/4\pi$.

Then the solution to the renormalization group equation is given by (Fig. 18.1):

$$\frac{1}{\alpha_i(\mu)} = \frac{1}{\alpha_i(M_X)} + \frac{b_i}{2\pi}\log\frac{M_X}{\mu} \tag{18.4}$$

Unfortunately, this shows that the unification of these three coupling constants takes place at an incredible energy scale of 10^{15} GeV, which is far beyond the ability of any accelerator to probe. Between 1 TeV and 10^{15} GeV, the simplest GUT theory predicts that there will be a large energy "desert" stretching across 12 orders of magnitude in which no new interactions will be found.

Although the existence of this enormous desert is one of the main criticisms of the theory, one attractive feature of this approach is that we have an experimental handle by which to verify such models, and this is proton decay. Since GUTs generically put leptons and quarks in the same multiplet, then the vector mesons of the theory will in general mix up these leptons and quarks, thereby mediating the decay of the quarks into leptons and hence producing proton decay. Since proton decay can be measured in the laboratory, this give us an experimental handle by which to accept or eliminate this approach.

18.2   *SU*(5)

One of the earliest GUT models was that of Pati and Salam.[2] Perhaps the most conservative choice for a model beyond the Standard Model is the "minimal" $SU(5)$ theory.[3] The Standard Model, with the gauge group $SU(3) \otimes SU(2) \otimes U(1)$, has four diagonal generators, corresponding to τ_3, τ_8 of color and τ_3 and Y of weak isospin. The minimal choice beyond the Standard Model is a rank 4 group. The complete set of rank 4 groups involving just one coupling constant can be easily written down. There are just nine of them (including products of identical Lie groups):

$$
\begin{array}{ccc}
SU(2)^4 & O(5)^2 & SU(3)^2 \\
(G_2)^2 & O(8) & O(9) \\
Sp(8) & F_4 & SU(5)
\end{array}
\tag{18.5}
$$

We also want groups with complex representations, because the complex conjugate of a field transforms differently from the field itself in the Standard Model. Of these, only $SU(5)$ contains the Standard Model's gauge group with the proper complex representations of quarks and leptons. There are phenomenological problems with all of these groups except for $SU(5)$.

In addition to being the minimal model compatible with complex representations, there are several other rather remarkable properties of $SU(5)$ that make it physically attractive:

1. It is free of anomalies.

2. It gives precisely the correct quantum number assignments for the 15 left-handed and right-handed quarks and leptons found in the Standard Model for one generation.

3. It gives, after radiative corrections, a reasonably good approximation to $\sin^2 \theta_W$.

4. It gives a scenario by which the model breaks down to the Standard Model via the Higgs mechanism:

$$SU(5) \rightarrow SU(3) \otimes SU(2) \otimes U(1) \rightarrow SU(3) \otimes U(1) \qquad (18.6)$$

Let us now study each of these features of the GUT theory.

18.3   Anomaly Cancellation

Very few groups and their representations give a cancellation of the chiral anomaly, but $SU(5)$ gives such a cancellation with precisely the correct number of quarks and leptons.

To analyze the representations of $SU(5)$, we first remind ourselves that the anomaly is proportional to:

$$\mathrm{Tr}\,[\{\tau^a, \tau^b\}\tau^c] \qquad (18.7)$$

All representations of $SU(N)$ can be found by tensoring the fundamental representation (hence the name) and then taking the various symmetric and anti-symmetric combinations of the indices found in the Young Tableaux. If we take the antisymmetric representations $[N, m]$ of $SU(N)$ (corresponding to a vertical stack of m boxes), then it is easy to see that their dimensionality is the number of ways we can take N things m at a time:

$$\dim [N, m] = \frac{N!}{m!(N - m)!} \qquad (18.8)$$

Furthermore, we can plug this fully antisymmetric representation of the generators of $SU(N)$ into the anomaly condition, and we arrive at:

$$A_{N,m} = \frac{(N - 3)!(N - 2m)}{(N - m - 1)!(m - 1)!} \qquad (18.9)$$

Now let $N = 5$. The fundamental representation is **5** with $m = 1$. If we multiply two of these together, then we can rearrange them in symmetric and antisymmetric combinations:

$$5 \otimes 5 = \mathbf{10} \oplus \mathbf{15} \qquad (18.10)$$

If we take $m = 1$ or 2 (corresponding to **10**), or 3, we find that they have the same anomaly contribution. Therefore, the anomaly contribution of a right-handed $\bar{\mathbf{5}}$ precisely cancels a left-handed **10**. For the fermion representation, anomaly cancellation demands that we take:

$$\text{Fermions}: \quad \bar{\mathbf{5}}_R + \mathbf{10}_L, \qquad (18.11)$$

The anomaly also cancels for the following combinations of $[N, m]$:

$$SU(5): \qquad [5, 1] \oplus [5, 3]$$

$$SU(6): \qquad 2[6, 1] \oplus [6, 4]$$

$$SU(7): \qquad [7, 2] \oplus [7, 4] \oplus [7, 6]$$

$$SU(8): \qquad [8, 1] \oplus [8, 2] \oplus [8, 5]$$

$$SU(9): \qquad [9, 2] \oplus [9, 5]$$

$$\text{or}: \qquad [9, 1] \oplus [9, 3] \oplus [9, 5] \oplus [9, 7]$$

$$SU(10): \qquad [10, 3] \oplus [10, 6] \qquad (18.12)$$

which, of course, does not exhaust all possible anomaly-free combinations.

18.4 Fermion Representation

Anomaly cancellation by itself is not so remarkable, since many other representations can achieve this. What is remarkable about this construction, however, is that the $\bar{\mathbf{5}} \oplus \mathbf{10}$ representation contains precisely the correct quantum numbers necessary to retrieve the Standard Model.

In the Standard Model, if we count the number of chiral fermion modes, we find that we have 12 modes from the u^i and d^i quark sectors, 2 modes from the electron field, and 1 from the massless neutrino, for a total of 15 modes.

However, since $SU(5)$ has no 15-dimensional representation, we must split up the fermions into two parts, the sum of a 5- and 10-dimensional representation. To accomplish this, we take the **5** to be right handed and the **10** to be left handed, so that $\bar{\mathbf{5}}$ and **10** are both left handed. But this, however, is precisely the anomaly-free combination that we just computed.

This 5-dimensional spinor, transforming as [5, 1], is given by:

$$\psi_5 = \begin{pmatrix} d^1 \\ d^2 \\ d^3 \\ e^c \\ \nu^c \end{pmatrix}_R \tag{18.13}$$

If we break this down into the representations of $SU(3) \otimes SU(2)$, we have:

$$\bar{\mathbf{5}} = (\bar{\mathbf{3}}, \mathbf{1}) \oplus (\mathbf{1}, \bar{\mathbf{2}}) \tag{18.14}$$

where the quarks correspond to $(\bar{\mathbf{3}}, \mathbf{1})$ and the leptons to $(\mathbf{1}, \bar{\mathbf{2}})$.

We will now take the left-handed $\mathbf{10}$ to consist of an antisymmetric [5, 2] with two antisymmetric indices:

$$\psi_{10} = \begin{pmatrix} 0 & u_3^c & -u_2^c & -u^1 & -d^1 \\ -u_3^c & 0 & u_1^c & -u^2 & -d^2 \\ u_2^c & -u_1^c & 0 & -u^3 & -d^3 \\ u^1 & u^2 & u^3 & 0 & -e^+ \\ d^1 & d^2 & d^3 & e^+ & 0 \end{pmatrix}_L$$

$$\mathbf{10} = (\mathbf{3}, \mathbf{2}) \oplus (\bar{\mathbf{3}}, \mathbf{1}) \oplus (\mathbf{1}, \mathbf{1}) \tag{18.15}$$

where the quarks correspond to $(\mathbf{3}, \mathbf{2}) \oplus (\bar{\mathbf{3}}, \mathbf{1})$ and the electron to $(\mathbf{1}, \mathbf{1})$.

Likewise, the gauge mesons transform according to the adjoint representation of $SU(5)$, which has 24 elements. The breakdown of these elements in terms of $SU(3) \otimes SU(2)$ is given by:

$$\mathbf{24} = (\mathbf{8}, \mathbf{1}) \oplus (\mathbf{1}, \mathbf{3}) \oplus (\mathbf{1}, \mathbf{1}) \oplus (\mathbf{3}, \mathbf{2}) \oplus (\bar{\mathbf{3}}, \mathbf{2}) \tag{18.16}$$

From this decomposition, we can identify the gauge mesons corresponding to the Standard Model. The $(\mathbf{8}, \mathbf{1})$ corresponds to the usual colored gauge bosons of $SU(3)_c$. The $(\mathbf{1}, \mathbf{3}) \oplus (\mathbf{1}, \mathbf{1})$ mesons correspond to the W_μ, Z_μ, and the electro-magnetic field. Finally, the $(\mathbf{3}, \mathbf{2}) \oplus (\bar{\mathbf{3}}, \mathbf{2})$ are new gauge mesons, which we call the X and Y vector mesons, which couple the quarks to the leptons and hence mediate proton decay.

Next, to extract any meaningful phenomenology from this model, we need a specific representation of the λ^a matrices, from which we can identify the charge and different isospin operators.

There exists, of course, an infinite number of ways in which we can choose these matrices. For convenience, we will take the following representation. Let us break up a 5×5 square matrix into four blocks, consisting of two smaller squares and two rectangles. The upper left-hand corner block will be a 3×3 submatrix. The lower right-hand corner block will be a 2×2 submatrix. In the diagonal upper right- and lower left-hand corners, we will place 3×2 and 2×3 rectangular submatrices.

We will adopt the normalization convention $\mathrm{Tr}\, L^a L^b = 2\delta^{ab}$. Then we can write:

$$L^a = \begin{pmatrix} \lambda^a & 0 \\ 0 & 0 \end{pmatrix}; \quad a = 1, 2, \ldots, 8 \tag{18.17}$$

where λ^a are the usual Gell-Mann matrices for $SU(3)$, and the 0 represents square and rectangular blocks that contain only zeros for entries.

For the 9th and 10th generators, we use two Pauli matrices $\sigma^{1,2}$ in the 2×2 block:

$$L^{9,10} = \begin{pmatrix} 0 & 0 \\ 0 & \sigma^{1,2} \end{pmatrix} \tag{18.18}$$

The 11th and 12th generators are taken to be diagonal, with the diagonal entries given by:

$$
\begin{aligned}
L^{11} &= \mathrm{diag}\,(0, 0, 0, 1, -1) \\
L^{12} &= \frac{1}{\sqrt{15}}\mathrm{diag}\,(-2, -2, -2, 3, 3)
\end{aligned}
\tag{18.19}
$$

To define the next set of matrices, it will be convenient to define rectangular matrices A and B:

$$A_1 = \begin{pmatrix} 1 & 0 \\ 0 & 0 \\ 0 & 0 \end{pmatrix}; \quad A_2 = \begin{pmatrix} 0 & 0 \\ 1 & 0 \\ 0 & 0 \end{pmatrix}; \quad \begin{pmatrix} 0 & 0 \\ 0 & 0 \\ 1 & 0 \end{pmatrix} \tag{18.20}$$

and:

$$B_1 = \begin{pmatrix} 0 & 1 \\ 0 & 0 \\ 0 & 0 \end{pmatrix}; \quad B_2 = \begin{pmatrix} 0 & 0 \\ 0 & 1 \\ 0 & 0 \end{pmatrix}; \quad B_3 = \begin{pmatrix} 0 & 0 \\ 0 & 0 \\ 0 & 1 \end{pmatrix} \tag{18.21}$$

Then the 13th through the 24th generators are given by (for $k = 1,2,3$):

$$L^{13,15,17} = L^{11+2k} = \begin{pmatrix} 0 & A_k \\ A_k^T & 0 \end{pmatrix}$$

$$L^{14,16,18} = L^{12+2k} = \begin{pmatrix} 0 & i\,A_k \\ -i\,A_k^T & 0 \end{pmatrix}$$

$$L^{19,21,23} = L^{17+2k} = \begin{pmatrix} 0 & B_k \\ B_k^T & 0 \end{pmatrix}$$

$$L^{20,22,24} = L^{18+2k} = \begin{pmatrix} 0 & -i\,B_k \\ i\,B_k^T & 0 \end{pmatrix} \tag{18.22}$$

Now let us identify the charge operator from this representation. If we analyze the **5** representation of the fermions, their charge assignment is given by:

$$Q = \frac{1}{2}\left(L^{11} + \sqrt{5/3}\,L^{12}\right) \tag{18.23}$$

Explicitly, the charge matrix is:

$$Q = \text{diag}\left(-1/3,\, -1/3,\, -1/3,\, 1,\, 0\right) \tag{18.24}$$

The quarks have fractional 1/3 charge compared to the electron. This is extremely important, because the charge assignments of the various quarks and leptons are now quantized; that is, GUT theory gives us *charge quantization*. This is different from the usual $U(1)$ Maxwell theory, where the charge e is a continuous parameter. Because the charge operator is now one of the generators of the group, its eigenvalues are quantized and we have a definite quantized charge assignment for the quarks and leptons. In fact, *the quarks have 1/3 charge relative to the leptons just because there are three colors within $SU(3)$.*

Similarly, the charge assignment of the **10** representation can be computed. The charge operator Q, acting on the mixed tensor $\mathbf{10} \equiv \psi_i^j$, yields:

$$Q(\psi_i^j) = Q_i - Q_j \tag{18.25}$$

From this, we can read off the charge assignments of the **10**, which are also experimentally correct. The relative ease with which we can generate the correct quantized quark and lepton assignments is one of the successes of the GUT approach.

We can now make the precise association between the vector mesons of $SU(5)$ and the vector mesons of the electroweak model ($W_\mu^\pm$, W_μ^3, B_μ) and the gluons G_μ^a of QCD: We find:

$$A_\mu^{1,2,\dots,8} \;\to\; G_\mu^a$$
$$A_\mu^{9\pm i10} \;\to\; W^\pm$$

$$A_\mu^{11} \quad \rightarrow \quad W_\mu^3$$

$$A_\mu^{12} \quad \rightarrow \quad B_\mu \tag{18.26}$$

The new vector mesons, which are only specific to $SU(5)$ and not the Standard Model, are given the names:

$$A_\mu^{13,14,\dots,18} \quad \rightarrow \quad X_\mu^i$$

$$A_\mu^{19,20,\dots,24} \quad \rightarrow \quad Y_\mu^i \tag{18.27}$$

Putting everything together, the 5×5 matrix $\mathbf{A}_\mu$ is given by:

$$\mathbf{A}_\mu = \sqrt{2} \sum_{i=1}^{24} A_\mu^i L^i \tag{18.28}$$

where this can be rewritten as:

$$\mathbf{A}_\mu^a = \begin{pmatrix} A & B \\ C & D \end{pmatrix} \tag{18.29}$$

where:

$$A = \mathbf{G}_\mu - \frac{2}{\sqrt{30}} B_\mu \mathbf{1}; \quad B = \begin{pmatrix} \bar{X}^1 & \bar{Y}^1 \\ \bar{X}^2 & \bar{Y}^2 \\ \bar{X}^3 & \bar{Y}^3 \end{pmatrix}$$

$$C = \begin{pmatrix} X_1 & X_2 & X_3 \\ Y_1 & Y_2 & Y_3 \end{pmatrix}; \quad D = \begin{pmatrix} \frac{W_3}{\sqrt{2}} + \frac{3B}{\sqrt{30}} & W^+ \\ W^- & \frac{-W^3}{\sqrt{30}} + \frac{3B}{\sqrt{30}} \end{pmatrix}$$

Next, we would like to calculate $\sin^2 \theta_W$. This is now easily accomplished by extracting out the coupling constant for $SU(2)$ and $U(1)$ from the coupling of gauge bosons.

The covariant derivative associated with W_μ^3 and B_μ can be related to the coupling of A_μ and Z_μ by extracting out the covariant derivative of the 11th and 12th generators:

$$
\begin{aligned}
D_\mu &= \partial_\mu - i(g/2)\left(W_\mu^3 L^{11} + B_\mu L^{12}\right) \\
&= \partial_\mu - i(g/2)\Big[A_\mu \left(\sin\theta_W L^{11} + \cos\theta_W L^{12}\right) \\
&\quad + Z_\mu \left(\cos\theta_W L^{11} - \sin\theta_W L^{12}\right)\Big] \\
&\equiv \partial_\mu - i(eQA_\mu + gQ_Z Z_\mu) \tag{18.30}
\end{aligned}
$$

Since $Q = (1/2)[L^{11} + \sqrt{5/3}L^{12}]$, we can take the ratio of 1 to $\sqrt{5/3}$, which equals the ratio of $\sin\theta_W$ and $\cos\theta_W$. Thus, we have:

$$\sin^2\theta_W = \frac{3}{8} = \frac{g_1^2}{g_1^2 + g_2^2} \tag{18.31}$$

This prediction, by itself, would be a disappointment, since experimentally the measured value is $\sin^2\theta_W \sim 0.23$. However, we must consider this to be a first-order value of the Weinberg angle θ_W that must be corrected by radiative corrections coming from the renormalization group.

Earlier, we analyzed how the renormalization group arguments give us a handle on the size of the GUT scale. Let us re-examine this calculation with $SU(5)$ in mind. For coupling constants g_1, g_2, g_3 for the groups within the Standard Model, we have the following solutions to the renormalization group equations:

$$\frac{g_1^2(\mu)}{4\pi} = (5/3)\frac{\alpha(\mu)}{\cos^2\theta_W}$$

$$\frac{g_2^2(\mu)}{4\pi} = \frac{\alpha(\mu)}{\sin^2\theta_W}$$

$$\frac{g_3^2(\mu)}{4\pi} = \alpha_3(\mu) \tag{18.32}$$

where $\alpha(\mu)$ describes the electromagnetic coupling.

By taking linear combinations of these equations, and using the fact that $g_i(M_X)$ are all equal, we find:

$$\log(M_X/\mu) \sim \frac{\pi}{11}\left(\frac{1}{\alpha(\mu)} - \frac{8}{3\alpha_3(\mu)}\right) \tag{18.33}$$

and:

$$\sin^2\theta_W \sim 3/8 - (55/24\pi)\alpha(\mu)\log(M_X/\mu) \tag{18.34}$$

A careful analysis of the parameters of the theory gives us:

$$M_X \sim 4 \times 10^{14}\,\text{GeV}$$
$$\sin^2\theta_W \sim 0.206^{+0.016}_{-0.004} \tag{18.35}$$

while the experimentally observed value is $\sin^2\theta_W \sim 0.2325 \pm 0.0008$.

18.5 Spontaneous Breaking of $SU(5)$

Next, we wish to use the Higgs mechanism to break down the theory to the Standard Model and eventually down to $SU(3)_c \otimes U(1)_{em}$ The first breaking down to the Standard Model is accomplished by taking a Higgs boson transforming in the **24** adjoint representation of the group. The second breaking down to $SU(3) \otimes U(1)$ is achieved via a **5** of Higgs:

$$\mathbf{24}: \qquad SU(5) \rightarrow SU(3) \otimes SU(2) \otimes U(1)$$

$$\mathbf{5}: \qquad SU(3) \otimes SU(2) \otimes U(1) \rightarrow SU(3) \otimes U(1) \qquad (18.36)$$

If we describe the breaking of the Lie algebra via the adjoint representation labeled by $(L^a)_{bc} = f^{abc}$, then the matrices would then be 24×24, which is unwieldy. To make things simpler, we note that:

$$\mathbf{5} \otimes \bar{\mathbf{5}} = \mathbf{24} \oplus \mathbf{1} \qquad (18.37)$$

which allows us to write down the Higgs as Φ_a^b, where a represents the $\bar{\mathbf{5}}$ index and b the **5** index. We then take the **24** Higgs to be represented by the product:

$$\Phi_a^b \equiv \phi_a \phi^b - \frac{1}{5} \phi_c \phi^c \delta_a^b \qquad (18.38)$$

where we have subtracted out the **1**.

We can, of course, reassemble these mesons in terms of the 24 members of the adjoint representation of $SU(5)$. In the adjoint representation, the Higgs meson becomes: $\Phi = \Phi^a L^a / 2$ for $a = 1, 2, \ldots, 24$. The kinetic term for the Higgs potential is given by:

$$L_0 = \frac{1}{2} \sum_{a=1}^{24} D_\mu \Phi^{a\dagger} D^\mu \Phi^a \qquad (18.39)$$

where:

$$D_\mu \Phi = \partial_\mu \Phi - ig \left[V_\mu^a \frac{L^a}{2}, \Phi \right] \qquad (18.40)$$

We now choose a potential $V(\Phi)$ such that the minimum is given by:

$$\langle \Phi \rangle = v \, \text{diag} \, (1, 1, 1, -3/2, -3/2) \qquad (18.41)$$

This is equal to the unit matrix in the $SU(3)$ subspace and equal to the unit matrix times $-3/2$ in the $SU(2)$ subspace. The only generators that have nonzero commutators with this diagonal matrix are $L^{13,\ldots,24}$. This tells us that only the X and Y vector bosons get mass, while preserving the massless nature of the vector bosons for $SU(3) \otimes SU(2) \otimes U(1)$. Thus, this successfully breaks $SU(5)$ down to $SU(3) \otimes SU(2) \otimes U(1)$.

The mass matrix for the vector bosons is given by:

$$\frac{1}{2}g^2 \operatorname{Tr}[\mathbf{A}_\mu, \langle \Phi \rangle]^2] \equiv m_{ab}^2 A_\mu^a A^{b,\mu} \tag{18.42}$$

Then we get:

$$m_X^2 = m_Y^2 = \frac{25}{8}g^2 v^2 \tag{18.43}$$

To arrive at this assignment of quantum numbers, we will take the Higgs potential to be:

$$-\mu^2 \operatorname{Tr}(\Phi^2) + \frac{1}{4}a\left(\operatorname{Tr}\Phi^2\right)^2 + \frac{1}{2}b \operatorname{Tr}\left(\Phi^4\right) \tag{18.44}$$

We can now shift the minimum and get an expression for v in terms of the Higgs potential parameters:

$$\mu^2 = \frac{15}{2}av^2 + \frac{7}{2}bv^2 \tag{18.45}$$

The second stage of Higgs breaking is mediated by a **5** Higgs boson, transforming as a doublet under $SU(2)$. From our discussion of the electroweak model, we know that a suitable choice for the Higgs meson is given by:

$$H = \begin{pmatrix} h^1 \\ h^2 \\ h^3 \\ h^+ \\ -h^0 \end{pmatrix} = (\mathbf{3}, \mathbf{1}) \oplus (\mathbf{1}, \mathbf{2}) \tag{18.46}$$

The potential for H is the same as for the electroweak model:

$$V(H) = -\frac{1}{2}v^2|H|^2 + \frac{1}{4}\lambda\left(|H|^2\right)^2 \tag{18.47}$$

As usual, the breaking of $SU(2)$ is performed by having an expectation value along the h^0 direction:

$$\langle H \rangle = \langle -h^0 \rangle = v_0 \tag{18.48}$$

where $v^2 = \lambda v_0^2$. Then, as expected, we will have:

$$M_W^2 = M_Z^2 \cos^2 \theta_W = \frac{1}{4} g_2^2 v_0^2 \qquad (18.49)$$

Finally, we look at the fermion masses of the theory. They are generated via the generic coupling $\bar{\psi} \psi \Phi$, where Φ acquires a vacuum expectation value.

Since fermion masses appear via the coupling of two fermions to the Higgs boson, we expect them to arise from interactions involving the combination $(\bar{\mathbf{5}} \oplus \mathbf{10}) \otimes (\bar{\mathbf{5}} \oplus \mathbf{10})$. This decomposition is given by:

$$\bar{\mathbf{5}} \otimes \mathbf{10} = \mathbf{5} \oplus \overline{\mathbf{45}}$$

$$\mathbf{10} \otimes \mathbf{10} = \bar{\mathbf{5}} \oplus \mathbf{45} \oplus \mathbf{50}$$

$$\bar{\mathbf{5}} \otimes \bar{\mathbf{5}} = \overline{\mathbf{10}} \oplus \overline{\mathbf{15}} \qquad (18.50)$$

Fermion masses are thus generated by the Higgs in the following representations: $\mathbf{5}, \mathbf{10}, \mathbf{15}, \mathbf{45}, \mathbf{50}$. Since the $\mathbf{5}$ Higgs was used in the minimal $SU(5)$ model to break the electroweak interactions, no new Higgs need be added to the minimal theory. Nonminimal $SU(5)$, involving more parameters, can be constructed using the $\mathbf{10}, \mathbf{15}, \mathbf{45}, \mathbf{50}$, which appear in the tensor product decomposition, meaning that these Higgs can couple to two fermions and generate masses.

The nice feature of this construction, however, is that $\mathbf{24}$ does not appear in the tensor product decomposition. Thus, the $\Phi_{\mathbf{24}}$ Higgs meson does not couple to two fermions, and hence the fermion masses cannot be of the GUT scale M_X. This is gratifying, since we do not want any of the quarks and leptons to have GUT scale masses. Notice that the $\mathbf{5}$ does appear in the tensor product decomposition, which means that the fermion masses can be of the order of M_W.

Although minimal $SU(5)$ seems to unify the strong and electroweak interactions in a surprisingly tight fashion, we should point out that the current experimental limits on proton decay seem to rule it out. For example, the theoretical decay rate of the proton into an electron can proceed via:

$$\Gamma^{-1}(p \to e^+ + \pi^0) = 4.5 \times 10^{29 \pm 1.7} \text{ yr} \qquad (18.51)$$

which is much too fast. Experiment has pushed the proton lifetime to:

$$\Gamma^{-1}(p \to e^+ + \pi^0), \ \Gamma^{-1}(n \to e^+ + \pi^-) > 6 \times 10^{31} \text{ yr} \qquad (18.52)$$

Furthermore, electroweak measurement of the Z mass are apparently precise enough to cast doubt on the value of $\sin^2 \theta_W \sim 0.206$, predicted by the minimal model [see Eq. (18.35)]. This, of course, does not rule out more complicated, nonminimal $SU(5)$ and models with more complicated GUT groups and couplings.

18.6   Hierarchy Problem

The most important theoretical challenge facing GUT theory is the *hierarchy problem*.[4] The origin of this problem is easy to isolate. The $SU(5)$ theory, for example, has two Higgs bosons, which introduce the mass scales M_X and M_W. Because of the vast difference between these two mass scales, it is important to keep the two scales apart, so there is no mixing between them. We must, therefore, "tune" our two mass scales so that we preserve the vanishingly small ratio between them:

$$\frac{M_W}{M_X} \sim 10^{-12} \tag{18.53}$$

However, it easy to show that the loop corrections lead to interactions connecting two Φ fields with two H fields. This, in turn, means that we must introduce a term in the action that corresponds to this new graph:

$$V(\Phi, H) = \alpha |H|^2 \operatorname{Tr} \Phi^2 + \beta \bar{H} \Phi^2 H \tag{18.54}$$

But introducing $\Phi^2 H^2$-type terms into the Higgs potential has mixed these two mass scales. Thus, the vast ratio between these two mass scales has been destroyed.

We can, at this one-loop order, retune the parameters within the Higgs potential so that we once again re-establish the hierarchy. The explicit calculation yields:

$$v^2 - (15\alpha + \frac{9}{2}\beta)v^2 \sim 10^{-24} v^2 \tag{18.55}$$

Although we can now tune α and β to one part in 10^{24}, we will find that the two-loop result reintroduces mixing between these two mass scales, and the hierarchy is again ruined.

We can always retune our parameters at the two-loop level, but then this retuning will not survive at the third-loop order. Clearly, we have a problem. It is difficult to imagine a more clumsy way in which to unify the known interactions than continually to perform a retuning of parameters to incredible accuracy at each order in perturbation theory.

18.7 $SO(10)$

Let us leave the minimal $SU(5)$ model and go to the next model, $SO(10)$,[5–6] which incorporates many of the attractive "accidents" of the $SU(5)$ model and explains their origin group theoretically. In general, the series $SO(N)$ is attractive because of anomaly cancellation. We know from Eq. (2.68) that the generators of $SO(N)$ can be written in terms of the matrix $(M^{ij})_{ab}$, which is antisymmetric in a, b and i, j. If we insert this value into the anomaly:

$$\text{Tr}\left[\{M^{ij}, M^{kl}\}M^{mn}\right] \tag{18.56}$$

we find that this number cannot, in general, be written as a constant tensor with the indices i, j, k, l, m, and n (except for $N = 6$, where a constant tensor with all the symmetry properties is given by ϵ^{ijklmn}). Thus, all $SO(N)$ theories are anomaly free except for $N = 6$.

Furthermore, we are interested in complex representations of the Lie group. However, not all $SO(N)$ groups have complex representations. In particular, the requirement of complex representations restricts us to the groups $SO(4n + 2)$. Therefore, the smallest orthogonal group of rank ≥ 4, with complex representations, is given by $SO(10)$.

The representation in which we are interested is the **16**-dimensional spinor representation of $SO(10)$. This is because the adjoint representation has 45 elements, which is too many, while the vector representation has too few, only 10 elements.

$SO(10)$ includes $SU(5)$ as a subgroup, therefore, all representations of $SO(10)$ can be described by giving its $SU(5)$ quantum numbers. The essential feature of the **16** is that, under $SU(5)$, it transforms as:

$$\mathbf{16} = \bar{\mathbf{5}} \oplus \mathbf{10} \oplus \mathbf{1} \tag{18.57}$$

where the **1** refers to the right-handed neutrino, which is missing in the minimal $SU(5)$ scheme. In one stroke, we see why the $\bar{\mathbf{5}} \oplus \mathbf{10}$ representation worked so well, and this is because they are actually part of the **16** representation of $SO(10)$.

The group $SO(10)$ has 45 generators, which can be broken down under $SU(5)$ as follows:

$$\mathbf{45} = \mathbf{24} \oplus \mathbf{1} \oplus \mathbf{10} \oplus \overline{\mathbf{10}} \tag{18.58}$$

where the **24** is the same multiplet of gauge bosons that we encountered earlier for $SU(5)$.

Symmetry breaking can proceed in numerous ways because of the large number of subgroups contained within $SO(10)$. The simplest route to symmetry breaking is given by:

$$SO(10) \rightarrow SU(5) \rightarrow SU(3) \otimes SU(2) \otimes U(1) \rightarrow SU(3) \otimes U(1) \tag{18.59}$$

The first breaking down to $SU(5)$ can be mediated with a **16** of Higgs bosons. The second breaking down to the Standard Model needs a Higgs in the **45** representation. And the last breaking is accomplished via **10** Higgs:

$$\mathbf{16}: \qquad SO(10) \rightarrow SU(5)$$

$$\mathbf{45}: \qquad SU(5) \rightarrow SU(3) \otimes SU(2) \otimes U(1)$$

$$\mathbf{10}: \qquad SU(3) \otimes SU(2) \otimes U(1) \rightarrow SU(3) \otimes U(1) \qquad (18.60)$$

Yet another favored route is given by:

$$
\begin{aligned}
SO(10) \quad &\rightarrow \quad SU(4) \otimes SU(2)_L \otimes SU(2)_R \\
&\rightarrow \quad SU(3) \otimes SU(2)_L \otimes SU(2)_R \otimes U(1)_{B-L} \\
&\rightarrow \quad SU(3) \otimes SU(2) \otimes U(1) \rightarrow SU(3) \otimes U(1) \qquad (18.61)
\end{aligned}
$$

This sequence of breakings is initiated by the following representations of Higgs bosons: **54**, **45**, **16**, **10**:

$$\mathbf{54}: \qquad SO(10) \rightarrow SU(4) \otimes SU(2)_L \otimes SU(2)_R$$

$$\mathbf{45}: \qquad SU(4) \otimes SU(2)_L \otimes SU(2)_R$$

$$\qquad\qquad \rightarrow SU(3) \otimes SU(2)_L \otimes SU(2)_R \otimes U(1)_{B-L}$$

$$\mathbf{16}: \qquad SU(3) \otimes SU(2)_L \otimes SU(2)_R \otimes U(1)_{B-L}$$

$$\qquad\qquad \rightarrow SU(3) \otimes SU(2) \otimes U(1)$$

$$\mathbf{10}: \qquad SU(3) \otimes SU(2) \otimes U(1) \rightarrow SU(3) \otimes U(1) \qquad (18.62)$$

Likewise, the fermion masses can also be analyzed. The fermion masses are generated by two fermion fields; therefore, we find:

$$\mathbf{16} \otimes \mathbf{16} = \mathbf{10} \oplus \mathbf{126} \oplus \mathbf{120} \qquad (18.63)$$

Thus, fermion masses can only be generated through Yukawa couplings via the **10**, **126**, or **120** representations for the Higgs particle. To see the relationship with the $SU(5)$ model, we can decompose these representations as:

$$10 = 5 \oplus \bar{5}$$

$$126 = 1 \oplus 5 \oplus 10 \oplus \overline{15} \oplus 45 \oplus \overline{50}$$

$$120 = 5 \oplus \bar{5} \oplus 10 \oplus \overline{10} \oplus 45 \oplus \overline{45} \tag{18.64}$$

To perform calculations with $SO(N)$ models, it is necessary to have a specific representation of the spinors. We can use a simple recursive technique, generating the spinor representation of $SO(2n+2)$ from the spinor representation of $SO(2n)$.

Let $\Gamma_i^{(n)}$ for $i = 1, \ldots, 2n$ form a Clifford algebra for $SO(2n)$. From these elements, we can generate the Clifford algebra for $SO(2n + 2)$, with elements $\Gamma_i^{(n+1)}$ for $i = 1, \ldots, 2n + 2$. The recursion relation is:

$$\Gamma_i^{(n+1)} = \begin{pmatrix} \Gamma_i^{(n)} & 0 \\ 0 & -\Gamma_i^{(n)} \end{pmatrix}; \quad i = 1, 2, \cdots, 2n$$

$$\Gamma_{2n+1}^{(n+1)} = \begin{pmatrix} 0 & 1 \\ 1 & 0 \end{pmatrix}$$

$$\Gamma_{2n+2}^{(n+1)} = \begin{pmatrix} 0 & -i \\ i & 0 \end{pmatrix} \tag{18.65}$$

For $SO(2n + 1)$, its Clifford algebra is formed from $\Gamma_i^{(n+1)}$ for $i = 1, \ldots, 2n + 1$, that is, by omitting the last element of the Clifford algebra.

Then the generators of the group, in terms of the spinors, are given by:

$$M^{ab} = \frac{i}{4}[\Gamma^a, \Gamma^b] \tag{18.66}$$

For models beyond $SO(10)$, it is useful to write down generic values of the predictions of the various theories. For example, we can compute $\sin^2 \theta_W$ with the simple observation that, before symmetry breaking, all couplings of the various subgroups of the gauge group are equal. Therefore, we know that $g_i^2 \text{Tr}(T_i^2)$ for the various subgroups is the same. Setting them equal, and solving for the ratio of the coupling constants, we can show that:

$$\sin^2 \theta_W = \frac{e^2}{g^2} = \frac{\text{Tr}(T_3^2)}{\text{Tr}(Q^2)}$$

$$\frac{\alpha_c}{\alpha_{em}} = \frac{\text{Tr}(Q^2)}{\text{Tr}\, T_c^2} \tag{18.67}$$

where c refers to the $SU(3)$ color subgroup.

There are, of course, many GUT models beyond the $SO(10)$ theory. Let us list a few of the restrictions on these models. First, we must have complex representations (unless we want to have "mirror" fermions with opposite handedness to the usual ones, but then we have another problem of explaining why their masses are so heavy and hence are not seen). The only complex representations occur for

1. $SU(N)$ for $N > 2$

2. E_6

3. $SO(4N + 2)$.

Second, we want anomaly cancellations, which must be checked by hand. For example, $SO(N)$ is anomaly free if $N \neq 6$.

Of these, the exceptional groups[7] look attractive, since there are only a finite number of them and not an infinite series. Of the exceptional groups, E_6 is attractive since it has complex representations and can be broken in many ways, including:

$$E_6 \quad \rightarrow \quad SO(10) \rightarrow SU(5)$$
$$\rightarrow SU(3)_c \otimes SU(2) \otimes U(1) \tag{18.68}$$

The fermions can be accommodated in the **27** of E_6. E_7 can accomodate the fermions in a **56** representation (except for the t quark), but has an unacceptable structure for the weak currents. E_8 has also been studied. Its lowest dimension representation is the adjoint with **248**.

One interesting sequence of breakings is given by:

$$E_8 \rightarrow SO(16) \rightarrow SO(10) \otimes SO(6) \rightarrow SU(5) \times SU(3) \tag{18.69}$$

E_8 and E_6 have also been seriously examined from the viewpoint of the superstring, where they are some of the preferred intermediate steps in symmetry breaking.

18.8 Beyond GUT

The GUT theories, although they have many compelling features, are a speculative step beyond the Standard Model. There are other speculative steps. We mention just a few of them.

18.8.1 Technicolor

Technicolor[8] is based on the philosophy that fundamental scalars are unattractive and undesirable features of an electroweak model. They can be eliminated if they emerge as bound states of some fermions. In order to drive the symmetry breaking to give the W boson a mass of around 80 GeV, we must postulate a new color-type interactions called hypercolor or technicolor, which becomes strong at 1 TeV.

One advantage of this approach is that the hierarchy problems seems to be avoided. The hierarchy problem emerged when mixing between fundamental scalars created radiative corrections that forced us to retune the parameters to preserve the mass difference between ordinary energies and GUT energies. In the technicolor picture, there are no such scalar couplings, because there are no scalars.

Unfortunately, the simplest versions of technicolor have been ruled out because they have problems with flavor-changing neutral currents. A possibility exists of avoiding this by proliferating technifamilies, but then these theories also have problems with the various counterparts of the Nambu–Goldstone boson, the technipions, which have not been discovered.

18.8.2 Preons or Subquarks

Everytime we have probed deeper into the structure of matter, we have seen a new layer of constituents, from molecules, to atoms, to nuclei, to subatomic particles, and to the quarks. It may not be such a leap of logic, therefore, to suppose that the quarks themselves are composite objects.

Several problems face subquark theories. First, there are technical ones, such as eliminating anomalies via the 't Hooft anomaly matching conditions. But there are also more physical questions, such as the lack of any guiding principle by which to construct subquark theories. Nature gives us few signals as to which direction to take in generating subquark models, of which there are many. One criterion is that these subquark theories must explain why the electron and neutrino seem point-like with a small or zero mass, even though the energy of their constituents is quite large by comparison. Naively, one would expect that, if the electron were a composite particle, its mass would be comparable to the energy scale of the composite particles, which would be enormous.

18.8.3 Supersymmetry and Superstrings

Nature, however, does give us very strong signals, at least in one direction, and this is the existence of gravitation. There is no question that gravitation exists, and that it is the basic glue holding much of the universe together. Ironically, although gravity was the very first force to have its fundamental classical equations revealed with the work of Newton three centuries ago, it resists unification with the other forces for a very fundamental reason: It is not renormalizable. Gravitation is a gauge theory of great sophistication, requiring new ideas in order to marry it to the other three fundamental interactions.

We now turn to the theory of general relativity and to the two theories that give us the only known nontrivial extensions of Einstein's theory: supergravity and superstrings. Not only does supersymmetry give us a plausible solution to the hierarchy problem, it also gives us theories of gravity in which the divergences are partially or even completely cancelled.

No one knows what will be the ultimate outcome to the vigorous theoretical pursuit of a quantum theory of gravity. However, from the standpoint of quantum field theory, it has already given us an incredibly rich laboratory by which to test old ideas and generate entirely new ones.

In summary, GUT theories give us the first nontrivial extension of the Standard Model. GUT theories based on gauge groups such as $SU(5)$, $O(10)$, and E_6 have the advantage that they are elegant and have fewer coupling constants than the Standard Model. Although the unification of the various forces takes place at approximately 10^{15} GeV, GUT theories can still be tested if the proton decays. Minimal $SU(5)$ has now been ruled out experimentally, but theories with more complicated groups and couplings are still consistent with experiments.

We now turn out attention to the problems raised by the GUT theories, such as the presence of gravity, the hierarchy problem, and the renormalization of quantum gravity.

18.9   Exercises

1. Show that, if we simply drop the massive X and Y mesons in the action for $SU(5)$ GUT theory, that the resulting theory (for the fermions and vector mesons) becomes essentially the Standard Model action with gauge group $SU(3) \otimes SU(2) \otimes U(1)$.

2. Take the spinor representation of $SO(10)$. Explicitly extract the generators of $SU(5)$ from the generators of $SO(10)$. Explicitly decompose this spinor into the $SU(5)$ representations in terms of quarks and leptons. In this way, show how the $SO(10)$ model decomposes into the Standard Model (for the leptons, quarks, and vector mesons only).

3. In minimal $SU(5)$, construct the explicit expression for the coupling of Higgs bosons to fermions to form the Yukawa potential. Show explicitly how the **24** representation of the Higgs can break this down to the Standard Model.

4. Isolate which graphs would contribute to proton decay in the minimal $SU(5)$ model. By dimensional arguments, do a quick order of magnitude calculation of the decay rate.

5. Let Γ_i generate a Clifford algebra. Define:

$$a_i = \frac{1}{2}\left(\Gamma_{2i-1} - i\Gamma_{2i}\right) \tag{18.70}$$

Show that a_i and $a_j^\dagger$ form a set of anticommuting annihilation and creation operators; that is:

$$\{a_i, a_j^\dagger\} = \delta_{ij} \tag{18.71}$$

with all other anticommutators being zero.

6. Let τ^a be a set of traceless Hermitian $n \times n$ matrices, which generate the algebra of $SU(n)$. Show that T^a, defined by:

$$T^a \equiv \sum_{jk} a_j^\dagger \left(\tau^a\right)_{jk} a_k \tag{18.72}$$

also generates the algebra of $SU(n)$. Now show that any bilinear $a_j^\dagger a_k$ can be expressed as a combination of generators M_{ij} of the group $SO(2n)$:

$$
\begin{aligned}
a_j^\dagger a_k &= \frac{1}{2}\delta_{jk} + \frac{i}{2}M_{2j-1,2k-1} + \frac{1}{2}M_{2j-1,2k} \\
&\quad - \frac{1}{2}M_{2j,2k-1} + \frac{i}{2}M_{2j,2k}
\end{aligned} \tag{18.73}
$$

Show that this proves that:

$$SU(n) \subset SO(2n) \tag{18.74}$$

This shows one way in which to embed $SU(5)$ GUT into $SO(10)$ GUT.

7. Let the $a_i^\dagger$ be five anticommuting creation operators acting on a vacuum $|0\rangle$. For $SO(10)$, show that a 32 dimensional spinor $|\psi\rangle$ can be decomposed as:

$$
\begin{aligned}
|\psi\rangle \;=\; & |0\rangle \psi_0 + a_i^\dagger |0\rangle \psi_i + \frac{1}{2} a_j^\dagger a_k^\dagger |0\rangle \psi_{jk} + \frac{1}{12} \epsilon^{ijklm} a_k^\dagger a_l^\dagger a_m^\dagger |0\rangle \bar\psi_{ji} \\
& + \frac{1}{24} \epsilon^{jklmn} a_k^\dagger a_l^\dagger a_m^\dagger a_n^\dagger |0\rangle \bar\psi_j + a_1^\dagger a_2^\dagger a_3^\dagger a_4^\dagger a_5^\dagger |0\rangle \bar\psi_0 \qquad (18.75)
\end{aligned}
$$

where ψ_{ij} is antisymmetric. Show that these form the $SU(5)$ representations of **1**, **5**, **10**, and their conjugates. Show that this generates the irreducible **16**-dimensional spinor and its conjugate. Now generalize your results for $SU(n)$ and $SO(2n)$, decomposing a $SO(2n)$ spinor into $SU(n)$ multiplets.

8. The breakdown of $SO(10)$ down to $SU(5)$ leaves us with an extra $U(1)$ symmetry. Show explicitly how this extra quantum number can be associated with $B - L$, where B is the baryon number and L is the lepton number. Explicitly construct the operator which generates $B - L$.

9. In Eq. (18.60), there are several ways in which $SO(10)$ may be broken down, with various representations of Higgs. Construct explicitly the Higgs potential for each of these breaking mechanisms. Analyze their strengths and weaknesses.

10. In a model with E_6 symmetry, we have $E_6 \supset SO(10) \otimes U(1)$. Thus, the **27** and adjoint **78** of E_6 can be broken down into:

$$
\begin{aligned}
\mathbf{27} \;&=\; \mathbf{16} \oplus \mathbf{10} \oplus \mathbf{1} \\
\mathbf{78} \;&=\; \mathbf{45} \oplus \mathbf{1} \oplus \mathbf{16} \oplus \overline{\mathbf{16}} \qquad (18.76)
\end{aligned}
$$

Rewrite this decomposition strictly in terms of $SU(5)$ representations. From this, describe the physical quark/lepton content of the **27** and the vector mesons of **78**. How many new particles must be postulated?

11. A Weyl neutrino cannot have a mass, since the mass term couples left- and hand-handed fermion fields. However, consider a theory in which the neutrino is a Majorana fermion, which obeys $\psi = \psi^c$. Then it is possible to construct a mass term for this field:

$$
\bar\psi_R \psi_L = (\psi_R^c) C \psi_L = \psi_L^T C \psi_L \qquad (18.77)
$$

Notice that this Majorana mass term is now defined totally in terms of ψ_L. Show that this quantity is Lorentz invariant. Show that, in contrast to the

Weyl neutrino action [which is invariant under $\nu \rightarrow e^{i\theta}\nu$, which generates a $U(1)$ symmetry, or lepton number], the Majorana mass term violates lepton number by two units.

12. Show that a Majorana neutrino cannot be generated in the Standard Model. (Hint: show that a term like $\nu_L^T C \nu_L$ transforms like $I = 1$, $I_3 = 1$, and see if such a term can be generated by the Higgs mechanism.)

13. Show that a Majorana neutrino mass cannot be generated in a minimal $SU(5)$ theory, even though there is a $I = 1$ Higgs field. Can it be generated, if the neutrino is a $SU(5)$ singlet? What about $SO(10)$?

14. Prove that the spinor matrices of $O(N)$ presented in Eq. (18.65) do, in fact, satisfy the correct Dirac algebra.

15. Prove Eq. (18.67).

Chapter 19
Quantum Gravity

I was sitting in a chair in the patent office at Bern when all of a sudden a thought occurred to me: "If a person falls freely he will not feel his own weight." I was startled. This simple thought made a deep impression on me. It impelled me toward a theory of gravitation. —A. Einstein

19.1 Equivalence Principle

One of the great physical problems of this century is to unify general relativity and quantum mechanics. Together they can explain a vast storehouse of physical knowledge, from the subatomic realm to the large-scale structure of the universe. However, attempts to unify quantum mechanics with general relativity have all met with frustration. General relativity has a negative dimensional coupling constant (Newton's constant) and hence is not renormalizable in the usual sense. To renormalize gravity, one must necessarily make a radical departure from quantum field theory as we know it.

To see the origin of the problems with quantum gravity, let us first describe the classical theory of general relativity. General relativity, like special relativity before it, can be reduced to a few postulates.[1]

Equivalence Principle: *The laws of physics in a gravitational field are identical to those in a local accelerating frame.*

Einstein stumbled upon this deceptively simple principle and its consequences when he noticed that a person in a freely falling elevator would experience no apparent weight. He called this "the happiest thought of my life." He generalized this to say that no physical experiment could differentiate a freely falling elevator from a frame without any gravity. In particular, it meant that in any gravitating system, one can at any point choose a new set of coordinates such that the gravitational field disappears. This new set of coordinates is the freely falling "elevator frame," in which space appears locally to resemble ordinary Lorentzian space. We wish, therefore, to construct a theory that is invariant under general coordinate transformations, that is, a theory in which one can choose coordinates such that the gravitational field vanishes locally.

Following our discussion of gauge invariance, we will begin our discussion of general relativity by proceeding in three steps:

1. First, we will write down the transformation properties of scalar, vector, and tensor fields under general coordinate transformations.

2. Then we will construct covariant derivatives of these fields by introducing connection fields (Christoffel symbols).

3. Finally, we will construct the action for general relativity and its coupling to matter fields.

Since we need to express the physical consequences of the equivalence principle mathematically, one needs a mathematical language by which we can easily transform from one frame to another, that is, tensor calculus. We will define a general coordinate transformation as an arbitrary reparametrization of the coordinate system:

$$\bar{x}^\mu = \bar{x}^\mu(x) \tag{19.1}$$

Unlike Lorentz transformations, which are *global* space–time transformations, general coordinate transformations are local and hence much more difficult to incorporate into a theory. A general coordinate transformation therefore describes a distinct reparametrization at every point in space–time. (Historically, the local nature of general coordinate transformations was one of the original inspirations that led Yang and Mills to postulate local gauge theories.)

Under reparametrizations, a scalar field transforms simply as follows:

$$\bar{\phi}(\bar{x}) = \phi(x) \tag{19.2}$$

Vectors transform like dx^μ or ∂_μ. Using ordinary calculus, we can construct two types of vectors under general coordinate transformations: *covariant vectors*, like ∂_μ, and *contravariant vectors*, like dx^μ:

$$\frac{\partial}{\partial \bar{x}^\mu} = \frac{\partial x^\nu}{\partial \bar{x}^\mu} \frac{\partial}{\partial x^\nu}$$

$$d\bar{x}^\mu = \frac{\partial \bar{x}^\mu}{\partial x^\nu} dx^\nu \tag{19.3}$$

(It is important to notice that x^μ is not a genuine tensor under general coordinate transformations. Not all fields with indices $\mu, \nu, \ldots$ are genuine tensors.)

Given these transformation laws, we can now give the abstract definition of covariant tensors, with lower indices, and contravariant tensors, with upper indices, depending on their transformation properties:

$$\bar{A}_\mu(\bar{x}) = \frac{\partial x^\nu}{\partial \bar{x}^\mu} A_\nu(x)$$

$$\bar{B}^\mu(\bar{x}) = \frac{\partial \bar{x}^\mu}{\partial x^\nu} B^\nu(x) \tag{19.4}$$

Since we have arbitrary coordinate transformations, these vectors transform under $GL(4)$, that is, arbitrary real 4×4 matrices.

Similarly, we can construct tensors of arbitrary rank or indices. They transform as the product of a series of first-rank tensors (vectors):

$$\bar{A}^{\bar{\nu}_1 \bar{\nu}_2 \cdots}_{\bar{\mu}_1 \bar{\mu}_2 \cdots}(\bar{x}) = \prod_{i=1}^{m} \left(\frac{\partial x^{\mu_i}}{\partial \bar{x}^{\bar{\mu}_i}} \right) \prod_{j=1}^{n} \left(\frac{\partial \bar{x}^{\bar{\nu}_i}}{\partial x^{\nu_i}} \right) A^{\nu_1 \nu_2 \cdots}_{\mu_1 \mu_2 \cdots}(x) \tag{19.5}$$

We can also construct an invariant under general coordinate transformations by contracting contravariant tensors with covariant ones:

$$A'_\mu B'^\mu = A_\mu B^\mu = \text{invariant} \tag{19.6}$$

We now introduce a *metric tensor* $g_{\mu\nu}$ that allows us to calculate distances on our space. The infinitesimal invariant distance between two points separated by dx^μ is given by:

$$ds^2 = dx^\mu g_{\mu\nu} dx^\nu \tag{19.7}$$

If $g_{\mu\nu}$ is defined to be a second-rank covariant tensor, then this distance ds^2 is a genuine invariant.

The metric tensor transforms as:

$$\bar{g}_{\bar{\mu}\bar{\nu}}(\bar{x}) = \left(\frac{\partial x^\mu}{\partial \bar{x}^{\bar{\mu}}} \right) \left(\frac{\partial x^\nu}{\partial \bar{x}^{\bar{\nu}}} \right) g_{\mu\nu}(x) \tag{19.8}$$

From this, we can deduce how the metric tensor changes under an infinitesimal general coordinate transformation $\delta x^\mu = \xi^\mu$. By expanding the previous transformation rule, we find that the variation of the metric tensor under an infinitesimal coordinate reparametrization is given as follows:

$$\delta g_{\mu\nu} = \partial_\mu \xi^\rho g_{\rho\nu} + \partial_\nu \xi^\sigma g_{\mu\sigma} - \xi^\rho \partial_\rho g_{\mu\nu} \tag{19.9}$$

One essential point is that it is always possible to find a local coordinate system in which we can diagonalize the metric tensor, so that $g_{\mu\nu}$ becomes the usual Lorentzian metric at a point. Then the space becomes "flat" at that point. (We emphasize that it is impossible to gauge an arbitrary metric tensor so that the space is flat at all points in space.) This is the mathematical expression of Einstein's original observation, that one should always be able to jump into the "elevator" frame at any single point in space–time, such that things look locally flat.

Now that we have defined how scalar, vector, and tensor fields transform under reparametrizations, the next step is to write down derivatives of these fields that are also covariant. The derivative of a scalar field is a genuine tensor under general coordinate transformations:

$$\frac{\partial}{\partial \bar{x}^\mu} \bar{\phi}(\bar{x}) = \frac{\partial x^\nu}{\partial \bar{x}^\mu} \frac{\partial}{\partial x^\nu} \phi(x) \tag{19.10}$$

However, as in the case of gauge transformations, we find that the derivative of a vector is not a genuine tensor under general coordinate transformations. Under this transformation, the derivative can act on the factor $\partial_\mu \bar{x}^\nu$, spoiling general covariance. The solution to this problem, as we know from gauge theory, is to introduce new fields, called connections, that absorb these unwanted terms. The connection field for general relativity is called the Christoffel symbol $\Gamma^\lambda_{\mu\nu}$. We introduce the symbol ∇_μ, which is a covariant derivative:

$$\begin{aligned} \nabla_\mu A_\nu &\equiv \partial_\mu A_\nu + \Gamma^\lambda_{\mu\nu} A_\lambda \\ \nabla_\mu A^\nu &\equiv \partial_\mu A^\nu - \Gamma^\nu_{\mu\lambda} A^\lambda \end{aligned} \tag{19.11}$$

We will define the transformation properties of the connection such that the derivative of a vector becomes a genuine tensor, paralleling the situation in gauge theory:

$$(\nabla_\mu A_\nu)' \equiv \left(\frac{\partial x^\lambda}{\partial \bar{x}^\mu} \frac{\partial x^\rho}{\partial \bar{x}^\nu} \right) \nabla_\lambda A_\rho \tag{19.12}$$

Given this transformation law, we can, as in gauge theories, extract the transformation law for the Christoffel symbol:

$$\bar{\Gamma}^\lambda_{\mu\nu} = \frac{\partial \bar{x}^\lambda}{\partial x^\tau} \frac{\partial x^\rho}{\partial \bar{x}^\mu} \frac{\partial x^\sigma}{\partial \bar{x}^\nu} \Gamma^\tau_{\rho\sigma} + \frac{\partial^2 \bar{x}^\lambda}{\partial x^\rho \partial x^\sigma} \frac{\partial x^\rho}{\partial \bar{x}^\mu} \frac{\partial x^\sigma}{\partial \bar{x}^\nu} \tag{19.13}$$

We find that the Christoffel symbol is not a genuine tensor, but has an inhomogeneous piece. [We recall that the gauge field A^a_μ also has an inhomogeneous piece in its transformation under $SU(N)$.]

Covariant derivatives can be constructed for increasingly complicated tensors by simply adding more and more Christoffel symbols:

$$\nabla_\rho A^{\nu_1 \nu_2 \cdots}_{\mu_1 \mu_2 \cdots} \equiv \partial_\rho A^{\nu_1 \nu_2 \cdots}_{\mu_1 \mu_2 \cdots}$$
$$+ \sum_{\text{perm}} \Gamma^\lambda_{\rho \mu_1} A^{\nu_1 \nu_2 \cdots}_{\lambda \mu_2 \mu_3 \cdots} - \sum_{\text{perm}} \Gamma^{\nu_1}_{\rho \lambda} A^{\lambda \nu_2 \nu_3 \cdots}_{\mu_1 \mu_2 \cdots} \qquad (19.14)$$

where we sum over all permutations of the indices. Notice that ∇_μ depends on the tensor it acts on. More and more Christoffel symbols are required if it acts upon increasingly mixed tensors.

At this stage, we have placed no restrictions on the connection, other than its transformation properties. The connection field, at this point, is an independent field. We would, however, like to construct a theory in which all fields, including the connection, are written in terms of the metric tensor. We thus need a constraint on the connection. From the equivalence principle, we know that we can always choose the "elevator frame" where the the metric tensor becomes the Lorentz metric; that is, the derivative of the metric tensor vanishes in this inertial frame. The covariant generalization of this statement is that the covariant derivative of the metric tensor in any frame vanishes:

$$\nabla_\mu g_{\nu\lambda} = \partial_\mu g_{\nu\lambda} + \Gamma^\rho_{\mu\nu} g_{\lambda\rho} + \Gamma^\rho_{\mu\lambda} g_{\nu\rho} = 0 \qquad (19.15)$$

The number of independent equations in this constraint ($4 \times 10 = 40$) is exactly equal to the number of independent components of the connection if we assume that the connection is symmetric in its lower indices. Thus, we can eliminate the connection field totally in terms of the metric tensor. To do this, we first write the equations in terms of the connection with only lower indices: $\Gamma_{\mu\nu,\sigma} = g_{\sigma\lambda} \Gamma^\lambda_{\mu\nu}$.

Now let us rewrite the vanishing of the covariant derivative of the metric tensor in terms of $\Gamma_{\mu\nu,\lambda}$. Written out explicitly (and cyclically rotating the indices), we find:

$$\partial_\mu g_{\nu\lambda} + \Gamma_{\mu\nu,\lambda} + \Gamma_{\mu\lambda,\nu} = 0$$
$$\partial_\nu g_{\lambda\mu} + \Gamma_{\nu\lambda,\mu} + \Gamma_{\lambda\nu,\mu} = 0$$
$$\partial_\lambda g_{\mu\nu} + \Gamma_{\lambda\mu,\nu} + \Gamma_{\nu\mu,\lambda} = 0 \qquad (19.16)$$

These three equations are identical. But by adding the first two equations and subtracting the last (and remembering that the Christoffel symbol is symmetric in the lower indices), we then find:

$$\Gamma_{\mu\nu,\sigma} = -\frac{1}{2}(\partial_\mu g_{\nu\sigma} + \partial_\nu g_{\mu\sigma} - \partial_\sigma g_{\mu\nu}) \qquad (19.17)$$

19.2 Generally Covariant Action

Now that we have defined the transformation properties of the fields and constructed covariant derivatives, the last step is to write down the action for general
relativity and couple it to other fields. To construct the action, we will need to
take the commutator between two covariant derivatives. In flat space, this commutator vanishes. However, for general coordinate transformations, we find that
this commutator does not vanish. By explicit construction, we find:

$$
\begin{aligned}
[\nabla_\mu, \nabla_\nu] A_\lambda &= R^\rho_{\mu\nu\lambda} A_\rho \\
R^\rho_{\mu\nu\lambda} &= \partial_\mu \Gamma^\rho_{\nu\lambda} - \partial_\nu \Gamma^\rho_{\mu\lambda} - \Gamma^\rho_{\mu\sigma} \Gamma^\sigma_{\nu\lambda} + \Gamma^\rho_{\nu\sigma} \Gamma^\sigma_{\mu\lambda}
\end{aligned}
\tag{19.18}
$$

We call $R^\rho_{\mu\nu\lambda}$ the *Riemann curvature tensor*. (Alternatively, we could have derived
the curvature tensor by taking a vector A_λ and then moving it around a closed
circle using parallel transport. After completing the circuit, the vector has rotated
by the amount $R^\rho_{\mu\nu\lambda} A_\rho \Delta^{\mu\nu}$, where $\Delta^{\mu\nu}$ is the area tensor of the closed path.)
From this, we can see the close analogy between the elements of gauge theory and
general relativity. This close correspondence can be symbolically represented as
follows:

$$
\begin{aligned}
A^a_\mu &\rightarrow \Gamma^\lambda_{\mu\nu} \\
D_\mu &\rightarrow \nabla_\mu \\
F^a_{\mu\nu} &\rightarrow R^\rho_{\mu\nu\lambda}
\end{aligned}
\tag{19.19}
$$

By suitably contracting the indices in the curvature tensor, we can reduce it to
tensors of smaller rank. Contracting ρ and ν gives us a second-rank curvature
tensor:

$$
R_{\mu\lambda} = R^\rho_{\mu\nu\lambda} \delta^\nu_\rho
\tag{19.20}
$$

This is called the *Ricci curvature tensor*.

Finally, we can construct a genuine invariant by contracting all the indices:

$$
R^\rho_{\mu\nu\lambda} \delta^\nu_\rho g^{\mu\lambda} = R
\tag{19.21}
$$

Using ordinary calculus, we can also construct the transformation properties
of the volume element:

$$
d^4 \bar{x} = \det \left(\frac{\partial \bar{x}^\mu}{\partial x^\nu} \right) d^4 x
\tag{19.22}
$$

It is also easy to calculate the transformation properties of the determinant of the metric tensor g. Because $\det(ABC) = \det A \det B \det C$, we can easily show:

$$\sqrt{-\bar{g}(\bar{x})} = \det\left(\frac{\partial x^{\mu}}{\partial \bar{x}^{\nu}}\right)\sqrt{-g(x)} \tag{19.23}$$

An object that transforms like this is not a scalar in the usual sense. We call it a scalar density.

The point is that now the product of these two is a genuine invariant:

$$\sqrt{-g}\, d^4x = \text{invariant} \tag{19.24}$$

From this we can construct actions.

In order to write down an action, we wish to fulfill a few key conditions:

1. The action must contain no more than two derivatives, or else there are ghosts in the theory that threaten unitarity.

2. The action must be invariant under general coordinate transformations.

Surprisingly, we find that there are only *two* solutions to these constraints, given by:

$$S = -\frac{1}{2\kappa^2}\int d^4x\sqrt{-g}\,R \tag{19.25}$$

(We can also add the cosmological term, which is proportional to $\Lambda\sqrt{-g}$, although experimentally Λ is very close to zero.) This is the celebrated Einstein–Hilbert action, which is the starting point for all calculations in general relativity.

We can also calculate the equations of motion from this action. By making a small variation in the metric $\delta g_{\mu\nu}$, we can compute:

$$\delta g = gg^{\mu\nu}\delta g_{\mu\nu}$$
$$\delta\sqrt{-g} = -\frac{1}{2}\sqrt{-g}\,g_{\mu\nu}\delta g^{\mu\nu}$$
$$\delta R_{\mu\nu} = \nabla_{\nu}\delta\Gamma^{\rho}_{\mu\rho} - \nabla_{\rho}\delta\Gamma^{\rho}_{\mu\nu} \tag{19.26}$$

Taking the variation of the action, we then find the equations of motion:

$$R_{\mu\nu} - \frac{1}{2}g_{\mu\nu}R = 0 \tag{19.27}$$

(The term $\delta R_{\mu\nu}$ does not contribute because it turns into a total derivative.)

In the presence of matter fields, we must alter this equation. We know that scalar matter couples to gravity via the interaction $\frac{1}{2}\sqrt{-g}g^{\mu\nu}\partial_\mu\phi\partial_\nu\phi \sim g^{\mu\nu}T_{\mu\nu}$; therefore, the right-hand side of the previous equation should contain the energy–momentum tensor.

One should mention that this equation reduces to the usual Newtonian potential equation in the limit that $c \to \infty$. In this limit, the metric tensor becomes the Lorentz metric, except for the term g_{00}:

$$g_{00} - 1 \sim \phi \tag{19.28}$$

Then the ϕ field becomes the scalar potential, and Einstein's equation reduces to Poisson's equation:

$$R_{\mu\nu} - \frac{1}{2}g_{\mu\nu}R = -\frac{8\pi\kappa}{c^2}T_{\mu\nu} \to \nabla^2\phi = 4\pi\kappa\rho \tag{19.29}$$

where ρ is the source term. From this, one can derive Newton's original universal law of gravitation, that the gravitational force is proportional to the product of the masses and inversely proportional to the distance of separation squared.

19.3   Vierbeins and Spinors in General Relativity

The coupling of the gravitational field to other fields is also straightforward. The generally covariant action for scalar and Yang–Mills fields is given by:

$$
\begin{aligned}
\mathscr{L} &= \frac{1}{2}\sqrt{-g}\left(g^{\mu\nu}\partial_\mu\phi\partial_\nu\phi - m^2\phi^2\right) \\
\mathscr{L} &= -\frac{1}{4}\sqrt{-g}g^{\mu\sigma}g^{\nu\rho}F_{\mu\nu}^a F_{\sigma\rho}^a
\end{aligned}
\tag{19.30}
$$

However, the coupling of gravity to spinor fields leads to an immediate difficulty: *There are no finite dimensional spinorial representations of GL(4).* This prevents a naive incorporation of spinors into general relativity. There is, fortunately, a trick that we may use to circumvent this problem. Although spinor representations do not exist for general covariance, there are, of course, spinorial representations of the Lorentz group. We utilize this fact and construct a flat *tangent space* at every point in the space. Imagine space–time as a rolling hill. Then the tangent space would correspond to placing a flat plane on each point of the hill. Spinors can then be defined at any point on the curved manifold only if they transform within the flat tangent space.

We will label the flat tangent space indices with Latin letters $a, b, c, \ldots$, while tensors under general coordinate transformations are labeled by Greek letters: $\mu, \nu, \rho, \ldots$.

In order to marry the two sets of indices, we will introduce the *vierbein* or *tetrad*, which is a mixed tensor:

$$\text{Vierbein}: \quad e_\mu^a(x) \qquad (19.31)$$

The inverse of this matrix is given by e_a^μ.

The vierbein can be viewed as the "square root" of the metric tensor via the following:

$$
\begin{aligned}
e_\mu^a e_\nu^a &= g_{\mu\nu} \\
e^{a\mu} &= g^{\mu\nu} e_\nu^a \\
e_\mu^a e^{b\mu} &= \delta^{ab}
\end{aligned}
\qquad (19.32)
$$

Since the Lorentz group acts on the tangent space indices, we can define spinors on the tangent space. The Dirac matrices γ^a can now be contracted onto vierbeins:

$$\gamma^a e^{a\mu} = \gamma^\mu(x) \qquad (19.33)$$

It is easy to show that the commutator between two of these matrices yields the metric tensor:

$$\{\gamma^\mu, \gamma^\nu\} = 2g^{\mu\nu}(x) \qquad (19.34)$$

Our goal is to construct the generally covariant Dirac equation. We introduce a spinor $\psi(x)$ that is defined to be a *scalar* under general coordinate transformations (and an ordinary spinor under flat tangent space Lorentz transformations):

$$
\begin{aligned}
\text{Coordinate transformations}: \quad &\psi \;\rightarrow\; \psi \\
\text{Lorentz transformations}: \quad &\psi \;\rightarrow\; e^{i\epsilon^{ab}(x)\sigma_{ab}} \psi
\end{aligned}
\qquad (19.35)
$$

It is important to note that we have introduced local Lorentz transformations on the flat tangent space, so ϵ_{ab} is a function of space–time.

This, of course, means that the derivative of a spinor is no longer a genuine tensor. As before, we must introduce a connection field ω_μ^{ab} that allows us to *gauge the Lorentz group*. The covariant derivative for gauging the Lorentz group is therefore:

$$\nabla_\mu \psi = (\partial_\mu - \frac{i}{4} \omega_\mu^{ab} \sigma_{ab}) \psi \qquad (19.36)$$

The generally covariant Dirac equation is therefore given by:

$$(i\gamma^\mu \nabla_\mu - m)\psi = 0 \qquad (19.37)$$

and hence the action for a Dirac particle interacting with gravity is given by:

$$\mathcal{L} = -\frac{1}{2\kappa^2} \sqrt{-g} R + e\bar{\psi}(i\gamma^\mu \nabla_\mu - m)\psi \qquad (19.38)$$

where $e \equiv \det e_\mu^a = \sqrt{-g}$.

This new connection field gives us an alternative way to construct the Riemann curvature tensor. By taking the commutator of two covariant derivatives, we can construct a new version of the curvature tensor:

$$[\nabla_\mu, \nabla_\nu]\psi = -\frac{i}{4} R_{\mu\nu}^{ab} \sigma^{ab} \psi \qquad (19.39)$$

Written out, this curvature tensor is generally covariant in μ, ν, but flat in a, b:

$$R_{\mu\nu}^{ab} = \partial_\mu \omega_\nu^{ab} - \partial_\nu \omega_\mu^{ab} + \omega_\mu^{ac} \omega_\nu^{cb} - \omega_\nu^{ac} \omega_\mu^{cb} \qquad (19.40)$$

At this point, the connection field ω_μ^{ab} is still an independent field. We can eliminate it in favor of the vierbein by placing an external constraint on the theory:

$$\nabla_\mu e_\nu^a = \partial_\mu e_\nu^a + \Gamma_{\mu\nu}^\lambda e_\lambda^a + \omega_\mu^{ab} e_\nu^b = 0 \qquad (19.41)$$

Again, the number of independent equations in the constraint ($4 \times 6 = 24$) equals the number of independent components of the connection field, so we have eliminated the connection field entirely as an independent field.

The connection field can be calculated in much the same way as the $\Gamma_{\mu\nu}^\rho$ was calculated, by rotating the various indices and then adding and subtracting them. The result is:

$$\omega_\mu^{ab} = \frac{1}{2} e^{av}(\partial_\mu e_\nu^b - \partial_\nu e_\mu^b) + \frac{1}{4} e^{a\rho} e^{b\sigma} (\partial_\sigma e_\rho^c - \partial_\rho e_\sigma^c) e_\mu^c - (a \leftrightarrow b) \qquad (19.42)$$

19.4 GUTs and Cosmology

With this elementary introduction, we can now make qualitative statements concerning the impact of GUT theory on cosmology. Any study of the origins of the universe, of course, must be prefaced with a clear statement of assumptions and prejudices, since the origin of the universe is not a reproducible event and cannot be duplicated in the laboratory.

However, general relativity has given us a theoretical and experimental framework in which to explain a large body of observational information. The scenario emerging from general relativity, that the universe started with a cataclysmic explosion 10-20 billion years ago, is supported by three strong pieces of information:

1. *Red shift.* The far-away stars and galaxies are receding from us, as measured by the Doppler shift. We do not see a blue shift in the heavens. General relativity is in agreement with Hubble's law,[2] which states that the farther away a galaxy is, the faster it is moving away from us. Experimentally, this linear relationship between distance and velocity is summarized in Hubble's constant, measured to be $H \sim 15$ km/sec per mega light year.

2. *Nucleosynthesis.* The theory predicts that about a quarter of all hydrogen in the heavens should have fused into helium by the Big Bang. It also correctly predicts the abundance of many other elements.

3. 3^0 *Background radiation.* The "echo" from the Big Bang, as predicted by Gamow,[3,4] should behave like blackbody radiation and should now have cooled down to the microwave range. The observed temperature of the background microwave radiation, measured by satellite to be $2.736 \pm 0.01°$, fits well with Gamow's original prediction.

More specifically, the Big Bang can be viewed as a solution to Einstein's equations in the presence of matter.

Let us assume an *ansatz* for the metric tensor that solves Einstein's equations, for example, the Robertson–Walker metric. We assume that the metric tensor is radially symmetric, with all angular dependence omitted, and that the time dependence of the metric is represented by a single function $R(t)$, which sets the scale of the universe and acts as an effective "radius" of the universe. We assume the ansatz:

$$
\begin{aligned}
ds^2 &= dx^\mu \, g_{\mu\nu} \, dx^\nu \\
&= dt^2 - R^2(t) \left(\frac{dr^2}{1 - kr^2} + r^2 \, d\Omega^2 \right)
\end{aligned}
\tag{19.43}
$$

where $d\Omega^2$ is the usual solid angle differential, and k is a constant. Now let us assume a highly idealized model of the universe, consisting of a fluid of galactic clusters, with an average density $\rho(t)$ and average internal pressure $p(t)$. In this idealized frame, the energy–momentum tensor becomes: $T_0^0 = \rho$, $T_i^i = -p$, and all other components are zero.

Because all angular dependence has been explicitly eliminated, we find that, after inserting this ansatz into Einstein's equations, these equations collapse into only two two equations for $R(t)$:

$$\left(\frac{\dot{R}}{R}\right)^2 = \frac{8\pi}{3}G_N\rho - \frac{k}{R^2}$$

$$\frac{\ddot{R}}{R} = -4\pi G_N\left(p + \frac{\rho}{3}\right) + \frac{\Lambda}{3} \tag{19.44}$$

We can always rescale R so that k is +1 (closed universe), 0 (flat universe), or -1 (open universe).

These two relativistic equations actually have a simplified Newtonian interpretation. Imagine a point particle at the surface of a sphere of radius R. The kinetic energy of the particle is $\frac{1}{2}\dot{R}^2$. Its potential energy is $G(M/R) = (4\pi R^3/3)\rho(G/R)$. Then conservation of energy states:

$$\frac{d}{dt}\left[\frac{1}{2}\dot{R}^2 - \left(\frac{4\pi R^3}{3}\rho\frac{G}{R}\right)\right] = 0 \tag{19.45}$$

which yields our first relativistic equation.

The second relativistic equation can also be seen as the conservation of energy. Imagine a sphere of radius R filled with a fluid, such that the conservation of energy yields $dU = -p\,dV$ for the kinetic theory of gases. Then this becomes:

$$\frac{d}{dt}\left(\frac{4\pi R^3\rho}{3}\right) = -p\frac{d}{dt}\left(\frac{4\pi R^3}{3}\right) \tag{19.46}$$

If we assume the cosmological constant is zero, $\Lambda = 0$, we can eliminate ρ and obtain one equation:

$$2R\ddot{R} + \dot{R}^2 + k = 0 \tag{19.47}$$

We assume that $k = 0$. For sufficiently large times, we find that the radius of the universe expands in time as a power law:

$$R = \left(\frac{9}{2GM}\right)^{1/3}t^{2/3} \tag{19.48}$$

This power law describes the *expanding Friedman universe*,[5,6] and was one of the first cosmological models to be found from Einstein's equations. This model, in turn, can explain the three experimental features of the Big Bang.

These general features, however, do not go far enough in terms of explaining precisely how the universe cooled down since the initial Big Bang. The general consensus is that the theory of elementary particles will ultimately play a decisive role in this respect. From the point of view of GUT theory, the Big Bang can be studied via a series of stages in the cooling of the universe. The boundary between each stage corresponds to the energy scale at which spontaneous symmetry breaking occurred.

A rough sequence of events, supported by the general features of any GUT theory, is as follows:

1. 10^{-44} sec. At the Planck energy 10^{19} GeV, all the symmetries of gauge theory were supposedly united into a single force. Gravitational effects were strong and, in fact, were unified with the GUT forces.

2. 10^{-36} sec. At the energy scale $M_X = 10^{15}$ GeV, the GUT gauge group broke apart into $SU(3) \otimes SU(2) \otimes U(1)$ of the Standard Model.

3. 10^{-10} sec. At 10^2 GeV, the electro-weak symmetry $SU(2) \otimes U(1)$ broke down into $U(1)_{em}$.

4. 10^{-6} sec. At 1 GeV, the quarks bound together to form hadrons. Shortly thereafter, nuclei slowly began to condense without being torn apart.

5. 10^{12} sec. At 10^{-9} GeV, atoms condensed without being ionized. Photons could now move through space without being easily absorbed, so space became black. Before this, space was full of ionized plasma and hence was opaque to light.

6. 10^{16} sec. Galaxies began to condense about 1 billion years after the Big Bang.

7. 10^{17} sec. The present day era, about 10 to 20 billion years after the Big Bang.

Given this rough sequence of events for the beginning of the universe, we can begin to ask what implications this has for the GUT theories. We find that the GUT theories give us a clue to the solution to two long-standing cosmological problems: the matter–antimatter asymmetry in the universe, and the flatness–horizon problems.

It is a fairly well established fact that our visible universe is composed primarily of matter, rather than antimatter. Although one may speculate that, at the Big Bang, there were equal quantities of matter and antimatter present, we find that our universe is quite asymmetric. In fact, a rough estimate of the baryon–antibaryon asymmetry is that the number N_B of baryons dominates over the number of $N_{\bar{B}}$ of antibaryons by a factor of 10^{-9}; that is:

$$\delta = \frac{N_B - N_{\bar{B}}}{N_B + N_{\bar{B}}} \sim 10^{-9} \tag{19.49}$$

(In fact, this small asymmetry between matter and antimatter is probably the reason we exist in the first place to ponder this question.)

Unfortunately, the Standard Model gives us no clue as to why this asymmetry between matter and antimatter exists. In the Standard Model, we must impose an initial asymmetry at $t = 0$. However, even if we put C and CP violating terms in the Standard Model at the origin of time, the CPT theorem can wash out baryon asymmetry. This is because, at equilibrium, baryons and antibaryons will have the same Boltzmann distribution because, by the CPT theorem, they must have the same mass.

Thus, in order to explain baryon asymmetry, we must have two features:

1. Breaking of C and CP symmetry[7,8] and baryon number at the origin of time.

2. A cosmological phase when these C and CP violating processes were out of equilibrium.

Fortunately, GUT theories can accomodate both these desirable features. The first criterion can be satisfied by GUT theory in a number of ways. The second criterion is also satisfied if we analyze the cooling of the early universe.

Assume that, at GUT times, there was an X particle that decayed into quarks and leptons and violated these symmetries. For very high temperatures, on the order of $kT > M_X$, the X particle existed in thermal equilibrium with other particles, and the decay of this particle could create a net baryon asymmetry. Normally, such a baryon asymmetry is cancelled by the inverse decays of the particles at equilibrium, so the net baryon asymmetry does not survive. However, as the temperature of the universe decreased and $kT < M_X$, one can show, by examining decay rates and cross sections, that the X was no longer in thermal equilibrium, and any net baryon asymmetry was frozen permanently. The population of X particles and the number of inverse decays was suppressed by the Boltzmann factor:

$$\exp\left(-M_X/kT\right) \tag{19.50}$$

To be more specific, consider the following reaction rates:

$$\begin{aligned}
\gamma_a &= \Gamma(X \to l^c q^c); \quad \gamma_b = \Gamma(X \to qq) \\
\bar{\gamma}_a &= \Gamma(X \to lq); \quad \bar{\gamma}_b = \Gamma(X \to q^c q^c)
\end{aligned} \tag{19.51}$$

where the superscript c refers to the charge conjugated particle. The CPT theorem demands that these reaction rates obey the following relations:

$$\gamma_a + \gamma_b = \bar{\gamma}_a + \bar{\gamma}_b \tag{19.52}$$

Furthermore, at the Born level, CPT enforces the following conditions:

$$\gamma_a = \bar{\gamma}_a; \quad \gamma_b = \bar{\gamma}_b \tag{19.53}$$

However, beyond the Born term, higher-order interactions can destroy this relation. The presence of C and CP violating higher-order processes can produce the following relations:

$$\gamma_a - \bar{\gamma}_a = \bar{\gamma}_b - \gamma_b \neq 0 \tag{19.54}$$

without violating the CPT theorem.

For example, in the minimal GUT theory, the first baryon asymmetric term enters in at the 10th level in perturbation theory. (Unfortunately, this is many orders of magnitude too small to explain the observed 10^{-9} asymmetry. More complicated GUT theories, however, can obtain the observed asymmetry.)

19.5 Inflation

There are two puzzles that, within the framework of classical general relativity, cannot be solved: the *flatness problem* and the *horizon problem*. A plausible explanation for both, however, can be given if we add the effects of gauge theories to general relativity.

The flatness problem arises because the universe appears much flatter than it should be. We know that there is a critical density ρ_c, such that if $\rho < \rho_c$, the gravitational pull of the matter in the universe is too weak to reverse the expansion, and the universe expands forever. For $\rho > \rho_c$, the gravitational pull is strong enough to force the expansion to stop and eventually reverse itself. However, the density of the universe today seems to be fairly close to $\rho \sim \rho_c = 3H^2/8\pi G \sim 5 \times 10^{-30}$ g/cm^3. If we define:

$$\Omega = \frac{\rho}{\rho_c} \tag{19.55}$$

then we find that $\Omega \sim 0.1 - 10$.

Now assume that we extrapolate Ω backwards in time, so that we compute Ω near the beginning of the universe. Ω rapidly becomes close to one as we go back in time, meaning that Ω was fine tuned in the early universe. For example, if we extrapolate back to the GUT universe, we find:

$$\Omega = 1 \pm O(10^{-55}) \text{ at } T = 10^{15} \text{ GeV} \qquad (19.56)$$

This means that, near the beginning of time, Ω was fine tuned to be 1 to one part in 10^{55}, which is difficult to believe.

The horizon problem has a similar origin. In general relativity, the horizon refers to the farthest distance that we can see. If we look in distant parts of the heavens, we find that the universe is quite isotropic. In fact, the universe seems to be much more isotropic than it should be. In particular, the background 2.7^0 radiation appears to be very uniform, no matter where we look in outer space. But this is difficult to understand. For distant parts of the universe to be isotropic, they had to be in causal contact with each other in the distant past. Because of the limitation imposed by the speed of light, one can show that distant parts of the visible universe could not be in causal contact with each other. Hence, the universe should not be so isotropic.

Although the classical theory of general relativity has difficulty explaining the flatness and horizon problem, one byproduct of gauge theory, Guth's *inflationary universe*,[9-11] has a plausible explanation for both.

Whenever we have spontaneous symmetry breaking in the Higgs sector coupled to gravity, we generate a constant term, which corresponds to increasing the energy density of the vacuum. Normally, we throw this away. However in general relativity, this constant is multiplied by $\sqrt{-g}$, so that it contributes to the cosmological constant.

If we have a large cosmological constant Λ in the Einstein equations, then we must use what is called de Sitter's solution. Like the standard Big Bang solution, the de Sitter solution is found by assuming spherical symmetry; so the metric is a function of the radius and time. Then Einstein's equations reduce to a simple equation that can be solved with an *exponential* expansion, rather than a standard power law expansion. The de Sitter solution, with the cosmological constant, therefore yields an exponential expansion rather than a power law expansion:

$$R(t) \sim e^{ct} \qquad (19.57)$$

where:

$$c = \left(\frac{8\pi\rho}{3M_P^2} \right)^{1/2} \sim \frac{T_c^2}{M_P} \qquad (19.58)$$

where T_c is the critical temperature at which inflation begins. This exponential expansion, which naturally emerges whenever a symmetry is broken spontaneously, might be large enough to solve the flatness and horizon problem if it were on the order of 10^{30}.

The flatness problem may be solved because the visible universe that we can observe is only a tiny fraction of the total universe. Thus, our universe appears to be flat only because the radius of the universe is so large.

The horizon problem may be solved because our present-day universe, extrapolated back in time, was only a tiny speck in the original primordial nucleus within which points were in thermal equilibrium. Thus, it is not surprising that distant points in today's universe can have the same uniform temperature. Near the beginning of time, our universe was small enough so that all points could be in causal contact with other points.

As attractive as the inflation theory is, only detailed experimental observations, for example, of the radiation left over from the early universe, will ultimately determine whether the inflation theory holds up with time. (There are, of course, problems with inflation. There is no unique way to introduce the potential necessary to yield an expansion of 10^{30}. There are several alternatives, but we often wind up reintroducing some form of fine-tuning back into the problem, which is undesirable.)

19.6 Cosmological Constant Problem

Although a naive application of GUT theory to cosmology seems to generate experimentally reasonable results, we should mention a serious problems with this (as well as any other) approach. This is the celebrated *cosmological constant problem*. Experimentally, we can measure the possible presence of the cosmological constant Λ by measuring exponential deviations from the standard $R(t) \sim t^{2/3}$ expansion. Experimentally, we find that it is consistent with zero to a remarkable degree:

$$\Lambda < 10^{-120} M_P^2 = 10^{-84} \text{ GeV}^2 \qquad (19.59)$$

However, every time we break a symmetry spontaneously, we generate a vacuum energy proportional to:

$$\Lambda_{\text{GUT}} \rightarrow \left(\frac{8\pi}{3} G_N V(\phi_i) \right)_{\phi_i = \langle \phi_i \rangle} \qquad (19.60)$$

Putting in the value of the $SU(5)$ potential minimum with order M_X^4, we find that Λ_{GUT} is 10^{100} times too big.

In all of physics, nowhere do we find a greater divergence between theory and experiment than in the cosmological constant problem. The addition of new symmetries (such as supersymmetry, which we discuss in the next chapter) can reduce this discrepancy, but only down to about 10^{50}.

The problem, at present, seems intractable. Even if we could somehow put $\Lambda \equiv 0$ at early times (which is one byproduct of supersymmetry), we still have new contributions to the cosmological constant when we break supersymmetry and approach present-day energies. These, too, must be set to zero by a mechanism that is yet unknown.

19.7 Kaluza–Klein Theory

Perhaps the most theoretically clumsy feature of GUT theory is that general relativity is spliced onto the theory by brute force. Ideally, we would like to see gravitational interactions and GUT theory emerge from a higher unified field theory from geometrical or group theoretical arguments, rather than being put in by hand. The search for a more sophisticated theory embracing both gauge theory and general relativity has led to a re-examination of the old theory of Kaluza–Klein, which is perhaps one of the most ingenious extensions of the theory of gravity. Kaluza[12] originally proposed uniting both Maxwell's theory of electromagnetism and Einstein's theory of general relativity by embedding both theories into a generally covariant *five-dimensional* space–time. When first proposed in 1919, the theory lacked an answer to the question: what happened to the fifth dimension? Seventy years later, we are still grappling with this question.

Kaluza assumed that the fifth dimension was curled up into a tiny ring so small that it could not be experimentally observed by any instrument. Thus, although space–time may actually be five dimensional, experiments designed to determine the size of the fifth dimension would be too crude to detect this. Klein[13] then assumed that quantum corrections caused the fifth dimension to curl up. In quantum gravity, there is only one dimensionful parameter, which is the Planck length, or 10^{-33} cm. Since this sets the scale for quantum gravity, it means that the fifth dimension might have curled up with approximately this radius, which is too small for any instrument to detect.

Since the fifth spatial dimension is periodic, if we move in this direction, eventually one returns to the starting point. The fifth dimension has the topology of a circle:

$$\phi(x_5) = \phi(x_5 + 2\pi r) \tag{19.61}$$

where r is the radius of the fifth dimension. If we expand the field $\phi(x)$ in this periodic space as:

$$\phi(x) = \sum_n \phi_n e^{ipx} \qquad (19.62)$$

we find that $p = n/r$ and the momentum conjugate is quantized in terms of the integer n. These higher modes ϕ_n correspond to particles of mass 10^{19} GeV. To analyze the low energy limit of the theory, we can safely ignore these higher mass particles and take only the $n = 0$ mode of the power expansion. This means that $\phi(x)$, in this approximation, loses all dependence on the fifth coordinate:

$$\partial_5 \phi(x) \sim 0 \qquad (19.63)$$

This, in turn, allows us to decompose five-dimensional general relativity into its four-dimensional fields.

Let $A, B, C, \ldots$ represent five-dimensional space–time indices. Let us define a new field, called $A_\mu = g_{5\mu}$. The metric tensor now decomposes as follows:

$$g_{AB} = \begin{pmatrix} g_{\mu\nu} + \kappa^2 A_\mu A_\nu & \kappa A_\nu \\ \kappa A_\mu & \phi \end{pmatrix} \qquad (19.64)$$

Einstein's action in five-dimensional space, with the four-dimensional fields separated out, now reads:

$$\sqrt{-\det g_{AB}} \, g^{AB} R_{AB} = \sqrt{-\det g_{\mu\nu}} \left(g^{\mu\nu} R_{\mu\nu} - \frac{1}{4} F_{\mu\nu} F_{\rho\sigma} g^{\mu\rho} g^{\nu\sigma} \right) + \cdots \quad (19.65)$$

We have decomposed a five-dimensional theory into a four-dimensional theory, yielding the usual Maxwell theory coupled to general relativity.

We can also see how Maxwell's equations emerges by analyzing the gauge symmetry. The metric tensor $g_{\mu 5} = A_\mu$ transforms as follows:

$$\delta g_{\mu 5} = \delta A_\mu = \partial_5 \xi_\mu + \partial_\mu \xi_5 \sim \partial_\mu \xi_5 \qquad (19.66)$$

By taking the fifth coordinate sufficiently small, we retrieve the gauge variation of the Maxwell field: $\delta A_\mu = \partial_\mu \Lambda$.

Since the Maxwell field emerges as a byproduct of dimensional reduction, one should be able to derive a relationship between the electric charge, Newton's constant, and the radius of the fifth coordinate. Consider, for example, the Dirac equation in a gravitational and electromagnetic field:

$$\mathcal{L} = \sqrt{-g} \, \bar{\psi} i \gamma^\mu \left(\partial_\mu + i e A_\mu \right) \psi \qquad (19.67)$$

The coupling of the fermion to the vector potential is given by:

$$ie A_\mu \bar{\psi} \gamma^\mu \psi \tag{19.68}$$

Now perform dimensional reduction on the fifth coordinate, using the fact that $\partial_5 \sim 1/r$, where r is the radius of the fifth coordinate. After dimensional reduction, we have:

$$\frac{\kappa}{r} \bar{\psi} \gamma_\mu \psi A^\mu \tag{19.69}$$

Equating the coefficients, we then have:

$$e \sim \kappa/r \tag{19.70}$$

For the electric charge to be $\sim 1/137$, this means that r is a bit larger than the Planck length.

19.8   Generalization to Yang–Mills Theory

The Kaluza–Klein method has a straightforward generalization to Yang–Mills theory. In fact, its first published announcement came as a homework problem in 1963 at the Les Houches Summer School.[14,15]

We now work in $(4 + N)$-dimensional space, which is decomposed as the product of flat Minkowski space M_4 and another N-dimensional manifold G. We will thus work with the space $M_4 \otimes G$. We use A, B, C indices to represent this larger space; μ, ν to represent the four-dimensional space; m, n to represent the N-dimensional space; and a, b to represent the adjoint representation of a gauge group.

To distinguish the metric tensors in various spaces, we will define γ_{AB} to be the metric tensor in the larger $4 + N$ space. Let μ, ν be the indices describing four-dimensional space and let $g_{\mu\nu}$ to be the metric tensor for this dimensional space. Correspondingly, let m, n be the indices describing the N-dimensional space, with metric γ_{mn}. Let x parametrize four-dimensional space, and let y parametrize N-dimensional space.

To isolate the Yang–Mills field, let us now reparametrize the metric tensor. There are many ways to do this, but a convenient choice involves introducing a new field B_μ^n, which is a mixed tensor. We will choose the following:

$$\gamma_{AB} = \begin{pmatrix} g_{\mu\nu} + \gamma_{mn} B_\mu^m B_\nu^m & \gamma_{mn} B_\mu^n \\ B_\nu^n \gamma_{nm} & \gamma_{mn} \end{pmatrix} \tag{19.71}$$

where $g_{\mu\nu}$ is only a function of x, and γ_{nm} is only a function of y.

By a direct calculation, we can show that the inverse metric is given by:

$$
\gamma^{AB} = \begin{pmatrix} g^{\mu\nu} & -B^m_\lambda g^{\lambda\mu} \\ -B^n_\lambda g^{\lambda\nu} & \gamma^{mn} + B^m_\lambda B^n_\sigma g^{\lambda\sigma} \end{pmatrix}
\tag{19.72}
$$

Our task is to now insert the value of the metric, parametrized in this way, into the Riemann curvature tensor defined over the larger $(4 + N)$-dimensional space. The calculation is straightforward, yielding:

$$
\begin{aligned}
\sqrt{\det g_{AB}} \; R_{AB} g^{AB} = {} & \sqrt{\det g_{\mu\nu} \det \gamma_{mn}} \left[R_4(x) + R_N(y) \right. \\
& \left. + \frac{1}{4} \gamma_{mn}(y) \tilde{F}^m_{\mu\nu}(x) \tilde{F}^n_{\lambda\rho}(x) g^{\mu\lambda}(x) g^{\nu\rho}(x) + \cdots \right]
\end{aligned}
\tag{19.73}
$$

where $R_4(x)$ and $R_N(y)$ are the respective four- and N-dimensional curvature scalars, but $\tilde{F}^n_{\mu\nu}$ is *not* the usual Yang–Mills field tensor. Instead, it equals:

$$
\tilde{F}^m_{\mu\nu} = \partial_\mu B^m_\nu - B^n_\nu \partial_n B^m_\mu - (\mu \leftrightarrow \nu)
\tag{19.74}
$$

Clearly, this is not the Yang–Mills tensor, and therefore B^m_μ cannot be the Yang–Mills field. Notice also that the structure constant of the gauge group f^a_{bc} appears nowhere in our discussion, so we are missing some essential element.

At this point, we must make one more assumption that is not so obvious at first. We will make the assumption that the manifold has a symmetry associated with it; that is, we say that the manifold has an "isometry." On manifolds with isometry, we can also extract a "Killing vector" ζ_μ that mathematically expresses the effect of this isometry.

For example, if a manifold loses all its dependence on the kth coordinate, then it has a symmetry that is mathematically expressed as $\partial_k g_{\mu\nu} = 0$. The generator of this symmetry is labeled by $L_\mu = \delta^\mu_k \partial_\mu = \partial_k$. The Killing vector is then defined as $\zeta^\mu = \delta^\mu_k$. Covariantly, it satisfies the equation:

$$
\nabla_\mu \zeta_\nu + \nabla_\nu \zeta_\mu = 0
\tag{19.75}
$$

which is sometimes taken to be the definition of a Killing vector.

One example of a manifold with a isometry and a Killing vector is a two-dimensional torus. Its isometry is the set of rotations in the azimuthal angle ϕ about its vertical axis. Its Killing vector is ∂_ϕ. Another example is the two-dimensional sphere S_2. Its isometries consist of rotations in three dimensions about its center. The set of motions generated by these rotations is, of course, the Lie group $SO(3)$.

If we have an arbitrary manifold G with a set of isometries associated with it, then these isometries will in general generate a Lie algebra associated with these symmetries. Let us say that the generators of this symmetry are described by:

$$L_a \equiv \zeta_a^m \partial_m \qquad (19.76)$$

such that they, by definition, generate a Lie algebra:

$$[L_a, L_b] = f_{bc}^a L_c \qquad (19.77)$$

where f_{bc}^a is the structure constant of a Lie algebra. Inserting the value of L_a into this equation, then we find:

$$\zeta_a^m \partial_m \zeta_b^n - \zeta_b^m \partial_m \zeta_a^n = f_{bc}^a \zeta_a^n \qquad (19.78)$$

With this Killing vector, we can now define:

$$B_\mu^m = \zeta_a^m A_\mu^a \qquad (19.79)$$

This is the redefinition we were seeking, where A_μ^a is the true Yang–Mills vector. Inserting this back into the $\tilde{F}_{\mu\nu}^m$ tensor, we get:

$$\tilde{F}_{\mu\nu}^m = \zeta_a^m F_{\mu\nu}^a + \cdots \qquad (19.80)$$

where $F_{\mu\nu}^a$ is the true Yang–Mills tensor, and the higher dimensional action contains the Yang–Mills action. It is now straightforward to show that the original action in $(4 + N)$-dimensional space splits up into two parts, the usual four-dimensional theory of Einstein and the standard Yang–Mills theory.

Now let us return to the expression we previously derived for the dimensional reduction of R_{AB}. The key idea is that now we can perform the integration over y, yielding:

$$\int d^N y \sqrt{\det \gamma_{mn}(y)} \; \gamma_{mn}(y) \zeta_a^n(y) \zeta_b^m(y) \to \Omega_N \delta_{ab} \qquad (19.81)$$

where Ω_N is the volume over the y space. This is the last step in the construction of the Yang–Mills action from the Einstein–Hilbert action in $4 + N$ dimensions.

The lesson learned from this exercise is that we cannot simply take a $(4 + N)$-dimensional manifold $M_4 \otimes G$ and expect the Yang–Mills theory to emerge. The extra assumption that we need is that the manifold G has a set of symmetries associated with it which generate a Lie algebra. Then the Yang–Mills field emerges as a function of the Killing fields of the isometries.

All this, of course, is formal, but let us see whether any possible phenomenology is possible with Kaluza–Klein Yang–Mills theories. Several questions come to mind immediately:

1. Can the Standard Model gauge group be included in this scenario?

2. Can complex representations of fermions be included?

3. Is the theory renormalizable?

4. Why should higher-dimensional space compactify?

5. What about the cosmological constant?

To answer the first, we will use the fact that the Yang–Mills theory is generated by isometries of the space–time manifold. Our task is to find the manifold that has the group of the Standard Model as its isometry group.[16]

The isometry of the circle S_1 is easy to find; it is represented by a simple rotation about its axis, which can be obtained via $SO(2)$ or $U(1)$. The isometries of the ordinary sphere S_2 can be obtained via rotations, labeled by $SO(3)$ or $SU(2)$. In general, the isometry group of S_n is given by $SO(n + 1)$. This is easy to see, because the defining equation of S_n is given by:

$$\sum_{i=1}^{n+1} x_i^2 = 1 \tag{19.82}$$

which is invariant under $SO(n + 1)$ rotations on x_i.

Likewise, the isometry group $SU(3)$ can be obtained via CP_2. [CP_n is the complex space spanned by $n + 1$ complex coordinates z_i, such that the point $\{z_1, z_2, \cdots, z_n\}$ is identified with the point $\{\lambda z_1, \lambda z_2, \cdots, \lambda z_n\}$ for nonzero complex λ. Notice that this definition is invariant under $SU(n + 1)$ rotations.]

We are therefore interested in the isometries of the $4 + 2 + 1 = 7$ dimensional manifold:

$$CP_2 \otimes S_2 \otimes S_1 \tag{19.83}$$

Thus, $4 + 7 = 11$ is the minimal number of total dimensions that we must have in order to have a Standard Model gauge group.

Now that we have successfully shown that a class of seven-dimensional manifolds exists that can reproduce the isometry group of the Standard Model, our next step is to ask whether this formalism can reproduce the complex fermions of the Standard Model.

Here, however, our formalism fails. There are powerful mathematical theorems that, in fact, forbid complex representations of fermions in this approach. This is disappointing, because it means that the Kaluza–Klein approach is not rich enough to support the fermionic representations of the Standard Model.

To see this, we first study the Dirac operator defined on a $(4 + N)$-dimensional product manifold, which splits up into two pieces:

$$i\Gamma^A D_A = i\gamma^\mu D_\mu(x) + i\gamma^m D_m(y) \tag{19.84}$$

where each covariant derivative depends crucially on the structure of the manifold (via the vierbein and the connection). In general, we are looking at the eigenvalues of the Dirac operator. If the Dirac operator on the B manifold has eigenvalue m, then we have:

$$i\Gamma^A D_A = i\gamma^\mu D_\mu(x) + m \tag{19.85}$$

However, the mass m is of the order of the Planck mass, which is much larger than the experimentally observed lepton and quark masses. Therefore, we must set $m = 0$, meaning that we must look at the zero eigenvalue of the Dirac operator.

However, there is the Atiyah–Hirzebruch index theorem, which states that manifolds that have zero eigenvalues of the Dirac operator can only have real representations of fermions.

This theorem leaves us with only a few options. Either we adopt complicated modifications of Riemannian manifolds in order to avoid this theorem, or we drop Riemannian manifolds entirely, and study supersymmetric and superstring-type theories.

But perhaps the most serious problem with quantum gravity and quantum Kaluza–Klein theory is that they are all nonrenormalizable. We now turn to this problem, which has baffled physicists for over half a century. Over the years, a number of alternative approaches have been proposed to renormalize gravity, none of them very successful. For example, let us assume that general relativity is an "effective theory," and assume that we introduce counterterms to cancel divergences at each order. Since we wish to preserve general covariance, the counterterms will be of the generic form R^2, R^3, R^4, ..., where R is composed of the Riemann curvature tensor. Because these higher terms contain higher derivatives, a theory of this type can be shown to converge sufficiently rapidly to be renormalizable. However, the modified theory is no longer unitary. R^2 has four derivatives in it, which leads to a theory with unitarity ghosts. (This is not surprising, since the higher R terms act like a Pauli–Villars cutoff, which we know introduces ghost states.) In other words, we gain renormalizability but lose unitarity.

19.9 Quantizing Gravity

To see why general relativity is not renormalizable, it is first important to explain how to quantize the theory. We begin the process of quantization by power expanding the metric tensor around some classical solution $g_{\mu\nu}^{(0)}$ of the equations of motion:

$$g_{\mu\nu}(x) = g_{\mu\nu}^{(0)} + \kappa h_{\mu\nu} \tag{19.86}$$

where $h_{\mu\nu}$ is the graviton field and $\kappa \sim \sqrt{G_N}$. The classical metric $g_{\mu\nu}^{(0)}$ is usually taken to be the Lorentz metric. Given this expansion, we can also expand the Christoffel symbols, and hence the entire action, in a power series in $h_{\mu\nu}$. Each term of the power series contains two derivatives and an increasing number of $h_{\mu\nu}$ fields and powers of the coupling constant. The action is nonpolynomial. The existence of a dimensional Newton's constant, then, is the origin of the problem of the nonrenormalizability of gravity.

Although the theory is not renormalizable, one can still study its Feynman rules and scattering matrices to lowest-order. The Feynman rules for the graviton propagator can be obtained by extracting the lowest order term quadratic in the graviton $h_{\mu\nu}$ field.

The Lagrangian, in this approximation, reduces to:

$$\mathscr{L}_0 = \frac{1}{4}\left[-(\partial_\nu h_{\rho\sigma})^2 + (\partial_\mu h_\rho^\rho)^2 - 2\partial_\sigma h_\rho^\rho \partial_\mu h^{\sigma\mu} + 2\partial_\rho h_{\nu\sigma}\partial^\nu h^{\rho\sigma}\right] \tag{19.87}$$

(where raising and lowering of indices is now performed by the flat Minkowski metric). If we make a gauge choice, we can simplify this a bit. We can, of course, add a term:

$$-\frac{1}{2}C_\mu^2; \quad C_\mu = \partial_\nu h_\mu^\nu - \frac{1}{2}\partial_\mu h_\nu^\nu \tag{19.88}$$

to the action to break the gauge.

The sum of both the original Lagrangian and the gauge part simplifies the total action to:

$$\mathscr{L}_0 = -\frac{1}{2}\partial_\lambda h_{\rho\sigma} V^{\rho\sigma\mu\nu}\partial^\lambda h_{\mu\nu}$$

$$V_{\rho\sigma\mu\nu} = \frac{1}{2}\delta_{\rho\mu}\delta_{\sigma\nu} - \frac{1}{4}\delta_{\rho\sigma}\delta_{\mu\nu} \tag{19.89}$$

We can now invert the matrix $V_{\rho\sigma\mu\nu}$ to obtain the final propagator. (One can check that the propagator is singular if the gauge-breaking part is missing.) We find the result for the propagator:

$$\frac{\delta_{\mu\rho}\delta_{\nu\sigma} + \delta_{\mu\sigma}\delta_{\nu\rho} - \delta_{\mu\nu}\delta_{\rho\sigma}}{k^2 + i\epsilon} \tag{19.90}$$

The calculations for the higher vertices, however, are prohibitively difficult. We must use special gauges and special tricks in order to reduce the number of possible interaction terms.

19.10   Counterterms in Quantum Gravity

Although quantum gravity is formally nonrenormalizable, we can still hope that (by a series of miracles) the divergences of the quantum loops cancel, leaving us with a finite theory. Usually, miracles occur because of a local symmetry, leading to Ward identities that cancel certain unwanted graphs. For quantum gravity, however, we have no more symmetries by which to cancel the higher-loop graphs.

For cancellations to happen, higher-loop counterterms must be forbidden by some unknown mechanism. If we can show that these higher-loop counterterms cannot exist, then the theory might have a chance at being finite. Let us first enumerate the total number of one-loop counterterms that are invariant. The total number of counterterms that are invariant is just three, given by the set: $R^2_{\mu\nu\rho\sigma}$, $R^2_{\mu\nu}$, and R^2.

In the background field method (see Exercise 14.17), the counterterms are gauge invariant, and we are allowed to eliminate some of them via the equations of motion. If we set $T_{\mu\nu} = 0$, then $R_{\mu\nu} - \frac{1}{2}g_{\mu\nu}R = 0$, which implies $R_{\mu\nu} = 0$. Thus, we are left with only one possible counterterm: $R^2_{\mu\nu\rho\sigma}$. The question is: Can some unforeseen identity or symmetry prevent this invariant from appearing as a counterterm? If so, then general relativity would be one-loop finite even without computing a single Feynman diagram.

It turns out that the answer is, indeed, yes. There is an identity, the Gauss–Bonnet identity, that allows us to eliminate this last invariant as a possible counterterm.

To see this, we first note that, as in Yang–Mills theory, there is a topological invariant corresponding to the square of curvature tensors:

$$\text{Total derivative} = \epsilon^{abcd}\epsilon^{\mu\nu\rho\sigma} R^{ab}_{\mu\nu} R^{cd}_{\rho\sigma} \tag{19.91}$$

We know how to reduce out the product of two antisymmetric constant tensors.

We find:

$$\epsilon^{abcd}\epsilon^{\mu\nu\rho\sigma} = e \sum_P (-1)^P e^{a\mu} e^{b\nu} e^{c\rho} e^{d\sigma} \tag{19.92}$$

where e is the determinant of the vierbein and we sum over the permutations in the indices, which preserves the antisymmetries of the antisymmetric tensor. (The left-hand side of this expression is a pure constant, while the right-hand side is a function of x. However, one can show that the x dependence of the right-hand side drops out.)

Plugging this expression into the original one, we find:

$$\text{Total derivative} \sim 4e(R_{\mu\nu\rho\sigma}R^{\mu\nu\rho\sigma} - 4R_{\mu\nu}R^{\mu\nu} + R^2) \tag{19.93}$$

This means that any counterterm that may appear at the one-loop level can be eliminated. The second and third tensors are eliminated by the equations of motion, and the first tensor is eliminated by the Gauss–Bonnet identity.

This is a truly remarkable result, indicating that quantum gravity is less divergent than previously expected. However, this fortuitous cancellation is actually an accident that does not generalize to higher loops. For example, at the two-loop level, it has been shown by computer that the following term cannot be cancelled by the equations of motion or any known identity[17,18]:

$$-\frac{1}{\epsilon}\frac{209}{2880}\frac{1}{(16\pi^2)^2} e \, C^{\mu_3\mu_4}_{\mu_1\mu_2} C^{\mu_5\mu_6}_{\mu_3\mu_4} C^{\mu_1\mu_2}_{\mu_5\mu_6} \tag{19.94}$$

(where $1/\epsilon$ represents the usual divergence found in quantum field theory, and where the $C_{\mu\nu\alpha\beta}$ tensor is the Weyl curvature tensor, which is composed of Riemann curvature tensors). The fact that this term does not cancel indicates that perturbative quantum gravity, by itself, is not a finite theory. This is a great disappointment, which has retarded progress in quantum gravity. The final answer will require essentially new ideas to remedy this defect.

Several approaches may be taken to this problem. First, one can still hope that the inclusion of matter fields will render the theory less divergent. Unfortunately, it can be shown that if we couple spin 0, 1/2, and 1 fields, then the theory becomes even more divergent. Even the first loop cancellation via the Gauss–Bonnet identity is spoiled, and quantum gravity becomes a divergent theory when coupled to matter.

Second, one might hope that coupling gravity to a spin 3/2 field may render the theory less divergent. In the next chapter, we will see that a miracle does, in fact, occur for this theory, called supergravity, at the first- and second-loop level if we couple quantum gravity to a spin 3/2 field. As one might expect, the cancellations occur because of a new symmetry in the theory, supersymmetry.

The Ward–Takahashi identities are sufficient to cancel a large class of divergences. Unfortunately, these identities are not powerful enough; supergravity appears to diverge at the third-loop level.

Third, one may observe that the nonzero coefficient appearing in the divergent two-loop term is 209. This factorizes into $(26 - D) \times 19/2$ for $D = 4$ dimensions. However, in 26 dimensions, this term might vanish exactly. The study of theories defined in $D = 26$ dimensions takes us into superstring theory, which we will study in Chapter 21.

In summary, we have seen that the equivalence principle naturally leads to a generally covariant description of gravity in terms of curved manifolds. When general relativity is combined with GUT theory, we find the theory of inflation, which gives a plausible but not conclusive solution to the flatness and horizon problems. Attempts to go beyond general relativity have led to renewed interest in Kaluza–Klein theories, which unfortunately are neither renormalizable nor do they accomodate chiral fermions. Next, we will study perhaps the most nontrivial extension of quantum gravity, the supergravity theory and finally the superstring theory, which holds the promise of successfully uniting all interactions into one finite framework.

19.11   Exercises

1. Let the $\Gamma^\lambda_{\mu\nu}$ be independent fields, along with $g_{\mu\nu}$. Take the usual Lagrangian, $\sqrt{-g}R(\Gamma)$, except keep the Christoffel symbols as independent fields, not related to the metric (this is called the Palatini form of the action). Prove that the equations of motion for the Christoffel symbols yields the usual identity Eq. (19.17), and hence the Palatini action is identical to the usual one, at least classically.

2. Do the same for ω^{ab}_μ. Prove that if the connection is an independent field and the action is taken as $\det e \, e^{a\mu} e^{b\nu} R^{ab}_{\mu\nu}(\omega)$, then the equations of motion for the connection are identical to its usual definition, given by Eq. (19.42). Unlike the Christoffel symbol, the connection ω^{ab}_μ is a generally covariant vector. Prove this.

3. To lowest order in κ, show that the lowest-order quadratic term in $h_{\mu\nu}$ arising from the linearized Einstein action equals Eq. (19.87). Prove that it is not invertible, so that a propagator does not exist unless we fix the gauge.

4. Prove Eq. (19.13).

5. Prove that, as $c \to \infty$, that the Einstein equations of motion reduce to the usual Poisson equations for a gravitational potential ϕ in the presence of a source ρ.

6. By varying the Einstein–Hilbert action, show explicitly that the equations of motion for Einstein's equations are $R_{\mu\nu} = 0$ (without matter fields). To show this, you must prove that the terms containing $\delta R_{\mu\nu}$ in Eq. (19.26) can be dropped when calculating the equations of motion.

7. Choose the harmonic gauge, where:

$$\frac{1}{2\alpha}(\partial_\mu h^{\mu\nu})^2 \tag{19.95}$$

is added to the action for arbitrary α. Calculate the graviton propagator, and the Faddeev–Popov ghost term. Compare the result with Eq. (19.90).

8. Prove that $\epsilon^{abcd}\epsilon^{\mu\nu\lambda\rho} R^{ab}_{\mu\nu}(\omega) R^{cd}_{\lambda\rho}(\omega)$ is a total derivative.

9. Starting with Maxwell's and Dirac's equations coupled to gravity, show that the metric tensor couples to the energy–momentum tensor of the Maxwell field and the Dirac field.

10. For the Kaluza–Klein theory, show explicitly that the five-dimensional Einstein–Hilbert action reduces to the usual four-dimensional Einstein–Hilbert action coupled to the Maxwell action, in the limit that the radius of the fifth dimension becomes large.

11. Construct explicitly the Kaluza–Klein decomposition of a theory where the isometry group is $O(N)$ and extract the Yang–Mills theory.

12. Insert a cosmological term. Show that the radially symmetric solution of Eq. (19.44) necessarily gives an exponential expansion (i.e., de Sitter space).

13. Prove, for an arbitrary matrix M:

$$\delta(\det M) = (\det M)(M^{-1})^{ij}\delta M_{ij} \tag{19.96}$$

14. Using the expansion $g_{\mu\nu} = \eta_{\mu\nu} + \kappa h_{\mu\nu}$, find the exact relationship between Newton's constant G and κ.

15. Prove that the metric in Eq. (19.72) is the inverse of the metric in Eq. (19.71).

16. Prove that the action in Eq. (19.73), with the proper Killing vectors, yields the usual Yang–Mills theory after dimensional reduction.

17. Power expand the Einstein–Hilbert action and explicitly derive all cubic terms in $h_{\mu\nu}$ in the harmonic gauge. Also, for the quartic and quintic terms, count the total number of ways in which four- and five-graviton fields and two derivatives may be contracted onto each other. From this, one can appreciate the complexity of doing calculations in quantum gravity.

Chapter 20
Supersymmetry and Supergravity

Supersymmetry is an answer looking for a problem.

—Anonymous

20.1 Supersymmetry

Supersymmetry has a long and interesting history. Apparently, the first known mention of a supersymmetric group was by Myazawa,[1] who discovered the supergroup $SU(M/N)$ in 1966. His motivation was to find a Master Group that could combine both internal groups and noncompact space–time groups in a nontrivial fashion. Supergroups, in fact, are the only known way in which to avoid the Coleman-Mandula theorem, which forbids naive unions of compact and noncompact groups. Unfortunately, this important work was largely ignored by the physics community.

Supersymmetry was rediscovered in 1971, from two entirely different approaches. In the first, the Neveu–Schwarz–Ramond superstring [1–2] was found to possess a new anticommuting gauge symmetry. From this, Gervais and Sakita [3] then wrote down the first supersymmetric action, the two-dimensional superstring action. The second approach was that of Gol'fand and Likhtman, [4] who were looking for a generalization of the usual space–time algebra and found the super Poincaré algebra.

In 1972, Volkov and Akulov [5] found a nonlinear supersymmetric theory. And finally in 1974, Wess and Zumino [6] wrote down the first four-dimensional point-particle field theory action.

Although a wide variety of supersymmetric actions were then discovered in the 1970s, for many years supersymmetry was considered a mathematical oddity, since none of the known subatomic particles had supersymmetric partners. However, its possible application to quantum physics came when attempts were made to iron out the inconsistencies of GUT theories.

In the previous chapter, we saw that one of the theoretical problems facing the GUT theory was the hierarchy problem; that is, renormalization effects will inevitably mix the two mass scales in the theory, the GUT scale M_X^2 and the electroweak energy scale M_W^2. Thus, even if we fine-tune the theory at the beginning to one part in 10^{12}, they will still mix, ruining the separation between these two mass scales. This means that we have to perform an infinite number of distinct fine-tunings for each order in perturbation theory, which is undesirable.

One appealing solution to the hierarchy problem is to include supersymmetry, both local and global. There are powerful nonrenormalization theorems in super-symmetric theories that show that higher order interactions do not renormalize the mass scale; that is, we do not have to fine-tune these parameters to each order in perturbation theory. One fine-tuning at the beginning is enough. This does not ex-plain where this original fine tuning came from; it only explains why higher-loop graphs do not mix the two mass scales.

There are, however, many other reasons for examining supersymmetric theo-ries. One of the main problems in building unified field theories is the inability to find a gauge group that can combine the particle spectrum with quantum gravity. The problem is the no-go theorem, which states that a group that nontrivially combines both the Lorentz group and a compact Lie group cannot have finite dimensional, unitary representations. This means that attempts to build a "master group" that combines both gravity and the particle spectrum face an insurmount-able difficulty.

There is, however, a way to evade the Coleman–Mandula theorem, and that is to use supersymmetry. Since anticommuting Grassmann numbers were never contemplated in the original derivation, the no-go theorem breaks down. The Coleman–Mandula theorem never analyzed a nontrivial symmetry that mixes bosonic and fermionic fields and places both in the same multiplet:

$$\text{Bosons} \leftrightarrow \text{Fermions} \tag{20.1}$$

Thus, there exists a supersymmetry operator Q that converts boson states $|B\rangle$ into fermion states:

$$Q|B\rangle = |F\rangle \tag{20.2}$$

As a consequence, electrons can appear in the same multiplet as the Maxwell field. In fact, there is the possibility of placing all the known particles found in nature into the same multiplet.

Perhaps one of the most remarkable aspects of supersymmetry is that it yields field theories that are finite to all orders in perturbation theory. In particular, we will outline the proof that the $N = 4$ super Yang–Mills theory, and certain versions of the $N = 2$ super Yang–Mills theory, are finite to all orders; that is, $Z = 1$ for all renormalization constants. [7-12] This is a surprising result, which indicates the power of supersymmetry in eliminating many, if not all, of the divergences of certain quantum field theories.

Yet another attractive feature of the theory is that once supersymmetry becomes a local gauge symmetry, it inevitably becomes a theory of gravity. This new theory, called supergravity, [13-15] has a new set of Ward identities that render the theory much more convergent than ordinary gravity. In fact, the largest supergravity theory, which has $SO(8)$ symmetry, is almost big enough to accomodate all the elementary particles.

We should caution the reader, however, about the limitations of supergravity as well. Although supergravity is not as divergent as ordinary gravity, the theory still is not finite. Local supersymmetry, by itself, is not powerful enough to cancel all divergences of the theory. Second, the group $SO(8)$ cannot (without extra bound states) include all the particles of the Standard Model.

To remedy some of these problems, we will have to go to yet another, more powerful theory, the superstring theory.

20.2 Supersymmetric Actions

We would first like to show that supersymmetry forces us to have equal numbers of bosons and fermions. The simplest example is the Hamiltonian:

$$H = \omega_a a^\dagger a + \omega_b b^\dagger b \tag{20.3}$$

where we have bosonic and fermionic harmonic operators that obey:

$$[a, a^\dagger] = \{b, b^\dagger\} = 1 \tag{20.4}$$

The supersymmetric operator Q is defined as:

$$Q \equiv b^\dagger a + a^\dagger b \tag{20.5}$$

If $a^\dagger|0\rangle$ is a one boson state, then $Qa^\dagger|0\rangle$ becomes a one fermion state, and vice versa. Q obeys the following identity:

$$[Q, H] = (\omega_a - \omega_b)Q \tag{20.6}$$

If $\omega_a = \omega_b = \omega$, then the supersymmetric operator Q commutes with the Hamiltonian and:

$$\{Q, Q^\dagger\} = \frac{2}{\omega} H \tag{20.7}$$

These identities show that Q and $Q^\dagger$ form a closed algebra with the Hamiltonian if the fermions and bosons have equal energy. The unusual feature of these identities is that the supersymmetric generator Q, in some sense, is the "square root" of the Hamiltonian. Furthermore, this highlights the fact that supersymmetry closes on space–time transformations. In this sense, it is radically different from the other symmetries that we have studied so far, which have treated space–time and isospin as entirely unrelated.

Another unusual consequence of this simple exercise is that the energy of the vacuum must be zero in order to have supersymmetry. To see this, take the vacuum expectation value of both sides of the previous equation. In order to have a supersymmetric vacuum, we must have:

$$Q|0\rangle = 0 \tag{20.8}$$

However, this implies that:

$$\langle 0|H|0\rangle = 0 \tag{20.9}$$

so that the vacuum must have zero energy. (This will have important implications later, when we discuss supersymmetry breaking. In broken theories, we will find that the vacuum energy becomes positive.)

To use symmetry to construct new actions, let us examine the very first and simplest supersymmetric action that was discovered in 1971. This is the action found by Gervais and Sakita that describes the Neveu–Schwarz–Ramond superstring:

$$\mathscr{L} = \bar{\psi}^a (i\gamma^\mu \partial_\mu)\psi^a + \partial_\mu \phi^a \partial^\mu \phi^a \tag{20.10}$$

which is defined in two dimensions for real, Majorana spinors (and where a is an additional vector index that does not concern us here). The action is invariant under:

$$
\begin{aligned}
\delta\psi^a &= -i\gamma^\mu \partial_\mu \phi^a \epsilon \\
\delta\phi^a &= \bar{\epsilon}\psi
\end{aligned}
\tag{20.11}
$$

There are several usual features of this action. We first notice that the supersymmetric parameter ϵ is a anticommuting spinor. This means that many of the classical theorems concerning Lie groups and Lie algebras no longer hold. Second, the fermions and bosons have the same index a; that is, they must transform under the same representation of some group. (This will have important implications for the theory of super GUTs, because the fermions usually transform under the fundamental representation, while the Yang–Mills field transforms under the adjoint representation. Super GUT theories, therefore, cannot easily place the quarks and gauge particles in the same representation.)

Third, we notice that if we anticommute the fields a second time, we find:

$$[\delta_1, \delta_2]\phi^a = \bar{\epsilon}_1 \gamma^\mu \epsilon_2 P_\mu \phi^a - (1 \leftrightarrow 2) \tag{20.12}$$

These commutation relations mean that there exists a spinor operator Q_α whose anticommutation relations with itself yield the translation operator P_μ. This generalizes the discussion we found earlier, where Q formed a closed algebra with H. Now, we find that the supersymmetric generator forms a closed algebra with the vector P_μ.

The previous action was written down in only two dimensions. To obtain a four-dimensional theory, we now study the free Wess–Zumino action, where we again have Majorana spinors:

$$S = \frac{1}{2} \int d^4x \left[(\partial_\mu A)^2 + (\partial_\mu B)^2 + i\bar{\psi}\gamma^\mu \partial_\mu \psi + F^2 + G^2 \right] \tag{20.13}$$

This action is invariant under:

$$
\begin{aligned}
\delta A &= \bar{\epsilon}\psi \\
\delta B &= \bar{\epsilon}\gamma_5 \psi \\
\delta F &= i\bar{\epsilon}\gamma_\mu \partial^\mu \psi \\
\delta G &= i\bar{\epsilon}\gamma_5 \gamma_\mu \partial^\mu \psi \\
\delta \psi &= -i\gamma^\mu \left(\partial_\mu A + \gamma_5 \partial_\mu B \right) \epsilon - (F + \gamma_5 G)\epsilon
\end{aligned}
\tag{20.14}
$$

This action contains equal numbers of fermions and bosons, as desired. There are four components within the off-shell Majorana field ψ, and four boson fields A, B, F, and G. It is easy to show that repeated variations of these $4 + 4$ off-shell fields close linearly among themselves. The supersymmetric algebra is thus linear. However, because F and G are auxiliary fields, we can eliminate them from the action from the very start. After this seemingly trivial elimination, the resulting action no longer has equal numbers of fermions and bosons. The action is still invariant under a modified form of supersymmetry, although it is no longer linear. By taking two such nonlinear supersymmetric variations, we find that the algebra does *not* close. This may seem disturbing, until we realize that the term which breaks the closure of the algebra is proportional to the equations of motion. The algebra then closes on-shell; that is, we must use the on-shell equations of motion in order to close the nonlinear supersymmetric relations.

It is more convenient therefore to retain these auxiliary fields in order to maintain the complete off-shell, linear algebra. In fact, one of the most pressing and unsolved problems in developing higher supersymmetric actions is to find all the auxiliary fields that will linearize the supersymmetric gauge transformations. The problem of writing down higher supersymmetric actions, in fact, often boils down to the highly nontrivial task of finding all auxiliary fields that linearize the supersymmetry algebra.

(It is also instructive to perform the on-shell counting of states for this action, to confirm that we have the same number of fermions and bosons on-shell. The Majorana fermion, which had four components off-shell, now only has two components on-shell. Likewise, on-shell the F and G fields vanish, leaving us with 2 + 2 fermion and boson states on-shell, as desired.)

Supersymmetry also generalizes to gauge theories. For example, the following gauge action with a Maxwell field and a Majorana spinor is invariant under global supersymmetry. It is the supersymmetric counterpart of QED:

$$S = \int d^4x \left(-\frac{1}{4} F_{\mu\nu}^2 + \frac{i}{2} \bar{\psi} \gamma^\mu \partial_\mu \psi + \frac{1}{2} D^2 \right) \qquad (20.15)$$

The fields transform under:

$$
\begin{aligned}
\delta A_\mu &= i\bar{\epsilon}\gamma_\mu \psi \\
\delta \psi &= \left(-\frac{i}{2} \sigma^{\mu\nu} F_{\mu\nu} - \gamma_5 D \right) \epsilon \\
\delta D &= i\bar{\epsilon}\gamma_5 \gamma^\mu \partial_\mu \psi \qquad (20.16)
\end{aligned}
$$

Once again, we have equal numbers of fermions and bosons off-shell. Off-shell, the Majorana field has 4 components, while the A_μ field has 3 components (because one is eliminated by gauge fixing), and the D field has one component. We therefore have 4 + 4 fermions and bosons. On-shell, we also have the same number of fermions and bosons. The ψ field now only has two components, the D field disappears, and the A_μ field has two components, so we are left with 2 + 2 fields on-shell. When we generalize this to non-Abelian gauge transformations, we will find that the fermionic ψ must transform in the *adjoint* representation, the same representation as the gauge fields, since they all belong in the same multiplet.

So far, we have been exploring actions written totally in terms of their component fields. This, however, becomes prohibitively difficult as we go to non-Abelian and gravitational theories. In order to systematically generate new supersymmetric actions, we now turn to a new formalism, the superspace formalism.

20.3 Superspace

Unfortunately, the number of fields rapidly escalates for higher supersymmetric actions. Perhaps one of the most beautiful ways in which to compress the blizzard of indices that often appears in supersymmetric theories is through *superspace*. [16] This construction postulates the existence of four antisymmetric coordinates θ_α that form the superpartner of the usual space–time coordinate:

$$\{x^\mu, \theta_\alpha\} \tag{20.17}$$

Supersymmetry, acting on the superspace coordinates, makes the following transformation:

$$x^\mu \quad \rightarrow \quad x^\mu + i\bar\epsilon\gamma^\mu\theta$$

$$\theta_\alpha \quad \rightarrow \quad \theta_\alpha + \epsilon_\alpha \tag{20.18}$$

In practice, the use of complex Dirac spinors leads to *reducible* representations of supersymmetry. In order to find irreducible representations, we will find it more convenient to use Majorana or Weyl spinors. We will therefore split the four-component spinor into two smaller spinors as follows:

$$\theta_\alpha \equiv \begin{pmatrix} \theta^a \\ \bar\theta_{\dot a} \end{pmatrix} \tag{20.19}$$

(Because of this split of four-component spinors down to two-component spinors, we will, unfortunately, find that the number of indices for irreducible representations proliferates considerably.)

In this formalism, we will take a modified Weyl representation of the Dirac matrices:

$$\gamma^\mu = \begin{pmatrix} 0 & \sigma^\mu \\ \bar\sigma^\mu & 0 \end{pmatrix}; \quad \gamma_5 = \begin{pmatrix} -i & 0 \\ 0 & i \end{pmatrix} \tag{20.20}$$

where:

$$\sigma^\mu = (\mathbf{1}, \sigma^i) = (\sigma^\mu)_{ab}; \quad \bar\sigma^\mu = (\mathbf{1}, -\sigma^i) = (\bar\sigma^\mu)^{\dot a b} = (\sigma^\mu)^{b\dot a} \tag{20.21}$$

Then the typical spinor breaks up as follows:

$$\psi \quad = \quad \begin{pmatrix} \phi_a \\ \bar\chi^{\dot a} \end{pmatrix}$$

$$\bar{\psi} = \psi^\dagger \gamma^0 = \left(\chi^a \ \bar{\phi}_{\dot{a}} \right) \tag{20.22}$$

In this representation, the spinors become reducible; that is, the four-spinor ψ has now been broken up into two two-spinors ϕ and $\bar{\chi}$, each of which forms a two-dimensional representation of the Lorentz group. The Lorentz group generators can be obtained by multiplying the old generators $M_{\mu\nu} = \sigma_{\mu\nu}/2$ by the chiral projection operators $P_\pm = \frac{1}{2}(1 \pm i\gamma_5)$. In this way, we can split the original 4×4 Lorentz generators into two distinct 2×2 blocks. Each two-spinor then transforms under a 2×2 complex representation of the Lorentz group, which we can show is $SL(2, C)$, the set of 2×2 complex matrices with unit determinant. We use the fact that:

$$O(3, 1) \sim SL(2, C) \tag{20.23}$$

Let the two-spinor θ^a transform as the fundamental representation of $SL(2, C)$, where $a = 1, 2$. The complex conjugate of these matrices generates an *inequivalent* representation of $SL(2, C)$. We will label these two-spinors as $\bar{\theta}_{\dot{a}}$, where $\dot{a} = 1, 2$ and the dot reminds us that the two-spinors transforms under the inequivalent complex conjugate representation of $SL(2, C)$.

We take the conjugation of spinors as follows:

$$\left(\theta^a \right)^* = \bar{\theta}^{\dot{a}}; \quad \left(\theta_a \right)^* = \bar{\theta}_{\dot{a}} \tag{20.24}$$

Raising and lowering in this two-dimensional space is done via:

$$\epsilon_{12} = \epsilon^{12} = -\epsilon_{\dot{1}\dot{2}} = -\epsilon^{\dot{1}\dot{2}} = +1 \tag{20.25}$$

so that:

$$\psi^a = \epsilon^{ab}\psi_b; \quad \bar{\psi}^{\dot{a}} = \bar{\psi}_{\dot{b}}\epsilon^{\dot{b}\dot{a}}$$

$$\psi_a = \psi^b\epsilon_{ba}; \quad \bar{\psi}_{\dot{a}} = \epsilon_{\dot{a}\dot{b}}\bar{\psi}^{\dot{b}} \tag{20.26}$$

Invariants under each of the two groups are given by:

$$\phi\chi = \phi^a \chi_a = -\phi_a \chi^a$$

$$\bar{\phi}\bar{\chi} = \bar{\phi}_{\dot{a}}\bar{\chi}^{\dot{a}} = -\bar{\phi}^{\dot{a}}\bar{\chi}_{\dot{a}} \tag{20.27}$$

and:

$$\theta^2 = \theta^a\theta_a; \quad \bar{\theta}^2 = \bar{\theta}_{\dot{a}}\bar{\theta}^{\dot{a}} \tag{20.28}$$

In this way, the standard invariant $\bar{\psi}\psi$ for four-spinors decomposes as:

$$
\bar{\psi}\psi = (\bar{\phi}_{\dot{a}} \; \chi^a)
\begin{pmatrix} 0 & 1 \\ 1 & 0 \end{pmatrix}
\begin{pmatrix} \phi_a \\ \bar{\chi}^{\dot{a}} \end{pmatrix}
$$

$$
= \bar{\phi}\bar{\chi} + \chi\phi \tag{20.29}
$$

We also have:

$$
\bar{\psi}_1 \gamma^\mu \psi_2 = \chi_1 \sigma^\mu \bar{\chi}_2 + \bar{\phi}_1 \bar{\sigma}^\mu \phi_2 \tag{20.30}
$$

In two-spinor notation, the supersymmetric transformation in superspace in Eq. (20.18) is written as:

$$
x^\mu \;\rightarrow\; x^\mu + i\epsilon\sigma^\mu\bar{\theta} - i\theta\sigma^\mu\bar{\epsilon}
$$

$$
\theta^a \;\rightarrow\; \theta^a + \epsilon^a
$$

$$
\bar{\theta}_{\dot{a}} \;\rightarrow\; \bar{\theta}_{\dot{a}} + \bar{\epsilon}_{\dot{a}} \tag{20.31}
$$

Given this superspace transformation, it is now a simple matter to extract the operators that generate this transformation:

$$
Q_a = i\frac{\partial}{\partial\theta^a} - \left(\sigma^\mu\bar{\theta}\right)_a \partial_\mu
$$

$$
\bar{Q}_{\dot{a}} = -i\frac{\partial}{\partial\bar{\theta}^{\dot{a}}} + (\theta\sigma^\mu)_{\dot{a}} \partial_\mu \tag{20.32}
$$

The supersymmetric algebra now reads:

$$
\{Q_a, \bar{Q}_b\} = 2(\sigma^\mu)_{ab} P_\mu
$$

$$
\{Q_a, Q_b\} = \{\bar{Q}_{\dot{a}}, \bar{Q}_b\} = 0
$$

$$
[Q_a, M_{\mu\nu}] = \frac{1}{2}\left(\sigma_{\mu\nu}\right)_a^b Q_b; \quad [\bar{Q}_{\dot{a}}, M_{\mu\nu}] = -\frac{1}{2}\bar{Q}_{\dot{b}}\left(\bar{\sigma}_{\mu\nu}\right)_{\dot{a}}^{\dot{b}}
$$

$$
[Q_a, P_\mu] = 0; \quad [\bar{Q}_{\dot{a}}, P_\mu] = 0 \tag{20.33}
$$

where $\sigma^{\mu\nu} = \frac{i}{2}[\gamma^\mu, \gamma^\nu]$.

Using superspace methods, let us now construct a few representations of supersymmetry. We begin by constructing a vector *superfield* $V(x, \theta, \bar{\theta})$ that is a function of superspace. Under a supersymmetric transformation, it transforms as:

$$
\delta V(x, \theta) = i[\epsilon Q + \bar{Q}\bar{\epsilon}, V] \tag{20.34}
$$

(Since the supersymmetry generator has spin $\frac{1}{2}$, this means that the supersymmetric partner of any particle must differ by only spin $\frac{1}{2}$. Supersymmetric multiplets can then be grouped into collections of particles, each differing from the other by spin $\frac{1}{2}$.)

These superfields have many nice properties. The most important is that the product of two superfields is again a superfield:

$$V_1 V_2 = V_3 \tag{20.35}$$

This can be simply checked by examining the transformation properties of both sides of the equation. Although this product rule is simple in superspace, written in component form it is highly nontrivial. Superspace thus gives a way of generating new representations of supersymmetry from old ones.

By power expanding $V(x, \theta, \bar{\theta})$ in a power series in θ and $\bar{\theta}$, we find that the series terminates after reaching the fourth power of the spinor because of its Grassmann nature. Since V is real, the most general parametrization is given by a Taylor expansion in the Grassmann variables:

$$
\begin{aligned}
V(x, \theta, \bar{\theta}) \;=\; & C - i\theta\chi + i\bar{\chi}\bar{\theta} - \frac{i}{2}\theta^2(M - iN) + \frac{i}{2}\bar{\theta}^2(M + iN) \\
& - \theta\sigma_\mu\bar{\theta}A^\mu - i\theta^2\bar{\theta}\left(\bar{\lambda} - \frac{i}{2}\bar{\sigma}_\mu\partial^\mu\chi\right) \\
& + i\bar{\theta}^2\theta\left(\lambda - \frac{i}{2}\sigma_\mu\partial^\mu\bar{\chi}\right) - \frac{1}{2}\theta^2\bar{\theta}^2\left(D + \frac{1}{2}\partial_\mu^2 C\right) \tag{20.36}
\end{aligned}
$$

This is called the vector superfield because it contains an ordinary vector field A_μ (and not because the superfield has a vector index on it). The vector superfield has 8 fermionic fields contained within λ and χ as well as 8 bosonic fields contained within C, D, M, N, A_μ; so we have an equal number of fermions and bosons, as desired.

Under a supersymmetric transformation, we have:

$$V(x, \theta, \bar{\theta}) \rightarrow V(x + i\epsilon\sigma_\mu\bar{\theta} - i\theta\bar{\sigma}^\mu\bar{\epsilon}, \theta + \epsilon, \bar{\theta} + \bar{\epsilon}) \tag{20.37}$$

By power expanding the previous equation and then equating coefficients, we can calculate the variation of all the fields within the vector superfield:

$$
\begin{aligned}
\delta C &= \bar{\epsilon}\gamma_5\chi \\
\delta\chi &= (M + \gamma_5 N)\epsilon - i\gamma^\mu(A_\mu + \gamma_5\partial_\mu C)\epsilon \\
\delta M &= \bar{\epsilon}(\lambda - i\slashed{\partial}\chi)
\end{aligned}
$$

$$\delta N = \bar{\epsilon}\gamma_5(\lambda - i\slashed{\partial}\chi)$$

$$\delta A_\mu = i\bar{\epsilon}\gamma_\mu\lambda + \bar{\epsilon}\partial_\mu\chi$$

$$\delta\lambda = -i\sigma^{\mu\nu}\epsilon\partial_\mu A_\nu - \gamma_5\epsilon D$$

$$\delta D = -i\bar{\epsilon}\slashed{\partial}\gamma_5\lambda \tag{20.38}$$

We can also introduce a new derivative operator:

$$D_a = \frac{\partial}{\partial\theta^a} - i\left(\sigma^\mu\bar{\theta}\right)_a \partial_\mu$$

$$\bar{D}_{\dot{a}} = -\frac{\partial}{\partial\bar{\theta}^{\dot{a}}} + i\left(\theta\sigma^\mu\right)_{\dot{a}} \partial_\mu \tag{20.39}$$

The importance of this derivative operator is that it anticommutes with the supersymmetric generators. We list a few useful identities of this operator that we will use extensively in this chapter:

$$\{D_a, D_b\} = 0; \quad \{\bar{D}_{\dot{a}}, \bar{D}_{\dot{b}}\} = 0$$

$$\{D_a, \bar{D}_{\dot{b}}\} = 2i(\sigma^\mu)_{a\dot{b}}P_\mu$$

$$D_a D_b D_c = \bar{D}_{\dot{a}}\bar{D}_{\dot{b}}\bar{D}_{\dot{c}} = 0$$

$$D^a\bar{D}^2 D_a = \bar{D}_{\dot{b}}D^2\bar{D}^{\dot{b}}$$

$$D^2\bar{D}^2 D^2 = -16\partial_\mu^2 D^2$$

$$\bar{D}^2 D^2\bar{D}^2 = -16\partial_\mu^2\bar{D}^2$$

$$[\bar{D}^2, D^2] = -16\partial_\mu^2 - 8i D\slashed{\partial}\bar{D} \tag{20.40}$$

[Proving these formulas is not as formidable as one might expect. Once one establishes the anticommutator between D and $\bar{D}$, the other relations follow. For example, the commutator between $\bar{D}^2$ and D^2 can be evaluated by pushing all $\bar{D}_{\dot{b}}$ to the right. Each time they pass a D_a, we pick up $2i(\sigma^\mu)_{a\dot{b}}\partial_\mu$. Thus, after pushing all $\bar{D}_{\dot{b}}$ to the right, we have:

$$[\bar{D}^2, D^2] = [\bar{D}_{\dot{a}}\epsilon^{\dot{a}\dot{b}}\bar{D}_{\dot{b}}, D_a\epsilon^{ab}D_b]$$

$$= 2(2i)(2i)\epsilon^{\dot{a}\dot{b}}\epsilon^{ab}(\sigma^\mu)_{a\dot{b}}(\sigma^\nu)_{b\dot{a}}\partial_\mu\partial_\nu + \cdots$$

$$= 8\,\mathrm{Tr}\left(\sigma^\mu\epsilon\sigma^{\nu T}\epsilon\right)\partial_\mu\partial_\nu + \cdots$$

$$= -16\partial_\mu^2 + \cdots \tag{20.41}$$

where ϵ is a 2×2 matrix $\begin{pmatrix} 0 & 1 \\ -1 & 0 \end{pmatrix}$. Finally, the identities involving six derivative operators are proven by multiplying $[\bar{D}^2, D^2]$ by $\bar{D}^2$ or D^2. In this way, all identities can be proved.]

Because D_a anticommutes with the supersymmetric generator, we can apply it at will on any superfield to form a constraint, such as:

$$\bar{D}_{\dot{a}}\phi = 0 \tag{20.42}$$

This constraint does not spoil the transformation of ϕ under supersymmetry because D_a anticommutes with the supersymmetry generator. A field that satisfies this constraint is called a *chiral superfield*. (An antichiral superfield satisfies $D_a\bar{\phi} = 0$.)

It is simple to write down the solution to the chiral constraint equation:

$$\phi(x, \theta, \bar{\theta}) = \exp\left(-i\theta \not{\partial}\bar{\theta}\right) \phi(x, \theta); \quad \bar{\phi}(x, \theta, \bar{\theta}) = \exp\left(i\theta \not{\partial}\bar{\theta}\right) \bar{\phi}(x, \bar{\theta}) \tag{20.43}$$

The problem of finding chiral superfields then reduces to the simpler problem of power expanding $\phi(x, \theta)$, which terminates after only three terms:

$$\phi(x, \theta) = A + 2\theta\psi - \theta^2 F \tag{20.44}$$

Once again, the number of fermion and boson fields are equal. ψ contains four components, while A and F are complex scalar fields with four components in all.

Written out explicitly, the variation of the fields is given by:

$$\begin{aligned} \delta A &= 2\epsilon\psi \\ \delta\psi &= -\epsilon F - i\partial_\mu A\sigma^\mu\bar{\epsilon} \\ \delta F &= -2i\partial_\mu\psi\sigma^\mu\bar{\epsilon} \end{aligned} \tag{20.45}$$

Given these vector and chiral superfields, we can now construct superfield actions that are manifestly supersymmetric. There are two ways in which supersymmetric invariant actions can be constructed, one for vector fields and the other for chiral fields. For vector fields, we simply integrate over all eight x, θ, and $\bar{\theta}$ indices. This integration selects out the "D" term that appears in the variation of the vector field. [The variation of the D term in Eq. (20.38) is a total derivative, and hence always integrates to zero. That is why D terms are always invariant.] For chiral fields, we only integrate over six variables, x and θ. This selects out the "F" term. (The variation of the F term is also a total derivative from the above equation.)

The simplest action based on superspace is the Wess–Zumino action, given as:

$$S = \int d^8x \, \frac{1}{2}\bar{\phi}\phi - \left[\int d^6x \left(\mu\phi + \frac{1}{2}m\phi^2 + \frac{\lambda}{3!}\phi^3\right) + \text{h.c.}\right] \qquad (20.46)$$

By integrating out the Grassmann variables, we retrieve the free Wess–Zumino action in Eq. (20.13) we wrote down earlier.

Not only can we find a superspace formulation of the Wess–Zumino model, we can also find the superspace formulation of gauge theory. To introduce gauge invariance, we first notice that the vector field V contains the field A_μ, which will form the basis of a gauge theory. The variation of V, in turn, looks very much like a chiral superfield Λ, which contains the combination $\partial_\mu A$.

What we want, therefore, is a real vector field V that transforms as:

$$\delta V = -\frac{i}{2}(\Lambda - \bar{\Lambda}) \qquad (20.47)$$

where Λ is a chiral superfield. It is easy to show that this variation contains the $U(1)$ symmetry transformation $\delta A_\mu \sim \partial_\mu A$.

Now we wish to construct the counterpart of the Maxwell tensor $F_{\mu\nu}$, which is invariant under this transformation. Let us define a chiral superfield W_a:

$$W_a \equiv \frac{i}{4}\bar{D}^2 D_a V$$

$$\bar{D}_{\dot{a}} W_a = 0 \qquad (20.48)$$

where the last identity is important because it shows that W_a is a chiral superfield. One can show that the Maxwell tensor is contained within W_a. Then it is easy to show that:

$$\delta W_a = 0 \qquad (20.49)$$

where we have used the fact that both Λ and W_a are chiral superfields.

Our gauge-invariant action is therefore:

$$\frac{1}{16}\int d^4x \, d^2\theta \, W^a W_a \qquad (20.50)$$

which is invariant under both gauge and supersymmetry.

Next, we must show that this yields the correct $U(1)$ action when we perform the integration. In general, this integration is rather lengthy; so we will use a trick.

We will use up many of the degrees of freedom in the gauge transformation so that the theory is defined only in terms of the important fields.

Since A and F within the chiral field Λ are complex, we will find it convenient to redefine these fields as $A \rightarrow A + iB$ and $F \rightarrow F + iG$. Then under this gauge transformation, we have:

$$
\begin{aligned}
C &\rightarrow C + B \\
\chi &\rightarrow \chi + \psi \\
M + iN &\rightarrow M + iN - F - iG \\
A_\mu &\rightarrow A_\mu + \partial_\mu A \\
\lambda &\rightarrow \lambda \\
D &\rightarrow D
\end{aligned}
\tag{20.51}
$$

so that the A field is the gauge parameter associated with the gauge field A_μ. We can obviously use B, ψ, F, and G to eliminate C, χ, M, and N. This leaves us with a reduced vector multiplet:

$$
V = (0, 0, 0, 0, A_\mu, \lambda, D)
\tag{20.52}
$$

We call this the *Wess–Zumino gauge*, [17] where we have partially used up the gauge degree of freedom within the chiral superfields, leaving us with only the gauge multiplet that includes the Maxwell field A_μ. Placing the chiral superfield W_a into the action, we obtain the original super Maxwell theory of Eq. (20.15).

The generalization of this construction to the full non-Abelian theory is also straightforward. Let us define $V = V^a \tau^a$, where the τ^a matrices generate some Lie group. The V transforms as:

$$
\delta V = ig(\Lambda - \Lambda^\dagger)
\tag{20.53}
$$

for some chiral superfield Λ. Then we also have:

$$
e^{-2V} \rightarrow e^{-i\bar{\Lambda}} e^{-2V} e^{i\Lambda}
\tag{20.54}
$$

Then define:

$$
W_a = -\frac{i}{8} \bar{D} \bar{D} e^{2V} D_a e^{-2V}
\tag{20.55}
$$

If we include matter fields within a chiral superfield, then:

$$
\phi \rightarrow e^{-i\Lambda} \phi; \quad \bar{\phi} \rightarrow \bar{\phi} e^{i\bar{\Lambda}}
\tag{20.56}
$$

Then the coupling to the matter fields arises through:

$$\bar{\phi}e^{-2V}\phi \tag{20.57}$$

Let us now put everything together. The most general coupling between a superfield V and a matter superfield ϕ is given by:

$$\mathcal{L} = \frac{1}{16}\text{Tr}\int d^4x\, d^2\theta\; W^a W_a + \text{h.c.}$$

$$+ \frac{1}{8}\int d^4x d^4\theta\; \bar{\phi}e^{-2V}\phi \tag{20.58}$$

After performing the Grassmann integrations, the action equals:

$$\mathcal{L} = \int d^4x\, \text{Tr}\left(-\frac{1}{4}F_{\mu\nu}F^{\mu\nu} + \frac{1}{2}i\bar{\lambda}\gamma^\mu\nabla_\mu\lambda + \frac{1}{2}D^2\right.$$

$$+ \frac{1}{2}\nabla_\mu A\nabla^\mu A + \frac{1}{2}\nabla_\mu B\nabla^\mu B + \frac{1}{2}i\bar{\psi}\bar{\sigma}^\mu\nabla_\mu\psi + \frac{1}{2}F^2 + \frac{1}{2}G^2$$

$$\left. - iA[B, D] - i\bar{\psi}[\lambda, A] - i\bar{\psi}\gamma_5[\lambda, B]\right) \tag{20.59}$$

where the matter field is in the adjoint representation of the group.

We also have the freedom of adding the most general renormalizable self-interaction of the ϕ field, which is at most cubic:

$$\int d^4x\, d^2\theta\; \left(\lambda_i\phi_i + \frac{1}{2}m_{ij}\phi_i\phi_j + \frac{1}{3}g_{ijk}\phi_i\phi_j\phi_k + \text{h.c.}\right) \tag{20.60}$$

where the terms in the interaction must be gauge invariant.

At this point, we can make a few remarks about supersymmetry breaking. There are two known ways in which we can break supersymmetry spontaneously. We can simply add the gauge superfield V (from which we constructed the supersymmetric gauge action) directly to the Lagrangian:

$$\mathcal{L} \rightarrow \mathcal{L} + kV \tag{20.61}$$

The integration of this V field, of course, generates a "D" term (Fayet–Iliopoulos term).[18] Since the variation of this D term is a total derivative, we still have a supersymmetric theory.

In general, this action creates a number of terms containing the D and F fields. Since they appear in this action as nonpropagating fields, we can eliminate them by their equations of motion. After this elimination, we are left with quartic terms in A, which generates a new effective potential. We then treat this effective potential in the usual way: We hunt for new vacua that allow us to break supersymmetry or gauge symmetry by shifting the vacuum $\langle 0|A|0\rangle \neq 0$.

The difficulty with this procedure, however, is that V in general transforms in the adjoint representation of the group, and hence cannot simply be added into the action. This means that we must have extra $U(1)$ symmetries so that V is invariant and can be added freely in the action. However, this is often not desirable phenomenologically.

Another more promising way to break supersymmetry is to add a term, called the "F" or O'Raifeartaigh term, [19] to the action, which is also supersymmetric.

Let us add a chiral term W, called the superpotential, into the action. As an example, consider the superpotential in Eq. (20.60). After performing the $d^2\theta$ integration, we are left with:

$$\int d^4x \left[F_i^* F_i + \lambda_i F_i + m_{ij}(A_i F_j - \frac{1}{2}\psi_i\psi_j) \right.$$
$$\left. + g_{ijk}(A_i A_j F_k - \psi_i\psi_j A_k) + \text{h.c.} \right] \tag{20.62}$$

Now let us solve for the equations of motion for the auxiliary F_i field:

$$- F_k = \lambda_k^* + m_{ik}^* A_i^* + g_{ijk} A_i^* A_j^* \tag{20.63}$$

Now substitute this value for F_k and F_k^* back into the action. The superpotential has now changed into the term:

$$-\frac{1}{2}m_{ik}\psi_i\psi_k - g_{ijk}\psi_i\psi_j A_k + \text{h.c.} - V(A_i, A_i^*) \tag{20.64}$$

where the potential $V(A_i, A_i^*)$ is given by:

$$V = \sum_k F_k^* F_k \tag{20.65}$$

As before, the elimination of the auxiliary fields F_i and F_i^* has generated an effective potential $V(A_i, A_i^*)$ can shift the vacuum. This, in turn, generates a fermion mass via the Yukawa term.

This simple exercise can be generalized for an arbitrary superpotential $W(\phi)$, where ϕ_i are the scalar fields within W. By repeating the same steps, we can show that the elimination of F_i and F_i^* from the action generates the following potential term:

$$\frac{1}{2}\sum_{i,j} \left(\frac{\partial^2 W(\phi)}{\partial\phi_i\partial\phi_j}\psi_i\psi_j + \text{h.c.} \right) - \sum_i \left| \frac{\partial W(\phi)}{\partial\phi_i} \right|^2 \tag{20.66}$$

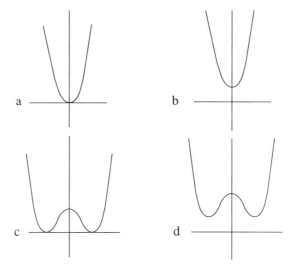

Figure 20.1. In (a), the potential respects both supersymmetry and gauge symmetry. In (b), the potential breaks supersymmetry (because the vacuum state no longer has zero energy) but gauge symmetry is still unbroken. In (c), the potential preserves supersymmetry but breaks gauge invariance. In (d), the potential breaks both supersymmetry and gauge invariance.

Pictorially, we can analyze how to break supersymmetry and gauge invariance with this potential. We recall from our previous discussion that supersymmetry is preserved if the vacuum has zero energy. This can also be generalized to show that if the vacuum has nonzero energy, then supersymmetry must be broken. From the previous expression, the potential V is positive definite. Thus, to have supersymmetry breaking, we need only show that some of the terms in V do not vanish, thereby giving the vacuum nonzero energy.

In Figure 20.1, we see various possibilities for the effective potential. In general, there are four possibilities. Potentials can be generated in which gauge symmetry and supersymmetry are broken together or independently of each other.

When this mechanism for spontaneous symmetry breaking is applied to model building, one problem is that we cannot put the gauge fields and matter fields in the same multiplet since they transform differently under the isospin group; that is, the fermions belong to the fundamental representation, while the gauge fields belong to the adjoint representation. Therefore, we must introduce superpartners for both the gauge fields and matter fields. We must therefore introduce the supersymmetric partners of the familiar particles: bosonic "squarks" and "sleptons" transforming in the fundamental representation, fermionic "gauginos" transforming in the adjoint representation, as well as "Higgsinos" and "Goldstinos." Another problem is that, since we do not see supersymmetry experimentally among the subatomic particles, we must be able to break supersymmetry at a sufficiently high mass scale so that these superpartners do not violate known experimental results.

In addition, there are stringent mass relations that must be obeyed whenever we use the "F" to break supersymmetry. Using the form of the potential in Eq. (20.66), a careful analysis of the most general superpotential shows that the spin 0, spin $\frac{1}{2}$, and spin 1 mass matrix M_i for spin i must obey the following relation:

$$\text{Tr}\,(M_0^2 - M_{1/2}^2 + 3M_1^2) = 0 \qquad (20.67)$$

This relation, unfortunately, is badly broken phenomenologically. It shows that the boson masses cannot be sufficiently heavier than the fermion masses as required in model building. Even if "D" terms are added to action, they are unable to lift this requirement. The only known way to avoid this mass condition is to add soft breaking terms to the action (which is undesirable) or move on to a more sophisticated theory, supergravity, which we will discuss shortly.

20.4   Supersymmetric Feynman Rules

There are several advantages to deriving the Feynman rules [20] for supersymmetric theories using the superspace formalism. First, the large number of component fields found within the superfield can be easily manipulated as a single block. Before superfields were introduced, calculations with the component fields were often long and tedious. Second, in the component formalism, the cancellations of certain divergent graphs are rather miraculous. In the superspace formalism, it is easy to see the origin of these cancellations. Finally, the usual rules of functional integration generalize naturally to the superspace formalism.

To begin, we wish to find an expression for:

$$Z(J, \bar{J}) = \int D\phi\, D\bar{\phi}\, \exp\left(iS + i\int d^4x\, d^2\theta\, J\phi + i\int d^4x\, d^2\bar{\theta}\, \bar{J}\bar{\phi}\right) \quad (20.68)$$

where J is the source term for a chiral field.

There are two types of Grassmann integrations found in the action, over $d^2\theta$ and over $d^4\theta$. To derive the Feynman rules, we will find it convenient to convert the chiral integral over $d^4x\, d^2\theta$ into an integral over $d^4x\, d^4\theta$. This is accomplished by remembering that taking an integral over a Grassmann number is the same as taking a derivative. We can show:

$$\int d^2\theta \;\; = \;\; \int d\theta^2\, d\theta^1 = \frac{1}{2}D^2$$

$$\int d^2\bar{\theta} \;=\; \int d\bar{\theta}^{\dot{1}} d\,\bar{\theta}^{\dot{2}} = \frac{1}{2}\bar{D}^2 \tag{20.69}$$

which hold when applied directly onto chiral superfields.

Using the formulas in Eq. (20.40), we find:

$$\bar{D}^2 D^2 \phi = -16\partial_\mu^2 \phi \tag{20.70}$$

since $\bar{D}_{\dot{a}}\phi = 0$. Then we write the chiral integral as:

$$\int d^4x \, d^2\theta \; \phi \;=\; -\int d^4x d^2\theta \; \left(\frac{\bar{D}^2 D^2}{16\partial_\mu^2}\right)\phi$$

$$=\; -\int d^4x \, d^4\theta \; \left(\frac{D^2}{8\partial_\mu^2}\right)\phi \tag{20.71}$$

The chiral integral over the mass term $m\phi^2$, for example, can be converted to an eight-dimensional integral as follows:

$$\int d^4x \, d^2\theta \; \phi^2 \;=\; -\int d^4x \, d^2\theta \; \phi\left[\frac{\bar{D}^2 D^2}{16\partial_\mu^2}\right]\phi$$

$$=\; -\int d^4x \, d^2\theta \frac{\bar{D}^2}{2} \left(\phi\frac{D^2}{8\partial_\mu^2}\phi\right)$$

$$=\; -\int d^4x \, d^4\theta \; \phi\left(\frac{D^2}{8\partial_\mu^2}\phi\right) \tag{20.72}$$

Putting everything together, we have, for the free Wess–Zumino action with an external chiral source J:

$$S \;=\; \int d^4x \, d^4\theta \;\left[\frac{1}{2}\bar{\phi}\phi - \frac{1}{2}m\phi\left(-\frac{D^2}{8\partial_\mu^2}\phi\right) - \frac{1}{2}m\bar{\phi}\left(-\frac{\bar{D}^2}{8\partial_\mu^2}\bar{\phi}\right)\right.$$

$$\left. + J\left(-\frac{D^2}{8\partial_\mu^2}\phi\right) + \bar{J}\left(-\frac{\bar{D}^2}{8\partial_\mu^2}\bar{\phi}\right)\right]$$

$$=\; \int d^4x \, d^4\theta \;\left(\frac{1}{4}\psi^T A\psi + \psi^T B\right) \tag{20.73}$$

where $\psi \equiv \begin{pmatrix} \phi \\ \bar{\phi} \end{pmatrix}$ and:

$$A = \begin{pmatrix} m D^2/4\partial_\mu^2 & 1 \\ 1 & m\bar{D}^2/4\partial_\mu^2 \end{pmatrix} ; \quad B = \begin{pmatrix} -(D^2/8\partial_\mu^2)J \\ -(\bar{D}^2/8\partial_\mu^2)\bar{J} \end{pmatrix} \qquad (20.74)$$

The functional integral can now be performed, leaving us with:

$$\begin{aligned}
Z(J, \bar{J}) &= \exp\left(-i \int d^8z\, d^8z'\, B^T(z')A^{-1}B(z)\right) \\
&= \exp\left(-\frac{i}{2}\int d^8z\, d^8z'\, \left[\frac{1}{2}J(z)\Delta_0(z, z')J(z')\right.\right. \\
&\quad \left.\left. + J(z)\Delta_1(z, z')\bar{J}(z') + \frac{1}{2}\bar{J}(z)\bar{\Delta}_0(z, z')\bar{J}(z')\right]\right) \quad (20.75)
\end{aligned}$$

where $d^8z = d^4x\, d^4\theta$. The propagators are given by:

$$\Delta_0 = -\frac{i}{4}\frac{m D^2}{p^2(p^2 - m^2)}$$

$$\Delta_1 = \frac{-i}{p^2 - m^2}\delta^4(\theta_1 - \theta_2) \qquad (20.76)$$

and where:

$$A^{-1} = \begin{pmatrix} -m\bar{D}^2/4(\partial_\mu^2 + m^2) & 1 + m^2\bar{D}^2 D^2/16\partial_\mu^2(\partial_\mu^2 + m^2) \\ 1 + m^2 D^2 \bar{D}^2/16\partial_\mu^2(\partial_\mu^2 + m^2) & -m D^2/4(\partial_\mu^2 + m^2) \end{pmatrix}$$

$$(20.77)$$

From these equations, we can write down the Feynman rules for a superfield theory. These rules will become crucial when we discuss nonrenormalization theorems.

20.5 Nonrenormalization Theorems

One truly remarkable feature of supersymmetry is the nonrenormalization theorem. This makes supersymmetric field theories a laboratory in which to test ideas about quantum field theory that have more sophisticated renormalization properties than ordinary ones.

The reason why supersymmetric theories have better convergence properties than ordinary field theories is that the fermion and boson loops of quantum field theory appear with opposite signs and hence cancel. For example, one can show that the quadratic divergences of simple supersymmetric theories cancel among themselves, leaving only logarithmic divergences. Furthermore, one can show that the mass and coupling constant corrections are actually finite to all orders in perturbation theory, and hence only the wave-function renormalization constant Z_ϕ appears.

One direct application of this is that the superpotential is not renormalized by higher-loop corrections, and hence any fine tuning of the potential will not receive any contributions from renormalization. There is no necessity of retuning the parameters at each order in perturbation theory. This gives us a potential solution to the hierarchy problem in GUT theory.

The proof that these cancellations persist to all orders in perturbation theory is prohibitive in the component formalism, where a series of miracles occur that cancel the divergent corrections to the mass and coupling constant. However, the proof to all orders in perturbation theory can be easily performed using the superspace method.

We recall that the propagators in the superspace formalism can be given by:

$$\langle \bar{\phi}(1)\phi(2) \rangle = \frac{1}{p^2 - m^2} \delta^4(\theta - \theta'); \quad \langle V(1)V(2) \rangle = -\frac{1}{p^2} \delta^4(\theta - \theta')$$

$$\langle \phi(1)\phi(2) \rangle = \frac{1}{4} \frac{mD^2}{p^2(p^2 - m^2)} \delta^4(\theta - \theta') \tag{20.78}$$

The vertices can be read off the Lagrangian, with the additional insertion of a factor of $-(1/4)\bar{D}^2$ [or $-(1/4)D^2$] acting on the propagator for each ϕ (or $\bar{\phi}$) line that leaves the vertex. There is also a factor of $\int d^4\theta$ at each vertex. Since we are integrating over a series of delta functions, we find that all θ integrations can be performed exactly. What is interesting is that, by simply counting Grassmann variables, we can show that $\int d^2\theta$ or $\int d^2\bar{\theta}$ *cannot* be the end product of all the θ integration. Only $\int d^4\theta\, f(\theta, \bar{\theta})$ survives the integration process.

However, we know that the masses and coupling constants all appear in the action via $m^2 \int d^2\theta\, \phi^2$ or $\lambda \int d^2\theta\, \phi^3$. Thus, corrections to these terms are finite. There is only the wave-function renormalization Z_ϕ function, which contributes to the mass and coupling constant renormalization via:

$$m \to Z_\phi^{-1} m; \quad g \to Z_\phi^{-3/2} g \tag{20.79}$$

Let us now analyze the terms that can be renormalized in gauge theory, that is, have the form $\int d^4\theta\, f(\theta, \bar{\theta})$. The degree of divergence of any graph (excluding the superpotential and terms that contain $\int d^2\theta$ or $\int d^2\bar\theta$) can be given as:

$$D = 2 - E - I \tag{20.80}$$

where E is the number of external chiral lines and I is the number of massive internal chiral propagators. From this, one can show that the possibly divergent contributions are given by:

$$\int d^4\theta \,\bar\phi\phi; \quad \int d^4\theta \,\bar\phi \,V\phi; \quad \int d^4\theta \,V; \quad \int d^4\theta \,VV; \quad \int d^4\theta \,VVV \tag{20.81}$$

We have used the fact that the dimensions of the fields and operators are given as:

$$[V] = 0; \quad [\phi] = 1; \quad [D_a] = \frac{1}{2}; \quad [d^4\theta] = 2 \tag{20.82}$$

Thus, all contributions (except for $\int d^4\theta \,V$), are logarithmically divergent. The nonrenormalization theorem states that we only have wave-function renormalization and gauge coupling renormalization constants Z_ϕ and Z_g, and that they are logarithmically divergent.

We should note that the quantity $\int d^4\theta \,V$, which is quadratically divergent, is gauge invariant only for $U(1)$. Thus, for non-Abelian theories, this term is absent. (Also, if $\text{Tr}\,Q = 0$, i.e., the trace of the charges of the scalar particles is zero, then this term also vanishes.)

20.6 Finite Field Theories

One of the most remarkable properties of supersymmetry is that supersymmetric field theories can be finite to all orders in perturbation theory, which was once thought to be impossible. [7-12] In some sense, these theories answer Dirac's old objections to quantum field theory, that renormalization theory was in some sense contrived and artificial.

We will now construct the global $SO(4)$ super Yang–Mills theory, which will turn out to be finite to all orders in perturbation theory, and then we will discuss the supergravity theory.

The $SO(4)$ Yang–Mills theory can be constructed by coupling the $N = 1$ super Yang–Mills theory, with the multiplet containing spins $(1, \frac{1}{2})$, to three copies of supersymmetric matter, containing the multiplet with spins $(\frac{1}{2}, 0)$. To construct the theory, we start with the usual Yang–Mills multiplet (A_μ, ψ) and then add three more fermion fields, which generalize ψ into four fields ψ_i. We must also, of course, add the corresponding scalar fields, which we choose to be self-dual and anti-self-dual:

$$A_{ij} = \frac{1}{2}\epsilon_{ijkl}A_{kl}; \quad B_{ij} = -\frac{1}{2}\epsilon_{ijkl}B_{kl} \tag{20.83}$$

Then the $SO(4)$ action becomes:

$$
\begin{aligned}
\mathcal{L} = \; & \mathrm{Tr}\Big(-\frac{1}{4}F_{\mu\nu}F^{\mu\nu} + \frac{1}{2}i\bar{\psi}_i\gamma^\mu D_\mu \psi_i + \frac{1}{8}D_\mu A_{ij}D^\mu A_{ij} + \frac{1}{8}D_\mu B_{ij}D^\mu B_{ij} \\
& + \frac{1}{2}i\bar{\psi}_i[\psi_j, A_{ij}] + \frac{1}{2}i\bar{\psi}_i\gamma_5[\psi_j, B_{ij}] + \frac{1}{32}[A_{ij}, B_{kl}][A_{ij}, B_{kl}] \\
& + \frac{1}{64}[A_{ij}, A_{kl}][A_{ij}, A_{kl}] + \frac{1}{64}[B_{ij}, B_{kl}][B_{ij}, B_{kl}] \Big) \tag{20.84}
\end{aligned}
$$

which is invariant under:

$$
\begin{aligned}
\delta A_\mu &= i\bar{\epsilon}_i\gamma_\mu \psi_i \\
\delta A_{ij} &= \bar{\epsilon}_i\psi_j - \bar{\epsilon}_j\psi_i + \epsilon_{ijkl}\bar{\epsilon}_k\psi_l \\
\delta B_{ij} &= \bar{\epsilon}_i\gamma_5\psi_j - \bar{\epsilon}_j\gamma_5\psi_i - \epsilon_{ijkl}\bar{\epsilon}_k\gamma_5\psi_l \\
\delta \psi_i &= -\frac{1}{2}i\sigma^{\mu\nu}\epsilon_i F_{\mu\nu} + i\gamma^\mu D_\mu(A_{ij} + \gamma_5 B_{ij})\epsilon_j \\
& \quad + \frac{1}{2}i[A_{ij} - \gamma_5 B_{ij}, \; A_{jk} + \gamma_5 B_{jk}]\epsilon_k \tag{20.85}
\end{aligned}
$$

The first indication that this action possessed remarkable renormalization properties came from the renormalization group, where it was noticed that β vanished to the first-, second- and even third-loop level.

For the single-loop β function, we have a slight modification of the result found for Dirac fermions:

$$\beta = -\frac{g^3}{16\pi^2}\frac{C_2(G)}{6}[22 - 4\nu(M) - \nu(R)] \tag{20.86}$$

where ν are the number of Majorana fermions or real scalar fields. (There is one Majorana fermion in the gauge multiplet. For each chiral superfield, there are two Majorana fermions and two real scalar fields.) This function vanishes for $N = 4$ super Yang–Mills theory, since there are three chiral superfields; therefore, $\nu(M) = 1 + 3 = 4$ and $\nu(R) = 2 \times 3 = 6$.

For $N = 1$ supersymmetry coupled to n chiral multiplets, the two-loop result is:

$$\beta = -\frac{g^3}{16\pi^2}\frac{C_2(G)}{6}(18 - 6n) - \frac{g^5}{(16\pi^2)^2}C_2(G)^2(6 - 10n + 8n) \tag{20.87}$$

For $N = 4$ super Yang–Mills theory, we have $n = 3$; therefore, β also cancels at the two-loop level.

Since then, there have been several different proofs showing that the theory is actually finite to all orders in perturbation theory. We will only summarize some of the arguments. One proof of the theorem rests on the fact that the $N = 4$ theory has the symmetry of the $N = 1$ and $N = 2$ theories as subsets. This yields a large number of constraints among the various renormalization constants, eventually giving us $Z = 1$ for all of them. Let us begin, for the moment, with the $N = 1$ super Yang–Mills theory coupled to supersymmetric matter fields. Let us couple enough $N = 1$ multiplets so that we have the same number of fields as the $N = 4$ super Yang–Mills theory. However, although we have enough fields to construct the $N = 4$ theory, assume that the coupling constants are *not* correlated with each other as in the $N = 4$ theory, so we only have $N = 1$ supersymmetry.

Our strategy will be to show that, as we gradually change the coupling constants so we recover $N = 2$ and then $N = 4$ supersymmetry, the resulting renormalization constants also change, until they all reduce to $Z_i = 1$ at the end. To construct the action, we will introduce a vector multiplet V and a matter multiplet Ψ_a^b in the adjoint representation. Also, we introduce chiral matter multiplets ϕ^a as well as ϕ_b. Then the most general superpotential one can write down for these fields is given by:

$$b^a \phi_a + b'_a \phi^a + b_a^b \Psi_b^a - m\phi^a \phi_a + g\phi^a \Psi_a^b \phi_b \qquad (20.88)$$

At this point, there is no correlation between the various coupling constants.

If we now turn on interactions, then we have the following possible renormalization constants:

$$V^2 \;\rightarrow\; Z_V V^2; \quad g_V \rightarrow Z_g g_V$$

$$\phi_a^2 \;\rightarrow\; Z_\phi \phi_a^2; \quad g_\phi \rightarrow Z'_g g_\phi$$

$$(\phi^a)^2 \;\rightarrow\; Z'_\phi (\phi^a)^2; \quad m \rightarrow m + \delta m$$

$$(\Psi_a^b)^2 \;\rightarrow\; Z_\Psi (\Psi_a^b)^2 \qquad (20.89)$$

We can also show (because of the nonrenormalization of e^{2V} and because of the properties of chirally supersymmetric interactions):

$$Z_g^2 Z_V \;=\; 1; \quad (Z'_g)^2 Z_\Psi Z_\phi Z'_\phi = 1$$

$$(1 + \delta m/m)^2 Z_\phi Z'_\phi \;=\; 1 \qquad (20.90)$$

which come from the nonrenormalization theorems for e^{2V} and for chiral superspace integrals.

Now we impose an additional constraint. We will change the coupling constants so that the matter multiplets and the $N = 1$ super Yang–Mills theory form $N = 2$ multiplets. This means that the gaugino from V and the spinor from Ψ form a doublet; so we have a new constraint on the coupling constants. Since the original coupling constants were arbitrary, we have the freedom to choose $N = 2$ symmetry, which gives us the additional restriction:

$$Z_g = Z_g'; \quad Z_V = Z_\Psi \tag{20.91}$$

which in turn implies:

$$Z_\phi Z_\phi' = 1; \quad \delta m = 0 \tag{20.92}$$

At this point, we have eliminated all but three renormalization constants, one for each of the three matter multiplets. However, we still have the freedom to change the remaining coupling constants so that the matter multiplets and the super Yang–Mills theory form $N = 4$ super multiplets. This further restriction on the coupling constants implies a symmetry between all the matter multiplets, so that:

$$Z_\phi = Z_\phi' = Z_\Psi \tag{20.93}$$

However, from Eq. (20.90), this also implies that:

$$Z = 1 \tag{20.94}$$

for all renormalization constants in the theory. *Thus, the $N = 4$ theory is finite to all orders in perturbation theory.*

By somewhat similar arguments, one can show that the $N = 2$ theory is finite for all higher loop levels beyond the one-loop level (where it diverges). If we relax $N = 4$ supersymmetry but keep $N = 2$ symmetry, then we lose the condition that $Z_\phi = Z_\phi'$ and hence lose finiteness. Hence an ordinary $N = 2$ theory is not finite. However, we still may be able to salvage this proof if we can find another way in which to make $Z_\phi = Z_\phi'$.

The way to patch up the proof of finiteness for $N = 2$ theories is to use a modification of the $N = 2$ theory, with real representations, in which a $SU(2)$ symmetry emerges that rotates ϕ_a into ϕ^a. Because of this additional symmetry, we can now equate Z_ϕ with Z_ϕ', and then the only divergences in this $N = 2$ theory come from Z_V, which is one-loop divergent. Thus, we have proved that the $N = 2$ theory, with real representations, is divergent only at the one loop-level and finite at all higher orders.

However, we can modify the theory still further by adding more multiplets to eliminate the one-loop divergence, rendering the modified theory completely finite to all orders. Once the one-loop divergence is eliminated, the modified $N = 2$ theory is finite to all orders.

In summary, *the N=4 super Yang–Mill theory and certain modified N=2 super Yang–Mills theories are finite to all orders.* These are remarkable results that are totally unexpected. However, this property does not persist when we build a supersymmetric theory of gravity. Although supergravity is much better behaved than ordinary Einstein gravity, it is divergent at the three-loop level.

20.7 Super Groups

Now that we have accumulated some practice in constructing supersymmetric theories, let us now analyze more systematically the group theoretic structure of super groups. We know from the classical works of Lie and Cartan that a complete classification of compact Lie groups is possible. Kac has generalized this result and given us a complete classification of the super groups.

Although there are many possible super groups, the only ones in which we are interested are the ones that generalize the standard space–time groups found in physics, that is, the Poincaré group and (for massless theories) the conformal group, $SU(2, 2) = O(4, 2)$. Each group, in turn, is part of an infinite series of super groups, which are called the orthosymplectic $Osp(N/M)$ and the superconformal $SU(N/M)$ groups, respectively. [21]

To see how these super groups are constructed, we recall that the orthogonal group $O(N)$ is the set of all real orthogonal $N \times N$ matrices that leave the following form invariant:

$$O(N): \quad x_i x_i = \text{invariant} \tag{20.95}$$

Likewise, the *symplectic group* is the set of $N \times N$ matrices that leave the following form invariant:

$$Sp(N): \quad \theta_m C_{mn} \theta_n = \text{invariant} \tag{20.96}$$

where C_{mn} is a real, antisymmetric matrix and the θ_n are anticommuting numbers.

The orthosymplectic group $Osp(N/M)$ is the set of matrices that leave the following form invariant:

$$Osp(N/M): \quad x_i x_i + \theta_m C_{mn} \theta_n = \text{invariant} \tag{20.97}$$

for $i = 1, 2, \ldots, N$ and $n = 1, 2, \ldots, M$. Not surprisingly, the algebra of the orthosymplectic group can be decomposed into blocks that contain the matrices of $Sp(M)$ and $O(N)$:

$$Osp(N/M) = \begin{pmatrix} O(N) & A \\ B & Sp(M) \end{pmatrix} \tag{20.98}$$

where A and B are determined by the commutation and anticommutation relations of the Jacobi identity. $O(N)$ and $Sp(M)$ are therefore subgroups of the orthosymplectic group:

$$O(N) \otimes Sp(M) \subset Osp(N/M) \tag{20.99}$$

More concretely, we are interested in the group $Osp(1/4)$, which is the gauge group of supergravity. It contains the symplectic group, $Sp(4)$, which is isomorphic to the de Sitter group, and contains the same number of generators as the Poincaré group, P_μ and $M_{\mu\nu}$. Its commutation relations differ only slightly from those of the Poincaré group; that is, the commutator of two translations $[P_\mu, P_\nu]$ is proportional to the Lorentz generator divided by the square of a length, which is called the de Sitter radius. In the limit of infinite de Sitter radius, two translations commute, and hence we have the same commutation relations as the Poincaré group (see Exercise 14.7).

We also point out that the second physically interesting super group is the superconformal group. The group $SU(N)$, of course, is the set of unitary $N \times N$ matrices with unit determinant. They leave the following form invariant:

$$SU(N): \quad (u^i)^* u^j \delta_{ij} \tag{20.100}$$

Not surprisingly, the superconformal group $SU(N/M)$ is the group with elements that leave the following form invariant:

$$(u^i)^* u^j \delta_{ij} + (\theta^m)^* g_{mn} \theta^n \tag{20.101}$$

where $i = 1, 2, \ldots, N$, $m = 1, 2, \ldots, M$, and where $g_{mn} = \pm\delta_{mn}$.

$SU(N/M)$ can naturally be decomposed into the following form:

$$SU(N) \otimes SU(M) \otimes U(1) \subset SU(N/M) \tag{20.102}$$

Let us now be concrete about the generators of $Osp(N/4)$. We know that this orthosymplectic group must contain the generators of $O(N)$, which we call T_i, as well as the generators of the symplectic group $Sp(4)$, which can be represented by the usual Poincaré generators P_μ and $M_{\mu\nu}$. We also have the supersymmetric generator Q_{ai}, which now carries the $O(N)$ index i, as well as the two-spinor index a. Then the generators of the super group are:

$$[T_i, T_j] = ic_{ij}^k T_k; \quad \{Q_{ai}, \bar{Q}_{bj}\} = 2\delta_{ij}(\sigma^\mu)_{ab} P_\mu$$

$$\{Q_{ai}, Q_{bj}\} = 2\epsilon_{ab} Z_{ij}; \quad \{\bar{Q}_{\dot{a}i}, \bar{Q}_{\dot{b}j}\} = -2\epsilon_{\dot{a}\dot{b}} Z_{ij}$$

$$[Q_{ai}, M_{\mu\nu}] = \frac{1}{2}(\sigma_{\mu\nu})_a^b Q_{bi}; \quad [\bar{Q}_{\dot{a}i}, M_{\mu\nu}] = -\frac{1}{2}\bar{Q}_{\dot{b}i}(\bar{\sigma}_{\mu\nu})_{\dot{a}}^{\dot{b}}$$

$$[Q_{ai}, T_j] = (b_j)_i^k Q_{ak}; \quad [\bar{Q}_{\dot{a}i}, T_j] = -(b_j)_i^k \bar{Q}_{\dot{a}k}$$

$$[T_i, P_\mu] = [T_i, M_{\mu\nu}] = 0$$

$$[Z_{ij}, \text{anything}] = 0 \tag{20.103}$$

where c_{ij}^k are the structure constants for $O(N)$, where the Z_{ij} are certain linear combinations of the generators T_i:

$$Z_{ij} = a_{ij}^k T_k \tag{20.104}$$

for the constant matrices a_{ij}^k, and the matrices b_i^k are Hermitian.

Using simple arguments, one can show that $SO(4)$ and $SO(8)$ are the largest possible groups for super Yang–Mills theory and for supergravity, respectively. The proof of this important fact is rather simple. We know that the supersymmetric generator for $SO(N)$ supergravity is given by Q_α^i, where α is a spinor index and i is an $SO(N)$ index, where $i = 1, \ldots, N$. We also know from group theoretical arguments that the spectrum of supergravity states can be generated by hitting the lowest helicity state $|-\rangle$ successively with the Q_α^i operator. For the super Yang–Mills theory, the field with the lowest helicity is the spin 1 vector particle, while for supergravity it is the spin-2 graviton.

If we act with this operator once, we have:

$$Q_\alpha^i |-\rangle \rightarrow \psi_\alpha^i \tag{20.105}$$

In the super Yang–Mills theory, this means that the partner of the Yang–Mills field is a spin $\frac{1}{2}$ field with isospin index i. In the supergravity theory, it means that the superpartner of the graviton is a spin 3/2 gravitino that also has an isospin index i.

Similarly, we can hit the lowest helicity state with two supersymmetric generators:

$$Q_\alpha^i Q_\beta^j |-\rangle \tag{20.106}$$

For the super Yang–Mills theory, this means that the spin 0 particle has two indices i, j that are antisymmetric. Thus, there are $N(N-1)/2$ such scalar particles. For the supergravity theory, this state corresponds to a vector particle that transforms as $A_\mu^{i,j}$, where i, j are antisymmetric.

This procedure can obviously be repeated an arbitrary number of times, each time generating spinning particles with isospin indices that are antisymmetric in $i, j, k, \ldots$. However, there is an important restriction. For the super Yang–Mills theory, we want the highest spin in the theory to be the Yang–Mills field. Thus, we can only hit the lowest helicity state $|-\rangle$ four times with $Q_{\alpha i}$ until we arrive at $|+\rangle$, which is the highest helicity state corresponding to the Yang–Mills field. Therefore, the maximum orthogonal group that we can accomodate without going to higher spins is $SO(4)$, because there are four half-steps in spin between the lowest and highest helicity state of the vector particle.

Counting antisymmetric indices, it is easy to see that the helicities and number of states for the $N = 4$ multiplet are given by:

$$
\begin{array}{|c|c|c|c|c|c|}
\hline
\text{Helicity :} & -1 & -\frac{1}{2} & 0 & \frac{1}{2} & 1 \\
\hline
\text{States :} & 1 & 4 & 6 & 4 & 1 \\
\hline
\end{array}
\tag{20.107}
$$

There are $1 + 6 + 1 = 8$ bosonic states and $4 + 4$ fermionic states, so we have equal numbers of bosons and fermions, as expected in any supersymmetric theory.

Similarly, we can only hit the graviton state with the lowest helicity $|-\rangle$ eight times with $Q_{\alpha i}$ until we arrive at $|+\rangle$, the highest helicity state that also corresponds to the graviton. We must stop at this point, because an interacting massless spin 3 theory is known to be inconsistent. Thus, this procedure must not generate spins beyond 2, or else we lose self-consistency. Since there are 8 half-steps between $|-\rangle$ and $|+\rangle$, the maximum number of generators Q_α^i must be $N = 8$. Thus, the highest symmetry group must be $SO(8)$. This is rather unfortunate, because this group is too small to accomodate the Standard Model.

If we count the helicity states as before, then the $N = 8$ multiplet has the following number of states:

$$
\begin{array}{|c|c|c|c|c|c|c|c|c|c|}
\hline
\text{Helicity :} & -2 & -\frac{3}{2} & -1 & -\frac{1}{2} & 0 & \frac{1}{2} & 1 & \frac{3}{2} & 2 \\
\hline
\text{States :} & 1 & 8 & 28 & 56 & 70 & 56 & 28 & 8 & 1 \\
\hline
\end{array}
\tag{20.108}
$$

Counting helicity states as before, we find that the number of antisymmetric indices $i, j, k, \ldots$ in a p-rank tensor is equal to:

$$
\binom{8}{p}
\tag{20.109}
$$

The total number of fields in the $SO(8)$ theory is therefore given by $1 + 8 + 28 + 56 + 70 + 56 + 28 + 8 + 1 = 256$. The number of bosonic states is given by $1 + 28 + 70 + 28 + 1 = 128$. Likewise, the total number of fermionic states is given by $8 + 56 + 56 + 8 = 128$. So we have an equal number of boson and fermion fields (on-shell).

20.8 Supergravity

Up to now, we have only considered global supersymmetry. However, the real beauty of this approach emerges when we consider gauging the supersymmetric group to produce a gauge theory of a new type. In this way, we will see that supergravity necessarily emerges when we gauge the super group.

There are many ways in which to formulate supergravity. The approach we will take will mimic the Yang–Mills approach as much as possible. We begin with the super group $Osp(1/4)$, which has 14 generators. In addition to the four supersymmetric operators Q^α, there are also the 10 generators of $Sp(4)$, which can be arranged as in the Poincaré group, consisting of the Lorentz generators M_{ab} and the translations P_a. As in Yang–Mills theory, we will introduce a separate connection field for each of the generators of $Osp(1/4)$.

Let M_A collectively refer to all the generators of $Osp(1/4)$. They satisfy:

$$[M_A, M_B\} = f_{AB}^C M_C \qquad (20.110)$$

where f_{AB}^C are the structure constants of the group, and we have both commutators and anticommutators in the algebra.

Let ω_μ^A collectively refer to all the connection fields. The fields transform as:

$$\delta\omega_\mu^A = f_{BC}^C \epsilon^B \omega_\mu^C \qquad (20.111)$$

where:

$$
\begin{aligned}
\omega_\mu^A &= \{e_\mu^a, \omega_\mu^{ab}, \bar\psi_\mu^\alpha\} \\
M_A &= \{P_a, -i M_{ab}, Q^\alpha\}
\end{aligned}
\qquad (20.112)
$$

The covariant derivative is therefore:

$$
\begin{aligned}
\nabla_\mu &= \partial_\mu + \omega_\mu^A M_A \\
&= \partial_\mu + e_\mu^a P_a - i\omega_\mu^{ab} M_{ab} + \bar\psi_{\mu,\alpha} Q^\alpha
\end{aligned}
\qquad (20.113)
$$

[Notice that gauging the translations P_a necessarily introduces a connection field e^a_μ, which is the vierbein field. Thus, gauging the super group $Osp(1/4)$ necessarily introduces the graviton field. There is no other choice. Local supersymmetry necessarily creates a theory of quantum gravity.]

The commutator of two covariant derivatives yields the curvature tensor:

$$[\nabla_\mu, \nabla_\nu] = R^A_{\mu\nu} M_A \qquad (20.114)$$

where:

$$R^A_{\mu\nu} = \partial_\mu \omega^A_\nu - \partial_\nu \omega^A_\mu + \omega^B_\nu \omega^C_\mu f^A_{CB} \qquad (20.115)$$

In component form, this reduces to:

$$
\begin{aligned}
R^a_{\mu\nu}(P) &= \partial_\mu e^a_\nu + \omega^{ab}_\mu e^b_\nu - (\mu \leftrightarrow \nu) \\
R^{ab}_{\mu\nu}(M) &= \partial_\mu \omega^{ab}_\nu + \omega^{ac}_\mu \omega^{cb}_\nu + e^a_\mu e^b_\nu - (\mu \leftrightarrow \nu) \\
R^a_{\mu\nu}(Q) &= \partial_\mu \bar\psi^\alpha_\nu + \bar\psi_\nu \omega^{ab}_\mu \sigma^{ab} + \bar\psi_\nu e^a_\mu \gamma^a - (\mu \leftrightarrow \nu)
\end{aligned}
$$

$$(20.116)$$

The variation of the curvature is equal to:

$$\delta R^A_{\mu\nu} = R^B_{\mu\nu} \epsilon^C f^A_{CB} \qquad (20.117)$$

The action for supergravity is now given by contracting the curvature tensors via the antisymmetric invariant tensors:

$$\mathcal{L} = \epsilon^{\mu\nu\rho\sigma} \left[R_{\mu\nu}(M)^{ab} R^{cd}_{\rho\sigma} \epsilon_{abcd} + R_{\mu\nu}(Q)^\alpha R_{\rho\sigma}(Q)^\beta (\gamma_5 C)_{\alpha\beta} \right] \qquad (20.118)$$

where C is the charge conjugation matrix. If we make a variation of the action, we find that the action is not invariant unless we enforce the condition:

$$R^a_{\mu\nu}(P) = 0 \qquad (20.119)$$

The action, at first, appears to be a R^2–type action, which is not unitary (because of ghosts). However, this is an illusion. The R^2 term is actually a total derivative and topologically invariant. Hence, it can be dropped from the action. The cross terms are linear in R and give us the supergravity action:

$$\mathcal{L} = -\frac{1}{2\kappa^2} eR - \frac{1}{2} \bar\psi_\mu \gamma_\nu \gamma^5 D_\sigma \psi_\rho \epsilon^{\mu\nu\rho\sigma} \qquad (20.120)$$

(There is also a term proportional to e^4, which comes from a generalization of the cosmological constant term divided by the de Sitter radius to the fourth power. We can drop this cosmological term if we set the de Sitter radius to infinity.)

In a similar fashion, we can also introduce the curvatures for the group $SU(2, 2/1)$, the superconformal group. By contracting products of these curvatures, one can write down a higher derivative theory that is locally superconformally symmetric, called conformal supergravity. [22]

Although the $N = 1$ supergravity action was relatively easy to construct, there are severe complications when generalizing this to $N = 8$ supersymmetry. To construct the $SO(8)$ action in the component formalism is prohibitively difficult. Even the superfield method is prohibitive in this case. Instead, we will construct the $SO(8)$ action by using a trick. We will formulate the $N = 1$ supergravity theory in 11 dimensions. Since the symmetry group for this higher-dimensional theory is only $N = 1$ supersymmetry, the action is easier to write down. Then we will use dimensional reduction or compactification to yield an $N = 8$ action in four dimensions.

To see that the number of states formally is the same, let us count the number of states within an 11-dimensional supergravity. The counting of states proceeds as follows:

$$e_M^A \quad \rightarrow \quad \frac{1}{2} 9 \times 10 - 1 = 44$$

$$\psi_M \quad \rightarrow \quad \frac{1}{2}(9 \times 32 - 32) = 128$$

$$A_{MNP} \quad \rightarrow \quad \binom{9}{3} = 84 \qquad\qquad (20.121)$$

where M, N represent 11-dimensional curved space indices, A, B represent flat space 11-dimensional indices, and the spinors are 32 dimensional. e_M^A represents the 11-dimensional vierbein linking the base manifold with the tangent space. ψ_M is a graviton field, and A_{MNP} is an antisymmetric tensor field. The total number of boson fields is 128, which equals the number of fermion fields, as it should. The total number of boson and fermion fields is thus 256, which is precisely the number of fields in the 11-dimensional $N = 8$ model. Thus, not only do we have equal numbers of fermions and bosons, we also have the same number that appear in the 11-dimensional action.

With some work, one can show that the action for 11-dimensional supergravity is [23]:

$$L = -\frac{1}{2\kappa^2} eR - \frac{1}{2} e\bar{\psi}_M \Gamma^{MNP} D_N [\frac{1}{2}(\omega + \hat{\omega})]\psi_P - \frac{1}{48} e F_{MNPQ}^2$$

$$- \frac{\sqrt{2}\kappa}{384} e \left(\bar{\psi}_M \Gamma^{MNPQRS} \psi_S + 12 \bar{\psi}^N \Gamma^{PQ} \psi^R \right) (F - \hat{F})_{NPQR}$$

$$- \frac{\sqrt{2}\kappa}{3456} \epsilon^{M_1 \cdots M_{11}} F_{M_1, \cdots, M_4} F_{M_5, \cdots, M_8} A_{M_9, M_{10}, M_{11}} \qquad (20.122)$$

which is invariant under:

$$\delta e_M^A = \frac{1}{2} \kappa \bar{\eta} \Gamma^A \psi_M$$

$$\delta A_{MNP} = -\frac{\sqrt{2}}{8} \bar{\eta} \Gamma_{[MN} \psi_{P]}$$

$$\delta \psi_M = \kappa^{-1} D_M(\hat{\omega}) \eta + \frac{\sqrt{2}}{288} (\Gamma_M^{PQRS} - 8\delta_M^P \Gamma^{QRS}) \eta \hat{F}_{PQRS}$$

$$(20.123)$$

and where:

$$\hat{\omega}_{MAB} = \omega_{MAB} + \frac{1}{8} \bar{\psi}^P \Gamma_{PMABQ} \psi^Q \qquad (20.124)$$

As mentioned earlier, the $SO(8)$ action is too small to include the Standard Model. If we go to higher supergravity theories beyond $SO(8)$, then we have interacting massless spin 3 fields, which are known to be inconsistent. This is disappointing.

One alternative is to couple supergravity to supersymmetric Yang–Mills fields with the gauge group given by the Standard Model. The addition of supergravity gives us nontrivial corrections to the effective potential in Eq. (20.66), which allow us to relax the stringent condition in Eq. (20.67). Supergravity coupled to super Yang–Mills theory thus has interesting phenomenology. However, this coupled theory diverges at the one-loop level, making it less convergent than supergravity (which diverges at the third-loop level) or even ordinary quantum gravity (which diverges at the second loop level). Thus, supergravity coupled to a super Standard Model has phenomenologically good properties, except that it is highly divergent.

In summary, we have seen that supersymmetric theories give us a theoretical laboratory to study field theories with radically different properties.

First, they mix isospin and space–time symmetries in a nontrivial way, thereby evading the Coleman–Mandula theorem. This gives us the hope of eventually putting all subatomic particles in the same irreducible representation.

Second, they can cancel enough divergences to render the $N = 4$ and certain $N = 2$ super Yang–Mills theories finite to all orders in perturbation theory. This realizes Dirac's original dream of field theories that do not require renormalizations.

Third, they can, in principle, solve the hierarchy problem. Supersymmetric theories have powerful nonrenormalization theorems that prove that the mass separation between GUT scale and low-energy physics is not renormalized.

Fourth, their local version necessarily contains gravity. Supersymmetric Ward identities can remove the first- and second-loop divergences, although they fail at the third loop level. Unfortunately, the maximum supergravity has $SO(8)$ symmetry, which is too small to include the Standard Model. In addition, the theory is not renormalizable.

Faced with the divergence of supergravity, one is forced to enlarge the gauge group, hoping to generate enough Ward identities that can cancel all possible counterterms. The only known nontrivial generalization of supergravity is the superstring theory, to which we turn in the next chapter.

20.9 Exercises

1. In quantum field theory, we must throw away, by hand, the infinite zero point energy of the scalar and fermionic fields. Show that in a simple supersymmetric theory, the zero-point energies of the bosonic and fermionic fields cancel by themselves.

2. Perform the integration over the Grassmann variables in the Wess–Zumino action in Eq. (20.46). Show that its free part is equivalent to the action written in terms of components in Eq. (20.13).

3. Perform the integration over the action (20.50) in the Wess–Zumino gauge and show that we recover the supersymmetric Yang–Mills theory in Eq. (20.15).

4. By direct computation, prove that the supersymmetric Yang–Mills theory in Eq. (20.15) is invariant under a supersymmetric transformation.

5. Prove (only to lowest order) that the $SO(4)$ Yang–Mills action in Eq. (20.84) is invariant under Eq. (20.85).

6. Consider the constraint $R^a_{\mu\nu}(P) = 0$ in Eq. (20.119). Show that this constraint is equivalent to the vanishing of the covariant derivative of the vierbein.

7. Prove all the relations in Eq. (20.40).

8. Write down an expression for the Noether (super) current for the Wess–Zumino model.

9. In supergravity, show that the anticommutator of two supersymmetry variations of the gravitino does not close properly, showing that the theory does not really form a group structure in the usual sense. (Show the presence of a few

terms that do not close, not the whole expression.) Show that the noninvariant terms are, in fact, proportional to the equations of motion, so that they vanish on-shell.

10. e_μ^a has 10 degrees of freedom, and ψ_μ has 16 degrees of freedom, yet we know that, in the canonical formalism of supergravity, we are only left with two helicities for both. Show how, using gauge invariance, we can remove all degrees of freedom down to two helicities.

11. In supergravity, the basic fields are the e_μ^a and the gravitino ψ_μ. Perform the counting of states both on-shell and off-shell. Show that off-shell, there is mismatch of 6 fields, requiring auxiliary fields.

12. To compensate for these 6 missing boson fields, let us add nonpropagating fields A_μ and S and P to the supergravity action:

$$L = -\frac{1}{2} eR - \frac{1}{2} \epsilon^{\mu\nu\rho\sigma} \bar{\psi}_\mu \gamma_5 \gamma_\nu D_\rho \psi_\sigma - \frac{e}{3}(S^2 + P^2 - A_\mu^2) \qquad (20.125)$$

Show (to lowest order) that this new action is invariant under:

$$\delta e_\mu^a = \frac{1}{2} \bar{\epsilon} \gamma^a \psi_\mu$$

$$\delta \psi_\mu = (D_\mu + \frac{i}{2} A_\mu \gamma_5)\epsilon - \frac{1}{2} \gamma_\mu \eta \epsilon$$

$$\delta S = \frac{1}{4} \bar{\epsilon} \gamma \cdot R^{\text{cov}}$$

$$\delta P = -\frac{i}{4} \bar{\epsilon} \gamma_5 \gamma \cdot R^{\text{cov}}$$

$$\delta A_\mu = \frac{3i}{4} \bar{\epsilon} \gamma_5 (R_\mu^{\text{cov}} - \frac{1}{3} \gamma_\mu \gamma \cdot R^{\text{cov}})$$

$$\eta = -\frac{1}{3}(S - i\gamma_5 P - i A\!\!\!/ \gamma_5) \qquad (20.126)$$

where:

$$R^{\mu,\text{cov}} = \epsilon^{\mu\nu\rho\sigma} \gamma_5 \gamma_\nu (D_\rho \psi_\sigma - \frac{i}{2} A_\sigma \gamma_5 \psi_\rho + \frac{1}{2} \gamma_\sigma \eta \psi_\rho) \qquad (20.127)$$

13. Show that the supergravity algebra with these auxiliary fields closes properly off-shell, without having to invoke the equations of motion. Calculate the closure of the algebra for the gravitino field. (Only calculate the lowest-order terms. Show that the terms that previously destroyed the closure of the algebra cancel.)

14. Show that, if we eliminate ω_μ^{ab} in Eq. (20.120), we find that the connection field picks up a contribution from the gravitino field. Show that this new term is proportional to the torsion (i.e., $\Gamma_{\mu\nu}^\lambda - \Gamma_{\nu\mu}^\lambda$).

15. Write down the supersymmetric version of $SU(5)$ GUT, introducing separate superfields for the various representations in the theory. Do not spontaneously break the theory.

16. Using superfield methods, show that the simplest one-loop graph in the Wess-Zumino model, after Grassmann integrations, is of the form $\int d^4\theta f(\theta, \bar\theta)$.

Chapter 21
Superstrings

But the creative principle resides in mathematics. In a certain sense, therefore, I hold it true that pure thought can grasp reality, as the ancients dreamed.

— A. Einstein

21.1 Why Strings?

At present, there is only one finite theory of quantum gravity, and this is the superstring theory. In this sense, the theory has no rivals. In addition, the theory can apparently reproduce all the known particle interactions found in nature. The fact that one can, in principle, construct solutions that include both gravity and all known interactions from such a simple physical picture, the string, is rather remarkable.

The desirable properties of string theory, as usual, derive from its powerful gauge groups. The gauge groups of the superstring include:

1. *Conformal and superconformal invariance.* These are the symmetries defined on the two-dimensional surfaces swept out by the string.

2. *General coordinate transformation.* Being a theory of quantum gravity, it possesses space–time reparametrization invariance.

3. $E_8 \otimes E_8$. This gauge group emerges when we compactify some of the higher dimensions of the theory.

4. *Space–time supersymmetry.* This symmetry helps to solve the hierarchy problem and cancel some of the potential infinities of the theory.

The symmetries found in particle physics and general relativity therefore emerge as a tiny subset of the symmetries of the superstring. In addition to being a finite theory of gravity, the theory also has definite phenomenological advantages over other theories:

1. The group $E_8 \otimes E_8$ is large enough to contain GUT theory. Supergravity, by contrast, was limited by its isospin group $SO(8)$, which was too small for phenomenology.

2. The superstring can accomodate complex fermion representations like those found in the Standard Model because it is not based on Riemannian manifolds. Kaluza–Klein theory, being Riemannian, cannot accomodate these fermion representations.

3. The superstring theory is completely free of anomalies. Gravity theory in higher dimensions, by contrast, has problems with anomalies once we have chiral fermions.

4. The superstring theory gives a plausible explanation of the generation problem in the Standard Model in terms of certain topological invariants that exist on six-dimensional manifolds. GUT theory, by contrast, cannot explain the presence of fermion generations.

5. The superstring theory has no hierarchy problem because of powerful non-renormalization theorems. The GUT theory, however, has a hierarchy problem.

We caution, however, that as with all theories defined at the Planck energy 10^{19} GeV, like quantum gravity, the superstring theory is subject to the severe criticism that it cannot be tested with present technology. Predictions of the theory, for example, that space–time was actually ten dimensional at the instant of the Big Bang, are beyond experimental verification. Unlike GUT theory, which yields testable predictions in the form of proton decay, it is difficult to find an experiment that can rule out (or in) superstring theory in the coming years.

Our philosophy, as we said before, is to treat superstring theory as a theoretical tool, as an example of a field theory which has highly nontrivial features that can probe the limits of quantum field theory. Underlying the superstring theory is a genuine quantum field theory; from its Lagrangian, we can derive the standard quantization rules and find the spectrum of states and the Feynman-like rules. This quantum field theory of strings apparently satisfies all the nontrivial constraints postulated for the S matrix. This quantum field theory also contains perhaps the most sophisticated, self-consistent Lagrangian to appear in physics, and hence deserves serious study. If we take particular subsets of this action, then we can find the usual actions describing quantum gravity, supergravity, gauge theory, and GUTs.

With a few rather mild assumptions, one can also find classical solutions to the string equations that come surprisingly close to the Standard Model, including three generations of fermions with a stable hierarchy. However, the theory also has an additional problem: It has millions of other solutions. It is not known how to select the true vacuum among the millions that have been found. A nonperturbative analysis is probably required to find the true vacuum of the theory, which is beyond our current calculational ability. The main problem facing superstring theory is thus theoretical, rather than experimental. If the true vacuum of the theory could be found theoretically, it should be possible to make a direct comparison with experiment. At that point, one can decide whether or not it correctly describes all quantum forces. But until the true vacuum is found, the theory does not have true predictive power. Until then, the superstring theory will remain a highly sophisticated quantum field theory without direct physical application.

Because of the mathematical complexity of string theory, we will only sketch some highlights of the theory in this chapter. The reader is referred to the literature for a more detailed explanation of the theory.

21.2   Points versus Strings

String theory, at first appearance, seems strange because it was historically formulated as a first quantized theory, rather than as a second quantized field theory. Therefore, it will be instructive to examine the simplest first quantized system, the relativistic point particle, and later develop the second quantized theory of points and strings.

Let the coordinate $x^\mu(\tau)$ represent a vector that points from the origin of our system to the location of a point particle. As the particle moves, it sweeps out a line, called the world-line, parametrized by τ. The action is proportional to the invariant length swept out by the world-line:

$$S = -m \int d\tau \sqrt{(\dot{x}^\mu)^2} \sim \text{length} \tag{21.1}$$

This action is invariant under reparametrizations of τ:

$$\tau \rightarrow \tilde{\tau}(\tau) \tag{21.2}$$

To quantize the action, we first compute the momenta:

$$p_\mu = \frac{\delta \mathcal{L}}{\delta \dot{x}^\mu}$$

$$= -m \frac{\dot{x}_\mu}{\sqrt{(\dot{x}^\mu)^2}} \tag{21.3}$$

Because of the last equation, the momenta are not all independent, signaling the presence of a gauge invariance. To find the precise dependence of the momenta, we take its square:

$$p_\mu^2 \equiv m^2 \tag{21.4}$$

Finally, we now apply this constraint directly onto the state vectors $|\phi\rangle$ of the theory:

$$\left(p_\mu^2 - m^2\right)|\phi\rangle = 0 \tag{21.5}$$

as in the Gupta–Bleuler formalism. At this point, we recognize this as the wave equation for the usual Klein–Gordon equation. This equation, in turn, can be derived from the standard covariant second quantized action:

$$S = \frac{1}{2} \int d^4x \, \phi(x) \left(\partial_\mu \partial^\mu + m^2\right) \phi(x) \tag{21.6}$$

In this way, we have made the transition from the first to the second quantized formalism for free point particles.

Alternatively, the reparametrization invariance of the theory allows us to select the following gauge choice:

$$x_0 = \tau \tag{21.7}$$

In this gauge, the action assumes the familiar nonrelativistic, ghost-free form in the limit of small velocities:

$$\begin{aligned} S &= -m \int d\tau \sqrt{1 - v_i^2} \\ &\sim \int d\tau \, \frac{1}{2} m v_i^2 \end{aligned} \tag{21.8}$$

Unfortunately, the first quantized formalism treats interactions in a rather clumsy fashion. To introduce scattering, we do *not* add an explicit interaction term, as in the second quantized formalism. Instead, we define the scattering amplitude by taking the path integral over a space–time configuration that has the topology of a Feynman graph. By summing over all such topologies, we arrive at the complete scattering amplitude:

$$A_N(p_1, p_2, \cdots, p_N) = \sum_{\text{topologies}} \int Dx_\mu(\tau) \exp \left(i \int d\tau \, \mathscr{L} + i \sum_{j=1}^{N} p_j \cdot x_j \right)$$

$$(21.9)$$

By explicitly evaluating the integral, we find the product of a series of propagators Δ_F and vertices. In this way, we can reproduce the usual Feynman series. By specifying the topology of the graph, we can reproduce the Feynman amplitude for any ϕ^n theory. To see this, we go to the Hamiltonian formulation, where the constraint in Eq. (21.4) is enforced by a Lagrange multiplier:

$$\mathscr{L} = p_\mu \dot{x}^\mu - \lambda(p_\mu^2 - m^2) \qquad (21.10)$$

By functionally eliminating p_μ and λ, we can retrieve our original Lagrangian appearing in Eq. (21.1).

Because we have reparametrization invariance, we have the freedom to fix the gauge by setting $\lambda = 1$. The new Hamiltonian is therefore $H(p, x) = p_\mu^2 - m^2$. The propagator in the Hamiltonian formalism is easily calculated. Between asymptotic states, it is given by the integral:

$$\int_0^\infty d\tau \, e^{-\tau H} = \frac{1}{p_\mu^2 - m^2} = \Delta_F(p) \qquad (21.11)$$

This reproduces the usual Feynman propagator.

If we now perform the path integration over the entire graph, then the path integral yields the product of these propagators joined together according to the topology of the Feynman diagram. In this way, the Feynman rules for any ϕ^n theory can be reproduced in the first quantized approach.

This simple example demonstrates that the Klein–Gordon theory can be formulated as a first quantized theory, but it is rather clumsy. In particular, the sum over the topologies of all Feynman graphs must be inserted by hand, which is undesirable. Also, unitarity is not obvious at higher orders. By contrast, the second quantized formalism is cleaner and can be derived from a single action.

Now let us make the transition from the point particle to the string, repeating the same steps as before. When a point moves in space–time, it sweeps out a one-dimensional world-line. When a string moves in space–time, it sweeps out a two-dimensional sheet, called the *world-sheet*.

Let $X_\mu(\sigma, \tau)$ represent a vector defined in D-dimensional space–time that begins at the origin of our coordinate system and ends at some point along the two-dimensional string world-sheet, as in Figure 21.1. We can place coordinates along the world-sheet labeled by $\xi^a = \{\sigma, \tau\}$.

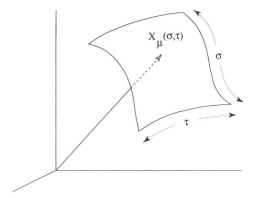

Figure 21.1. X^μ is a vector that goes from the origin to a point on a two-dimensional world-sheet swept out by a string.

Let the metric $\eta_{\mu\nu} = (+, -, -, - \cdots)$ be the flat metric in D-dimensional space–time, where $\mu = 0, 1, 2, \ldots, D-1$. Let g^{ab} represent a two-dimensional metric defined on the surface.

Our action can be written as[1]:

$$S = -\frac{1}{4\pi\alpha'} \int d^2\xi \sqrt{g}\, g^{ab}\, \partial_a X_\mu\, \partial_b X_\nu\, \eta^{\mu\nu} \tag{21.12}$$

where we define $\alpha' = 1/2$ for open strings and $\alpha' = 1/4$ for closed strings. α' is equal to the Regge slope, which we will define shortly. The action is manifestly reparametrization invariant. If we reparametrize the two-dimensional world sheet according to:

$$\sigma \to \bar{\sigma}(\sigma, \tau), \qquad \tau \to \bar{\tau}(\sigma, \tau) \tag{21.13}$$

then the action is invariant under this two-dimensional general co-ordinate transformation if:

$$\bar{g}^{ab}(\bar{\xi}) = \left(\frac{\partial \bar{\xi}^a}{\partial \xi^c}\right)\left(\frac{\partial \bar{\xi}^b}{\partial \xi^d}\right) g^{cd}(\xi) \tag{21.14}$$

Under this transformation, the action is manifestly co-ordinate invariant (because the transformation of $\sqrt{g}$ cancels against the transformation of the two-dimensional measure).

Under an infinitesimal transformation, the transformation of the fields becomes:

$$\delta g^{ab} = \epsilon^c \partial_c g^{ab} - g^{ac} \partial_c \epsilon^b - g^{bc} \partial_c \epsilon^a$$

$$\delta X_\mu \;=\; \epsilon^a \, \partial_a X_\mu \tag{21.15}$$

The action is also trivially invariant under local scale transformations:

$$g^{ab} \rightarrow e^\phi g^{ab} \tag{21.16}$$

Since the two-dimensional metric in the action does not propagate, we may eliminate it via its equations of motion. Then, we find:

$$g_{ab} \sim \partial_a X_\mu \, \partial_b X^\mu \tag{21.17}$$

Reinserting this value of the metric tensor back into the action, we find the original Nambu–Goto action[2–4]:

$$S \;=\; \frac{1}{2\pi\alpha'} \int d^2\xi \, \sqrt{(\dot X^\mu)^2 (X^{\mu\prime})^2 - (\dot X_\mu X^{\mu\prime})^2}$$

$$\sim \quad \text{surface area} \tag{21.18}$$

where $\dot X^\mu$ equals $\partial_\tau X^\mu$ and X'^μ equals $\partial_\sigma X^\mu$.

It is remarkable that string theory, which provides a comprehensive scheme in which to unite general relativity with quantum mechanics and all known physical forces, begins with this simple statement: the first quantized action is proportional to the area of the string world-sheet.

21.3   Quantizing the String

To calculate the spectrum of the string and its properties, we will quantize the free theory using three different methods:

1. The Gupta–Bleuler formalism in the conformal gauge.

2. The light-cone gauge.

3. The BRST formalism.

21.3.1   Gupta–Bleuler Quantization

The gauge degree of freedom allows us to choose the conformal gauge:

$$\text{Conformal gauge}: \quad g^{ab} = \delta^{ab} \tag{21.19}$$

Then, our Lagrangian linearizes to the following[5]:

$$\mathcal{L} = \frac{1}{4\pi\alpha'}\left[(\dot{X}_\mu)^2 + (X'_\mu)^2\right] = \frac{1}{2\pi}\,\partial_z X_\mu\,\partial_{\bar{z}} X^\mu \tag{21.20}$$

where, after a Wick rotation, we have introduced the complex variable z:

$$z = \sigma + i\tau \tag{21.21}$$

The equations of motion are:

$$\left(\frac{\partial^2}{\partial^2\sigma} + \frac{\partial^2}{\partial^2\tau}\right) X_\mu = 0 \tag{21.22}$$

(In deriving these equations of motion, we had to eliminate a surface term; so we must also set $X'_\mu = 0$ at the ends of the string.) The gauge-fixed action is no longer locally reparametrization invariant, but it is still globally invariant under a subgroup of reparametrization, *conformal transformations*:

$$z \to f(z) \tag{21.23}$$

Under conformal transformations, the string transforms as:

$$\delta X_\mu(z, \bar{z}) = \epsilon\partial_z X_\mu + \bar{\epsilon}\partial_{\bar{z}} X_\mu \tag{21.24}$$

To quantize the system, we first introduce the canonical conjugate:

$$\begin{aligned}
\left[P_\mu(\sigma), X_\nu(\sigma')\right] &= -i\eta_{\mu\nu}\delta(\sigma - \sigma') \\
P_\mu &= \frac{\delta\mathcal{L}}{\delta\dot{X}^\mu}
\end{aligned} \tag{21.25}$$

We can always decompose the string variable via the Fourier series:

$$\begin{aligned}
X^\mu(\sigma) &= x^\mu + i\sum_{n=1}^{\infty}\frac{1}{\sqrt{n}}(a_n^\mu - a_{-n}^\mu)\cos n\sigma \\
P^\mu(\sigma) &= \frac{1}{\pi}\left(p^\mu + \sum_{n=1}^{\infty}\sqrt{n}\,(a_n^\mu + a_{-n}^\mu)\cos n\sigma\right)
\end{aligned} \tag{21.26}$$

where the commutation relations between the string variable and its momentum conjugate are satisfied if:

$$[a_{n\mu}, a_{m\nu}] = \delta_{n,-m}\eta_{\mu\nu} \tag{21.27}$$

To calculate the spectrum of states, we calculate the Hamiltonian:

$$H = \int_0^\pi d\sigma (P_\mu \dot{X}^\mu - \mathcal{L}) = \sum_{n=1}^\infty n a_{-n\mu} a_n^\mu + \alpha' p_\mu^2 \qquad (21.28)$$

Fortunately, the Hamiltonian is the simplest possible operator for an extended object: the sum over an infinite set of independent harmonic oscillators. The eigenfunctions of the Hamiltonian are therefore just the products of free creation operators $a_{-n,\mu}$ acting on the vacuum:

$$\prod_{n,\mu} \{a_{-n,\mu}\} |0\rangle \qquad (21.29)$$

This allows us to display the spectrum of states, which correspond to an infinite tower of point particles of arbitrary spin. The lowest states include a tachyon and a massless vector meson (the Maxwell field or, if we include isospin, the Yang–Mills field):

$$
\begin{array}{rcl}
\text{Tachyon} & = & |0\rangle \\
\text{Massless vector} & = & a^\mu_{-1} |0\rangle
\end{array} \qquad (21.30)
$$

(Historically, the tachyon was viewed as troublesome feature of the string model. However, one can also view it as a blessing in disguise, because it signals the presence of spontaneous symmetry breaking to a new, perhaps more physical vacuum. Also, the tachyon disappears when we generalize the theory to the supersymmetric string.)

The series continues indefinitely. The next few states include a massive spin-2 field and massive vector field:

$$
\begin{array}{rcl}
\text{Massive spin} - 2 \text{ field} & = & a^\mu_{-1} a^\nu_{-1} |0\rangle \\
\text{Massive vector field} & = & a^\mu_{-2} |0\rangle
\end{array} \qquad (21.31)
$$

In Figure 21.2, we plot the resonances on a chart, with mass squared on the x axis and spin on the y axis. The linearly rising trajectories are called "Regge trajectories" with Regge slope α'. The point where the leading Regge trajectory hits the y axis is called the "intercept." The important point is to observe that the massless Maxwell and Yang–Mills fields (with intercept one) are necessarily included as part of the string spectrum. For closed strings, the intercept is equal to two, so we necessarily have a theory of massless gravitons. (In the limit of zero slope, we see that only the massless particles remain. Thus, the zero-slope limit is a convenient limit in which we may retrieve point-particle field theory.)

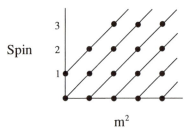

Figure 21.2. Linearly rising Regge trajectories for open strings. The resonances of string theory are states of arbitrarily high spin and mass. In the massless closed string sector, the theory necessarily includes quantum gravity.

Next, we calculate the energy–momentum tensor of the system:

$$T_{ab} = -4\pi\alpha' \frac{1}{\sqrt{g}} \frac{\delta\mathscr{L}}{\delta g^{ab}} \tag{21.32}$$

This, in turn, can be shown to equal:

$$T_{ab} = \partial_a X_\mu \partial_b X^\mu - \frac{1}{2} g_{ab} g^{cd} \partial_c X^\mu \partial_d X_\mu \tag{21.33}$$

We notice several important features of the energy–momentum tensor; that is, it satisfies:

$$\partial_b T^{ab} = 0; \qquad \text{Tr } T_{ab} = 0 \tag{21.34}$$

Notice that the energy–momentum tensor forms a closed algebra. The Fourier modes of the energy–momentum tensor form the *Virasoro generators* L_m[6]:

$$
\begin{aligned}
L_m &= \frac{1}{4\pi\alpha'} \int_0^\pi d\sigma \left[e^{im\sigma}(T_{00} + T_{01}) + e^{-im\sigma}(T_{00} - T_{01}) \right] \\
&= \frac{1}{8\pi\alpha'} \int_{-\pi}^\pi d\sigma \left[e^{im\sigma}(\dot{X} + X')^2 \right] \\
&= \frac{1}{2} \sum_{n=-\infty}^\infty \alpha_{m-n}\alpha_n
\end{aligned}
\tag{21.35}
$$

where $\alpha_n = \sqrt{|n|}a_n$ for $n \neq 0$ and $\alpha_0 = \sqrt{2\alpha'}p$. They obey the algebra:

$$[L_n, L_m] = (n-m)L_{n+m} + \frac{c}{12}\delta_{n,-m}n(n^2 - 1) \tag{21.36}$$

where c is the central charge and equals the dimension of space–time.

(Another way to derive this algebra is to start with the Nambu–Goto action, and then construct P_μ as $\delta \mathcal{L}/\delta \dot{X}_\mu$. We find that the momenta are not all independent, but instead satisfy an additional set of constraints:

$$P_\mu^2 + \frac{1}{(2\pi\alpha')^2} X_\mu'^2 \equiv 0$$

$$P_\mu X'^\mu \equiv 0 \tag{21.37}$$

The moments of these constraints also form the Virasoro generators.)

In the Gupta–Bleuler quantization scheme, the ghosts that propagate in the system (corresponding to the longitudinal modes of a_{-n}^μ) can be eliminated by applying the gauge constraints directly on the Fock space. Thus, we apply:

$$L_n|R\rangle = 0, \quad n > 0$$

$$(L_0 - 1)|R\rangle = 0 \tag{21.38}$$

where the second condition is the mass-shell condition. After a rather tedious calculation, one can show that these conditions are sufficient to eliminate all unphysical states from the physical spectrum. However, there is an unexpected result: the spectrum is ghost-free only if the dimension of space–time is 26.

21.3.2 Light-Cone Gauge

As in ordinary field theory, we can alternatively formulate the system in the light-cone gauge[7] where the unphysical longitudinal modes are eliminated from the very start.

We will define the light-cone coordinates as:

$$X^+ = \frac{1}{\sqrt{2}}\left(X^0 + X^{D-1}\right)$$

$$X^- = \frac{1}{\sqrt{2}}\left(X^0 - X^{D-1}\right) \tag{21.39}$$

and fix the gauge as:

$$X^+(\sigma, \tau) = p^+\tau \tag{21.40}$$

We can use this light-cone gauge to eliminate all nonphysical modes. By taking the constraints in Eq. (21.37), we can eliminate unwanted longitudinal vibrations by solving for:

$$P^-(\sigma) = \frac{\pi}{2p^+}\left(P_i^2 + \frac{X_i'^2}{\pi^2}\right)$$

$$X^-(\sigma) = \int_0^\sigma d\sigma' \frac{\pi}{p^+}(P_i X_i') \tag{21.41}$$

The Hamiltonian in the light-cone gauge reduces to:

$$H = \frac{\pi}{2}\int_0^\pi \left(P_i^2 + \frac{X_i^2}{\pi^2}\right) d\sigma \tag{21.42}$$

Notice that the physical Fock space consists of transverse harmonic oscillators, which are ghost-free. Of course, we still must check that the theory is Lorentz invariant. We do this by rewriting the Lorentz generators in terms of the independent transverse modes. This is a bit awkward, but straightforward:

$$M^{\mu\nu} = \int_0^\pi d\sigma \left(X^\mu P^\nu - X^\nu P^\mu\right)$$

$$= x^\mu p^\nu - x^\nu p^\mu - i\sum_{n=1}^\infty \frac{1}{n}\left(\alpha_{-n}^\mu \alpha_n^\nu - \alpha_{-n}^\nu \alpha_n^\mu\right) \tag{21.43}$$

The surprising feature of the Lorentz generators is that, in general, they fail to close properly unless we impose one more constraint:

$$\left[M^{-i}, M^{-j}\right] = -\frac{1}{p^{+2}}\sum_{n=1}^\infty \left(\alpha_{-n}^i \alpha_n^j - \alpha_{-n}^j \alpha_n^i\right)\Delta_n = 0 \tag{21.44}$$

where:

$$\Delta_n = \frac{n}{12}(26 - D) + \frac{1}{n}\left(\frac{D - 26}{12} + 2 - 2a\right) \tag{21.45}$$

where a is the intercept. In order to have Lorentz invariance, we must set Δ_n equal to zero, that is,

$$D = 26; \quad a = 1 \tag{21.46}$$

This is consistent with the result found in the conformal gauge, that self-consistency of the string theory forces the dimension of space–time to be 26.

21.3.3 BRST Quantization

Likewise, the BRST method should also reproduce this result. To BRST quantize the string, we start with the invariance of the metric tensor:

$$\delta g_{ab} = g_{ac}\, \partial_b \delta v^c + \partial_a \delta v^c\, g_{cb} - \delta v^c\, \partial_c g_{ab} = \nabla_a\, \delta v_b + \nabla_b\, \delta v_a \qquad (21.47)$$

The BRST procedure begins with the construction of the Faddeev–Popov determinant, which is the determinant of the variation of the constraint. This, however, is just the determinant of the operator ∇_a. The Faddeev–Popov determinant can thus be rewritten as:

$$\Delta_{FP} = \det(\nabla_a) = \det \nabla_z \, \det \nabla_{\bar{z}} \qquad (21.48)$$

We now introduce Faddeev–Popov ghosts by exponentiating this determinant:

$$\Delta_{FP} = \int Db\, D\bar{b}\, Dc\, D\bar{c}\, e^{i \int \mathscr{L}^{bc}\, d^2 z} \qquad (21.49)$$

where:

$$\mathscr{L}^{bc} = \frac{1}{\pi}\left(b\, \partial_{\bar{z}} c + \bar{b}\, \partial_z \bar{c}\right) \qquad (21.50)$$

As usual, adding this ghost term to the original conformal gauge action in Eq. (21.20) yields a residual global symmetry, called the BRST symmetry, which can be generated by the BRST charge, which is computable from the Noether current. A straightforward calculation of the Noether current yields[8]:

$$
\begin{aligned}
Q &= \sum_{n=-\infty}^{\infty} : c_{-n}\left(L_n^X + \frac{1}{2}L_n^{bc} - a\delta_{n,0}\right) \\
&= c_0(L_0 - a) + \sum_{n=1}^{\infty}\left(c_{-n}L_n + L_{-n}c_n\right) \\
&\quad - \frac{1}{2}\sum_{n,m=-\infty}^{\infty} : c_{-m}c_{-n}b_{n+m} : (m-n)
\end{aligned} \qquad (21.51)
$$

where:

$$\{c_n, b_m\} = \delta_{n,-m} \qquad (21.52)$$

and:

$$Q^2 = \frac{1}{2}\sum_{m=-\infty}^{\infty}\left(\frac{D}{12}(m^3 - m) + \frac{1}{6}(m - 13m^3) + 2am\right)c_m c_{-m} \qquad (21.53)$$

which vanishes only if $D = 26$ and $a = 1$, as before.

The new Fock space now consists of all possible products of all possible creation oscillators, including the ghost oscillators:

$$\prod_{n,m,p,\mu} \{a^{\mu}_{-n}\}\{b_{-m}\}\{c_{-p}\}|0\rangle \tag{21.54}$$

Although the Fock space now has a vastly increased number of ghost states, we can eliminate all of them by applying the BRST operator onto the Fock space:

$$Q|\psi\rangle = 0 \tag{21.55}$$

Thus, all three quantization programs can be shown to have the same physical spectrum if the dimension of space–time is 26. Also, the intercept condition forces us to incorporate spin-1 Maxwell fields for the open string and spin-2 gravitational fields for the closed string.

21.4   Scattering Amplitudes

Interactions are introduced by postulating that the string can break and reform an arbitrary number of times. The world-sheet corresponding to this is therefore the set of all two-dimensional complex surfaces with g holes or "handles," as shown in Figure 21.3. (Two-dimensional complex surfaces are called Riemann surfaces.) In this way, we introduce Feynman-like diagrams in a first quantized formalism.[9]

These simple Feynman-like diagrams conceal a large amount of information. For example, if we carefully extract out the zero mass, spin-2 sector from these Feynman diagrams, we will reproduce all of Einstein's theory of general relativity power expanded around flat space.

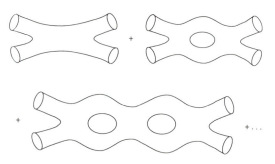

Figure 21.3. Strings can break and reform, thereby sweeping out two-dimensional Riemann surfaces of genus g. In this way, the string reproduces Feynman-like diagrams.

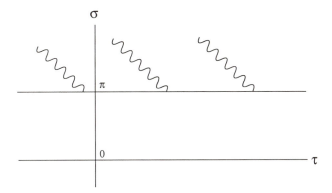

Figure 21.4. The world-sheet for an N-point tree amplitude. For calculational purposes, we have conformally mapped the world-sheet onto this surface. Momenta from tachyons enter from the top of the world-sheet.

Let $\mathscr{R}_{[N,g]}$ represent all conformally inequivalent Riemann surfaces with genus g and N "punctures" (external strings) located at infinity. Then the complete N point tachyon amplitude is therefore given by summing over all functional integrals defined over $\mathscr{R}_{[N,g]}$:

$$
\begin{aligned}
A_N(k_1, k_2, \cdots, k_N) &= \sum_g \int_{\mathscr{R}_{[N,g]}} DX_\mu \int d\mu \\
&\quad \times \exp\left[i \int d^2z \, \mathscr{L}(z) + \sum_{i=1}^{N} i k_{i,\nu} X^\nu(z_i) \right] \\
&= \sum_g \int d\mu \left\langle \prod_{i=1}^{N} e^{ik_i X_i} \right\rangle_{\mathscr{R}_{[N,g]}}
\end{aligned}
\tag{21.56}
$$

where $d\mu$ is a conformal measure on the Riemann surface. This is a generalization of the first quantized point-particle path integral that we analyzed in Eq. (21.9).

Fortunately, for tree diagrams this amplitude is easily calculable. Because the theory is conformally invariant, we will find it convenient to perform the functional integral over the world-sheet corresponding to a long horizontal strip with momenta entering the world-sheet at points along the top (Fig. 21.4). To obtain the amplitude, we will then conformally map this strip onto the upper half-plane. The external tachyon lines will then lie on the real axis.

To solve this functional integral, let us shift the integration variable by a solution to the classical equation:

$$
X_\mu \to X_{\mu,\text{classical}} + X_\mu
\tag{21.57}
$$

where the classical solution is determined via the Green's function for Laplace's equation on the upper half-plane:

$$
\begin{aligned}
X_{\mu,\text{classical}} &= -i\alpha' \int G(z, z') J(z') \, dz' \\
G(z, z') &= \log|z - z'| + \log|z - z^{*'}| \tag{21.58}
\end{aligned}
$$

(This Neumann function is easily calculated by the method of images. The electrostatic potential at a point z in the upper half-plane is the sum of the contributions from a point charge placed at z' and also the image charge placed in the lower half-plane at $z^{*'}$.)

With this Neumann function, the Gaussian integral can be performed, leaving us with the N-point tree amplitude[10-14]:

$$
A_N = \int \prod_{i=3}^{N-1} dz_i \prod_{2 \leq i < j \leq N} |z_i - z_j|^{-k_i k_j} \tag{21.59}
$$

where the z_i are on the real axis and obey: $\infty = z_1 > z_2 = 1 \geq z_3 \cdots z_{N-1} \geq z_N = 0$. This is the scattering amplitude that describes the scattering of N tachyons.

For $N = 4$, this expression reduces down to the celebrated Veneziano formula:[15,16]

$$
B_4(s, t) = \int_0^1 dx \, x^{-s/2-2}(1 - x)^{-t/2-2} = \frac{\Gamma(-\alpha(s))\Gamma(-\alpha(t))}{\Gamma(-\alpha(s) - \alpha(t))} \tag{21.60}
$$

where $\alpha(s) = 1 + \frac{1}{2}s$, $\alpha(t) = 1 + \frac{1}{2}t$, $s = (k_1 + k_2)^2$, and $t = (k_2 + k_3)^2$.

The accidental discovery of this formula in 1968 by Veneziano and Suzuki, who were trying to describe the scattering matrix for hadronic interactions, marked the birth of what eventually became superstring theory. (They were originally trying to find a formula for the scattering of pions, using S matrix theory and finite-energy sum rules, when they stumbled across the Euler beta function, which satisfied almost all the properties of the S matrix except unitarity.)

In practice, it is often more convenient to work with the operator formalism. To convert the path integral to the operator formalism, we need to make the transition from the Lagrangian formalism to the Hamiltonian formalism on the world-sheet. Then the path integral will be defined in terms of X^μ as well as its conjugate momentum P^μ, which is an operator. This transition is easily done, since the Lagrangian describes an infinite set of noninteracting harmonic oscillators.

When we make the transition to the Hamiltonian formalism, the vertex $e^{ik \cdot X}$ appearing in the path integral now becomes the operator expression:

$$
V(k) = : e^{ik_\mu X^\mu} := \exp\left(k \cdot \sum_{n=1}^{\infty} \frac{\alpha_{-n}}{n}\right) \exp\left(-k \cdot \sum_{n=1}^{\infty} \frac{\alpha_n}{n}\right) e^{ik \cdot x} \tag{21.61}
$$

(We must normal order this vertex. This is because in the path integral for the N-point function we deleted the sum over $i = j$, which diverges. To eliminate this divergent quantity in the operator formalism, we normal order the oscillators.)

In the Hamiltonian formalism, the transition element between two string states is given by $\langle X | e^{itH} | X' \rangle$. If we make a Wick rotation, then the integrated propagator between two states becomes:

$$D = \int_0^\infty e^{-\tau(L_0-1)} d\tau = \frac{1}{L_0 - 1} \tag{21.62}$$

sandwiched between any two string states. The Hamiltonian on the world-sheet is given by $L_0 - 1$.

For the path integral describing the N-point amplitude, the transition to the Hamiltonian formalism gives us an expression for the N-point function[17]:

$$A_N = \langle 0, k_1 | V(k_2) D V(k_3) \cdots V(k_{N-1}) | 0, k_N \rangle \tag{21.63}$$

where $|0, k\rangle = |0\rangle e^{ik \cdot x}$, where x^μ is the center-of-mass variable describing X^μ. To contract these oscillators, which are all written in terms of exponentials, we use the coherent state formalism. We define a coherent state by:

$$|\lambda\rangle \equiv \sum_{n=0}^\infty \frac{\lambda^n}{n!} (a^\dagger)^n |0\rangle = e^{\lambda a^\dagger} |0\rangle \tag{21.64}$$

Then we have the identities:

$$
\begin{aligned}
\langle \mu | \lambda \rangle &= e^{\mu^* \lambda} \\
x^{a^\dagger a} | \lambda \rangle &= | x\lambda \rangle \\
e^{\mu a^\dagger} | \lambda \rangle &= | \lambda + \mu \rangle
\end{aligned}
\tag{21.65}
$$

By contracting the oscillators, we reproduce the N-point amplitude in Eq. (21.59), as desired.

From a physical point of view, the more interesting theory is the closed string theory, which includes Einstein's theory of general relativity as a subset. Closed strings can also be quantized in much the same way. The only major difference mathematically is that the closed string contains two independent sets of oscillators, not just one.

We can decompose the string variable in terms of two sets of commuting harmonic oscillators:

$$X_\mu(\sigma) = x_\mu + \left(\frac{\alpha'}{2}\right)^{1/2} \sum_{n=1}^\infty \frac{1}{\sqrt{n}} \left(a_n e^{-in\sigma} + \tilde{a}_n e^{in\sigma}\right)$$

$$
+ a_n^\dagger e^{in\sigma} + \tilde{a}_n^\dagger e^{-in\sigma}\Big)_\mu
$$

$$
P_\mu(\sigma) = \frac{p_\mu}{2\pi} + \frac{1}{2\pi\sqrt{2\alpha'}} \sum_{n=1}^{\infty} \sqrt{n}\left(- i\tilde{a}_n e^{-in\sigma} - i a_n e^{in\sigma}\right.
$$

$$
\left. + i a_n^\dagger e^{in\sigma} + i\tilde{a}_n^\dagger e^{-in\sigma}\right)_\mu \tag{21.66}
$$

where $X_\mu(0) = X_\mu(2\pi)$.

The Hamiltonian now also has doubled the number of oscillators:

$$
H = \pi \int_0^{2\pi} d\sigma \left(\alpha' P_\mu^2 + \frac{X_\mu'^2}{4\pi^2\alpha'}\right) = \sum_{n=1}^{\infty}(n a_n^\dagger a_n + n\tilde{a}_n^\dagger \tilde{a}_n) + \alpha' p_\mu^2 \tag{21.67}
$$

The graviton naturally emerges as the massless state with spin 2:

$$
\begin{aligned}
\text{Tachyon} &= |0\rangle \\
\text{Graviton} &= \left(a_{-1}^\mu \tilde{a}_{-1}^\nu + \tilde{a}_{-1}^\mu a_{-1}^\nu\right)|0\rangle
\end{aligned} \tag{21.68}
$$

This, in fact, is perhaps the most attractive, and most mysterious, feature of string theory, that general relativity is necessarily part of the theory. While other point-particle theories try to avoid including the graviton, string theory views gravity as an inseparable part of its formulation.

The propagator for closed strings is similar to the open string propagator, except for one difference: There is an extra rotation factor P that guarantees that the final result is not dependent on the origin of the parametrization. Thus, the propagator is:

$$
\frac{1}{L_0 + \tilde{L}_0 - 2} P \tag{21.69}
$$

where:

$$
P = \int_0^{2\pi} d\theta \, e^{i\theta(L_0 - \tilde{L}_0)} \tag{21.70}
$$

where $L_0 - \tilde{L}_0$ is the operator that rotates the closed string. The propagator can be written in an equivalent way:

$$
D = \frac{1}{2\pi} \int_{|z|\leq 1} z^{L_0-2}\bar{z}^{\tilde{L}_0-2} \, d^2 z = \frac{\sin \pi(L_0 - \tilde{L}_0)}{\pi(L_0 - \tilde{L}_0)} \frac{1}{L_0 + \tilde{L}_0 - 2} \tag{21.71}
$$

The N-point amplitude then becomes[18,19]:

$$A_N = \sum_{\text{perm}} \langle 0, k_1 | V(k_2) D V(k_3) \cdots V(k_{N-1}) | 0, k_N \rangle \qquad (21.72)$$

where the permutation is taken over all possible orderings of the external legs. When expanded out, the resulting N-point amplitude for closed strings is almost identical to the one for open strings (except the z_i variables are now integrated over all complex space, not just the real axis).

The Virasoro constraints can also be written for the theory, which now become:

$$
\begin{aligned}
L_n | \phi \rangle &= \tilde{L}_n | \phi \rangle = 0 \\
(L_0 + \tilde{L}_0 - 2) | \phi \rangle &= 0 \\
(L_0 - \tilde{L}_0) | \phi \rangle &= 0
\end{aligned}
\qquad (21.73)
$$

where the last constraint is due to the fact that the states should be independent of where we chose the origin of our parametrization.

21.5 Superstrings

To make the spectrum more realistic, we must now turn to the superstring, which introduces a new symmetry: supersymmetry. In fact, supersymmetry, as a symmetry of an action, was first discovered in 1971 in string theory, and only later was adapted to four-dimensional point particle theories.

Let us introduce a new fermion field ψ_μ, the counterpart of X_μ, which is a vector in space–time but transforms as a two-dimensional spinor in the two-dimensional world-sheet. Then, the Neveu–Schwarz–Ramond (NSR) model[20,21] can be expressed as a two-dimensional action. Gervais and Sakita introduced the following Lagrangian[22]:

$$\mathcal{L} = -\frac{1}{2\pi} \left(\partial_a X_\mu \, \partial^a X^\mu - i \bar{\psi}^\mu \rho^a \, \partial_a \psi_\mu \right) \qquad (21.74)$$

where:

$$\rho^0 = \begin{pmatrix} 0 & -i \\ i & 0 \end{pmatrix}, \qquad \rho^1 = \begin{pmatrix} 0 & i \\ i & 0 \end{pmatrix} \qquad (21.75)$$

and:

$$\psi^\mu = \begin{pmatrix} \psi_0^\mu \\ \psi_1^\mu \end{pmatrix}, \qquad \bar{\psi}^\mu = \psi^\mu \rho^0 \tag{21.76}$$

with the metric $\{\rho^a, \rho^b\} = -2\eta^{ab}$, where η is diagonal and given by $(-1, +1)$.
    Written explicitly, this equals:

$$\mathscr{L} = \frac{1}{2\pi} \left[\dot{X} \cdot \dot{X} - X' \cdot X' + i\psi_0(\partial_\tau + \partial_\rho)\psi_0 + i\psi_1(\partial_\tau - \partial_\rho)\psi_1 \right] \tag{21.77}$$

The important feature of this action is that it is explicitly invariant under the
following supersymmetry:

$$\delta X^\mu = \bar{\epsilon}\psi^\mu, \qquad \delta\psi^\mu = -i\rho^a \partial_a X^\mu \epsilon \tag{21.78}$$

The energy–momentum tensor can be written as:

$$T_{ab} = \partial_a X_\mu \partial_b X^\mu + \frac{i}{4}\bar{\psi}^\mu \rho_a \partial_b \psi_\mu + \frac{i}{4}\bar{\psi}^\mu \rho_b \partial_a \psi_\mu - \text{(trace)} \tag{21.79}$$

By Noether's method, we can derive the conserved supercurrent:

$$J_a = \frac{1}{2}\rho^b \rho_a \psi^\mu \partial_b X_\mu \tag{21.80}$$

We can rewrite the superconformal current J_a as:

$$T_F = -\frac{1}{2}\psi_\mu \partial X^\mu \tag{21.81}$$

and its Fourier moments as:

$$G_n = 2 \oint \frac{dz}{2\pi i} z^{n+(1/2)} T_F(z) \tag{21.82}$$

We quantize the fermionic oscillators in the usual way:

$$\{\psi_a^\mu(\sigma, \tau), \psi_b^\nu(\sigma', \tau)\} = \pi \delta_{ab}\delta(\sigma - \sigma')\eta^{\mu\nu} \tag{21.83}$$

Because we have more fields, there are actually two different boundary conditions we may take on the theory, either periodic (R) or antiperiodic (NS) boundary
conditions. The ψ_i fields are equal to each other at $\sigma = 0$, but at $\sigma = \pi$ they obey:

$$\begin{aligned} \text{R}: \quad \psi_0(\pi, \tau) &= \psi_1(\pi, \tau) \\ \text{NS}: \quad \psi_0(\pi, \tau) &= -\psi_1(\pi, \tau) \end{aligned} \tag{21.84}$$

With these boundary conditions, the harmonic oscillator decomposition is given by:

$$
\text{R}: \qquad \psi_{0,1}^{\mu} = \frac{1}{\sqrt{2}} \sum_{n=-\infty}^{\infty} d_n^{\mu} e^{-in(\tau \pm \sigma)}
$$

$$
\text{NS}: \qquad \psi_{0,1}^{\mu} = \frac{1}{\sqrt{2}} \sum_{r \in \mathbf{Z}+1/2}^{\infty} b_r^{\mu} e^{-ir(\tau \pm \sigma)} \tag{21.85}
$$

where we associate the lower index 0 (1) with the $+$ ($-$) sign appearing in the exponential, where the R states are integral moded, while NS states are half-integral moded, and where we have the anticommutation relation among oscillators:

$$
\text{R} \quad : \quad \{d_n^{\mu}, d_m^{\nu}\} = \eta^{\mu\nu} \delta_{n,-m}
$$

$$
\text{NS} \quad : \quad \{b_r^{\mu}, b_s^{\nu}\} = \eta^{\mu\nu} \delta_{r,-s} \tag{21.86}
$$

The Fock space of the theory now describes either an infinite tower of bosonic fields, or fermionic ones:

$$
\text{R}: \qquad \prod_{n,r} \{a_{-n}^{\mu}\} \{d_{-r}^{\nu}\} |0\rangle u_{\alpha}
$$

$$
\text{NS}: \qquad \prod_{n,r} \{a_{-n}^{\mu}\} \{b_{-r}^{\nu}\} |0\rangle \tag{21.87}
$$

where u_{α} is a 10-dimensional (32-component) spinor.

The commutators and anticommutators of the energy–momentum tensor and the supercurrent now form a closed algebra, called the *superconformal algebra*:

$$
[L_m, L_n] = (m-n)L_{m+n} + \frac{\hat{c}}{8}(m^3 - m)\delta_{m+n,0}
$$

$$
[L_m, G_r] = \left(\frac{m}{2} - r\right)G_{m+r}
$$

$$
\{G_r, G_s\} = 2L_{r+s} + \frac{\hat{c}}{2}\left(r^2 - \frac{1}{4}\right)\delta_{r+s,0} \tag{21.88}
$$

where $\hat{c} = 2c/3$.

An explicit representation of the NS superconformal operators is given by:

$$
L_m = \frac{1}{2} \sum_{n=-\infty}^{\infty} : \alpha_{-n}\alpha_{m+n} : + \frac{1}{2} \sum_{r=-\infty}^{\infty} \left(r + \frac{1}{2}m\right) : b_{-r}b_{m+r} :
$$

$$
G_r = \sum_{n=-\infty}^{\infty} \alpha_{-n}b_{r+n} \tag{21.89}
$$

For the R sector, the generators are given by:

$$L_m = \frac{1}{2} \sum_{n=-\infty}^{\infty} : \alpha_{-n}\alpha_{m+n} : + \frac{1}{2} \sum_{n=-\infty}^{\infty} \left(n + \frac{1}{2}m\right) : d_{-n}d_{m+n} :$$

$$G_m = \sum_{n=-\infty}^{\infty} \alpha_{-n}d_{m+n} \tag{21.90}$$

Finally, let us define the operator Q_{BRST}. We find that the Faddeev–Popov ghost factor can be written in terms of two commuting ghosts β, γ as:

$$L = \frac{1}{\pi}\left(\beta\,\partial_{\bar{z}}\gamma + \text{c.c.}\right) \tag{21.91}$$

where c.c. is the complex conjugate.

The complete superconformal generators must also include the presence of the b, c and β, γ ghosts:

$$L_m^{\text{ghost}} = \sum_{n=-\infty}^{\infty}(m+n) : b_{m-n}c_n + \sum_{n=-\infty}^{\infty}\left(\frac{1}{2}m+n\right) : \beta_{m-n}\gamma_n :$$

$$G_m^{\text{ghost}} = -2\sum_{n=-\infty}^{\infty} b_{-n}\gamma_{m+n} + \sum_{n=-\infty}^{\infty}\left(\frac{1}{2}n - m\right)c_{-n}\beta_{m+n} \tag{21.92}$$

Finally, Q can be written as:

$$Q = \sum_{n=-\infty}^{\infty}(L_{-n}c_n + G_{-n}\gamma_n) - \frac{1}{2}\sum_{m,n=-\infty}^{\infty}(m-n) : c_{-m}c_{-n}b_{m+n} :$$

$$+ \sum_{m,n=-\infty}^{\infty}\left(\frac{3n}{2}+m\right)c_{-n}\beta_{-m}\gamma_{m+n} + \sum_{m,n=-\infty}^{\infty}\gamma_{-m}\gamma_{-n}b_{m+n} - ac_0 \tag{21.93}$$

As usual, we can check for the vanishing of Q^2, and we find the constraints:

$$D = 10, \qquad a = \begin{cases} \frac{1}{2} & \text{(NS)} \\ 0 & \text{(R)} \end{cases} \tag{21.94}$$

Although the NSR formulation is quite simple and easy to work with, one disadvantage is that ten-dimensional space–time supersymmetry (not to be confused with the two-dimensional superconformal symmetry of the NSR model) is not manifest. There exists another reformulation of this model, called the Green–Schwarz model,[23] which introduces two genuine ten-dimensional spinor fields S^1

and S^2 (which have $2^{10/2} = 32$ components each). In this model, ten-dimensional space–time supersymmetry is manifest as a symmetry of an action. (However, a detailed discussion of this model is beyond the scope of this book.) The advantage of introducing these spinors S^i is that we can construct various superstring theories from them.

21.6 Types of Strings

At this point, we may ask what are the various types of string theories one can write that are supersymmetric, ghost free, and anomaly free. The easiest way to catalog the various possibilities is through the light-cone quantization of the GS string, since all ghosts have been removed and the theory is globally supersymmetric in space–time.

The list of totally self-consistent superstring theories consists of:

1. Type I.

2. Type IIA.

3. Type IIB.

4. Heterotic.

(At present, the leading superstring theory is the heterotic string. When we refer to the superstring theory, we are therefore implicitly referring to the heterotic string.)

It may seem surprising that there are so few self-consistent string theories, while there are an infinite number of point particle theories. The reason for this is that the Feynman diagrams of a point particle are based on one-dimensional *graphs*, upon which we can impose any number of Lorentz covariant vectors and spinors with arbitrary isospin indices in our Feynman's rules. However, the Feynman diagrams of string theory are two-dimensional *manifolds*, obeying strict self-consistency constraints; so it is not surprising that we only find four self-consistent string theories.

21.6.1   Type I

The first string theory is called type I, which contains both open strings and closed strings. The two spinor fields S^1 and S^2 of the GS model have the same chirality. (Because the closed string emerges as a bound state of open string graphs, we must add the closed string sector to the open string in order to maintain unitarity.)

Gauge invariance can be added into the theory by multiplying the N-point function with appropriate traces over the generators of some Lie algebra (called Chan–Paton factors). The gauge group must be $SO(32)$ in order to cancel all anomalies.

21.6.2   Type IIA

For closed strings, there are two ways to choose the chiralities of S^1 and S^2. If we choose them to be of opposite chirality, then we have the type IIA string. Type IIA closed string theory is appealing because it has no chiral anomalies from the very beginning (since the two chiral sectors cancel against each other). In the zero-slope limit, when only the massless sector of the theory survives, the theory reduces to the point particle $N = 2$, $D = 10$ supergravity theory.

21.6.3   Type IIB

For closed strings, if S^1 and S^2 have the same chirality, then we have the type IIB superstring. However, in the zero-slope limit, when we analyze the massless sector, we find that there does not exist any known covariant version of this theory. Its light-cone reduction is well defined, but its covariant precursor apparently cannot be written. (This may be because of our limited understanding of how to construct point particle supersymmetric theories in ten dimensions.)

At present, it seems, however, that the type II string cannot describe the physical $SU(3) \otimes SU(2) \otimes U(1)$ symmetry of our low-energy universe. By compactifying from ten dimensions to four dimensions, the type II string can introduce a wide array of symmetries, but none of them seems to fit the description of our world.

21.6.4   Heterotic String

The string theory that holds the most promise of describing the physical world is the heterotic string.[24] While the type I string uses multiplicative Chan–Paton factors to introduce isospin symmetry, the heterotic string introduces isospin in an unorthodox fashion. We recall that the closed string has two sets of operators, a_n and $\tilde{a}_n$, which do not interact; that is, as the closed string propagates, it has right-moving and left-moving oscillator modes. The heterotic string splits these modes apart. The left-moving modes are purely bosonic and live in a 26-dimensional space labeled by X^μ which has been compactified to ten dimensions, leaving us with a compact 16-dimensional space. If we use the symbol X^i (X^I) to represent the 10 (16)-dimensional space, then we have:

$$X^\mu \rightarrow \{X^i, X^I\} \tag{21.95}$$

We will choose the compactified 16-dimensional string, labeled by X^I, to live on the root lattice space of an $E_8 \otimes E_8$ isospin symmetry. Since E_8 is a rank-eight Lie group, the heterotic string can be compactified so that its spectrum is $E_8 \otimes E_8$ [or Spin(32)/Z_2], which is certainly large enough to permit a serious phenomenological investigation.

However, the right-moving modes only live in a ten-dimensional space and contain the supersymmetric GS or NSR theory. When the left-moving half (containing the isospin) and the right-moving half (containing the supersymmetry) are put together, they produce a self-consistent, ghost-free, anomaly-free, one-loop finite theory, the heterotic string (meaning "hybrid vigor").

The action for the heterotic string is therefore:

$$S = -\frac{1}{4\pi\alpha'} \int d\tau \int_0^{2\pi} d\sigma \left[\partial_a X^i \, \partial^a X^i + \sum_{I=1}^{16} \partial_a X^I \, \partial^a X^I + i\bar{S}\gamma^-(\partial_\tau + \partial_\sigma)S \right] \tag{21.96}$$

where $I = 1, 2, \ldots, 16$ and is an isospin index and where we enforce the constraints:

$$(\partial_\tau - \partial_\sigma)X^I = 0, \qquad \gamma^+ S = \frac{1}{2}(1 + \gamma_{11})S = 0 \tag{21.97}$$

where $\gamma^+ = 2^{-1/2}(\gamma^0 + \gamma^9)$.

In the zero-slope limit, this theory yields ten-dimensional supergravity coupled to a super Yang–Mills gauge multiplet with $E_8 \otimes E_8$ local gauge symmetry. Clearly, we have enough symmetry to include the Standard Model and extract interesting phenomenology.

21.7 Higher Loops

There are three main aspects to superstring theory:

1. Superstring perturbation theory.

2. Superstring compactification and phenomenology.

3. Nonperturbative approaches and string field theory.

We will discuss each of them separately.

From the point of view of quantum field theory, superstring perturbation theory gives us entirely new, unexpected mechanisms by which to cancel the divergences found in quantum gravity. We find, for example, that the higher loops are not ultraviolet divergent at all (because the presence of the infinite tower of resonances acts, in some sense, like a Pauli–Villars regulator). The only problem comes from infrared divergences, which in turn can be controlled by using symmetry.

To see this, we will only sketch the calculation of the single-loop amplitude, omitting many details. We will, as expected, obtain the Neumann function defined over a disc with a hole (defined in terms of Jacobi θ functions). To calculate the first loop amplitude for N external tachyons, we will simply trace over a series of vertices and propagators, using the coherent state formula:

$$\frac{1}{\pi} \int d^2\lambda \, e^{-|\lambda|^2} |\lambda\rangle\langle\lambda| = 1 \tag{21.98}$$

Using Eqs. (21.65) and (21.98), it is now a simple, although tedious, matter to take the trace over a string of vertices and propagators:

$$
\begin{aligned}
A_N &= \int d^{26}p \; \mathrm{Tr} \; [V(k_1)DV(k_2)\cdots DV(k_N)D] \\
&= \int d^{26}p \int \prod_{i=1}^{N} |z_i|^{-k_i^2/4} \frac{1}{\pi^2} \int d^2\lambda \, d^2\tilde{\lambda} \\
&\quad \times \langle\lambda|\langle\tilde{\lambda}|V(k_1)z_1^R V(k_2)z_2^R \cdots V(k_N)z_N^R|\lambda\rangle|\tilde{\lambda}\rangle \\
&= \int \prod_{i=1}^{N} d^2 z_i |w|^{-4} |f(w)|^{-48} \left(\frac{-4\pi}{\log|w|}\right)^{13} \prod_{i<j} \left[\chi_{ij}(c_{ji}, w)\right]^{k_i \cdot k_j/2}
\end{aligned}
\tag{21.99}
$$

where:

$$
\begin{aligned}
\nu_j &= (2\pi i)^{-1} \log z_1 z_2 \cdots z_j \\
\nu_{ji} &= \nu_j - \nu_i \\
\tau &= (2\pi i)^{-1} \log w \\
w &= z_1 z_2 \cdots z_N \\
c_{ji} &= z_{i+1} z_{i+2} \cdots z_j
\end{aligned}
\tag{21.100}
$$

and where:

$$
\chi(z, w) = \exp\left(\frac{\log^2 |z|}{2 \log|w|}\right) \left|z^{-1/2} \prod_{m=1}^{\infty} \frac{(1 - w^m z)(1 - w^m/z)}{(1 - w^m)^2}\right|
$$

$$= 2\pi \exp\left(\frac{-\pi (\mathrm{Im}\, \nu_{ji})^2}{\mathrm{Im}\, \tau}\right) \left|\frac{\theta_1(\nu_{ji}|\tau)}{\theta_1'(0|\tau)}\right| \tag{21.101}$$

This, in turn, can be written as:

$$A_N = \int_F d^2\tau\ (\mathrm{Im}\, \tau)^{-2}\, C(\tau) F(\tau) \tag{21.102}$$

where:

$$C(\tau) = 4(\frac{1}{2}\mathrm{Im}\, \tau)^{-12} e^{4\pi \mathrm{Im}\, \tau} |f(e^{2\pi i\tau})|^{-48}$$

$$F(\tau) = \pi^N (\mathrm{Im}\, \tau) \int \prod_{i=1}^{N-1} d^2\nu_i \prod_{i<j} (\chi_{ij})^{-k_i \cdot k_j/2} \tag{21.103}$$

The important point is that ultraviolet divergences are completely missing in this amplitude, which is astonishing because it contains the one-loop contribution from the graviton and an infinite tower of massive particles. However, the theory is infrared divergent, which corresponds to $w \to 1$, or to the interior hole shrinking to a point. This infrared divergence can be eliminated when we go to the superstring theory. To see this, we will analyze the superstring single-loop amplitude. We simply present the result:

$$A_N \sim \int \prod_{i=1}^{N} d^2 z_i\ |w|^{-2} \left(\frac{-4\pi}{\log |w|}\right)^5 \prod_{i<j} (\chi_{ij})^{-k_i \cdot k_j/2}$$

$$= \int_F d^2\tau (\mathrm{Im}\, \tau)^{-2} F_S(\tau) \tag{21.104}$$

where:

$$F_S(\tau) = (\mathrm{Im}\, \tau)^{-3} \int \prod_{i=1}^{2} d^2\nu_i \prod_{i<j} (\chi_{ij})^{-(1/2)k_i \cdot k_j} \tag{21.105}$$

This amplitude appears to diverge as $\tau \to 0$. However, there is a symmetry that is protecting the amplitude from diverging. This symmetry is *modular invariance*,[25] which is a global symmetry, a subset of conformal invariance. A modular transformation on the τ variable is generated by:

$$\tau' = \frac{a\tau + b}{c\tau + d} \tag{21.106}$$

where a, b, c, d are *integers*.

The divergence of string theory is similar to the divergence of gauge theory found by non-Abelian gauge theory. As in gauge theory, the path integral diverges because of an infinite overcounting of the gauge symmetry, which is eliminated by slicing the gauge orbit once. In string theory, the counterpart of slicing the orbit is to take one "fundamental" region of the complex plane. We can do this by taking the fundamental region to be:

$$\text{Fundamental region} = \begin{cases} -\frac{1}{2} \le \operatorname{Re} \tau \le \frac{1}{2} \\ \operatorname{Im} \tau \ge 0 \\ |\tau| \ge 1 \end{cases} \tag{21.107}$$

Under a modular transformation, the fundamental region can be mapped into all other points in the complex plane. Thus, the divergence is removed by taking one fundamental region and throwing away the rest.

Multiloops have also be calculated in the string formalism.[26−29] The integrands of the multiloop amplitudes correspond to the Neumann functions defined over Riemann surfaces of genus g. Because of this close analogy with Riemann surfaces, we can see intuitively that the rather miraculous cancellation of all divergences at the first-loop level persists to all loop levels. We know, by conformal invariance, that we can isolate the divergence of each loop by "pinching" each hole separately. Thus, the same arguments we used in the single-loop cancellation can be used to show that the divergence of each "pinch" can be eliminated.

Once we have eliminated the divergences associated with each hole separately, we still have to consider the subtle divergences associated with the multiple deformation of the topology of the surface, that is, when several holes collapse together. This is easiest to study in the light-cone gauge, where Mandelstam has eliminated all divergences of the superstring.

21.8 Phenomenology

One of the main problems in superstring research has been to find the true vacuum of the theory, either perturbatively or nonperturbatively. Therefore, intense research over the years has been spent trying to catalog the various possible four-dimensional compactified strings.

A few classes of these solutions include:

1. Calabi–Yau manifolds,[30] which are highly nonlinear, nontrivial manifolds studied by mathematicians.

2. Orbifolds,[31] which are certain manifolds which have fixed points on them (e.g., a cone is an orbifold).

3. Free fermion/free boson solutions.[32−34]

Unfortunately, we now know millions upon millions of possible string vacua. In fact, it is conjectured that the complete set of all possible string vacua is the totality of possible conformal field theories.

Although there are an enormous number of possible four-dimensional string vacua, the surprising feature of string theory is that, with a few rather mild assumptions, one can come fairly close to describing the physical universe. Earlier, we saw that Kaluza–Klein theory was too restrictive to describe the physical universe. In particular, the Standard Model's gauge group and complex fermion representations could not be accommodated. However, the string model, because it is not based on Riemannian space, does not suffer from these problems. To begin, let us make the following assumptions:

1. The string has compactified down to a four-dimensional Minkowski space times a six-dimensional space:

$$M_{10} \rightarrow M_4 \otimes K_6 \qquad (21.108)$$

where M_4 is a maximally symmetric space; that is,

$$R_{\mu\nu\alpha\beta} = \frac{R}{12}(g_{\mu\alpha}g_{\nu\beta} - g_{\mu\beta}g_{\nu\alpha}) \qquad (21.109)$$

2. $N = 1$ local supersymmetry has survived the compactification down to four dimensions.

3. Some of the bosonic fields in ten-dimensional superstring theory can be set to zero.

The second assumption, in particular, yields very stringent constraints on the possible string vacua. The variation of a fermion ψ_i transforming under $N = 1$ supergravity is given by:

$$\delta\psi_i = [\bar{\epsilon}Q, \psi_i] \sim D_i\epsilon \qquad (21.110)$$

If supersymmetry is preserved, then the vacuum is annihilated by Q, and therefore:

$$Q|0\rangle = 0 \rightarrow \langle 0|\delta\psi_i|0\rangle = 0 \qquad (21.111)$$

In the classical limit, this means that $\delta\psi_i$ itself must vanish:

$$\delta\psi_i \sim D_i\epsilon = 0 \tag{21.112}$$

This deceptively simple statement is quite restrictive, because it means that ϵ is a *generally covariant constant spinor*. This, in turn, is only possible on a highly specialized set of six-dimensional manifolds. To find these manifolds, we must study the covariant derivative D_i, which has the physical interpretation of being a covariant displacement operator on the K manifold. If we travel in closed loops in K space, then the effect of this is equivalent to taking multiple variations of the fermion, so we arrive at:

$$\epsilon \rightarrow \epsilon + \Delta^{ij}[D_i, D_j]\epsilon \tag{21.113}$$

where Δ^{ij} is the area tensor of the loop.

The statement that ϵ is a covariantly constant spinor means that it is invariant under multiple displacements in K space, so that:

$$[D_i, D_j]\epsilon \sim R_{ijkl}\Gamma^{kl}\epsilon = 0 \tag{21.114}$$

In other words, this means that:

$$R_{ij} = 0 \tag{21.115}$$

that is, the manifold K is *Ricci-flat*.

On the manifold K, the displacement operator D_i contains a connection field, which is an $O(6)$ gauge field. However, we also know that $O(6) = SU(4)$. Normally, a $O(6)$ spinor has eight elements. However, this eight-component spinor can be decomposed according to $SU(4)$ as:

$$\mathbf{8} = \mathbf{4} \oplus \mathbf{4} \tag{21.116}$$

that is, the eight-component spinor transforms as the sum of two four-spinors of opposite chirality. We will take ϵ to have positive chirality, so it transforms as one $\mathbf{4}$. The fact that ϵ is a covariantly constant spinor now reduces to the simple question: What is the largest group that will leave a constant spinor invariant? The answer is easy to see if we write the spinor as:

$$\epsilon = \begin{pmatrix} 0 \\ 0 \\ 0 \\ \epsilon_0 \end{pmatrix} \tag{21.117}$$

Clearly, $SU(3)$ rotations, which do not affect the first three rows of ϵ, are the largest group that can leave ϵ invariant.

In summary, we have shown that, with rather mild assumptions (namely, that $N = 1$ supersymmetry survives the compactification), the manifold K is both Ricci-flat and has $SU(3)$ holonomy. We call such manifolds *Calabi–Yau* manifolds.

Next, we must check that the Bianchi identities are satisfied for the theory. Usually, these identities are trivially satisfied. However, for our case this is no longer true, especially when we invoke the third condition, that certain fields vanish. The Bianchi identities become:

$$\text{Tr } R \wedge R = \frac{1}{30}\text{Tr } F \wedge F \tag{21.118}$$

This is a highly nontrivial constraint, because the Riemann curvature sits on the left-hand side, while the Yang–Mills field for the exceptional group sits on the right-hand side.

However, there is a nontrivial solution to this constraint. This constraint essentially forces us to make a link between the Yang–Mills connection field and the connection field of Riemannian K space. Since we know that K is a Calabi–Yau manifold with $SU(3)$ holonomy, we can insert the connection field of K into the connection field of $E_8 \otimes E_8$. We know that E_8 contains $SU(3) \otimes E_6$ as a subgroup. By preserving the $SU(3)$ contained within E_8, we achieve a breaking of the original exceptional group symmetry, so that:

$$E_8 \otimes E_8 \rightarrow SU(3) \otimes E_6 \otimes E_8 \tag{21.119}$$

The original fermions of $E_8 \otimes E_8$, which formed a representation **248**, transform under $SU(3) \otimes E_6$ as follows:

$$\mathbf{248} = (\mathbf{3}, \mathbf{27}) \oplus (\bar{\mathbf{3}}, \overline{\mathbf{27}}) \oplus (\mathbf{8}, \mathbf{1}) \oplus (\mathbf{1}, \mathbf{78}) \tag{21.120}$$

We can now place yet another restriction on the theory. The Bianchi identity cannot be satisfied with any choice of fermions in four dimensions. A careful analysis of the constraint shows that the fermions of the low-energy spectrum must belong to the **27**, which is precisely the favored GUT representation for E_6.

Finally, one great advance of this construction over previous ones is that we can determine the number of generations from purely topological reasons. In standard GUT theory, we recall, there is no compelling reason to introduce three exact copies of the theory. In superstring theory, we have an additional constraint coming from topology. By analyzing which manifolds allow fermions to propagate on them, this gives us a determination of how many generations of fermions are allowed. In particular, there are several manifolds that allow precisely three generations of fermions on them.

In summary, with very mild assumptions, we have found a vast number of solutions to the string equations of motion that mimic many of the features of the physical universe.

21.9   Light-Cone String Field Theory

So far, we have developed string theory in the first quantized formalism, where we postulated a large number of ad hoc rules to derive the S matrix. In this language, we could not prove unitarity or fix the weights of the various diagrams. In this section, we will derive the second quantized field theory of strings,[35] where all the Feynman rules are derived from a single action. We will first discuss the light-cone string field theory, where unitarity is manifest, and then the BRST string field theory, where Lorentz covariance is manifest.

The field theory of strings is based on $\Phi(X)$, which is a functional; that is, it is a function of every point $X_\mu(\sigma)$ along the string for all possible values of σ:

$$\Phi(X_\mu) = \Phi\left(X_\mu(\sigma_1), X_\mu(\sigma_2), \ldots, X_\mu(\sigma_N)\right) \qquad (21.121)$$

where σ_i are the points along the string and we let $N \to \infty$.

We can also decompose this string functional in any basis we wish. In the harmonic oscillator basis, the string field has a particularly simple form:

$$\Phi(X) = \langle X | \Phi(x_0) \rangle \qquad (21.122)$$

where:

$$|\Phi(x_0)\rangle = \phi(x_0)|0\rangle + A_\mu(x_0)a_1^{\mu\dagger}|0\rangle + g_{\mu\nu}(x_0)a_1^{\mu\dagger}a_1^{\nu\dagger}|0\rangle + \cdots \qquad (21.123)$$

where x_0 is the usual four-vector representing ordinary space–time. Here, we see the explicit decomposition of the field functional in terms of the tachyon field $\phi(x_0)$, the Maxwell field $A_\mu(x_0)$, a massive graviton field $g_{\mu\nu}(x_0)$, etc.

We can repeat our discussion in Section 8.3, where we made the transition from first to second quantization for point particles. We find that the free light-cone action is given by:

$$S_0 = \int d\tau \, DX_i \, dp^+ \left[\Phi_{p^+}(X_i, \tau)(i\partial_\tau - H)\Phi_{p^+}(X_i, \tau)\right] \qquad (21.124)$$

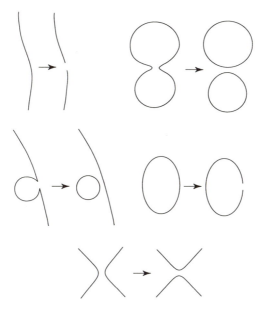

Figure 21.5. In light-cone string field theory, there are five ways in which open and closed strings may interact. Notice that all interactions take place locally along the string at certain points. Closed string field theory has only a cubic interaction.

where the light-cone Hamiltonian is given in Eq. (21.42) and $\Phi_{p^+}(X_i, \tau)$ is the Fourier transform of $\Phi(X^-, X^+, X_i)$ with respect to X_- after we have taken the light-cone gauge $X^+ = p^+\tau$.

The measure DX_i is equal to:

$$DX_i \equiv \prod_i \prod_n dX_{ni} = \prod_i \prod_\sigma dX_i(\sigma) \tag{21.125}$$

Next, we will only sketch how to write the interaction Lagrangian for the light-cone string field theory. In Figure 21.5, we list the five possible interactions that are required to describe open string field theory. (For a purely closed string field theory, only the cubic term is necessary.) For open strings, an examination of Figure 21.5 shows that strings can join at their endpoints (or break at an interior point). Among other terms, the interaction Lagrangian contains a Φ^3 term, with a Dirac delta function sandwiched in between:

$$S_3 = \int \prod_{i=1}^{3} dp_r^i \, DX_i \, \delta \left(\sum_{r=1}^{3} p^{+r} \right) \delta_{123} \, \Phi_1(X_1)\Phi_2(X_2)\Phi_3^\dagger(X_3) + \text{h.c.} \tag{21.126}$$

where:

$$\delta_{123} = \prod_{i=1}^{2} \prod_{0<\sigma_i<\pi\alpha_i} \delta\big(X_3(\sigma_3) - \theta(\pi\alpha_1 - \sigma)X_1(\sigma_1) - \theta(\sigma - \pi\alpha_1)X_2(\sigma_2)\big) \quad (21.127)$$

where the string variables are defined as:

$$\begin{aligned}
\sigma_1 &= \sigma, & 0 \le \sigma \le \pi\alpha_1 \\
\sigma_2 &= \sigma - \pi\alpha_1, & \pi\alpha_1 \le \sigma \le \pi(\alpha_1 + \alpha_2) \\
\sigma_3 &= \pi(\alpha_1 + \alpha_2) - \sigma, & 0 \le \sigma \le \pi(\alpha_1 + \alpha_2) \quad (21.128)
\end{aligned}$$

with the condition $\sum_i \alpha_i = 0$. To calculate all N-point functions and loops diagrams, it is necessary to use all five interactions shown in Figure 21.5. If we let Φ (Ψ) represent open (closed) string fields, we can symbolically represent the interactions for the open and closed strings:

$$\begin{aligned}
\mathscr{L}_{\text{open}} &= \Phi^3 + \Phi^4 + \Phi^2\Psi + \Psi^3 + \Phi\Psi \\
\mathscr{L}_{\text{closed}} &= \Psi^3 \quad (21.129)
\end{aligned}$$

In other words, the open string vertex function by itself cannot generate all string amplitudes, so we must necessarily include closed strings as well. Thus, even if we started out with an open string theory without any gravitons, we find that gravitons necessarily creep back into the theory. There is no choice: String theory is by its very nature a theory of quantum gravity.

The purely closed string action, by contrast, is cubic. This is rather remarkable: The theory of quantum gravity, which is highly nonlinear, coupled to an infinite tower of spinning fields, is cubic. The Ψ^3 interaction is sufficient to generate all the interactions of gravity coupled to matter fields.

21.10   BRST Action

There is also a second quantized formalism in which gauge invariance and Lorentz invariance is manifest. Let us choose the BRST action:

$$S = \int DX\, Db\, Dc\, D\bar{b}\, D\bar{c}\ \Phi Q\Phi \quad (21.130)$$

where Φ is defined to have ghost number $-\frac{1}{2}$, and Q is the usual BRST operator.

The advantage of this second approach is that one can see explicitly the gauge invariance of the theory. The theory is invariant under:

$$\delta \Phi = Q \Lambda \tag{21.131}$$

because Q is nilpotent.

As we mentioned, the first quantized string theory required a sum over the set of all conformally inequivalent topologies. This conveniently concealed many difficult questions concerning how to place coordinates (or moduli) on arbitrary Riemann surfaces. The principal problem is that, until recently, mathematicians have been unable to triangulate moduli space successfully for genus g Riemann surfaces, even after a century of experience with these surfaces. Remarkably, string field theory gives an exact triangulation of moduli space, thus solving a long-standing mathematical problem.

Let us begin our discussion by first requiring that open string field theory be a gauge theory that satisfies the axioms of gauge theory. Specifically, we need to postulate the existence of a derivative Q and a product operation $*$. We postulate the following five axioms:

1. The existence of nilpotent derivative Q such that $Q^2 = 0$.

2. The associativity of the $*$ product:

$$(A * B) * C = A * (B * C) \tag{21.132}$$

3. The Leibnitz rule:

$$Q(A * B) = QA * B + (-1)^{|A|} A * QB \tag{21.133}$$

4. The product rule:

$$\int A * B = (-1)^{|A||B|} \int B * A \tag{21.134}$$

5. The integration rule:

$$\int QA = 0 \tag{21.135}$$

where $(-1)^{|A|}$ is -1 if A is Grassmann odd and $+1$ if A is Grassmann even.

We postulate that the field A has the following transformation rule:

$$\delta A = Q\Lambda + A * \Lambda - \Lambda * A \tag{21.136}$$

Figure 21.6. The symmetric interaction of Witten's covariant open string field theory.

Then we can construct a curvature form given by:

$$F = QA + A * A \tag{21.137}$$

such that:

$$\delta F = F * \Lambda - \Lambda * F \tag{21.138}$$

It is easy therefore to show that the following is a total derivative:

$$\int F * F = \int Q \left(A * QA + \frac{2}{3} A * A * A \right) \tag{21.139}$$

Then the Witten action[36] is given as a Chern–Simons form:

$$\mathscr{L} = A * QA + \frac{2}{3} A * A * A \tag{21.140}$$

(The Chern–Simons form is preferable to the usual F^2 form found in ordinary gauge theory, because Q already has two derivatives contained within it.)

Our task is to find a multiplication operation that satisfies the postulates of the $*$ product. Then, the gauge invariance of the theory is automatic, without any more work. We notice, first of all, that the $*$ operation is symmetric in all three strings. There is only one unique configuration that is symmetrical in all three fields, and that is given in Figure 21.6, where the midpoint of the strings has been singled out.

The multiplication operation:

$$|X_3\rangle = |X_1\rangle * |X_2\rangle \tag{21.141}$$

simply means that we have exchanged the Fock spaces of strings 1 and 2 for string 3, such that the points along 1 and 2 have been identified with points along string 3. The triple product (without ghosts) can be defined as a delta function:

$$\Phi * \Phi * \Phi \;=\; \int DX_1 \, DX_2 \, DX_3 \, \Phi(X_1)\Phi(X_2)\Phi(X_3)$$

$$\times \prod_{r=1}^{3} \prod_{0 \le \sigma_r \le \pi/2} \delta\left(X_{\mathbf{r},\mu}(\sigma_{\mathbf{r}}) - X_{\mathbf{r}-1,\mu}(\pi - \sigma_{r-1})\right)$$

$$(21.142)$$

(where we omit the ghost delta functions).

Let us now write the ghost number for all the operators in the theory. The c ghost has ghost number 1, the b ghost has ghost number -1, so that Q has ghost number 1. This, in turn, fixes the ghost number of the A field to be $-\frac{1}{2}$, since the action contains a term $\langle A|Q|A \rangle$, which must have total ghost number 0.

The ghost number of the gauge parameter Λ and the $*$ operation can be fixed by observing the gauge variation of the A field. In order for the left-hand side (with ghost number $-\frac{1}{2}$) to equal the ghost number of the right-hand side, the ghost number of Λ must be $-\frac{3}{2}$ and the ghost number of the $*$ operation must be $+\frac{3}{2}$.

Similarly, we can fix the ghost number of the $\int$ operation by demanding that the action have total ghost number zero. Putting everything together, we have the following set of ghost numbers:

$$
\begin{array}{llll}
c: & 1; & *: & \frac{3}{2} \\[2mm]
b: & -1; & \int: & -\frac{3}{2} \\[2mm]
Q: & 1; & \Lambda: & -\frac{3}{2} \\[2mm]
A: & -\frac{1}{2}; & &
\end{array}
$$

$$(21.143)$$

What is more interesting, of course, is a covariant closed string field theory. Unfortunately, it is more complicated, requiring a nonpolynomial action where the closed string interactions have the topology of polyhedra.[37-39] In this short chapter we are unable to present this action or other interesting features of the superstring theory. We could only sketch the highlights. The interested reader is therefore urged to consult the literature concerning the many fascinating properties of superstrings that are beyond the scope of this book.

In summary, the advantages of the superstring theory are:

1. The theory is finite to all orders in perturbation theory. It requires no renormalization.

2. The theory necessarily includes quantum gravity and gauge theory as subsets. Dropping quantum gravity from the action, in fact, destroys the properties of the theory.

3. The theory contains all the symmetries so far found in quantum field theory as a subset, yet it is totally free of anomalies.

4. With only a few assumptions, one can obtain the chiral fermion spectrum contained within the **27** of E_6, which includes all the known fermions.

5. The generation problem can be formally solved by analyzing the topological invariants of a six-dimensional manifold.

6. The model is so tightly constrained that only a handful of self-consistent string theories is possible.

However, we should also mention the formidable problems facing superstring theory:

1. As with any theory of quantum gravity, the superstring theory is defined at the Planck energy, and hence testing the superstring theory becomes problematic, if not impossible.

2. Millions of vacua for the theory have been found, some of which have three generations of fermions and can reproduce many of the features of the Standard Model. However, the outstanding problem is finding which one, if any, is the true vacuum of the theory.

3. Experimentally, the theory cannot explain why the cosmological constant is extremely close to zero. Supersymmetry, before symmetry breaking, is powerful enough to fix the cosmological constant to be zero. However, once supersymmetry is broken, it is not known how to keep the cosmological constant zero.

Of these various problems, the most fundamental is perhaps the second. Until the true, nonperturbative vacuum of the theory can be isolated among the millions that have been discovered, the theory has no real predictive power. However, since the superstring equations are perfectly well defined, the true nonperturbative vacuum solution can, in principle, be found. Thus, the main problem facing superstring theory at present is theoretical, to isolate the true vacuum of the theory, rather than experimental. Until this solution is found, our attitude is to treat the superstring theory as a highly sophisticated theoretical laboratory in which to test the limits of quantum field theory.

21.11 Exercises

1. Show that Veneziano amplitude in Eq. (21.60) satisfies all the properties of an S matrix, except for one; that is, show that it is analytic, Lorentz invariant,

CPT invariant, crossing symmetric, and Regge behaved:

$$A(s, t) \rightarrow s^{\alpha(t)} \tag{21.144}$$

for large s and fixed t. Why does the Veneziano formula not satisfy the last remaining constraint of the S matrix, unitarity?

2. Using harmonic oscillators, expand out the N-point amplitude in Eq. (21.63) using coherent state methods and show it to be equivalent to Eq. (21.59).

3. Express the N-point Veneziano formula so that it is manifestly invariant under a real projective transformation performed on the integration variables:

$$z' = \frac{az + b}{cz + d} \tag{21.145}$$

where $ad - bc = 1$.

4. For the Nambu–Goto string, calculate the momenta P_μ and prove that it satisfies Eq. (21.37).

5. Given the BRST operator Q, show by direct calculation that it is nilpotent only in 26 dimensions; that is, prove Eq. (21.53).

6. Show that the condition $Q|\Phi\rangle = 0$ is sufficient to eliminate the longitudinal mode of the Maxwell field.

7. For the string field given in Eq. (21.123), show that the variation:

$$\delta|\Phi\rangle = L_{-1}|\Lambda\rangle \tag{21.146}$$

contains within it the gauge variation of the Maxwell field: $\delta A_\mu = \partial_\mu \Lambda$.

8. Show that the field variation:

$$\delta|\Phi\rangle = \overline{L_{-1}}|\Lambda\rangle + L_{-1}|\bar{\Lambda}\rangle \tag{21.147}$$

yields, for the spin-2 field:

$$\delta h_{\mu\nu} = \partial_\mu \Lambda_\nu + \partial_\nu \Lambda_\mu \tag{21.148}$$

9. In the commutation relation for the Virasoro algebra in Eq. (21.36), explicitly show where the c-number term comes from. (Hint: take the vacuum expectation value of the Virasoro algebra.)

10. Consider the state:

$$|\psi\rangle = \left(L_{-2} + aL_{-1}^2\right)|\phi\rangle \tag{21.149}$$

where $|\phi\rangle$ satisfies the Virasoro constraint. Show that this state does not couple to real states that satisfy the Virasoro constraint; that is, show that it is spurious. Now demand that this state also be real, that it also satisfy the Virasoro constraint:

$$L_1|\psi\rangle = L_2|\psi\rangle = 0 \tag{21.150}$$

Show that this fixes $D = 26$ and $a = 3/2$. At first, this may seem to be a disaster: We have constructed a real state that is also spurious. But show that this state has zero norm, and hence the theory still makes sense at $D = 26$.

11. Calculate the four-point function for the scattering of four tachyons in the Neveu–Schwarz model. Show that:

$$
\begin{aligned}
A_4(s,t) &= \langle 0; k_1 | k_1 \cdot b_{1/2} V(k_2) \frac{1}{L_0 - 1} V(k_3) k_4 \cdot b_{-1/2} | 0; k_4 \rangle \\
&= \frac{\Gamma\left(1 - \alpha(s)\right) \Gamma\left(1 - \alpha(t)\right)}{\Gamma\left(1 - \alpha(s) - \alpha(t)\right)}
\end{aligned} \tag{21.151}
$$

where $V = k_\mu \psi^\mu V_0$, where $\alpha(s) = 1 + \alpha's$ and $\alpha'k^2 = 1$.

12. Prove:

$$
\begin{aligned}
&\exp\left(\frac{1}{2}a_i A_{ij} a_j\right) \exp\left(\frac{1}{2}a_i^\dagger B_{ij} a_j^\dagger\right)|0\rangle \\
&= \det^{-1/2}(1 - AB) \exp\left[\frac{1}{2}a_i^\dagger B_{ij}\left(\frac{1}{1 - AB}\right)_{jk} a_k^\dagger\right]|0\rangle
\end{aligned} \tag{21.152}
$$

[Hint: use Eqs. (21.65) and (21.98) by contracting onto coherent states.]

13. Prove the L_{-1}, L_0, and L_1 generate the group $SL(2, R)$ (the set of 2×2 real matrices with unit determinant). This is also called the projective group.

14. The modular group, which is the symmetry of the one-loop string amplitude, is generated by:

$$
\begin{aligned}
\tau &\rightarrow \tau + 1 \\
\tau &\rightarrow -\frac{1}{\tau}
\end{aligned} \tag{21.153}
$$

Show that the group generated by these mappings is equivalent to the group of transformations:

$$\tau \longrightarrow \frac{a\tau + b}{c\tau + d} \tag{21.154}$$

where a, b, c, d are arbitrary *integers*.

15. For the open bosonic string, prove:

$$(L_n - L_0 - n + 1)V_0 = V_0(L_n - L_0 + 1)$$

$$(L_n - L_0 + 1)\frac{1}{L_0 - 1} = \frac{1}{L_0 + n - 1}(L_n - L_0 - n + 1) \tag{21.155}$$

From this, prove that:

$$(L_n - L_0 - n + 1)V_0 D \cdots D V_0 |0\rangle = 0 \tag{21.156}$$

Show that this means that ghost states do not couple to trees, although they can couple to loops. How does this compare with the way Yang–Mills ghosts couple to trees and loops?

16. Prove that Eq. (21.10) is equivalent to the original Lagrangian in Eq. (21.1) by functionally eliminating p_μ and λ.

Appendix

A.1 $SU(N)$

From the work of Lie and Cartan, we have a complete classification of the various compact Lie groups. If we restrict ourselves to compact, real forms, then the complete set is given by the infinite series, labeled:

$$
\begin{aligned}
A_n &= SU(n+1) \\
B_n &= SO(2n+1) \\
C_n &= Sp(2n) \\
D_n &= SO(2n)
\end{aligned}
\tag{A.1}
$$

as well as the exceptional groups, labeled by E_6, E_7, E_8, F_4, and G_2.

Of special interest to physicists is the Lie group $SU(N)$, which is the set of all special, unitary $N \times N$ complex matrices. If U is a member of $SU(N)$, then it satisfies:

$$
\begin{aligned}
UU^\dagger &= 1 \\
\det U &= 1
\end{aligned}
\tag{A.2}
$$

By counting the constraints in this equation, we know that the matrix has $N^2 - 1$ unknowns, or parameters.

Any unitary matrix can be represented by the exponential:

$$
U = e^{iH}
\tag{A.3}
$$

where H is Hermitian:

$$
H^\dagger = H
\tag{A.4}
$$

(This can be proved by taking the conjugate of both sides of the equation.)

Since there are $N^2 - 1$ independent Hermitian $N \times N$ matrices, we can also write:

$$U = \exp\left(i \sum_{i=1}^{N^2-1} \theta^a \tau^a\right) \tag{A.5}$$

where τ^a are independent Hermitian matrices, the generators of the group, satisfying:

$$[\tau^a, \tau^b] = i f^{abc} \tau^c \tag{A.6}$$

To create irreducible representations of $SU(N)$, we first postulate the existence of N complex fields ϕ^i that transform as:

$$\phi'^i = U^i_j \phi^j \tag{A.7}$$

We also introduce a new field ψ that transforms as $\psi^* \rightarrow \psi^* U^\dagger$. Then an invariant is given by:

$$\text{Invariant}: \quad \psi_i^* \phi^i \tag{A.8}$$

This is easily shown to be an invariant, because the transformed object contains $U^\dagger U$ sandwiched between the two fields. Since $U^\dagger U = 1$, we see that $\psi_i^* \phi^i$ is an invariant.

In fact, we can use this as an alternative definition of the group; that is, $SU(N)$ consists of all complex transformations with unit determinant that leave $\psi_i^* \phi^i$ invariant. That is,

$$
\begin{aligned}
\psi'^*_i \phi'^i &= \left(\psi_k^* (U^\dagger)^k_i\right)\left(U^i_l \phi^l\right) \\
&= \psi_k^* \left(U^{\dagger k}_i U^i_l\right) \phi^i \\
&= \psi_i^* \phi^i
\end{aligned}
\tag{A.9}
$$

Notice that, unlike the case of $O(N)$, the placement of the indices in the upstairs or downstairs position is extremely important, because the location of the indices indicates whether the vector transforms under U or under $U^\dagger$.

The ϕ^i transform according to the *fundamental representation* of the group. The name is appropriate because we can derive the higher representations by taking tensor products of the fundamental representation.

Higher tensors transform exactly like the product of various fundamental representations:

$$T^{j_1 j_2 \cdots j_M}_{i_1 i_2 \cdots i_N} \tag{A.10}$$

where it is important to keep track of the upstairs and downstairs indices.

A.2 Tensor Products

In general, such tensors are reducible. To find the irreducible representations, we must take symmetric and antisymmetric combinations of the indices.

This tedious process of taking symmetric and antisymmetric combinations is made simpler by noticing that there are two genuine constant tensors for the theory:

$$\delta^j_i; \quad \epsilon^{i_1 i_2 \cdots i_N} = \epsilon_{i_1 i_2 \cdots i_N} \tag{A.11}$$

To prove that these are genuine tensors, we simply act on these tensors with U matrices. As in the case of the group $O(N)$, δ^{ij} can be shown to be a constant tensor because U is unitary. Also, $\epsilon^{i_1 i_2 \cdots i_N}$ is a genuine tensor because the determinant of U is equal to one.

For example, the tensor product $A^i B^j$, composed of two vectors, is reducible. To create smaller subsets that transform among themselves, let us take the symmetric or the antisymmetric combinations of $A^i B^j$. We can write:

$$A^i B^j = \frac{1}{2} A^{[i} B^{j]} + \frac{1}{2} A^{(i} B^{j)}$$
$$A^{[i} B^{j]} \equiv A^i B^j - A^j B^i$$
$$A^{(i} B^{j)} \equiv A^i B^j + A^j B^i \tag{A.12}$$

For $SO(2)$, let the symbol **2** represent the two elements of a vector:

$$\mathbf{2} = A^i \tag{A.13}$$

In Eq. (A.12), there is only one element in the antisymmetric combination, which we represent symbolically as **1**, and there are three elements in the symmetric combination (one of which can be separated out as the trace). Then a shorthand notation for Eq. (A.12) is given by:

$$\mathbf{2} \otimes \mathbf{2} = \mathbf{2} \oplus \mathbf{1} \oplus \mathbf{1} \tag{A.14}$$

It is easy to show that the symmetric and antisymmetric tensors, by themselves, form a separate representation of $O(2)$, thereby proving that the tensor product $A^i B^j$ is reducible.

To construct irreducible representations of $SO(3)$, it is useful to know that δ^{ij} and ϵ^{ijk} are covariant tensors, and that we can take reducible representations and extract out irreducible tensors from them.

As before, we can take the product of two vectors A^i and B^i, each of which transforms as a triplet **3**, and extract irreducible tensors. For example, we can extract the singlet $A^i B^i$ and the triplet $\epsilon^{ijk} A^j B^k$ from the product.

In general, the product of two triplets can be reduced according to whether they are symmetric or antisymmetric. The symmetric combination is represented as a **5** plus the trace **1**, while the antisymmetric combination is represented as a **3**, so we have two equivalent ways of representing this:

$$A^i B^j = \frac{1}{2} A^{(i} B^{j)} + \frac{1}{2} A^{[i} B^{j]}$$

$$\mathbf{3} \otimes \mathbf{3} = \mathbf{5} \oplus \mathbf{1} \oplus \mathbf{3} \tag{A.15}$$

Similarly, we can construct the irreducible representations of $SU(3)$ by taking tensor products of the fundamental representation:

$$\mathbf{3} \otimes \mathbf{3} = \mathbf{6} \oplus \bar{\mathbf{3}} \tag{A.16}$$

We can also take the combination ψ_i^* times ϕ^j, which reduces to:

$$\mathbf{3} \otimes \bar{\mathbf{3}} = \mathbf{8} \oplus \mathbf{1} \tag{A.17}$$

where **1** is represented by $\psi_i^* \phi^i$.

For $SU(N)$, this identity can be written as:

$$\mathbf{N} \otimes \bar{\mathbf{N}} = \left(\mathbf{N}^2 - \mathbf{1}\right) \oplus \mathbf{1} \tag{A.18}$$

For more complicated tensor products, taking tensor products becomes rather tedious, so we use the method of Young tableaux. Let the box symbol represent ϕ^i. If we have the product of two vectors and take the symmetric product, we have $\phi^{(i} \phi^{j)}$, which is represented by two horizontal boxes.

In general, n horizontal boxes means that we have an n rank tensor such that the indices are symmetrized.

The number of independent components within this horizontal array of boxes is given by:

$$\binom{N+n-1}{n} = \frac{N(N+1)\cdots(N+n-1)}{n!} \tag{A.19}$$

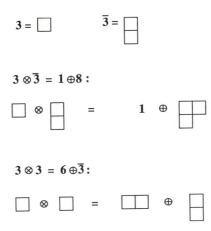

$$3 \otimes \bar{3} = 1 \oplus 8 :$$

$$3 \otimes 3 = 6 \oplus \bar{3} :$$

Figure A.1. In this diagram, we see that the product of a quark and an antiquark gives us an octet and singlet, and that two quarks give us an antitriplet and a sextet.

When we have two boxes stacked vertically, this means that we are taking a second-rank tensor and then taking the antisymmetric combination of the two indices. In general, m boxes stacked vertically means that m indices are antisymmetrized. The number of independent elements in such a vertical array is given by N elements taken m at a time. Thus, the dimension of m boxes stacked vertically

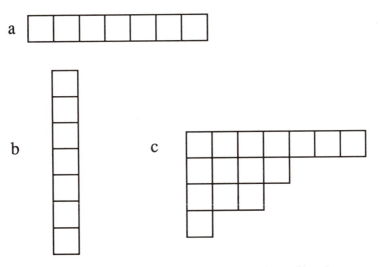

Figure A.2. In (a), a series of horizontal boxes corresponds to taking the symmetrized tensor product of n vectors. In (b), a series of vertical boxes corresponds to taking the antisymmetrized tensor product. In (c), we have a mixed tensor, with symmetrization for horizontal boxes and antisymmetrization for vertical ones.

is given by:

$$\binom{N}{m} = \frac{N(N-1)\cdots(N-m+1)}{m!} \tag{A.20}$$

For example, in Figure A.1 we the Young tableaux for Eqs. (A.16) and (A.17). In Figure A.2, we see more general types of young tableaux.

Of course, this process cannot be continued indefinitely. For $SU(3)$, for example, three boxes stacked vertically only have one element:

$$\mathbf{1} = \epsilon_{ijk} T^{ijk} \tag{A.21}$$

Therefore, for $SU(N)$, N boxes stacked vertically corresponds to a tensor with only one independent element, denoted by $\mathbf{1}$.

Also, notice that $N-1$ boxes stacked vertically has N elements. This state corresponds to ϕ_i^*, which also has N elements. If we add one more vertical box to $N-1$ vertical boxes, then we get a scalar. Similarly, if we contract ϕ_i^* with ϕ^i, we also get a scalar.

By convention, an arbitrary mixed tensor consists of a series of boxes stacked both vertically and horizontally. Let f_i equal the number of boxes stacked horizontally in the i row. We take the convention that $f_i > f_{i+1}$; so the number of horizontal rows diminishes as we go down the Young tableaux (see Fig. A.2). An arbitrary Young tableaux can therefore be designated by a series of numbers $(f_1, f_2, \ldots f_k)$, with each number representing the number of horizontal boxes in each row.

For example, a series of n horizontal boxes is designed by $(n, 0, 0 \ldots)$. A series of m boxes stacked vertically is given by $(1, 1, \ldots, 1)$ with m entries.

Then there is a classical theorem from group theory that the dimensionality or number of independent elements in the mixed tensor $(f_1, f_2, \ldots, f_k)$ is given by:

$$
\begin{aligned}
D(f_1, f_2, \ldots, f_k) &= (1 + f_1 - f_2)(1 + f_2 - f_3)\cdots(1 + f_k) \\
&\times \left(1 + \frac{f_1 - f_3}{2}\right)\left(1 + \frac{f_2 - f_4}{2}\right)\cdots\left(1 + \frac{f_{k-1}}{2}\right) \\
&\times \left(1 + \frac{f_1 - f_4}{2}\right)\left(1 + \frac{f_2 - f_5}{2}\right)\cdots\left(1 + \frac{f_{k-2}}{2}\right) \\
&\times \left(1 + \frac{f_1 - f_5}{2}\right)\cdots\left(1 + \frac{f_{k-3}}{2}\right) \\
&\cdots \\
&\times \left(1 + \frac{f_1}{2}\right) \tag{A.22}
\end{aligned}
$$

A.3   *SU*(3)

We can repeat many of the same steps for $SU(3)$ using ladder operators. Notice that there are two generators that commute among each other (because they are diagonal):

$$T_3 = F_3; \quad Y = \frac{2}{\sqrt{3}} T_8 \qquad (A.23)$$

This means that we can simultaneously diagonalize both operators, and that the eigenstates of these operators are indexed by two numbers, the ordinary isospin and the hypercharge.

We therefore have states labeled by their eigenvalues:

$$
\begin{aligned}
T_3 |t_3, y\rangle &= t_3 |t_3, y\rangle \\
Y |t_3, y\rangle &= y |t_3, y\rangle
\end{aligned}
\qquad (A.24)
$$

As with $SU(2)$, we will now introduce the ladder operators of $SU(3)$:

$$T_\pm = F_1 \pm i F_2; \quad U_\pm = F_6 \pm i F_7; \quad V_\pm = F_4 \pm i F_5 \qquad (A.25)$$

The new commutation relations now become:

$$
\begin{array}{ll}
[T_3, T_\pm] = \pm T_\pm & [Y, T_\pm] = 0 \\
[T_3, U_\pm] = \mp U_\pm/2 & [Y, U_\pm] = \pm U_\pm \\
[T_3, V_\pm] = \pm V_\pm/2 & [Y, V_\pm] = \pm V_\pm \\
[T_+, V_-] = -U_- & [T_+, U_+] = V_+ \\
[U_+, V_-] = T_- & [T_3, Y] = 0 \\
[T_+, T_-] = 2T_3 & [U_+, U_-] = (3/2)Y - T_3 \\
[V_+, V_-] = (3/2)Y + T_3 & [T_+, V_+] = 0 \\
[T_+, U_-] = 0 & [U_+, V_+] = 0
\end{array}
\qquad (A.26)
$$

By examining the commutators carefully, we see that T_+ raises the eigenvalue t_3 by one unit, and T_- lowers it by one unit. Since $T_\pm$ commute with Y, they leave y the same. We also see that U_+ lowers t_3 by one-half unit, and raises y by one unit. Likewise, V_+ raises t_3 by one-half unit and raises y by one unit.

Graphically, we can represent this in eigenvalue space by plotting t_3 horizontally and y vertically. Then the action of the ladder operators is to raise or lower

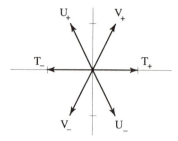

Figure A.3. The ladder operators $T_\pm$, $U_\pm$, and $V_\pm$ change the eigenvalues of a state in the direction of the arrows shown in this chart.

the various eigenvalues along the horizontal, vertical, and diagonal lines, as in Figure A.3.

By hitting an eigenstate $|t_3, y\rangle$ with these operators $U_\pm$, $V_\pm$, and $T_\pm$, this eigenstate is converted into an eigenstate that lies one step removed from the original state, according to the prescription given.

In general, all the Lie groups can be analyzed in this fashion via the ladder operators. Given the generators of an algebra, we can divide them into two types of generators:

1. *The Cartan subalgebra*, consisting of the generators H_i, which all mutually commute among themselves:

$$[H_i, H_j] = 0 \tag{A.27}$$

The number of such generators in the Cartan subalgebra is called the *rank* r of the group. [For example, $SU(3)$ is a rank-two group because its Cartan subalgebra consists of T_3 and Y.]

Then we can simultaneously diagonalize the members of the Cartan algebra, so that an eigenvector of these operators is given by:

$$|l_1, l_2, \ldots, l_r\rangle \tag{A.28}$$

2. *The ladder operators*, which move the various eigenvalues of the eigenvector by various amounts:

$$L_\pm^i |l_1, \ldots, l_i, \ldots\rangle = C|l_1, \ldots, l_i \pm 1, \ldots\rangle \tag{A.29}$$

Since each ladder operator changes the eigenvalue of the state, we can label each ladder operator by a vector α in the space of eigenvalues, which is called root space.

Then, by taking successive products of the various ladder operators acting on a state of highest weight, we can fill out any representation of the group.

In this way, we can systematically exhaust all possible representations of all possible Lie groups.

A.4   Lorentz Group

Because $O(4) = SU(2) \otimes SU(2)$, we can also categorize the irreducible representations of the Lorentz group using two-component, complex spinors belonging to $SU(2)$.

We can decompose the four-spinor as:

$$\psi = \begin{pmatrix} \psi_R \\ \psi_L \end{pmatrix} \tag{A.30}$$

We can then take the two-spinors as:

$$(1/2, 0) = \frac{1 + \gamma_5}{2} \psi$$

$$(0, 1/2) = \frac{1 - \gamma_5}{2} \psi \tag{A.31}$$

We can then construct higher spin fields by taking tensor products between the spinors. For example, vectors can be constructed by taking the product of two spinors:

$$\text{Vector}: \quad (1/2, 0) \otimes (0, 1/2) = (1/2, 1/2) \tag{A.32}$$

A spin 3/2 field can be represented in several ways, but the most common is to take the product of a vector and a spinor:

$$(1/2, 1/2) \otimes (1/2, 0) = (1, 1/2) \oplus (0, 1/2) \tag{A.33}$$

In more familiar language, this corresponds to constructing a four-spinor with a vector index attached:

$$\text{Spin 3/2}: \quad \psi_\mu = \begin{pmatrix} \psi_{\mu R} \\ \psi_{\mu L} \end{pmatrix} \tag{A.34}$$

Then the $(0, 1/2)$ spinor corresponds to contracting the spin 3/2 field with a gamma matrix:

$$(0, 1/2) = \gamma^\mu \psi_\mu \tag{A.35}$$

The $(1, 1/2)$ field then corresponds to a spin 3/2 field that has zero contraction on a gamma matrix:

$$(1, 1/2) = \psi_\mu - (1/4)\gamma_\mu \gamma^\nu \psi_\nu \tag{A.36}$$

Similarly, a spin-2 field can be represented as the product of two vectors:

$$\text{Spin 2}: \ (1/2, 1/2) \otimes (1/2, 1/2) = [(0, 0) \oplus (1, 1)]_S \oplus [(0, 1) \oplus (1, 0)]_A \tag{A.37}$$

where $S(A)$ represents a symmetric (antisymmetric) combination.

In more familiar language, this means that we can take the symmetric or antisymmetric combination of a second-rank tensor:

$$g_{\mu\nu} = \frac{1}{2} g_{(\mu\nu)} + \frac{1}{2} g_{[\mu\nu]} \tag{A.38}$$

where the parentheses (brackets) represent taking the symmetric (antisymmetric) combinations.

Then we can extract out the trace part of the symmetric tensor:

$$(1, 1) = g_{\mu\nu} - \frac{1}{4} \delta_{\mu\nu} g_\nu^\nu$$
$$(0, 0) = g_\mu^\mu \tag{A.39}$$

Thus, $(1, 1)$ corresponds to a traceless, symmetric second-rank tensor, which we adopt as our definition of the spin-2 field.

We can go to higher and higher representations, but there is eventually a problem: A theory of interacting massless spin 3 particles does not seem to be consistent.

Finally, we remark that it is customary to decompose the Lorentz group into various pieces, depending on the sign of certain parameters.

For example, we saw that we could take $\det \Lambda = \pm 1$. If we only take the $\det \Lambda = 1$, we have the *proper* Lorentz transformations, forming the subgroup $SO(3, 1)$; that is, the Lie group is special; the determinant is equal to 1. The group is called the *improper* Lorentz group if $\det \Lambda = -1$.

From the definition of the metric, we know that:

$$g_{\mu\nu} = \Lambda^\rho_{\ \mu} g_{\rho\sigma} \Lambda^\sigma_{\ \nu} \tag{A.40}$$

Taking the $0 - 0$ component of this equation, we arrive at:

$$1 = (\Lambda^0_0)^2 - (\Lambda^i_0)^2 \tag{A.41}$$

so that:

$$(\Lambda^0_0)^2 \geq 1 \tag{A.42}$$

We thus have the *orthochronous* Lorentz group, with $\Lambda^0_0 \geq 1$, or the *nonorthochronous* Lorentz transformation, with $\Lambda^0_0 \leq -1$.

Thus, there are four ways in which we can decompose the Lorentz group, depending on the sign of $\det \Lambda$ and Λ^0_0:

$$\begin{array}{|llll|}
\hline
\text{Proper orthochronous :} & \det \Lambda = 1 & \Lambda^0_0 \geq 1 & 1 \\
\text{Improper orthochronous :} & \det \Lambda = -1 & \Lambda^0_0 \geq 1 & P \\
\text{Improper nonorthochronous :} & \det \Lambda = -1 & \Lambda^0_0 \leq 1 & T \\
\text{Proper nonorthochronous :} & \det \Lambda = 1 & \Lambda^0_0 \leq 1 & PT \\
\hline
\end{array} \tag{A.43}$$

For example, ordinary rotations and boosts (which can be smoothly deformed back to the identity) are part of the proper orthochronous Lorentz group.

A parity transformation $x^i \rightarrow -x^i$ belongs to the improper orthochronous Lorentz group. Time inversion $t \rightarrow t$ belongs to the improper non-orthochronous Lorentz group. Full inversion $x^\mu \rightarrow -x^\mu$, which is the product of a parity and time inversion, belongs to the proper nonorthochronous Lorentz group.

A.5   Dirac Matrices

Independent of any representation, the Dirac matrices obey a number of identities that follow from the definitions:

$$\begin{aligned}
\{\gamma^\mu, \gamma^\nu\} &= \gamma^\mu \gamma^\nu + \gamma^\nu \gamma^\mu = 2g^{\mu\nu} \\
\gamma^0 &= \beta; \gamma^i = \beta \alpha^i \\
\gamma^5 &= \gamma_5 = i\gamma^0 \gamma^1 \gamma^2 \gamma^3 = -\frac{i}{4!}\epsilon_{\mu\nu\sigma\rho}\gamma^\mu \gamma^\nu \gamma^\sigma \gamma^\rho \\
\gamma^2_5 &= 1 \\
\sigma^{\mu\nu} &= \frac{i}{2}[\gamma^\mu, \gamma^\nu]
\end{aligned}$$

$$\gamma^5 \sigma^{\mu\nu} = \frac{i}{2} \epsilon^{\mu\nu\sigma\rho} \sigma_{\sigma\rho}$$

$$\not{a}\not{b} = a \cdot b - i\sigma_{\mu\nu} a^\mu b^\nu$$

$$\gamma^\mu \gamma_\mu = 4$$

$$\gamma^\mu \gamma^\nu \gamma_\mu = -2\gamma^\nu$$

$$\gamma^\mu \gamma^\nu \gamma^\lambda \gamma_\mu = 4g^{\nu\lambda}$$

$$\gamma^\mu \gamma^\nu \gamma^\lambda \gamma^\sigma \gamma_\mu = -2\gamma^\sigma \gamma^\lambda \gamma^\nu$$

$$\gamma^\mu \gamma^\nu \gamma^\lambda \gamma^\sigma \gamma^\rho \gamma_\mu = 2(\gamma^\rho \gamma^\nu \gamma^\lambda \gamma^\sigma - \gamma^\sigma \gamma^\lambda \gamma^\nu \gamma^\rho)$$

$$\gamma^\mu \sigma^{\nu\lambda} \gamma_\mu = 0$$

$$\gamma^\mu \sigma^{\nu\lambda} \gamma^\sigma \gamma_\mu = 2\gamma^\sigma \sigma^{\nu\lambda} \qquad (A.44)$$

They also obey the following trace identities:

$$\mathrm{Tr}(\gamma^{\mu_1} \cdots \gamma^{\mu_n}) = 0; \quad n \text{ odd}$$

$$\mathrm{Tr}(\gamma^\mu \gamma^\nu) = 4g^{\mu\nu}$$

$$\mathrm{Tr}(\gamma^\mu \gamma^\nu \gamma^\rho \gamma^\sigma) = 4(g^{\mu\nu} g^{\rho\sigma} - g^{\mu\rho} g^{\nu\sigma} + g^{\mu\sigma} g^{\nu\rho})$$

$$\mathrm{Tr}(\gamma^5 \gamma^\mu \gamma^\nu \gamma^\rho \gamma^\sigma) = -4i\epsilon^{\mu\nu\rho\sigma}$$

$$\mathrm{Tr}(\not{a}_1 \not{a}_2 \cdots \not{a}_{2n}) = a_1 \cdot a_2 \, \mathrm{Tr}(\not{a}_3 \cdots \not{a}_{2n}) - a_1 \cdot a_3 \, \mathrm{Tr}(\not{a}_2 \not{a}_4 \cdots \not{a}_{2n})$$

$$+ \quad \cdots + a_1 \cdots a_{2n} \, \mathrm{Tr}(\not{a}_2 \cdots \not{a}_{2n-1}) \qquad (A.45)$$

Under Hermitian conjugation and charge conjugation, the Dirac matrices obey:

$$\gamma^{0\dagger} = \gamma^0; \quad \gamma^{i\dagger} = -\gamma^i$$

$$\gamma^0 \gamma^\mu \gamma^0 = \gamma^{\mu\dagger}$$

$$\gamma^0 \gamma_5 \gamma^0 = -\gamma_5^\dagger = -\gamma_5$$

$$\gamma^0 \gamma_5 \gamma^\mu \gamma^0 = (\gamma_5 \gamma^\mu)^\dagger$$

$$\gamma^0 \sigma^{\mu\nu} \gamma^0 = (\sigma^{\mu\nu})^\dagger$$

$$C^T = C^\dagger = -C; \quad C^2 = 1; \quad CC^\dagger = C^\dagger = 1;$$

$$C\gamma_\mu C^{-1} = -\gamma_\mu^T$$

$$C\gamma_5 C^{-1} = \gamma_5^T$$

$$C\sigma_{\mu\nu} C^{-1} = -\sigma_{\mu\nu}^T$$

$$C\gamma_5\gamma_\mu C^{-1} = (\gamma_5\gamma_\mu)^T \qquad (A.46)$$

Let us now specialize to specific representations. The most common is the *Dirac representation*, which has four complex components:

$$\gamma^0 = \beta = \begin{pmatrix} 1 & 0 \\ 0 & -1 \end{pmatrix}; \quad \gamma^5 = \gamma_5 = \begin{pmatrix} 0 & 1 \\ 1 & 0 \end{pmatrix}$$

$$\gamma^i = \beta\alpha^i = \begin{pmatrix} 0 & \sigma^i \\ -\sigma^i & 0 \end{pmatrix}$$

$$\sigma^{0i} = i\alpha^i = i\begin{pmatrix} 0 & \sigma^i \\ \sigma^i & 0 \end{pmatrix}; \quad \sigma^{ij} = \epsilon_{ijk}\begin{pmatrix} \sigma^k & 0 \\ 0 & \sigma^k \end{pmatrix}$$

$$C = i\gamma^2\gamma^0 = \begin{pmatrix} 0 & -i\sigma^2 \\ -i\sigma^2 & 0 \end{pmatrix} \qquad (A.47)$$

Under the Lorentz group, the Dirac representation is reducible. Each Dirac representation can be split up into two smaller representations. We can take the chiral projection, which gives us the *Weyl representation* for left-handed and right-handed spinors:

$$\gamma^0 = \beta = \begin{pmatrix} 0 & -1 \\ -1 & 0 \end{pmatrix}; \quad \gamma^5 = \gamma_5 = \begin{pmatrix} 1 & 0 \\ 0 & -1 \end{pmatrix}$$

$$\gamma^i = \begin{pmatrix} 0 & \sigma^i \\ -\sigma^i & 0 \end{pmatrix}$$

$$\sigma^{0i} = i\begin{pmatrix} \sigma^i & 0 \\ 0 & -\sigma^i \end{pmatrix}; \quad \sigma^{ij} = \epsilon_{ijk}\begin{pmatrix} \sigma^k & 0 \\ 0 & \sigma^k \end{pmatrix}$$

$$C = \begin{pmatrix} -i\sigma^2 & 0 \\ 0 & i\sigma^2 \end{pmatrix} \qquad (A.48)$$

We can also take a purely imaginary representation of the spinors, given by the *Majorana representation*:

$$\gamma^0 \;=\; \beta = \begin{pmatrix} 0 & \sigma^2 \\ \sigma^2 & 0 \end{pmatrix}; \gamma^5 = \gamma_5 = \begin{pmatrix} \sigma^2 & 0 \\ 0 & -\sigma^2 \end{pmatrix}$$

$$\gamma^1 \;=\; \begin{pmatrix} i\sigma^3 & 0 \\ 0 & i\sigma^3 \end{pmatrix}$$

$$\gamma^2 \;=\; \begin{pmatrix} 0 & -\sigma^2 \\ \sigma^2 & 0 \end{pmatrix}$$

$$\gamma^3 \;=\; \begin{pmatrix} -i\sigma^1 & 0 \\ 0 & -i\sigma^1 \end{pmatrix}$$

$$C \;=\; \begin{pmatrix} 0 & -i\sigma^2 \\ -i\sigma^2 & 0 \end{pmatrix} \tag{A.49}$$

We define conjugate spinors by:

$$\bar{\psi} \;=\; \psi^\dagger \gamma^0$$

$$\bar{u} \;=\; u^\dagger \gamma^0$$

$$\bar{v} \;=\; v^\dagger \gamma^0 \tag{A.50}$$

On-shell, the spinors u and v represent the electron and positron wave function. They obey:

$$(\not{p} - m)u(p) \;=\; 0$$

$$(\not{p} + m)v(p) \;=\; 0$$

$$\bar{u}(p)(\not{p} - m) \;=\; 0$$

$$\bar{v}(p)(\not{p} + m) \;=\; 0 \tag{A.51}$$

The spinors u and v also obey a number of normalization and completeness relations. They are normalized as follows:

$$\bar{u}(p, s)u(p, s) \;=\; 1$$

$$\bar{v}(p, s)v(p, s) \;=\; -1 \tag{A.52}$$

These spinors obey certain completeness relations:

$$\sum_s u_\alpha(p, s)\bar{u}_\beta(p, s) - v_\alpha(p, s)\bar{v}_\beta(p, s) = \delta_{\alpha\beta} \tag{A.53}$$

$$u_\alpha(p, s)\bar{u}_\beta(p, s) = \left[\frac{\not{p} + m}{2m}\frac{1 + \gamma_5 \not{s}}{2}\right]_{\alpha\beta} \tag{A.54}$$

and:

$$v_\alpha(p, s)\bar{v}_\beta(p, s) = -\left[\frac{m - \not{p}}{2m}\frac{1 + \gamma_5 \not{s}}{2}\right]_{\alpha\beta} \tag{A.55}$$

If we sum over the helicity s, we have two projection operators:

$$[\Lambda_+(p)]_{\alpha\beta} = \sum_{\pm s} u_\alpha(p, s)\bar{u}_\beta(p, s) = \left(\frac{\not{p} + m}{2m}\right)_{\alpha\beta} \tag{A.56}$$

$$[\Lambda_-(p)]_{\alpha\beta} = -\sum_{\pm s} v_\alpha(p, s)\bar{v}_\beta(p, s) = \left(\frac{-\not{p} + m}{2m}\right)_{\alpha\beta} \tag{A.57}$$

These are projection operators, and hence they satisfy:

$$\Lambda_\pm^2 = \Lambda_\pm; \quad \Lambda_+\Lambda_- = 0; \quad \Lambda_+ + \Lambda_- = 1 \tag{A.58}$$

A.6 Infrared Divergences to All Orders

Although we have proved that infrared divergences can be eliminated to lowest order by adding the bremsstrahlung diagram to the vertex correction diagram, we would now like to generalize our result to all orders in perturbation theory. At first, this may seem like an impossible task, since there are an infinite number of ways in which the infrared divergence enters into various Feynman diagrams.

However, the problem is actually tractable for two reasons. First, only a small subset of all possible Feynman diagrams actually contributes to the infrared divergence. We will therefore only concentrate on those diagrams where the emitted real photon is attached to the initial or final electron leg, which contributes to the infrared divergence when they are on the mass shell. If the photon has a small momentum q_i and is emitted from an on-shell electron with momentum p, then the Feynman propagator contains a factor:

$$\frac{1}{(p + q_i)^2 - m^2 + i\epsilon} \sim \frac{1}{2p \cdot q_i + i\epsilon} \qquad (A.59)$$

For small q_i, we see that we have an infrared problem. (Photons attached to internal electron lines, or electron lines which are far off the mass shell, will not contribute.)

Second, there are remarkable identities that make it possible to show that all these divergences cancel exactly. The calculation to all orders is not difficult once we realize that it is possible to sum the contribution of the real and virtual photons into an exponential function. Let the contribution of the emission of each real photon contribute a factor R, while the contribution of integrating over each virtual photon contributes V. Then the contribution of summing over arbitrary numbers of real and virtual photons, we will show, conveniently sums up to an exponential, given by:

$$\frac{d\sigma}{d\Omega} = \left(\frac{d\sigma}{d\Omega}\right)_0 \exp R \, \exp 2V \qquad (A.60)$$

Before, we found that the integration over the real photon contribution is taken from μ to some detector sensitivity energy E_0, and hence yields a factor of $\log(E^2/\mu^2)$. The integration over the virtual photon contribution is given by an integration over the four-momenta, which yields $\log(-q^2/\mu^2)$. Since we are taking the exponentials of these two divergent factors, the $\log \mu^2$ cancels perfectly, and the final result is convergent.

To begin the process of summing over photon lines, let us analyze a process where we have an electron coming in with momentum p and scattering off with momentum p'. If there were no infrared divergence problems to worry about, the contribution of this diagram would be of the form $\bar{u}(p')Ou(p)$. However, because of the infrared corrections, we must calculate the contribution due to the emission of real photons and the integration over virtual photons.

To perform the calculation, it will be convenient to insert a large number of photons radiating from the electron line with momenta q_i, as in Figure A.4.

Our job will be to calculate how to attach these various photon lines in various ways in order to perform the summation over real and virtual photons.

To calculate V, the contribution of the virtual photons, we will pair off these photon lines, in arbitrary order, and then perform the integration over the virtual photon's momenta. Then we must perform the summation over all possible pairings. To calculate R, by contrast, the contribution from the emission of a real photon, we will sum over the photon polarization and integrate over the photon momenta.

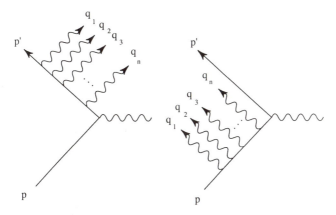

Figure A.4. N-point graph for the emission of soft photons, which has an infrared divergence.

If we examine, using Feynman's rules, the sequence of propagators in the figure (with photon legs attached near the top of the diagram), we find:

$$\bar{u}(p')\cdots(-i\gamma^{\mu_i})\frac{[i\gamma\cdot(p'+q_1+\cdots+q_i)+m]}{(p'+q_1+\cdots q_i)^2-m^2+i\epsilon}(-i\gamma^{\mu_{i+1}})$$

$$\times\ \frac{[i\gamma\cdot(p'+q_1+\cdots q_{i+1})+m]}{(p'+q_1+\cdots+q_{i+1})^2-m^2+i\epsilon}\cdots u(p) \qquad (A.61)$$

This Feynman diagram is not as hopeless as it may seem, especially when we assume that each q_i is small. First, consider the photon line with momenta q_i that is near the emitted electron, with momentum p'. We can significantly simplify the numerator by shoving all momenta p' to the left, where they hit $\bar{u}(p')$, and then we can use the Dirac equation. Since all p' can be reduced to m, the only possible tensor left-over is p'^{μ_i}; so the numerator simply becomes the product of p'^{μ_i}. Similarly, the denominators can be simplified.

For small q_i, the product of Feynman propagators becomes:

$$\bar{u}(p')\prod_i^n\left(\frac{p'^{\mu_i}}{p'\cdot\left(\sum_{k=1}^i q_k\right)}\right)\cdots \qquad (A.62)$$

The next task is to sum over all permutations of the q_i appearing in the product. Although this may seem difficult, we can use a formula that simplifies this calculation enormously:

$$\sum_{\text{perm}} \frac{1}{p' \cdot q_1} \frac{1}{p' \cdot (q_1 + q_2)} \cdots \frac{1}{p' \cdot (q_1 + \cdot + q_n)}$$

$$= \frac{1}{p' \cdot q_1} \frac{1}{p' \cdot q_2} \cdots \frac{1}{p' \cdot q_n} \tag{A.63}$$

where we sum over all permutations of the q_i appearing on the left hand side of the equation. (The proof of this formula can be carried out by induction.)

In summary, we are now left with a very simple expression, with each photon contributing a factor of $p'^{\mu_i}/p' \cdot q_i$ to the product. Not surprisingly, the same process can be carried out for photon legs attached to the bottom half of the diagram. Then we shove all $\not{p}$ to the right and use the Dirac equation. Then the Feynman diagram reduces to a product of $p^{\mu_i}/p \cdot q_i$.

Now let us sum up the contribution of all such diagrams, in any possible order. The photon with momentum q_i can be attached to the initial or the final electron. Thus, the photon with momentum q_i contributes two factors, depending on whether it attached to the initial or final electron leg. Since this photon can be attached to either leg, the correction factor for all the various photons is simply:

$$e^n \bar{u}(p') O u(p) \prod_{i=1}^{n} Q^{\mu_i}(q_i) \tag{A.64}$$

where we define:

$$Q^{\mu_i}(q_i) \equiv \left(\frac{p'^{\mu_i}}{p' \cdot q_i} - \frac{p^{\mu_i}}{p \cdot q_i} \right) \tag{A.65}$$

At this point, we must now begin the contraction process on the various photon lines. Let us say that there is a total of $2N + M$ photon lines. We will contract and integrate over N pairs of photon lines in order to calculate the contribution of virtual photons. The remaining M photons will be emitted as real photons, contributing to bremsstrahlung.

To calculate the virtual photon contribution, we must pair off the photon legs, insert a photon propagator for each of the N pairs, and then perform the integration over d^4q. Each contraction of a pair of virtual photon legs contributes the following factor:

$$V \equiv -\frac{ie^2}{2} \int \frac{d^4q}{(2\pi)^4} Q(q_i) \cdot Q(-q_i) \tag{A.66}$$

But we must also sum over N such contractions (and divide by a factor of $N!$, which represents the number of ways that we can permute these lines). This gives us the following correction factor for virtual photons:

$$\sum_{N=0}^{\infty} \frac{V^N}{N!} = \exp V \tag{A.67}$$

Similarly, we must now calculate the contribution of M emitted real photons to the scattering cross section. This means inserting the photon polarizations, summing over these polarizations, squaring the matrix element, and integrating over the photon's phase space. Fortunately, the summation over the polarizations just gives us a delta function, so the contribution of each emitted photon gives us a scalar product between $Q(q)$ and $Q(-q)$, as before. The value of R is therefore:

$$R \equiv -e^2 \int \frac{d^3q}{(2\pi)^3} \frac{1}{2k} Q(q) \cdot Q(-q) \tag{A.68}$$

By the same logic as before, we must sum over M of these factors, emitted photons, giving us R^M, and then divide by $M!$. As before, this gives us a factor of $\exp R$.

The net effect of summing over all possible permutations of the $2N + M$ photon lines, which generate both the real emitted photons and the virtual photon loops, is therefore the product of two exponentials. We can now summarize the contribution of both the real and the virtual photons by the formula:

$$\frac{d\sigma}{d\Omega} = \left(\frac{d\sigma}{d\Omega}\right)_0 \exp(2V) \exp R \tag{A.69}$$

The last and final step then involves inserting the actual value of V and R into the above formula. These values were already computed for the one photon bremsstrahlung process and vertex corrections that were computed earlier. We find:

$$2V = -\frac{\alpha}{\pi} f(q^2) \log\left(\frac{-q^2}{\mu^2}\right)$$

$$R = \frac{\alpha}{\pi} f(q^2) \log\left(\frac{E^2}{\mu^2}\right) \tag{A.70}$$

As expected, we find a cancellation between the two factors, yielding an expression that is finite and independent of μ. We can now safely take the limit as μ goes to zero, therefore obtaining the correct result without any infrared divergences.

A.7 Dimensional Regularization

The following formulas can be derived by taking the derivative of the formulas presented in Chapter 7:

$$
\int \frac{d^n k \, k_\mu}{(k^2 + 2k \cdot q - m^2)^\alpha} = \frac{-i\pi^{n/2}}{\Gamma(\alpha)(-q^2 - m^2)^{\alpha-n/2}} q_\mu \Gamma(\alpha - n/2)
$$

$$
\int \frac{d^n k \, k_\mu k_\nu}{(k^2 + 2k \cdot q - m^2)^\alpha} = \frac{i\pi^{n/2}}{\Gamma(\alpha)(-q^2 - m^2)^{\alpha-n/2}} \left[q_\mu q_\nu \Gamma(\alpha - n/2) \right.
$$
$$
\left. + \frac{1}{2} g_{\mu\nu} (-q^2 - m^2) \Gamma(\alpha - 1 - n/2) \right]
$$

$$
\int \frac{d^n k \, k_\mu k_\nu k_\lambda}{(k^2 + 2k \cdot q - m^2)^\alpha} = \frac{i\pi^{n/2}}{\Gamma(\alpha)(-q^2 - m^2)^{\alpha-n/2}} \left[-q_\mu q_\nu q_\lambda \Gamma(\alpha - n/2) \right.
$$
$$
- \frac{1}{2} \left(g_{\mu\nu} q_\lambda + g_{\nu\lambda} q_\mu + g_{\lambda\mu} q_\nu \right)
$$
$$
\left. \times (-q^2 - m^2) \Gamma(\alpha - 1 - n/2) \right]
$$

$$
\int \frac{d^n k \, k_\mu k_\nu k_\lambda k_\rho}{(k^2 + 2k \cdot q - m^2)^\alpha} = \frac{i\pi^{n/2}}{\Gamma(\alpha)(-q^2 - m^2)^{\alpha-n/2}} \left[q_\mu q_\nu q_\lambda q_\rho \Gamma(\alpha - n/2) \right.
$$
$$
+ \frac{1}{2} \left(q_\mu q_\nu g_{\lambda\rho} + \text{perm} \right) (-q^2 - m^2) \Gamma(\alpha - 1 - n/2)
$$
$$
\left. + \frac{1}{4} \left(g_{\mu\nu} g_{\lambda\rho} + \text{perm} \right) (-q^2 - m^2) \Gamma(\alpha - 2 - n/2) \right]
$$

$$(A.71)$$

A general formula is given as follows:

$$
\int \frac{d^n k \, k_{\mu_1} k_{\mu_2} \cdots k_{\mu_p}}{(k^2 + 2k \cdot q - m^2)^\alpha} = \frac{i\pi^{n/2}}{\Gamma(\alpha)(-q^2 - m^2)^\alpha} T_{\mu_1 \mu_2 \cdots \mu_p} \qquad (A.72)
$$

where:

$$
T_{\mu_1 \mu_2 \cdots \mu_p} = (-1)^p \left(q_{\mu_1} q_{\mu_2} \cdots q_{\mu_p} \Gamma(\alpha - n/2) \right.
$$
$$
+ \frac{1}{2} \sum_{\text{perm}} \left(g_{\mu_1 \mu_2} q_{\mu_3} \cdots q_{\mu_p} \right) (-q^2 - m^2) \Gamma(\alpha - 1 - n/2)
$$
$$
+ \frac{1}{4} \sum_{\text{perm}} \left(g_{\mu_1 \mu_2} g_{\mu_3 \mu_4} q_{\mu_5} \cdots q_{\mu_p} \right) (-q^2 - m^2)^2 \Gamma(\alpha - 2 - n/2)
$$

$$\vdots$$

$$+ (2)^{-p/2} \sum_{\text{perm}} \left(g_{\mu_1 \mu_2} g_{\mu_3 \mu_4} \cdots g_{\mu_{p-1} \mu_p} \right)$$

$$\times (-q^2 - m^2)^{p/2} \Gamma(\alpha - p/2 - n/2) \Big) \tag{A.73}$$

for p even. For p odd, the last term should be:

$$+ (2)^{-[p/2]} \sum_{\text{perm}} \left(g_{\mu_1 \mu_2} \cdots g_{\mu_{p-2} \mu_{p-1}} q_{\mu_p} \right)$$

$$\times (-q^2 - m^2)^{[p/2]} \Gamma\left(\alpha - [p/2] - n/2\right) \tag{A.74}$$

where $[m]$ means taking the largest integer not greater than m.

By contracting the various k_μ, we can also derive a succession of related formulas involving k^2.

Notes

Chapter 1. Why Quantum Field Theory?

1. Dirac, P.A.M., 1927, *Proc. Roy. Soc. Lon.* **A114**, 243.

2. Tomonaga, S., 1946, *Prog. Theor. Phys* **1**, 27; 1948, *Phys. Rev.* **74**, 224.

3. Schwinger, J., 1949, *Phys. Rev.* **75**, 651; **76**, 790.

4. Feynman, R. P., 1949, *Phys. Rev.* **76**, 749, 769.

5. For a historical review and more complete references, see: Pais, A., 1986, *Inward Bound.* Oxford: Oxford University Press.; Crease, R. P., and Mann, C. C., 1986, *Second Creation.* New York: Macmillan.

6. Yukawa, H., 1935, *Prog. Phys. Math. Soc. of Japan*, **17**, 48.

7. Goldberger, M. L., 1955, *Phys. Rev.* **97**, 508; **99**, 979; Gell-Mann, M., Goldberger, M. L., and Thirring, W., 1954, *Phys. Rev.* **95**, 1612; **96**, 1428.

8. Chew, G. F., 1962, *S-Matrix Theory of Strong Interactions.* Reading: Benjamin.

9. Gell-Mann, M., 1961 (unpublished); 1962, *Phys. Rev.*, **125**, 1067; 1964, *Phys. Lett.* **8**, 214.

10. Ne'eman, Y., 1961, *Nucl. Phys.* **26**, 222; Gell-Mann, M., and Ne'eman, Y., 1964, *The Eightfold Way.* Reading: Benjamin.

11. Zweig, C., CERN Rep. 8419/TH 412.

12. Sakata, S., 1956, *Prog. Theor. Phys.* **16**, 686.

13. Ikeda, M., Ogawa, S., and Ohnuki, Y., 1959, *Prog. Theor. Phys.* **22**, 715.

14. Lee, T. D., and Yang, C. N., 1956, *Phys. Rev.* **104**, 254.

15. Wu, C. S., Ambler, E., Hayward, R. W., Hoppes, D. D., and Hudson, R. P., 1957, *Phys. Rev.* **105**, 1413.

16. Garwin, R. L., Lederman, L. M., Weinrich, M., 1957, *Phys. Rev.* **105**, 1415.

17. 't Hooft, G., 1971, *Nucl. Phys.* **B33**, 173; **B35**, 167.

18. Weinberg, S., 1967, *Phys. Rev. Lett.* **19**, 1264.

19. Salam, A., 1968, *Elementary Particle Theory*, ed. N. Svaratholm, Stockholm: Almquist and Forlag.

20. Gross, D. J., and Wilczek, F., 1973, *Phys. Rev.* **D8**, 3497.

21. Politzer, H. D., 1973, *Phys. Rev. Lett.* **26**, 1346.

22. 't Hooft, G., 1972, Conference on Lagrangian Field Theory, Marseille (unpublished).

23. Wilson, K.G., 1974, *Phys. Rev.* **D10**, 2445.

Chapter 3. Spin 0 and $\frac{1}{2}$ Fields

1. Gordon, W., 1926, *Z. Physik* **40**, 117

2. Klein, O., 1927, *Z. Physik* **41**, 407.

3. Fock, V., 1926, *Z. Physik* **38**, 242.

4. Schrödinger, E., 1926, *Ann. der Phys.* **81**, 109.

5. de Donder, Th. and van den Dungen, H., 1926, *Comptes Rundus* **183**, 22.

6. Kudar, J., 1926, *Ann. der Phys.* **81**, 632.

7. Pauli, W., and Weisskopf, V., 1934, *Helv. Phys. Act.* **7**, 709.

8. Dirac, P.A.M., 1928, *Proc. Roy. Soc. Lon.* **A117**, 610; 1930, **A126**, 360.

9. Majorana, E., 1937, *Nuovo Cim.* **14**, 171.

10. Weyl, H., 1929, *Z. Physik* **56**, 330.

Chapter 4. Quantum Electrodyanics

1. Darwin, C. G., 1928, *Proc. Roy. Soc. Lon.* **A118**, 654.

2. Gordon, W., 1928, *Z. Physik,* **48**, 11.

3. Gupta, S. N., 1950, *Proc. Phys. Soc. Lon.* **A63**, 681.

4. Bleuler, K., 1950, *Helv. Phys. Acta* **23**, 567.

5. Schwinger, J., 1951, *Phys. Rev.* **82**, 914; **91**, 713.

6. Lüders, G., 1954, *Kgl. Dansk. Vidensk. Selsk. Mat.-Fys. Medd.* **28**, 5.

7. Pauli, W., 1955, *Niels Bohr and the Development of Physics.* New York: McGraw-Hill.

Chapter 5. Feynman Rules and Reduction

1. Schwinger, J., 1949, *Phys. Rev.* **75**, 651; **76**, 790.

2. Tomonaga, S., 1946, *Prog. Theor. Phys* **1**, 27; 1948, *Phys. Rev.* **74**, 224.

3. Feynman, R. P., 1949, *Phys. Rev.* **76**, 749, 769.

4. Mott, N. F., 1929, *Proc. Roy. Soc. Lon.* **A124**, 425.

5. Lehmann, H., Symanzik, K., and Zimmermann, W., 1957, *Nuovo Cim.* **6**, 319.

6. Wick, G. C., 1950, *Phys. Rev.* **80**, 268.

7. Furry, W. H., 1937, *Phys. Rev.* **81**, 115.

8. Källén, G. 1952, *Helv. Phys. Acta* **52**, 417.

9. Lehmann, 1954, *Nuovo Cim.* **11**, 342.

Chapter 6. Scattering Processes and the S-Matrix

1. Klein, O., and Nishina, Y., 1929, *Z. Physik* **52**, 853.

2. Dirac, P.A.M., 1930, *Proc. Cam. Phil. Soc.* **26**, 361.

3. Møller, C., 1932, *Ann. Phys.* **14**, 531.

4. Bhabha, H. J., 1935, *Proc. Roy. Soc. Lon.* **A154**, 195,

5. Heitler, W., 1954, *The Quantum Theory of Radiation.* Oxford: Clarendon Press.

6. Schwinger, J., 1949, *Phys. Rev.* **75**, 651; **76**, 790.

7. Yennie, D. R., Frautschi, S. C., and Surra, H., 1961, *Ann. Phys.* **13**, 379.

8. Bloch, F., and Nordsieck, A., 1937, *Phys. Rev.* **52**, 54.

9. Pauli, W., and Villars, F., 1949, *Rev. Mod. Phys.* **21**, 433.

10. Schwinger, J., 1948, *Phys. Rev.* **73**, 1256.

11. Combley, F. H., 1979, *Rep. Prog. Phys.* **42**, 1889.

12. Lamb, W. E. and Retherford, R. C., 1947, *Phys. Rev.* **72**, 241.

13. Bethe, H. A., 1947, *Phys. Rev.* **72**, 339.

14. Uehling, E. A., 1935, *Phys. Rev.* **48**, 55.

15. Serber, R., 1935, *Phys. Rev.* **48**, 49.

16. Chew, G. F., 1962, *S-Matrix Theory of Strong Interactions.* Reading, Benjamin. See also: Eden, R. J., Landshoff, P. V., Olive, D. I., and Polkinghorne, J. C., 1966, *The Analytic S-Matrix.* Cambridge: Cambridge University Press.

17. Gell-Mann, M., Goldberger, M. L., and Thirring, W., 1954, *Phys. Rev.* **95**, 1612; **96**, 1428.

18. Goldberger, M. L., 1955, *Phys. Rev.* **97**, 508; **99**, 979.

19. Landau, L. D., 1959, *Nucl. Phys.* **13**, 181.

20. Nambu, Y., 1958, *Nuovo Cim.* **9**, 610.

Chapter 7. Renormalization of QED

1. Dyson, F. J., 1949, *Phys. Rev.* **75**, 486, 1736.

2. Mills, R. L., and Yang, C. N., 1966, *Prog. of Theor. Phys. Suppl.* **37-38**, 507.

3. Wu, T. T., 1961, *Phys. Rev.* **125**, 1436.

4. Salam, A., 1951, *Phys. Rev.* **82**, 217; **84**, 426.

5. Steuckelberg, E.C.G., and Petermann, A., 1953, *Helv. Phys. Acta* **5**, 499.

6. Gell-Mann, M., and Low, F.E., 1954, *Phys. Rev.* **95**, 1300.

7. Bollini, C. G., and Giambiagi, J. J., 1964, *Nuovo Cim.* **31**, 550.

8. Bollini, C. G., and Giambiagi, J. J., 1972, *Phys. Lett.* **40B**, 566.

9. 't Hooft, G., and Veltman, M., 1972, *Nucl. Phys.* **B44**, 189.

10. 't Hooft, G., 1973, *Nucl. Phys.* **B62**, 444.

11. 't Hooft, G., and Veltman, M., 1973, CERN Report 73-9, *Diagrammar*.

12. 't Hooft, G., 1973, *Nucl. Phys.* **B61**, 455.

13. Cicuta, G. M., and Montaldi, E., 1972, *Lett. Nuovo Cim.* **4**, 392.

14. Butera, P., Cicuta, G. M., and Montaldi, E., 1974, *Nuovo Cim.* **19A**, 513.

15. Ashmore, J. F., 1972, *Lett. Nuovo Cim.* **4**, 289.

16. Speer, E. R., 1968, *J. Math. Phys.* **9**, 1404.

17. Speer, E. R., 1971, *Comm. Math. Phys.* **23**, 23.

18. Ward, J. C., 1950, *Phys. Rev.* **78**, 182.

19. Takahashi, Y., 1957, *Nuovo Cim.* **6**, 371.

20. Mills, R. L., and Yang, C. N., 1966, *Prog. Theor. Phys. Suppl.* **37-8**, 507.

21. Wu, T. T., 1962, *Phys. Rev.* **125**, 1436.

22. Velo, G., and Wightman, A. S., 1976, *Renormalization Theory*. Dordrecht: Riedel.

23. Bjorken, J. D., and Drell, S. D., 1965, *Relativistic Quantum Fields*. New York: McGraw-Hill, pp. 283–363.

24. Weinberg, S., 1960, *Phys. Rev.* **118**, 838.

Chapter 8. Path Integrals

1. Feynman, R. P., 1948, *Rev. Mod. Phys.* **20**, 267.

2. Feynman, R. P., and Hibbs, A. R., 1965, *Quantum Mechanics and Path Integrals*. New York: McGraw-Hill.

3. Dirac, P.A.M., 1933, *Physik Z. Sov. Union* **3**, 64.

4. Nambu, Y., 1968, *Phys. Lett.* **26B**, 626.

Chapter 9. Gauge Theory

1. Klein, O., 1938, *New Theories in Physics*, 77, Intern. Inst. of Intellectual Co-operation, League of Nations.

2. Yang, C. N., and Mills, R. L., 1954, *Phys. Rev.* **96**, 191.

3. Shaw, R., 1954, *The Problem of Particle Types and Other Contributions to the Theory of Elementary Particles*, Cambridge University Ph.D. thesis (unpublished).

4. Utiyama, R., 1956, *Phys. Rev.* **101**, 1597.

5. 't Hooft, G., 1971, *Nucl. Phys.* **B33**, 173; **B35**, 167.

6. Faddeev, L. D., and Popov, V. N., 1967, *Phys. Lett.* **25B**, 29; See also: Mandelstam, S., 1962, *Ann. Phys.* **19**, 1.

7. Feynman, R. P., 1963, *Acta Physica Polonica* **24**, 697.

8. Gribov, V. N., 1978, *Nucl. Phys.* **B139**, 1.

Chapter 10. The Weinberg–Salam Model

1. Nambu, Y., 1960, *Phys. Rev. Lett.* **4**, 380.

2. Goldstone, J., 1961, *Nuovo Cim.* **19**, 15.

3. Goldstone, J., Salam, A., and Weinberg, S., 1962, *Phys. Rev.* **127**, 965.

4. Higgs, P. W., 1964, *Phys. Lett.* **12**, 132.

5. Higgs, P. W., 1966, *Phys. Rev.* **145**, 1156.

6. Kibble, T.W.B., 1967, *Phys. Rev.* **155**, 1554.

7. Fermi, E., 1935, *Z. Physik* **88**, 161.

8. Sudarshan, E.C.G., and Marshak, R. E., 1958, *Phys. Rev.* **109**, 1860.

9. Feynman, R. P., and Gell-Mann, M., 1958, *Phys. Rev.* **109**, 193.

10. Weinberg, S., 1967, *Phys. Rev. Lett.* **19**, 1264.

11. Salam, A., 1968, *Elementary Particle Theory*, ed. N. Svaratholm. Stockholm: Almquist and Forlag.

12. Langacker, P., Luo, M., and Mann, A., 1992, *Rev. Mod. Phys.* **64**, 87.

13. Aguilar–Benitez, M. et al., 1992, *Review of Particle Properties, Phys. Rev.* **D45**, 1.

14. Renton, P., 1990, *Electroweak Interactions*. Cambridge: Cambridge University Press.

15. Sirlin, A., 1984, *Phys. Rev.* **D29**, 89.

16. Lee, T. D., and Yang, C. N., 1955, *Phys. Rev.* **98**, 1501.

17. 't Hooft, G., 1971, *Nucl. Phys.* **B33**, 173; **B35**, 167.

18. Coleman, S., and Weinberg, E., 1973, *Phys. Rev.* **D7**, 1888.

19. Nambu, Y., and Jona-Lasiniao, G., 1961, *Phys. Rev.* **124**, 246.

Chapter 11. The Standard Model

1. Sakata, S., 1956, *Prog. Theor. Phys.* **16**, 686.

2. Ikeda, M., Ogawa, S., and Ohnuki, Y., 1959, *Prog. Theor. Phys.* **22**, 715.

3. Gell-Mann, M., 1961 (unpublished); 1962, *Phys. Rev.* **125**, 1067.

4. Gell-Mann, M., and Ne'eman, Y., 1964, *The Eightfold Way*. Reading: W. Benjamin.

5. Ne'eman, Y., 1961, *Nucl. Phys.* **26**, 222.

6. Gell-Mann, M., 1964, *Phys. Lett.* **8**, 214.

7. Zweig, C., CERN Rep. 8419/TH 412.

8. Nishijima, K., 1955, *Prog. Theor. Phys.* **13**, 285.

9. Gell-Mann, M., 1956, *Nuovo Cim. Supp.* **4**, 848.

10. Okubo, S., 1962, *Prog. Theor. Phys.* **27**, 949; **28**, 24.

11. Sakita, B., 1964, *Phys. Rev.* **136B**, 1756.

12. Gürsey, F., and Radicati, L., 1964, *Phys. Rev. Lett.* **13**, 173.

13. Zweig, G., 1965, in *Symmetries in Elementary Particle Physics*, ed. A. Zichichi. New York: Academic Press.

14. Tarjanne, P., and Teplitz, V. L., 1963, *Phys. Rev. Lett.* **11**, 447; Krolikowski, W., 1964, *Nucl. Phys.* **52**, 342; Hara, Y., 1963, *Phys. Rev.* **134B**, 701; Bjorken, B. J., and Glashow, S. L., 1964, *Phys. Lett.* **11**, 255; Maki, Z., and Ohnuki, Y., 1964, *Prog. Theor. Phys.* **32**, 144; Amati, D., Bacry, H., Nuyts, J., and Prentki, J., 1964, *Nuovo Cim.* **34**, 1732; Okun, L. B., 1964, *Phys. Lett.* **12**, 250.

15. Han, M. Y., and Nambu, Y., 1965, *Phys. Rev.* **139B**, 1006. See also the para-statistics formulation of Greenberg, O. W., 1964, *Phys. Rev. Lett.* **13**, 598.

For reviews of current algebra, see Refs. 16 and 17:

16. Adler, S., and Dashen, R., 1968, *Current Algebras*. New York: Benjamin.

17. de Alfaro, V., Fubini, S., Furlan, G., and Rossetti, C., 1973, *Currents in Hadron Physics*. Amsterdam: North-Holland.

18. Feynman, R. P., and Gell-Mann, M., 1958, *Phys. Rev.* **109**, 193.

19. Nambu, Y., 1960, *Phys. Rev. Lett.* **4**, 380.

20. Gell-Mann, M., and Levy, M., 1960, *Nuovo Cim.* **16**, 705.

21. Chou, K. C., 1961, *Soviet Phys., JETP*, **12**, 492.

22. Goldberger, M. L. and Treiman, S. B., 1958, *Phys. Rev.* **109**, 193.

23. Adler, S. L., 1965, *Phys. Rev. Lett.* **14**, 1051.

24. Weisberger, W. I., 1965, *Phys. Rev. Lett.* **14**, 1047.

25. Cabibbo, N., 1963, *Phys. Rev. Lett.* **10**, 531.

26. Glashow, S. L., Iliopoulos, J., and Maiani, L., 1970, *Phys. Rev.* **D2**, 1285.

27. Kobayashi, M., and Maskawa, K., 1973, *Prog. Theor. Phys.* **49**, 652.

28. Christensen, J. H., Cronin, J. W., Fitch, V. L., and Turlay, R., 1964, *Phys. Rev. Lett.* **13**, 138.

Chapter 12. Ward Identities, BRST, and Anomalies

1. Ward, J. C., 1950, *Phys. Rev.* **78**, 182.

2. Takahashi, Y., 1957, *Nuovo Cim.* **6**, 371.

3. Taylor, J. C., 1971, *Nucl. Phys.* **B33**, 436.

4. Slavnov, A. A., 1972, *Theor. and Math. Phys.* **10**, 99.

5. Becchi, C., Rouet, A., and Stora, R., 1975, *Comm. Math. Phys.* **52**, 55.

6. Kugo, T., and Ojima, I., 1978, *Phys. Lett.* **73B**, 459.

7. Adler, S. L., 1969, *Phys. Rev.* **177**, 2426.

8. Bell, J. S., and Jackiw, R., 1969, *Nuovo Cim.* **60A**, 47.

9. Bardeen, W. A., 1969, *Phys. Rev.* **184**, 1848.

10. Fujikawa, K., 1979, *Phys. Rev. Lett.* **42**, 1195.

Chapter 13. BPHZ Renormalization of Gauge Theories

1. Bogoliubov, N. N., and Parasiuk, O., 1957, *Acta Math.* **97**, 227.

2. Hepp, K., 1966, *Comm. Math. Phys.* **2**, 301.

3. Zimmerman, W., 1968, *Comm. Math. Phys.* **11**, 1; 1969, **15**, 208.

Chapter 14. QCD and the Renormalization Group

1. Bjorken, J. D., 1969, *Phys. Rev.,* **179**, 1547.

2. Feynman, R. P., 1969, *Phys. Rev. Lett.* **23**, 1415.

3. Bjorken, J. D., and Paschos, E.A., 1969, *Phys. Rev.* **185**, 1975.

4. Callan, C. G., and Gross, D., 1969, *Phys. Rev. Lett.* **22**, 156.

5. Adler, S., 1966, *Phys. Rev.* **143**, 1144.

6. Gross, D., and Llewellyn Smith, C. H., 1969, *Nucl. Phys.* **B14**, 337.

7. Wilson, K. G., 1969, *Phys. Rev.* **179**, 1499.

8. For a complete set of references, see: Frishman, Y., 1974, *Phys. Rep.* **13C**, 1.

9. Stueckelberg, E.C.G., and Petermann, A., 1953, *Helv. Phys. Acta* **26**, 499.

10. Gell-Mann, M., and Low, F. E., 1954, *Phys. Rev.* **95**, 1300.

11. Callan, C. G., 1970, *Phys. Rev.* **D2**, 1541.

12. Symanzik, K., 1970, *Comm. Math. Phys.* **18**, 227.

13. Gross, D. J., and Wilczek, F., 1973, *Phys. Rev.* **D8**, 3497.

14. Politzer, H. D., 1973, *Phys. Rev. Lett.* **26**, 1346.

15. 't Hooft, G., 1972, Conference on Lagrangian Field Theory, Marseille (unpublished).

16. 't Hooft, G., 1973, *Nucl. Phys.* **B61**, 455.

17. Blaer, A., and Young, K., 1974, *Nucl. Phys.* **B83**, 493.

18. Callan, C. G., 1976, in *Methods in Field Theory*, ed. R. Galian, and J. Zinn-Justin. Amsterdam: North-Holland/World Scientific.

Chapter 15. Lattice Gauge Theory

1. Wilson, K. G., 1974, *Phys. Rev.* **D10**, 2445.

2. For more complete references, see: Kogut, J. B., 1983, *Rev. Mod. Phys.* **55**, 775.

3. Creutz, M., 1979, *Phys. Rev. Lett.* **43**, 553.

4. See: Rebbi, C., 1982, in *Non-Perturbative Aspects of Quantum Field Theory*, ed. J. Julve, and M. Ramon-Medrano. Singapore: World Scientific.

5. Kogut, J., and Susskind, L., 1975, *Phys. Rev.* **D11**, 395.

Chapter 16. Solitons, Monopoles, and Instantons

1. Russel, J. S., 1844, *Rep. 14th Meet. Brit. Assoc. Adv. Sci.*, 311. London: John Murray.

2. For more references, see: Rajaraman, R., 1989, *Solitons and Instantons*. Amsterdam: North-Holland.

3. Dirac, P.A.M., 1931, *Proc. Roy. Soc.* **A133**, 60.

4. 't Hooft, 1974, *Nucl. Phys.* **B79**, 276.

5. Polyakov, A. M., 1974, *JETP Lett.* **20**, 194.

6. Belavin, A. A., Polyakov, A. M., Schwartz, A. S., and Tyupkin, Yu. S., 1975, *Phys. Lett.* **59B**, 85.

7. For more references on instantons, see: Coleman, S., 1985, *Aspects of Symmetry*. Cambridge: Cambridge University Press; 't Hooft, G., 1976, *Phys. Rev. Lett.* **37**, 8.

8. 't Hooft, G., 1976, *Phys. Rev. Lett.* **37**, 8.

9. Jackiw, R., and Rebbi, C., 1976, *Phys. Rev. Lett.* **37**, 172.

10. Pecci, R. D., and Quinn, H. R., 1977, *Phys. Rev.* **D16**, 1791.

11. Dine, M., Fischler, W., Srednicki, M., 1981, *Phys. Lett.* **104B**, 199.

Chapter 17. Phase Transitions and Critical Phenomena

1. Baxter, R. J., 1982, *Exactly Solved Models in Statistical Mechanics*. San Diego: Academic Press.

2. For further references, see: Domb, C., and Lebowitz, J. L., eds., 1986, *Phase Transitions and Critical Phenomena* **10**. San Diego: Academic Press.

3. Ising, E., 1925, *Z. Physik* **31**, 253.

4. Onsager, L., 1944, *Phys. Rev.* **65**, 117.

5. McCoy, B. M. and Wu, T. T., 1973, *The Two Dimensional Ising Model.* Cambridge: Harvard University Press.

6. Yang, C. N., 1952, *Phys. Rev.* **85**, 808.

7. Landau, L. D., 1937, *Phys. Zurn. Sowjetunion* **11**, 26, 545.

8. Ginzburg, V. L., and Landau, L. D., 1950, *JETP* **20**, 1064.

9. Kadanoff, L. P., 1965, *Physics* **2**, 263.

10. Wilson, K. G., and Kogut, J., 1974, *Phys. Rep.* **12C**, 76.

11. Brezin, E., Le Guillou, J. C., Zinn-Justin, J., and Nickel, B. G., 1974, *Phys. Lett.* **44A**. 227.

Chapter 18. Grand Unified Theories

1. Georgi, H. M., Quinn, H. R., and Weinberg, S., 1974, *Phys. Rev. Lett.* **33**, 451.

2. Pati, J. C., and Salam, A., 1973, *Phys. Rev. Lett.* **31**, 275.

3. Georgi, H., and Glashow, S. L., 1974, *Phys. Rev. Lett.* **32**, 438.

4. Gildener, E., 1976, *Phys. Rev.* **D14**, 1667.

5. Fritzsch, H., and Minkowski, P., 1975, *Ann. Phys.* **93**, 193.

6. Georgi, H., 1975, in *Particles and Fields—1974*, ed. C. E. Carlson. New York: AIP Press.

7. Gürsey, F., Ramond, P., and Sikivie, P., 1976, *Phys. Lett.* **60B**, 177.

8. Farhi, E., and Susskind, L., 1981, *Phys. Rep.* **74**, 277.

Chapter 19. Quantum Gravity

1. Einstein, A., 1915, *Sitzungsber. Preuss. Akad. Wiss.* 778, 779, 844.

2. Hubble, E. P., 1936, *Astrophys. J.* **84**, 270.

3. Gamow, G., 1946, *Phys. Rev.* **70**, 572.

4. Alpher, R. A., Bethe, H., and Gamow, G., 1948, *Phys. Rev.* **73**, 803.

5. Friedman, A., 1922, *Z. Physik* **10**, 377.

6. Robertson, H. P., 1935, *Astrophys. J.* **82**, 284.

7. Sakharov, A. D., 1967, *Zh, Ek. Teor. Fiz.* **5**, 24.

8. Yoshimura, M., 1978, *Phys. Rev. Lett.* **41**, 281.

9. Guth, A. H., 1981, *Phys. Rev.* **D23**, 347.

10. Linde, A. D., 1982, *Phys. Lett.* **108B**, 389.

11. Linde, A. D., 1984, *Rep. Prog. Phys.* **47**, 925.

12. Kaluza, Th., 1921, *Sitz. Preuss. Akad. Wiss* **K1**, 966.

13. Klein, O., 1926, *Z. Phys.* **37**, 895.

14. DeWitt, B. S., 1963, in *Dynamical Theory of Groups and Fields*, 1963 Les Houches Summer School.

15. For more references, see: Appelquist, T., Chodos, A., and Freund, P.G.O., 1987, *Modern Kaluza–Klein Theories*. Reading: Addison-Wesley.

16. Witten, E., 1981, *Nucl. Phys.* **B186**, 412.

17. Goroff, M. H., and Sagnotti, A., 1985, *Phys. Rev.* **160B**, 81; 1986, *Nucl. Phys.* **B266**, 709.

18. van de Ven, A.E.M., DESY-91-115.

Chapter 20. Supersymmetry and Supergravity

1. Myazawa, H., *Prog. Theor. Phys.* 1966, **36**, 1266; 1968, *Phys. Rev.* **170**, 1586.

2. Neveu, A., and Schwarz, J. H., 1971, *Nucl Phys.* **B31**, 86; Ramond, P., 1971, *Phys. Rev.* **D3**, 2415.

3. Gervais, J. L., and Sakita, B., 1971, *Nucl. Phys.* **B34**, 632.

4. Gol'fand, Yu. A., and Likhtman, E. P., 1971, *Sov. Phys.: JETP Lett.* **13**, 323.

5. Volkov, D. V., and Akulov, V. P., 1972, *JETP Lett.* **16**, 438.

6. Wess, J., and Zumino, B., 1974, *Nucl. Phys.* **B70**, 34.

7. Mandelstam,S., 1982, *Proc. 21st. Int. Conf. on High Energy Physics*, ed. P. Petiau, and J. Pomeuf. *J. Phys.* **12**, 331; 1983, *Nucl. Phys.* **B213**, 149.

8. Brink, L., Lindgren, O., and Nilsson, B., 1983, *Phys. Lett.* **123B**, 323.

9. Howe, P., Stelle, K., and Townsend, P., 1983, *Nucl. Phys.* **B212**, 401.

10. Grisaru, M., and Siegel, W., 1982, *Nucl. Phys.* **B236**, 125.

11. Sohnius, M., and West, P., 1981, *Phys. Lett.* **100B**, 45.

12. Flume, R., 1983, *Nucl. Phys.* **B217**, 531.

13. Freedman, D. Z., van Nieuwenhuizen, P., and Ferrara, S., 1976, *Phys. Rev.* **D13**, 3214.

14. Deser, S., and Zumino, B., 1976, *Phys. Lett.* **62B**, 335.

15. For another approach based on supermetric tensors, see: Arnowitt, R., and Nath, P., 1975, *Phys. Lett.* **56B**, 177.

16. Salam, A., and Strathdee, J., 1974, *Phys. Lett.* **51B**, 353.

17. Wess, J., and Zumino, B., 1974, *Nucl. Phys.* **B70**, 39.

18. Fayet, P., and Illiopoulos, J., *Phys. Lett.* **51B**, 461.

19. O'Raifeartaigh, L., 1975, *Nucl. Phys.* **B96**, 331.

20. See: Gates, S. J., Grisaru, M. T., Rocek, M., and Siegel, W., 1983, *Superspace: Or One Thousand and One Lessons in Supersymmetry*. Reading: Benjamin/Cummings.

21. Haag, R., Lopuszanski, J. T., and Sohnius, M. F., 1975, *Nucl. Phys.* **B88**, 257.

22. Kaku, M., Townsend, P., and van Nieuwenhuizen, P., 1978, *Phys. Rev.* **D17**, 3179.

23. Cremmer, E., Julia, B., and Scherk, J., 1978, *Phys. Lett.* **76B**, 409.

Chapter 21. Superstrings

1. Polyakov, A. M., 1981, *Phys. Lett.* **103B**, 207, 211.

2. Nambu, Y., 1970, *Lectures at the Copenhagen Summer Symposium.*

3. Goto, T., 1971, *Prog. Theor. Phys.* **46**, 1560.

4. See also the earlier work of: Susskind, L., 1970, *Nuovo Cim.* **69A**, 457; Nielsen, H. B., 1970, *15th Int. Conf. of High Energy Phys.*, Kiev.

5. Hsue, C. S., Sakita, B., and Virasoro, M. B., 1970, *Phys. Rev.* **D2**, 2857.

6. Virasoro, M. A., 1969, *Phys. Rev. Lett.* **22**, 37.

7. Goddard, P., Goldstone, J., Rebbi, C., and Thorn, C. B., 1973, *Nucl. Phys.* **B56**, 109.

8. Kato, M., and Ogawa, K., 1983, *Nucl. Phys.* **B212**, 443.

9. Kikkawa, K., Sakita, B., and Virasoro, M. B., 1969, *Phys.Rev.* **184**, 1701.

10. Bardakci, K., and Ruegg, H., 1969, *Phys. Rev.* **181**, 1884.

11. Virasoro, M. A., 1969, *Phys. Rev. Lett.* **22**, 37.

12. Goebel, C. J., and Sakita, B., *Phys. Rev. Lett.* **22**, 257.

13. Chan, H. M., *Phys. Lett.* **28B**, 425.

14. Koba, Z. J., and Nielsen, H. B., 1969, *Nucl. Phys.* **B12**, 517.

15. Veneziano, G., 1976, *Nucl. Phys.* **B117**, 519.

16. Suzuki, M. (unpublished).

17. Fubini, S., Gordon, D., and Veneziano, G., 1969, *Phys. Lett.* **29B**, 679.

18. Virasoro, M. A., 1969, *Phys. Rev.* **177**, 2309.

19. Shapiro, J., 1970, *Phys. Lett.* **33B**, 361.

20. Ramond, P., 1971, *Phys. Rev.* **D3**, 2415.

21. Neveu, A., and Schwarz, J. H., 1971, *Nucl Phys.* **B31**, 86.

22. Gervais, J. L., and Sakita, B., 1971, *Nucl. Phys.* **B34**, 632.

23. Green, M., and Schwarz, J. H., 1982, *Nucl Phys.* **B198**, 252, 441.

24. Gross, D. J., Harvey, J. A., Martinec, E., and Rohm, R., 1985, *Phys. Rev. Lett.* **54**, 502.

25. Shapiro, J., 1972, *Phys. Rev.* **D5**, 1945.

26. Kaku, M., and Yu, L. P., 1970, *Phys. Lett.* **33B**, 166; 1971, *Phys. Rev.* **D3**, 2992, 3007, 3020.

27. Lovelace, C., 1970, *Phys. Lett.* **32B**, 703; 1971, **34B**, 500.

28. Allesandrini, V., 1971, *Nuovo Cim.* **2A**, 321.

29. For more references, see: D'Hoker, E., and Phong, D. H., 1988, *Rev. Mod. Phys.* **60**, 917.

30. Candelas, P., Horowitz, G., Strominger, A., and Witten, E., 1985, *Nucl Phys.* **B258**, 46.

31. Dixon, L., Harvey, J., Vafa, C., and Witten, E., 1985, *Nucl. Phys.* **B261**, 678.

32. Kawai, H., Lewellen, D. C., and Tye, S.H.H., 1986, *Phys. Rev. Lett.* **57**, 1832.

33. Antoniadis, I., Bachas, C., and Kounnas, C., 1987, *Nucl. Phys.* **B289**, 87.

34. Lerche, W., Lust, D., and Schellekens, A. N., 1987, *Nucl. Phys.* **B287**, 477.

35. Kaku, M., and Kikkawa, K., 1974, *Phys. Rev.* **D10**, 1110, 1823.

36. Witten, E., 1986, *Nucl. Phys* **B268**, 253.

37. Kaku, M., 1990, *Phys. Rev.* **D41**, 3733.

38. Kugo, T., Kunitomo, H., and Suehiro, K., 1989, *Phys. Lett.* **B226**, 48.

39. Saadi, M., and Zwiebach, B., 1989, *Ann. Phys.* **192**, 213.

References

Field Theory

1. Bjorken, J. D., and Drell, S. D. 1964. *Relativistic Quantum Mechanics*. New York: McGraw-Hill.

2. Bjorken, J. D., and Drell, S. D. 1965. *Relativistic Quantum Fields*. New York: McGraw-Hill.

3. Boboliubov, N. N., and Shirkov, D. V. 1959. *Introduction to the Theory of Quantized Fields*. New York: Wiley.

4. Chang, S. J. 1990. *Introduction to Quantum Field Theory*. Singapore: World Scientific.

5. Collins, J. 1984. *Renormalization*. Cambridge: Cambridge University Press.

6. Itzykson, C., and Zuber, J-B. 1980. *Quantum Field Theory*. New York: McGraw-Hill.

7. Jauch, J. M., and Rohrlich, F. 1955. *The Theory of Photons and Electrons*. Reading: Addison-Wesley.

8. Mandl, F., and Shaw, G. 1984. *Quantum Field Theory*. New York: Wiley.

9. Ramond, R. 1989. *Field Theory: A Modern Primer*. Reading: Addison-Wesley .

10. Ryder, L. H. 1985. *Quantum Field Theory*. Cambridge: Cambridge University Press.

11. Sakurai, J. J. 1967. *Advanced Quantum Mechanics*. Reading: Addison-Wesley.

12. Schweber, S. S. 1961. *An Introduction to Relativistic Quantum Field Theory*. New York: Harper & Row.

13. Schwinger, J. 1958. *Quantum Electrodynamics*. New York: Dover.

14. Zinn-Justin, J. 1989. *Quantum Field Theory and Critical Phenomena*. Oxford: Oxford University Press.

Gauge Theories

1. Cheng, T.-P., and Li, L.-F. 1984. *Gauge Theory of Elementary Particle Physics*. Oxford: Oxford University Press.

2. Faddeev, L. D., and Slavnov, A. A. 1980. *Gauge Fields: Introduction to Quantum Theory*. Reading: Benjamin/Cummings.

3. Frampton, P. H. 1987. *Gauge Field Theories*. Reading: Benjamin/Cummings.

4. Muta T. 1987. *Foundations of Quantum Chromodynamics*. Singapore: World Scientific.

5. Pokorski, S. 1987. *Gauge Field Theories*. Cambridge: Cambridge University Press.

Particle Physics

1. Becher, P., Bohm, M., and Joos, H. 1984. *Gauge Theories of Strong and Electroweak Interaction*. New York: Wiley.

2. Gasiorowicz, S. 1966. *Elementary Particle Physics*. New York: Wiley.

3. Huang, K. 1982. *Quarks, Leptons, and Gauge Fields*. Singapore: World Scientific.

4. Lee, T. D. 1981. *Particle Physics and Introduction to Field Theory*. New York: Harwood Academic.

5. Renton, P. 1990. *Electroweak Interations*. Cambridge: Cambridge University Press.

6. Ross, G. G. 1985. *Grand Unified Theories*. Reading: Benjamin/Cummings.

Critical and Non-Perturbative Phenomena

1. Amit, D. J. 1978. *The Renormalization Group and Critical Phenomena*. New York: McGraw-Hill.

2. Creutz, M. 1983. *Quarks, Gluons and Lattices*. Cambridge: Cambridge University Press.

3. Ma, S.-K. 1976. *Modern Theory of Critical Phenomena*. Reading: Benjamin/Cummings.

4. Rajaraman, R. 1989. *Solitons and Instantons*. Amsterdam: North-Holland.

5. Rebbi, C. 1983. *Lattice Gauge Theories and Monte Carlo Simulations*, Singapore: World Scientific.

6. Sakita, B. 1985. *Quantum Theory of Many-Variable Systems and Fields*. Singapore: World Scientific.

Supergravity

1. Gates, S. J., Grisaru, M. T., Rocek, M., and Siegel, W. 1983. *Superspace*. Reading: Benjamin/Cummings.

2. Jacob, M., ed. 1986. *Supersymmetry and Supergravity*. Amsterdam: North-Holland and World Scientific.

3. Mohapatra, R. N. 1986. *Unification and Supersymmetry: The Frontiers of Quark–Lepton Physics.* New York: Springer-Verlag.

4. West, P. 1990. *Introduction to Supersymmetry and Supergravity.* Singapore: World Scientific.

Superstrings

1. Frampton, P. H. 1974. *Dual Resonance Models.* Reading: Benjamin/Cummings.

2. Green, M. B., Schwarz, J. H., and Witten, E. 1987. *Superstring Theory.* Vols. I and II. Cambridge: Cambridge University Press.

3. Jacob, M., ed. 1974. *Dual Theory.* Amsterdam: North-Holland.

4. Kaku, M. 1988. *Introduction to Superstrings.* New York: Springer-Verlag.

5. Kaku, M. 1991. *Strings, Conformal Fields, and Topology.* New York: Springer-Verlag.

6. Schwarz, J. H., ed. 1985. *Superstrings.* Vols. I and II. Singapore: World Scientific.

Index